RISK FACTORS FOR PSYCHOSIS

RISK FACTORS FOR PSYCHOSIS

PARADIGMS, MECHANISMS, AND PREVENTION

Edited by

ANDREW D. THOMPSON

Orygen, Parkville, VIC, Australia;
Centre for Youth Mental Health, The University of Melbourne, Parkville, VIC, Australia;
Warwick Medical School, University of Warwick, Coventry, United Kingdom

MATTHEW R. BROOME

Professor and Chair in Psychiatry and Youth Mental Health, Director of the Institute for Mental Health, School of Psychology, University of Birmingham Distinguished Research Fellow, Oxford Uehiro Centre for Practical Ethics, University of Oxford

ELSEVIER

ACADEMIC PRESS
An imprint of Elsevier

Academic Press is an imprint of Elsevier
125 London Wall, London EC2Y 5AS, United Kingdom
525 B Street, Suite 1650, San Diego, CA 92101, United States
50 Hampshire Street, 5th Floor, Cambridge, MA 02139, United States
The Boulevard, Langford Lane, Kidlington, Oxford OX5 1GB, United Kingdom

Library of Congress Cataloging-in-Publication Data
A catalog record for this book is available from the Library of Congress

British Library Cataloguing-in-Publication Data
A catalogue record for this book is available from the British Library

ISBN: 978-0-12-813201-2

For information on all Academic Press publications visit our website at https://www.elsevier.com/books-and-journals

Publisher: Nikki Levy
Acquisitions Editor: Joslyn T. Chaiprasert-Paguio
Editorial Project Manager: Leticia M. Lima
Production Project Manager: Nirmala Arumugam
Cover Designer: Miles Hitchen

Typeset by TNQ Technologies

Contents

6. Familial high risk and high-risk studies

LIANA ROMANIUK, STELLA W.Y. CHAN, ALIX MACDONALD, JESSIKA E. SUSSMANN, ANDREW M. MCINTOSH, HEATHER C. WHALLEY, AND STEPHEN M. LAWRIE

7. Psychotic-like experiences in the general population

COLM HEALY AND MARY CANNON

8. 22q11.2 deletion syndrome: a neurodevelopmental model of psychosis

CORRADO SANDINI, STEPHAN ELIEZ, MAUDE SCHNEIDER, AND MARCO ARMANDO

II

Specific areas and risk

9. Neuroimaging studies in people at clinical high risk for psychosis

GEORGE GIFFORD, ROB ERT MCCUTCHEON, AND PHILIP MCGUIRE

10. Genetic studies of psychosis

HANNAH J. JONES, STANLEY ZAMMIT, AND JAMES T.R. WALTERS

11. Immune processes and risk of psychosis

ADAM AL-DIWANI AND THOMAS A. POLLAK

12. Neurochemical models of psychosis risk and onset

DOMINIC OLIVER, GEMMA MODINOS, AND PHILIP MCGUIRE

13. Clinical risk factors for psychosis

ASWIN RATHEESH, JESSICA A. HARTMANN, AND BARNABY NELSON

14. Cognitive risk factors for psychosis

KELLY ALLOTT AND ASHLEIGH LIN

15. Society and risk of psychosis

CRAIG MORGAN, TESSA ROBERTS, BRIAN O. DONOGHUE, AND ANDREW D. THOMPSON

16. Is there sufficient evidence that cannabis use is a risk factor for psychosis?

MARCO COLIZZI AND SAGNIK BHATTACHARYYA

III

Interventions

17. The ethics of identifying and treating psychosis risk

PAOLO CORSICO AND ILINA SINGH

Contributors

Jean Addington Hotchkiss Brain Institute, Department of Psychiatry, University of Calgary, Calgary, AB, Canada

Adam Al-Diwani Wellcome Research Training Fellow, Department of Psychiatry, University of Oxford, Oxford, United Kingdom

Nikolai Albert Mental Health Center Copenhagen, Copenhagen University Hospital, Copenhagen, Denmark

Kelly Allott Orygen, Parkville, VIC, Australia; Centre for Youth Mental Health, The University of Melbourne, Parkville, VIC, Australia

Marco Armando Developmental Imaging and Psychopathology Lab, Department of Psychiatry, School of Medicine, University of Geneva, Geneva, Switzerland

Neus Barrantes-Vidal Universitat Autònoma de Barcelona, Bellaterra, Barcelona, Spain; Sant Pere Claver - Fundació Sanitària, Barcelona, Spain; CIBER Salud Mental, Instituto de Salud Carlos III, Barcelona, Spain

A. Bechdolf Orygen, The National Centre of Excellence in Youth Mental Health, University of Melbourne, Melbourne, VIC, Australia; Vivantes Klinikum Am Urban, Department of Psychiatry, Psychotherapy and Psychosomatic Medicine with Early intervention and recognition centre – FRITZ am Urban, Teaching hospital of Charité Medizin, Berlin, Germany; Vivantes Klinikum im Friedrichshain, Department of Psychiatry, Psychotherapy and Psychosomatic Medicine, Teaching hospital of Charité Medizin, Berlin; Universitätsklinikum Köln, Department of Psychiatry and Psychotherapy, Cologne, NRW, Germany

Sagnik Bhattacharyya Department of Psychosis Studies, Institute of Psychiatry, Psychology and Neuroscience, King's College London, London, United Kingdom

E. Burkhardt Orygen, The National Centre of Excellence in Youth Mental Health, University of Melbourne, Melbourne, VIC, Australia; Vivantes Klinikum Am Urban, Department of Psychiatry, Psychotherapy and Psychosomatic Medicine with Early intervention and recognition centre – FRITZ am Urban, Teaching hospital of Charité Medizin, Berlin, Germany

Mary Cannon Department of Psychiatry, Royal College of Surgeons in Ireland, Dublin, Leinster, Ireland; Trinity College Institute of Neuroscience, Trinity College Dublin, Dublin, Leinster, Ireland

Stella W.Y. Chan Section of Clinical Psychology, University of Edinburgh, Edinburgh, United Kingdom

Sherry K.W. Chan Department of Psychiatry, University of Hong Kong, Hong Kong

W.C. Chang Department of Psychiatry, University of Hong Kong, Hong Kong

Eric Y.H. Chen Department of Psychiatry, University of Hong Kong, Hong Kong

Marco Colizzi Department of Psychosis Studies, Institute of Psychiatry, Psychology and Neuroscience, King's College London, London, United Kingdom; Section of Psychiatry, Department of Neurosciences, Biomedicine and Movement Sciences, University of Verona, Verona, Italy

Paolo Corsico Centre for Social Ethics and Policy, Department of Law, School of Social Sciences, the University of Manchester, Manchester, United Kingdom; Department of Psychiatry, University of Oxford, Oxford, United Kingdom

Daniel J. Devoe Hotchkiss Brain Institute, Department of Psychiatry, University of Calgary, Calgary, AB, Canada

Brian O. Donoghue Orygen, the National Centre of Excellence in Youth Mental Health, VIC, Australia

Stephan Eliez Developmental Imaging and Psychopathology Lab, Department of Psychiatry, School of Medicine, University of Geneva, Geneva, Switzerland

Rahel Flückiger University Hospital of Child and Adolescent Psychiatry and Psychotherapy, University of Bern, Bern, Switzerland

George Gifford Department of Psychosis Studies, Institute of Psychiatry, Psychology & Neuroscience, King's College London, London, United Kingdom

Louise Birkedal Glenthøj Mental Health Center Copenhagen, Copenhagen University Hospital, Copenhagen, Denmark

Jessica A. Hartmann Orygen, Parkville, VIC, Australia; Centre for Youth Mental Health, The University of Melbourne, Parkville, VIC, Australia

Colm Healy Department of Psychiatry, Royal College of Surgeons in Ireland, Dublin, Leinster, Ireland

Christy L.M. Hui Department of Psychiatry, University of Hong Kong, Hong Kong

Hannah J. Jones Centre for Academic Mental Health, Department of Population Health Sciences, Bristol Medical School, University of Bristol, Bristol, United Kingdom

Peter B. Jones Department of Psychiatry, University of Cambridge, Cambridge, United Kingdom; CAMEO Early Intervention Services, Cambridgeshire and Peterborough NHS Foundation Trust, Cambridge, United Kingdom

Thomas R. Kwapil University of Illinois at Urbana-Champaign, Champaign, IL, United States

Stephen M. Lawrie Division of Psychiatry, University of Edinburgh, Edinburgh, United Kingdom

Edwin H.M. Lee Department of Psychiatry, University of Hong Kong, Hong Kong

K. Leopold Vivantes Klinikum Am Urban, Department of Psychiatry, Psychotherapy and Psychosomatic Medicine with Early intervention and recognition centre – FRITZ am Urban, Teaching hospital of Charité Medizin, Berlin, Germany; Vivantes Klinikum im Friedrichshain, Department of Psychiatry, Psychotherapy and Psychosomatic Medicine, Teaching hospital of Charité Medizin, Berlin

Ashleigh Lin Telethon Kids Institute, The University of Western Australia, Perth, WA, Australia

Alix Macdonald Division of Psychiatry, University of Edinburgh, Edinburgh, United Kingdom

Robert McCutcheon Department of Psychosis Studies, Institute of Psychiatry, Psychology & Neuroscience, King's College London, London, United Kingdom

Patrick D. McGorry Orygen, Parkville, VIC, Australia; Centre for Youth Mental Health, The University of Melbourne, Parkville, VIC, Australia

Philip McGuire Department of Psychosis Studies, Institute of Psychiatry, Psychology & Neuroscience, King's College London, London, United Kingdom; OASIS Service, South London and the Maudsley NHS National Health Service Foundation Trust, London, United Kingdom; National Institute for Health Research (NIHR) Biomedical Research Centre for Mental Health, Institute of Psychiatry, Psychology & Neuroscience, King's College London, London, United Kingdom

Andrew M. McIntosh Division of Psychiatry, University of Edinburgh, Edinburgh, United Kingdom

Cristina Mei Orygen, Parkville, VIC, Australia; Centre for Youth Mental Health, The University of Melbourne, Parkville, VIC, Australia

Chantal Michel University Hospital of Child and Adolescent Psychiatry and Psychotherapy, University of Bern, Bern, Switzerland; Developmental Clinical Psychology Research Unit, Faculty of Psychology and Educational Sciences, University of Geneva, Geneva, Switzerland

Gemma Modinos Department of Psychosis Studies, Institute of Psychiatry, Psychology & Neuroscience, King's College London, London, United Kingdom; Department of Neuroimaging, Institute of Psychiatry, Psychology & Neuroscience, King's College London, London, United Kingdom

Craig Morgan ESRC Centre for Society and Mental Health and Social Epidemiology Research Group, King's College London, London, United Kingdom

Barnaby Nelson Orygen, Parkville, VIC, Australia; Centre for Youth Mental Health, The University of Melbourne, Parkville, VIC, Australia

Merete Nordentoft Mental Health Center Copenhagen, Copenhagen University Hospital, Copenhagen, Denmark; University of Copenhagen, Copenhagen, Denmark

Dominic Oliver Department of Psychosis Studies, Institute of Psychiatry, Psychology & Neuroscience, King's College London, London, United Kingdom

Jesus Perez Department of Psychiatry, University of Cambridge, Cambridge, United Kingdom; CAMEO Early Intervention Services, Cambridgeshire and Peterborough NHS Foundation Trust, Cambridge, United Kingdom; Norwich Medical School, University of East Anglia, Norwich, United Kingdom

Danijela Piskulic Hotchkiss Brain Institute, Department of Psychiatry, University of Calgary, Calgary, AB, Canada

Thomas A. Pollak Department of Psychosis Studies, Institute of Psychiatry, Psychology and Neuroscience, King's College London, London, United Kingdom

Anna Racioppi Universitat Autònoma de Barcelona, Bellaterra, Barcelona, Spain

Aswin Ratheesh Orygen, Parkville, VIC, Australia; Centre for Youth Mental Health, University of Melbourne, Parkville, VIC, Australia

Tessa Roberts ESRC Centre for Society and Mental Health and Social Epidemiology Research Group, King's College London, London, United Kingdom

Liana Romaniuk Division of Psychiatry, University of Edinburgh, Edinburgh, United Kingdom

Corrado Sandini Developmental Imaging and Psychopathology Lab, Department of Psychiatry, School of Medicine, University of Geneva, Geneva, Switzerland

Olga Santesteban-Echarri Hotchkiss Brain Institute, Department of Psychiatry, University of Calgary, Calgary, AB, Canada

Maude Schneider Developmental Imaging and Psychopathology Lab, Department of Psychiatry, School of Medicine, University of Geneva, Geneva, Switzerland; Center for Contextual Psychiatry, Department of Neurosciences, Leuven, Belgium

Frauke Schultze-Lutter University Hospital of Child and Adolescent Psychiatry and Psychotherapy, University of Bern, Bern, Switzerland; Department of Psychiatry and Psychotherapy, Medical Faculty, Heinrich-Heine University, Düsseldorf, Germany

Ilina Singh Department of Psychiatry & Wellcome Centre for Ethics and Humanities, University of Oxford, Oxford, United Kingdom

Emma Soneson Department of Psychiatry, University of Cambridge, Cambridge, United Kingdom

Rachael Spooner Orygen, Parkville, VIC, Australia; Centre for Youth Mental Health, The University of Melbourne, Parkville, VIC, Australia

Jacqueline Stowkowy Hotchkiss Brain Institute, Department of Psychiatry, University of Calgary, Calgary, AB, Canada

Y.N. Suen Department of Psychiatry, University of Hong Kong, Hong Kong

Jessika E. Sussmann Division of Psychiatry, University of Edinburgh, Edinburgh, United Kingdom

Anastasia Theodoridou Department of Psychiatry, Psychotherapy and Psychosomatics, University Hospital of Psychiatry, Zurich, Switzerland

Andrew D. Thompson Orygen, the National Centre of Excellence in Youth Mental Health, VIC, Australia; Mental Health and Wellbeing Division, Warwick Medical School, University of Warwick, Coventry, United Kingdom

James T.R. Walters MRC Centre for Neuropsychiatric Genetics and Genomics, Division of Psychological Medicine and Clinical Neurosciences, School of Medicine, Cardiff University, Cardiff, United Kingdom

Heather C. Whalley Division of Psychiatry, University of Edinburgh, Edinburgh, United Kingdom

Alison R. Yung Centre for Youth Mental Health and Orygen, The University of Melbourne, Parkville, Victoria, Australia; Division of Psychology and Mental Health, School of Health Sciences, The University of Manchester, Manchester, United Kingdom

Stanley Zammit MRC Centre for Neuropsychiatric Genetics and Genomics, Division of Psychological Medicine and Clinical Neurosciences, School of Medicine, Cardiff University, Cardiff, United Kingdom

Foreword

This book is an up-to-date presentation of the state-of-the-art research findings by clinical academics and researchers from across the globe. One of the strengths of the book is that it emphasizes a multidisciplinary approach to increase the understanding of psychosis risk. The book begins with a historical overview of the concept of psychosis risk, followed by a description of the concepts of psychosis risk, screening, and prevention. Over the past 25 years, we have seen a proliferation of research dealing with the biopsychosocial mechanisms involved in the etiology and maintenance of psychosis risk and of psychosis. Nevertheless, gaps still confront our understanding of the individual factors that lead to psychosis risk and to psychosis. More research is needed to better understand what combination of factors can lead to transition to psychosis in one individual at risk, while another with a similar degree of psychosis risk at first presentation will develop a different mental health problem or indeed stabilize and recover.

The second part of the book comprises chapters devoted to the main types of risk paradigms. The discussion of each risk paradigm in a dedicated chapter presents the reader with a very informative overview of the evolution of the different paradigms (At risk mental state, Psychosis Risk Syndrome, Basic symptoms, Self-disturbance, Schizotypy, Familial high risk, Psychotic like symptoms, and 22q), as well as demonstrating where the different paradigms overlap or diverge from each other.

The third segment of the book explores emerging approaches to the research of specific areas of psychosis risk. The chapters in this section give an excellent illustration of the sheer complexity of factors that contribute to psychosis risk: an individual personal history, their personality traits, their neuropsychological profile, their social context, their use of substances, alongside the infinite number of possible interactions with the individual genetic predisposition, make a heavy demand on our ability to develop explanatory models of the mechanisms involved in the onset and maintenance of psychosis risk and can only be addressed in a multidisciplinary context.

In the final part of the book we read how this growing body of evidence has not only led to an improved understanding of the etiology of psychosis risk, but also, importantly, how it has been accompanied by the development of novel interventions, clinical implementation, and service reform across the world. Somewhat paradoxically, in the same period of time, this complexity has fueled a resistance in some areas of psychiatry to the concept of prevention of psychosis. This resistance can be difficult to comprehend, especially at a time when medicine and psychology are emphasizing prevention, early detection, and early intervention to improve short-term and long-term outcomes and decrease long-term economic and personal costs associated with delayed recognition of illness onset. However, the danger of labeling a young individual as being at risk of developing psychosis and the ethical issues of identifying and treating psychosis risk should not be underestimated and is addressed in the content of this section.

This book is written primarily for researcher and clinical academics working in early detection and early intervention for psychosis risk but will prove valuable to clinicians looking for a comprehensive overview of the most recent research on psychosis risk. The book will also be useful to anyone who is interested in increasing their knowledge about psychosis risk.

The editors and authors of *Risk Factors for Psychosis: Paradigms, Mechanisms, and Prevention* deserve our congratulations for writing what is undoubtedly a most important addition to the literature on psychosis risk, and it is an honor for me to introduce it to the reader.

Dr. Lucia Valmaggia
President of IEPA—Early Intervention in Mental Health
Reader in Clinical Psychology and Digital Mental Health, King's College London, Institute of Psychiatry, Psychology and Neuroscience

CHAPTER

1

Historical perspectives on psychosis risk

Cristina Mei[1,2], *Patrick D. McGorry*[1,2]

[1]**Orygen, Parkville, VIC, Australia;** [2]**Centre for Youth Mental Health, The University of Melbourne, Parkville, VIC, Australia**

Introduction

Early detection and prevention of potentially serious medical conditions, such as cancer and cardiovascular disease, have long been strongly advocated to improve patient outcomes and survival. Achieving this endeavor for psychotic disorders has lagged considerably in comparison, despite their potentially devastating impacts and high burden of disease (Rössler et al., 2005; Morgan et al., 2012). Although a preemptive psychiatry was aspired to almost a century ago (Sullivan, 1927), it was not until decades later that this potential gained momentum. A significant barrier to progress was the early conceptualization of psychotic disorders, particularly schizophrenia, which were viewed with pessimism, with an inevitable deteriorating course and largely palliative treatment focus that offered limited hope and opportunity for recovery. The current conceptualization of psychotic disorders is more optimistic, with the course of illness understood to be modifiable and not restricted to an inevitable poor prognosis and decline in social and functional outcomes (Henry et al., 2010; McGorry et al., 1990; McGorry, 1992; Killackey et al., 2019; Anderson et al., 2018; Correll et al., 2018).

This more positive outlook is largely due to the research and early intervention efforts in the 1980s and 1990s that focused on the timely recognition and phase-specific treatment of first-episode psychosis (McGorry, 2015). This early intervention model represented a form of secondary prevention, with the target being psychosis (rather than schizophrenia) across three stages: prepsychotic, first-episode, and recovery. These stages represented key differences in the timing and duration of antipsychotic medication as well as the underlying risk of chronicity (McGorry et al., 2008). The establishment of an early psychosis paradigm led to increased research and clinical efforts to identify and intervene during the earliest or prodromal stage of psychosis to prevent or at least delay the onset of psychosis. Well before this paradigm shift, a prolonged period of signs and symptoms conceptualized as a "prodrome" were known to commonly precede the onset of a frank psychotic episode (Bleuler, 1911; Kraepelin, 1919). A number of seminal works attempted to reconstruct the prodromal period to identify the earliest symptoms of psychotic disorders, particularly schizophrenia. However, these research attempts faced numerous conceptual limitations and challenges in prospectively identifying those in the

Risk Factors for Psychosis
https://doi.org/10.1016/B978-0-12-813201-2.00001-6

prodromal phase of illness. In this chapter, we summarize the conceptual origins of the prodrome that have led to contemporary at-risk approaches.

The prodrome concept

Within clinical medicine, the prodrome refers to the early signs of an illness that precede the emergence of specific diagnostic symptoms that denote a fully fledged illness. The development of measles is a prime example of the concept; its earliest prodromal signs and symptoms are nonspecific (fever, cough, conjunctivitis, and coryza), with a definitive diagnosis of measles possible following the appearance of a specific rash (Yung and McGorry, 1996b).

The prodrome for psychotic disorders has long been recognized. Since the early 1900s, early signs of schizophrenia had been observed prior to the onset of a clinically diagnosable psychotic disorder (Sullivan, 1927; Cameron, 1938b; Meares, 1959). This prodromal phase was characterized by a range of nonspecific signs and symptoms (e.g., mood changes, anxiety, sleep disturbance, impaired functioning) and attenuated or subthreshold psychotic symptoms that represented a change in premorbid functioning (Beiser et al., 1993; Yung and McGorry, 1996b; Loebel et al., 1992). This makes the prodrome conceptually distinct to the premorbid phase, which represents the stage prior to the onset of prodromal symptoms and functional decline. While these distinct phases exist, identifying clear-cut boundaries between the various stages of a psychotic disorder (i.e., pre-morbid, prodromal, first-episode) is challenging and often blurred (Yung and McGorry, 1996b). For instance, as the early symptoms of psychotic disorders are typically nonspecific and emerge gradually, it can be difficult to identify the precise point in which an individual's typical behavior or symptoms transition to the point of a prodrome.

Clinical implications of the prodrome

Arising from the prodrome concept was the idea of early identification and prevention of psychosis. One of the earliest notions of indicated prevention in psychiatry was described by Sullivan (1927, p. 106–107): "I feel certain that many incipient cases might be arrested before the efficient contact with reality is completely suspended, and a long stay in institutions made necessary." This challenged deterministic 19th century notions of psychotic disorders, particularly schizophrenia, that were derived from degeneration theory, which instilled despair and stigma and were entrenched with minimal challenge for decades. This is exemplified through the phenomenon of dementia praecox, later termed schizophrenia, which connoted inevitable chronicity and deterioration, and embedded deep stigmatization. Its subsequent effect on the care of individuals with psychosis was devastating, with a largely palliative treatment focus adopted, even after effective treatments were discovered. Authors in the early 1900s questioned this early notion of schizophrenia, recognizing that treatment was often offered too late: "the psychiatrist sees too many end states and deals professionally with too few of the pre-psychotic" (Sullivan, 1927, p.106). This was echoed in the lament of McGlashan (1996, p. 198), who stated: "I remain convinced that with them [patients with schizophrenia] I came upon the scene too late; most of the damage was already done." The prodrome was viewed as a potential solution and was identified as a key target for early detection to prevent the onset of psychosis and the loss of human potential (Cameron, 1938b; Meares, 1959; Sullivan, 1927). This was supported by the fact that much of the disability associated with psychotic disorders developed during this prepsychotic phase (Hafner et al., 1995) and was difficult to reverse even when remission of the psychotic episode was achieved (Hafner et al., 2003).

Conceptual limitations

Although intervening within the prodrome represented a potential opportunity for prevention of psychotic disorders, early obstacles, which included challenges in prospectively identifying the prodrome and the lack of effective treatments, limited the realization of this aspiration. A key component of prevention is the accurate identification of at-risk individuals. However, for psychotic disorders, the nonspecific nature of prodromal symptoms and the difficulty in differentiating these from other psychopathology (e.g., major depression) (Hafner et al., 2005) meant that the prodrome could not be accurately identified. That is, the prodrome is a retrospective concept that can only be applied after meeting diagnostic criteria for a full-threshold psychotic disorder (Yung et al., 1996). Although the prodrome concept assumes an inevitable progression to a psychotic disorder (unless prevented through intervention), prodromal symptoms have only limited predictive power in relation to schizophrenia (Jackson et al., 1995). Prior to a diagnosis of psychosis, there is no certainty that those who are suspected to be within the prodromal phase will indeed later develop a psychotic disorder. Even in the general population, attenuated or isolated psychotic symptoms are common and may not evolve into a full-threshold psychotic disorder (Tien, 1991; McGorry and Hickie, 2019; van Os et al., 2001). We have elsewhere addressed this issue in light of the past 25 years of new research (McGorry et al., 2018; see Chapter 7 in this book for further discussion).

The above factors had implications on the utility of the prodrome concept in enhancing early detection and treatment. Firstly, the assumption that prodromal features represent the beginning of an inevitable progressive disorder would result in unnecessary labeling and treatment, with patients being treated as if they already had schizophrenia, an issue highlighted by Falloon (1992). Secondly, a reliance on nonspecific prodromal symptoms to identify those at imminent onset of psychosis would result in a substantial number of false positives. The obvious need was to be able to enrich the sample with predictors that conferred greater specificity and more accurate predictive power. Such predictors could be historical or premorbid, clinical features, neurobiological markers or risk factors. This alternative approach began in Melbourne from 1991 as the Early Psychosis Prevention and Intervention Centre (EPPIC) program came into being and led to the creation of new criteria and new clinical platforms (McGorry et al., 1996).

Early research characterizing the psychosis prodrome

These conceptual limitations of the prodrome became apparent following a series of studies attempting to identify the signs and symptoms that precede a diagnosis of schizophrenia. Early researchers in the field sought to reconstruct the prodromal phase through clinical case notes or interviews conducted with individuals who had passed the diagnostic threshold for a psychotic disorder (Yung and McGorry, 1996a). A key limitation of retrospective accounts was the possibility of recall bias. A review of early studies found that the most commonly reported prodromal features were nonspecific and included reduced concentration, attention and motivation, depressed mood, sleep disturbance, anxiety, social withdrawal, suspiciousness, decline in functioning, and irritability (Yung and McGorry, 1996b). Although these early studies lacked methodological rigor and were largely anecdotal reports, the prodromal features described are largely in agreement with more recent findings (Iyer et al., 2008). In their detailed review, Yung and McGorry (1996b) identified a number of prodromal symptoms that had been suggested as having some specificity for psychosis. These symptoms included disorders of selective attention and perception, suspiciousness, change in the sense of self and the world, certain cognitive basic symptoms, and cenesthesias (Bowers and Freedman, 1966; Ebel et al., 1989; Conrad, 1958; Chapman, 1966). However, it is now evident that individual prodromal symptoms are not pathognomonic of schizophrenia.

An important finding emerging from these early works was the prolonged period between the earliest behavioral changes observed and the onset of a frank psychotic episode. In one of the first accounts, Cameron (1938b) highlighted the variability of the prodromal phase, with the duration ranging from days to years for both nonspecific and prepsychotic symptoms. The average duration of symptoms prior to the first hospital admission ranged from 6 months or less (32.4%), 6 months to 2 years (17.5%), and over 2 years (48.1%) (Cameron, 1938a). This variability is also evident in more recent ultra high risk samples where the risk of transition to psychosis can extend to 10 years, with the peak period between 2 and 3 years (Fusar-Poli et al., 2012; Nelson et al., 2013a).

Characterizing illness course

A number of initial theories were proposed describing the illness course leading to psychosis (Yung and McGorry, 1996b). A key notion was that psychotic disorders were preceded by a period of nonspecific behavioral changes, which were subsequently followed by specific prepsychotic symptoms (Cameron, 1938b; Docherty et al., 1978; Meares, 1959). Early conceptualizations separated the initial nonspecific symptoms of schizophrenia into hyperactive (e.g., nervousness, sleep disturbance, mood changes) and hypoactive behavioral changes (e.g., withdrawn, fatigue, depression) (Cameron, 1938b). Cameron (1938b) detailed how these initial disturbances then progressed to a less varied constellation of symptoms (e.g., confusion, hallucinations) that were identified as early indications of clinically recognizable schizophrenia.

Chapman (1966) provided an alternative conceptualization of the symptomatology. He suggested that although "every kind of neurotic symptom was encountered in the early stages" (p. 245) of schizophrenia, these were "superficial" and represented emotional reactions to the illness rather than underlying symptoms. Chapman (1966) proposed that specific subjective experiences, which included disturbances in attention, perception, memory, mobility, thinking, and speech, preceded neurotic symptoms and were more relevant to the illness.

A staged approach to describing the course of schizophrenia was also proposed. Docherty et al. (1978) termed the stages of schizophrenia as (I) overextension (e.g., anxiety, confusion, irritability), (II) restricted consciousness (e.g., depressed mood, fatigue, lack of energy, social withdrawal), (III) disinhibition (e.g., poor impulse control), and (IV) psychotic disorganization (e.g., hallucinations and distortions). These stages were proposed to be sequentially ordered, with the prodrome phase initiating at stage I and ending at stage III. In Conrad's (1958) model, the initial early phase of schizophrenia was characterized by a period of nonspecific features such as depression, anxiety, and irritability, which was then followed by stages characterized by hallucinations, delusions, and thought disorder. This pattern of emergence preceded the onset of a full-blown psychotic illness. In more recent years, the utility of these unidirectional stage-based models of psychosis has come into question, with evidence failing to support both models (Hafner et al., 2003).

The schizotypy construct offered an alternative perspective on the development of schizophrenia and related disorders. Based on the early observations of Bleuler (1911) and Kraepelin (1971), a vulnerability to schizophrenia was highlighted, which was termed "latent schizophrenia" and referred to schizophrenia-like traits in patients and their nonpsychotic relatives. This was described as a mild form of the illness, which included characteristics such as atypical personality, irritability, mood changes, suspiciousness, and social withdrawal, although these cases rarely sought professional care (Kendler, 1985). These observations were later extended by Rado (1953) and Meehl (1962) to create the construct of schizotypy, which was viewed as representing a latent liability for schizophrenia. In Meehl's model, gene–environment interactions were thought to be associated with schizophrenia, with only 10% of "schizotypes" transitioning to schizophrenia. This led to the development of a fully dimensional model (Claridge and Beech, 1995), which proposed that schizotypy, representing a variety of personality traits, was distributed within the general population on a continuum from normal variation to pathological (e.g., schizophrenia) (Grant et al., 2018). This view was supported by genetic research based on family, adoption and twin designs, which revealed that schizotypy in healthy populations and in those with psychotic disorders have shared underpinnings (Nelson et al., 2013b). The role of schizotypy in the emergence of psychotic disorders remains an area of ongoing research, particularly in regard to its association with transition in high-risk samples (i.e., individuals in the putative prodromal phase) (Debbane et al., 2015). While no link between schizotypy and transition has been reported (Brucato et al., 2017), a number of studies have found an association between the negative dimension of schizotypy (particularly anhedonia) and transition to psychosis in high-risk samples (Michel et al., 2019; Fluckiger et al., 2016; Kotlicka-Antczak et al., 2019; Debbane et al., 2015). For further discussion on schizotypy, see Chapter 5.

Identifying individuals in the prodromal phase

Despite early attempts to characterize the prodrome, the prospective identification of individuals at imminent onset of a psychotic disorder was not achieved. It was not until the 1980s that the endeavor of early detection became plausible. During this time, Huber and Gross (1989) conducted the first prospective study on the early identification of schizophrenia. They described the basic symptoms of schizophrenia; early, subjectively experienced disturbances in a range of domains (e.g., cognition, affect, perception, and movement) that occurred prior to the onset of frank psychosis. These formed the basis of two distinct precursor syndromes: "prodromes" that develop into a psychotic disorder, and "outpost" syndromes that are fully remitting without immediate transition to a frank psychotic episode (Gross, 1997). The symptoms are described as "basic" since they are viewed as representing the "immediate symptomatic expression of the neurobiological processes underlying psychosis and the earliest form of self-experienced symptoms" (Schultze-Lutter and Theodoridou, 2017, p. 104). Within this perspective, attenuated and frank psychotic symptoms are presumed to develop later due to an inability to cope with initial symptoms. As described in Chapter 4, the basic symptoms approach created a new strategy for identifying individuals at risk of schizophrenia (Klosterkotter et al., 1997, 2001). In a study examining the prognostic accuracy of basic symptoms, it was found that during a mean follow-up period of 9.6 years, transition to psychosis was experienced by 49.4% of patients and the presence of basic symptoms predicted schizophrenia with a probability of 70% (Klosterkotter et al., 2001).

Further progress was made through Hafner and colleagues' attempt to map the transition from prodrome to psychosis (Hafner et al., 1992). They described two prephases of early psychosis: first, a prepsychotic prodromal stage of negative and nonspecific symptoms commencing from the first sign of illness until the first positive psychotic symptoms, with a mean duration of approximately 5 years; and second, a psychotic prephase, commencing from the first positive symptom until the first admission, with a mean duration of approximately 1 year (Hafner and Maurer, 2006; Hafner et al., 1995). Hafner and colleagues' work emphasized the need for early intervention, with the vast majority of negative social consequences occurring during the prodromal phase, typically coinciding with early adulthood (Hafner et al., 1995). There was, however, a failure to distinguish between subthreshold and full-threshold psychosis. Despite the recognition of early intervention, intervening during the prodrome remained a challenge as a prospective approach to examining the onset of psychosis was considered impossible (Hafner et al., 1995).

The introduction of Mrazek and Haggerty's (1994) framework for preventive interventions, with its inclusion of indicated preventive interventions that target high-risk individuals such as those with subthreshold symptoms, provided a genuine possibility for averting a full-threshold psychotic disorder. Around this time, Eaton et al. (1995) proposed the notion of "precursor signs and symptoms," while McGorry and Singh (1995) introduced the ***"at-risk mental state"*** concept. These terms differed to "prodrome" in that they indicated *risk* of progression to a psychotic disorder without the assumption that all cases will transition to psychosis, thus addressing the false positives issue associated with the prodrome. In the mid 1990s, based on the at-risk mental state concept, Yung et al. (1996) introduced operationalized criteria to prospectively identify individuals at ultra high risk (UHR) of psychosis and who could therefore be considered to be in the putative prodromal phase or at incipient risk of psychosis (see Chapter 3). Given the relative nonspecificity of prodromal symptoms, these criteria utilized a "close-in" strategy (Bell, 1992) to enhance prediction of those at high risk of psychosis. This involved a sequential screening process where a number of screening measures were created to concentrate the level of risk in the sample, that is, an individual had to meet a number of conditions to qualify as UHR of psychosis. The criteria incorporated the risk factor of age (adolescence and young adulthood) with clinical features (attenuated or transient psychotic symptoms), and genetic risk (including schizotypal personality disorder) combined with a decline in functioning. The term "close-in" was chosen because the risk was captured close to onset of the fully fledged syndrome.

This was a departure from the traditional method of identifying high-risk cohorts based on family history of a psychotic disorder, particularly schizophrenia. The genetic high-risk model offered limited scope for a broad-based early detection strategy given that most cases emerged in the absence of a positive family history of schizophrenia and only a low proportion of individuals with a positive family history were diagnosed with schizophrenia or a psychotic disorder, typically after a long latent period (McGorry et al., 2003).

Initial validation supported the UHR approach, with 40.8% of individuals who met the UHR criteria developing a psychotic disorder within 12 months (Yung et al., 2003). Based on meta-analytic data, rates of transition to psychosis have been estimated at 22% and 36% within 1 and 3 years, respectively (Fusar-Poli et al., 2012). The UHR approach has been influential, leading to specialized clinical services for individuals at risk of psychotic disorders, with an integrated research program that has generated new knowledge in the field (Yung et al., 1995, 1996; McGorry et al., 2003). Specifically, it shifted the focus of psychosis and schizophrenia research to the earlier stages of illness, which has spearheaded an exponential growth in research on this topic. The impact of research findings over the last 2 decades has been significant and has guided the formulation of new diagnostic strategies, including transdiagnostic and clinical staging approaches (McGorry et al., 2018; see Chapter 23).

Conclusion: challenges and future directions

From its beginnings with the prodrome concept, the endeavor of predicting and preventing psychosis has been complex and remains an area of refinement. Almost a century ago, early writers in the field recognized the potential to intervene during the prodromal or prepsychotic phase. Key learnings from these early retrospective studies revealed that the prodrome of psychosis was characterized by a range of mostly nonspecific signs and symptoms that had limited power to accurately predict the onset of a psychotic disorder. This has understandably raised a number of ethical dilemmas over the years, including overtreatment and stigmatization (see Chapter 18 for full details). While Sullivan alluded to the potential negative effects of providing unnecessary treatment to those suspected to be experiencing the prodrome of schizophrenia (see discussion of Cameron, 1938b), the issue of false positives was highlighted in Falloon's (1992) influential study that attempted to identify and treat schizophrenia during the prodromal phase. Although results from this study were encouraging and indicated a reduction in the incidence of schizophrenia, the challenges of the prodrome concept were recognized in that some cases (i.e., false positives) may have been unnecessarily labeled as having schizophrenia and exposed to treatment even though they may not have been truly experiencing the prodrome of schizophrenia. However, this does not imply that treatment should be withheld until a frank episode of psychosis. A genuine need for care exists prior to this point (Yung et al., 1996) and treatment should be proportionate to current need and risk of illness progression (McGorry et al., 2018).

The wide adoption of the at-risk mental state approach produced a major new wave of research and service reform. Despite leading to significant progress in the early detection and treatment of individuals at risk of a psychotic disorder (van der Gaag et al., 2013), there are limitations to the UHR approach. This includes the modest transition rates to psychosis and the associated heterogeneous outcomes of the UHR criteria (e.g., nonpsychotic disorders) and pathways to first-episode psychosis (Addington et al., 2011; Shah et al., 2017; Lin et al., 2015). This has necessitated a broader and more ambitious strategy to predicting psychosis as well as other mental disorders (McGorry, 2013). A transdiagnostic stage-based approach has recently been proposed (Hartmann et al., 2019; McGorry and Nelson, 2016; McGorry et al., 2018), which has the potential to provide the statistical power needed to accurately predict low-incidence mental disorders such as schizophrenia (Cuijpers, 2003). Although this approach requires validation, which is currently underway, it offers substantial potential and could lead to new avenues in understanding the mechanisms underlying psychosis risk that can then be targeted via preventive interventions.

References

Addington, J., Cornblatt, B.A., Cadenhead, K.S., Cannon, T.D., McGlashan, T.H., Perkins, D.O., Seidman, L.J., Tsuang, M.T., Walker, E.F., Woods, S.W., Heinssen, R., 2011. At clinical high risk for psychosis: outcome for nonconverters. Am. J. Psychiatry 168 (8), 800–805.

Anderson, K.K., Norman, R., MacDougall, A., Edwards, J., Palaniyappan, L., Lau, C., Kurdyak, P., 2018. Effectiveness of early psychosis intervention: comparison of service users and nonusers in population-based health administrative data. Am. J. Psychiatry 175 (5), 443–452.

Beiser, M., Erickson, D., Fleming, J.A., Iacono, W.G., 1993. Establishing the onset of psychotic illness. Am. J. Psychiatry 150 (9), 1349–1354.

Bell, R.Q., 1992. Multiple-risk cohorts and segmenting risk as solutions to the problem of false positives in risk for the major psychoses. Psychiatry 55 (4), 370–381.

Bleuler, E., 1911. Dementia Praecox or the Group of Schizophrenias. International Universities Press, New York, 1950.

Bowers Jr., M.B., Freedman, D.X., 1966. "Psychedelic" experiences in acute psychoses. Arch. Gen. Psychiatr. 15 (3), 240–248.

Brucato, G., Masucci, M.D., Arndt, L.Y., Ben-David, S., Colibazzi, T., Corcoran, C.M., Crumbley, A.H., Crump, F.M., Gill, K.E., Kimhy, D., Lister, A., Schobel, S.A., Yang, L.H., Lieberman, J.A., Girgis, R.R., 2017. Baseline demographics, clinical features and predictors of conversion among 200 individuals in a longitudinal prospective psychosis-risk cohort. Psychol. Med. 47 (11), 1923–1935.

Cameron, D.E., 1938a. Early diagnosis of schizophrenia by the general practitioner. N. Engl. J. Med. 218 (5), 221–224.

Cameron, D.E., 1938b. Early schizophrenia. Am. J. Psychiatry 95 (3), 567–582.

Chapman, J., 1966. The early symptoms of schizophrenia. Br. J. Psychiatry 112 (484), 225–251.

Claridge, G., Beech, T., 1995. Fully and quasi-dimensional constructions of schizotypy. In: Raine, A., Lencz, T., Mednick, S.A. (Eds.), Schizotypal Personality. Cambridge University Press, Cambridge.

Conrad, K., 1958. Die Beginnende Schizophrenic. Thieme, Stuttgart, Germany.

Correll, C.U., Galling, B., Pawar, A., et al., 2018. Comparison of early intervention services vs treatment as usual for early-phase psychosis: a systematic review, meta-analysis, and meta-regression. JAMA Psychiatry 75 (6), 555–565.

Cuijpers, P., 2003. Examining the effects of prevention programs on the incidence of new cases of mental disorders: the lack of statistical power. Am. J. Psychiatry 160 (8), 1385–1391.

Debbane, M., Eliez, S., Badoud, D., Conus, P., Fluckiger, R., Schultze-Lutter, F., 2015. Developing psychosis and its risk states through the lens of schizotypy. Schizophr. Bull. 41 (Suppl. 2), S396–S407.

Docherty, J.P., Van Kammen, D.P., Siris, S.G., Marder, S.R., 1978. Stages of onset of schizophrenic psychosis. Am. J. Psychiatry 135 (4), 420–426.

Eaton, W.W., Badawi, M., Melton, B., 1995. Prodromes and precursors: epidemiologic data for primary prevention of disorders with slow onset. Am. J. Psychiatry 152 (7), 967–972.

Ebel, H., Gross, G., Klosterkotter, J., Huber, G., 1989. Basic symptoms in schizophrenic and affective psychoses. Psychopathology 22 (4), 224–232.

Falloon, I.R., 1992. Early intervention for first episodes of schizophrenia: a preliminary exploration. Psychiatry 55 (1), 4–15.

Fluckiger, R., Ruhrmann, S., Debbane, M., Michel, C., Hubl, D., Schimmelmann, B.G., Klosterkotter, J., Schultze-Lutter, F., 2016. Psychosis-predictive value of self-reported schizotypy in a clinical high-risk sample. J. Abnorm. Psychol. 125 (7), 923–932.

Fusar-Poli, P., Bonoldi, I., Yung, A.R., Borgwardt, S., Kempton, M.J., Valmaggia, L., Barale, F., Caverzasi, E., McGuire, P., 2012. Predicting psychosis: meta-analysis of transition outcomes in individuals at high clinical risk. Arch. Gen. Psychiatr. 69 (3), 220–229.

Grant, P., Green, M.J., Mason, O.J., 2018. Models of schizotypy: the importance of conceptual clarity. Schizophr. Bull. 44 (Suppl. 2), S556–S563.

Gross, G., 1997. The onset of schizophrenia. Schizophr. Res. 28 (2–3), 187–198.

Hafner, H., Maurer, K., 2006. Early detection of schizophrenia: current evidence and future perspectives. World Psychiatry 5 (3), 130–138.

Hafner, H., Maurer, K., Loffler, W., an der Heiden, W., Hambrecht, M., Schultze-Lutter, F., 2003. Modeling the early course of schizophrenia. Schizophr. Bull. 29 (2), 325–340.

Hafner, H., Maurer, K., Löffler, W., Bustamante, S., van der Heiden, W., Riecher-Rössler, A., Nowotny, B., 1995. Onset and early course of schizophrenia. In: Häfner, H., Gattaz, W.F. (Eds.), Search for the Causes of Schizophrenia. Springer, Berlin, Heidelberg.

Hafner, H., Maurer, K., Trendler, G., an der Heiden, W., Schmidt, M., Konnecke, R., 2005. Schizophrenia and depression: challenging the paradigm of two separate diseases — a controlled study of schizophrenia, depression and healthy controls. Schizophr. Res. 77 (1), 11–24.

Hafner, H., Riecher-Rossler, A., Hambrecht, M., Maurer, K., Meissner, S., Schmidtke, A., Fatkenheuer, B., Loffler, W., van der Heiden, W., 1992. IRAOS: an instrument for the assessment of onset and early course of schizophrenia. Schizophr. Res. 6 (3), 209–223.

Hartmann, J.A., Nelson, B., Spooner, R., Paul Amminger, G., Chanen, A., Davey, C.G., McHugh, M., Ratheesh, A., Treen, D., Yuen, H.P., McGorry, P.D., 2019. Broad clinical high-risk mental state (CHARMS): methodology of a cohort study validating criteria for pluripotent risk. Early Interv. Psychiatry 13 (3), 379–386.

Henry, L.P., Amminger, G.P., Harris, M.G., Yuen, H.P., Harrigan, S.M., Prosser, A.L., Schwartz, O.S., Farrelly, S.E., Herrman, H., Jackson, H.J., McGorry, P.D., 2010. The EPPIC follow-up study of first-episode psychosis: longer-term clinical and functional outcome 7 years after index admission. J. Clin. Psychiatry 71 (6), 716–728.

Huber, G., Gross, G., 1989. The concept of basic symptoms in schizophrenic and schizoaffective psychoses. Recent. Prog. Med. 80 (12), 646–652.

Iyer, S.N., Boekestyn, L., Cassidy, C.M., King, S., Joober, R., Malla, A.K., 2008. Signs and symptoms in the pre-psychotic phase: description and implications for diagnostic trajectories. Psychol. Med. 38 (8), 1147–1156.

Jackson, H.J., McGorry, P.D., Dudgeon, P., 1995. Prodromal symptoms of schizophrenia in first-episode psychosis: prevalence and specificity. Compr. Psychiatr. 36 (4), 241–250.

Kendler, K.S., 1985. Diagnostic approaches to schizotypal personality disorder: a historical perspective, 11 (4), 538–553.

Killackey, E., Allott, K., Jackson, H.J., Scutella, R., Tseng, Y.P., Borland, J., Proffitt, T.M., Hunt, S., Kay-Lambkin, F., Chinnery, G., Baksheev, G., Alvarez-Jimenez, M., McGorry, P.D., Cotton, S.M., 2019. Individual placement and support for vocational recovery in first-episode psychosis: randomised controlled trial. Br. J. Psychiatry 214 (2), 76–82.

Klosterkotter, J., Hellmich, M., Steinmeyer, E.M., Schultze-Lutter, F., 2001. Diagnosing schizophrenia in the initial prodromal phase. Arch. Gen. Psychiatr. 58 (2), 158–164.

Klosterkotter, J., Schultze-Lutter, F., Gross, G., Huber, G., Steinmeyer, E.M., 1997. Early self-experienced neuropsychological deficits and subsequent schizophrenic diseases: an 8-year average follow-up prospective study. Acta Psychiatr. Scand. 95 (5), 396–404.

Kotlicka-Antczak, M., Karbownik, M.S., Pawelczyk, A., Zurner, N., Pawelczyk, T., Strzelecki, D., Urban-Kowalczyk, M., 2019. A developmentally-stable pattern of premorbid schizoid-schizotypal features predicts psychotic transition from the clinical high-risk for psychosis state. Compr. Psychiatr. 90, 95–101.

Kraepelin, E., 1919. Dementia Praecox and Paraphrenia. Krieger, New York.

Kraepelin, E., 1971. Dementia Praecox and Paraphrenia, eighth ed. Krieger, New York.

Lin, A., Wood, S.J., Nelson, B., Beavan, A., McGorry, P., Yung, A.R., 2015. Outcomes of nontransitioned cases in a sample at ultra-high risk for psychosis. Am. J. Psychiatry 172 (3), 249–258.

Loebel, A.D., Lieberman, J.A., Alvir, J.M., Mayerhoff, D.I., Geisler, S.H., Szymanski, S.R., 1992. Duration of psychosis and outcome in first-episode schizophrenia. Am. J. Psychiatry 149 (9), 1183–1188.

McGlashan, T.H., 1996. Early detection and intervention in schizophrenia: editor's introduction. Schizophr. Bull. 22 (2), 197–199.

McGorry, P., 2013. Early clinical phenotypes and risk for serious mental disorders in young people: need for care precedes traditional diagnoses in mood and psychotic disorders. Can. J. Psychiatr. 58 (1), 19–21.

McGorry, P., Hickie, I. (Eds.), 2019. Clinical Staging in Psychiatry: Making Diagnosis Work for Research and Treatment. Cambridge University Press, Cambridge.

McGorry, P., Nelson, B., 2016. Why we need a transdiagnostic staging approach to emerging psychopathology, early diagnosis, and treatment. JAMA Psychiatry 73 (3), 191–192.

McGorry, P.D., 1992. The concept of recovery and secondary prevention in psychotic disorders. Aust. N. Z. J. Psychiatr. 26 (1), 3–17.

McGorry, P.D., 2015. Early intervention in psychosis: obvious, effective, overdue. J. Nerv. Ment. Dis. 203 (5), 310–318.

McGorry, P.D., Copolov, D.L., Singh, B.S., 1990. Current concepts in functional psychosis. The case for a loosening of associations. Schizophr. Res. 3 (4), 221–234.

McGorry, P.D., Edwards, J., Mihalopoulos, C., Harrigan, S.M., Jackson, H.J., 1996. EPPIC: an evolving system of early detection and optimal management. Schizophr. Bull. 22 (2), 305–326.

McGorry, P.D., Hartmann, J.A., Spooner, R., Nelson, B., 2018. Beyond the "at risk mental state" concept: transitioning to transdiagnostic psychiatry. World Psychiatry 17 (2), 133–142.

McGorry, P.D., Killackey, E., Yung, A., 2008. Early intervention in psychosis: concepts, evidence and future directions. World Psychiatry 7 (3), 148–156.

McGorry, P.D., Singh, B.S., 1995. Schizophrenia: risk and possibility. In: Raphael, B., Burrows, G.D. (Eds.), Handbook of Studies on Preventive Psychiatry. Elsevier, Amsterdam.

McGorry, P.D., Yung, A.R., Phillips, L.J., 2003. The "close-in" or ultra high-risk model: a safe and effective strategy for research and clinical intervention in prepsychotic mental disorder. Schizophr. Bull. 29 (4), 771–790.

Meares, A., 1959. The diagnosis of prepsychotic schizophrenia. Lancet 1 (7063), 55–58.

Meehl, P.E., 1962. Schizotaxia, schizotypy, schizophrenia. Am. Psychol. 17 (12), 827–838.

Michel, C., Flückiger, R., Kindler, J., Hubl, D., Kaess, M., Schultze-Lutter, F., 2019. The trait-state distinction between schizotypy and clinical high risk: results from a one-year follow-up. World Psychiatry 18 (1), 108–109.

Morgan, V.A., Waterreus, A., Jablensky, A., Mackinnon, A., McGrath, J.J., Carr, V., Bush, R., Castle, D., Cohen, M., Harvey, C., Galletly, C., Stain, H.J., Neil, A.L., McGorry, P., Hocking, B., Shah, S., Saw, S., 2012. People living with psychotic illness in 2010: the second Australian national survey of psychosis. Aust. N. Z. J. Psychiatr. 46 (8), 735–752.

Mrazek, P.J., Haggerty, R.J., 1994. Reducing Risks for Mental Disorders: Frontiers for Preventive Intervention Research. National Academy Press, Washington.

Nelson, B., Yuen, H., Wood, S.J., et al., 2013a. Long-term follow-up of a group at ultra high risk ("prodromal") for psychosis: the pace 400 study. JAMA Psychiatry 70 (8), 793–802.

Nelson, M.T., Seal, M.L., Pantelis, C., Phillips, L.J., 2013b. Evidence of a dimensional relationship between schizotypy and schizophrenia: a systematic review. Neurosci. Biobehav. Rev. 37 (3), 317–327.

Rado, S., 1953. Dynamics and classification of disordered behaviour. Am. J. Psychiatry 110 (6), 406–416.

Rössler, W., Joachim Salize, H., van Os, J., Riecher-Rössler, A., 2005. Size of burden of schizophrenia and psychotic disorders. Eur. Neuropsychopharmacol. 15 (4), 399–409.

Schultze-Lutter, F., Theodoridou, A., 2017. The concept of basic symptoms: its scientific and clinical relevance. World Psychiatry 16 (1), 104–105.

Shah, J.L., Crawford, A., Mustafa, S.S., Iyer, S.N., Joober, R., Malla, A.K., 2017. Is the clinical high-risk state a valid concept? Retrospective examination in a first-episode psychosis sample. Psychiatr. Serv. 68 (10), 1046–1052.

Sullivan, H.S., 1927. The onset of schizophrenia. Am. J. Psychiatry 84 (1), 105–134.

Tien, A.Y., 1991. Distributions of hallucinations in the population. Soc. Psychiatry Psychiatr. Epidemiol. 26 (6), 287–292.

van der Gaag, M., Smit, F., Bechdolf, A., French, P., Linszen, D.H., Yung, A.R., McGorry, P., Cuijpers, P., 2013. Preventing a first episode of psychosis: meta-analysis of randomized controlled prevention trials of 12 month and longer-term follow-ups. Schizophr. Res. 149 (1), 56–62.

van Os, J., Hanssen, M., Bijl, R.V., Vollebergh, W., 2001. Prevalence of psychotic disorder and community level of psychotic symptoms: an urban-rural comparison. Arch. Gen. Psychiatr. 58 (7), 663–668.

Yung, A.R., McGorry, P.D., 1996a. The initial prodrome in psychosis: descriptive and qualitative aspects. Aust. N. Z. J. Psychiatr. 30 (5), 587–599.

Yung, A.R., McGorry, P.D., 1996b. The prodromal phase of first-episode psychosis: past and current conceptualizations. Schizophr. Bull. 22 (2), 353–370.

Yung, A.R., McGorry, P.D., McFarlane, C.A., Jackson, H.J., Patton, G.C., Rakkar, A., 1996. Monitoring and care of young people at incipient risk of psychosis. Schizophr. Bull. 22 (2), 283–303.

Yung, A.R., McGorry, P.D., McFarlane, C.A., Patton, G.C., 1995. The Pace Clinic: development of a clinical service for young people at high risk of psychosis. Australas. Psychiatr. 3 (5), 345–349.

Yung, A.R., Phillips, L.J., Yuen, H. sP., Francey, S.M., McFarlane, C.A., Hallgren, M., McGorry, P.D., 2003. Psychosis prediction: 12-month follow up of a high-risk ("prodromal") group. Schizophr. Res. 60 (1), 21–32.

CHAPTER

2

Principles of risk, screening, and prevention in psychiatry

Emma Soneson[1], *Jesus Perez*[1,2,3], *Peter B. Jones*[1,2]

[1]**Department of Psychiatry, University of Cambridge, Cambridge, United Kingdom; [2]CAMEO Early Intervention Services, Cambridgeshire and Peterborough NHS Foundation Trust, Cambridge, United Kingdom; [3]Norwich Medical School, University of East Anglia, Norwich, United Kingdom**

The value of a diagnostic test depends on the prevalence of the condition in the population tested.

"Well what does that mean?" asked the Minister.

"It means that my crystal ball will work best on people who already have a high chance of having a disease," answered the Wizard. "If I use it on lots of people who aren't very poorly then it won't give the right answers." ***From "The Gatekeeper and the Wizard: a fairy tale" Mathers and Hodgkin (1989).***

Introduction

In his seminal work on prevention in public health, epidemiologist Geoffrey Rose argues that any successful model of prevention must begin with an epidemiological understanding of the condition (Rose, 2001). Confronted with someone with a mental disorder (or any other health condition), Rose encourages us to ask ourselves "why did *this person* get *this disease* at *this time*?" In seeking to answer to this question, we begin to get an idea of the underlying risk factors for ill health, the understanding of which is crucial for anyone interested in the prevention and treatment of psychiatric disorders.

In this chapter, we shall discuss the concepts of risk, screening, and prevention as they apply in the field of psychiatry. Throughout the chapter, we shall illustrate these concepts through use of mathematical examples, national guidelines, and case studies from diverse branches of psychiatry.

Risk

In this section, we shall discuss the evolution of risk as a concept, review the mathematical underpinnings of risk, and present a series of common risk factors for mental disorders using a mediator and moderator approach to introduce epidemiological principles in the field of psychiatry.

https://doi.org/10.1016/B978-0-12-813201-2.00002-8

Risk, relative risk, and excess risk—the mathematical underpinnings

By the 19th century, risk became a quantitative measure to be calculated as a probability of something happening, and statistical methods began to be used to model it (Daston, 1987). In modern epidemiology, *risk* is defined as the probability of an event occurring over a given time interval (Last et al., 2001; Borson et al., 2013). It may be thought of as the likelihood of having an outcome of interest given exposure or nonexposure to certain factors. In essence, risk is simply a fraction.

In medicine, the concept of risk is particularly useful when used as a relative measure; relative risks can point us toward causes.

Relative risk (also known as the risk ratio) is defined as the ratio of two risks. Generally, it is calculated as the risk of the outcome given exposure divided by the risk of the outcome given nonexposure (Last et al., 2001). The relative risk may be interpreted in various ways. It could be said that the relative risk refers to *how many times* more likely is it that those who are exposed have the outcome, compared with those who were not exposed. It is also possible to express relative risk as the *percent increase* in likelihood for the outcome, given exposure. For example, with a relative risk of 1.5, it could be said that those who are exposed have a 1.5 times greater likelihood of the outcome of interest than those who are not exposed. Equally, this ratio could be interpreted as those exposed being 50% more likely to have the outcome of interest compared with those who were not exposed.

A related concept to the relative risk is the *odds ratio*. There are two common ways to think about odds ratios. The first is the *exposure-odds ratio*, which is used in case—control and cross-sectional studies. The exposure-odds ratio is a ratio of odds of exposure in those with a given condition (cases) to odds of exposure in those without the condition (noncases) (Last et al., 2001). The second is the *disease-odds ratio*, used in cross-sectional and cohort studies. This ratio is defined as a ratio of odds of having the condition (cases) given exposure to the odds of having the condition given nonexposure (Last et al., 2001). Importantly, the odds ratio calculates to the same value regardless of if conceptualized as an exposure- or disease-odds ratio. For populations where the prevalence of the condition of interest is less than approximately 5%, the odds ratio approximates the risk ratio. The risk ratio in turn equates to the *incidence rate ratio*, defined as the incidence rate given exposure divided by the incidence rate given nonexposure (Last et al., 2001). Odds ratios are more easily modeled in logistic and other approaches to modeling and are commonly used in the analysis of case—control studies and are a common means of expressing statistical association.

The commonly used term *risk factor* generally refers to a characteristic or exposure that generates a relative risk greater than one; where such an association is unlikely to be due to quirks of study design or alternative explanations, such a risk factor may point toward causation. In observational epidemiology, as opposed to evidence from randomized experiments, such a possibility requires judgment according to rules of inferential logic such as those developed by Sir Austin Bradford Hill in 1965 (Hill, 1965) where the size of the association is but one factor to be considered.

Another way to compare risks is through the *excess risk* (also known as the *absolute risk difference*). The excess risk is simply the difference between two risks, generally calculated as the risk in those exposed to a putative cause, minus the risk in those unexposed (Last et al., 2001). This individual excess risk can be used to calculate the *population excess risk* (or *population attributable risk*), defined as the incidence of the outcome that may be attributed to the risk factor of interest (Last et al., 2001). Population excess risk is calculated by multiplying the individual excess risk by the prevalence of the *exposure* in the population. Unlike relative risk, which may indicate causation, absolute risks and risk differences put these into context, indicating how important such an effect may be.

Risk calculations

To understand the concept of medical risk, 2 × 2 *contingency tables* are particularly useful (Fig. 2.1). These tables show measures of exposure and outcome, where (+) indicates presence of exposure or outcome and (−) represents absence of exposure or outcome. In Fig. 2.1, for example, "a" represents those who have both the exposure and outcome of interest.

Using this generic contingency table, we can calculate a number of measures related to the risk of having the outcome:

- *Risk given exposure* is calculated as $\frac{a}{a+b}$
- *Risk given no exposure* is calculated as $\frac{c}{c+d}$
- *Relative risk* is calculated as $\frac{\frac{a}{a+b}}{\frac{c}{c+d}}$
- *Odds ratio* is calculated as $\frac{ad}{bc}$
- *Excess risk (absolute risk difference)* is calculated as $\frac{a}{a+b} - \frac{c}{c+d}$
- *Population excess risk* is calculated by multiplying the excess risk by the population prevalence of the exposure, i.e., $\left(\frac{a}{a+b} - \frac{c}{c+d}\right)\left(\frac{a+b}{a+b+c+d}\right)$

Below is an example of how risk can be applied in psychiatry, using an example of mothers with postnatal depression (PND).

Example 1. Adapted from *Gender, Poverty, and Postnatal Depression: A Study of Mothers in Goa, India* (Patel et al., 2002)

270 expectant mothers in Goa, India were recruited to participate in a study that aimed to identify risk factors for PND. Overall, 59 mothers were diagnosed with PND over a 6-month period.

One of the risk factors measured was whether the mothers had a current score of 5 or more on the General Health Questionnaire (GHQ). A total of 113 women scored 5 or more on the GHQ. About 46 women who had a score ≥5 were diagnosed with PND, while 13 women who had a score <5 were diagnosed with PND.

a. Constructing a 2 × 2 contingency table yields the following:

		Outcome (PND)		
		+	-	
Exposure (≥5 on GHQ)	+	46	67	**113**
	-	13	144	**157**
		59	**211**	**270**

		Outcome (disease or no disease; case or control)		
		+	-	
Exposure	+	a	b	**a + b**
	-	c	d	**c + d**
		a + c	**b + d**	**Total (a + b + c + d)**

FIGURE 2.1 2 × 2 Contingency table.

b. The *risk* for PND over 6 months given a score of ≥5 on the GHQ is therefore equal to $\frac{46}{113}$ = 40.7%.

c. The *risk* for PND over 6 months given a score of <5 on the GHQ is equal to $\frac{13}{157}$ = 8.3%.

d. The *relative risk* for PND over 6 months can be calculated by $\frac{\frac{46}{113}}{\frac{13}{157}} = 4.9$, meaning that mothers whose prenatal scores on the GHQ were ≥5 had a 4.9 times greater risk for PND diagnosis than those whose scores were <5.

e. The *absolute (or excess) risk* over 6 months can be calculated by subtracting the two risks, i.e., $\frac{46}{113} - \frac{13}{157} = 0.324 = 32.4\%$.

f. The *population excess risk* over 6 months can be calculated by multiplying the excess risk by the prevalence of the exposure in the population, i.e., $\left(\frac{46}{113} - \frac{13}{157}\right)\left(\frac{113}{270}\right) = .135 = 13.5\%$.

Taken together, these measures of risk can help us better understand the epidemiology of psychiatric disorders, which in turn helps guide the provision of care and support.

Risk factors for mental ill health

Mental disorders, like many conditions in medicine, are complex in nature, and their development cannot be attributed to one single cause. Instead, mental disorders are influenced by genetic, biological, psychological, social, and environmental risk factors that interact with one another to influence psychiatric outcomes (Fig. 2.2) (Kraemer et al., 2001). Thus, it can be quite complicated to apply these concepts of risk within the field of psychiatry. To fully understand risk in psychiatry, therefore, one needs to have an understanding of each of these levels of risk (Dahlgren and Whitehead, 1991).

The Dahlgren and Whitehead (1991) model is useful for visualizing layers of risk, but it is often difficult to place a given risk factor within one of

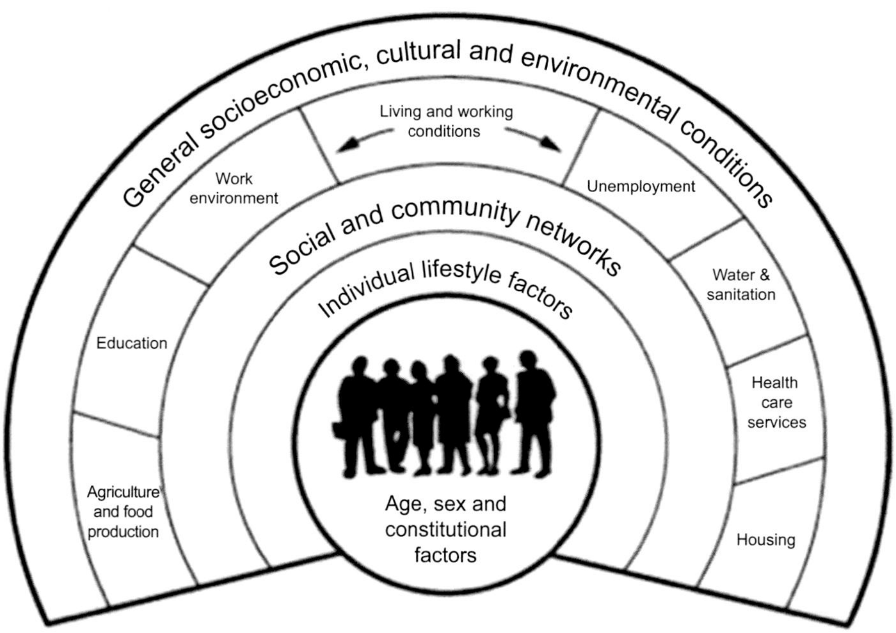

FIGURE 2.2 Dahlgren and Whitehead's model of the social determinants of health. *Reproduced with permission from Dahlgren, G., Whitehead, M., 1991. Policies and Strategies to Promote Social Equity in Health. Institute for Future Studies, Stockholm.*

these levels. This difficulty has led to the creation of risk categories including "psychosocial" and "socioenvironmental" risks, which operate in the crossover between different levels. To avoid these troublesome categorizations, this chapter will instead use a "mediator" and "moderator" approach to risk factors for mental ill health (Breitborde et al., 2010). *Moderators* may be thought of as context variables in the social and physical environment that can lead to negative mental health outcomes. *Mediators* are the pathways through which potential causal factors, such as stress or loss, "get under the skin" and lead to negative mental health outcomes.

In this way, moderators may explain differential susceptibility to common events (like "stress"), while mediating factors may, themselves, differ between people or over the life course, as discussed below. This naturally introduces the reciprocal concepts of protection, protective factors, and resilience (Van Harmelen et al., 2017). Protective factors are simply characteristics or exposures where the relative risk is computed as less than one. Resilience can be construed as sets of moderating or mediating factors that negate or in some way buffer against the adverse sequelae of conventional risk factors, thereby explaining differential effects and susceptibility (Rutter, 2008). While beyond the scope of this chapter, resilience is a concept of increasing interest in current mental health research.

Moderators

Moderators of mental ill health may be found throughout the life course, including from before birth. Below are some examples of moderators for mental ill health. The list is by no means exhaustive because the nature of mental ill health means that each mental disorder may have a complex etiology and several interacting contributing factors. Instead, the included examples are meant to help conceptualize some of the different ways in which biological, social, and environmental factors may influence mental health. The section will follow a life course approach, starting with risk factors early in life and progressing through to factors in adolescence and adulthood.

Early developmental

Pre- and perinatal risk factors

Risk for mental ill health begins even before birth, when maternal mental and physical health have an important influence on outcomes in the child. Poor maternal mental health before, during, and immediately after pregnancy may increase the chances of the offspring developing a lifetime mental disorder (Verbeek et al., 2012). Depression in the pregnancy period is one of the best-studied examples of how maternal health can affect offspring mental health. Children whose mothers have had depression during pregnancy are more likely to have negative affect as toddlers, while children whose mothers have experienced postpartum depression are more likely to have depression or other internalizing disorders as adolescents (Verbeek et al., 2012; Murray et al., 2011). Similarly, children of mothers with antenatal anxiety are more likely than their peers to have behavioral and emotional problems as toddlers (O'Connor et al., 2002).

Maternal physical health status during pregnancy can also affect offspring mental ill health, and several physical conditions are known to influence development and psychiatric outcomes of offspring. For example, exposure to prenatal infection or to obstetric complications are established risk factors for later development of schizophrenia and other psychotic outcomes in the offspring (Mäki et al., 2005; Buka et al., 2001; Khandaker et al., 2013; Cannon et al., 2002). Maternal nutritional status is also important; while antenatal famine has been shown to contribute to increased risk of schizophrenia (Mäki et al., 2005), high intake of unhealthy food during pregnancy is associated with increased risk for externalizing problems in the child (Jacka et al., 2013). Finally, maternal smoking, alcohol, and drug use during pregnancy lead to increased risk of attention deficit hyperactivity disorder (ADHD), externalizing disorders, and antisocial behavior (Wakschlag et al., 2002; Huizink and Mulder, 2006; Thapar et al., 2003; Button et al., 2007).

Gender

There are important differences in the prevalence of mental health disorders between men and women that make gender an important risk factor for the development of different mental disorders. In general, women are much more likely than men to experience common mental disorders (i.e., depressive and anxiety disorders). In England, one in five women has a common mental disorder, compared with one in eight men (McManus et al., 2016). Women are also more likely to develop eating disorders, and are about three times as likely as men to have diagnoses of anorexia nervosa or bulimia nervosa (Hudson et al., 2007).

Conversely, there is a set of mental disorders for which being male serves as a risk factor. While women are more prone to internalizing disorders, men are more likely to have antisocial behaviors (World Health Organization, 2002). Men are also more likely to engage in substance use than women across all age groups and are more than twice as likely as women to be dependent on illicit drugs (4.3% of men vs. 1.9% women in England demonstrate signs of dependence) (McManus et al., 2016). Similarly, the rates of hazardous drinking and alcohol dependence are nearly twice as high in men as in women, with over a quarter of men in England drinking at or above "hazardous" levels (McManus et al., 2016). Finally, male gender is a risk factor for psychotic disorders (Jongsma et al., 2017) and completed suicides (Mergl et al., 2015; Värnik, 2012).

Ethnicity and migration

Ethnicity may also influence lifetime mental health outcomes. Ethnic groups are complex constructs, but may broadly be viewed as self-identified groups with common ancestry or social background, culture, traditions, language, and history (Office for National Statistics, 2011; Senior and Bhopal, 1994). One of the clearest examples of the effect of ethnicity on mental health is visible ethnic minority status in psychotic disorders. For first-episode psychosis (FEP), a recent multinational study of five European countries and one South American country estimated that the incidence rate is 1.59 times higher in ethnic minority populations than ethnic majorities (Jongsma et al., 2017). These differences in incidence have been linked to immigration, and exposure to adversity in all stages of migration has been suggested as a contributor to the increased risk for psychotic disorders (Fig. 2.3) (Kirkbride et al., 2017; Bourque et al., 2011). Immigrating at a young age is another risk factor for psychotic disorder; in the United Kingdom, those who immigrated during childhood (5–12 years of age) have more than double the likelihood of developing FEP than their UK-born White British peers (Kirkbride et al., 2017). Evidence is mixed on whether first- or subsequent generation immigrants have higher risks of FEP, and risk varies by ethnic group (Kirkbride et al., 2017; Bourque et al., 2011).

The risks associated with different ethnicities, however, is not always so clear. For example, there is little evidence at the population level to suggest variation in prevalence of common mental disorders by ethnicity (Weich et al., 2004; Bhui et al., 2005). The EMPIRIC study of common mental disorders in England found significant differences only among Irish and Pakistani adult men, who had higher rates of

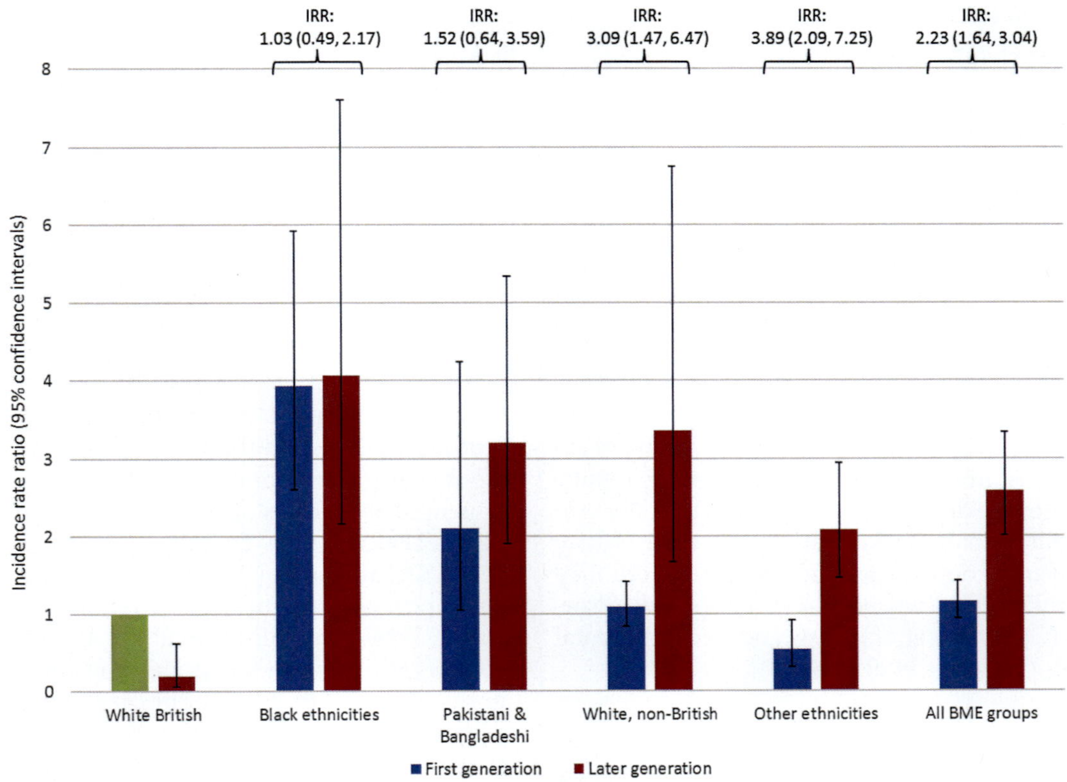

FIGURE 2.3 Incidence rate ratios for first-episode psychosis (FEP) by generation status and broad ethnic group. Overall there was evidence that FEP risk by generation status varied by ethnic group (LRT-$\chi^2$$P$-value for interaction between ethnicity and generation status on three degrees of freedom: $\chi^2 = 9.2$; $P = 0.03$). Thus, compared with the UK-born white British group, rates were raised to a similar extent for first- and later-generation black and Pakistani and Bangladeshi groups, with no statistically significant differences in risk by generation. For non-British white and other ethnic groups, excess rates were confined to later-generation groups. Foreign-born white British groups and first-generation "other" ethnic groups were at significantly reduced psychosis risk compared with the UK-born white British group. All incidence rate ratios are adjusted for age and sex. The white British (UK-born) reference population is shown in green, with the white British (born overseas) shown in red. *BME*, black and minority ethnic. *Reproduced with permission from Kirkbride, J.B., Hameed, Y., Ioannidis, K., Ankireddypalli, G., Crane, C.M., Nasir, M., et al., 2017. Ethnic Minority Status, Age-at-Immigration and Psychosis Risk in Rural Environments: Evidence From the SEPEA Study. Schizophr. Bull sbx010.*

common mental disorders than their White counterparts (Weich et al., 2004). Compared with White women, higher rates of disorder were found in Pakistani women and older Indian women and lower rates were found in Bangladeshi women (Weich et al., 2004). The researchers did, however, find important differences in mental disorder when they examined the effect of unfair treatment and racial insults (Bhui et al., 2005). Ethnic minority individuals who reported these forms of discrimination were twice as likely to have a common mental disorder than their White counterparts (Bhui et al., 2005).

Childhood and adolescence

Adverse childhood experiences

There are several events that may occur in childhood that can serve as risk factors for immediate or future psychopathology. *Adverse childhood experiences* comprise one group of early life events that can increase risk for mental disorder (St. Clair et al., 2015). These experiences include child maltreatment (emotional, physical, and sexual abuse), neglect, parental separation, domestic violence, household mental illness, household alcohol abuse or drug use, and incarceration of a family member (Bellis et al., 2014). In young adulthood, these adverse childhood experiences are associated with higher depressive symptoms, drug use, and antisocial behavior (Schilling et al., 2007). The effects of these early life events continue into adulthood, where they are linked to increased risk of depressive disorders (Chapman et al., 2004) and general mental health (Edwards et al., 2003).

Substance use

Illicit drug use can also serve as a risk factor for lifetime mental disorder. Perhaps the best-studied example of this relationship is that of substance use and psychotic outcomes. In populations with FEP, for example, lifetime substance use has been estimated at twice the rate of the general population (Barnett et al., 2007). Cannabis use in particular has been well studied within the realm of psychotic outcomes. Cohort studies have shown an increased risk for psychotic outcomes (including psychotic disorders or subthreshold psychotic symptoms) of 41% (95% CI: 20%–65%) among those who use cannabis compared with those who do not (Moore et al., 2007). Furthermore, these studies have shown a dose–response relationship, meaning that there is a positive linear relationship between amount of cannabis used and risk of psychotic outcome (Moore et al., 2007).

Lifetime

Social determinants of health

Another important, overarching category of risk concerns the *social determinants of health*, which refer to higher-level social and structural factors influencing health outcomes. While recent decades have seen a resurgence of emphasis on the social determinants of health, particularly in terms of the origins of major inequalities in health between communities and large groups, the idea of social and environmental factors influencing health outcomes is hardly new. Beginning with the campaigns to improve sanitation in the 19th century, the field of public health recognized that health could be influenced by factors acting above the level of the individual. The founders of the World Health Organization (WHO) clearly recognized the importance of the social determinants of health. They maintained that the WHO must work to address the social and environmental causes of ill health. In fact, this very focus was included within the constitution of WHO, whose definition of health includes "physical, mental, and social well-being" (Irwin and Scali, 2007).

For many years following the founding of the WHO, the focus of public health campaigns was centered around technological and biomedical advances and tended to be disease-specific. In recent decades, however, there has been a renewed importance given to the social determinants of health, and in 2014, the WHO released a report on the social determinants of mental health. The report emphasized that there are factors acting above the individual level that influence mental health and offered recommendations for interventions addressing these risk factors that are universal and proportionate to need (Fig. 2.4) (World Health Organization, 2014).

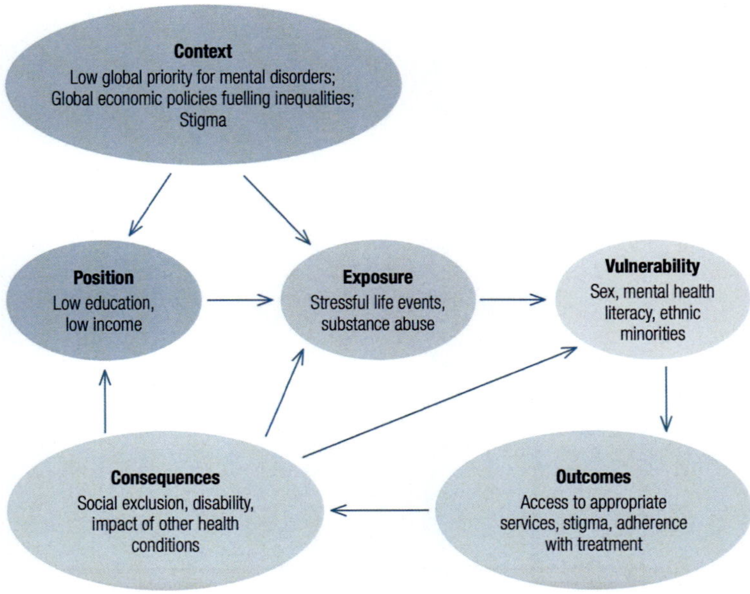

FIGURE 2.4 The vicious cycle of social determinants and mental disorders. *Reproduced with permission from Patel, V., Lund, C., Hatherill, S., Plagerson, S., Corrigall, J., Funk, M., et al., 2010. Mental Disorders: Equity and Social Determinants. Equity, Social Determinants and Public Health Programmes. vol. 115.*

Income inequality and socioeconomic disadvantage

Two important social determinants of mental health are *income inequality* (at a country level) and *socioeconomic disadvantage* (at an individual level) (Marmot and Bell, 2012; Allen et al., 2014). Within high-income countries, there is great variation in the burden of mental ill health, and high-income countries with greater income inequality suffer from increased prevalence of mental disorders (Fig. 2.5) (Pickett and Wilkinson, 2010). It has been proposed that this increased prevalence results from the lower quality of social relationships and the greater scale of status differentiation in these countries, each of which adds to the risk of developing a mental disorder (Pickett and Wilkinson, 2010). Thus, a potent risk factor may appear quite subtle, which is itself not an absolute concept but a relative one.

At an individual level, people with low socioeconomic status living in higher socioeconomic settings carry a disproportionate share of the burden of mental ill health in society (Allen et al., 2014). This difference begins early in the life course, whereby children 3–5 years old from low-income households have greater emotional and behavioral difficulties than their wealthier peers (Kelly et al., 2011). At the ages of 10–15 years, the prevalence of depressed mood and anxiety in young people of low socioeconomic status is 2.5 times higher than in young people of high socioeconomic status (Lemstra et al., 2008). In adulthood, there is a gradient effect of socioeconomic disadvantage, whereby those with greater disadvantage are at proportionally increased risk for developing a common mental disorder (McManus, 2009).

Mediators

Mediators of mental ill health are the mechanisms by which experiences with the social and physical environment "get under the skin" and lead to psychopathology.

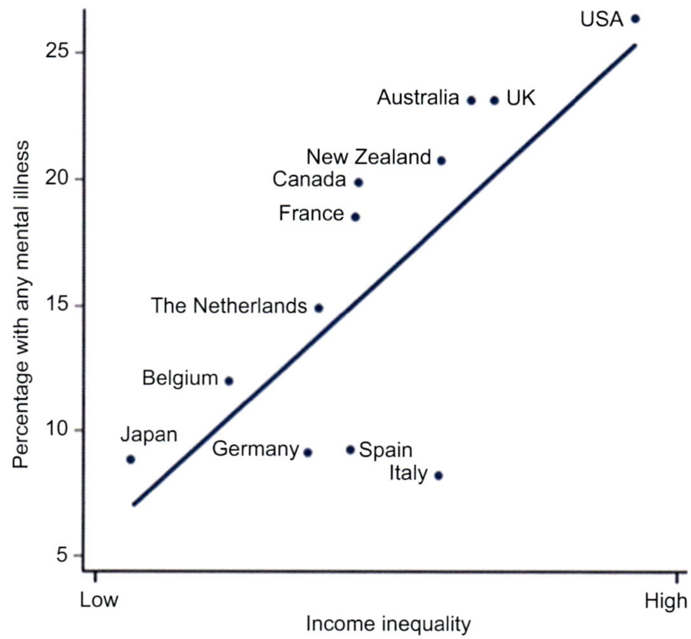

FIGURE 2.5 Income inequality and prevalence of mental disorders in twelve high income countries. *Reproduced with permission from Pickett, K.E., Wilkinson, R.G., 2010. Inequality: An Underacknowledged Source of Mental Illness and Distress. RCP.*

The heritability of mental disorders has long been recognized. While early geneticists tended to look for single causative genes, researchers now understand the underlying genetic complexity of mental disorders. As already discussed, disorders are now recognized as being *multifactorial*, meaning that they are caused by a combination of risk factors on multiple levels. Within this combination of risk factors, there may be several genes contributing to increased risk, and indeed most disorders now have several known *susceptibility genes*, or genetic variations that can lead to increased risk for disorder (Hyman, 2000). For example, a 2015 review of candidate genes for five common mental disorders (ADHD, anxiety disorders, bipolar disorder, major depressive disorder, and schizophrenia) reported a total of 134 significantly associated risk variants across these five disorders (Gatt et al., 2015).

Since 2005, *genome-wide association studies* have been used to map genetic risk factors for mental disorder (Fig. 2.6) (Psychiatric GWAS Consortium Coordinating Committee, 2009). By looking for *single-nucleotide polymorphisms* and *copy number variants*, two types of genetic variations, these studies have uncovered several candidate genes for mental disorder (Psychiatric GWAS Consortium Coordinating Committee, 2009). In some cases, risk loci may be shared among multiple mental disorders. A recent genome-wide association study including over 50,000 participants found cross-disorder genetic risk factors shared by autism, ADHD, clinical depression, bipolar disorder, and schizophrenia (Cross-Disorder Group of the Psychiatric Genomics Consortium, 2013).

Schizophrenia is a particularly good example of the role of genetics within mental disorder. Through use of family studies, twin studies

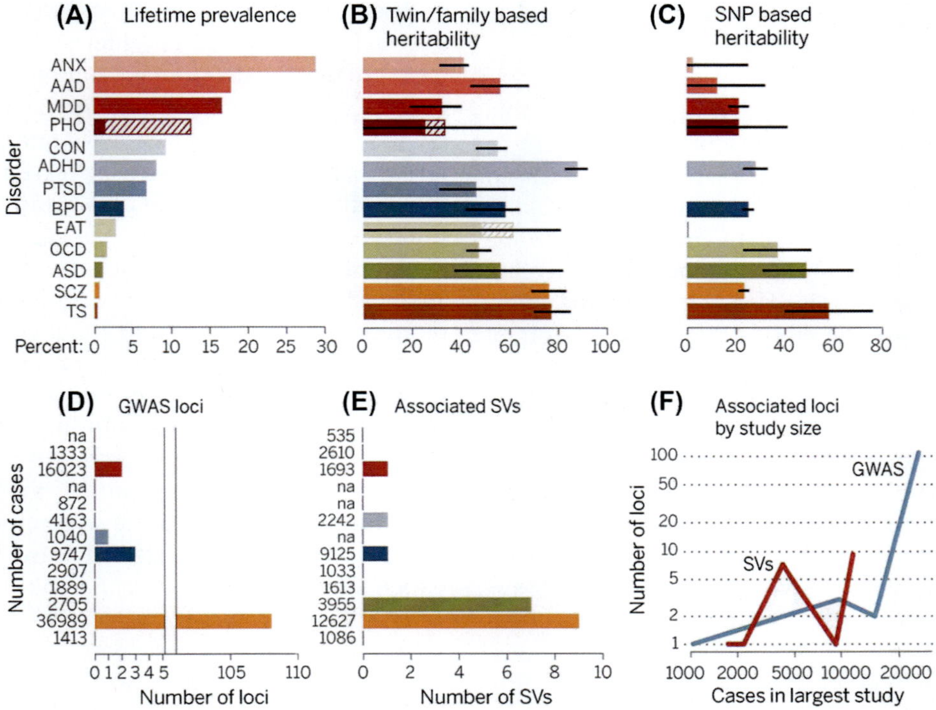

FIGURE 2.6 Summary of genetic analyses performed on 13 psychiatric disorders. (A) Highest lifetime (point for ASD) prevalence in percentages. The discontinuous bar in phobias represents the range in different forms. (B) Heritability estimates; bars, standard error (SE). (C) Single-nucleotide polymorphism–based heritability estimates; bars, SE. (D) Number of genome-wide significant loci. The x-axis is discontinuous because of the large difference of associated loci between disorders. (E) The number of associated structural variants (SVs) that either reach genome-wide significance or have been replicated with $P \leq 0.01$ in another study. (F) The y-axis shows associated genome-wide association study (GWAS) loci (blue) and SVs (red) by the number of cases (x-axis) in the largest study for that disorder. The number of cases in the largest study for GWAS (D) and SV studies (E) is reported next to each disorder. *AAD*, alcohol abuse disorder; *ANX*, any anxiety disorder; *CON*, conduct disorders; *EAT*, eating disorders; *MDD*, major depressive disorder; *PHO*, any phobia; *PTSD*, posttraumatic stress disorder; *TS*, Tourette syndrome. The order of disorders and their color coding are maintained throughout the bar plots. *Reproduced with permission from Geschwind, D.H., Flint, J., 2015. Genetics and genomics of psychiatric disease. Science 349 (6255), 1489–1494.*

(i.e., of monozygotic and dizygotic twins), and adoption studies, researchers discovered the high heritability of schizophrenia within families. In fact, familial risk has been identified as the single strongest risk factor for schizophrenia (Mäki et al., 2005).

Numerous studies also highlight the importance of the interaction between vulnerability genes and different environmental moderators/influences, such as adverse experiences, for the development of mental disorders, such as schizophrenia (Van Os and Kapur, 2009). This interaction may be responsible for changes in stress response that may be reflected in alterations of the hypothalamus–pituitary–adrenal axis. This may initiate a cascade of effects on several systems, affecting immunological and inflammatory responses, which may play an important part in the development of severe psychopathology (Walker et al., 2017).

Identification of mental disorders and screening

Understanding the mediators and moderators of mental disorder is particularly important in terms of identifying people who are at increased risk for poor psychiatric outcomes and linking them with appropriate care and support. While there is no single, "perfect" model for detecting cases of mental disorders, two of the most commonly used methods for identification of disorders in psychiatry are *screening* and *case-finding*. The basic definition of *screening* is "the process of identifying healthy people who may be at increased risk of a disease or condition" (Fig. 2.7) (UK National Screening Committee, 2013). In the realm of physical health, most screening programs look for signs of the disease in its latent phase. Screening for mental disorders, however, presents a unique situation. Unlike with many physical health conditions, there is no "latent" phase of mental ill health, and so the standard definition of screening does not apply. Instead, when we screen for mental disorders, we are effectively looking for fragments of extant disorders in terms of existing impairment and symptomology (Levitt et al., 2007).

The other major type of case identification is *case-finding*, whereby psychiatric measurement tools are used to determine the presence of a disorder. Despite its common use in the field of psychiatry, the definition of case-finding is not altogether clear. Some define case-finding as inherently patient-initiated and add that case-finding, unlike screening, has no guaranteed benefit for the patient (Sackett and Holland, 1975). However, this is not a universally held definition. Even Wilson and Jungner, who were the first to set out clear criteria for screening programs, did not make a clear differentiation between screening and case-finding (Wald and Morris, 1996). Indeed, in their report on the principles of screening programs, they defined case-finding as "that form of screening of which the main object is to detect disease and bring patients to treatment" (Wilson et al., 1968).

This section will focus on the concept of screening for mental disorders. It will provide formal definitions of two different types of screening, explain the mathematical basis of screening, present two different sets of criteria for screening programs, and give two examples of settings for screening programs.

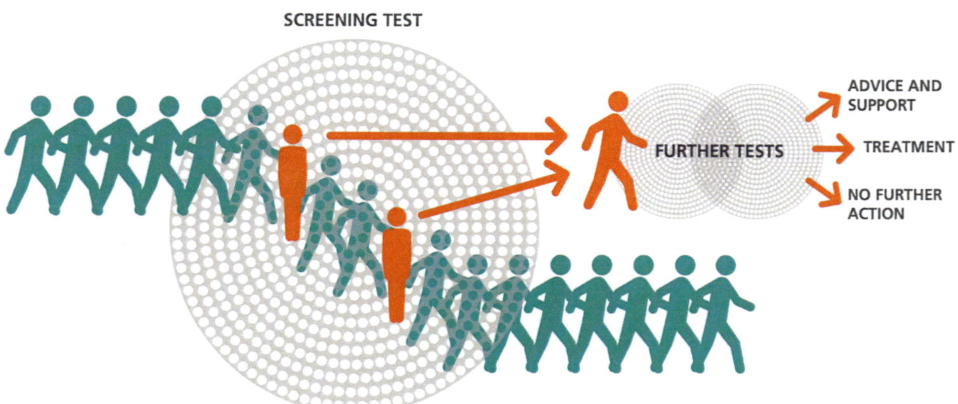

FIGURE 2.7 Screening programs: an overview. *Reproduced from UK National Screening Committee, 2013. NHS Population Screening Explained. Public Health England. Available from: https://www.gov.uk/guidance/nhs-population-screening-explained.*

Definitions—universal and selective

Screening programs may be divided into *universal* and *selective screening*. *Universal screening* refers to programs that screen for mental ill health for everyone in a given population and do not discriminate on the basis of risk factors. For example, if all children were screened for social-emotional difficulties upon entry to primary school, this would be considered a universal screening program.

Selective (or targeted) screening refers to programs that target specific groups known to be at increased risk for mental disorders (O'Doherty et al., 2014). Effectively, selective screening aims to test groups with greater risk of mental health problems so as to increase the positive predictive value (PPV) of the measurement tool and thereby limit the number of false positives identified in the process (i.e., those who are identified as having the condition who do not actually have the condition) (Ruf and Morgan, 2008). Although this method is similar in principle to case-finding, the use of the term *case-finding* may preclude the rigorous evaluation processes expected of screening programs (Wald and Morris, 1996). Conversely, *selective screening* implies that all evaluation processes that apply to universal screening continue to apply. An example of a selective screening process would be screening for depression among pregnant women and new mothers, as women are more vulnerable to depression in the pre-, peri-, and postnatal periods (Bennett et al., 2004).

The screening tool—psychometric properties

In any screening program, there must be a screening measure that is able to accurately distinguish those who have the condition from those who do not. In psychiatry, *psychometric testing* refers to the study of the properties of a screening measure. Fig. 2.8 presents a 2 × 2 table that gives generic psychometric properties of a screening tool. The table may be broken down into true positives, true negatives, false positives, and false negatives, defined as follows:

- *True positives*: people who have the condition and test positive.
- *True negatives*: people who do not have the condition and test negative.
- *False positives*: people who do not have the condition and test positive.
- *False negatives*: people who have the condition and test negative.

Using these categorizations, we can calculate a number of properties related to the screening measure:

- *Sensitivity*, calculated as $\frac{TP}{TP+FN}$, refers to the proportion of people who have the condition who test positive on the screening measure.

Truth (presence of outcome)

		+	-	
Test result	+	True positives (TP)	False positives (FP)	**TP + FP**
	-	False negatives (FN)	True negatives (TN)	**FN + TN**
		TP + FN	**FP + TN**	

FIGURE 2.8 Psychometric properties of a screening tool.

The sensitivity of a screening test refers to its ability to correctly refer to further testing those people with the condition.

- *Specificity*, calculated as $\frac{TN}{FP+TN}$, refers to the proportion of people who do not have the condition who test negative on the screening measure. The specificity of a screening test refers to its ability *not* to refer to further testing people who *do not* have the condition.
- *Positive predictive value (PPV)*, calculated as $\frac{TP}{TP+FP}$, refers to the proportion of those who test positive on the screening measure who actually have the condition.
- *Negative predictive value (NPV)*, calculated as $\frac{TN}{FN+TN}$, refers to the proportion of those who test negative on the screening measure who do not have the condition.

Both PPV and NPV depend on the prevalence of a given condition in the population being screened. In a population with high prevalence, the PPV of screening tools is increased, while NPV is decreased. For this reason, screening tools used in populations with high prevalence of mental disorders (e.g., psychiatric in-patients, people referred by primary care providers to psychiatric services) will appear to have higher PPVs than when used in the general population. (For an illustrative example of psychometric properties, please see "The Gatekeeper and the Wizard: a fairy tale" (Mathers and Hodgkin, 1989).)

Fig. 2.9 provides a visual of screening. In this example, the blue people have the condition of interest and the green people do not. The screening program is represented as a sieve through which the people are separated. A perfect screening measure, as represented by the picture on the right, would be one where all individuals with the condition were identified by the measure (100% sensitivity) and no individuals without the condition were identified (100% specificity). This perfect example would further have a value of 100% for PPV and NPV, as all those who tested positive have the condition and all those who tested negative do not have the condition.

In reality, no screening test is perfect. Taking psychosis as an example, we can examine the psychometric properties of the Community Assessment of Psychic Experiences-Positive

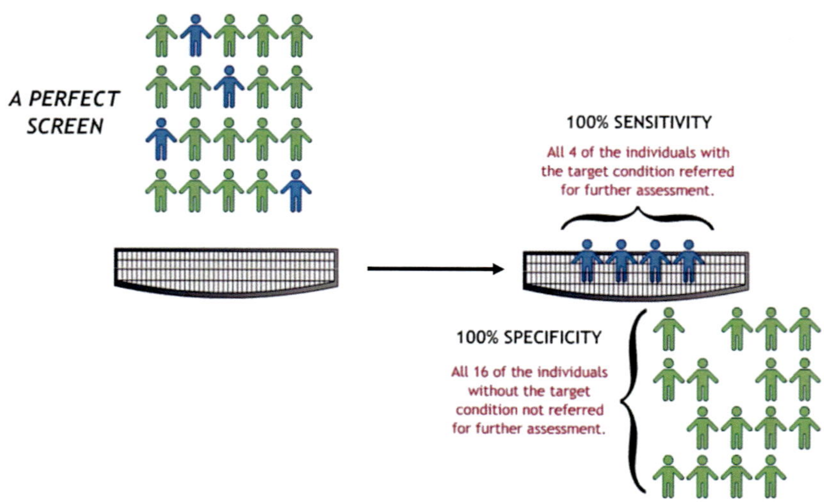

FIGURE 2.9 Screening programs: a perfect screen. *Adapted from UK National Screening Committee, 2013. NHS Population Screening Explained. Public Health England Available from: https://www.gov.uk/guidance/nhs-population-screening-explained.*

Scale (CAPE-P15) (Bukenaite et al., 2017), a scale developed to detect people at ultra high risk (UHR) of psychosis as measured using an accepted, gold standard definition generated by the Comprehensive Assessment of At-Risk Mental States (CAARMS) (Yung et al., 2005). With a cutoff score of 1.47, the scale has a sensitivity of 77%, specificity of 58%, PPV of 66%, and NPV of 71% (Bukenaite et al., 2017). In other words, 77% of those who test positive are actually at UHR, while 58% of those who screen negative are actually negative (i.e. *not* at UHR). Furthermore, 66% of UHR individuals are so identified by the CAPE and 71% of those who are *not at* UHR are so identified. These results indicate that there will be a concerning number of false positives and false negatives.

Furthermore, when used for population screening, screening measures must have known psychometric properties derived from testing among the general population. For many screening measures, psychometric properties have been determined by using the instrument among psychiatric patients. As the prevalence of mental disorder will be higher in these populations, the PPV will be lower in the general population compared with psychiatric patients. Therefore, caution must be applied when generalizing results from these special populations to the wider population, or vice versa, without being cognizant of this fact (Hartmann et al., 2016).

Screening criteria

The ability of a screening test to accurately identify people with the condition is not, however, the only criterion for a screening program to be implemented. In 1968, Wilson and Jungner set out criteria for screening programs that the WHO continues to champion to this day. According to these criteria, a justifiable screening program must meet 10 criteria (Wilson et al., 1968):

1. The condition is an important health problem.
2. There is an accepted treatment available for the condition.
3. There are sufficient facilities for diagnosis and treatment.
4. The condition may be recognized in a latent or early symptomatic stage.
5. There exists a suitable test for the condition.
6. The test for the condition is acceptable to the population.
7. The natural history of the condition is understood.
8. There is a clear cutoff for which screening participants will be treated as patients.
9. The program is economically balanced in terms of cost and benefit.
10. The screening program is a process repeated at regular intervals.

These criteria are still used to this day as standards for judging the effectiveness, feasibility, and acceptability of screening programs. In the field of psychiatry, it is often difficult to fulfill all 10 criteria. As an example of how these criteria may be applied to screening for mental disorders, we can use the example of psychotic disorders:

The condition is an important health problem: Psychotic disorders represent an important health problem globally and in the United Kingdom. For instance, globally, schizophrenia accounts for 7.4% of disability-adjusted life years associated with mental disorders (Whiteford et al., 2013). While the prevalence of psychotic disorders is lower than that of many other mental disorders (Whiteford et al., 2013), they have significant consequences for individuals, health systems, and societies that make them an important health problem.

There is an accepted treatment available for the condition: Pharmacological, psychological, and social interventions are available to treat these disorders. A wealth of evidence supports the use of antipsychotic medications and psychotherapy, mostly cognitive behavioral therapy, to treat these conditions (Leucht et al., 2013; Pilling et al., 2002).

There are sufficient facilities for diagnosis and treatment: Clinical facilities to treat psychotic disorders are available in both clinical and community settings. The diagnosis and treatment of psychotic disorders is a fundamental part of the training received by health care professionals worldwide.

The condition may be recognized in a latent or early symptomatic stage: Psychotic disorders may be recognized in their prodromal state, which comprises a period of decreased functioning before onset of frank disorder (Yung and McGorry, 1996). Early intervention in the prodromal period may serve to alter the course of illness (Yung et al., 2005; Fusar-Poli et al., 2013). However, it is important to note that there is a high rate of false positive results in the identification of patients in the prodromal stages of psychotic disorders (Yung et al., 2005), in part because of their low prevalence and the high sensitivity of currently available screening tools.

There exists a suitable test for the condition: There is a number of agreed criteria in current international diagnostic classifications, i.e., DSM-5 (American Psychiatric Association, 2013) and ICD-10 (World Health Organization, 1992) to diagnose psychotic disorders. Furthermore, clinicians have access to validated categorical and dimensional tools to diagnose and test the severity and evolution of these conditions over time, such as the Positive and Negative Syndrome Scale (PANSS) (Kay et al., 1987).

The test for the condition is acceptable to the population: Although population acceptability of screening for these conditions is not fully known, tools currently used to determine risk of developing psychotic disorders, such as the Comprehensive Assessment of At-Risk Mental States (CAARMS) (Yung et al., 2005) or the Community Assessment of Psychic Experiences (CAPE) (Stefanis et al., 2002), have been widely used not only in clinical but also in population settings, which suggests high acceptability and tolerability (Bukenaite et al., 2017; Perez et al., 2017; Stochl et al., 2015).

The natural history of the condition is understood: Psychotic disorders generally follow a relatively consistent natural history. For instance, the longitudinal course of schizophrenia is usually characterized by different stages: premorbid, prodromal, deterioration, and residual (Lieberman et al., 2001). FEP patients usually respond to treatment and may achieve symptom remission, whereas recurrent episodes tend to lead to more lasting cognitive and functional deterioration (Lieberman et al., 2001). It has also been demonstrated that the longer psychotic symptoms remain untreated, the poorer the outcomes (Marshall et al., 2005).

There is a clear cutoff for which screening participants will be treated as patients: As stated before, there are a number of diagnostic manuals and tools that set the number of criteria and thresholds to be met to determine caseness and, therefore, establish suitability for treatment.

The program is economically balanced in terms of cost and benefit: In England, the most recent estimates place the societal costs of schizophrenia at £11.8 billion per year, with £7.2 billion falling onto the public sector (Andrew et al., 2012). Specialized early psychosis programs are highly cost-effective compared to standard public mental health services (Mihalopoulos et al., 2009). Early detection services have the potential to save the UK National Health Service (NHS) £18.4 million per year (National Institute for Clinical Excellence, 2014).

The screening program is a process repeated at regular intervals: There is no reason to suggest that a screening program for schizophrenia and other psychotic disorders could not be repeatable, although it would be necessary to determine how regularly to screen.

In the United Kingdom, the *National Screening Committee* (*NSC*) oversees screening programs. The NSC guidelines, similar in scope to the guidelines provided by Wilson and Jungner in 1968, are presented in Box 2.1. They are divided into four main categories: the condition, the test, the intervention, and the screening program, with an additional fifth category for implementation criteria.

At the time of writing, there is no recommendation by the NSC to universally screen for psychiatric disorders (UK National Screening Committee). The NSC has produced reports and recommendations concerning screening programs for alcohol misuse, autism, dementia, depression, development and behavior, and perinatal maternal psychiatric illnesses. For many of the conditions, there is insufficient evidence to fulfill all criteria. The NSC does not currently, for example, recommend screening for depression in adults because of the low PPV of available tests and lack of evidence for improved long-term outcomes due to early intervention in subthreshold depression (UK National Screening Committee). The picture is similar for recommendations on dementia screening, with the NSC citing the large number of false positives (around two for every three positive tests) as a key reason for not recommending screening (Box 2.2) (UK National Screening Committee).

Settings for universal psychiatric screening

In considering the implementation of screening programs, it is important to consider the setting in which the program would take place. Below, we examine two of the most commonly proposed settings for universal and selective screening for mental disorders: schools and primary care settings.

Schools

In recent years, school have been championed as an ideal setting for universal screening programs for mental health difficulties in children (Weist et al., 2007; Humphrey and Wigelsworth, 2016; Williams, 2013). From a practical standpoint, schools reach the vast majority of children and young people in the United Kingdom (Levitt et al., 2007). Importantly, this reach includes children and young people from marginalized populations, who experience mental health problems at higher rates than their peers (Green et al., 2005). Furthermore, given that half of psychiatric disorders begin by the age of 14 (Kessler et al., 2005), the majority of onsets will occur during the schooling years, creating opportunity for early identification and intervention.

There are several examples of screening programs that have been trialed in school settings, including screenings for depression and anxiety, emotional and behavioral problems, eating disorders, substance abuse, and suicide risk (Anderson et al., 2019). The earliest of the programs are designed to screen for difficulties in preschoolers and primary school children, while others screen for risk in secondary school students. In addition to universal screening, the school setting also offers the opportunity for selective screening by targeting only groups of students at high risk for mental health problems. For example, by only screening students who have high truancy rates, poor academic performance, or family-level risk, these screening programs can better predict mental ill health and avoid false positive test results. While school-based screening programs are a good idea in principle, more evidence collected on the effectiveness, feasibility, and acceptability is necessary before their widespread adoption (Soneson et al., In press).

Primary care

Another widely suggested setting for mental health screening is primary care. The detection of mental disorders in this setting makes sense conceptually for a number of reasons. First, as of 2017, nearly 60 million people in the United Kingdom are registered with a general

BOX 2.1

UK NATIONAL SCREENING COMMITTEE CRITERIA FOR SCREENING PROGRAMS

The condition

1. The condition should be an important health problem as judged by its frequency and/or severity. The epidemiology, incidence, prevalence, and natural history of the condition should be understood, including development from latent to declared disease and/or there should be robust evidence about the association between the risk or disease marker and serious or treatable disease.
2. All the cost-effective primary prevention interventions should have been implemented as far as practicable.
3. If the carriers of a mutation are identified, as a result of screening the natural history of people with this status should be understood, including the psychological implications.

The test

4. There should be a simple, safe, precise, and validated screening test.
5. The distribution of test values in the target population should be known and a suitable cutoff level defined and agreed.
6. The test, from sample collection to delivery of results, should be acceptable to the target population.
7. There should be an agreed policy on the further diagnostic investigation of individuals with a positive test result and on the choices available to those individuals.
8. If the test is for a particular mutation or set of genetic variants, the method for their selection and the means through which these will be kept under review in the program should be clearly set out.

The intervention

9. There should be an effective intervention for patients identified through screening, with evidence that intervention at a presymptomatic phase leads to better outcomes for the screened individual compared with usual care. Evidence relating to wider benefits of screening, for example, those relating to family members, should be taken into account where available. However, where there is no prospect of benefit for the individual screened, then the screening program should not be further considered.
10. There should be agreed evidence-based policies covering which individuals should be offered interventions and the appropriate intervention to be offered.

The screening program

11. There should be evidence from high-quality randomized controlled trials that the screening program is effective in reducing mortality or morbidity. Where screening is aimed solely at providing information to allow the person being screened to make an "informed choice" (such as Down's syndrome or cystic fibrosis carrier screening), there must be evidence from high-quality trials that the test accurately measures risk. The information that is provided about the test and its outcome must be of value and readily understood by the individual being screened.
12. There should be evidence that the complete screening program (test, diagnostic procedures, treatment/intervention) is clinically, socially, and ethically acceptable to health professionals and the public.
13. The benefit gained by individuals from the screening program should outweigh any

BOX 2.1 (*cont'd*)

UK NATIONAL SCREENING COMMITTEE CRITERIA FOR SCREENING PROGRAMS

harms, for example, from overdiagnosis, overtreatment, false positives, false reassurance, uncertain findings, and complications.

14. The opportunity cost of the screening program (including testing, diagnosis and treatment, administration, training, and quality assurance) should be economically balanced in relation to expenditure on medical care as a whole (value for money). Assessment against these criteria should have regard to evidence from cost-benefit and/or cost-effectiveness analyses and have regard to the effective use of available resource.

Implementation criteria

15. Clinical management of the condition and patient outcomes should be optimized in all health care providers before participation in a screening program.

16. All other options for managing the condition should have been considered (such as improving treatment or providing other services) to ensure that no more cost-effective intervention could be introduced or current interventions increased within the resources available.

17. There should be a plan for managing and monitoring the screening program and an agreed set of quality assurance standards.

18. Adequate staffing and facilities for testing, diagnosis, treatment, and program management should be available before the commencement of the screening program.

19. Evidence-based information, explaining the purpose and potential consequences of screening, investigation, and preventative intervention or treatment, should be made available to potential participants to assist them in making an informed choice.

20. Public pressure for widening the eligibility criteria for reducing the screening interval and for increasing the sensitivity of the testing process should be anticipated. Decisions about these parameters should be scientifically justifiable to the public.

Reproduced with permission from Public Health England, 2015. Criteria for Appraising the Viability, Effectiveness and Appropriateness of a Screening Programme.

practitioner (GP) (National Health Service, 2017), meaning that GP practices have a wide reach within the population and across all age ranges. Second, a large proportion of people who attend primary care suffer from mental ill health. In the United Kingdom, it is estimated that 18.8% of men and 18.2% of women who attend primary care have at least one type of mental disorder (King et al., 2008). This means that there is a high need for detection and care among those going regularly to their GP. Third, the vast majority of mental disorders are managed within primary care, and so it makes sense that detection of mental disorders should also occur at this level (Bower and Gilbody, 2005). Detection and treatment of mental disorders at a primary care level can furthermore facilitate

TABLE 2.1 Total returns on investment (all years): economic payoffs per £1 expenditure[a].

	NHS	Other public sector	Nonpublic sector	Total
Early identification and intervention as soon as mental disorder arises				
Early intervention for conduct disorder	1.08	1.78	5.03	7.89
Health visitor interventions to reduce postnatal depression	0.4	–	0.4	0.8
Early intervention for depression in diabetes	0.19	0	0.14	0.33
Early intervention for medically unexplained symptoms[b]	1.01	0	0.74	1.75
Early diagnosis and treatment of depression at work	0.51	–	4.52	5.03
Early detection of psychosis	2.62	0.79	6.85	10.27
Early intervention in psychosis	9.68	0.27	8.02	17.97
Screening for alcohol misuse	2.24	0.93	8.57	11.75
Suicide training courses provided to all general practitioners (GPs)	0.08	0.05	43.86	43.99
Suicide prevention through bridge safety measures	1.75	1.31	51.39	54.45
Promotion of mental health and prevention of mental disorder				
Prevention of conduct disorder through social and emotional learning programs	9.42	17.02	57.29	83.73
School-based interventions to reduce bullying	0	0	14.35	14.35
Workplace health promotion programs	–	–	9.69	9.69

[a] *Returns on investment calculated as gross economic payoffs divided by expenditure on the intervention. Depending on the availability of data, these returns may be calculated over different time periods for different interventions. Returns and expenditures discounted back to present values, expressed in 2009/10 prices.*

[b] *For e-learning of GPs, plus cognitive behavioral therapy for all people with somatoform conditions (including subthreshold cases as well as those with full somatoform disorders).*

Reproduced with permission from Knapp, M., McDaid, D., Parsonage, M., 2011. Mental Health Promotion and Mental Illness Prevention: The Economic Case.

greater access to care for those who screen positive (World Health Organization, 2008).

However, although GPs are well placed to identify patients experiencing mental ill health, they often do not do so, leading to underdetection of those with disorders and misidentification of those without disorders (National Collaborating Centre for Mental Health, 2010; Kessler et al., 2002). Systematic, standardized screening programs are one way to improve identification within primary care. There are several available screening tools that can help primary care providers more accurately determine the presence of mental disorder in their patients (Sharp and Lipsky, 2002; Schmitz et al., 1999). While some countries recommend screening for common mental disorders in primary care (Calonge et al., 2009), the National Institute for Health and Care Excellence (NICE) in the United Kingdom currently supports no such recommendation. Instead, NICE supports a selective screening approach, focusing on screening only in high-risk groups, including those with chronic illness, alcohol use problems, or histories of depression (Gilbody et al., 2006). By restricting screening to these groups, the accuracy of screening measures can be much

BOX 2.2

A CAUTIONARY TALE—THE PRIME MINISTER'S CHALLENGE ON DEMENTIA

When David Cameron became Prime Minister of the United Kingdom, he announced that fighting dementia would be a "personal priority" for his time in office (Department of Health, 2015). As part of that promise, Mr. Cameron introduced "The Prime Minister's Challenge on Dementia" to expand upon the Department of Health's National Dementia Strategy (Department of Health, 2009). The strategy aimed to improve awareness, quality care, and research in the field of dementia (Department of Health, 2015). One of the main goals was to improve early identification of dementia. As part of the strategy, the Department of Health made a recommendation that NHS Health Checks for people aged 64–75 include giving information to raise awareness of dementia and related supportive services (Alzheimer's Society, 2013). Others, however, proposed universal screening as a mechanism for early detection of dementia in those aged 75 years and older (Ashford et al., 2006).

As is the case with any national-level screening program, it is necessary to apply the NSC criteria (see Table 2.1) to examine the potential harms and benefits of screening (Public Health England, 2015). When put to the test against these criteria, the concepts behind dementia screening did not hold. In their reports on universal screening for dementia, the NSC cited the poor psychometric properties of dementia screeners as a key reason why it did not currently recommend universal testing in older adults (UK National Screening Committee). The NSC argues that, using screeners currently available, the number of false negative and false positive results would be significant and could lead to undue harm for those who partook in the program.

Beyond the failure to meet the criterion of having a valid and reliable test, dementia screening also misses the mark on several other NSC criteria. Another requirement for implementing universal screening is that the screening test and program as a whole must be acceptable to the population and to health professionals (Public Health England, 2015). A recent systematic review by Martin and colleagues (2015) suggested that dementia screening "may not be acceptable to either the general public or health care professionals." One potential reason for this lack of acceptance is to do with the unavailability of evidence that treatment for dementia improves prognosis (Martin et al., 2015). Indeed, the NSC report lists the current lack of evidence-based treatment as a key reason why screening is not recommended (UK National Screening Committee). Without evidence that the potential risks of screening (e.g., false positive/false negative results) are outweighed by the potential benefits, it is not possible for screening to meet all NSC criteria.

Although reducing the burden of dementia continues to be a top priority in the United Kingdom and elsewhere, screening programs for it are not exempt from meeting the criteria required of all other programs. Before it will be realistic to recommend screening on a universal scale, dementia researchers must prove the effectiveness, acceptability, and feasibility of dementia screening programs.

improved and resources saved within the NHS and social sectors.

Pros and cons of screening for mental disorders

There are many reasons why screening for mental disorder is desirable. First, screening could help identify people suffering from previously unidentified mental disorders, which could lead to reduction in the unmet need for mental health care if there exist sufficient treatment services (Demyttenaere et al., 2004). Second, screening (especially within populations of children and young people) could increase opportunities for early intervention. In turn, early treatment of mental disorders generally results in better outcomes as compared with treatment once difficulties have developed into full psychiatric disorders (Kelly et al., 2007). Third, screening for mental disorders is likely to be economically advantageous. Because investing in early identification and treatment of mental health problems is efficient and cost-effective compared with treating already developed problems (Knapp et al., 2011; World Health Organization, 2003; Beddington et al., 2008), early detection of mental disorders can greatly reduce the social and economic costs of mental ill health over the lifetime (Beddington et al., 2008).

While there are many potential benefits to screening for mental disorders, it is also important to consider potential harms. One of the most important harms of screening is the effects of false positive and false negative results (Public Health England, 2015). Subpar sensitivity and specificity of psychiatric screening tools can mean high rates of identification of people who *do not* have the condition or nonidentification of people who *do* have the condition. For any screening test, there are significant psychological ramifications of incorrect results, and in psychiatric screening, there is the additional concern of stigmatization among those who screen positive (particularly for certain – more severe – disorders). Until tests with better psychometric properties exist, it will be crucial to consider these effects.

Another concern is the availability of effective and acceptable treatment after a positive screening result. For example, strong arguments can be made against autism or dementia screening programs on the basis that there is a substantial lack of evidence that early identification can lead to improved outcomes (Al-Qabandi et al., 2011; Brayne and Davis, 2012). Furthermore, there is a large cost associated with screening programs, which is a serious concern in any system with limited resources as investment in screening for subthreshold mental disorders may divert money from the treatment of those with more serious mental disorders (Thombs et al., 2012). Finally, and perhaps most importantly, the evidence of the ultimate benefit screening programs to patient outcomes is not well-established, and greater certainty would therefore be needed to recommend screening program implementation (Gilbody et al., 2006).

In recommending any psychiatric screening program at a population level, it is important to weigh the potential harms and benefits of the program. The Wilson and Jungner (1968) and NSC criteria (Public Health England, 2015) for screening programs are useful guidelines in determining whether a screening program will ultimately benefit the population.

Prevention

One of the purposes of screening programs is to serve as a preventive measure. *Prevention* in mental health refers to the reduction of risk factors and enhancement of protective factors linked to mental disorders (World Health Organization, 2004). More precisely, prevention of mental disorders refers to "reducing incidence, prevalence, recurrence of mental disorders, the time spent with symptoms, or the risk condition for a mental illness, preventing or delaying recurrences and also decreasing the impact of illness in the affected person, their families and the society" (Haggerty and Mrazek, 1994). In many areas of health, such as cancer and heart disease, there is a shared emphasis on cures for existing disease and prevention strategies to reduce incident cases of disease in the population (Insel and Scolnick, 2006). This stands in contrast with the field of psychiatry, where the majority of available resources are devoted to treating existing mental disorders rather than preventing the onset of new disorders (Insel and Scolnick, 2006). In fact, this dearth of preventive action has been flagged by leading experts in mental health as one of the six main goals among the grand challenges in the field (Collins et al., 2011).

To reduce the global burden of disease attributable to mental ill health, it is important to begin to intervene *before* the onset of disorder occurs. In this section, we shall set forth an economic argument for prevention in psychiatry, define the levels of prevention strategies, and give examples of successful interventions in each of the levels of prevention.

The economic argument for prevention

The financial burden associated with mental ill health is enormous. The World Economic Forum estimates the global economic burden of mental ill health to be US$2.5 trillion (approximately £1.86 trillion) and estimates that approximately two-thirds of these costs are indirect (e.g., impact through lost income due to foregone productivity) (Bloom et al., 2012). Lifetime mental ill health is linked with lower quality of life, decreased economic productivity, and potential destabilization of communities (Belfer, 2008), each of which significantly contributes to the global economic burden of mental disorders. In England alone, the estimated annual financial cost of mental ill health is £105 billion, with £21.3 billion spent on health and social care (Center for Mental Health, 2010). Additional societal costs include educational costs, burdens on the criminal justice system, and increased welfare usage among those with mental disorders (Belfer, 2008).

Despite the proven burden of mental disorders, government spending for the care and prevention of mental disorders is relatively low, especially when compared with physical health conditions with similar contributions to the burden of disease. While mental disorders account for 21.37% of disability-adjusted life years in high-income countries, the average mental health budget accounts for a mere 6.88% of the total health budget in these countries (Saxena et al., 2007). There is a strong economic case to be made for prevention of mental disorder, and indeed several prevention and early intervention measures have shown to be cost-effective (Table 2.1). Investing in prevention is much more cost-effective than intervening after the onset of mental disorder, and preventive measures greatly reduce the social and economic costs of mental ill health over the lifetime (Beddington et al., 2008).

Levels of prevention strategies

There are three main levels of prevention: primary, secondary, and tertiary (Fig. 2.10). *Primary prevention* refers to strategies that target people who are susceptible to mental ill health. This category includes *universal*, *selective*, and

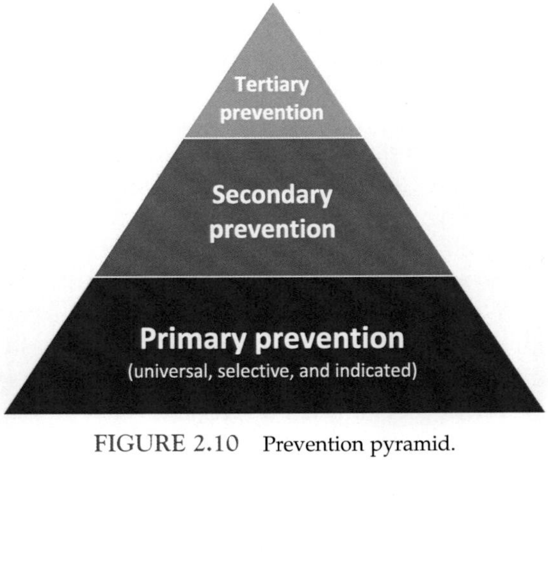

FIGURE 2.10 Prevention pyramid.

indicated prevention strategies. Classification of primary preventive interventions is accomplished by assessing the risk of individuals or subpopulations with regard to developing a mental disorder. Below, we further explore each level of prevention and discuss how each level applies in psychiatry.

Universal prevention strategies target entire populations, regardless of risk status. This population-centered approach is based on the idea that the majority of the burden of mental disorder in the population comes not from the small proportion of people at the highest risk, but rather from the larger proportion of people with moderate or slightly above-average risk. The use of universal preventive interventions, therefore, is an attempt to shift the distribution of risk factors, and ultimately mental disorder, within the population (Rose, 2001).

Some of the best universal approaches to prevention address the macrolevel risk factors for mental ill health. The WHO lists the following strategies as key in promoting positive mental health and preventing mental disorder: improving nutrition, improving housing, improving access to education, reducing economic insecurity, and strengthening community networks (World Health Organization, 2004). Within each of these categories, there are several examples of successful interventions. For example, through the Communities that Care Program, communities across the United Kingdom have improved the healthy psychosocial development of children through identifying and addressing risk factors for mental ill health, including youth delinquency, violence, and substance abuse (Hawkins et al., 2008, 2014). By addressing these social determinants of mental ill health, public health may "shift the curve" for these risk factors, leading to reduced burden.

In addition to community-level approaches, universal approaches may take place on smaller scales including in schools and workplaces. Universal preventive approaches in schools have shown success in reducing emotional and behavioral problems, depression, conduct problems, stress, and substance use in a range of age groups (Rones and Hoagwood, 2000; Durlak et al., 2011). Similarly, interventions that tackle mental health in the workplace have great potential for reaching the vast majority of adults in most high income countries. Interventions to improve mental health at work have aimed to reduce stress, promote better coping, and increase job satisfaction (Czabała et al., 2011). Workplace-based interventions can have benefits for both employees and employers by limiting negative effects of mental ill health, including absenteeism and "presenteeism" (attendance but reduced productivity at work) (Knapp et al., 2011).

Selective prevention strategies target individuals or subpopulations who have higher risk than the general population for onset of mental disorder. This risk may be assessed using a biopsychosocial model through the evaluation of biological, psychological, or social risk factors for mental ill health. Many selective prevention programs target children who have certain risk factors for poor mental health outcomes due to individual (e.g., ethnicity, cognitive functioning), family (e.g., parent socioeconomic status, history of mental disorder, family functioning), and social (e.g., neighborhood level factors) characteristics (Merikangas et al., 2009). Intervening early in the development process may serve to interrupt some of the pathways that lead from risk factors to mental disorder.

A notable example of an intervention that targets mental ill health through selective prevention is the *Incredible Years* program (Webster-Stratton and Reid, 2003; Webster-Stratton et al., 2008; Webster-Stratton, 2001). *Incredible Years* is a training program for parents and teachers meant to encourage social competence in children through improved parenting skills, parent–teacher relationships, and teacher classroom managerial skills. The program has been used as a selective prevention measure for children involved in Head Start programs in the United States. Head Start students come primarily from low-income families or else face other significant disadvantage that puts them at risk for emotional and behavioral problems. The intervention has consistently shown success in reducing conduct problems at home and in school for children whose parents and teachers participated in the program (Webster-Stratton et al., 2001; Reid et al., 2004).

Indicated prevention refers to interventions designed for high-risk populations who are already identified as having symptomatology of mental disorder, but whose symptoms are subthreshold for diagnosis. This type of prevention represents the highest tier of the three types of primary prevention strategies.

Indicated prevention strategies are particularly important in terms of reducing the incidence of psychosis. Patients who are at increased risk for developing a psychotic disorder may be identified through use of the CAARMS, which classifies patients as at-risk for psychosis if they have (1) attenuated psychotic symptoms, (2) brief limited intermittent psychotic symptoms (BLIPS), or (3) family history of psychosis along with decreased functioning (Yung et al., 2005). Once a person is identified as at-risk for psychotic disorder, there are preventive treatments that may reduce psychotic symptoms (Addington et al., 2011; Shi et al., 2017) and the likelihood of transitioning to frank psychosis (Bechdolf et al., 2012; Morrison et al., 2004). In the United Kingdom, the NICE guidelines for preventing onset of psychosis recommend cognitive behavioral therapy along with interventions for anxiety disorders, depression, emerging personality disorders, or substance misuse as needed (National Institute for Clinical Excellence, 2014). There is some evidence that psychological therapies for patients at UHR for psychosis are cost-saving due to prevention of First Episode Psychosis (FEP) (Ising et al., 2015; Phillips et al., 2009).

Secondary prevention in psychiatry refers to reducing the burden of mental ill health in the population through early detection and intervention of diagnosable mental disorders. Single programs may focus on using both of the strategies together (i.e., programs that include both early detection and treatment) or individually.

Suicide prevention is an important example of secondary prevention in psychiatry. With around 800,000 people dying by suicide every year worldwide (Prince et al., 2007), it is clear why suicide prevention strategies are a global priority. Many countries, including the United Kingdom, have developed complex strategies aimed at reducing the burden of suicide in society (HM Government/Department of Health, 2012). The United Kingdom's suicide prevention strategy focuses on six key areas: reducing suicide risk; improving mental health; reducing access to means of suicide; supporting those affected by suicide; providing guidance to media portrayal of suicide; and supporting research, data collection, and monitoring (HM Government/Department of Health, 2012). Since the publication of the strategy, the number of people dying by suicide has decreased, although only slightly (Office for National Statistics, 2017), signaling a need for the continued development of prevention strategies.

In practice, secondary prevention of suicide typically begins with the identification of suicide risk. Due to the high burden of suicide among young people (Prince et al., 2007), schools serve as an important setting for suicide screening. There are several screening programs in schools in particular that aim to detect risk for suicide by looking at distress, suicidal ideation, substance use, and self-harming behaviors (Mann et al., 2005). Some of these screening programs have been successful in increasing service use through providing referrals to further care for those who screen positive (Gould et al., 2009). Early intervention for those who are at increased risk for suicide is another form of secondary prevention. These programs and treatments target the risk factors for suicide early on so as to prevent them from causing additional distress. There are several examples of successful interventions for people at risk of suicide, including psychotherapy and pharmacotherapy interventions (Mann et al., 2005). Combined focus on detection and early intervention is a key secondary preventive strategy that can reduce the global burden of disease due to suicide and related behaviors.

Early intervention in FEP is another prominent example of secondary prevention in psychiatry. The amount of time between a first episode of psychosis and the beginning of effective treatment is known as the duration of untreated psychosis (DUP). Early intervention is particularly important in FEP because longer DUP is associated with poorer outcomes in terms of psychotic symptoms, depression, anxiety, and functioning (Harrigan et al., 2003; Marshall et al., 2005; Addington et al., 2004). One barrier in reducing DUP is identifying people with FEP. Lowering referral thresholds for psychosis services, increasing community awareness of psychosis, and improving help-seeking behaviors of patients and their families may all enhance early detection of untreated FEP (Johannessen et al., 2001). In terms of treatment, NICE recommendations for treatment of FEP include pharmacological, psychological, social, occupational, and educational interventions (National Institute for Health and Care Excellence, 2014). Early detection and intervention in psychosis is highly cost-effective (Knapp et al., 2011); early intervention services in the United Kingdom are estimated to save the NHS £5,536 per service user in the first year of psychosis and £15,862 over 3 years (Andrew et al., 2012). When societal benefits and increased productivity are taken into account, this 3-year figure rises to £21,512 in savings (Andrew et al., 2012).

Tertiary prevention occupies the highest level of prevention and is concerned with managing ongoing mental illness. Specifically, this level of prevention refers to strategies aimed at limiting disability associated with mental disorder and/or improving rehabilitation. Tertiary prevention can also refer to relapse prevention. As of 2013, mental health and substance use disorders account for nearly 23% of all years lived with disability (Whiteford et al., 2013), making the reduction of disability associated with psychiatric disorders a key strategy for reducing the global burden of mental ill health.

Tertiary prevention has a large role to play within the realm of preventing relapses in people who abuse substances. Among those diagnosed with substance use disorder, the relapse rate for 12 months after treatment completion is estimated at 70%–80% (Hunt et al., 1971; Bradizza et al., 2006). In reducing this relapse rate, the prevalence of substance abuse would decrease accordingly. One relatively new intervention that has shown promise for preventing relapse in substance use is Mindfulness-Based Relapse Prevention (Bowen et al., 2009; Witkiewitz et al., 2005). This 8-week intervention relies on a cognitive behavioral therapy model to teach mindfulness-based techniques that can be used to address negative emotions and cravings within the context of relapse (Witkiewitz et al., 2005). In a trial of this intervention, rates of substance use were significantly lower 4 months postintervention in the mindfulness group versus the treatment as usual group.

Tertiary preventive strategies may further be used in reducing relapse rates in psychotic disorders. Antipsychotic medication leads to a significant reduction in 1-year relapse rates as evidenced by relapse rates of 27% and 64% in patients on antipsychotics and placebo, respectively (risk ratio of 0.40 (95% CI 0.33–0.49)) (Leucht et al., 2012). For FEP patients, there is some evidence that tailored psychosocial interventions may also be effective in preventing second episodes of psychosis (Alvarez-Jiménez et al., 2009). However, the evidence for a beneficial effect of individual or family cognitive-based relapse prevention therapy is not clear (Alvarez-Jiménez et al., 2009; Lynch et al., 2010; Jones et al., 2012).

Conclusion

In this chapter, we have covered principles of risk, screening, and prevention in psychiatry. To understand the epidemiology of mental disorders and provision for care and support, it is necessary to have an understanding of risk. There are many ways to express risk, each with its own usefulness and applicability within certain situations. Risk in turn plays an important role in psychiatric screening, where it relates to the psychometric properties of screening tools. In addition to having valid and reliable tools, screening programs must be effective, acceptable, and feasible to be recommended for implementation. Ultimately, the practical value of a screening or diagnostic test will depend on the prevalence of the condition in the population tested; rare conditions are very difficult to predict.

Mental health screening programs are a good example of secondary prevention and can interact with the other tiers of preventive interventions to decrease the burden of mental disorders in the population. In the coming chapters, the concepts of risk, screening, and prevention will be keys to understanding the landscape of psychosis in society.

References

Addington, J., Van Mastrigt, S., Addington, D., 2004. Duration of untreated psychosis: impact on 2-year outcome. Psychol. Med. 34, 277–284.

Addington, J., Epstein, I., Liu, L., French, P., Boydell, K.M., Zipursky, R.B., 2011. A randomized controlled trial of cognitive behavioral therapy for individuals at clinical high risk of psychosis. Schizophr. Res. 125, 54–61.

Al-Qabandi, M., Gorter, J.W., Rosenbaum, P., 2011. Early autism detection: are we ready for routine screening? Pediatrics 128, e211–e217.

Allen, J., Balfour, R., Bell, R., Marmot, M., 2014. Social determinants of mental health. Int. Rev. Psychiatry 26, 392–407.

Alvarez-Jiménez, M., Parker, A.G., Hetrick, S.E., McGorry, P.D., Gleeson, J.F., 2009. Preventing the second episode: a systematic review and meta-analysis of psychosocial and pharmacological trials in first-episode psychosis. Schizophr. Bull. 37, 619–630.

Alzheimer's Society, 2013. The Prime Minister's Challenge on Dementia: Delivering Major Improvements in Dementia Care and Research by 2015. Annual report of progress 2013.

American Psychiatric Association, 2013. Diagnostic and Statistical Manual of Mental Disorders (DSM-5®). American Psychiatric Pub.

Anderson, J.K., Ford, T., Soneson, E., Coon, J.T., Humphrey, A., Rogers, M., Moore, D., Jones, P.B., Clarke, E., Howarth, E., 2019. A systematic review of effectiveness and cost-effectiveness of school-based identification of children and young people at risk of, or currently experiencing mental health difficulties. Psychol. Med. 49 (1), 9–19.

Andrew, A., Knapp, M., Mccrone, P.R., Parsonage, M., Trachtenberg, M., 2012. Effective Interventions in Schizophrenia: The Economic Case. London School of Economics and Political Science, London.

Ashford, J.W., Borson, S., O'Hara, R., Dash, P., Frank, L., Robert, P., Shankle, W.R., Tierney, M.C., Brodaty, H., Schmitt, F.A., 2006. Should older adults be screened for dementia? Alzheimer's Dement. 2, 76–85.

Barnett, J.H., Werners, U., Secher, S.M., Hill, K.E., Brazil, R., Masson, K.I.M., Pernet, D.E., Kirkbride, J.B., Murray, G.K., Bullmore, E.T., 2007. Substance use in a population-based clinic sample of people with first-episode psychosis. Br. J. Psychiatry 190, 515–520.

Bechdolf, A., Wagner, M., Ruhrmann, S., Harrigan, S., Putzfeld, V., Pukrop, R., Brockhaus-Dumke, A., Berning, J., Janssen, B., Decker, P., Bottlender, R., Maurer, K., Möller, H.J., Gaebel, W., Häfner, H., Maier, W., Klosterkötter, J., 2012. Preventing progression to first-episode psychosis in early initial prodromal states. Br. J. Psychiatry 200, 22–29.

Beddington, J., Cooper, C.L., Field, J., Goswami, U., Huppert, F.A., Jenkins, R., Jones, H.S., Kirkwood, T.B.L., Sahakian, B.J., Thomas, S.M., 2008. The mental wealth of nations. Nature 455, 1057.

Belfer, M.L., 2008. Child and adolescent mental disorders: the magnitude of the problem across the globe. J. Child Psychol. Psychiatry 49, 226–236.

Bellis, M.A., Hughes, K., Leckenby, N., Perkins, C., Lowey, H., 2014. National household survey of adverse childhood experiences and their relationship with resilience to health-harming behaviors in England. BMC Med. 12, 72.

Bennett, H.A., Einarson, A., Taddio, A., Koren, G., Einarson, T.R., 2004. Prevalence of depression during pregnancy: systematic review. Obstet. Gynecol. 103, 698–709.

Bhui, K., Stansfeld, S., Mckenzie, K., Karlsen, S., Nazroo, J., Weich, S., 2005. Racial/ethnic discrimination and common mental disorders among workers: findings from the EMPIRIC Study of Ethnic Minority Groups in the United Kingdom. Am. J. Public Health 95, 496–501.

Bloom, D., Cafiero, E., Jané-Llopis, E., Abrahams-Gessel, S., Bloom, L., Fathima, S., Feigl, A., Gaziano, T., Hamandi, A., Mowafi, M., 2012. The global economic burden of noncommunicable diseases. In: Program on the Global Demography of Aging.

Borson, S., Frank, L., Bayley, P.J., Boustani, M., Dean, M., Lin, P.-J., Mccarten, J.R., Morris, J.C., Salmon, D.P., Schmitt, F.A., 2013. Improving dementia care: the role of screening and detection of cognitive impairment. Alzheimer's Dement. 9, 151–159.

Bourque, F., van der Ven, E., Malla, A., 2011. A meta-analysis of the risk for psychotic disorders among first-and second-generation immigrants. Psychol. Med. 41, 897–910.

Bowen, S., Chawla, N., Collins, S.E., Witkiewitz, K., Hsu, S., Grow, J., Clifasefi, S., Garner, M., Douglass, A., Larimer, M.E., 2009. Mindfulness-based relapse prevention for substance use disorders: a pilot efficacy trial. Subst. Abus. 30, 295–305.

Bower, P., Gilbody, S., 2005. Managing common mental health disorders in primary care: conceptual models and evidence base. BMJ 330, 839–842.

Bradizza, C.M., Stasiewicz, P.R., Paas, N.D., 2006. Relapse to alcohol and drug use among individuals diagnosed with co-occurring mental health and substance use disorders: a review. Clin. Psychol. Rev. 26, 162–178.

Brayne, C., Davis, D., 2012. Making Alzheimer's and dementia research fit for populations. Lancet 380, 1441.

Breitborde, N.J.K., Srihari, V.H., Pollard, J.M., Addington, D.N., Woods, S.W., 2010. Mediators and moderators in early intervention research. Early Interv. Psychiatry 4, 143–152.

Buka, S.L., Tsuang, M.T., Torrey, E.F., Klebanoff, M.A., Bernstein, D., Yolken, R.H., 2001. Maternal infections and subsequent psychosis among offspring. Arch. Gen. Psychiatr. 58, 1032–1037.

Bukenaite, A., Stochl, J., Mossaheb, N., Schäfer, M.R., Klier, C.M., Becker, J., Schloegelhofer, M., Papageorgiou, K., Montejo, A.L., Russo, D.A., 2017. Usefulness of the CAPE-P15 for detecting people at ultra-high risk for psychosis: psychometric properties and cut-off values. Schizophr. Res. 189, 69–74.

Button, T.M.M., Maughan, B., McGuffin, P., 2007. The relationship of maternal smoking to psychological problems in the offspring. Early Hum. Dev. 83, 727–732.

Calonge, N., Petitti, D.B., Dewitt, T.G., Dietrich, A.J., Gordis, L., Gregory, K.D., Harris, R., Isham, G., Lefevre, M.L., Leipzig, R.M., 2009. Screening for depression in adults: US preventive services task force recommendation statement. Ann. Intern. Med. 151, 784–792.

Cannon, M., Jones, P.B., Murray, R.M., 2002. Obstetric complications and schizophrenia: historical and meta-analytic review. Am. J. Psychiatry 159, 1080–1092.

Centre for Mental Health, 2010. Economic and Social Costs of Mental Health Problems. Available from: https://www.centreformentalhealth.org.uk/economic-and-social-costs.

Chapman, D.P., Whitfield, C.L., Felitti, V.J., Dube, S.R., Edwards, V.J., Anda, R.F., 2004. Adverse childhood experiences and the risk of depressive disorders in adulthood. J. Affect. Disord. 82, 217–225.

Collins, P.Y., Patel, V., Joestl, S.S., March, D., Insel, T.R., Daar, A.S., Bordin, I.A., Costello, E.J., Durkin, M., Fairburn, C., 2011. Grand challenges in global mental health. Nature 475, 27–30.

Cross-Disorder Group Of The Psychiatric Genomics Consortium, 2013. Identification of risk loci with shared effects on five major psychiatric disorders: a genome-wide analysis. Lancet 381, 1371–1379.

Czabała, C., Charzyńska, K., Mroziak, B., 2011. Psychosocial interventions in workplace mental health promotion: an overview. Health Promot. Int. 26, i70–i84.

Dahlgren, G., Whitehead, M., 1991. Policies and Strategies to Promote Social Equity in Health. Institute for future studies, Stockholm.

Daston, L.J., 1987. The domestication of risk: mathematical probability and insurance 1650–1830. Probabilistic Revolut. 1, 237–260.

Demyttenaere, K., Bruffaerts, R., Posada-Villa, J., Gasquet, I., Kovess, V., Lepine, J., Angermeyer, M.C., Bernert, S.D., DE Girolamo, G., Morosini, P., 2004. Prevalence, severity, and unmet need for treatment of mental disorders in the world health organization world mental health surveys. JAMA 291, 2581–2590.

Department of Health, 2009. Living Well with Dementia: A National Dementia Strategy. Department of Health.

Department of Health, 2015. Prime Minister's Challenge on Dementia 2020. Department of Health London, UK.

Durlak, J.A., Weissberg, R.P., Dymnicki, A.B., Taylor, R.D., Schellinger, K.B., 2011. The impact of enhancing students' social and emotional learning: a meta-analysis of school-based universal interventions. Child Dev. 82, 405–432.

Edwards, V.J., Holden, G.W., Felitti, V.J., Anda, R.F., 2003. Relationship between multiple forms of childhood maltreatment and adult mental health in community respondents: results from the adverse childhood experiences study. Am. J. Psychiatry 160, 1453–1460.

Fusar-Poli, P., Borgwardt, S., Bechdolf, A., Addington, J., Riecher-Rössler, A., Schultze-Lutter, F., Keshavan, M., Wood, S., Ruhrmann, S., Seidman, L.J., 2013. The psychosis high-risk state: a comprehensive state-of-the-art review. JAMA Psychiatry 70, 107–120.

Gatt, J.M., Burton, K.L.O., Williams, L.M., Schofield, P.R., 2015. Specific and common genes implicated across major mental disorders: a review of meta-analysis studies. J. Psychiatr. Res. 60, 1–13.

Geschwind, D.H., Flint, J., 2015. Genetics and genomics of psychiatric disease. Science 349, 1489–1494.

Gilbody, S., Sheldon, T., Wessely, S., 2006. Should we screen for depression? BMJ 332, 1027–1030.

Gould, M.S., Marrocco, F.A., Hoagwood, K., Kleinman, M., Amakawa, L., Altschuler, E., 2009. Service use by at-risk youths after school-based suicide screening. J. Am. Acad. Child Adolesc. Psychiatry 48, 1193–1201.

Green, H., McGinnity, Á., Meltzer, H., Ford, T., Goodman, R., 2005. Mental Health of Children and Young People in Great Britain, 2004. Palgrave Macmillan, Basingstoke.

Haggerty, R.J., Mrazek, P.J., 1994. Reducing Risks for Mental Disorders: Frontiers for Preventive Intervention Research. National Academies Press.

Harrigan, S.M., Mcgorry, P.D., Krstev, H., 2003. Does treatment delay in first-episode psychosis really matter? Psychol. Med. 33, 97–110.

Hartmann, J.A., Yuen, H.P., McGorry, P.D., Yung, A.R., Lin, A., Wood, S.J., Lavoie, S., Nelson, B., 2016. Declining transition rates to psychotic disorder in "ultra-high risk" clients: investigation of a dilution effect. Schizophr Res.

Hawkins, J.D., Brown, E.C., Oesterle, S., Arthur, M.W., Abbott, R.D., Catalano, R.F., 2008. Early effects of Communities that Care on targeted risks and initiation of delinquent behavior and substance use. J. Adolesc. Health 43, 15–22.

Hawkins, J.D., Catalano, R.F., Kuklinski, M.R., 2014. Communities that care. In: Encyclopedia of Criminology and Criminal Justice. Springer.

Hill, A.B., 1965. The environment and disease: association or causation? Proc. R. Soc. Med. 58, 295–300.

HM Government/Department of Health, 2012. Preventing Suicide in England: A Cross-Government Outcomes Strategy to Save Lives.

Hudson, J.I., Hiripi, E., Pope, H.G., Kessler, R.C., 2007. The prevalence and correlates of eating disorders in the National Comorbidity Survey Replication. Biol. Psychiatry 61, 348–358.

Huizink, A.C., Mulder, E.J.H., 2006. Maternal smoking, drinking or cannabis use during pregnancy and neurobehavioral and cognitive functioning in human offspring. Neurosci. Biobehav. Rev. 30, 24–41.

Humphrey, N., Wigelsworth, M., 2016. Making the case for universal school-based mental health screening. Emot. Behav. Difficulties 21, 22–42.

Hunt, W.A., Barnett, L.W., Branch, L.G., 1971. Relapse rates in addiction programs. J. Clin. Psychol. 27, 455–456.

Hyman, S.E., 2000. The genetics of mental illness: implications for practice. Bull. World Health Organ. 78, 455–463.

Insel, T.R., Scolnick, E.M., 2006. Cure therapeutics and strategic prevention: raising the bar for mental health research. Mol. Psychiatry 11, 11–17.

Irwin, A., Scali, E., 2007. Action on the social determinants of health: a historical perspective. Glob. Public Health 2, 235–256.

Ising, H.K., Smit, F., Veling, W., Rietdijk, J., Dragt, S., Klaassen, R.M.C., Savelsberg, N.S.P., Boonstra, N., Nieman, D.H., Linszen, D.H., 2015. Cost-effectiveness of preventing first-episode psychosis in ultra-high-risk subjects: multi-centre randomized controlled trial. Psychol. Med. 45, 1435–1446.

Jacka, F.N., Ystrom, E., Brantsaeter, A.L., Karevold, E., Roth, C., Haugen, M., Meltzer, H.M., Schjolberg, S., Berk, M., 2013. Maternal and early postnatal nutrition and mental health of offspring by age 5 years: a prospective cohort study. J. Am. Acad. Child Adolesc. Psychiatry 52, 1038–1047.

Johannessen, J.O., McGlashan, T.H., Larsen, T.K., Horneland, M., Joa, I., Mardal, S., Kvebæk, R., Friis, S., Melle, I., Opjordsmoen, S., 2001. Early detection strategies for untreated first-episode psychosis. Schizophr. Res. 51, 39–46.

Jones, C., Hacker, D., Cormac, I., Meaden, A., Irving, C.B., 2012. Cognitive Behavioural Therapy versus Other Psychosocial Treatments for Schizophrenia. The Cochrane Library.

Jongsma, H.E., Gayer-Anderson, C., Lasalvia, A., Quattrone, D., Mulè, A., Szöke, A., Selten, J.-P., Turner, C., Arango, C., Tarricone, I., 2017. Treated incidence of psychotic disorders in the multinational EU-GEI study. JAMA Psychiatry 75 (1), 36–46. https://doi.org.10.1001/jamapsychiatry.2017.3554.

Kay, S.R., Fiszbein, A., Opfer, L.A., 1987. The positive and negative syndrome scale (PANSS) for schizophrenia. Schizophr. Bull. 13, 261.

Kelly, C.M., Jorm, A.F., Wright, A., 2007. Improving mental health literacy as a strategy to facilitate early intervention for mental disorders. Med. J. Aust. 187, S26.

Kelly, Y., Sacker, A., Del Bono, E., Francesconi, M., Marmot, M., 2011. What role for the home learning environment and parenting in reducing the socioeconomic gradient in child development? Findings from the Millennium Cohort Study. Arch. Dis. Child. 96 (9), 832–837.

Kessler, D., Bennewith, O., Lewis, G., Sharp, D., 2002. Detection of depression and anxiety in primary care: follow up study. BMJ 325, 1016–1017.

Kessler, R.C., Berglund, P., Demler, O., Jin, R., Merikangas, K.R., Walters, E.E., 2005. Lifetime prevalence and age-of-onset distributions of DSM-IV disorders in the national comorbidity survey replication. Arch. Gen. Psychiatr. 62, 593–602.

Khandaker, G.M., Zimbron, J., Lewis, G., Jones, P.B., 2013. Prenatal maternal infection, neurodevelopment and adult schizophrenia: a systematic review of population-based studies. Psychol. Med. 43, 239–257.

King, M., Nazareth, I., Levy, G., Walker, C., Morris, R., Weich, S., Bellón-Saameño, J.Á., Moreno, B., Svab, I., Rotar, D., 2008. Prevalence of common mental disorders in general practice attendees across Europe. Br. J. Psychiatry 192, 362–367.

Kirkbride, J.B., Hameed, Y., Ioannidis, K., Ankireddypalli, G., Crane, C.M., Nasir, M., Kabacs, N., Metastasio, A., Jenkins, O., Espandian, A., Spyridi, S., Ralevic, D., Siddabattuni, S., Walden, B., Adeoye, A., Perez, J., Jones, P.B., 2017. Ethnic minority status, age-at-immigration and psychosis risk in rural environments: evidence from the SEPEA study. Schizophr. Bull. 43 (6), 1251–1261.

Knapp, M., McDaid, D., Parsonage, M., 2011. Mental Health Promotion and Mental Illness Prevention: The Economic Case.

Kraemer, H.C., Stice, E., Kazdin, A., Offord, D., Kupfer, D., 2001. How do risk factors work together? Mediators, moderators, and independent, overlapping, and proxy risk factors. Am. J. Psychiatry 158, 848–856.

Last, J.M., Abramson, J.H., Freidman, G.D., 2001. A Dictionary of Epidemiology. Oxford University Press, New York.

Lemstra, M., Neudorf, C., D'Arcy, C., Kunst, A., Warren, L.M., Bennett, N.R., 2008. A systematic review of depressed mood and anxiety by SES in youth aged 10–15 years. Can. J. Public Health (Rev. Can. Sante'e Publique) 125–129.

Leucht, S., Tardy, M., Komossa, K., Heres, S., Kissling, W., Salanti, G., Davis, J.M., 2012. Antipsychotic drugs versus placebo for relapse prevention in schizophrenia: a systematic review and meta-analysis. Lancet 379, 2063–2071.

Leucht, S., Cipriani, A., Spineli, L., Mavridis, D., Örey, D., Richter, F., Samara, M., Barbui, C., Engel, R.R., Geddes, J.R., 2013. Comparative efficacy and tolerability of 15 antipsychotic drugs in schizophrenia: a multiple-treatments meta-analysis. Lancet 382, 951–962.

Levitt, J.M., Saka, N., Romanelli, L.H., Hoagwood, K., 2007. Early identification of mental health problems in schools: the status of instrumentation. J. Sch. Psychol. 45, 163–191.

Lieberman, J.A., Perkins, D., Belger, A., Chakos, M., Jarskog, F., Boteva, K., Gilmore, J., 2001. The early stages of schizophrenia: speculations on pathogenesis, pathophysiology, and therapeutic approaches. Biol. Psychiatry 50, 884–897.

Lynch, D., Laws, K.R., Mckenna, P.J., 2010. Cognitive behavioural therapy for major psychiatric disorder: does it really work? A meta-analytical review of well-controlled trials. Psychol. Med. 40, 9–24.

Mäki, P., Veijola, J., Jones, P.B., Murray, G.K., Koponen, H., Tienari, P., Miettunen, J., Tanskanen, P., Wahlberg, K.-E., Koskinen, J., 2005. Predictors of schizophrenia—a review. Br. Med. Bull. 73, 1–15.

Mann, J.J., Apter, A., Bertolote, J., Beautrais, A., Currier, D., Haas, A., Hegerl, U., Lonnqvist, J., Malone, K., Marusic, A., 2005. Suicide prevention strategies: a systematic review. JAMA 294, 2064–2074.

Marmot, M., Bell, R., 2012. Fair society, healthy lives. Public Health 126, S4–S10.

Marshall, M., Lewis, S., Lockwood, A., Drake, R., Jones, P., Croudace, T., 2005. Association between duration of untreated psychosis and outcome in cohorts of first-episode patients: a systematic review. Arch. Gen. Psychiatr. 62, 975–983.

Martin, S., Kelly, S., Khan, A., Cullum, S., Dening, T., Rait, G., Fox, C., Katona, C., Cosco, T., Brayne, C., 2015. Attitudes and preferences towards screening for dementia: a systematic review of the literature. BMC Geriatr. 15, 66.

Mathers, N., Hodgkin, P., 1989. The Gatekeeper and the Wizard: a fairy tale. BMJ 298, 172.

McManus, S.B., 2009. Enhancing Positive Early Childhood Mental Health Outcomes in Young Children. University of Oregon.

McManus, S., Bebbington, P., Jenkins, R., Brugha, T., 2016. Mental Health and Wellbeing in England: Adult Psychiatric Morbidity Survey 2014. NHS Digital, Leeds, pp. 39–40.

Mergl, R., Koburger, N., Heinrichs, K., Székely, A., Tóth, M.D., Coyne, J., Quintão, S., Arensman, E., Coffey, C., Maxwell, M., 2015. What are reasons for the large gender differences in the lethality of suicidal acts? An epidemiological analysis in four European countries. PLoS One 10, e0129062.

Merikangas, K.R., Nakamura, E.F., Kessler, R.C., 2009. Epidemiology of mental disorders in children and adolescents. Dialogues Clin. Neurosci. 11, 7.

Mihalopoulos, C., Harris, M., Henry, L., Harrigan, S., McGorry, P., 2009. Is early intervention in psychosis cost-effective over the long term? Schizophr. Bull. 35, 909–918.

Moore, T.H.M., Zammit, S., Lingford-Hughes, A., Barnes, T.R.E., Jones, P.B., Burke, M., Lewis, G., 2007. Cannabis use and risk of psychotic or affective mental health outcomes: a systematic review. Lancet 370, 319–328.

Morrison, A.P., French, P., Walford, L., Lewis, S.W., Kilcommons, A., Green, J., Parker, S., Bentall, R.P., 2004. Cognitive therapy for the prevention of psychosis in people at ultra-high risk: randomised controlled trial. Br. J. Psychiatry 185, 291–297.

Murray, L., Arteche, A., Fearon, P., Halligan, S., Goodyer, I., Cooper, P., 2011. Maternal postnatal depression and the development of depression in offspring up to 16 years of age. J. Am. Acad. Child Adolesc. Psychiatry 50, 460–470.

National Collaborating Centre for Mental Health, 2010. Depression: the treatment and management of depression in adults, updated edition. British Psychological Society.

National Health Service, December 2017. Patients Registered at a GP Practice. Available from: http://digital.nhs.uk/catalogue/PUB30171.

National Institute For Health And Care Excellence, 2014. Psychosis and Schizophrenia in Adults: Prevention and Management,. NICE Guidelines [CG178].

O'Connor, T.G., Heron, J., Golding, J., Beveridge, M., Glover, V., 2002. Maternal antenatal anxiety and children's behavioural/emotional problems at 4 years. Br. J. Psychiatry 180, 502–508.

Office For National Statistics, 2011. Ethnic Group, National Identity and Religion. London. Available from: www.ons.gov.uk/methodology/classificationsandstandards/measuringequality/ethnicgroupnationalidentityandreligion–ethnic-group.

Office For National Statistics, 2017. Statistical Bulletin: Suicides in Great Britain: 2016 Registrations. Available from: https://www.ons.gov.uk/peoplepopulationandcommunity/birthsdeathsandmarriages/deaths/bulletins/suicidesintheunitedkingdom/2016registration.

O'Doherty, L.J., Taft, A., Hegarty, K., Ramsay, J., Davidson, L.L., Feder, G., 2014. Screening women for intimate partner violence in healthcare settings: abridged Cochrane systematic review and meta-analysis. BMJ 348, g2913.

Patel, V., Rodrigues, M., DeSouza, N., 2002. Gender, poverty, and postnatal depression: a study of mothers in Goa, India. Am. J. Psychiatry 159, 43–47.

Patel, V., Lund, C., Hatherill, S., Plagerson, S., Corrigall, J., Funk, M., Flisher, A.J., 2010. Mental Disorders: Equity and Social Determinants. Equity, Social Determinants and Public Health Programmes, vol. 115.

Perez, J., Russo, D.A., Stochl, J., Clarke, J., Martin, Z., Jassi, C., French, P., Fowler, D., Jones, P.B., 2017. Common mental disorder including psychotic experiences: trailblazing a new recovery pathway within the Improving Access to Psychological Therapies programme in England. Early Interv. Psychiatry 12 (3), 497–504.

Phillips, L.J., Cotton, S., Mihalopoulos, C., Shih, S., Yung, A.R., Carter, R., McGorry, P.D., 2009. Cost implications of specific and non-specific treatment for young persons at ultra high risk of developing a first episode of psychosis. Early Interv. Psychiatry 3, 28–34.

Pickett, K.E., Wilkinson, R.G., 2010. Inequality: An Underacknowledged Source of Mental Illness and Distress. RCP.

Pilling, S., Bebbington, P., Kuipers, E., Garety, P., Geddes, J., Orbach, G., Morgan, C., 2002. Psychological treatments in schizophrenia: I. Meta-analysis of family intervention and cognitive behaviour therapy. Psychol. Med. 32, 763–782.

Prince, M., Patel, V., Saxena, S., Maj, M., Maselko, J., Phillips, M.R., Rahman, A., 2007. No health without mental health. Lancet 370, 859–877.

Psychiatric GWAS Consortium Coordinating Committee, 2009. Genomewide association studies: history, rationale, and prospects for psychiatric disorders. Am. J. Psychiatry 166 (5), 540–556.

Public Health England, 2015. Criteria for Appraising the Viability, Effectiveness and Appropriateness of a Screening Programme.

Reid, M.J., Webster-Stratton, C., Baydar, N., 2004. Halting the development of conduct problems in Head Start children: the effects of parent training. J. Clin. Child Adolesc. Psychol. 33, 279–291.

Rones, M., Hoagwood, K., 2000. School-based mental health services: a research review. Clin. Child Fam. Psychol. Rev. 3, 223–241.

Rose, G., 2001. Sick individuals and sick populations. Int. J. Epidemiol. 30, 427–432.

Ruf, M., Morgan, O., 2008. Differences Between Screening and – Diagnostic Tests, Case Finding. Public Health Acation Support Team (PHAST). Available from: https://www.healthknowledge.org.uk/public-health-textbook/disease-causation-diagnostic/2c-diagnosis-screening/screening-diagnostic-case-finding.

Rutter, M., 2008. Developing concepts in developmental psychopathology. In: Developmental Psychopathology and Wellness: Genetic and Environmental Influences, pp. 3–22.

Sackett, D., Holland, W., 1975. Controversy in the detection of disease. Lancet 306, 357–359.

Saxena, S., Thornicroft, G., Knapp, M., Whiteford, H., 2007. Resources for mental health: scarcity, inequity, and inefficiency. Lancet 370, 878–889.

Schilling, E.A., Aseltine, R.H., Gore, S., 2007. Adverse childhood experiences and mental health in young adults: a longitudinal survey. BMC Public Health 7, 30.

Schmitz, N., Kruse, J., Heckrath, C., Alberti, L., Tress, W., 1999. Diagnosing mental disorders in primary care: the general health Questionnaire (GHQ) and the symptom check list (SCL-90-R) as screening instruments. Soc. Psychiatry Psychiatr. Epidemiol. 34, 360–366.

Senior, P.A., Bhopal, R., 1994. Ethnicity as a variable in epidemiological research. BMJ 309, 327.

Sharp, L.K., Lipsky, M.S., 2002. Screening for depression across the lifespan: a review of measures for use in primary care settings. Am. Fam. Physician 66, 1001–1008.

Shi, J., Wang, L., Yao, Y., Zhan, C., Su, N., Zhao, X., October 2017. Systemic therapy for youth at clinical high risk for psychosis: a pilot study. Front. Psychiatry 8 (no pagination).

Soneson E., Howarth E., Ford T., Humphrey A., Jones P.B., Thompson Coon J., Rogers M. and Anderson J.K.,

Feasibility of school-based identification of children and adolescents experiencing, or at risk of developing, mental health difficulties: A systematic review. Prevention Science, In press.
St. Clair, M.C., Croudace, T., Dunn, V.J., Jones, P.B., Herbert, J., Goodyer, I.M., 2015. Childhood adversity subtypes and depressive symptoms in early and late adolescence. Dev. Psychopathol. 27, 885–899.
Stefanis, N.C., Hanssen, M., Smirnis, N.K., Avramopoulos, D.A., Evdokimidis, I.K., Stefanis, C.N., Verdoux, H., VAN Os, J., 2002. Evidence that three dimensions of psychosis have a distribution in the general population. Psychol. Med. 32, 347–358.
Stochl, J., Khandaker, G.M., Lewis, G., Perez, J., Goodyer, I.M., Zammit, S., Sullivan, S., Croudace, T.J., Jones, P.B., 2015. Mood, anxiety and psychotic phenomena measure a common psychopathological factor. Psychol. Med. 45, 1483–1493.
Thapar, A., Fowler, T., Rice, F., Scourfield, J., van den Bree, M., Thomas, H., Harold, G., Hay, D., 2003. Maternal smoking during pregnancy and attention deficit hyperactivity disorder symptoms in offspring. Am. J. Psychiatry 160, 1985–1989.
Thombs, B.D., Coyne, J.C., Cuijpers, P., de Jonge, P., Gilbody, S., Ioannidis, J.P.A., Johnson, B.T., Patten, S.B., Turner, E.H., Ziegelstein, R.C., 2012. Rethinking recommendations for screening for depression in primary care. Can. Med. Assoc. J. 184, 413–418.
UK National Screening Committee, 2017. Current UK NSC Recommendations. Available from: https://legacyscreening.phe.org.uk/screening-recommendations.php.
UK National Screening Committee, 2013. NHS Population Screening Explained. Public Health England. Available from: https://www.gov.uk/guidance/nhs-population-screening-explained.
van Harmelen, A.L., Kievit, R.A., Ioannidis, K., Neufeld, S., Jones, P.B., Bullmore, E., Dolan, R., Fonagy, P., Consortium, N.S.P.N., Goodyer, I., 2017. Adolescent friendships predict later resilient functioning across psychosocial domains in a healthy community cohort. Psychol. Med. 1–11.
van Os, J., Kapur, S., 2009. Schizophrenia. Lancet 374, 635–645.
Värnik, P., 2012. Suicide in the world. Int. J. Environ. Res. Public Health 9, 760–771.
Verbeek, T., Bockting, C.L.H., Van Pampus, M.G., Ormel, J., Meijer, J.L., Hartman, C.A., Burger, H., 2012. Postpartum depression predicts offspring mental health problems in adolescence independently of parental lifetime psychopathology. J. Affect. Disord. 136, 948–954.
Wakschlag, L.S., Pickett, K.E., Cook Jr., E., Benowitz, N.L., Leventhal, B.L., 2002. Maternal smoking during pregnancy and severe antisocial behavior in offspring: a review. Am. J. Public Health 92, 966–974.
Wald, N.J., Morris, J.K., 1996. What Is Case—Finding? SAGE Publications Sage UK, London, England.
Walker, F.R., Pfingst, K., Carnevali, L., Sgoifo, A., Nalivaiko, E., 2017. In the search for integrative biomarker of resilience to psychological stress. Neurosci. Biobehav. Rev. 74, 310–320.
Webster-Stratton, C., 2001. The Incredible Years: parents, teachers, and children training series. Resid. Treat. Child. Youth 18, 31–45.
Webster-Stratton, C., Reid, M.J., 2003. The Incredible Years Parents, Teachers and Children Training Series: A Multifaceted Treatment Approach for Young Children with Conduct Problems. Guilford Press.
Webster-Stratton, C., Reid, M.J., Hammond, M., 2001. Preventing conduct problems, promoting social competence: a parent and teacher training partnership in Head Start. J. Clin. Child Psychol. 30, 283–302.
Webster-Stratton, C., Jamila Reid, M., Stoolmiller, M., 2008. Preventing conduct problems and improving school readiness: evaluation of the incredible years teacher and child training programs in high-risk schools. J. Child Psychol. Psychiatry 49, 471–488.
Weich, S., Nazroo, J., Sproston, K., Mcmanus, S., Blanchard, M., Erens, B., Karlsen, S., King, M., Lloyd, K., Stansfeld, S., 2004. Common mental disorders and ethnicity in England: the EMPIRIC study. Psychol. Med. 34, 1543–1551.
Weist, M.D., Rubin, M., Moore, E., Adelsheim, S., Wrobel, G., 2007. Mental health screening in schools. J. Sch. Health 77, 53–58.
Whiteford, H.A., Degenhardt, L., Rehm, J., Baxter, A.J., Ferrari, A.J., Erskine, H.E., Charlson, F.J., Norman, R.E., Flaxman, A.D., Johns, N., 2013. Global burden of disease attributable to mental and substance use disorders: findings from the Global Burden of Disease Study 2010. Lancet 382, 1575–1586.
Williams, S.N., 2013. Bring in universal mental health checks in schools. BMJ 347, f5478.
Wilson, J.M.G., Jungner, G., World Health, Organization., 1968. Principles and Practice of Screening for Disease.
Witkiewitz, K., Marlatt, G.A., Walker, D., 2005. Mindfulness-based relapse prevention for alcohol and substance use disorders. J. Cogn. Psychother. 19, 211–228.
World Health Organization, 1992. The ICD-10 Classification of Mental and Behavioural Disorders: Clinical Descriptions and Diagnostic Guidelines. World Health Organization.
World Health Organization, 2002. Gender and Mental Health.
World Health Organization, 2003. Investing in Mental Health.

World Health Organization, 2004. Prevention of Mental Disorders: Effective Interventions and Policy Options: Summary Report.

World Health Organization, 2008. Integrating Mental Health into Primary Care: A Global Perspective. World Health Organization.

World Health Organization, 2014. Social Determinants of Mental Health. World Health Organization.

Yung, A.R., McGorry, P.D., 1996. The prodromal phase of first-episode psychosis: past and current conceptualizations. Schizophr. Bull. 22, 353.

Yung, A.R., Yuen, H.P., McGorry, P.D., Phillips, L.J., Kelly, D., Dell'Olio, M., Francey, S.M., Cosgrave, E.M., Killackey, E., Stanford, C., 2005. Mapping the onset of psychosis: the comprehensive assessment of at-risk mental states. Aust. N. Z. J. Psychiatr. 39, 964–971.

Risk paradigms

CHAPTER

3

At-risk mental states

Alison R. Yung[1,2]

[1]Centre for Youth Mental Health and Orygen, The University of Melbourne, Parkville, Victoria, Australia; [2]Division of Psychology and Mental Health, School of Health Sciences, The University of Manchester, Manchester, United Kingdom

Psychotic disorders such as schizophrenia have a lifetime prevalence of over 1% and are among the world's leading causes of disability. They have potentially high personal, social, and economic impact (Saha et al., 2005; van Os and Kapur, 2009). Their onset is usually in adolescence and young adulthood (Abel et al., 2010), meaning that the most productive years of a person's life are affected and could lead to living many years with the illness.

Given this background, there has been an increasing emphasis on early intervention to detect illness early and minimize disability. Early intervention services that aim to provide timely and high quality multidisciplinary care have now been established worldwide. Such early intervention, during the first episode of psychosis, can significantly improve quality of life, reduce the rate of relapse, risk of suicide, and number of hospital admissions (Correll et al., 2018). They can improve employment and education outcomes and wellbeing, are cost effective (Craig et al., 2004; Petersen et al., 2005; Garety et al., 2006; McCrone et al., 2010) and well-liked by service-users and carers (Schizophrenia Commission, 2012).

Following the success of early intervention in the first episode of psychosis, in the late 1990s, consideration was given to whether even earlier intervention was possible—before the first episode of psychosis. That is, we aimed to intervene during the lead up to the first psychotic episode in what is called the "prodrome" or "prodromal phase". The idea was that intervention in the prodromal phase of schizophrenia and related psychoses may result in attenuation, delay, or even prevention of the onset of psychosis in some individuals (Yung et al., 1996). (McGlashan et al., 2003; McGorry et al., 2002; Morrison et al., 2004).

Identifying the prodrome of psychotic disorder

There were several issues that we needed to deal with when considering the identification of the prodrome of psychotic disorder. The first was conceptual: the term prodrome is a retrospective concept and is used after the onset of fully fledged disorder. For example, once the characteristic rash of measles develops, it is

Risk Factors for Psychosis
https://doi.org/10.1016/B978-0-12-813201-2.00003-X

possible to look back and say that the cough and coryza that preceded the rash was the "measles prodrome." However, when only the cough and coryza are present, it is not be possible to label these symptoms as the measles prodrome as they may equally indicate a common cold, hay fever, or pneumonia. This example also highlights the second issue: that prodromal symptoms and signs are necessarily nonspecific, otherwise they would not be the prodrome of illness but would be the illness itself. The third issue was that we needed more information about characteristic features of the psychotic prodrome.

Coining the term "At-Risk Mental State" was a way of dealing with the fact that "prodrome" is a retrospective concept. This term does not imply inevitable progression to full disorder, unlike the term "prodrome". We then set about developing criteria for the "At-Risk Mental State" or "ARMS". The criteria were determined following two pieces of work. The first was a comprehensive review of the literature describing the signs and symptoms leading up to a first episode of psychosis or schizophrenia (Yung and McGorry, 1996). The second was a qualitative study exploring the experiences of individuals who had recently recovered after a first episode of psychosis (Yung and McGorry, 1996).

The synthesis of these two studies concluded that the first manifestations of a psychotic prodrome were likely to be nonspecific symptoms, such as sleep disturbance, anxiety, and low mood. More proximal to the onset of psychotic symptoms were subthreshold or attenuated versions of these symptoms. Examples include an overvalued idea that one might be being followed or watched which was held with some doubt, rather than a fully formed persecutory delusion, or hearing mumbling or whispering but not clear auditory hallucinations with the same quality as normal perceptions. This led to the ARMS category of attenuated psychotic symptoms. The literature review and study of recovered patients also revealed that short and self-limited episodes of frank psychotic symptoms could also occur in the lead up to a more prolonged psychotic episode. This led to the ARMS category of brief limited intermittent psychotic symptoms. Also included as an ARMS category was a group with both trait and state risk factors for psychotic disorder. The thinking was that a young person with a trait risk factor, such as a family history of schizophrenia, was at higher risk for development of psychosis than an individual without such family history, and that marked and prolonged decline in functioning may indicate a developing psychotic disorder. This led to the ARMS category of the trait and state risk factor group. We also made an arbitrary definition of frank psychosis as the outcome variable. This was based on the presence of clear threshold level psychotic symptoms (delusions, hallucinations, and formal thought disorder) occurring several times per week for at least 1 week. This threshold is that at which antipsychotic medication would commonly be commenced in clinical practice (Yung et al., 2003). The criteria are summarized in Box 3.1.

In developing the ARMS criteria (or Ultra High Risk Criteria (UHR)), we aimed to describe clinical features that would indicate *high* and *imminent* risk of psychotic disorder, such that identification and treatment of these individuals would reduce risk for psychotic disorder. The idea behind the criteria and the name "At-Risk Mental State" was that the syndrome indicated risk for psychosis at the time that the person was experiencing such symptoms. That is, it was considered a *state* risk factor for psychotic disorder. Risk is linked to the presence of symptoms. The converse is a *trait* risk factor, such as genetic risk, which places the individual at risk of psychotic disorder regardless of their current mental state.

BOX 3.1

CRITERIA FOR AN AT-RISK MENTAL STATE FOR PSYCHOSIS

(A) Age between 15 and 30 years
(B) Help-seeking at a mental health service, and
(C) Belonging to at least one of the following three groups:
- **(1)** *Trait and State Risk Factor Group* having a trait risk factor of either a first-degree relative with a psychotic disorder and/or schizotypal personality disorder plus a state risk factors of a significant decrease in functioning within the last year or chronic low functioning.
- **(2)** *Attenuated Psychotic Symptoms Group*—having attenuated positive psychotic symptoms, for at least 1 week, several times per week, within the last year
- **(3)** *Brief Limited Intermittent Psychotic Symptoms Group*—having experienced short episodes of frank psychotic symptoms that have resolved without treatment within the last year

True positives, false positives, and false false positives

ARMS individuals who transition to a psychotic disorder are known as "true positives." That is, the ARMS syndrome with which they presented was the prodrome to their first episode of psychosis. ARMS individuals who do not develop a psychotic disorder are known as "false positives." This means that their ARMS syndrome was not indicative of a psychosis prodrome, but rather symptoms may have been transient or related to some other disorder or experience. However, we must also consider the concept of "false false positives." These are individuals *who would have* developed a psychotic disorder, but did not during the period of monitoring. This may be due to an intervention that was provided, or some change in their circumstances (e.g., they stopped using cannabis). The problem we face is that it is difficult to differentiate "false positives" and "false false positives." Phenotypically they are the same. In theory, the "false false positives" should resemble the "true positives" in factors that predict transition to psychotic disorder (such as biomarkers). However, these groups should differ on factors that ultimately predicted development of psychotic disorder. For example, while both the "true positives" and the "false false positives" may have a high polygenic risk score for schizophrenia, the "true positives" may have high levels of perceived stress and substance use and low social support, while the "false false positives" may have lower levels of stress and substance use and strong social support, all of these factors being potentially relevant to development of a psychotic disorder. "False false positives" therefore continue to be at risk of psychotic disorder.

In 1994, the first clinical service for individuals meeting the UHR criteria, the Personal Assessment and Crisis Evaluation (PACE) clinic was established in Melbourne, Australia. The service was designed to be low stigma and so was located at a general adolescent health service. Although it was able to receive referrals from a range of services, including young people themselves and their families, the majority of referrals were from the nearby Early Psychosis Prevention and Intervention Centre (EPPIC), a service for people in the first few years after psychosis onset. Referrals from EPPIC to PACE were individuals who were referred to EPPIC but who were found to have subthreshold rather than threshold psychotic symptoms. Early treatment at the PACE clinic consisted of management of presenting complaints, supportive therapy and monitoring of mental state.

Operationalization of the ARMS criteria—The Comprehensive Assessment of At-Risk Mental States (CAARMS)

A new instrument was needed to assess emerging subthreshold and brief psychotic symptoms. This was because existing instruments such as the Brief Psychiatric Rating Scale (BPRS) (Overall and Gorham, 1962) and the Comprehensive Assessment of Symptoms and History (CASH) (Andreasen, 1987) were adequate to assess symptom intensity, but they were unable to measure the more "fine-grained" aspects of these symptoms such as their frequency, duration, and recency. The Comprehensive Assessment of At-Risk Mental States (CAARMS) was designed so that all relevant domains (intensity, frequency, duration, and recency) could be assessed with the one tool (Yung et al., 2005). Early publications referenced the BPRS and CASH, while the CAARMS instrument was being validated.

The CAARMS consists of a semistructured interview and rating system and was designed to be used by trained interviewers (Yung et al., 2005). It is able to detect change in symptoms over time and should be used as a repeated measure. No strict frequency of use is in place. Ideally an individual meeting ARMS criteria should be assessed every 3–4 months to monitor symptom progression or reduction and determine if a first episode of psychosis has developed.

There are three main aims of the CAARMS:

1. To determine if an individual meets the criteria for an "At-Risk Mental State,"
2. To rule out, or confirm, criteria for first episode psychosis,
3. To map a range of psychopathology and functioning factors over time.

For Aims 1 and 2, the positive symptoms scales are needed. For Aim 3, the CAARMS measures other symptoms thought to be indicative of imminent psychotic disorder. A literature review found that these experiences are present in the period leading up to a first episode of psychosis (Yung and McGorry, 1996). These include negative symptoms, depression, anxiety, obsessive-compulsive symptoms, dissociative symptoms, concentration and attention difficulties, and "basic" symptoms (Gross, 1989). Basic symptoms were first defined in the German psychiatric literature (Huber et al., 1980), and are subjective deficit features, the experiential equivalents of the behaviorally defined negative symptoms. They have been shown to occur in the prodromal phase of schizophrenia, as well as in the schizophrenia deficit state. Some basic symptoms, particularly cognitive, language, and perceptual disturbances, have been shown to predict first episode of psychosis (Klosterkoetter et al., 1996; Schultze-Lutter et al., 2007) (see Chapter 4). Instruments have been designed to measure basic symptoms in individuals considered at risk of psychotic disorder (Schultze-Lutter et al., 2012). Table 3.1 shows the CAARMS subscales and items.

TABLE 3.1 Full CAARMS subscales.

Subscale	Items
Positive symptoms	Unusual thought content
	Non-bizarre ideas
	Perceptual abnormalities
	Disorganized speech
Cognitive change, attention, concentration	Subjective experience
	Observed cognitive change
Emotional disturbance	Subjective emotional disturbance
	Observed blunted affect
	Observed inappropriate affect
Negative symptoms	Alogia
	Avolition/apathy
	Anhedonia
Behavioral change	Social isolation
	Impaired role function
	Disorganizing/odd/stigmatizing behavior
	Aggression/dangerous behavior
Motor/physical changes	Subjective complaints of impaired motor functioning
	Informant reported or observed changes in motor functioning
	Subjective complaints of impaired bodily sensation
	Subjective complaints of impaired autonomic functioning
General psychopathology	Mania
	Depression
	Suicidality and self harm
	Mood swings/lability
	Anxiety
	Obsessive compulsive symptoms
	Dissociative symptoms
	Impaired tolerance to normal stress

The original version of the CAARMS included three types of positive symptoms, disorders of thought content (e.g., delusional mood, overvalued ideas, and delusions), perceptual abnormalities (e.g., distortions, illusions, and hallucinations), and conceptual disorganization (e.g., subjectively experienced difficulties with forming thoughts and objective assessment of formal thought disorder). However, several factor analyses of psychotic symptoms suggested that this three-factor model did not adequately describe the different forms of positive symptoms. For example, there are three four-factor models (paranoia, first rank symptoms, hallucinations, and grandiosity (Stefanis et al., 2002)), (bizarre experiences, persecutory ideation, perceptual abnormalities, and magical thinking (Yung et al., 2009)), (bizarre experiences, persecutory ideation, perceptual abnormalities, and grandiosity (Armando et al., 2010)) and a five-factor model (hallucinations, delusions, paranoia, grandiosity, and paranormal beliefs (Dominguez et al., 2011)). These studies have all been conducted in nonclinical populations, and the factors are similar. Persecution, persecutory ideas, paranoia are all likely to be describing the same phenomena, as are first rank symptoms and bizarre experiences, and hallucinations and perceptual abnormalities. Paranormal beliefs and grandiosity may be equivalent to the dimension of magical thinking. In the light of these findings, the CAARMS disorders of thought content scale was split into two: unusual thought content (UTC) and non-bizarre ideas (NBI), with UTC describing clearly implausible symptoms, including the first rank symptoms of thought insertion, withdrawal, and broadcasting, and NBI including symptoms that could be plausible, such as persecutory and jealous ideas.

An abbreviated version of the CAARMS is also available which includes only the positive symptom subscales as these are the ones needed to confirm or rule out an At-Risk Mental State. This version is frequently used in clinical services.

Each CAARMS item is scored according to its intensity and frequency (from 0 (absent/no frequency) to 6 (severe and psychotic/continuous)). In addition, the positive symptom items include a distress scale, which is scored from 0 (no distress) to 100 (extremely distressing). The intensity, frequency, and distress scales of the positive symptoms have been shown to measure distinct aspects of the CAARMS (Wilson et al., under submission).

The CAARMS has been translated into 18 languages and is widely used worldwide. An alternative instrument, the Structured Interview for Prodromal Syndromes (SIPS) (Miller et al., 2003) is based on the CAARMS and similar in structure. It is widely used in the United States.

The CAARMS and the ARMS criteria have fairly good validity and reliability, meaning that the ARMS can be differentiated from psychotic disorder and from mental states below the ARMS threshold and that there tends to be agreement between clinicians about into which of these groups an individual falls (Woods et al., 2009). Further, the ARMS criteria are relatively specific for psychotic disorders, given their low incidence in the general population (Woods et al., 2009; Lin et al., 2015).

The predictive validity of the ARMS criteria

Early studies found that the ARMS criteria had high predictive ability for development of psychosis. The original study from the PACE clinic found that over 40% of ARMS individuals (20 out of 49) developed a psychotic disorder within 1 year, a rate several hundred-fold over expected rates for that age group in the general population (Yung et al., 2003). Similarly, in the United States, a high risk was found in the earliest study using the SIPS to define ARMS criteria, with 7 of 14 patients (50%) developing a psychotic disorder within 1 year (Miller et al., 2003). Subsequently, the criteria have shown predictive validity for psychotic disorders across different countries and service settings. A meta-analysis found that ARMS individuals have a risk of developing a full psychotic disorder of 15%–30% within 12 months, and over 36% after 3 years (Fusar-Poli et al., 2012). Most individuals who develop a psychotic disorder have a diagnosis of schizophrenia or a schizophrenia spectrum disorder.

However, several more recent studies have found that the risk of developing a psychotic disorder has decreased in the ARMS population. We explored reasons for this in a 2007 paper (Yung et al., 2007). This paper examined the reduction in transition rate in the PACE clinic over time between 1993 and 2000 and found that the decline was associated with a shorter duration of symptoms prior to being referred to the PACE clinic (Yung et al., 2007). That is, people were referred to PACE earlier. Thus, the relationship between early detection and decreased transition rate could have been due to a "lead time bias". This means that, due to earlier identification, a group will experience the event in question after a longer period following detection. It is a term frequently used in relation to cancer and survival times, and describes the situation in which a person diagnosed earlier may appear to survive longer, but in fact, early detection has no real effect on the time course of events. In ARMS studies, this means that there is no genuine reduction in transition rate, rather a longer time designated as being at ARMS before psychosis onset. That is, that there is an apparent decline in the transition rate but this is due to earlier detection and the rate will increase if longer follow up is provided.

Alternatively, the reduction in transition rate may be a reflection of more effective treatment, especially if that treatment is applied earlier in the course of illness. Several interventions have been used in the ARMS population (see below). It may be that if they are applied very early in the ARMS phase, perhaps before brain changes have occurred, then treatment may be more effective and perhaps preventive than if applied later. The reduction in transition rate would then be due to more "false false positives"—those who would have developed psychosis but were prevented from doing so by the intervention.

Another possibility that may explain the reduced transition rate is a "dilution effect". That is, as the work of the PACE clinic has become more well-known, potential referrers may be more likely to ask about psychotic like symptoms. For example, a young person may seek help from a general practitioner or counseling service due to depressive symptoms. He or she may then be questioned about psychotic like experiences, as the well-meaning clinician is keen to detect any hint of risk for psychosis. As transient psychotic like experiences are not uncommon in adolescents and young adults (Van Os et al., 2009), these may be detected and lead to referral. In other words, the number truly at risk may be "diluted" with false positives: those not at risk of psychotic disorder.

Intervention studies to prevent transition to psychotic disorder

Treatment of ARMS individuals has two aims: to manage current symptoms and problems and to reduce the risk of developing a psychotic disorder (Yung et al., 1996). Intervention trials have tended to have "transition to psychosis" as the primary outcome, with symptoms, level of functioning, and distress sometimes included as secondary outcome measures. A recent meta-analysis studied 10 randomized trials that reported effects on transition rates of low-dose antipsychotic medication, cognitive behavioral therapy (CBT), omega 3 fatty acid, and integrated treatment that included family therapy, cognitive remediation, social skills training, and CBT (van der Gaag et al., 2013). This study found that receipt of any specific intervention significantly reduced the risk of developing a first episode of psychosis both at 12 months and over the longer term (2–4 years), albeit with diminished effects over time. The reduced effect at long-term follow-up suggests that at least some ARMS individuals remain at risk and that interventions might delay, rather than prevent, onset of psychosis. Even so, such a delay could be of benefit, enabling people to, for example, finish education and develop supportive networks outside the family of origin.

Since publication of this meta-analysis, several negative trials have been published, including of omega 3 fatty acids (McGorry et al., 2017) and CBT (Addington et al., 2011a). Additionally, a network meta-analysis examining trials with transition to psychosis as the outcome found that no specific treatment was effective. However, this claim was based on a comparison of specific treatments to a combination of several different "needs-based interventions," many of which included active components.

Additionally individuals who develop a first episode of psychosis after having been treated in the prodromal phase have improved outcomes compared to their counterparts who did not receive such very early intervention (Valmaggia et al., 2015). Interventions for ARMS individuals are further discussed in detail in Chapters 18 and 19.

Broadening the outcomes of interest

As noted above, transition to psychosis has traditionally been the main outcome of interest in this ARMS group (Yung et al., 2003; Ruhrmann et al., 2010; Cannon et al., 2008). However, the majority of ARMS patients do not develop psychotic disorder even up to 10 years post identification (Nelson et al., 2013). Nonetheless, many remain symptomatic and disabled (Schlosser et al., 2012; Addington et al., 2011b; Lin et al., 2014; Cotter et al., 2014; Yung et al., 2010). This functional disability appears to be independent of positive symptoms (Cotter et al., 2014; Carrión et al., 2013; Meyer et al., 2014), although its relationship to transition and diagnosis of psychotic disorder is unclear. It is important, therefore, to study psychosocial functioning as an outcome in at-risk research. Other outcomes of interest are persistent subthreshold positive symptoms, persistent negative symptoms (Yung et al., 2018), and nonpsychotic disorders such as anxiety and depression (Lin et al., 2015).

Reflecting the need to include different outcomes in ARMS trials, recently some novel treatments have also been piloted in this group. These have had more targeted outcomes, based on hypothesized mechanisms of action of the intervention rather than global aims of reducing transition risk. For example, a small study of lithium hypothesized that it may have a neuroprotective effect and examined hippocampal MRI T2 relaxation time and proton magnetic resonance spectroscopy as outcomes testing this hypothesis (Berger et al., 2012). Glycine has been tested in two small pilot trials with outcomes of symptoms and neurocognitive functioning (Woods et al., 2013), a study of biofeedback measured anxiety and distress as outcomes (McAusland and Addington, 2016), and a trial of processing speed training examined improvement in processing speed and its correlation with social functioning (Choi et al., 2016). A trial of a family intervention measured caregiver warmth, family communication, and social functioning as outcomes (O'Brien et al., 2006). All studies showed feasibility and either significant results or trends to significance, indicating future avenues of research.

A recent systematic review and network meta-analysis examined treatments for negative symptoms in the ARMS group (Devoe et al., 2018). Psychosocial treatments and pharmacological treatments were examined. While *N*-methyl-D-aspartate-receptor (NMDAR) modulators (such as D-serine and glycine) ranked higher than all other treatments, no intervention significantly reduced negative symptoms. However, interpretation of this meta-analysis should take into account the relative paucity of studies in this area, small sample sizes and that the primary outcome for the majority of interventions was development of a psychotic disorder and not reduction in negative symptoms (Devoe et al., 2018). This meant that the psychosocial treatments did not include specific elements targeting negative symptoms. Further, as negative symptoms were not the focus of the studies, many ARMS participants had low levels of negative symptoms (e.g., (Nordentoft et al., 2006; Amminger et al., 2010)).

The attenuated psychosis syndrome

More recently, the psychotic disorders working group of the Diagnostic and Statistical Manual of Mental Disorders (DSM V) in the United States considered the possibility of introducing a "psychosis risk syndrome" or "attenuated psychosis syndrome" into the psychiatric classification system (Tsuang et al., 2013). The syndrome described is roughly equivalent to the category of attenuated psychotic symptoms described in the ARMS criteria. There were a number of arguments for and against including it in the main document (Woods et al., 2010; Yung et al., 2010), but the eventual decision was to introduce the clinical term "Attenuated Psychosis Syndrome" (APS) in to the Appendix section (Section 3) as a condition for further systematic study (Yung et al., 2012). Further research in the field in the next few years will also concentrate on individuals meeting this specific set of criteria for an attenuated psychosis syndrome.

The next wave of research—refining prediction

One problem in ARMS research is that we lack understanding of the factors that predict different outcomes, including underlying biological mechanisms. This means that some ARMS individuals are having unnecessary treatment, and others are having ineffective interventions that potentially delay initiation of effective treatment. There is a need for investigation into factors that predict different outcomes. The aim would be to stratify the ARMS group according to their underlying pathological processes and target treatment accordingly, moving the field toward personalized management. Several "risk prediction tools" have now been published, but validation studies of them are lacking. One difficulty is that most studies of ARMS individuals have small sample sizes and low numbers of individuals developing psychotic disorder. Variables predictive of transition in one study may not be predictive in another and may not be generalizable to the whole population of ARMS individuals (not just those who consent to research studies). The field would benefit from pooled resources enabling analysis of large sample sizes, multiple factors, and interaction between factors. This would allow predictors with even small effects to be examined.

References

Abel, K.M., Drake, R., Goldstein, J.M., 2010. Sex differences in schizophrenia. Int. Rev. Psychiatr. 22 (5), 417–428.

Addington, J., Epstein, I., Liu, L., French, P., Boydell, K., Zipursky, R.B., 2011a. A randomized controlled trial of cognitive behavioral therapy for individuals at clinical high risk of psychosis. Schizophr. Res. 125, 54–61.

Addington, J., Cornblatt, B., Cadenhead, K., et al., 2011b. At clinical high risk for psychosis: outcome for nonconverters. Am. J. Psychiatr. 168, 800–805.

Amminger, G.P., Schäfer, M., Papageorgiou, K., et al., 2010. Long-Chain omega-3 fatty acids for indicated prevention of psychotic disorders: a randomized, placebo-controlled trial. Arch. Gen. Psychiatr. 67 (2), 146–154.

Andreasen, N., 1987. The Comprehensive Assessment of Symptoms and History (CASH) Interview. University of Iowa.

Armando, M., Nelson, B., Yung, A.R., Ross, M., Birchwood, M., Girardi, P., Nastro, P.F., 2010. Psychotic-like experiences and correlation with distress and depressive symptoms in a community sample of adolescents and young adults. Schizophr. Res. 119, 258–265.

Berger, G.E., Wood, S.J., Ross, M., et al., 2012. Neuroprotective effects of low-dose lithium in individuals at ultra-high risk for psychosis. A longitudinal MRI/MRS study. Curr. Pharmaceut. Des. 18 (4), 570–575.

Cannon, T.D., Cadenhead, K., Cornblatt, B., et al., January 2008. Prediction of psychosis in youth at high clinical risk: a multisite longitudinal study in North America. Arch. Gen. Psychiatr 65 (1), 28–37.

Carrión, R.E., McLaughlin, D., Goldberg, T.E., Auther, A.M., Olsen, R.H., Olvet, D.M., Correll, C.U., Cornblatt, B., 2013. Prediction of functional outcome in individuals at clinical high risk for psychosis. JAMA Psychiatry 70, 1133–1142.

Choi, J., Corcoran, C., Fiszdon, J., et al., 2016. Pupillometer-based neurofeedback cognitive training to improve processing speed and social functioning in individuals at clinical high risk for psychosis. Psychiatr. Rehabil. J. 40 (1), 33–42.

Correll, C.U., Galling, B., Pawar, A., et al., 2018. Comparison of early intervention services vs treatment as usual for early-phase psychosis a systematic review, meta-analysis, and meta-regression. JAMA Psychiatr. 75, 555–565.

Cotter, J., Drake, R.J., Bucci, S., Firth, J., Edge, D., Yung, A.R., 2014. What drives poor functioning in the at-risk mental state? A systematic review. Schizophr. Res. in press.

Craig, T.K., Garety, P., Power, P., Rahaman, N., Colbert, S., Fornells-Ambrojo, M., Dunn, G., 2004. The Lambeth Early Onset (LEO) team: randomised controlled trial of the effectiveness of specialised care for early psychosis. BMJ 329 (7474), 1067.

Devoe, D., Peterson, A., Addington, J., 2018. Negative symptom interventions in youth at risk of psychosis: a systematic review and network meta-analysis. Schizophr. Bull. 44, 807–823.

Dominguez, M.D.G., Wichers, M., Lieb, R., Wittchen, H.-U., Van Os, J., 2011. Evidence that onset of clinical psychosis is an outcome of progressively more persistent subclinical psychotic experiences: an 8-year cohort study. Schizophr. Bull. 37, 84–93.

Fusar-Poli, P., Bonoldi, I., Yung, A.R., Borgwardt, S., Kempton, M., Barale, F., 2012. Predicting psychosis: a meta-analysis of evidence. Arch. Gen. Psychiatr. 69 (3), 220–229.

Garety, P.A., Craig, T.K., Dunn, G., Fornells-Ambrojo, M., Colbert, S., Rahaman, N., Read, J., Power, P., 2006. Specialised care for early psychosis: symptoms, social functioning and patient satisfaction Randomised controlled trial. Br. J. Psychiatr. 188 (1), 37–45.

Gross, G., 1989. The "basic" symptoms of schizophrenia. Br. J. Psychiatr. 155 (Suppl. 7), 21–25.

Huber, G., Gross, G., Schuttler, R., Linz, M., 1980. Longitudinal studies of schizophrenic patients. Schizophr. Bull. 6, 592–605.

Klosterkoetter, J., Ebel, H., Schultze-Lutter, F., Steimeyer, E.M., 1996. Diagnostic validity of basic symptoms. Eur. Arch. Psychiatry Clin. Neurosci. 246 (3), 147–154.

Lin, A., Wood, S.J., Nelson, B., Beavan, A., McGorry, P., Yung, A.R., 2014. Outcomes of nontransitioned cases in a sample at ultra-high risk for psychosis. Am. J. Psychiatry in press.

Lin, A., Wood, S.J., Nelson, B., Beavan, A., McGorry, P., Yung, A.R., 2015. Outcomes of non-transitioned cases in a sample at ultra-high risk for psychosis. Am. J. Psychiatr. 172 (3), 249–258.

McAusland, L., Addington, J., 2016. Biofeedback to treat anxiety in young people at clinical high risk for developing psychosis. Early Interv. Psychiatr. https://doi.org/10.1111/eip.12368 epub ahead of print.

McCrone, P., Craig, T., Power, P., Garety, P., 2010. Cost-effectiveness of an early intervention service for people with psychosis. Br. J. Psychiatr. 196 (5), 377–382.

McGlashan, T., Zipursky, R., Perkins, D., et al., May 2003. The PRIME North America randomized double-blind clinical trial of olanzapine versus placebo in patients at risk of being prodromally symptomatic for psychosis. I. Study rationale and design. Schizophr. Res. 61 (1), 7–18.

McGorry, P., Nelson, B., Markulev, C., Yuen, H.P., Verma, S., Thompson, A., Yung, A.R., Amminger, P., 2017. Effect of ω-3 polyunsaturated fatty acids in young people at ultra-high risk for psychotic disorders: the NEURAPRO randomized clinical trial. JAMA Psychiatr. 74, 19–27.

McGorry, P.D., Yung, A.R., Phillips, L.J., et al., October 2002. Randomized controlled trial of interventions designed to reduce the risk of progression to first-episode psychosis in a clinical sample with subthreshold symptoms. Arch. Gen. Psychiatr. 59 (10), 921–928.

Meyer, E.C., Carrión, R.E., Cornblatt, B.A., et al., 2014. The relationship of neurocognition and negative symptoms to social and role functioning over time in individuals at clinical high risk in the first phase of the North American Prodrome Longitudinal Study. Schizophr. Bull. in press.

Miller, T.J., McGlashan, T.H., Rosen, J.L., et al., 2003. Prodromal assessment with the structured interview for prodromal syndromes and the scale of prodromal symptoms: predictive validity, interrater reliability, and training to reliability. Schizophr. Bull. 29 (4), 703–715.

Morrison, A.P., French, P., Walford, L., Lewis, S.W., Kilcommons, A., Green, J., Parker, S., Bentall, R.P., 2004. Cognitive therapy for the prevention of psychosis in people at ultra-high risk: randomised controlled trial. Br. J. Psychiatr. 185 (4), 291–297.

Nelson, B., Yuen, H.P., Wood, S.J., et al., 2013. Long term follow up of a group at ultra high risk ("prodromal") for psychosis: the PACE 400 study. JAMA Psychiatr. 70 (8), 793–802.

Nordentoft, M., Thorup, A., Peterson, L., et al., 2006. Transition rates from schizotypal disorder to psychotic disorder for first-contact patients included in the OPUS trial. A randomized clinical trial of integrated treatment and standard treatment. Schizophr. Res. 83 (1), 29–40.

O'Brien, M.P., Gordon, J.L., Bearden, C.E., Lopez, S.R., Kopelowicz, A., Cannon, T.D., 2006. Positive family environment predicts improvement in symptoms and social functioning among adolescents at imminent risk for onset of psychosis. Schizophr. Res. 81, 269–275.

Overall, J.E., Gorham, D.R., 1962. The brief psychiatric rating scale. Psychol. Rep. 10, 799–812.

Petersen, L., Jeppesen, P., Thorup, A., et al., 2005. A randomised multicentre trial of integrated versus standard treatment for patients with a first episode of psychotic illness. BMJ 331 (7517), 602.

Ruhrmann, S., Schultze-Lutter, F., Salokangas, R.K.R., et al., 2010. Prediction of psychosis in adolescents and young adults at high risk: results from the prospective European Prediction of Psychosis Study. Arch. Gen. Psychiatr. 67 (3), 241–251.

Saha, S., Chant, D., Welham, J., McGrath, J.J., 2005. A systematic review of the prevalence of schizophrenia. PLoS Med. 2 (5), e141.

Schizophrenia Commission, 2012. Schizophrenia – The Abandoned Illness.

Schlosser, D.A., Jacobson, S., Chen, Q., Sugar, C.A., Niendam, T.A., Li, G., Bearden, C.E., Cannon, T.D., 2012. Recovery from an at-risk state: clinical and functional outcomes of putatively prodromal youth who do not develop psychosis. Schizophr. Bull. 38 (6), 1225–1233.

Schultze-Lutter, F., Ruhrmann, S., Fusar-Poli, P., Bechdolf, A., Schimmelmann, B.G., Klosterkötter, J., 2012. Basic symptoms and the prediction of first-episode psychosis. Curr. Pharmaceut. Des. 18, 351–357.

Schultze-Lutter, F., Klosterkotter, J., Picker, H., Steinmeyer, E.-M., Ruhrmann, S., February 2007. Predicting first-episode psychosis by basic symptom criteria. Clin. Neuropsychiatry J. Treat Eval. 4 (1), 11–22.

Stefanis, N.C., Hanssen, M., Smirnis, N.K., Avramopoulos, D.A., Evdokimidis, I.K., Stefanis, C.N., Verdoux, H., Van Os, J., 2002. Evidence that three dimensions of psychosis have a distribution in the general population. Psychol. Med. 32 (2), 347–358.

Tsuang, M.T., Van Os, J., Tandon, R., Barch, D.M., Bustillo, J., Gaebel, W., et al., 2013. Attenuated psychosis syndrome in DSM-5. Schizophr. Res. 150 (1), 31–35.

Valmaggia, L., Byrne, M., Day, F., et al., 2015. Duration of untreated psychosis and need for admission in patients who engage with mental health services in the prodromal phase. Br. J. Psychiatr. 207, 130–134.

van der Gaag, M., Smit, F., Bechdolf, A., French, P., Linszen, D.H., Yung, A.R., McGorry, P., Cuijpers, P., 2013. Preventing a first episode of psychosis: meta-analysis of randomized controlled prevention trials of 12 month and longer-term follow-ups. Schizophr. Res. 149, 56–62.

Van Os, J., Kapur, S., 2009. Schizophrenia. Lancet 374 (9690), 635–645.

Van Os, J., Linscott, R.J., Myin-Germeys, I., Delespaul, P., Krabbendam, L., 2009. A sytematic review and meta-analysis of the psychosis continuum: evidence for a psychosis proneness-persistence-impairment model of psychotic disorder. Psychol. Med. 39, 179–195.

Woods, S.W., Walsh, B.C., Saksa, J.R., McGlashan, T.H., 2010. The case for including Attenuated Psychotic Symptoms Syndrome in DSM-5 as a psychosis risk syndrome. Schizophr. Res. 123 (2–3), 199–207.

Woods, S.W., Walsh, B., Hawkins, K., et al., 2013. Glycine treatment of the risk syndrome for psychosis: report of two pilot studies. Eur. Neuropsychopharmacol. 23, 931–940.

Woods, S.W., Addington, J., Cadenhead, K.S., et al., September 1, 2009. Validity of the prodromal risk syndrome for first psychosis: findings from the North American Prodrome Longitudinal Study. Schizophr. Bull. 35 (5), 894–908.

Yung, A.R., McGorry, P.D., 1996a. The prodromal phase of first episode psychosis: past and current conceptualisations. Schizophr. Bull. 22 (2), 353–370.

Yung, A.R., McGorry, P.D., 1996b. The initial prodrome in psychosis: descriptive and qualitative aspects. Aust. N. Z. J. Psychiatr. 30 (5), 587–599.

Yung, A.R., McGorry, P.D., McFarlane, C.A., Jackson, H.J., et al., 1996. Monitoring and care of young people at incipient risk of psychosis. Schizophr. Bull. 22 (2), 283–303.

Yung, A.R., Phillips, L.J., Yuen, H.P., Francey, S.M., McFarlane, C.A., Hallgren, M., McGorry, P.D., 2003. Psychosis prediction: 12-month follow up of a high-risk ("prodromal") group. Schizophr. Res. 60 (1), 21–32.

Yung, A.R., Yuen, H.P., McGorry, P.D., et al., 2005. Mapping the onset of psychosis – the comprehensive assessment of at risk mental states (CAARMS). Aust. N. Z. J. Psychiatr. 39, 964–971.

Yung, A.R., Yuen, H.P., Berger, G., Francey, S., Hung, T., Phillips, L.J., McGorry, P.D., 2007. Declining transition rate in Ultra High Risk (prodromal) services: dilution or reduction of risk? Schizophr. Bull. 33, 673–681.

Yung, A.R., Nelson, B., Baker, K., Buckby, J., Baksheev, G., Cosgrave, E., 2009. Psychotic-like experiences in a community sample of adolescents: implications for the continuum model of psychosis and prediction of schizophrenia. Aust. N. Z. J. Psychiatr. 43 (2), 118–128.

Yung, A.R., Nelson, B., Thompson, A., Wood, S.J., 2010. The psychosis threshold in Ultra High Risk ("prodromal") research: is it valid? Schizophr. Res. 120, 1–6.

Yung, A.R., Nelson, B., Thompson, A.D., Wood, S.J., 2010. Should a "risk syndrome for psychosis" be included in the DSMV? Schizophr. Res. 120 (1–3), 7–15.

Yung, A.R., Woods, S.W., Ruhrmann, S., Addington, J., Schultze-Lutter, F., Cornblatt, B.A., et al., 2012. Whither the attenuated psychosis syndrome? Schizophr. Bull. 38 (6), 1130–1134.

Yung, A.R., Nelson, B., McGorry, P.D., Wood, S.J., Lin, A., 2018. Persistent negative symptoms in individuals at Ultra High Risk for psychosis. Schizophr. Res. https://doi.org/10.1016/j.schres.2018.10.019 in press.

CHAPTER

4

Subjective disturbances in emerging psychosis: basic symptoms and self-disturbances

Frauke Schultze-Lutter[1,2], *Chantal Michel*[1,3], *Rahel Flückiger*[1], *Anastasia Theodoridou*[4,a]

[1]University Hospital of Child and Adolescent Psychiatry and Psychotherapy, University of Bern, Bern, Switzerland; [2]Department of Psychiatry and Psychotherapy, Medical Faculty, Heinrich-Heine University, Düsseldorf, Germany; [3]Developmental Clinical Psychology Research Unit, Faculty of Psychology and Educational Sciences, University of Geneva, Geneva, Switzerland; [4]Department of Psychiatry, Psychotherapy and Psychosomatics, University Hospital of Psychiatry, Zurich, Switzerland

Subjective experiences in early psychoses

Kraepelin (1896) observed that mental disorders rarely have an acute onset but, commonly, an insidious onset in which subtle changes in affect, drive, and cognition, as well as sleep and appetite, occur. Already in the early 20th century, the observation that precursor states are the rule rather than the exception led some researchers to entertain the idea of detecting psychosis early, ideally even before the onset of the first psychotic episode (Häfner et al., 2003; Rzesnitzek, 2013). One of them was the German psychiatrist Wilhelm Mayer-Gross (1889–1961) who, in 1932, wondered "why hitherto one [the psychiatrist] has so infrequently made use of the impressive experience that is represented by the first irruption of a thought disturbance, a decrease in activity, an aberration in sympathy and other emotions into the healthy personality [of persons with emerging psychosis]." (Mayer-Gross, 1932, p. 296; translated by F.S.-L.). Next to the limited presentation of patients in the early phase of illness in psychiatric practice, he regarded the facts "that these slowly emerging changes in mental habitus often escape the observation by others and by the patient or are not paid attention to for their slightness" (Mayer-Gross, 1932, p. 296; translated by F.S.-L.) as reasons for their neglect in

[a] Senior author.

Risk Factors for Psychosis
https://doi.org/10.1016/B978-0-12-813201-2.00004-1

psychiatric practice. Yet, he argued, in particular, disturbances in cognitive processes would often be retrospectively reported in detail and might even lead to an increased self-observation. Mayer-Gross' detailed, illustrated description of these cognitive disturbances as well as other subtle symptoms of an insidious onset was one of the first comprehensive descriptions of prodromal symptoms of first-episode psychosis. It inspired the subsequent works on an early detection of Gerd Huber (1921–2012), who started developing the *basic symptom* (BS) concept in the 1960s (Huber, 1966, 1983; 1986, 1995; Gross and Huber, 1989; Huber et al., 1980).

Concept of basic symptoms

Preventative research in psychosis has renewed interest in subjective and subclinical psychopathology beyond positive and associated symptoms such as negative and disorganized symptoms since the mid-1990s. The main approach to a detailed description of such subtle disturbances in mental processes is the concept of BS (Huber, 1966, 1983, 1995; Gross and Huber, 1989; Schultze-Lutter, 2009, 2016; Schultze-Lutter and Theodoridou, 2017; Schultze-Lutter et al., 2012c, 2016).

Definition of basic symptoms and original basic symptom model

Although BSs are not per se specific to psychosis and might occur in other severe nonorganic mental disorders, in particular, affective disorders (Klosterkötter et al., 1996), they were considered as an integral part of psychosis and to appear throughout all stages of the disorder (Schultze-Lutter, 2009; Schultze-Lutter and Theodoridou, 2017). BSs are generally defined as subtle, subclinical disturbances in any kind of mental process such as stress tolerance, drive, affect, thinking, speech, (body) perception, motor action, and central vegetative functions that are self-experienced with immediate and full insight into their abnormal nature (Gross and Huber, 1989; Schultze-Lutter, 2009; Schultze-Lutter et al., 2012c, 2016). Consequently, the ability to experience BS with insight and to report them often attenuates with progressive illness, i.e., with loss of insight and emerging psychotic symptoms, but is commonly restored upon remission (Gross and Huber, 1989). The term BSs was chosen for their conceptualization as the earliest primarily subjective psychopathological correlates of the physiological disturbances of information processing underlying the development of psychosis—or other severe mental disorders—that develop on the basis of, and partly in reaction to, them (Huber, 1966; Schultze-Lutter et al., 2012c, 2016). Thus, by definition, BSs differ from what patients consider to be their *normal* mental self and functions; this makes them distinct from trait-like schizotypy features considered as part of the *normal* self (Schultze-Lutter, 2009, 2016; Debbanné et al., 2015; see Relation of basic symptoms to other psychosis-risk concepts section). For their self-experienced deviation from the normal self, a subgroup of BS in combination with selected attenuated psychotic symptoms (APSs) and related symptoms was recently conceptualized as *self-disturbances* (SDs) or *self-disorders* and a core schizophrenia vulnerability phenotype (Parnas et al., 2005b; see Concept of self-disturbances section). Because BSs are independent from thought contents in terms of (delusional) ideas or explanations of perceptions as well as fluctuating in their occurrence, they remain commonly private and are rarely directly observable to others. However, others might observe patients' self-initiated coping strategies (avoidance strategies, compensatory behaviors, and/or self-medication) and emotional reactions (such as feelings of helplessness, anxiety, tension, and/or confusion) in response to BS, e.g., as negative or affective symptoms, or suicidality (Schmidt et al., 2017).

The original BS concept distinguished between three permeable symptom dimensions—or stages—of different severity and specificity to psychoses (Gross and Huber, 1989; Schultze-Lutter, 2009, 2016). Upon debut at stage 1, first unspecific BS, mainly involving disturbances of drive, volition, affect, concentration, and memory, will occur. At stage 2, these will be followed by the occurrence of qualitatively peculiar and more specific BS, including disturbances of thinking, speech, (body) perception, and motor action, which will gradually increase in number and severity. In most cases and based on BS, attenuated and frank psychotic symptoms will ultimately develop at stage 3, although temporary improvements are possible (Huber, 1986; Gross and Huber, 1989; Schultze-Lutter, 2009, 2016). In some cases, BS will even spontaneously remit completely before reoccurring and developing into full-blown psychotic symptoms. Because these symptomatic phases that can mimic true prodromal stages potentially announce the subsequent prodrome, they were called *outpost syndromes* (Gross and Huber, 1989; Schultze-Lutter, 2009).

The BS model of the development of psychoses assumes that neurobiological aberrations impair information processing at a cerebral level. The resulting pathologically vulnerable information processing capacity will become more susceptible to negative effects of stress, and already everyday situations and demands will more quickly overstrain the remaining information processing capacity and trigger the development of characteristic level-2 BS and their conversion to (attenuated) psychotic symptoms (Gross and Huber, 1989; Schultze-Lutter, 2009, 2016). Yet, if environmental and personal conditions are favorable (e.g., if there is a supportive social network and the person possesses good social, problem-solving, and coping skills), BS can well be coped with at any state as long as their number and/or severity do not overstrain personal resources. Thus, in early phases, the emerging disorder will only become apparent to others when inadequate coping strategies are employed or when compensatory abilities are exhausted and, consequently, BSs start to interfere with behavior. They might then become apparent as functional deficits, disorganized communication, changes in affect, and emotional stability. Furthermore, they might become observable by others when insight cannot be maintained and inadequate explanatory models are developed, which might present as attenuated or frank delusional or hallucinatory psychotic symptoms (Gross and Huber, 1989; Klosterkötter, 1992).

Following the first episode of psychosis, BS will evolve into one of three different types of outcome or, in the nomenclature of the original BS concept (Gross and Huber, 1989), *postsymptomatic basic stages*. One is a good outcome persistent with full remission of BS and other symptoms; the second is a partially poor outcome with only temporary full or partial remission and relapse episodes; and the third is a poor outcome with persistence of BS, especially of disturbances in drive, stress tolerance, affect, and cognition (Gross and Huber, 1989).

Assessments of basic symptoms

Following a first unpublished initial assessment, the Heidelberg Checklist, the *Bonn Scale for the Assessment of Basic Symptoms* (BSABS; Gross et al., 1987; Huber, 1986) and the *Frankfurt Complaint Questionnaire* (FCQ; Süllwold, 1986, 1991) were developed in parallel—the BSABS as a semistructured clinical interview on 142 single BS and the FCQ as a 98-item self-report questionnaire with some BS assessed by multiple items. Both BSABS and FCQ rate only presence/absence but not severity of single BS. Good interrater reliability was reported for the BSABS after 10 training sessions (Gross et al., 1990).

Based on multidimensional analyses of BSABS data (Schultze-Lutter et al., 2008, 2012c), the *Schizophrenia Proneness Instruments* were developed in an adult version (SPI-A; Schultze-Lutter et al., 2007a) and a children and youth version (SPI-CY; Schultze-Lutter et al., 2012b) as economic BS assessments that also allow rating of severity in terms of frequency (Daneault et al., 2013). Both instruments possess good interrater reliability after five training sessions (Schultze-Lutter et al., 2007a, 2012b). Furthermore, the BSABS greatly informed the concept-driven development of the *Examination of Anomalous Self-Experience* (EASE; Parnas et al., 2005b). The EASE includes 40 BSABS items rated for severity based on their frequency and associated distress (see also Table 4.2 below), and, consequently, EASE total scores were highly correlated to risk criteria–relevant BS (Raballo et al., 2016). More recently, drawing on both BSABS and EASE, the *Examination of Anomalous World Experience* (EAWE; Sass et al., 2017) was developed as a complement to the EASE to explore subjective anomalies in a person's experience of the external world, language, and other people. Of the 48 BSs of the BSABS included in the EAWE, the majority, in particular perceptual BSs, had not been included in the EASE.

Despite the conjoint development of the BSABS and the FCQ, subsequent studies reported poor correspondence between interview-assessed BSs and their presumably corresponding counterpart in the FCQ, indicating poor convergent validity of the self-rating of BS in comparison to their gold standard interview assessment (Maβ et al., 1997; Michel et al., 2017). Only two visual perception items clearly showed beyond-chance agreement between questionnaire and interview assessment; all other BSs were commonly overestimated by the FCQ (Michel et al., 2017). The same was true for the two BS psychosis-risk criteria (see Basic symptom criteria section) whose presence was overestimated by the FCQ by 30%–60% (Michel et al., 2017). Reinforcing these reports of poor correspondence between interview- and questionnaire-assessed BS and contrary to interview-assessed cognitive and perceptual BS, the FCQ or single FCQ factors did not predict transition to psychosis in a recent study of a clinical sample of psychosis-risk patients (Uttinger et al., 2018). Thus, a questionnaire approach to the assessment of BS criteria using the FCQ must be regarded as unreliable.

Basic symptom criteria

Cross-sectional studies in the 1980s and 1990s indicated that most BSs are indeed not specific to psychosis and may occur in other, especially nonpsychotic affective disorders (Klosterkötter et al., 1996; Schultze-Lutter et al., 2012c). Yet, the first-ever long-term prospective early detection study initiated by Gerd Huber in the late 1980s (Gross and Huber, 1989) with an average follow-up of 9.6 years (Klosterkötter et al., 2001) found that altogether 14 BSs were specific to the development of first-episode schizophrenia (Table 4.1). These were employed in the following two BS criteria that resulted from

TABLE 4.1 Prognostic accuracy of the risk-relevant BS and of the two BS criteria in the criteria-generation study (Klosterkötter et al., 2001).

Basic symptom (BS)	SEN	SPE	PPV	NPV	PLR	NLR	OR	FPP%	FNP%
Inability to divide attention	0.20	0.91	0.67	0.54	2.22	0.88	2.53	4.4	39.4
Captivation by details of the visual field	0.09	1	1	0.53	#	0.91	#	0	50.6
Disturbance of abstract thinking	0.04	1	1	0.52	#	0.96	#	0	50.6
Disturbance of expressive speech	0.23	0.94	0.78	0.56	3.83	0.82	4.68	3.1	47.5
Disturbance of receptive speech	0.39	0.91	0.82	0.61	4.33	0.67	6.46	4.4%	30.0%
Thought interference	0.42	0.91	0.83	0.62	4.67	0.64	7.32	4.4%	28.8%
Thought pressure	0.38	0.96	0.91	0.62	9.50	0.65	14.7	1.9%	30.6%

TABLE 4.1 Prognostic accuracy of the risk-relevant BS and of the two BS criteria in the criteria-generation study (Klosterkötter et al., 2001).—cont'd

Basic symptom (BS)	SEN	SPE	PPV	NPV	PLR	NLR	OR	FPP%	FNP%
Thought blockages	0.34	0.86	0.71	0.57	2.43	0.77	3.16	6.9%	32.5%
Unstable ideas of reference	0.39	0.89	0.78	0.60	3.55	0.69	5.17	5.6%	30.0%
Thought perseveration	0.32	0.88	0.71	0.57	2.67	0.77	3.45	6.3%	33.8%
Decreased ability to discriminate between ideas/perception, phantasy/true memories	0.27	0.95	0.84	0.57	5.40	0.77	7.03	2.5%	36.3%
Derealization	0.28	0.90	0.73	0.56	2.80	0.80	3.50	5.0%	35.6%
Visual perception disturbances	0.46	0.85	0.75	0.62	3.07	0.64	4.83	7.5%	26.9%
Acoustic perception disturbances	0.29	0.89	0.72	0.53	2.64	0.80	3.30	5.6%	35.0%
At least 1 of the 10 last-listed BS (COPER)	0.87	0.54	0.65	0.82	1.89	0.23	7.86	23.1%	6.3%
At least 2 of the 9 first-listed BS (COGDIS)	0.67	0.83	0.79	0.72	3.9	0.40	9.91	8.8%	16.3%

#, cannot be calculated because of disallowed division by 0; *COGDIS (bold)*, cognitive disturbances criterion; *COPER (bold)*, cognitive-perceptive BS criterion; *FNP%*, percent false-negative predictions; *FPP%*, percent false-positive predictions; *NLR*, negative likelihood ratio; *NPV*, negative predictive value; *OR*, odds ratio; *PLR*, positive likelihood ratio; *PPV*, positive predictive value; *SEN*, sensitivity; *SPE*, specificity.

TABLE 4.2 Comparison of phenomena relevant to the three psychosis-risk approaches and self-disturbances with emphasis on psychosis-risk symptoms.

Assumed dimension	Schizotypy (according to WSS)	SPD	BS (according to SPI-A/SPI-CY)	APS/BLIPS (according to SIPS)	SD[a] (according to EASE)
Positive	Magical ideation (beliefs that are regarded as invalid and magical by the conventional standards), perceptual aberration (deviations and distortions in the perception of one's body and the environment as well as sensory hypersensitivity)	Odd beliefs or magical thinking; suspiciousness or paranoid ideation; ideas of reference; unusual perceptual experiences (scoring at APS level in the SIPS)	Derealization, decreased ability to discriminate between ideas/perceptions, phantasy/true memories, visual and acoustic perception disturbances, SPI-CY main item and SPI-A subscale "body perception disturbances," unstable ideas of reference (scoring below APS level in the SIPS)	P1 unusual thought content/delusional ideas; P2 suspiciousness/persecutory ideas; P3 grandiose ideas; P4 perceptual abnormalities/hallucinations; D1 bizarre thinking	EASE subscales "bodily experiences" and "existential reorientation," derealization, diminished presence, inability to discriminate modalities of intentionality, morphological change, mirror-related phenomena, disturbance in experience of time, perceptualization of inner speech or thought, identity confusion, sense of change in relation to chronological age, sense of change in relation to gender, confusion with the other, passivity mood

(*Continued*)

TABLE 4.2 Comparison of phenomena relevant to the three psychosis-risk approaches and self-disturbances with emphasis on psychosis-risk symptoms.—cont'd

Assumed dimension	Schizotypy (according to WSS)	SPD	BS (according to SPI-A/SPI-CY)	APS/BLIPS (according to SIPS)	SD[a] (according to EASE)
Negative	Social anhedonia (diminished pleasure in social or interpersonal situations), physical anhedonia (deficits to experience pleasure in different sensory domains)	Lack of close friends; excessive social anxiety associated with paranoid fears; inappropriate or constricted affect	Most BS of the SPI-CY subscale "adynamia" and the SPI-A subscales "affective-dynamic disturbances" and "cognitive-attentional impediments," autopsychic depersonalization, difficulties to hold things in mind for less than an hour	N1 social withdrawal N2 avolition N3 expression of emotion N4 experience of emotion and self N6 occupational functioning D3 trouble with focus and attention	Most phenomena of the EASE subscale "self-awareness and presence," distorted first-person perspective, dissociative depersonalization, other states of depersonalizations, confusion with one's own specular image, disorder of short-term memory, discontinuous awareness of own action
Disorganized	Cognitive slippage (speech deficits and disorganized, confused thinking), schizotypal ambivalence (tendency to simultaneously experience divergent emotions)	Odd thinking and speech (scoring at APS level in the SIPS); behavior or appearance that is odd, eccentric, or peculiar	Disturbances of abstract thinking, (vaguely in WSS: thought pressure, thought blockage)	P5 disorganized communication D1 odd behavior and appearance D4 impairment in personal hygiene N5 ideational richness	
Cognitive			Thought interference, blockage, pressure, or perseveration; disturbances in abstract thinking or in receptive or expressive speech; inability to divide attention; captivation of attention		Thought interference, block, pressure or rumination, silent thought echo, disturbance of expressive language function, attentional disturbances

APS, attenuated psychotic symptom; *BLIPS*, brief intermittent psychotic symptom; *BS*, basic symptom; *EASE*, Examination of Anomalous Self-Experience; *SD*, self-disturbance; *SIPS*, Structured Interview for Psychosis-Risk Syndromes; *SPD*, schizotypal personality disorder; *SPI-A/SPI-CY*, Schizophrenia Proneness Instrument, Adult/Child & Youth version; *WSS*, Wisconsin Schizotypy Scales. The 14 criteria-relevant BS and 5 positive items of the SIPS are underlined.

[a] *For lack of empirical studies, the Examination of Anomalous World Experience, EAWE, is not considered here.*

different analyses and share five BSs (Klosterkötter et al., 2001; Schultze-Lutter et al., 2006, 2007b):

- *Cognitive-Perceptive Basic Symptoms* (COPER) requires the presence of at least one of the following 10 BSs: *thought interference; thought perseveration; thought pressure; thought blockages; disturbance of receptive speech; disturbance in discriminating ideas and perception, phantasy, and true memory content; unstable ideas of reference; derealization; visual perception disturbances* (except hypersensitivity to light and blurred vision); and *acoustic perception disturbances* (except hypersensitivity to sound). The respective BS has to be first noted at least 12 months previous to assessment and, for any period of time within the previous 3 months, has to have occurred at a frequency of at least once per week (corresponding to a severity rating on SPI-A/SPI-CY of at least "3").
- *Cognitive Disturbances* (COGDIS) requires the presence of at least two of the following nine BSs: *thought interference; thought pressure; thought blockages; disturbance of receptive speech; disturbance of expressive speech; disturbance in abstract thinking; unstable ideas of reference; inability to divide attention;* and *captivation of attention by details of the visual field.* The respective BSs have to have occurred at a frequency of at least once per week (corresponding to a severity rating on the SPI of at least "3") for any period of time within the previous 3 months, irrespective of time of first notice.

Recently, a community study of 2683 16- to 40-year-old citizens of the Swiss canton Bern reported that any 1 of the 14 BSs included in COPER and or COGDIS was reported by 9.8% of the participants within the 3 months prior to the telephone interview conducted by trained psychologists (Schultze-Lutter et al., 2018a). In doing so, 8.3% of the participants reported at least any 1 of the 10 BSs included in COPER and 6.3% at least any 1 of the 9 BSs included in COGDIS. Yet, these BSs commonly occurred infrequently so that only 2.01% met any one of the two BS psychosis-risk criteria: 1.94% met COPER, 0.56% in addition to COGDIS, while COGDIS was met exclusively only in 0.07% (Fig. 4.1; Schultze-Lutter et al., 2018a). When the association of BS criteria with the presence of a nonpsychotic axis-I disorder and with a functional deficit in terms of a score below 71 on the Social And Occupational Functioning Assessment Scale (American Psychiatric Association, 1994) was examined as a proxy measure of clinical significance, the exclusive presence of BS criteria in the absence of APS-related risk criteria was significantly related to both. The association with a functional deficit (OR = 15.890; 95% CIs: 8.763–28.813) was considerably stronger than that with a mental disorder (OR = 5.350; 95% CIs: 3.002–9.534), indicating that BS criteria are not merely an indicator of mental illness but of clinical significance by themselves (Schultze-Lutter et al., 2018a).

Concept of self-disturbances

SDs, also known as anomalous self-experiences or self-disorders, have long been described as playing a central role in the concept of schizophrenia. Starting from Eugen Bleuler's claim that schizophrenia is essentially characterized by a "loss of the unity of personality" (1911), and drawing on the phenomenological tradition in philosophy, i.e., the works of Edmund Husserl or Maurice Merleau-Ponty, SDs have been continually explored further and conceptually refined (Blankenburg, 1971; Kronfeld, 1930; Minkowski, 1927; Laing, 1959; Sass and Parnas, 2003; Sass, 2014). The *self,* in these theories, is commonly understood as the center of a subject's conscious life that imparts a sense of *mineness* to all of someone's conscious experiences (Zahavi, 2005; Guillot, 2017). Yet, a consensus on the meaning of self is still lacking, and the construct of (minimal) self or aspects is even questioned completely by narrative theory, "no self-conceptions," or some neuroscience

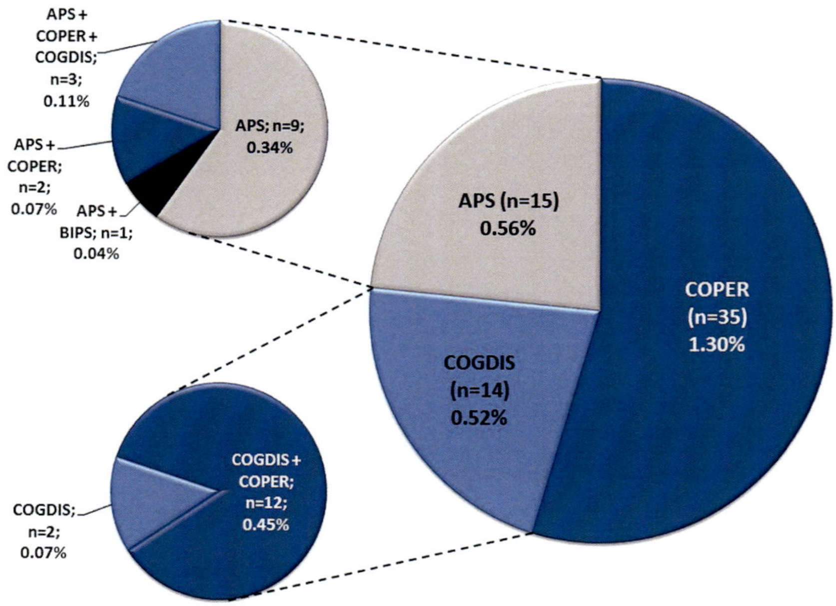

FIGURE 4.1 **Prevalence and co-occurrence of interview-assessed psychosis-risk criteria in a random representative general population sample of the canton Bern, age 16–40 years (N = 2683).** *According to Schultze-Lutter, F., Hubl, D., Schimmelmann, B.G., Michel, C., 2017. Age effect on prevalence of ultra high risk for psychosis symptoms: replication in a clinical sample of an early detection of psychosis service. Eur. Child Adolesc. Psychiatry 26, 1401–1405.*

approaches. Yet, from a psychopathological perspective, the sense of self in schizophrenia is disturbed, thus, for example, resulting in feelings of self-alienation ("as if" feelings), solipsistic grandiosity, or bizarre attitudes. In clinical psychiatry, interest in SD—just as in BS—was renewed by the emergence of the field of early recognition and intervention in psychosis. Attempting to detect the earliest signs of a psychotic development, the attention is turned to subtle changes in subjective experiences. Similar to BSs, SDs are conceptualized as a constellation of interrelated anomalies of subjective experience. Yet, in contrast to BS, in the SD concept, these anomalies are thought to gravitate around pervasive distortions of the "minimal self" or "core self." The minimal self is a common concept in contemporary cognitive sciences, phenomenology, and philosophy of mind that refers to the primary sense of *mineness*, *myness*, and *for-me-ness* (Sass and Parnas, 2003). Parnas and Sass (2011, p. 532) described the disturbed minimal self as "fragile and unstable, as if it loses its automaticity."

As with BS (Klosterkötter et al., 1996, 1997), and despite variable sensitivity and specificity of single SD items for psychotic disorders (Haug et al., 2012; Nordgaard and Parnas, 2014; Møller et al., 2011), SDs were shown to aggregate in schizophrenia (Mishara et al, 2014; Henriksen and Nordgaard, 2014) and to agglomerate as a rather specific index of vulnerability to schizophrenia-spectrum disorders (independently of the concomitant presence of overt psychotic symptoms), therefore also occurring in schizotypal conditions (Raballo and Parnas, 2011). After the initial validation of the SD construct (conducted through proxy scores extrapolated from the BSABS), the EASE was developed on mainly theoretical grounds as a specific assessment tool for SD (Parnas et al., 2005b). Despite some natural item overlaps with the BSABS (although even for analog items, definitions are usually slightly broader in the EASE—see Parnas et al., 2005b), the EASE mainly focuses on *gestalt* transformations of the normal articulation of the stream of consciousness and incorporates also micropsychotic experiential variants that phenomenologically overlap with APS as assessed with the *Structured Interview for Psychosis-Risk Syndromes* (SIPS; McGlashan et al., 2010) or the *Comprehensive Assessment of At-Risk Mental States* (Yung et al., 2006) (see also Table 4.2 below). Thus, SD also aggregate in psychosis-risk states defined by BS or ultra high risk (UHR) criteria (Raballo et al., 2016). Both state-like BS and supposedly trait-like SD manifest aspects of disturbed subjective experience that may (or may not) progress into surface-level psychotic symptoms. However, while the BS concept and related assessment tools (see Assessments of basic symptoms section) were progressively refined for the prediction of imminent risk of psychosis, SDs have been conceived as a core phenotype for the construct validity of the broad schizophrenia spectrum (in its clinical and nonclinical counterpart, e.g., schizotypy) and less studied for their psychosis-predictive value so far (see Prediction of psychosis by self-disturbances section).

Recently, a revision of phenomenological theory provided a biophenosocial model for schizophrenia trying to integrate experiential, neurocognitive, psychosocial, and neurobiological aspects (Sass et al., 2018). This approach proposes a model incorporating primary (e.g., trait-like features) and secondary factors (consequential, defensive, compensatory features) on phenolevel (e.g., dissolution of first-person perspective/dissociative symptoms) and neurolevel (e.g., aberrant salience, source-monitoring deficits). As a phenomenological SD concept, this approach aims to account for the temporal aspects of the symptoms, to disentangle overlapping concepts such as depersonalization and derealization in the context of some nonschizophrenic syndromes and, finally, to integrate other aspects such as the influence of trauma or culture (Sass et al., 2018). Thus, in the search for an adequate model of schizophrenia, the BS concept might provide the required tools for an empirical approach accounting for the legitimacy of the notion of self and the potential distinct pathways of the shared disturbances of self-experiences in psychosis.

Relation of basic symptoms to other psychosis-risk concepts

Next to the BS approach, two other main approaches are currently followed in the endeavor to detect psychosis early (Fusar-Poli et al., 2013b; Debbané et al., 2015): the UHR approach (see Chapter 3) and the schizotypy approach (see Chapter 5). UHR criteria consist of brief (limited) intermittent psychotic symptoms (B(L)IPS) and APS that account for majority of patients in UHR samples (Schultze-Lutter et al., 2015). They also consist of a trait-state criterion, i.e., a genetic risk due to a first degree relative with a psychotic disorder or due to the presence of a schizotypal personality disorder (SPD), indicating the trait, along with a significant recent but persistent functional decline, indicating the state. While both the UHR and the BS approach are mainly followed in clinical samples, the schizotypy approach is mainly followed in nonclinical samples, frequently college students (Debbané et al., 2015).

Phenomenological differences and similarities between concepts

Schizotypy is conceptualized as a multidimensional *trait* construct representing a not necessarily pathological (genetic) predisposition or liability to schizophrenia resembling the schizophrenia (spectrum) symptom dimensions with a negative (interpersonal), positive (cognitive perceptual), and disorganization dimension (Nelson et al., 2013; Lenzenweger, 2015; see Chapter 5). Using self-report-questionnaires, schizotypy is typically assessed psychometrically, of which the *Wisconsin Schizotypy Scales* (WSS) is a commonly used instrument (Chapman et al., 1994). These comprise of four scales: the *Perceptual Aberration* Scale (Chapman et al., 1978) and the *Magical Ideation* Scale (Eckblad and Chapman, 1983) that, together, form the positive schizotypy dimension (Kwapil et al., 2008) as well as the *Physical Anhedonia* Scale (Chapman et al., 1976) and the *Social Anhedonia* Scale (Eckblad et al., 1982) that, together, form the negative schizotypy dimension (Kwapil et al., 2008). The third, frequently neglected schizotypy dimension of disorganization can be assessed by the *Cognitive Slippage Scale* (Miers and Raulin, 1987) and the *Schizotypal Ambivalence Scale* (Kwapil et al., 2002). The characterization of schizotypy scales, in particular, the WSS, as psychosis liability found support in several studies applied in (nonclinical) community samples, where high scores on the WSS, i.e., both positive scales and *Social Anhedonia,* were predictive of psychosis conversion (Chapman et al., 1994; Kwapil, 1998; Kwapil et al., 2013). In contrast, a study examining the psychosis-predictive value of the WSS in a clinical sample showed exclusively the *Physical Anhedonia* Scale to be predictive but only in conjunction with a positive CHR status (Flückiger et al., 2016a).

As initially described (see Definition of basic symptoms and original basic symptom model section), BS as well as APS and BLIPS and schizotypy differ in their conceptualization as well as their course as states and traits, respectively (Michel et al., 2019). Furthermore, BS and schizotypy greatly differ in the definition of their symptoms or features (Table 4.2). For example, only 4 of the 41 items of *Perceptual Aberration* and *Magical Ideation* vaguely relate to perceptive BS in terms of a "changed perception of patient's own face (or body)," "derealization," and "changed quality of acoustic stimuli"; and only 6 of the 35 items of the *Cognitive Slippage Scale* distantly relate to cognitive BS, 5 of them to "thought pressure" and 1 to "thought blockage." In contrast, the concept of schizotypy and, relatedly, the definition of SPD had significantly informed the development of UHR criteria, in particular of APS (Table 4.2). APS comprises positive symptoms not yet held or experienced with full conviction as initially described as clinician-assessed *psychotic-like experiences* (Chapman and Chapman, 1980) and the five positive features of SPD (Ruhrmann et al., 2010a; Schultze-Lutter et al., 2012a).

Structural relationships between concepts

Only few studies exist about the relationship between schizotypy features and BS. Yon et al. (2009) examined the relationship between schizotypy, assessed by the four WSS, the *Cognitive Slippage Scale* and the *Schizotypal Ambivalence Scale*, and the FCQ. In a student sample, a two-factor solution was yielded, with the negative factor containing only Social and Physical Anhedonia and all other five scales forming an extended positive factor. Yon et al. (2009) concluded that questionnaire-assessed BSs were associated with the positive rather than with the negative dimension of schizotypy. Yet, because FCQ-assessed BSs hardly correspond to interview-assessed BSs (see Assessments of basic symptoms section), these first results convey, at best, a very preliminary estimation of the relationship between schizotypy measures and risk-relevant interview-assessed BS.

Extending the findings of Yon et al. (2009) to interview-assessed risk-relevant BS and UHR symptoms, factor analyses and structural equation modeling of 4 WSS, the 14 predictive BS of the SPI-A/SPI-CY, and 5 positive items of SIPS (McGlashan et al., 2010) were carried out in a large patient sample ($N = 277$; aged 8–40, 22 ± 6 years) from 2 early detection services (Flückiger et al., 2016b). Initial calculation of Spearman's Rho (ρ) revealed that three items, each rarely reported by $\leq 5\%$ of patients, had to be excluded from further analyses due to $\rho < 0.3$: the BS "captivation of attention by details of the visual field," "disturbance of abstract thinking", and the positive item "grandiosity." As expected, one negative dimension consisted of two anhedonia scales of the WSS, and another positive dimension consisted of the two WSS forming the positive schizotypy dimension, the remaining three delusional and hallucinatory positive items, and the three BS included in these as below-APS phenomena (see Table 4.2; a fourth BS, derealization, failed to link to the positive dimension in the structural equation model). The eight cognitive BS and the positive item "disorganized communication" formed a third cognitive dimension (Fig. 4.2). The structural equation model revealed that negative and positive dimensions were not related to each other but both significantly to the cognitive dimension (Fig. 4.3). The cognitive dimension explained the largest amount of variance (58%) and acted as a higher-order factor in the model. Its superior position indicates that cognitive BS exacerbate or provoke symptoms of the other two dimensions, thus indicating the importance of subjective cognitive deficits in the early detection and treatment of psychoses (Flückiger et al., 2016b).

Current evidence for the basic symptom model

Basic symptoms and their association to neurobiological aberrations

Despite the fact that BSs are considered to be a direct manifestation of brain abnormalities in patients with schizophrenia, the nature of such abnormalities appears subtle and heterogeneous, requiring sophisticated methods of analyses to be detected. A recent review of the so far limited number of neurobiological studies on BS provides preliminary evidence for various different neurobiological mechanisms underlying BS (Schultze-Lutter et al., 2016a). It highlighted that diverse anatomical, pharmacological, and functional correlates may be involved in the manifestation of BSs in psychotic and psychosis-risk patients. These include structural alterations, changes in event-related potentials and neural oscillations, aberrations in neurotransmitter systems, and evidence for changes in large-scale networks as assessed with fMRI. Yet, many of these studies either involved samples selected by COPER and/or COGDIS or studied the totality of BS without consideration of their various natures. Thus, the review concluded that different and more differentiated strategies should be employed to gain more detailed and possibly mechanistic insights into the neurobiology of BS. These insights may not only be useful for an improved general understanding of BS but may ultimately give critical insights into the development of psychosis, which could be crucial for early diagnosis and intervention. Furthermore, because BSs are present throughout the course of the illness, studies of the neurobiology of BS may also give new insights into the neurobiological determinants of unfavorable outcomes including functional deterioration (Schultze-Lutter et al., 2016; Schultze-Lutter and Theodoridou, 2017).

Developmental aspects of basic symptoms and attenuated psychotic symptoms in the general population

Another recent finding that was interpreted as linking BS to neurobiology in terms of brain maturation was reported from epidemiological studies (Schultze-Lutter et al., 2018b). In a combined sample of two Swiss general population studies (Schimmelmann et al., 2015), which together comprised individuals aged 8–40 years, the influence of age on the prevalence of APS as assessed with the SIPS and BS employed in COPER and COGDIS as assessed with SPI-A/SPI-CY was studied. Since high prevalence rates had been reported for hallucinatory experiences in children and young adolescents of the community (Schimmelmann et al., 2013; Schultze-Lutter and Schmidt, 2016), perceptual and non-perceptual (delusion-like or cognitive) psychosis-risk symptoms were distinguished.

FIGURE 4.2 **The development of the Examination of Anomalous Self-Experience (EASE) from the Bonn Scale for the Assessment of Basic Symptoms (BSABS).** *Modified from Schultze-Lutter, F., Debbané, M., Theodoridou, A., Wood, S.J., Raballo, A., Michel, C., Schmidt, S.J., Kindler, J., Ruhrmann, S., Uhlhaas, P., 2016a. Revisiting the basic symptom concept: towards translating risk symptoms for psychosis into neurobiological targets. Front. Psychiatry 7, 9.*

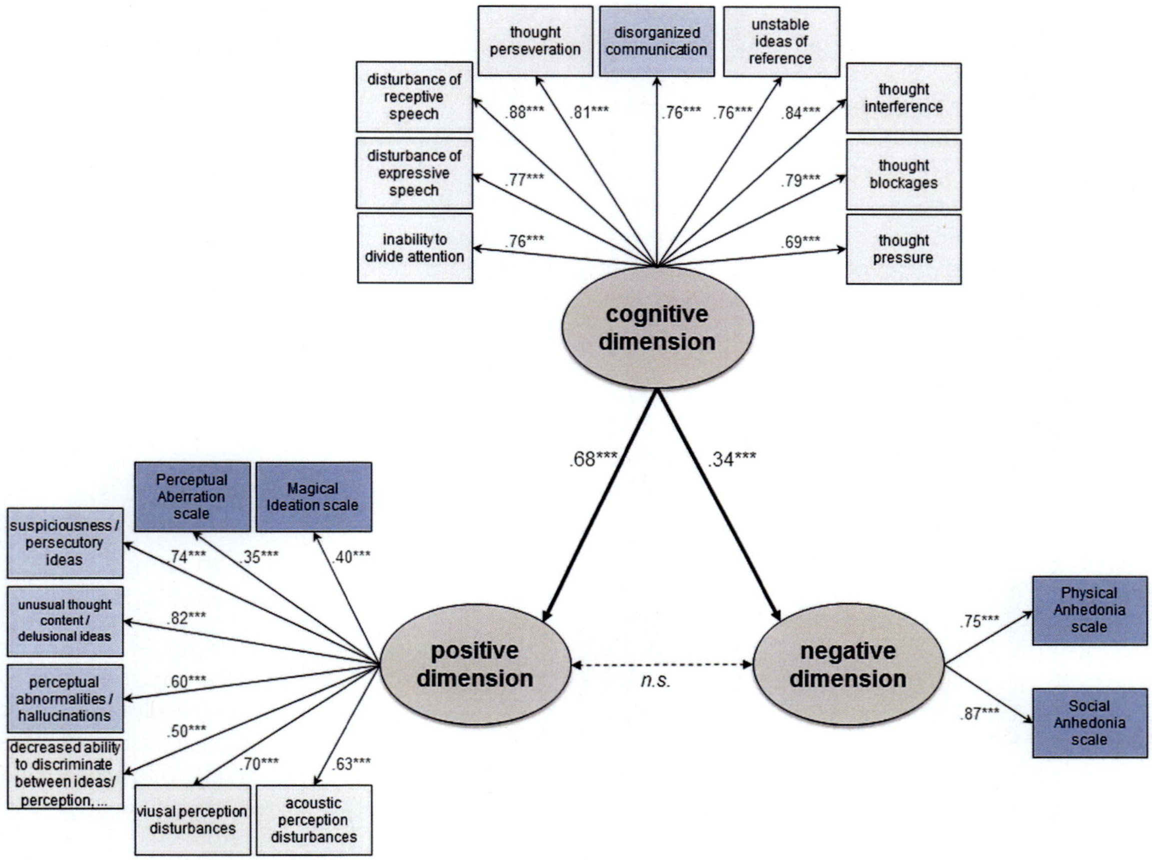

FIGURE 4.3 **Structural equation model of the common dimensions of the three psychosis-risk concepts.** Initial model fit was adequate ($\chi^2_{(149)} = 251.283$, $P < 0.001$, CFI = 0.95, RMSEA = 0.050 [90% CIs = 0.039–0.060], WRMR = 1.002); it improved and became good following the elimination of the insignificant path between the positive and negative from the final model ($\chi^2_{(150)} = 248.851$, $P < 0.001$; CFI = 0.95, RMSEA = 0.049 [90% CIs = 0.038–0.059], WRMR = 1.004). Observed variables are indicated by *squares*, and bright blue indicates positive items of the SIPS, blue indicates WSS, and gray indicates BS.

To estimate the clinical relevance of these symptoms and their association with age, a symptom-independent current global level of psychosocial functioning, in particular, an indication toward a functional deficit, and current mental disorders according to DSM-IV criteria were also assessed (Schimmelmann et al., 2015).

With regard to the prevalence of the 14 BSs included in psychosis-risk criteria, any 1 BS was reported by 18% of the participants within the 3 months prior to the interview: 13% reported at least any 1 of the 12 cognitive BS, 3% any 1 of the 2 perceptual BS, and 2% both (Schultze-Lutter et al., 2018b). Any one of the two BS psychosis-risk criteria was met by 3%, whereby the cognitive-perceptive BS criterion COPER was reported approximately 2.5 times more often than the cognitive disturbances criterion COGDIS. Perceptual and cognitive BSs were differently affected by age, whereby age thresholds were indicated for perceptual BS in late adolescence (around age 18) and for cognitive BS in young adulthood (early twenties)—in each case, with higher prevalence rates, but a lesser association with functional deficits and the presence of mental disorder in the below-threshold groups. Thereby, interaction effects between age and BS on functioning and mental disorder were commonly stronger than the individual effects of age and BS. Thus, with an age threshold around the age of 18 years for perceptual BS and within the early twenties for cognitive BS, effects of perceptual and cognitive BS seem to follow normal brain maturation processes (Houston et al., 2014; Klein et al., 2014), in which they might occur as infrequent and temporary nonpathological disturbances (Fig. 4.4). Thus, their persistence beyond and/or occurrence after the conclusion of main brain maturation processes were discussed as possibly signifying aberrant maturation and neurodegenerative processes, respectively (Schultze-Lutter et al., 2018a,b). Interestingly, within the same sample, the age threshold for APS was lower, around the age of 16 years (Schimmelmann et al., 2015). At point prevalence rates of any one APS of 10% and of the APS-related risk criterion of slightly over 1%, 8- to 15-year-olds reported more perceptual APS compared to older individuals that were generally unrelated to functional impairment, regardless of age. Conversely, nonperceptual APS were related to functional deficit, in particular, in those of age 16 years or older (Schimmelmann et al., 2015). This age threshold was largely confirmed recently in a clinical sample (Schultze-Lutter et al., 2017). In light of the BS concept and the above findings on BS, APS were regarded as being affected primarily by the development of cognitive abilities, which is largely concluded by late adolescence (Vijayakumar et al., 2014; Schultze-Lutter et al., 2018b). Thus, APS is more frequent, but less clinically significant, before development of main cognitive abilities is completed, while the same might hold true for BSs, in particular cognitive ones, before the conclusion of the maturation of more frontal brain regions (Schultze-Lutter et al., 2018b, Fig. 4.4). If future studies support this proposed theory, it will provide new important insight in the pathogenesis of psychosis. Further, it will also have important therapeutic implications that would support the search for effective neuroprotective interventions in particular for BS, while emphasizing on cognitive-behavioral interventions for APS (Bentall et al., 2007; French and Morrison, 2004). The combination of both interventions, however, might be indicated in the group with the highest risk for psychosis, i.e., help seekers with both APS and COGDIS (Schmidt et al., 2015; see Conversion to psychosis in samples meeting basic symptom criteria section).

Prediction of psychosis

Conversion to psychosis in samples meeting basic symptom criteria

A recent metaanalysis (Schultze-Lutter et al., 2015a) conducted as part of the Guidance project of the European Psychiatric Association reported pooled conversion to psychosis rates in samples meeting COGDIS—exclusively or in combination with other CHR criteria between 25% (95% CIs: 23%–28%) at 12 months and 61% (95% CIs: 44%–79%) at follow-ups beyond 4 years (Fig. 4.5). Because COPER has been less frequently studied, pooled conversion rates could only be calculated for 12- and 24-month follow-ups; these were similarly high as those of COGDIS: 14% (95% CIs: 7%–22%) at 12 months and 21% (95% CIs: 11%–32%) at 24 months (Schultze-Lutter et al., 2015).

Pooled conversion rates between the three UHR criteria differed significantly, mainly due to higher conversion rates in BLIPS and/or lower in the genetic risk and functional decline criterion (GRFD) samples (Fig. 4.5). COPER and COGDIS samples did not significantly differ from each other or from UHR samples—irrespective of their potential co-occurrence—in 1- and 2-year pooled conversion rates but in long-term pooled conversion rates, which were

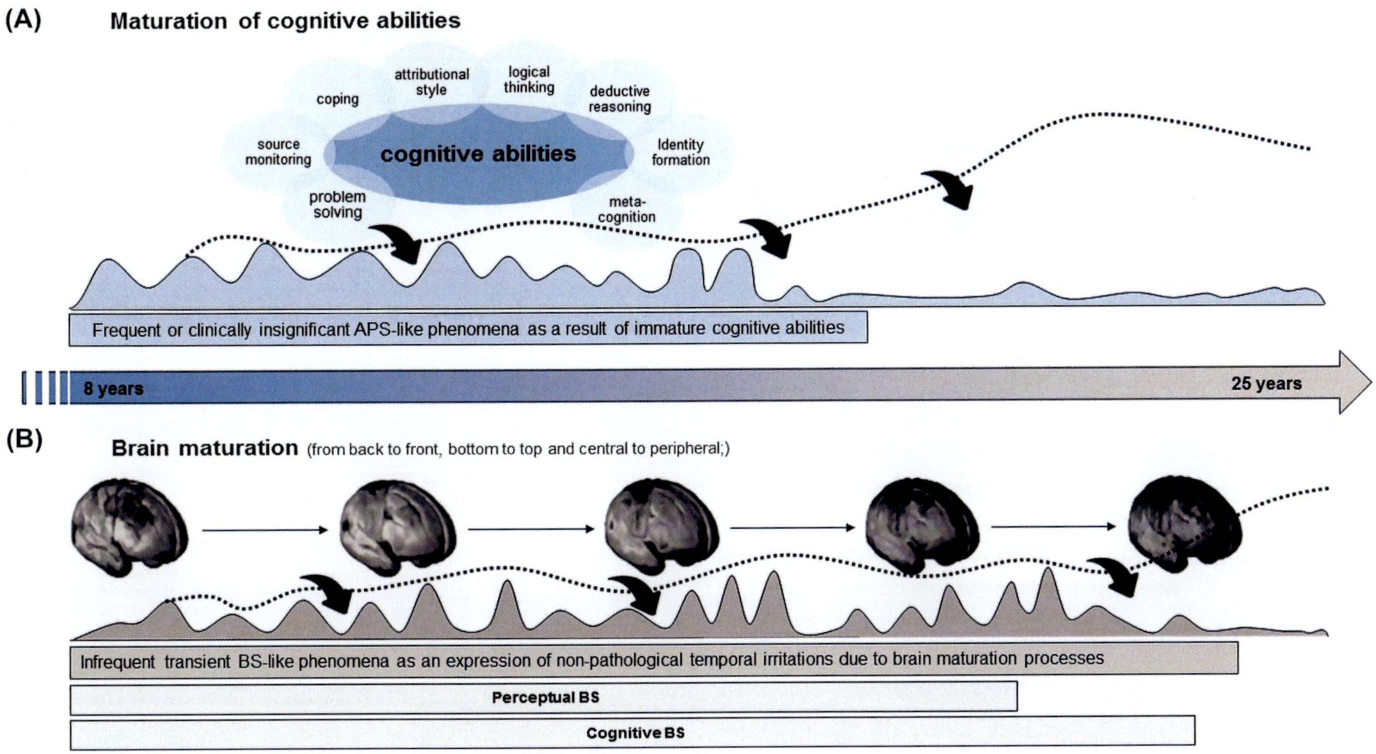

FIGURE 4.4 **Illustration of the assumed association between basic symptoms (BSs) and brain maturation and attenuated psychotic symptoms (APSs) and maturation of cognitive abilities according to Schultze-Lutter et al. (2018b).** This model assumes on the one hand that (A) unusual thought contents and unusual perceptual experiences identical to APS might occur during childhood and early adolescence as an expression of not yet fully matured cognitive abilities (*blue-shaded curve*). If the maturation of cognitive abilities is impaired by risk factors or stressors (*arrows*) or neurodevelopmental disturbances in information processing, which might be signaled by the occurrence of BS, APS might persist or progress (*upper dotted line*) as a result of poor coping, externalization biases, or poor source monitoring, potentially leading to schizotypal traits, schizotypal personality disorder, and/or psychosis. On the other hand, the model assumes that (B) subjective subclinical disturbances in cognition and perception phenomenologically identical to BS might occur during childhood and adolescence as infrequent temporary expressions of minor transient dysfunctions in the wake of brain maturation processes (*gray-shaded curve*). However, if these BSs occur more frequently and are persistent (*lower dotted line*), they might indicate disturbances in brain maturation that, in *line* with a neurodevelopmental model of psychosis, predispose the person to the development of psychosis. A genetic predisposition as well as childhood adversities, other risk factors, and stressful life events (*arrows*) and cognitions promoting the development of APS might further fuel the development and persistence of information processing disturbances. Thus, when co-occurring, BS and APS might reciprocally amplify each other, significantly enhancing the risk of developing psychosis, whereby the effect of BS on APS might be greater than the reverse effect (see Fig. 4.3).

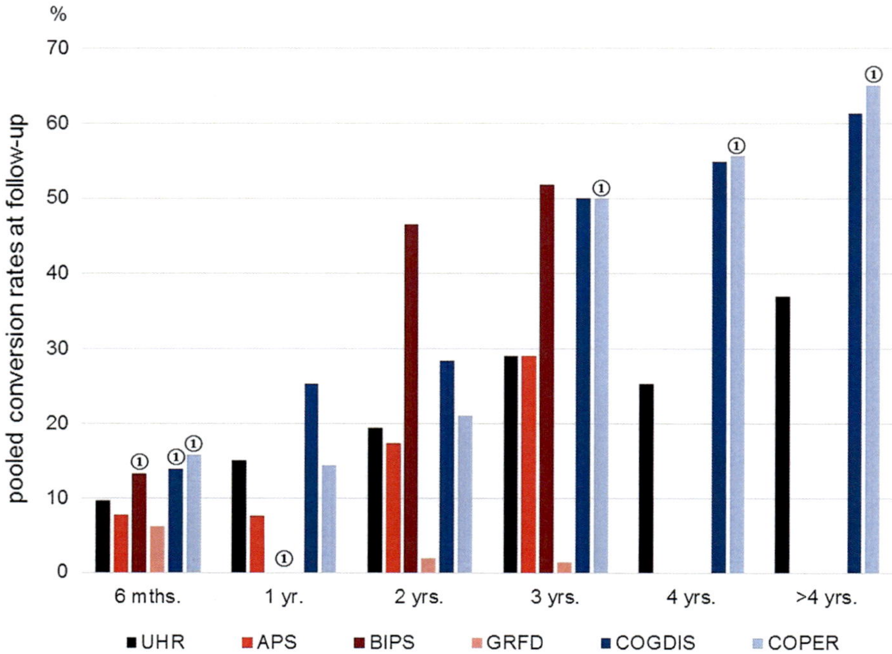

FIGURE 4.5 **Pooled conversion rates to psychosis associated with single psychosis-risk criteria irrespective of each other at different follow-ups according to Schultze-Lutter et al. (2015).** ① indicates not pooled but single-study conversion rates; *APS*, attenuated psychotic symptom criterion of the UHR criteria; *BIPS*, brief intermittent psychotic symptom criterion of the UHR criteria (equals BLIPS); *COGDIS*, cognitive disturbances criterion; *COPER*, cognitive-perceptive basic symptoms criterion; *GRFD*, genetic risk and functional decline (i.e., trait-state) criterion of the UHR criteria; *UHR*, any one ultra high risk criterion.

higher in COGDIS samples as well as in the one long-term study of COPER (Klosterkötter et al., 2001) as compared to UHR samples (Schultze-Lutter et al., 2015a). Higher long-term conversion rates in samples meeting BS criteria compared to UHR samples and the low conversion rates of GRFD samples were confirmed by a subsequent replication study of pooled conversion rates associated with different psychosis-risk criteria that, however, failed to distinguish between COPER and COGDIS (Fusar-Poli et al., 2016). Furthermore, an earlier metaanalysis on the type of psychosis upon conversion in psychosis-risk samples had linked BS criteria to a higher likelihood of developing schizophrenia compared to UHR criteria, which were more frequently followed by affective psychoses (Fusar-Poli et al., 2013a).

With regard to the co-occurrence of UHR and BS criteria, most studies employing both UHR and BS criteria, predominately COGDIS (Marshall et al., 2013; Ruhrmann et al., 2010b; Schultze-Lutter et al., 2014; Ziermans et al., 2011), indicated considerably higher conversion rates in subsamples meeting criteria of both approaches, both compared to subsamples meeting criteria of only one approach at short- and long-term follow-ups. In contrast, 3-year hazard rates of psychosis were lower in combined UHR and BS subsamples compared to UHR subsamples with and without BS criteria in a sample with a high attrition rate, although specificity and positive predictive power improved by the combination (Hengartner et al., 2017). Yet, when only APSs were considered, the combination of APS with any BS criterion again increased hazard rates in comparison to only considering APS, COPER, or COGDIS by themselves and irrespective of each other (Hengartner et al., 2017).

With regard to the course of psychosis-risk symptoms in nonconverters, a recent study reported high overall remission rates over an average 4-year course (43%; Michel et al., 2018). In doing so, a higher maintenance rate of BS of COGDIS (18%) compared to APS/BLIPS (12%) was found, in particular when the exclusive presence of COGDIS (29%) and APS/BLIPS (4%) was considered (Michel et al., 2018). Together with the higher conversion rates of COGDIS samples in the long term (Fig. 4.5; Schultze-Lutter et al., 2015a), this might indicate that BS criteria are indeed able to detect psychosis earlier than the UHR criteria that had been explicitly developed to detect psychosis within the 12 months before its onset (Schultze-Lutter, 2006).

Prediction of psychosis by self-disturbances

Within the framework of the SD concept, psychosis-risk criteria have not been defined so far. Yet, two studies examined the psychosis-predictive power of subscales of SD measures (Nelson et al., 2012; Parnas et al., 2011). Based on a BSABS version that was expanded with additional items concerning self-experience and arranged in 7 subscales (Parnas et al., 2005a), the 5-year follow-up of a prospective, observational study of first admission patients in putative state of beginning psychosis ($N = 151$) reported that high baseline scores of the subscales *perplexity* (11 BSs, 5 of them included in COPER/COGDIS) and *self-disorders* (10 BSs, 5 of them included in COPER/COGDIS, plus 5 additional items) yielded the best prediction of the subsequent development of schizophrenia-spectrum disorders with $OR = 6.11$ (95% CIs: 1.34–27.96) and $OR = 12.00$ (2.15–67.07), respectively (Parnas et al., 2011).

Furthermore, examination of the EASE in a small sample of 49 UHR patients followed for a mean of 19 months (Nelson et al., 2012) indicated that total EASE score significantly predicted time to transition when other significant variables (duration of symptoms prior to baseline examination and functioning levels) were controlled for. This was mainly due to the effect of two domains of the EASE, *Cognition and Stream of Consciousness* and *Self Awareness and Presence*, which were both found to predict transition, after controlling for the same variables (Nelson et al., 2012). While 12 (70%) of the 17 main items of Cognition and Stream of Consciousness are BSs taken from the BSABS, 8 of them included in COPER/COGDIS, only half of the 18 main items of Self-awareness and Presence are BSs, and of that only one, *derealization*, is part of COPER.

Conclusion

Self-experienced subtle disturbances in mental processes, in particular, in cognition, as initially described by the BS concept and as later included in the SD concept seem to aggregate in psychotic disorders, especially of the schizophrenia spectrum. Yet, these self-experienced subtle disturbances might also occur as infrequent, temporary phenomena as an expression of the significant neurodevelopmental reorganization during brain maturation, thus supporting their assumed close relationship to the neurobiological aberrations underlying psychotic disorders. Both their psychosis-predictive value and their neurobiological foundation make them valuable targets in early detection and etiological research.

List of acronyms and abbreviations

APS Attenuated psychotic symptom
BLIPS Brief limited intermittent psychotic symptom

BS Basic symptom
BSABS Bonn Scale for the Assessment of Basic Symptoms
CFI Comparative fit index
CI Confidence interval
COGDIS Cognitive disturbances
COPER Cognitive-perceptive basic symptoms
EASE Examination of Anomalous Self-Experiences
EAWE Examination of Anomalous World Experience
RMSEA Root mean square error of approximation
SD Self-disturbances (also called self-disorders)
SIPS Structured Interview for Psychosis-Risk Syndromes
SPD Schizotypal personality disorder
SPI-A Schizophrenia Proneness Instrument, Adult version
SPI-CY Schizophrenia Proneness Instrument, Child and Youth version
UHR Ultra high risk
WLSMV Weighted least squares means and variance adjusted estimation
WRMR Weighted root mean square residual
WSS Wisconsin Schizotypal Scales

References

American Psychiatric Association, 1994. Diagnostic and Statistical Manual of Mental Disorders, fourth ed. American Psychiatric Association, Washington.

Bentall, R.P., Fernyhough, C., Morrison, A.P., Lewis, S., Corcoran, R., 2007. Prospects for a cognitive-developmental account of psychotic experiences. Br. J. Clin. Psychol. 46 (Pt 2), 155–173.

Blankenburg, W., 1971. Der Verlust der natürlichen Selbstverständlichkeit. Ein Beitrag zur Psychopathologie symptomarmer Schizophrenien. Enke, Stuttgart (in German).

Bleuler, E., 1911. Dementia praecox oder Gruppe der Schizophrenien. In: Aschaffenburg, G. (Ed.), Handbuch der Psychiatrie. Franz Deuticke, Leipzig (in German).

Chapman, L.J., Chapman, J.P., 1980. Scales for rating psychotic and psychotic-like experiences as continua. Schizophr. Bull. 6 (3), 477–489.

Chapman, L.J., Chapman, J.P., Kwapil, T.R., Eckblad, M., Zinser, M.C., 1994. Putatively psychosis-prone subjects 10 years later. J. Abnorm. Psychol. 103 (2), 171–183.

Chapman, L.J., Chapman, J.P., Raulin, M.L., 1976. Scales for physical and social anhedonia. J. Abnorm. Psychol. 85, 374–382.

Chapman, L.J., Chapman, J.P., Raulin, M.L., 1978. Body-image aberration in schizophrenia. J. Abnorm. Psychol. 87, 399–407.

Daneault, J.G., Stip, E., Refer-O-Scope Group, 2013. Genealogy of instruments for prodrome evaluation of psychosis. Front. Psychiatry 4, 25.

Debbané, M., Eliez, S., Badoud, D., Conus, P., Flückiger, R., Schultze-Lutter, F., 2015. Developing psychosis and its risk states through the lens of schizotypy. Schizophr. Bull. 41 (Suppl. 2), S396–S407.

Eckblad, M., Chapman, L.J., 1983. Magical ideation as an indicator of schizotypy. J. Clin. Consult. Psychol. 51, 215–225.

Eckblad, M., Chapman, L.J., Chapman, J.P., Mishlove, M., 1982. The Revised Social Anhedonia Scale. Unpublished test. University of Wisconsin, Wisconsin.

Flückiger, R., Ruhrmann, S., Debbané, M., Michel, C., Hubl, D., Schimmelmann, B.G., Klosterkötter, J., Schultze-Lutter, F., 2016a. Psychosis-predictive value of self-reported schizotypy in a clinical high risk sample. J. Abnorm. Psychol. 125 (7), 923–932.

Flückiger, R., Schmidt, S.J., Ruhrmann, S., Debbané, M., Michel, C., Hubl, D., Schultze-Lutter, F., 2016b. Dimensional structure among schizotypy measures and clinical high risk for psychosis criteria in patients from an early detection service. Early Interv. Psychiatry 10 (Suppl. 1), 117.

French, P., Morrison, A., 2004. Early Detection and Cognitive Therapy for People at High Risk of Developing Psychosis: A Treatment Approach. John Wiley & Sons, Ltd, West Sussex, London.

Fusar-Poli, P., Bechdolf, A., Taylor, M.J., Bonoldi, I., Carpenter, W.T., Yung, A.R., McGuire, P., 2013a. At risk for schizophrenic or affective psychoses? A meta-analysis of DSM/ICD diagnostic outcomes in individuals at high clinical risk. Schizophr. Bull. 39 (4), 923–932.

Fusar-Poli, P., Borgwardt, S., Bechdolf, A., Addington, J., Riecher-Rössler, A., Schultze-Lutter, F., Keshavan, M., Wood, S.J., Ruhrmann, S., Seidman, L.J., Valmaggia, L., Cannon, T., Velthorst, E., De Haan, L., Cornblatt, B.A., Bonoldi, I., Birchwood, M., McGlashan, T.H., Carpenter, W.T., McGorry, P.D., Klosterkötter, J., McGuire, P., Yung, A.R., 2013b. The psychosis high-risk state: a comprehensive state-of-the-art review. JAMA Psychiatry 70 (1), 107–120.

Fusar-Poli, P., Cappucciati, M., Borgwardt, S., Woods, S.W., Addington, J., Nelson, B., Nieman, D.H., Stahl, D.R., Rutigliano, G., Riecher-Rössler, A., Simon, A.E., Mizuno, M., Lee, T.Y., Kwon, J.S., Lam, M.M., Perez, J., Keri, S., Amminger, P., Metzler, S., Kawohl, W., Rössler, W., Lee, J., Labad, J., Ziermans, T., An, S.K., Liu, C.C., Woodberry, K.A., Braham, A., Corcoran, C., McGorry, P.D., Yung, A.R., McGuire, P., 2016. Heterogeneity of psychosis risk within individuals at clinical high risk: a meta-analytical stratification. JAMA Psychiatry 73 (2), 113–120.

Gross, G., Huber, G., 1989. Das Basissymptomkonzept idiopathischer psychosen. Zentralblatt Neurol. Psychiatr. 252, 655–673 (in German).

Gross, G., Huber, G., Klosterkötter, J., Linz, M., 1987. Bonner Skala für die Beurteilung von Basissymptomen (BSABS; Bonn Scale for the Assessment of Basic Symptoms). Springer-Verlag, Berlin (in German).

Gross, G., Stassen, H.H., Huber, G., Klosterkötter, J., 1990. Reliability of the psychopathological documentation scheme BSABS. In: Stefanis, C.N., Rabavilas, A.D., Soldatos, C.R. (Eds.), Psychiatry: A World Perspective. Excerpta Medica, Amsterdam, New York, Oxford, pp. 199–203.

Guillot, M., 2017. I Me Mine: on a confusion concerning the subjective character of experience. Rev. Philos. Pychol. 8, 23–53.

Haug, E., Lien, L., Raballo, A., Bratlien, U., Oie, M., Andreassen, O.A., 2012. Selective aggregation of self-disorders in first-treatment DSM-IV schizophrenia spectrum disorders. J. Nerv. Ment. Dis. 200 (7), 632–636.

Häfner, H., Maurer, K., Löffler, W., an der Heiden, W., Hambrecht, M., Schultze-Lutter, F., 2003. Modeling the early course of schizophrenia. Schizophr. Bull. 29 (2), 325–340.

Hengartner, M.P., Heekeren, K., Dvorsky, D., Walitza, S., Rössler, W., Theodoridou, A., 2017. Checking the predictive accuracy of basic symptoms against ultra high-risk criteria and testing of a multivariable prediction model: evidence from a prospective three-year observational study of persons at clinical high-risk for psychosis. Eur. Psychiatry 45, 27–35.

Henriksen, M.G., Nordgaard, J., 2014. Schizophrenia as a disorder of the self. J. Psychopathol. 20, 435–441.

Houston, S.M., Herting, M.M., Sowell, E.R., 2014. The neurobiology of childhood structural brain development: conception through adulthood. Curr. Top. Behav. Neurosci. 16, 3–17.

Huber, G., 1966. Reine Defektsyndrome und Basisstadien endogener Psychose. Fortschr. Neurol. Psychiatr. 34, 409–426 (in German).

Huber, G., 1983. Das Konzept substratnaher Basissymptome und seine Bedeutung für Theorie und Therapie schizophrener Erkrankungen. Der Nervenarzt 54, 23–32 (in German).

Huber, G., 1986. Psychiatrische Aspekte des Basisstörungskonzepts. In: Süllwold, L., Huber, G., Hrsg (Eds.), Schizophrene Basisstörungen. Springer, Berlin Heidelberg New York, pp. 39–143 (in German).

Huber, G., 1995. Prodrome der Schizophrenie. Fortschr. Neurol. Psychiatr. 63 (4), 131–138 (in German).

Huber, G., Gross, G., Schüttler, R., Linz, M., 1980. Longitudinal studies of schizophrenic patients. Schizophr. Bull. 6 (4), 592–605.

Klein, D., Rotarska-Jagiela, A., Genc, E., Sritharan, S., Mohr, H., Roux, F., Han, C.E., Kaiser, M., Singer, W., Uhlhaas, P.J., 2014. Adolescent brain maturation and cortical folding: evidence for reductions in gyrification. PLoS One 9 (1), e84914.

Klosterkötter, J., 1992. The meaning of basic symptoms for the genesis of the schizophrenic nuclear syndrome. Psychiatry Clin. Neurosci. 46 (3), 609–630.

Klosterkötter, J., Ebel, H., Schultze-Lutter, F., Steinmeyer, E.M., 1996. Diagnostic validity of basic symptoms. Eur. Arch. Psychiatry Clin. Neurosci. 246, 147–154.

Klosterkötter, J., Gross, G., Huber, G., Wieneke, A., Steinmeyer, E.M., Schultze-Lutter, F., 1997. Evaluation of the 'Bonn Scale for the Assessment of Basic Symptoms – BSABS' as an instrument for the assessment of schizophrenia proneness: a review of recent findings. Neurol. Psychiatry Brain Res. 5, 137–150.

Klosterkötter, J., Hellmich, M., Steinmeyer, E.M., Schultze-Lutter, F., 2001. Diagnosing schizophrenia in the initial prodromal phase. Arch. Gen. Psychiatr. 58, 158–164.

Kraepelin, E., 1896. Psychiatrie – Ein Lehrbuch für Studierende und Ärzte (5. Aufl.). Barth Verlag, Leipzig (in German).

Kronfeld, A., 1930. Perspektiven der Seelenheilkunde. Thieme, Leipzig (in German).

Kwapil, T.R., 1998. Social anhedonia as a predictor of the development of schizophrenia-spectrum disorders. J. Abnorm. Psychol. 107 (4), 558–565.

Kwapil, T.R., Barrantes-Vidal, N., Silvia, P.J., 2008. The dimensional structure of the Wisconsin schizotypy scales: factor identification and construct validity. Schizophr. Bull. 34 (3), 444–457.

Kwapil, T.R., Gross, G.M., Silvia, P.J., Barrantes-Vidal, N., 2013. Prediction of psychopathology and functional impairment by positive and negative schizotypy in the Chapmans' ten-year longitudinal study. J. Abnorm. Psychol. 122 (3), 807–815.

Kwapil, T.R., Mann, M.C., Raulin, M.L., 2002. Psychometric properties and concurrent validity of the schizotypal ambivalence scale. J. Nerv. Ment. Dis. 190, 290–295.

Laing, R.D., 1959. The Divided Self. An Existential Study in Sanity and Madness. Travistock Publications, London, UK.

Lenzenweger, M.F., 2015. Thinking clearly about schizotypy: hewing to the schizophrenia liability core, considering interesting tangents, and avoiding conceptual quicksand. Schizophr. Bull. 41 (Suppl. 2), S483–S491.

Marshall, M., Johnson, C., Neelam, K., Drake, R., 2013. Predicting psychosis in clinical practice: outcome of patients seen in an NHS Prodrome Clinic. Eur. Arch. Psychiatry Clin. Neurosci. 63 (Suppl. 1), S14.

Maß, R., Hitschfeld, K., Wall, E., Wagner, H.B., 1997. Validität der Erfassung schizophrener Basissymptome. Der Nervenarzt 68 (3), 205–211 (in German).

Mayer-Gross, W., 1932. Die Klinik. In: Bumke, O. (Ed.), Handbuch der Geisteskrankheiten. Fünfter Teil. Die Schizophrenie. Julius Springer, Berlin, pp. 293–337 (in German).

McGlashan, T.H., Walsh, B., Woods, S., 2010. The Psychosis-Risk Syndrome. Handbook for Diagnosis and Follow-Up. Oxford University Press, New York.

Michel, C., Flückiger, R., Kindler, J., Hubl, D., Kaess, M., Schultze-Lutter, F., 2019. The trait-state distinction of schizotypy and clinical high risk: results from a 1-year follow-up. World Psychiatry 18 (1), 108–109.

Michel, C., Kutschal, C., Schimmelmann, B.G., Schultze-Lutter, F., 2017. Convergent and concurrent validity of the Frankfurt Complaint Questionnaire as a screener for psychosis risk. J. Risk Res. 20 (11), 1480–1496.

Michel, C., Ruhrmann, S., Klosterkötter, J., Schimmelmann, B.G., Schultze-Lutter, F., 2018. Course of clinical high risk states for psychosis beyond conversion. Eur. Arch. Psychiatry Clin. Neurosci. 268, 39–48.

Miers, T.C., Raulin, M.L., 1987. Cognitive slippage scale. In: Corcoran, K.J., Fischer, J. (Eds.), Measures for Clinical Practice: A Sourcebook. The Free Press, New York, NY, pp. 125–127.

Minkowski, E., 1927. La schizophrènie: Psychopathologie des schizoïdes et des schizophrènes. Payot, Paris (in French).

Mishara, A.L., Lysaker, P.H., Schwartz, M.A., 2014. Self-disturbances in schizophrenia: history, phenomenology, and relevant findings from research on metacognition. Schizophr. Bull. 40 (1), 5–12.

Møller, P., Haug, E., Raballo, A., Parnas, J., Melle, I., 2011. Examination of anomalous self-experience in first-episode psychosis: interrater reliability. Psychopathology 44 (6), 386–390.

Nelson, B., Thompson, A., Yung, A.R., 2012. Basic self-disturbance predicts psychosis onset in the ultra high risk for psychosis "Prodromal" population. Schizophr. Bull. 38 (6), 1277–1287.

Nelson, M.T., Seal, M.L., Pantelis, C., Phillips, L.J., 2013. Evidence of a dimensional relationship between schizotypy and schizophrenia: a systematic review. Neurosci. Biobehav. Rev. 37 (3), 317–327.

Nordgaard, J., Parnas, J., 2014. Self-disorders and the schizophrenia spectrum: a study of 100 first hospital admissions. Schizophr. Bull. 40 (6), 1300–1307.

Parnas, J., Handest, P., Jansson, L., Sæbye, D., 2005a. Anomalous subjective experience among first-admitted schizophrenia spectrum patients: empirical investigation. Psychopathology 38 (5), 259–267.

Parnas, J., Møller, P., Kircher, T., Thalbitzer, J., Jansson, L., Handest, P., Zahavi, D., 2005b. EASE: examination of anomalous self-experience. Psychopathology 38 (5), 236–258.

Parnas, J., Raballo, A., Handest, P., Jansson, L., Vollmer-Larsen, A., Sæbye, D., 2011. Self-experience in the early phases of schizophrenia: 5-year follow-up of the Copenhagen Prodromal Study. World Psychiatry 10 (3), 200–204.

Parnas, J., Sass, L., 2011. The structure of self-consciousness in schizophrenia. In: Gallagher, S. (Ed.), The Oxford Handbook of the Self. Oxford University Press, Oxford, UK, pp. 521–576.

Raballo, A., Pappagallo, E., Dell' Erba, A., Lo Cascio, N., Patane', M., Gebhardt, E., Boldrini, T., Terzariol, L., Angelone, M., Trisolini, A., Girardi, P., Fiori Nastro, P., 2016. Self-disorders and clinical high risk for psychosis: an empirical study in help-seeking youth attending community mental health facilities. Schizophr. Bull. 42 (4), 926–932.

Raballo, A., Parnas, J., 2011. The silent side of the spectrum: schizotypy and the Schizotaxic self. Schizophr. Bull. 37 (5), 1017–1026.

Rzesnitzek, L., 2013. "Early psychosis" as a mirror of biologist controversies in post-war German, Anglo-Saxon, and Soviet psychiatry. Front. Psychol. 4, 481.

Ruhrmann, S., Schultze-Lutter, F., Klosterkötter, J., 2010a. Probably at-risk, but certainly ill – advocating the introduction of a psychosis spectrum disorder in DSM-V. Schizophr. Res. 120 (1–3), 23–37.

Ruhrmann, S., Schultze-Lutter, F., Salokangas, R.K.R., Heinimaa, M., Linszen, D., Dingemans, P., Birchwood, M., Patterson, P., Juckel, G., Heinz, A., Morrison, A., Lewis, S., Graf von Reventlow, H., Klosterkötter, J., 2010b. Prediction of psychosis in adolescents and young adults – results from a prospective European Multicenter study (EPOS). Arch. Gen. Psychiatr. 67 (3), 241–251.

Sass, L., Borda, J.P., Madeira, L., Pienkos, E., Nelson, B., 2018. Varieties of self disorder: a bio-pheno-social model of schizophrenia. Schizophr. Bull. 44 (4), 720–727.

Sass, L., Pienkos, E., Skodlar, B., Stanghellini, G., Fuchs, T., Parnas, J., Jones, N., 2017. EAWE: examination of anomalous world experience. Psychopathology 50, 10–54.

Sass, L.A., 2014. Self-disturbance and schizophrenia: structure, specificity, pathogenesis (Current issues, New directions). Schizophr. Res. 152 (1), 5–11.

Sass, L.S., Parnas, J., 2003. Schizophrenia, consciousness, and the self. Schizophr. Bull. 29, 427–444.

Schimmelmann, B.G., Michel, C., Martz-Irngartinger, A., Linder, C., Schultze-Lutter, F., 2015. Age matters in the prevalence and clinical significance of ultra-high-risk for psychosis symptoms and criteria in the general population: findings from the BEAR and BEARS-Kid studies. World Psychiatry 14, 189–197.

Schimmelmann, B.G., Walger, P., Schultze-Lutter, F., 2013. The significance of at-risk symptoms for psychosis in children and adolescents. Can. J. Psychiatr. 58, 32–40.

Schmidt, S.J., Schultze-Lutter, F., Bendall, S., Groth, N., Inderbitzin, N., Michel, C., Schimmelmann, B.G., Hubl, D., Nelson, B., 2017. Psychological and symptomatic mediators linking childhood adversities and trauma to suicidal behavior in adolescents and adults at clinical high risk for psychosis. Front. Psychiatry 8, 242.

Schmidt, S.J., Schultze-Lutter, F., Schimmelmann, B.G., Maric, N.P., Salokangas, R.K.R., Riecher-Rössler, A., van der Gaag, M., Meneghelli, A., Nordentoft, M., Marshall, M., Morrison, A., Raballo, A., Klosterkötter, J., Ruhrmann, S., 2015. EPA guidance on the early intervention in clinical high-risk states of psychoses. Eur. Psychiatry 30, 388–404.

Schultze-Lutter, F., 2009. Subjective symptoms of schizophrenia in research and the clinic: the basic symptom concept. Schizophr. Bull. 35, 5–8.

Schultze-Lutter, F., 2016. First signs of emerging psychosis. In: Riecher-Rössler, A., McGorry, P.D. (Eds.), Early Detection and Intervention in Psychosis – State of the Art and Future Perspectives, Key Issues in Mental Health, vol. 181. Karger, Basel, New York, pp. 29–41.

Schultze-Lutter, F., Addington, J., Ruhrmann, S., Klosterkötter, J., 2007a. Schizophrenia Proneness Instrument, Adult Version (SPI-a). Giovanni Fioriti Editore s.r.l, Rome.

Schultze-Lutter, F., Debbané, M., Theodoridou, A., Wood, S.J., Raballo, A., Michel, C., Schmidt, S.J., Kindler, J., Ruhrmann, S., Uhlhaas, P., 2016a. Revisiting the basic symptom concept: towards translating risk symptoms for psychosis into neurobiological targets. Front. Psychiatry 7, 9.

Schultze-Lutter, F., Hubl, D., Schimmelmann, B.G., Michel, C., 2017. Age effect on prevalence of ultra-high risk for psychosis symptoms: replication in a clinical sample of an early detection of psychosis service. Eur. Child Adolesc. Psychiatry 26, 1401–1405.

Schultze-Lutter, F., Klosterkötter, J., Michel, C., Winkler, K., Ruhrmann, S., 2012a. Personality disorders and accentuations in at-risk persons with and without conversion to first-episode psychosis. Early Interv. Psychiatry 6, 389–398.

Schultze-Lutter, F., Klosterkötter, J., Picker, H., Steinmeyer, E.M., Ruhrmann, S., 2007b. Predicting first-episode psychosis by basic symptom criteria. Clin. Neuropsychiatry 4 (1), 11–22.

Schultze-Lutter, F., Klosterkötter, J., Ruhrmann, S., 2014a. Improving the clinical prediction of psychosis by combining ultra-high risk criteria and cognitive basic symptoms. Schizophr. Res. 154 (1–3), 100–106.

Schultze-Lutter, F., Marshall, M., Koch, E., 2012b. Schizophrenia Proneness Instrument, Child and Youth Version; Extended English Translation (SPI-CY EET). Giovanni Fioriti Editore s.r.l, Rome.

Schultze-Lutter, F., Michel, C., Ruhrmann, S., Schimmelmann, B.G., 2018a. Prevalence and clinical relevance of interview-assessed psychosis risk symptoms in the young adult community. Psychol. Med. 48, 1167–1178.

Schultze-Lutter, F., Michel, C., Schmidt, S.J., Schimmelmann, B.G., Maric, N.P., Salokangas, R.K.R., Riecher-Rössler, A., van der Gaag, M., Nordentoft, M., Raballo, A., Meneghelli, A., Marshall, M., Morrison, A., Ruhrmann, S., Klosterkötter, J., 2015a. EPA guidance on the early detection of clinical high risk states of psychosis. Eur. Psychiatry 30, 405–416.

Schultze-Lutter, F., Ruhrmann, S., Fusar-Poli, P., Bechdolf, A., Schimmelmann, B.G., Klosterkötter, J., 2012c. Basic symptoms and the prediction of first-episode psychosis. Curr. Pharmaceut. Des. 18 (4), 351–357.

Schultze-Lutter, F., Ruhrmann, S., Michel, C., Kindler, J., Schimmelmann, B.G., Schmidt, S.J., 2018b. Age effects on basic symptoms in the community: a route to gain new insight into the neurodevelopment of psychosis? Eur. Arch. Psychiatry Clin. Neurosci. https://doi.org/10.1007/s00406-018-0949-4.

Schultze-Lutter, F., Ruhrmann, S., Klosterkötter, J., 2006. Can schizophrenia be predicted phenomenologically? In: Johannessen, J.O., Martindale, B., Cullberg, J. (Eds.), Evolving Psychosis. Different Stages, Different Treatments. Routledge, London, New York, pp. 104–123.

Schultze-Lutter, F., Schmidt, S.J., 2016. Not just small adults – the need for developmental considerations in psychopathology. Austin Child Adolesc Psychiatry 1 (1), 1001.

Schultze-Lutter, F., Steinmeyer, E.M., Ruhrmann, S., Klosterkötter, J., 2008. The dimensional structure of self-reported 'prodromal' disturbances in schizophrenia. Clin. Neuropsychiatry 5 (3), 140–150.

Schultze-Lutter, F., Theodoridou, A., 2017. The concept of basic symptoms: its relevance in research and the clinic. World Psychiatry 16 (1), 104–105.

Süllwold, L., 1986. Die Selbstwahrnehmung defizitärer Störungen. Psychologische Aspekte des Basisstörungskonzepts. In: Süllwold, L., Huber, G., Hrsg (Eds.), Schizophrene Basisstörungen. Springer, Berlin Heidelberg New York, pp. 1–38 (in German).

Süllwold, L., 1991. Manual Zum Frankfurter Beschwerde-Fragebogen (FBF). Springer, Berlin, Heidelberg, New York (in German).

Uttinger, M., Studerus, E., Ittig, S., Heitz, U., Schultze-Lutter, F., Riecher-Rössler, A., 2018. The Frankfurt Complaint Questionnaire for self-assessment of basic symptoms in the early detection of psychosis – factor structure, reliability and predictive validity. Int. J. Methods Psychiatr. Res. 27 (2), e1600.

Vijayakumar, N., Whittle, S., Yücel, M., Dennison, M., Simmons, J., Allen, N.B., 2014. Prefrontal structural correlates of cognitive control during adolescent development: a 4-year longitudinal study. J. Cogn. Neurosci. 26 (5), 1118–1130.

Yon, V., Loas, G., Monestès, J.L., 2009. Relationships between schizotypy and subjective experiences in a sample of 399 university students. Compr. Psychiatr. 50 (2), 142–150.

Yung, A.R., Phillips, L.J., Yuen, H.P., McGorry, P.D., 2006. Comprehensive Assessment of at Risk Mental State. The PACE Clinic, ORYGEN Research Centre, University of Melbourne, Department of Psychiatry, Parkville, Australia.

Zahavi, D., 2005. Subjectivity of Selfhood: Investigating the First-Person Perspective. MIT Press, Camebridge, Mass.

Ziermans, T.B., Schothorst, P.F., Sprong, M., van Engeland, H., 2011. Transition and remission in adolescents at ultra-high risk for psychosis. Schizophr. Res. 126 (1–3), 58–64.

CHAPTER

5

Schizotypy, schizotypal personality, and psychosis risk

Neus Barrantes-Vidal[1,2,3], *Anna Racioppi*[1], *Thomas R. Kwapil*[4]

[1]Universitat Autònoma de Barcelona, Bellaterra, Barcelona, Spain; [2]Sant Pere Claver - Fundació Sanitària, Barcelona, Spain; [3]CIBER Salud Mental, Instituto de Salud Carlos III, Barcelona, Spain; [4]University of Illinois at Urbana-Champaign, Champaign, IL, United States

Schizotypy as a framework to study psychosis risk

The term schizotypy was coined by Rado (1953) and theoretically elaborated by Meehl (1962) to represent a continuous phenotype of schizophrenia-like psychopathology and impairment reflecting the inherited vulnerability to schizophrenia. Previously, both Kraepelin (1919) and Bleuler (1950) described schizophrenic-like traits and symptoms in patients before illness onset, as well as in their nonpsychotic relatives. This line of work crystallized in the construct of schizophrenia-spectrum disorders, which assumes that several phenotypes representing more transient or mild forms of schizophrenia, such as schizotypy and schizotypal personality disorder (SPD), share some common etiological factors with schizophrenia, which is the most extreme and rare manifestation of this spectrum. This conceptualization entails that schizotypy serves as a risk marker for schizophrenia-spectrum disorders (Kwapil and Barrantes-Vidal, 2012). Schizotypy traits are heterogeneous, comprising odd beliefs and perceptual disturbances, paranoia, thought poverty, avolition and anhedonia, social disinterest, and disturbances in the capacity to organize and express thoughts, speech, affect, and behavior. As noted by Kwapil and Barrantes-Vidal (2015), schizotypy offers a useful and unifying construct for understanding schizophrenia-spectrum psychopathology from relatively mild subclinical presentations to severe psychosis. Thus, schizophrenia, spectrum disorders, the prodrome, and subclinical manifestations can all be understood as expressions of the schizotypy continuum.

When these schizotypy traits have a transsituational, chronic, dysfunctional, and impairing nature, they merit a clinical diagnosis of SPD. SPD stands at a unique crossroads in the characterization of psychopathology, as it is conceptualized both as stable personality pathology and a milder manifestation of schizophrenia (Kwapil and Barrantes-Vidal, 2012). This idiosyncrasy has become fully explicit in DSM-5 (American Psychiatric Association, 2013), in which SPD is

Risk Factors for Psychosis
https://doi.org/10.1016/B978-0-12-813201-2.00005-3

both listed in the category of schizophrenia-spectrum and other psychotic disorders and in the category of personality disorders. SPD is defined as "a pervasive pattern of social and interpersonal deficits, including reduced capacity for close relationships; cognitive or perceptual distortions; and eccentricities of behavior, usually beginning by early adulthood but in some cases first becoming apparent in childhood and adolescence. Abnormalities of beliefs, thinking, and perception are below the threshold for the diagnosis of a psychotic disorder" (p. 89).

The construct of schizotypy was also developed in the context of individual differences. Kretschmer (1925) was one of the first proponents of continuity between personality and psychopathology in psychosis, suggesting a continuum of schizothymia-cyclothymia with schizophrenia and affective psychosis, respectively, at its end points. Drawing on this work, Claridge (1997) proposed that schizotypy reflects normal genetic and temperamental variation, thus constituting both healthy variation in cognitive personality organization and vulnerability to, predominantly, psychotic disorders. He noted that this model entails a *fully* dimensional conceptualization, as continuity extends from normal variation in personality dimensions to disorder; in contrast, the schizotypy model proposed by Meehl, that is dominant in the medical tradition, conceives schizotypy as taxonic and a *forme frustre* of psychosis, as continuity exists only within the illness domain as an analogue of severity. These theoretical issues are relevant for the study of schizotypy in the context of high-risk research (see Grant et al., 2018). The fully dimensional perspective promotes a broader approach by suggesting that the assessment of schizotypy serves both as a risk indicator and as a psychological profile that can aid, for example, tailoring treatment designs in those identified at high risk. Also, it provides a framework to understand that schizotypy can be expressed as healthy and positive psychological manifestations, such as heightened creativity (e.g., Mohr and Claridge, 2015).

The psychometric high-risk strategy involves the identification of psychosis-prone or schizotypic individuals on the basis of scores on self-reported schizotypy questionnaires. This should ease unraveling developmental pathways to psychosis in individuals who do not present the typical confounding factors associated with schizophrenia-spectrum disorders, such as medication, hospitalization, symptom severity, and high comorbidity. The study of continuities and discontinuities between schizotypy and schizophrenia-spectrum disorders should improve our knowledge of the heterogeneity in terms of pathways toward both nonclinical and clinical outcomes, helping to identify compensating or protective factors (Barrantes-Vidal et al., 2015).

Schizotypy, and by extension, schizophrenia, is heterogeneous at etiological phenotypic and developmental levels. Both constructs share a common multidimensional structure, consisting of, at least, positive, negative, and disorganization dimensions (Kwapil and Barrantes-Vidal, 2015). Positive schizotypy is characterized by odd beliefs (including full-blown delusions), unusual perceptual experiences (including illusions and hallucinations), and suspiciousness/paranoia. Negative schizotypy involves diminished functioning including anhedonia, flattened affect, social disinterest, avolition, and anergia. The disorganization dimension involves disruptions in thought, behavior, and communication. These dimensions are associated with differential patterns of impairment and symptoms. Therefore, taking into consideration this multidimensionality is critical to map differential developmental routes of risk and resilience for psychosis (Debbané and Barrantes-Vidal, 2015).

Developmentally, signs of schizotypy will mostly emerge in adolescence, consistent with the timing of the initial expression of symptoms of schizophrenia, and could be considered the most distal marker of psychosis risk (Debbané and Barrantes-Vidal, 2015). As noted, the expression is variable in terms of types of traits and intensity and also in terms of developmental trajectories. Some people with schizotypic features will transition into schizophrenia-spectrum disorders; however, it is expected that the majority will not—although they may continue to demonstrate mild and transient symptoms and impairment. In terms of SPD, DSM-5 (American Psychiatric Association, 2013) indicates that it "may be first apparent in childhood and adolescence with solitariness, poor peer relationships, social anxiety, under-achievement in school, hypersensitivity, peculiar thoughts and language, and bizarre fantasies" (p. 657).

Most research using schizotypy as a risk indicator has been performed in nonclinical samples, although it has also been assessed in samples of individuals at high genetic risk, in which schizotypy has been examined as a potential marker to refine the identification of true risk carriers. Other research has analyzed the developmental course of schizophrenia-spectrum personality disorders (PDs), mostly SPD. More recently, as the field of early detection and intervention has developed, there has been emerging interest in the potential value of studying schizotypy and spectrum disorders in the context of clinical high risk (CHR) samples. As Debbané et al. (2015) pointed out, schizotypy has traditionally been considered as a *trait* indicator of vulnerability for schizophrenia-spectrum disorders, whereas the field of CHR has put the emphasis on "state" indicators of an imminent transition to overt psychosis (ultra high risk; UHR) as well as the detection of the earliest possible prodromal specific signs of prepsychosis (basic symptoms). Thus, UHR criteria are defined by a *state* status and entail an onset/worsening requirement (Schultze-Lutter et al., 2013), which exclude schizotypal manifestations that present with a "trait" character unless meeting a diagnostic status associated with functional decline. However, it appears that state CHR and UHR indicators can best be understood as manifestations along the schizotypy continuum. The following section will review research examining schizotypy and SPD in these four types of populations.

Schizotypy and schizotypal personality disorder as predictors of psychosis spectrum psychopathology

Schizotypy as a marker of psychosis proneness in nonclinical samples

A substantial body of research has employed schizotypy measures in nonclinical samples to test its associations with schizophrenia-spectrum phenomenology and functional impairment. The Wisconsin Schizotypy Scales (WSS) have been one of the most extensively used schizotypy measures in the last 40 years (for complete summaries and reviews of available schizotypy measures, please see Chapman et al., 1995; Kwapil and Chun, 2015; Mason, 2015; Mason et al., 1997). The WSS comprise four scales. Two of them tap positive features: the Perceptual Aberration Scale (PerAb; Chapman et al., 1978) assesses psychotic-like bodily distortions and perceptual experiences, and the Magical Ideation Scale (MagId; Eckblad and Chapman, 1983) taps belief in invalid causation. The remaining two scales predominately assess negative schizotypy: the Physical Anhedonia Scale (PhysAnh; Chapman et al., 1976) assesses deficits in sensory and esthetic pleasure, and the Revised Social Anhedonia Scale (SocAnh; Eckblad et al., 1982) measures schizoid asociality.

The majority of studies have used *cross-sectional designs*, either comparing groups of extreme scorers with control participants or analyzing the correlates of schizotypy dimensions. For example, Horan et al. (2007) compared a group of high scorers on SocAnh and MagId scales with control participants. The group with high scores on SocAnh presented high levels of schizoid, schizotypal, and paranoid symptoms, as well as poor social coping style and poor social support, whereas the high MagId group specifically reported high levels of schizotypal symptoms. On the other hand, using a dimensional approach, Barrantes-Vidal et al. (2013a) examined the correlates of conceptually driven positive and negative schizotypy dimensions derived from the WSS. Consistent with the dimensional heterogeneity reported within clinical psychosis, these dimensions showed specific as well as common patterns of associations. As hypothesized, positive schizotypy was associated with psychotic-like, paranoid, and borderline symptoms, whereas negative schizotypy was uniquely associated with both self-report and interview negative and schizoid symptoms. Both dimensions were associated with schizotypal and avoidant personality symptoms, which are consistent with the mixture of positive and negative features characterizing SPD and the social dysfunction captured in avoidant personality. Also, both dimensions were associated with ratings of suspiciousness and impaired functioning. Congruent with the robust association of positive and affective symptoms, only positive schizotypy was associated with measures of anxiety, depression, and low self-esteem. Moreover, in terms of interpersonal schemas, positive schizotypy was associated with a negative view of self and others, whereas negative schizotypy was associated with diminished positive views of self and others. This pattern highlights a differential role of affect in positive and negative dimensions that holds across nonclinical and clinical levels, such that positive schizotypy is associated with affect dysregulation and high negative affect, whereas negative schizotypy tends to be characterized by diminished positive affect. Note that other studies have reported that negative schizotypy is associated with heightened negative affect (e.g., Fonseca-Pedrero et al., 2011). However, such findings seem to reflect the use of problematic measures of negative schizotypy that are saturated with depression and neuroticism (Barrantes-Vidal et al., 2013a; Gross et al., 2014, 2018; Kwapil et al., 2018). Depression and neuroticism share phenotypic similarities with negative symptoms but are not conceptualized to be part of the construct (as negative schizotypy is presumed to involve flattened or diminished affect, not increased negative affect or affective reactivity). The abovementioned findings offer insight into the long-term trajectories of these dimensions, such as social anxiety for positive and schizoid withdrawal for negative schizotypy.

Overall, cross-sectional research has shown that there is a consistent and meaningful pattern of associations of positive and negative schizotypy measures with schizophrenia-spectrum symptoms and impairment—patterns that have been distinctively associated with positive and negative symptom dimensions across various psychosis spectrum disorders (e.g., Barrantes-Vidal et al., 2009, 2010; Blanchard et al., 2011; Bolinskey et al., 2015; Bolinskey and Gottesman, 2010; Ettinger et al., 2015; Horton et al., 2014; Kaczorowski et al., 2009; Kwapil et al., 2008). More recently, a few studies have examined whether schizotypy measures are also valid to detect the real-life expression of behavioral patterns either characterizing psychosis or those associated with an increased risk for psychosis in genetic risk or epidemiological cohorts. This work employs experience sampling methodology (ESM), a within-day self-assessment method in which individuals are repeatedly prompted at random intervals to complete brief questionnaires in the moment about their current experiences as they occur in their daily life environment. This method offers several advantages to traditional assessment procedures (Myin-Germeys et al., 2009), as it allows researchers to study the context of experiences, their prospective and dynamic association with symptoms, and potential mechanistic pathways. Barrantes-Vidal et al. (2013b) examined the real-life expression of psychometric schizotypy dimensions using ESM for 1 week in 206 young adults with a mean age of 20 years. Consistent with findings from the psychometric research just described, positive schizotypy was associated with daily life reports of psychotic-like experiences (PLEs) and paranoid symptoms, whereas negative schizotypy was associated with a subset of these symptoms and with negative symptoms such as diminished thoughts, emotions, and social contact in daily life. Examination of the hypothesis that positive symptoms in particular are related to an abnormal pattern of stress sensitivity revealed that ESM appraisals of stress in the moment were associated with psychotic-like and paranoid symptoms, but only in those with high *positive* schizotypy. Strikingly, time-lagged analyses showed that stress at the preceding signal predicted psychotic-like symptoms at the current assessment, but only for individuals high in positive schizotypy. In an extension of this study, Chun et al. (2017) obtained schizotypy dimensions based on dimensional *interview* ratings of cluster A PDs and replicated the same findings: positive (as well as paranoid and disorganized) schizotypy was associated with elevated stress reactivity, whereas negative schizotypy was related with diminished reactivity. Finally, Kwapil et al. (2012) found that both positive and negative schizotypy were associated with the desire to be alone when being with others in daily life but that this social discomfort resulted from excessive anxiety in positive schizotypy and from reduced positive affect in negative schizotypy. This finding indicates that the same abnormality (social disinterest) results from distinct psychological mechanisms, which implies that such social discomfort may require different preventive or therapeutic strategies. People high in negative schizotypy consistently reported that they wanted to be alone when with others and that they wanted to remain alone when they were by themselves. People high in positive schizotypy wanted to be alone when with others, but desired social contact when alone (suggesting that feeling socially comfortable is difficult for people high in positive schizotypy). Overall, these studies demonstrate the ecological validity of ESM as a method able to detect precursors of schizophrenia-spectrum psychopathology by examining the daily life experiences of schizotypic individuals.

Regarding the *prospective* association of schizotypy with clinical outcomes in nonclinical samples, Chapman et al. (1994) reported a pioneering 10-year follow-up study of college students. They selected four high-risk groups based on the following WSS: PerAb or MagId (PerMag), Impulsive Nonconformity (Chapman et al., 1984), and PhysAnh, as well as a combined-risk group and a control group. Of the 534 subjects interviewed at the first assessment, 95% were reassessed 10 years later. At the cross-sectional assessment, Chapman et al. noted that high-risk individuals, and especially those high on PerMag group, reported elevated rates of PLEs and schizotypal symptoms (note that none was psychotic at baseline). At the 10-year reinterview, PerMag participants exceeded the control group in rates of psychotic disorders, psychotic-like, paranoid, and schizotypal symptoms, and mood and substance use disorders. Furthermore, when Chapman et al. examined a combination of multiple predictors, the rate of psychosis increased dramatically at the 10-year follow-up: 14% of magical ideation subjects who also reported PLEs at the initial assessment developed psychotic disorders; participants who were deviant on MagId and scored above the mean on SocAnh had a 21% rate of psychosis at follow-up (note that the Social Anhedonia Scale was not used to select a high-risk group). Finally, the study showed that PLEs were an excellent indicator of psychosis proneness, since the rate of psychosis increased to 40% in the MagId-SocAnh subjects who exhibited moderate PLEs at the initial assessment. Kwapil (1998) investigated in this sample whether SocAnh scores independently predicted the development of schizophrenia-spectrum disorders 10 years later. He reported that 24% of participants identified by elevated scores on the SocAnh suffered from schizophrenia-spectrum disorders at follow-up. Furthermore, individuals with high scores on the scale who did not develop such disorders still reported elevated schizophrenia-spectrum symptoms and impaired functioning 10 years later. These findings suggest that SocAnh appears to identify a specific group of at-risk individuals with a latent vulnerability for schizophrenia-spectrum disorders. In 2013, Kwapil and colleagues used the Chapman et al. (1994) longitudinal dataset to test the predictive validity of conceptually driven positive and negative dimensions. They assigned positive and negative schizotypy dimensional scores based on the WSS to participants in the 10-year longitudinal study. It was found that positive schizotypy predicted the development of psychotic disorders (i.e., schizophrenia and psychosis not otherwise specified), whereas both positive and negative schizotypy dimensions predicted schizophrenia-spectrum disorders (including psychotic disorders and schizophrenia-spectrum PDs). Moreover, positive schizotypy was associated with major depressive and manic/hypomanic episodes, substance use disorders, and mental health treatment. In contrast, negative schizotypy was associated with schizoid personality symptoms and social impairment at follow-up. These findings are especially striking given that participants were functioning well enough to enroll in a major university at the start of the study, they were only part way into the window of greatest risk for developing psychosis at the time of the follow-up assessments, schizotypy ratings predicted schizophrenia-spectrum symptoms even after omitting participants with psychotic and spectrum disorders, and the rates of transition into schizophrenia-spectrum disorders were as high or higher than typically seen in genetic high-risk studies (Kwapil et al., 2013). Overall, these findings supported the predictive validity of the WSS and showed that schizotypy is a useful phenotype to detect developmental risk for schizophrenia and related disorders.

A similar study conducted by Gooding et al. (2005) identified two high-risk groups based on the WSS: a group of high scorers on the PerAb and/or MagId scales and high scorers on the SocAnh. At a 5-year follow-up, these groups displayed more frequent and severe PLEs compared to a control group. However, in contrast to Chapman et al. (1994) findings, none of the baseline PLEs was an indicator of schizophrenia-spectrum psychopathology at the 5-year follow-up, and no participants met diagnostic criteria for a psychotic disorder. The authors argued that this difference was probably related to the fact that Chapman's 10-year follow-up covered a greater proportion of the lifetime risk of psychosis expression compared to their 5-year follow-up. Furthermore, participants in the Gooding et al. (2005) study had a better baseline functioning (as indicated by GAF scores) than those in the Chapman study, even if all of them were college students. Consistent with Kwapil (1998), they indicated that high scorers on SocAnh were at high risk for the development of schizophrenia-spectrum disorders rather than psychopathology in general.

Bolinskey et al. (2017) reported a 2-year follow-up of a group of high schizotypy scorers on MagId, PerAb, or SocAnh and a matched control group. The high schizotypy group met more criteria for avoidant, paranoid, schizoid, and SPDs than the comparison group with medium to large effect sizes, both at the baseline assessment and the 2-year follow-up, indicating a consistent temporal stability of these traits over time. As part of an ongoing longitudinal study on psychosis risk and resilience factors, Racioppi et al. (2018) reported the 3-year follow-up outcome of participants described earlier in the study by Barrantes-Vidal (2013a). As mentioned, they used conceptually driven schizotypy dimensions rather than single-scale or mixed-scale predictors. Positive schizotypy predicted interview ratings of PLEs, depression, general psychopathology, and low self-esteem, and negative schizotypy predicted schizoid personality traits, emotional disturbances, and mental health treatment during the past year. As expected, both schizotypy dimensions predicted schizotypal, paranoid, and avoidant personality traits, suspiciousness, and impaired functioning. Despite the fact that the studies used different outcome measures in different cultures, the longitudinal findings for the schizotypy dimensions appear strikingly consistent.

The studies described so far have been conducted in samples of college students, which entail testing hypotheses in a highly conservative manner as participants have more protective factors and less likelihood of developing schizophrenia-spectrum disorders. However, they limit the generalization of findings, so research from community samples is necessary to complement this research approach. Cohen et al. (2012) reported on the 3-year follow-up of a group with high SocAnh and a control group. The former had greater schizotypal, schizoid, and paranoid ratings, as well as poorer functioning and more prevalent history of psychiatric treatment and major depressive disorder diagnosis than controls. Using the prospective Northern Finland 1966 birth cohort, Miettunen et al. (2011) examined the predictive validity of various psychological scales including the PerAb, PhysAnh, and SocAnh in an 11-year longitudinal study in an unselected general population sample of 4926 participants. The PerAb had the best concurrent validity with psychotic diagnoses at a cross-sectional assessment and overall presented the highest predictive validity. Bogren et al. (2010) created a semistructured interview to assesses premorbid symptoms and behaviors, such as cluster A personality traits, in an unselected general population sample ($n = 1797$) and examined the association of these traits with the incidence of psychosis in a 50-year follow-up. Schizotypal-paranoid traits, as well as anxiety-proneness and affective/cognitive blunting, were significantly associated with a diagnosis of psychosis (not schizophrenia) 50 years later.

Schizotypy in genetically at-risk samples

In addition to using schizotypy measures as primary predictors of schizophrenia-spectrum psychopathology, researchers have also examined schizotypy in genetic high-risk studies. The New York High-Risk Project followed up the offspring of persons with schizophrenia, affective disorders, and controls from childhood to adulthood. PhysAnh was administered to 161 adolescents and was found to be a precursor of psychotic outcomes and social isolation in females, but not males (Erlenmeyer-Kimling et al., 1993). In the Edinburgh High Risk Study of Schizophrenia (EHRS), Miller et al. (2002a) examined the predictive validity of schizotypy among young individuals with genetic high risk for psychosis, with a first episode of psychosis meeting criteria for schizophrenia, and without genetic risk. A total of 212 individuals were classified as control, high-risk individuals who did not report psychotic symptoms, high-risk individuals who displayed psychotic symptoms but did not meet criteria for psychosis, and as first-episode patients with schizophrenia 4.5 years later. A principal components analysis of the Structured Interview for Schizotypy (SIS) baseline scores revealed that four schizotypy features (social withdrawal, positive symptoms, socioemotional dysfunction, and odd behavior) accounted for the 47.8% of the variance in outcome. Patients with schizophrenia reported the highest scores on social withdrawal, whereas high-risk individuals with psychotic symptoms reported the highest scores on the odd behavior. Both patients and high-risk individuals with psychotic symptoms reported similar high scores on positive schizotypy. Social withdrawal was the best single predictor of transition, although a combination of odd behavior, social withdrawal, socioemotional dysfunction, and positive schizotypy features had the best predictive power. In a subsequent study of the EHRS project, Johnstone et al. (2005) examined factors predicting transition into schizophrenia at 2.5 years in a sample of 163 young adults with 2 relatives with schizophrenia as well as in control participants and individuals with risk for other psychiatric disorders. The strongest predictors of transition to psychosis were the Rust Inventory of Schizotypal Cognition (Miller et al., 2002b), which mainly taps positive schizotypy such as bizarre and eccentric ideas, and SIS (Miller et al., 2002a) at study entry, with social withdrawal and oddness factors being the most powerful predictors—providing further evidence that schizotypy is able to detect developmental vulnerability for schizophrenia. In a 10-year follow-up report of this study, Tijms et al. (2015) reported that there was a strong association between schizotypal cognitions and gray matter network alterations for genetic risk individuals who later developed schizophrenia before illness onset. More recently, Zarogianni et al. (2017) reported that introducing schizotypal cognitions in predictive models improved outcome classification, reaching a predictive accuracy of 94% when combining schizotypal cognitions, declarative memory, and structural MRI data.

In a study with 96 young first- and second-degree relatives of persons with schizophrenia, Shah et al. (2012) analyzed the predictive power of a multivariate model integrating socioenvironmental, ecological, neurodevelopmental, and clinical etiological and risk factors. The three schizotypy scales administered (MagId, PerAb, and SocAnh) were a direct predictor of transition to psychosis at 2.3 years, acting as mediators of the indirect effect of distal (e.g., familial, biological, and socioenvironmental) and cognitive risk factors on psychosis risk.

A novel approach was recently used by the Genetic Risk and Outcome in Psychosis consortium (van Os et al., 2017) by examining the association of a polygenic risk score (PRS) with measures of schizotypy in a sample of first-degree relatives (siblings and parents) of patients with psychotic disorder and healthy controls in a longitudinal study. The PRS was associated with a total summary score of schizotypy as assessed with the Community Assessment of Psychic Experiences (Stefanis et al., 2002) and with the Structured Interview for Schizotypy-Revised (Kendler et al., 1989), as well as with the positive dimension of both measures and presence of any lifetime affective disorder, both in relatives and controls. They suggested that in individuals at elevated genetic risk, the emerging expression of phenotypic alterations may yield floor effects, which would obscure the detection of associations.

Schizotypy and schizotypal personality disorder in clinical at-risk samples

In the Cologne Early Recognition study (Klosterkötter et al., 2001), the predictive validity of basic symptoms was tested in a sample of 160 patients who were followed up for approximately 10 years. In addition, the value of PDs assessed at baseline was examined, and it was found that only SPD was a predictor of transition to schizophrenia regardless of the presence of basic symptom criteria.

Mason et al. (2004) examined the predictive power of a wide variety of factors (family history, perinatal complications, premorbid social functioning and personality, recent life events, and current symptoms) for transition to psychosis in a sample of help-seeking CHR young individuals (mean age = 17.3) in New South Wales. One year after the baseline assessment, half of the sample developed a psychotic disorder. Among the premorbid factors studied, the best scale-based predictor of transition to psychosis were SPD ratings, although the score difference between those who transitioned and those who did not was small, which suggested that SPD scores would have limited clinical value in terms of improving the identification of at-risk individuals. When analyzing individual items, SPD items such as odd beliefs/magical thinking, auditory hallucinations, blunted or inappropriate affect, and anhedonia/asociality, together with impairment in role functioning, were good predictors of transition 1 year later. Importantly, the authors acknowledged that SPD symptoms were rated as present or absent in the last month (along with the minimum of a 5-year presence) and that this measure was used in the study. Therefore, they can actually be considered as current prodromal symptoms at study entry and hardly possible to distinguish from premorbid personality.

In the context of the Early Detection and Intervention Evaluation study with CHR individuals, Morrison et al. (2002) reported that high-risk youngsters receiving primary care showed elevated self-reported positive, negative, and disorganized schizotypy as measured with the Oxford-Liverpool Inventory of Feelings and Experiences (O-LIFE; Mason et al., 1995) as compared to published data from sociodemographically comparable samples of healthy controls.

Cannon et al. (2008) examined the predictive power of a comprehensive set of predictors for transition to psychosis at 2.5 years in a sample of 291 CHR individuals enrolled in the North American Prodrome Longitudinal Study (NAPLS). SPD ratings did not prove to be significant predictors. Brucato et al. (2017) also did not find SPD ratings (based on a checklist following DSM-IV; APA, 1994) to predict transition to psychosis among a large set of variables in a prospective study of 200 CHR individuals followed up for 2 years at the Center of Prevention and Evaluation. Of note, analyzing data from the NAPLS, Woods et al. (2009) reported that a group of patients with SPD who did not meet CHR criteria reported worse social functioning compared with the CHR group (which encompassed the three CHR syndromes, including participants with deterioration in functioning within the last year meeting either a diagnosis of SPD or a family history risk criterion). In terms of transition, 40% of putatively prodromal patients transitioned into psychosis (schizophrenia-spectrum psychosis, affective psychosis, delusional disorder, brief psychotic disorder, and psychosis not otherwise specified), and 36% of the SPD group transitioned to undifferentiated schizophrenia, schizoaffective disorder, and depression with psychosis during the course of the study. The authors argued that SPD may identify a more gradual progression of illness than UHR criteria among adolescents and young adults and suggested that SPD may capture an independent risk syndrome for psychosis. This is not entirely surprising given that UHR criteria largely focus on acute PLEs, whereas SPD involves trait-like positive, negative, interpersonal, and disorganized symptoms. A similar conclusion was made by Parnas et al. (2011) with data from the Copenhagen Prodromal Study. It was found that 25% of patients with SPD transitioned to schizophrenia over a 5-year period, although none of their various sociodemographic and psychopathological variables predicted transition to psychosis in these patients. Similar to Woods et al. (2009), Parnas and colleagues suggested that SPD seems to be a subpsychotic condition, highly similar to schizophrenia, but with a different degree of severity.

Schultze-Lutter et al. (2012) examined schizophrenia-spectrum PDs and single PD features in a sample of CHR with and without transition to psychosis with a self-report questionnaire. Cluster A PDs were not frequent (14%), despite the fact that PDs were generally frequent (46%). At baseline, paranoid (9%) was the most prevalent PD among cluster A PDs, whereas schizotypal (7%) and schizoid (3%) PDs were less frequent. Unexpectedly, they found that only schizoid, but not schizotypal PD, was a predictor of transition to psychotic disorders. Among the schizoid personality features, "lack of close friends or confidants other than first-degree relatives" and "emotional detachment observed by others" were most prevalent in converted patients.

Only a few studies have employed self-reported measures in early psychosis samples. In the European Prediction of Psychosis Study, both schizotypy and SPD have been investigated a cohort of 245 CHR individuals meeting criteria for either basic symptoms (cognitive disturbances, COGDIS) or UHR criteria followed up up to 18 months. Ruhrmann et al. (2010) found that defined SPD (which requires its presence for only at least 1 year) was part of the prediction model of transition to psychosis with positive symptoms, bizarre thinking, sleep disturbances, functioning, and years of education among patients who met UHR and COGDIS (positive predictive value of 83.3%). They suggested that this algorithm might be used as a second step in the identification of at-risk individuals to improve the predictive power of UHR and COGDIS criteria. Salokangas et al. (2013) reported on the predictive value of two schizotypal features assessed at baseline with the Schizotypal Personality Questionnaire (SPQ; Raine, 1991), ideas of reference, and lack of close friends. Both schizotypal traits were predictive of transition to psychosis, and their co-occurrence greatly enhanced the risk of transition to psychosis, even when the effect of SPD was considered. However, other traits, such as suspiciousness, odd beliefs, magical thinking, and unusual perceptual experiences, were not associated with transition to psychosis. Salokangas et al. (2013) suggested that the predictive power of the combination of ideas of reference and lack of close friends can be interpreted in the context of the differential relationship of schizotypal features with affect, whereas ideas of reference align with high negative affectivity, lack of close friends maps on to diminished positive affectivity, and this combination is associated in schizophrenia-spectrum disorders. Similarly, the former is a core component of positive schizotypy, whereas the latter is a central feature of negative schizotypy. Of note, this pattern has also been found in nonclinical individuals. In a reanalysis of the Chapman's longitudinal study, Kwapil et al. (1997) found that the co-occurrence of high rates of MagId and SocAnh increased the predictive power to identify individuals with psychosis proneness. Authors of both studies suggested that negative schizotypy traits such as social anhedonia and lack of close friends would strengthen positive schizotypy traits by restricting the possibility of reality testing.

Flückiger et al. (2016) employed the MagId, PerAb, and PhysAnh scales in a sample of 128 young adults self-seeking for help. They were classified as CHR+ or CHR− based on the UHR and/or basic symptoms criteria. The CHR+ group scored higher on positive schizotypy measures and the PhysAnh, but the latter was the only predictor of the CHR state. This was an unexpected finding, and authors suggested that they may been unable to tap predictive value for positive schizotypy due to the characteristics of the sample. The mean scores of the positive WSS in their sample were superior to most nonclinical and genetic high-risk samples, so they argued that they may have captured individuals positioned in the upper level of the continuum and so close to psychosis that the WSS positive schizotypy has no sufficient discriminatory power (and likely had a restricted range and distribution). Moreover, the interaction between PhysAnh and CHR status was predictive of transition to psychosis, such that those classified as CHR+ who developed psychosis had elevated PhysAnh scores, whereas low scores seemed to predict transition in the CHR group. Data on SocAnh were missing for almost half of the sample, so it was difficult to firmly conclude on the negative dimension (although estimated statistical models seemed to confirm the role of PhysAnh and discarded a role for SocAnh). In light of their findings, Flückiger et al. (2016) proposed the need to further explore the potential of PhysAnh as a valid screening method in clinical populations to detect a CHR status and for the prediction of psychosis among these patients.

Recently, Bang et al. (2019) examined whether the presence and interaction of "schizophrenia-specific" basic symptoms and schizotypy predicted transition to psychosis in UHR. At baseline, UHR individuals showed higher levels of basic symptoms and schizotypy, with higher rates of SocAnh and PhysAnh compared to the controls. Transition over the course of 25.8 months was better predicted by the interaction of basic symptoms with PhysAnh than basic symptoms alone. In addition, basic symptoms severity increased the risk of transition in UHR individuals with high PhysAnh, but not in low scorers. These findings are consistent with those reported by Salokangas et al. (2013) and Flückiger et al. (2016) indicating that PhysAnh is a significant predictor of transition in CHR and concur with the latter study in showing that PhysAnh increases the predictive power of basic symptoms and improves the discrimination of CHR individuals who will develop psychosis.

The above studies demonstrate that positive, negative, and disorganized schizotypy dimensions are differentially distributed in CHR individuals and exhibited differential patterns of association with psychosis transition. The assessment of schizotypy offers a valid screening method that increased the predictive power of CHR criteria in some of the studies reviewed. Furthermore, schizotypy offers a useful phenotype that is associated with the predisposition to schizophrenia and the developmental trajectory of vulnerability for psychosis.

Psychosis spectrum personalities as predictors of psychosis in samples with schizophrenia-spectrum personality disorders

In the prospective Chestnut Lodge follow-up study, 17% of the 105 participants who met at last three DSM-III criteria for SPD developed schizophrenia after a 15-year period (Fenton and McGlashan, 1989). Some specific traits (social isolation, suspiciousness, and magical thinking) individually predicted transition to schizophrenia. Similarly, Schulz and Soloff (1987) indicated a 25% transition over a 2-year period. Wolff et al. (1991) reassessed at age 27 a sample of 32 10-year old children diagnosed as "schizoid" (although this diagnosis significantly overlapped with current SPD conceptualizations) and found that 75% met criteria for SPD at the reassessment and 6% developed schizophrenia. In a study with adolescents with either SPD or paranoid PD, Weiser et al. (2001) indicated an odds ratio of 21.5 (95% CI 12.6–36.6) for the development of schizophrenia over a 4- to 8-year period.

In the context of the Danish OPUS project, a randomized clinical trial examining the effect of specialized early intervention in a sample of 79 SPD patients, Nordentoft et al. (2006) reported a transition rate to psychosis of 65% at a 2-year follow-up, specifically 25% of those exposed to an integrated treatment and 48.3% to the standard treatment. In a multivariate model, the integrated treatment reduced the risk of transition, whereas male gender and negative symptoms increased the risk. Recently, Albert et al. (2017) extended these findings by examining data from the OPUS II longitudinal study. Of the entire sample, 59 patients with SPD were reassessed 5 years, and 32.2% of them developed psychosis. There were no differences in transition rates between the two groups of intervention, which authors suggested that might relate to a delivery of the intervention later than in prior studies. Among predictors, and consistent with most previous research, level of functioning was the most important one. These results are in line with longitudinal studies investigating psychometric schizotypy—that includes features of SPD—that reported a significant prediction of schizophrenia-spectrum disorders or psychotic disorders in nonclinical samples over a 5-year

(Gooding et al., 2005) or a 10-year period (Chapman et al., 1994; Kwapil et al., 2013) follow-up. As noted by Woods et al. (2009) and Parnas et al. (2011), these findings suggest that SPD and schizotypy are not different types of phenotypes in relation to schizophrenia but best conceptualized as different expressions of severity degree along the schizophrenia continuum.

Conclusion

Schizophrenia and schizophrenia-spectrum disorders present a major public health problem across the globe. Although these disorders are relatively rare, they are among the leading causes of disability worldwide. Schizophrenia tends to have a chronic and episodic course with many patients experiencing marked impairment in between acute episodes. Schizophrenia and spectrum disorders result in marked health, social, and economic burden for patients, their families, and society (see recent review by Chong et al., 2016). Traditional medical models of schizophrenia conceptualized the disorder as categorical. However, more recent formulations have viewed schizophrenia as the most extreme manifestation of a continuum clinical and subclinical symptoms and impairment (Kwapil and Barrantes-Vidal, 2015; Tandon et al., 2009). Psychiatric formulations that have broadened the conceptualization of schizophrenia such as the Diagnostic and Statistical Manual of Mental Disorders (e.g., American Psychiatric Association, 2013) have done so by adding additional intermediate categorical disorders (e.g., schizophreniform disorder and other specified disorders). However, the expansion of the schizophrenia spectrum to a continuum of schizotypy appears more in line with current dimensional approaches to psychopathology such as the recent Research Domain Criteria proposed by the National Institute of Mental Health (e.g., Kozak and Cuthbert, 2016).

The construct of schizotypy offers a promising framework for conceptualizing the continuum of schizophrenic symptoms and impairment and a promising platform for research. Schizotypy unifies multiple related constructs such as psychosis proneness, the prodrome, PDs, and psychotic disorders with the assumption that these conditions share comparable underlying etiologies and that their differences reflect a matter of degree or severity, not qualitative differences. The schizotypy continuum is presumed to be dynamic. This dynamic nature is readily understood at the pathological end, as patients decompensate into psychotic episodes, followed by recovery into partial or full remission of symptoms. However, the entire continuum is presumed to be dynamic as some individuals who exhibit mild forms of schizotypy will transition into prodromal symptoms and others will continue to develop clinical disorders. It is also important to highlight that a large proportion of individuals who present with elevated traits of schizotypy will not develop any clinical psychotic manifestation or reach a subclinical, prodromal threshold. High "trait" schizotypy can be compared to high levels of neuroticism: it is both a specific constellation of characteristics defining individual differences as well as a risk factor for the experience of psychotic manifestations (or, in the case of neuroticism, affective psychopathology). In other words, as with any other dimension of personality, schizotypy has a dual nature, with both trait and state manifestations (Claridge, 1997). Schizotypy has traditionally been used to describe the "stable" personality structure, whereas PLEs or subclinical symptoms refer to the temporal, symptomatic states that fluctuate depending on developmental transient states (e.g., neurodevelopmental reorganizations) that are influenced by endogenous and exogenous factors (Barrantes-Vidal et al., 2015). Therefore, a critical difference between the schizotypy and clinical high-risk paradigm is that the latter focuses on a narrow fraction of this dynamic trait-state continuum by focusing on the "state" aspects that almost reach a clinical threshold (be it positive symptoms or a time-defined, "state," functional decline in the case of SPD or family history of psychosis). Schizotypy, thus, provides a promising framework for exploring the etiology and development of schizophrenia-spectrum disorders, the identification of risk and protective factors, and the trajectory of schizotypic symptoms and impairment. Importantly, the reliable identification of schizotypic individuals and factors that influence their trajectory is essential for the development of preventative interventions.

Despite its promise, the conceptualization of schizotypy has suffered from a number of limitations that have restricted its utility. As noted, the construct developed from different historic pathways primarily represented by Meehl's taxonic model and Claridge's fully dimensional personality model. As a result, clear and consistent operationalizations of schizotypy have not been agreed upon (Kwapil and Barrantes-Vidal, 2015). Likewise, multiple measures have been put forth, which often differ in terms of their factor structure and their basic conceptualization of schizotypy. Thus, at times in the literature, it appears that schizotypy is defined more by the particular measure used to assess the construct than by an a priori model.

Perhaps the most important issue in terms of conceptualizing schizotypy is considering its multidimensional structure. Schizotypy and by extension schizophrenia are characterized by considerable heterogeneity in terms of etiology, development, symptoms, impairment, and expression. This heterogeneity appears to be best captured by a multidimensional structure. Support for this multidimensional structure comes from the descriptive psychopathology literature (e.g., Dollfus and Lyne, 2017), clinical assessment (e.g., Andreasen et al., 1990), psychiatric nosology (e.g., American Psychiatric Association, 2013), and factor analytic studies (e.g., Fonseca-Pedrero et al., 2018). The strongest support appears to be for positive (psychotic-like), negative (deficit), and disorganized schizotypy dimensions. As reviewed above, questionnaire measures have reliably assessed these dimensions. Furthermore, cross-sectional and longitudinal studies have demonstrated the validity of these dimensions in that they have unique patterns of associations with measures of clinical symptoms, personality traits, daily life experiences, neurocognitive deficits, biobehavioral markers, genetic indicators, and neural assessments. It is essential to note that failure to consider and to adequately assess schizotypy as a multidimensional construct risks obscuring important findings. For example, Sahakyan and Kwapil (2016) reported that episodic memory deficits were exhibited by a high negative schizotypy group, but not by high positive schizotypy or control groups. Reanalysis of the data indicated the group differences were eliminated when the positive and negative schizotypy participants were combined into a single "high schizotypy" group that was compared to the control group. In addition, a comprehensive multidimensional mapping would facilitate the increasingly perceived need to move beyond unidimensional, highly specific links between personality traits, risk states, and very narrow psychopathology phenotypes. For example, it has been suggested that positive schizotypy might bear the risk for the expression of psychosis in both the schizophrenia spectrum and mood psychoses, which may explain their strong association. Similarly, negative schizotypy may account for the still poorly understood overlap between schizophrenia and autistic spectrum disorders. Thus, future studies should offer specific hypotheses for the schizotypy dimensions (not for schizotypy as an omnibus construct) and should carefully assess the individual dimensions.

Although we strongly advocate against treating schizotypy as a homogenous construct, it appears that the combination of elevations on multiple schizotypy dimensions may predict elevated impairment and symptoms. For example, in both nonclinical and CHR samples, the joint presence of positive and negative features predicted risk for psychosis (e.g., Kwapil, 1998; Salokangas et al., 2013). This includes both additive and interactive effects. For example, Barrantes-Vidal et al. (2010) employed cluster analysis and reported that the high positive and negative schizotypy cluster exhibited the most severe impairment and symptoms relative to the high positive schizotypy cluster, high negative schizotypy cluster, and the low schizotypy cluster. Although these effects are often additive, we have repeatedly found that interaction of positive and negative schizotypy significantly predicts paranoid symptoms (e.g., Kwapil et al., 2013). This highlights the importance of assessing premorbid schizotypy traits in CHR, as obtaining a comprehensive profile may help to tailor psychological and pharmacological interventions and address features that may be obscured by prominent attenuated psychotic symptoms. This is comparable to assessing affective temperaments in individuals with an emerging or established mood disorders. The presence of hyperthymic temperament before the emergence of a depressive episode would alert a clinician to consider the risk of bipolar disorder. Similarly, the presence of high trait negative or disorganized schizotypy should inform clinicians working with CHR populations on the developmental routes and mechanisms preceding the current prepsychosis state and relevant treatment strategies.

Although the positive, negative, and disorganized dimensions appear to capture much of the heterogeneity of schizotypy and schizophrenia, this does not mean that they are perfectly operationalized or measured or that there are not additional relevant dimensions. For example, many of the current measures of schizotypy were not developed based upon rich conceptual descriptions of the schizotypy dimensions. The WSS positive and negative schizotypy dimensions were derived from factor analysis of scales measuring characteristics described in Meehl's (1964) schizotypy checklist. Although the measures appear to capture variance relevant to these dimensions, the positive schizotypy dimension does not include items assessing paranoia, and the negative schizotypy dimension is largely limited to tapping social and physical anhedonia (not to mention the WSS do not include a measure of disorganized schizotypy). The SPQ was designed to be a measure of SPD, rather than a measure of schizotypy, per se. As a result, the interpersonal factor is used as a proxy for negative schizotypy, despite the fact that it loads highly on neuroticism and social anxiety (which are not typically considered part of negative schizotypy). In terms of assessing disorganized schizotypy, the O-LIFE cognitive disorganization factor largely taps social anxiety and the SPQ disorganization factor appears to be more related to oddness and eccentricity characteristic of positive schizotypy, than cognitive and behavioral disorganization. In contrast, the Multidimensional Schizotypy Scale (Kwapil et al., 2018; Gross et al., 2018) was specifically developed to assess positive, negative, and disorganized schizotypy overcoming many of the limitations of previous measures. Future research using new generation measures as the one just mentioned should enhance the predictive validity of schizotypy and refine its capacity to map developmental routes to psychosis spectrum conditions.

Thus, schizotypy provides a promising model for expanding the schizophrenia phenotype. The identification of nondisordered schizotypes should enhance the search for relevant etiological factors. For example, genetic investigations and neuroscience studies have often been limited by classifying participants as affected or nonaffected based solely on the presence of clinical disorders. As a result, nondisordered schizotypes are often erroneously classified as nonaffected, despite the fact that they may have putative genetic risk alleles or neurophysiological deficits. The study of putative etiological factors and task performance in patients with schizophrenia is often confounded by the consequences of the disorder such as medication, stigma, hospitalization, and severe symptom presentation. For example, the performance of patients with schizophrenia on a measure of sustained attention may be grossly impaired due to paranoid ideation regarding the laboratory equipment, preoccupation with hallucinatory voices, marked loss of motivation related to negative symptoms, or disruptions in understanding the task instructions due to cognitive disorganization—regardless of their actual ability on the task. The study of nondisordered schizotypes allows for the investigation of relevant etiological factors and task performance relatively untainted by the consequences of schizophrenia-spectrum disorders. Finally, schizotypy provides a powerful method for identifying individuals at risk for developing schizophrenia-spectrum disorders.

Glossary

Disorganized schizotypy It is characterized by features such as disorganized thought and behavior, confusion, racing thoughts, loose associations, disrupted speech, difficulty following conversations, and slowness of thought.

Negative schizotypy It is characterized by social disinterest, flat affect, anhedonia, alogia, anergia, and avolition.

Paranoid personality disorder It is one of the so-called schizophrenia-spectrum or cluster A personality disorders. Its main feature is a pervasive and persistent suspiciousness of others and the world

Positive schizotypy It is characterized by disruptions in content of thought (ranging from magical ideation to delusional ideas), perceptual oddities (including illusions and hallucinations), and suspiciousness/paranoia.

Schizoid personality disorder It is one of the so-called schizophrenia-spectrum or cluster A personality disorders. It is characterized by a persistent pattern of social disinterest and a limited range of emotional expression.

Schizophrenia-spectrum disorders This term generally refers to schizophrenia, schizoaffective disorder, schizophreniform disorder, brief psychotic disorder, delusional disorder, and schizotypal personality disorder.

Schizotypal personality disorder It is one of the so-called schizophrenia-spectrum or cluster A personality disorders It is characterized by odd thinking and behavior and by a profound discomfort in social and interpersonal settings.

Schizotypy It is considered both a dimension of personality and the phenotypic manifestation of the underlying vulnerability for schizophrenia-spectrum psychopathology (expressed across a broad range from subclinical expression to the prodrome to schizophrenia-spectrum personality disorders to full-blown psychosis).

List of acronyms and abbreviations

ARMS At risk mental states
CAPE Community Assessment of Psychic Experiences
CHR Clinical high risk
CHR+ Presence of a clinical high risk state
CHR− Absence of a clinical high risk state
COGDIS Cognitive disturbances
COPE Center of Prevention and Evaluation
DSM-III Diagnostic and Statistical Manual of Mental Disorders, Third Edition
DSM-IV Diagnostic and Statistical Manual of Mental Disorders, Fourth Edition
EHRS Edinburgh High Risk Study of Schizophrenia
EPOS European Prediction of Psychosis Study
ESM Experience sampling methodology
GROUP Genetic Risk and Outcome in Psychosis
MagId Magical Ideation Scale
MRI Magnetic resonance imaging
NAPLS North American Prodrome Longitudinal Study
O-LIFE Oxford-Liverpool Inventory of Feelings and Experiences
PDs Personality disorders
PerAb Perceptual Aberration Scale
PerMag Perceptual Aberration Scale or Magical Ideation Scale
PhysAnh Physical Anhedonia Scale
PLEs Psychotic-like experiences
PRS Polygenic risk score
RISC Rust Inventory of Schizotypal Cognition
SIPS Structured Interview for Prodromal Syndromes
SIS Structured Interview for Schizotypy
SIS-R Structured Interview for Schizotypy-Revised
SocAnh Revised Social Anhedonia Scale
SPD Schizotypal personality disorder
SPQ Schizotypal Personality Questionnaire
UHR Ultra high risk
WSS Wisconsin Schizotypy Scales

Acknowledgments

Authors are supported by the Spanish Ministerio de Economía y Competitividad (PSI2017-87512-C2-00) and Comissionat per a Universitats i Recerca of the Generalitat de Catalunya (2017SGR1612). N. Barrantes-Vidal is supported by the Institució Catalana de Recerca i Estudis Avançats (ICREA) Academia Award and by Centro de Investigación Biomédica en Red de Salud Mental (CIBERSAM), Instituto de Salud Carlos III, Barcelona, Spain. A. Racioppi is supported by the Spanish Ministerio de Economía y Competitividad (BOE-A-2015-6508) and by the European Social Found (ESF).

References

Albert, N., Glenthøj, L.B., Melaua, M., Jensena, H., Hjorthøj, C., Nordentoft, M., 2017. Course of illness in a sample of patients diagnosed with a schizotypal disorder and treated in a specialized early intervention setting. Findings from the 3.5 year follow-up of the OPUS II study. Schizophr. Res. 182, 24–30.

American Psychiatric Association, 1994. Diagnostic and Statistical Manual of Mental Disorders, fourth ed. American Psychiatric Association, Washington.

American Psychiatric Association, 2013. Diagnostic and Statistical Manual of Mental Disorders, fifth ed. American Psychiatric Association, Washington.

Andreasen, N.C., Flaum, M., Swayze, V.W., Tyrrell, G., Arndt, S., 1990. Positive and negative symptoms in schizophrenia. A critical reappraisal. Arch. Gen. Psychiatr. 47, 615–621.

Bang, M., Park, J.Y., Kim, K.R., Lee, S.Y., Song, Y.Y., Kang, J.I., Lee, E., An, S.K., 2019. Psychotic conversion of individuals at ultra-high risk for psychosis: the potential roles of schizotypy and basic symptoms. Early Interv. Psychiatry 13 (3), 546–554.

Barrantes-Vidal, N., Chun, C.A., Myin-Germeys, I., Kwapil, T.R., 2013b. Psychometric schizotypy predicts psychotic-like, paranoid, and negative symptoms in daily life. J. Abnorm. Psychol. 122, 1077–1087.

Barrantes-Vidal, N., Grant, P., Kwapil, T.R., 2015. The role of schizotypy in the study of the etiology of schizophrenia spectrum disorders. Schizophr. Bull. 41 (Suppl. 2), S408–S416.

Barrantes-Vidal, N., Gross, G.M., Sheinbaum, T., Mitjavila, M., Ballespí, S., Kwapil, T.R., 2013a. Positive and negative schizotypy are associated with prodromal and schizophrenia-spectrum symptoms. Schizophr. Res. 145, 50–55.

Barrantes-Vidal, N., Lewandowski, K.E., Kwapil, T.R., 2010. Psychopathology, social adjustment and personality correlates of schizotypy clusters in a large nonclinical sample. Schizophr. Res. 122, 219–225.

Barrantes-Vidal, N., Ros-Morente, A., Kwapil, T.R., 2009. An examination of neuroticism as a moderating factor in the association of positive and negative schizotypy with psychopathology in a nonclinical sample. Schizophr. Res. 115, 303–309.

Blanchard, J.J., Collins, L.M., Aghevli, M., Leung, W.W., Cohen, A.S., 2011. Social anhedonia and schizotypy in a community sample: the Maryland Longitudinal Study of Schizotypy. Schizophr. Bull. 37, 587–602.

Bleuler, E.P., 1950. Dementia Praecox or the Group of Schizophrenias. J., Zinkin, Trans. International Universities Press, New York. Original work published in 1911.

Bogren, M., Mattisson, C., Tambs, K., Horstmann, V., Munk-Jørgensen, P., Nettelbladt, P., 2010. Predictors of psychosis: a 50-year follow-up of the Lundby population. Eur. Arch. Psychiatry Clin. Neurosci. 260, 113–125.

Bolinskey, P.K., Gottesman, I.I., 2010. Premorbid personality indicators of schizophrenia-related psychosis in a hypothetically psychosis-prone college sample. Scand. J. Psychol. 51, 68–74.

Bolinskey, P.K., James, A.V., Cooper-Bolinskey, D., Novi, J.H., Hunter, H.K., Hudak, D.V., Schuder, K.M., Myers, K.R., Iati, C.A., Lenzenweger, M.F., 2015. Revisiting the blurry boundaries of schizophrenia: spectrum disorders in psychometrically identified schizotypes. Psychiatry Res. 225, 335–340.

Bolinskey, P.K., Smith, E.A., Schuder, K.M., Cooper-Bolinskey, D., Myers, K.R., Hudak, D.V., James, A.V., Hunter, H.K., Novi, J.H., Guidi, J.P., Gonzalez, Y., McTiernan, E.F., Arnold, K.M., Iati, C.A., Gottesman, I.I., 2017. Schizophrenia spectrum personality disorders in psychometrically identified schizotypes at two-year follow-up. Psychiatry Res. 252, 289–295.

Brucato, G., Masucci, M.D., Arndt, L.Y., Ben-David, S., Colibazzi, T., Corcoran, C.M., Crumbley, A.H., Crump, F.M., Gill, K.E., Kimhy, D., Lister, A., Schobel, S.A., Yang, L.H., Lieberman, J.A., Girgis, R.R., 2017. Baseline demographics, clinical features and predictors of conversion among 200 individuals in a longitudinal prospective psychosis-risk cohort. Psychol. Med. 47, 1923–1935.

Cannon, T.D., Cadenhead, K., Cornblatt, B., Woods, S.W., Addington, J., Walker, E., Seidman, L.J., Perkins, D., Tsuang, M., McGlashan, T., Heinssen, R., 2008. Prediction of psychosis in youth at high clinical risk: a multisite longitudinal study in North America. Arch. Gen. Psychiatr. 65, 28–37.

Chapman, L.J., Chapman, J.P., Kwapil, T.R., Eckblad, M., Zinser, M., 1994. Putatively psychosis-prone subjects 10 years later. J. Abnorm. Psychol. 103, 171–183.

Chapman, J.P., Chapman, L.J., Kwapil, T.R., 1995. Scales for the measurement of schizotypy. In: Raine, A., Lencz, T., Mednick, S. (Eds.), Schizotypal Personality Disorder. Cambridge University Press, Cambridge, pp. 79–106.

Chapman, L.J., Chapman, J.P., Numbers, J.S., Edell, W.S., Carpenter, B.N., Beckfield, D., 1984. Impulsive nonconformity as a trait contributing to the prediction of psychotic-like and schizotypal symptoms. J. Nerv. Ment. Dis. 172, 681–691.

Chapman, L.J., Chapman, J.P., Raulin, M.L., 1976. Scales for physical and social anhedonia. J. Abnorm. Psychol. 85, 374–382.

Chapman, L.J., Chapman, J.P., Raulin, M.L., 1978. Body image aberration in schizophrenia. J. Abnorm. Psychol. 87, 399–407.

Chong, H.Y., Teoh, S.L., Wu, D.B.C., Kotirum, S., Chiou, C.F., Chaiyakunapruk, N., 2016. Global economic burden of schizophrenia: a systematic review. Neuropsychiatric Dis. Treat. 12, 357–373.

Chun, C.A., Barrantes-Vidal, N., Sheinbaum, T., Kwapil, T.R., 2017. Expression of schizophrenia-spectrum personality traits in daily life. Personal. Disord. 8, 64–74.

Claridge, G., 1997. Theoretical background issues. In: Claridge, G. (Ed.), Schizotypy: Implications for Illness and Health. Oxford University Press, Oxford, pp. 3–18.

Cohen, A.S., Couture, S.M., Blanchard, J.J., 2012. Neuropsychological functioning and social anhedonia: three year follow-up data from a longitudinal community high risk study. J. Psychiatr. Res. 46 (7), 898–904.

Debbané, M., Barrantes-Vidal, N., 2015. Schizotypy from a developmental perspective. Schizophr. Bull. 41 (Suppl. 2), S386–S395.

Debbané, M., Eliez, S., Badoud, D., Conus, P., Flückiger, R., Schultze-Lutter, F., 2015. Developing psychosis and its risk states through the lens of schizotypy. Schizophr. Bull. 41 (Suppl. 2), S396–S407.

Dollfus, S., Lyne, J., 2017. Negative symptoms: history of the concept and their position in diagnosis of schizophrenia. Schizophr. Res. 186, 3–7.

Eckblad, M., Chapman, L.J., 1983. Magical ideation as an indicator of schizotypy. J. Consult. Clin. Psychol. 51, 215–225.

Eckblad, M.L., Chapman, L.J., Chapman, J.P., Mishlove, M., 1982. The Revised Social Anhedonia Scale (Unpublished Test Copies), 27402. University of North Carolina, Department of Psychology, Greensboro, NC.

Erlenmeyer-Kimling, L., Cornblatt, B.A., Rock, D., Roberts, S., Bell, M., West, A., 1993. The New York High-Risk Project: anhedonia, attentional deviance, and psychopathology. Schizophr. Bull. 19, 141–153.

Ettinger, U., Mohr, C., Gooding, D.C., Cohen, A.S., Rapp, A., Haenschel, C., Park, S., 2015. Cognition and brain function in schizotypy: a selective review. Schizophr. Bull. 41 (Suppl. 2), S417–S426.

Fenton, W.S., McGlashan, T.H., 1989. Risk of schizophrenia in character disordered patients. Am. J. Psychiatry 146, 1280–1284.

Flückiger, R., Ruhrmann, S., Debbané, M., Michel, C., Hubl, D., Schimmelmann, B.G., Klosterkötter, J., Schultze-Lutter, F., 2016. Psychosis-predictive value of self-reported schizotypy in a clinical high-risk sample. J. Abnorm. Psychol. 125 (7), 923–932.

Fonseca-Pedrero, E., Debbané, M., Ortuño-Sierra, J., Chan, R.C.K., Cicero, D.C., Zhang, L.C., Brenner, C., Barkus, E., Linscott, R.J., Kwapil, T.R., Barrantes-Vidal, N., Cohen, A., Raine, A., Compton, M.T., Tone, E.B., Suhr, J., Muñiz, J., Fumero, A., Giakoumaki, S., Tsaousis, I., Preti, A., Chmielewski, M., Laloyaux, J., Mechri, A., Lahmar, M.A., Wuthrich, V., Larøi, F., Badcock, J.C., Jablensky, A., 2018. The structure of schizotypal personality traits: a cross-national study. Psychol. Med. 48, 451–462.

Fonseca-Pedrero, E., Paino, M., Lemos-Giráldez, S., Muñiz, J., 2011. Schizotypal traits and depressive symptoms in nonclinical adolescents. Compr. Psychiatr. 52, 293–300.

Gooding, D.C., Tallent, K.A., Matts, C.W., 2005. Clinical status of at-risk individuals 5 years later: further validation of the psychometric high-risk strategy. J. Abnorm. Psychol. 114, 170–175.

Grant, P., Green, M.J., Mason, O.J., 2018. Models of schizotypy: the importance of conceptual clarity. Schizophr. Bull. 44 (2), S556–S563.

Gross, G.M., Mellin, J., Silvia, P.J., Barrantes-Vidal, N., Kwapil, T.R., 2014. Comparing the factor structure of the Wisconsin schizotypy scales and the Schizotypal Personality Questionnaire. Personal. Disord. 5, 397–405.

Gross, G.M., Kwapil, T.R., Raulin, M.L., Silvia, P.J., Barrantes-Vidal, N., 2018. The multidimensional schizotypy scale-brief: scale development and psychometric properties. Psychiatry Res. 261, 7–13.

Horan, W.P., Brown, S.A., Blanchard, J.J., 2007. Social anhedonia and schizotypy: the contribution of individual differences in affective traits, stress, and coping. Psychiatry Res. 149, 147–156.

Horton, L.E., Barrantes-Vidal, N., Silvia, P.J., Kwapil, T.R., 2014. Worries about being judged versus being harmed: disentangling the association of social anxiety and paranoia with schizotypy. PLoS One 9, e96269.

Johnstone, E.C., Ebmeier, K.P., Miller, P., Owens, D.G., Lawrie, S.M., 2005. Predicting schizophrenia: findings from the Edinburgh high-risk study. Br. J. Psychiatry 186, 18–25.

Kaczorowski, J.A., Barrantes-Vidal, N., Kwapil, T.R., 2009. Neurological soft signs in psychometrically identified schizotypy. Schizophr. Res. 115, 293–302.

Kendler, K.S., Lieberman, J.A., Walsh, D., 1989. The structured interview for Schizotypy (SIS): a preliminary report. Schizophr. Bull. 15, 559–571.

Klosterkötter, J., Hellmich, M., Steinmeyer, E.M., Schultze-Lutter, F., 2001. Diagnosing schizophrenia in the initial prodromal phase. Arch. Gen. Psychiatr. 58 (2), 158–164.

Kozak, M.J., Cuthbert, B.N., 2016. The NIMH research domain criteria initiative: background, issues, and pragmatics. Psychophysiology 53, 286–297.

Kraepelin, E., 1919. Dementia Praecox and Paraphrenia. Livingstone, Edinburgh. Original work published in 1913.

Kretschmer, E., 1925. Physique and Character. W.J.H., Sprott, Trans. Kegan, Trench and Trubner, London.

Kwapil, T.R., 1998. Social anhedonia as a predictor of the development of schizophrenia-spectrum disorders. J. Abnorm. Psychol. 107, 558–565.

Kwapil, T.R., Barrantes-Vidal, N., 2012. Schizotypal personality disorder: an integrative review. In: Widiger, T.A. (Ed.), The Oxford Handbook of Personality Disorders. Oxford University Press, New York, pp. 437–477.

Kwapil, T.R., Barrantes-Vidal, N., 2015. Schizotypy: looking back and moving forward. Schizophr. Bull. 41 (Suppl. 2), S366–S373.

Kwapil, T.R., Barrantes-Vidal, N., Silvia, P.J., 2008. The dimensional structure of the Wisconsin Schizotypy Scales: factor identification and construct validity. Schizophr. Bull. 34, 444–457.

Kwapil, T.R., Brown, L.H., Silvia, P.J., Myin-Germeys, I., Barrantes-Vidal, N., 2012. The expression of positive and negative schizotypy in daily life: an experience sampling study. Psychol. Med. 42, 2555–2566.

Kwapil, T.R., Chun, C.A., 2015. The psychometric assessment of schizotypy. In: Mason, O., Claridge, G. (Eds.), Schizotypy. Oxford University Press, New York, pp. 7–32.

Kwapil, T.R., Gross, G.M., Silvia, P.J., Barrantes-Vidal, N., 2013. Prediction of psychopathology and functional impairment by positive and negative schizotypy in the Chapmans' ten-year longitudinal study. J. Abnorm. Psychol. 122 (3), 807–815.

Kwapil, T.R., Gross, G.M., Silvia, P.J., Raulin, M.L., Barrantes-Vidal, N., 2018. Development and psychometric properties of the Multidimensional Schizotypy Scale: a new measure for assessing positive, negative, and disorganized schizotypy. Schizophr. Res. 193, 209–217.

Kwapil, T.R., Miller, M.B., Zinser, M.C., Chapman, J., Chapman, L.J., 1997. Magical ideation and social anhedonia as predictors of psychosis proneness: a partial replication. J. Abnorm. Psychol. 106 (3), 491–495.

Mason, O., 2015. The assessment of schizotypy and its clinical relevance. Schizophr. Bull. 41 (Suppl. 2), S374–S385.

Mason, O., Claridge, G., Jackson, M., 1995. New scales for the assessment of schizotypy. Personal. Individ. Differ. 18, 7–13.

Mason, O., Claridge, G., Williams, L., 1997. Questionnaire measurement. In: Claridge, G. (Ed.), Schizotypy: Implications for Illness and Health. Oxford University Press, New York, pp. 19–37.

Mason, O., Startup, M., Halpin, S., Schall, U., Conrad, A., Carr, V., 2004. Risk factors for transition to first episode psychosis among individuals with "at-risk mental states". Schizophr. Res. 71, 227–237.

Meehl, P.E., 1962. Schizotaxia, schizotypy, schizophrenia. Am. Psychol. 17, 827–838.

Meehl, P.E., 1964. Manual for Use with Checklist of Schizotypic Signs. Unpublished manuscript.

Miettunen, J., Veijola, J., Isohanni, M., Paunio, T., Freimer, N., Jääskeläinen, E., Taanila, A., Ekelund, J., Järvelin, M.R., Peltonen, L., Joukamaa, M., Lichtermann, D., 2011. Identifying schizophrenia and other psychoses with psychological scales in the general population. J. Nerv. Ment. Dis. 199, 230–238.

Miller, P., Byrne, M., Hodges, A., Lawrie, S.M., Owens, D.G., Johnstone, E.C., 2002a. Schizotypal components in people at high risk of developing schizophrenia: early findings from the Edinburgh High-Risk Study. Br. J. Psychiatry 180, 179–184.

Miller, P.M., Lawrie, S.M., Byrne, M., Cosway, R., Johnstone, E.C., 2002b. Self-rated schizotypal cognitions, psychotic symptoms and the onset of schizophrenia in young people at high risk of schizophrenia. Acta Psychiatr. Scand. 105, 341–345.

Mohr, C., Claridge, G., 2015. Schizotypy—do not worry, it is not all worrisome. Schizophr. Bull. 41 (Suppl. 2), S436–S443.

Morrison, A.P., Bentall, R.P., French, P., Walford, L., Kilcommons, A., Knight, A., Kreutz, M., Lewis, S.W., 2002. Randomised controlled trial of early detection and cognitive therapy for preventing transition to psychosis in high-risk individuals. Study design and interim analysis of transition rate and psychological risk factors. Br. J. Psychiatry 181 (Suppl. 43), S78–S84.

Myin-Germeys, I., Ooschort, M., Collip, D., Lataster, J., Delespaul, P., van Os, J., 2009. Experience sampling research in psychopathology: opening the black box of daily life. Psychol. Med. 39, 1533–1547.

Nordentoft, M., Thorup, A., Petersen, L., Ohlenschlaeger, J., Melau, M., Christensen, T.O., Krarup, G., Jorgensen, P., Jeppesen, P., 2006. Transition rates from schizotypal disorder to psychotic disorder for first-contact patients included in the OPUS trial: a randomized clinical trial of integrated treatment and standard treatment. Schizophr. Res. 83 (1), 29–40.

Parnas, J., Raballo, A., Handest, P., Jansson, L., Vollmer-Larsen, A., Saebye, D., 2011. Self-experience in the early phases of schizophrenia: 5-year follow-up of the Copenhagen prodromal study. World Psychiatry 10, 200–204.

Racioppi, A., Sheinbaum, T., Gross, G.M., Ballespí, S., Kwapil, T.R., Barrantes-Vidal, N., Prediction of prodromal symptoms and schizophrenia-spectrum personality disorder traits by positive and negative schizotypy: a 3-year prospective study. PLoS ONE 13 (11), e0207150.

Rado, S., 1953. Dynamics and classification of disordered behavior. Am. J. Psychiatry 110, 406–416.

Raine, A., 1991. The SPQ: a scale for the assessment of schizotypal personality based on DSM III–R Criteria. Schizophr. Bull. 17 (4), 555–564.

Ruhrmann, S., Schultze-Lutter, F., Salokangas, R.K., Heinimaa, M., Linszen, D., Dingemans, P., Birchwood, M., Patterson, P., Juckel, G., Heinz, A., Morrison, A., Lewis, S., von Reventlow, H.G., Klosterkötter, J., 2010. Prediction of psychosis in adolescents and young adults at high risk: results from the prospective European prediction of psychosis study. Arch. Gen. Psychiatr. 67, 241–251.

Sahakyan, L., Kwapil, T.R., 2016. Positive and negative schizotypy are associated with differential patterns of episodic memory impairment. Schizophr. Res. Cogn. 5, 35–40.

Salokangas, R.K., Dingemans, P., Heinimaa, M., Svirskis, T., Luutonen, S., Hietala, J., Ruhrmann, S., Juckel, G., Graf von Reventlow, H., Linszen, D., Birchwood, M., Patterson, P., Schultze-Lutter, F., Klosterkotter, J., 2013. Prediction of psychosis in clinical high-risk patients by the Schizotypal Personality Questionnaire. Results of the EPOS project. Eur. Psychiatry 28, 469–475.

Schultze-Lutter, F., Klosterkötter, J., Michel, C., Winkler, K., Ruhrmann, S., 2012. Personality disorders and accentuations in at-risk persons with and without conversion to first-episode psychosis. Early Interv. Psychiatry 6, 389–398.

Schultze-Lutter, F., Schimmelmann, B.G., Ruhrmann, S., Michel, C., 2013. 'A rose is a rose is a rose', but at-risk criteria differ. Psychopathology 46, 75–87.

Schulz, P.M., Soloff, P.H., 1987. Still borderline after all these years. In: Proceedings of the 140th Annual Meeting of the American Psychiatric Association, Chicago, IL, May 12, 1987.

Shah, J., Eack, S.M., Montrose, D.M., Tandon, N., Miewald, J.M., Prasad, K.M., Keshavan, M.S., 2012. Multivariate prediction of emerging psychosis in adolescents at high risk for schizophrenia. Schizophr. Res. 14, 189–196.

Stefanis, N.C., Hanssen, M., Smirnis, N.K., Avramopoulos, D.A., Evdokimidis, I.K., Stefanis, C.N., Verdoux, H., van Os, J., 2002. Evidence that three dimensions of psychosis have a distribution in the general population. Psychol. Med. 32 (2), 347–358.

Tandon, R., Nasrallah, H.A., Keshavan, M.S., 2009. Schizophrenia, "just the facts" 4. Clinical features and conceptualization. Schizophr. Res. 110, 1–23.

Tijms, B.M., Sprooten, E., Job, D., Johnstone, E.C., Owens, D.G., Willshaw, D., Seriès, P., Lawrie, S.M., 2015. Grey matter networks in people at increased familial risk for schizophrenia. Schizophr. Res. 168, 1–8.

van Os, J., van der Steen, Y., Islam, M.A., Gülöksüz, S., Rutten, B.P., Simons, C.J., GROUP Investigators, 2017. Evidence that polygenic risk for psychotic disorder is expressed in the domain of neurodevelopment, emotion regulation and attribution of salience. Psychol. Med. 47 (14), 2421–2437.

Weiser, M., Reichenberg, A., Rabinowitz, J., Kaplan, Z., Mark, M., Bodner, E., Nahon, D., Davidson, M., 2001. Association between nonpsychotic psychiatric diagnoses in adolescent males and subsequent onset of schizophrenia. Arch. Gen. Psychiatr. 58 (10), 959–964.

Wolff, S., Townshend, R., McGuire, R., Weeks, D., 1991. 'Schizoid' personality in childhood and adult life. II: adult adjustment and the continuity with schizotypal personality disorder. Br. J. Psychiatry 159, 620–629.

Woods, S.W., Addington, J., Cadenhead, K.S., Cannon, T.D., Cornblatt, B.A., Heinssen, R., Perkins, D.O., Seidman, L.J., Tsuang, M.T., Walker, E.F., McGlashan, T.H., 2009. Validity of the prodromal risk syndrome for first psychosis: findings from the North American Prodrome Longitudinal Study. Schizophr. Bull. 35, 894–908.

Zarogianni, E., Storkey, A.J., Johnstone, E.C., Owens, D.G., Lawrie, S.M., 2017. Improved individualized prediction of schizophrenia in subjects at familial high risk, based on neuroanatomical data, schizotypal and neurocognitive features. Schizophr. Res. 181, 6–12.

CHAPTER

6

Familial high risk and high-risk studies

Liana Romaniuk[1], Stella W.Y. Chan[2], Alix Macdonald[1], Jessika E. Sussmann[1], Andrew M. McIntosh[1], Heather C. Whalley[1], Stephen M. Lawrie[1,]*

[1]Division of Psychiatry, University of Edinburgh, Edinburgh, United Kingdom; [2]Section of Clinical Psychology, University of Edinburgh, Edinburgh, United Kingdom

Introduction

The observation that major mental illness is highly heritable offers several opportunities. It suggests a means by which we might learn more about the molecular mechanisms underlying various diagnoses. It also allows us to focus on the individuals at highest risk, to identify which factors distinguish them from the general population, and what in turn keeps some people well, while others are less fortunate.

Gottesman (1991) summarized the 20th century literature on the relative risks of schizophrenia in family members of various probands depending upon their genetic propinquity. In summary, people with a first degree relative with schizophrenia have a roughly 10-fold increased risk. This, obviously, includes nonidentical twins. Identical twins have a 50-fold increased risk, but this still means that one generally gets a roughly equal proportion of discordant and concordant monozygotic twin pairs for schizophrenia. Thus, differential epigenetic and environmental risk factors are implicated. Overall, the heritability of schizophrenia is around 80% and bipolar disorder (BD) slightly higher at 85%, which includes both genetic and gene−environment interactions.

Familial high-risk (HR) studies come in several forms, each having varying abilities to disentangle the genetic from the shared environmental contributions to risk. Twin studies allow for an informative estimation of genetic heritability, as well as portioning out the remaining variance to shared versus unique environmental factors. However, they necessarily involve small sample sizes, which limit their statistical power, and are especially hard to perform longitudinally if the goal is illness prediction. As rare as twin studies are, adoption studies are rarer still. However, studies of those with an affected parent or sibling expand the potential sample size at the expense of a less clear delineation of

* Senior author.

Risk Factors for Psychosis
https://doi.org/10.1016/B978-0-12-813201-2.00006-5

genetic and environmental risk. Cross-sectional approaches provide a snapshot in time, which can help identify potential endophenotypes, but cannot clear the fog of previous environmental factors. The longitudinal approach allows deeper insights into causality and prediction. This is especially important in illnesses such as schizophrenia and BD, which both tend to involve a prodrome well in advance of the index episode precipitating diagnosis (Duffy et al., 2017). Some studies adopt an alternative approach whereby they use this prodrome as a means to define a population as being at clinical (as opposed to familial) high risk. Here, the rates of transition to full illness tend to be much higher and likewise the population's heterogeneity. This means they may better capture the variance seen in patient groups, but at the expense of gaining a clearer mechanistic understanding of the illness (Bois et al., 2015c).

The nature of prodromal changes and how they alter with illness progression, symptom exposure, and age is only just beginning to be elucidated. It is unclear, however, which changes occur before the onset of symptoms and whether they correlate with the severity of those symptoms once expressed. Once an illness becomes established, the degenerative effects of a multiepisodic long-term illness become confounded by the presence of medication. Disentangling these complex interactions poses a significant challenge, and any inconsistencies that exist in the literature may relate to the cross-sectional nature of the studies involved and the variation in age of their samples. To truly determine predictive markers of illness and how neural abnormalities change over the course of illness development, prospective longitudinal studies of young descendants of people with a well-characterized illness commencing before typical ages of onset are critical (Schneider et al., 2012). The identification of specific vulnerability markers could promote a mechanistic understanding of disease etiology, as well as advancing prevention and early intervention strategies.

In this chapter we summarize findings from two prospective longitudinal familial HR studies: the Edinburgh High Risk Study of schizophrenia (EHRS) and the Bipolar Family Study (BFS).

Schizophrenia

In Edinburgh, we started recruiting and examining participants with at least two first- and/or second-degree relatives with schizophrenia in 1995. A total of 229 potentially suitable HR subjects were identified, 162 provided some clinical, neuropsychological, and/or imaging data, and 150 had one or more structural magnetic resonance imaging (sMRI) scans by 1999. Smaller groups of similarly aged healthy controls (CON) and first-episode schizophrenia (FES) patients were also examined. Most of the HR subjects and controls returned for at least one sMRI and functional MRI (fMRI) scan between 2000 and 2005.

Baseline findings/trait effects on structural magnetic resonance imaging

Preliminary findings on sMRI, in the first 100 HR subjects (Lawrie et al., 1999), indicated that semi-automated "region-of-interest" traced volumes of the amygdala–hippocampal complex (AHC) were significantly smaller (by about 4%) than in controls, but about 4% larger than in FES patients (effect size about 0.3 of a standardized difference). Reductions in the volume of the thalamus were only evident in the high risk versus controls contrast. These main findings were replicated on voxel-based morphometry (VBM) analysis (Job et al., 2002, 2003), which also showed anterior cingulate and medial prefrontal reductions in gray matter density with an FES < HR < CON pattern. More recent reanalysis with FreeSurfer has, however, only partially replicated these findings (Sprooten et al., 2013b; Bois et al., 2015a). This may be because these data were acquired with a relatively weak magnet (1.0T) by contemporary standards, with the resulting images perhaps not having the required spatial resolution and gray/white matter contrast to allow for surface-based parcellation methods such as FreeSurfer to detect these subtle differences.

In the full HR sample, prefrontal lobe and thalami volumes and whole brain volumes, but not those of the AHC, were associated with a matrix-derived measure of the strength of genetic liability to schizophrenia (Lawrie et al., 2001). Ventricular volumes at baseline showed associations with self-reported alcohol and cannabis use, with significant dose—response effects (Welch et al., 2011a), while thalamus volumes were smaller in the HR subgroup who indicated that they smoked cannabis regularly (Welch et al., 2011b). Using FreeSurfer, both AHC volumes and cortical thickness were related to proxy measures for childhood adversity, namely social work involvement and/or the child having been referred to the Children's Panel, neither of which was simply attributable to having a parent with schizophrenia (Barker et al., 2016a,b).

Psychotic symptom associations/state effects on structural magnetic resonance imaging

Psychotic symptoms were elicited with the Present State Examination (PSE) (Wing et al., 1974), and isolated or transient symptoms were found in many more HR subjects than we had expected (66 out of the 146 subjects who had completed at least one PSE). In those with psychotic symptoms at any point in the first 5 years of the study, the whole brain volume at study entry was reduced (after controlling for age, sex, paternal social class, height, and handedness) compared to those that did not have psychotic symptoms over this time (Lawrie et al., 2001). No other brain volumes at intake were related to a liability to psychotic symptoms. Those with psychotic symptoms, and two or more sMRI scans, had nonsignificant reductions in whole brain and (left) AHC volume over almost 2 years, but did have significant reductions in the (right) temporal lobes over time (Lawrie et al., 2002). Regular cannabis consumption between two scans was associated with reduced gray matter density in the thalamus (Welch et al., 2013).

Functional magnetic resonance imaging

Our fMRI studies were able to demonstrate a compatible picture of trait-related deficits in prefrontal cortex, thalamus, and cerebellum on a sentence completion task (Whalley et al., 2004), alongside increased parietal lobe activity in those with isolated or transient symptoms at the time of the scan. The latter was accompanied by an increased frontoparietal connectivity, presumably to compensate for genetically related prefrontal dysfunction (Whalley et al., 2005). Functional and effective connectivity analyses showed a common pattern of reduced frontothalamic connectivity, with greater effects in those with symptoms on the latter (Dauvermann et al., 2013). A comparable picture of symptom-related exaggerations of trait deficits in prefrontal cortex and cerebellum was evident on another two tasks (Marjoram et al., 2006; Whyte et al., 2006). HR subjects who developed ICD-10 schizophrenia demonstrated increased activation of the parietal lobe, decreased activation of the anterior cingulate, and smaller increases in activation with increasing task difficulty in the right lingual gyrus and bilateral temporal regions. Although these findings have to be considered cautiously, as only four subjects who had an fMRI scan subsequently became ill, they suggest functional abnormalities are present in HR subjects who later became ill, which distinguish them not only from normal control subjects but also from those at high risk who had not developed the disorder (Whalley et al., 2006).

Predicting schizophrenia

About 21 subjects (13%), 20 with some data, developed schizophrenia by the close of the study (Johnstone et al., 2005). As far as we are aware, none more have done since then. Sixteen of these participants had at least one sMRI scan, eight had two or more, and four also had an fMRI scan, before becoming ill. Preliminary results suggested that relatively large brains (and relatively small thalami in women) predicted the early onset of psychosis (within the first 5 years of the study) in the HR subjects (Johnstone et al., 2002). Subsequent, complete analyses have, however, found little more than weak predictive utility of reduced AHC or thalamus volume (Job et al., 2005; Johnstone et al., 2005), and any predictive effect of regional brain volumes for the subsequent development of schizophrenia was at most weak (Lawrie, 2007).

Baseline measures of increased cortical folding, particularly in prefrontal cortex, were relatively strong predictors of schizophrenia, whether analyzed by hand-tracing (Harris et al., 2004a, b), an in-house automated measure (gyrification index, GI) (Harris et al., 2007), or using FreeSurfer (Bois et al., 2015b; Neilson et al., 2017). None of these approaches had sufficient specificity so as to be of potential clinical utility. This may reflect the largely developmental nature of gyral folding. The major gyri are mainly formed during the second trimester in utero and the GI measure is itself relatively resistant to the subsequent impact of any cortical thinning. In the HR subjects with both sMRI and fMRI data, significant positive associations were found between prefrontal GI and prefrontal lateral—medial functional connectivity, while a negative correlation was found between prefrontal GI and prefrontal-thalamic connectivity (Dauvermann et al., 2012). These associations were more pronounced than in control subjects and suggest that increased prefrontal gyrification may therefore relate to increased local short-range prefrontal connectivity and reduced long-range connectivity.

Job et al. (2005) extended the earlier findings of reductions in temporal lobe volume in those with psychotic symptoms with VBM, both by showing distributed changes in prefrontal and temporal lobes and by revealing additional reductions over approximately 18 months in gray matter density in left (para)hippocampal uncus, fusiform gyrus, and right cerebellar cortex in the eight individuals at high risk who developed schizophrenia on average 2.5 years after the first of the two scans. This finding, despite the very small sample, makes it clear that such changes may occur years before diagnosis and cannot be attributable to medication as none of the participants were medicated until after their diagnosis. We have since evaluated the diagnostic properties of these reductions in gray matter density and shown positive predictive values (PPVs) of around 60% for these regional reductions individually and about 70% in combination (Job et al., 2006). These PPVs can be contrasted with the much lower PPVs seen when using psychotic symptoms and even the most predictive clinical or behavioral measures. These latter measures, of declarative memory impairment and schizotypy (Johnstone et al., 2005), were most strongly correlated with gray matter density in thalamic nuclei and parts of the medial temporal lobe, with additional schizotypal symptom severity correlations in the superior temporal gyrus (Lymer et al., 2006). Note that here we are talking of psychotic symptoms elicited at PSE interview rather than at risk mental state (ARMS) criteria, although most of the HR subjects with such symptoms would likely have met ARMS criteria for brief or attenuated symptoms.

Taking all 434 sMRI scans acquired in the EHRS, in a longitudinal analysis of changes over time, we were able to demonstrate striking reductions in overall prefrontal and temporal lobe extracted volumes, which were strongly correlated with increased severity of the delusions, hallucinations, and passivity experiences, which are the hallmark symptoms of schizophrenia (McIntosh et al., 2011).

Finally, most recently, we have used machine learning approaches to optimize the predictive utility of brain structural findings. Taking only the baseline sMRI scan, an average of 2.5 years before diagnosis, we were able to distinguish those HR participants with psychotic symptoms who would go on to develop schizophrenia from those who would not with an impressive 88% overall accuracy—rising to 94% if we included the two most powerful clinical predictors of memory function and schizotypy (Zarogianni et al., 2017b). Furthermore, this exact sMRI predictive algorithm had an overall accuracy of 74% in a completely independent sample who were recruited at elevated symptomatic rather than genetic risk (Zarogianni et al., 2017a).

Summary

The overall pattern of results from the EHRS therefore suggests genetically mediated structural and functional deficits in prefrontal cortex, medial temporal lobe, and thalamus, with additional volume decrements in the prefrontal and temporal lobes being related to the onset of more severe symptomatology and frank schizophrenia. These results are very much in keeping with studies of clinical HR populations. Several research groups around the world have gone on to show that these findings are sufficiently powerful to be of clinical utility in what might be referred to as "proof-of-concept" studies.

Bipolar disorder

BD is characterized by episodes of emotion dysregulation, potentially driven by disturbances in neural systems mediating emotion processing. Neuroimaging studies in patient groups have found structural and functional abnormalities in the corticolimbic network, particularly the amygdala, hippocampus, cingulate cortex, and dorsal and ventrolateral prefrontal regions (Phillips et al., 2003; Hajek et al., 2009; Savitz and Drevets, 2009; Elliott et al., 2010; Bora et al., 2012; Sacher et al., 2012). Such models propose either a reduction of prefrontal higher-order cognitive control over limbic regions or conversely a hyperactive limbic network overriding cortical control (Phillips et al., 2003).

BD is highly heritable, with likely complex genetic architecture of a likely polygenic nature. Estimates of heritability are believed to be as much as 85%, with approximately one-third of the genetic variance being shared between depression and mania (McGuffin et al., 2003). In pursuit of potential neural vulnerability markers for illness, several studies have examined cohorts of unaffected individuals at increased familial risk for BD (see reviews in Fusar-Poli et al. (2012); Piguet et al. (2015)). Here, we focus on the Scottish BFS, which is a prospective longitudinal neuroimaging investigation examining young individuals (16–25 years) at high familial risk of mood disorder (HR), compared with a group of healthy controls with no family history of major psychiatric disorder. There were three waves of follow-up assessments, 2 years apart (Fig. 6.1) (Whalley et al., 2015).

The majority of HR individuals (>90%) had a first-degree relative with BD, and the remainder had >2 second-degree relatives. All were well at the time of recruitment and had no personal history of any psychiatric disorder. Imaging assessments took place at baseline (time 1) and

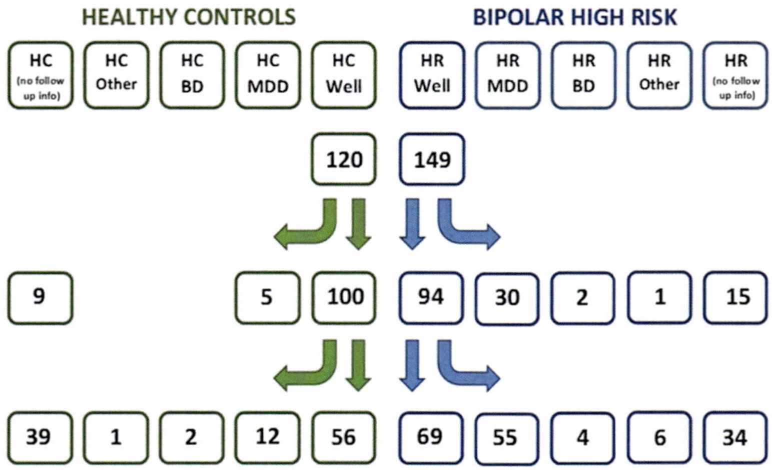

FIGURE 6.1 Flow diagram of numbers of Bipolar Family Study individuals in each group category. *BD*, bipolar disorder; *HC*, healthy controls; *HR*, high risk; *MDD*, major depressive disorder. Top *line* represents initial recruitment (time 1, all individuals well), subsequent follow-up assessments every 2 years represented as *lines* 2 (time 2) and 3 (time 3). Depicts total number of individuals on whom any form of data was collected, not just individuals providing imaging data.

the first round of follow-up (time 2), whereas clinical assessments took place at all three waves. This included a battery of neuropsychological tests; measures of other risk factors such as childhood trauma; trait markers such as neuroticism and extraversion; and clinical symptom severity measures.

Over the course of the BFS, a number of individuals were reported to have developed a mood disorder, predominantly major depressive disorder (MDD), with a smaller number developing BD.

Structural imaging

The first structural longitudinal imaging study from this cohort measured cortical thickness from 111 initially unaffected HR subjects, of whom 20 developed MDD by time 2, versus 93 healthy control individuals (Papmeyer et al., 2015). Considering the main effect of group, HR participants who developed MDD (HR-MDD) demonstrated reduced cortical thickness in the right parahippocampal and fusiform gyri compared to HR who remained well (HR-Well), who in turn had reduced thickness compared to healthy controls. The parahippocampal gyrus was reported on average to be ~4% thinner in HR-Well individuals versus controls and ~11% thinner in the HR-MDD group across both time points. However, although the HR-MDD group was significantly different from the controls, formal statistical post hoc pairwise comparisons only revealed a trend level of significance for the comparison of the two HR groups themselves, which would have indicated a stronger predictive capacity of the findings.

A second similar study from the same cohort examined changes over time in several subcortical regions, including the basal ganglia, thalamus, and limbic areas. However, this revealed no changes in the 2-year follow-up period between baseline and time 2 using FreeSurfer-based volumetric analysis (Papmeyer et al., 2016).

Another structural study of this sample, this time employing longitudinal VBM analysis and using clinical information gathered at the third study assessment to categorize groups, revealed significant structural decreases over time in the amygdala in those who became ill (Nickson et al., 2016). These greater reductions over time were seen in the HR-MDD individuals versus both HR individuals who remained well and healthy controls. The findings therefore indicated that dynamic structural changes in the amygdala were associated with the development of MDD in this young sample. It was also reported in this paper that structural change in this region was not associated with measures of childhood trauma, symptom severity, or neuroticism, although these latter measures were significantly different between the groups at baseline assessment.

In addition to structural volume, white matter integrity—as measured by fractional anisotropy during diffusion tensor imaging—is known to be generally reduced in those at familial risk of BD (Sprooten et al., 2011). A genome-wide association study comparing HR-Well and controls isolated a number of single-nucleotide polymorphisms key to white matter development, neuronal plasticity, and cell adhesion (Sprooten et al., 2013a). However, the trajectories of change in white matter integrity over time are similar when comparing HR-MDD, HR-Well, and controls (Ganzola et al., 2017).

Functional imaging

Using a sentence completion fMRI task designed to probe executive language function, we examined whether it was possible to distinguish at baseline those HR participants who became ill (defined as a first manic or major depressive episode) from those who remained well (Whalley et al., 2013). The study reported that HR subjects that later became ill demonstrated increased activation of the insula cortex and inferior frontal regions compared to both the control subjects and those HR who remained well. A follow-up study used an emotional memory task acquired at time 2 and clinical data from time 3 to determine whether this finding could be replicated in a separate subsample of HR individuals who became ill (Whalley et al., 2015). Consistent with the above study, it also reported increased activation of the insula, along with thalamus and anterior cingulate cortex, in those HR individuals who subsequently became ill versus those who remained well. This provides partial replication of the finding, as they are based on two separate subsamples; however, there is still a need to achieve replication in a truly independent population. The findings in insula, thalamus, and anterior cingulate cortex also speak to the potential consistency across imaging paradigms utilizing different cognitive domains. However, as both these tasks were acquired at a single time point, they cannot comment on changes over time and the impact these could have on symptoms expressed.

Summary

The results of the BFS have reinforced the important role of limbic regions in the development of BD, with parahippocampal and insula reductions in those that become unwell, above and beyond any influence of childhood trauma. Beyond genetic vulnerability, hippocampal size is also reduced in those who have experienced many adverse life events (Hoy et al., 2012). The hippocampus therefore stands as a structure whose size potentially reflects accumulated insults both genetic and environmental.

Cortical regions involved in emotions representing internal states (insula) and regulation (inferior frontal gyrus, anterior cingulate) demonstrate altered functional activation across two task domains. The amygdala, prefrontal cortex, and wider limbic networks respond to the emotional salience of stimuli and are fundamental in the formation and stability of affective states (LeDoux, 2000). In individuals with mood disorders, the amygdala consistently demonstrates abnormally increased activation in response to negative or nonemotional stimuli (Drevets et al., 2008) and impaired prefrontal connectivity (Matthews et al., 2008), perhaps underpinning the emotion regulation difficulties experienced in psychotic disorders with a strong affective component and in those at risk of them (Phillips et al., 2003). Previous cross-sectional studies report abnormal corticolimbic structure and function in unaffected relatives, including the amygdala (e.g., Mannie et al. (2014)), and in relation to cumulative genetic risk (Whalley et al., 2012). Corticolimbic abnormalities are therefore not restricted to established illness or the confounding effects of medication. They may relate to the overall burden of genetic risk, and these findings add to the suggestion of dynamic change concurrent with illness development.

White matter integrity is globally reduced in those that become unwell, although the rate of change over time does not appear to differ from HR-Well or controls. Findings within the bipolar literature to date are somewhat equivocal regarding longitudinal changes with illness duration. The suggestion has been made that white matter abnormalities are relatively stable and may represent an at-risk state or onset of illness at its earliest stages (Schneider et al., 2012).

Discussion

A transdiagnostic approach?

Schizophrenia and BD share 44% of their heritability (Song et al., 2015), so some overlap in their neurobiology is to be expected. Both show reductions in medial temporal volume, with genetic risk being associated with a degree of reduction, and those that go on to develop psychosis having smaller volumes still. Within those at risk of schizophrenia, the rate of parahippocampal volume decline was partly able to predict who would go on to become unwell.

Findings that were more specific to schizophrenia included frontothalamic deficits, with those at risk demonstrating reduced thalamic volumes and functional responsivity to task demands; reduced prefrontal cortex volume, midline gray matter density, and functional activation; and impairments in effective connectivity between the two, especially in those who were symptomatic. This may relate to the observation that so-called "cold" cognition tasks such as problem solving and working memory tend to be impaired in schizophrenia (Snitz et al., 2006), as opposed to the "hot" cognition deficits seen in BD, such as emotional Stroop and facial emotion processing (Miskowiak et al., 2017).

Cerebellum showed lower functional activation and longitudinally decreasing gray matter density that predicted future illness. Temporal and frontal lobe volume reductions were also a feature of schizophrenia risk, with changes over time showing a correlation with increasing transient psychotic symptoms. Parietal lobe hyperactivation was also seen in those with transient symptoms, together with increased frontoparietal connectivity.

Risk of BD was associated with decreased volume, but increased functional activation of insula cortex. Nearby inferior frontal cortex, as well as anterior cingulate, also showed increased activation. In contrast to what was seen in the schizophrenia group, they demonstrated thalamic hyperactivation. Generalized white matter deficits were also a strong feature. Affective instability is believed to be in part dependent on inferior frontal cortex's overinvolvement in the brain's salience network and altered cingulate function in governing behavior (Broome et al., 2015).

Within the schizophrenia HR sample, we found that those who had been experiencing transient psychotic symptoms already demonstrated reduced whole brain volumes and that their right temporal lobes continue to decrease. This highlights the potential fluidity in time for reaching a diagnosis and when to draw the line between a prodromal state and a generally agreed transition to a categorical psychosis. The same applies to the observed changes in gross anatomy, which may represent a risk state or the very early phase of a formal illness. This issue is further clouded by another feature shared by both illnesses: their clinical heterogeneity. Although it is hoped that some of that heterogeneity can be accounted for by identifying clinical subtypes with associated biomarkers, it will arguably remain the case that accurate diagnosis cannot be reached solely by noting a series of neurobiological changes, but must be taken alongside the individual's context of symptoms, psychology, and social functioning.

Tracking environmental factors

Although the presence of a family history of mental illness is one of the highest risk factors for psychosis, it by no means accounts for all the variance. For example, the Brain Center Rudolf Magnus in Utrecht, the Netherlands, has recently begun publishing on a longitudinal twin study into BD. By comparing concordant and discordant mono/dizygotic twins, they found that baseline total cortical surface area/volume and subcortical volumes were highly heritable, whereas regional measures were more sensitive to unique environmental contributions (Bootsman et al., 2015b). In addition, longitudinal changes over time (mean 7.5 years) were not found to have a strong heritable component (Bootsman et al., 2015a, 2016).

Here, the EHRS of schizophrenia highlighted illness-relevant brain changes imposed by both cannabis and alcohol, as well as childhood adversity. Childhood trauma has a now well-established relationship with the later development of psychosis, with meta-analysis suggesting a threefold increase in risk (van Winkel et al., 2013). The additional environmental challenge of having a parent with psychosis has also been noted: Doucette and colleagues find an association between perceived maternal neglect and subsequent likelihood of illness in their 16-year longitudinal study of bipolar offspring (Doucette et al., 2016). A 12-year study by Iacono et al. of children with one bipolar parent found that parental control over children mediates the relationship between having a parent with BD and subsequently becoming unwell (Iacono et al., 2017). Disentangling the means by which a parent with an illness confers risk on their children likely requires studies involving fostered and adopted children, for example, the Finnish Adoptive Family Study of Schizophrenia (Wahlberg et al., 2004), which present significant methodological challenges. Alternatively, animal models can allow understanding of gene × environment interactions that are much more genetically, experimentally, and developmentally constrained, but at the expense of interspecies applicability (Cash-Padgett and Jaaro-Peled, 2013). The impact of the environment on gene expression will be felt by the individual over their life course; however, over the longer term, genes will also become adapted to prevailing environmental circumstances. It is important to note that this process clearly does not operate in a linear manner: an adaptation in one gene can impact multiple associated functional pathways and resultant traits.

Whatever resilience one's genome confers can be overridden by environmental factors such as childhood physical abuse (Cicchetti and Rogosch, 2012). However, in addition to individuals becoming unwell, they can also be rescued, for example, through targeted therapeutic interventions (Fusar-Poli et al., 2017). One means of quantifying the impact of environmental factors against a background of implied genetic risk is by assessing genome methylation, i.e., epigenetics (Henikoff and Matzke, 1997). A twin study involving people with either schizophrenia or BD found both shared and more illness-specific methylation differences affecting networks governing neural development and function

(Dempster et al., 2011). However, there was significant variation within and between families, and some methylation sites showed changes in opposite directions for the two diseases. There is early evidence that certain epigenetic changes may herald the shift into acute psychosis (Kebir et al., 2017), whereas others report sustained changes following early life adversity (Tomassi and Tosato, 2017). This suggests that epigenetics presents a promising but complex insight into gene–environment interactions.

Neurodevelopmental models, particularly of schizophrenia, might suggest that you can never start a longitudinal study early enough. The evidence of the suggested impact of childhood adversity, assessed indirectly and well after the fact, would support this notion. However, the practicalities of such an undertaking increase the appeal of any available alternative methods. Population-wide studies, based on excellent digitized record keeping, are one option (for example, see Cantor-Graae and Pedersen (2007)). One could build on such systems to recruit biologically related people of interest for additional brain imaging. Having access to reliable neonatal, developmental, educational, and socioeconomic information, combining this with large-scale longitudinal neuroimaging, and overlaying this on top of well-characterized genetic risk would be the familial HR paradigm at its most powerful.

Such factors will be essential in informing wider models of risk prediction for psychosis. The EHRS highlighted how taking individual well-circumscribed neural differences did not make for good generalized prediction. Instead it was the combination of these with clinical measures of memory function and schizotypy that was able to achieve a predictive accuracy of 74% in an independent sample. However, the word "combination" is somewhat opaque in this case, as the machine learning methods underpinning it do not permit a clear view of the neuroanatomical measures that contribute to its success.

Limitations

The Edinburgh studies have focused primarily on combining clinical, neuropsychological trait characteristics, and neuroimaging measures in their investigations of familial risk of psychosis. It is important to note that there are other important measures of interest, some of which may be amenable to producing more clinically convenient biomarkers. The Dutch Bipolar Offspring Study (Haarman et al., 2014) has followed 140 12–21-year olds with at least one bipolar parent over a 12-year period, focusing on inflammation and the genes that underpin it. Ellenbogen and colleagues have undertaken a longitudinal study of 77 adolescent bipolar offspring in Montreal, Canada, examining the hypothalamic–pituitary axis and stress reactivity (Ellenbogen et al., 2006). The Pittsburgh Bipolar Offspring Study followed 412 young offspring up for at least 9 years, leading to the development of a risk calculator based on severity of mania, depression, anxiety, mood lability, psychosocial functioning, and parental age at mood disorder (Hafeman et al., 2017). Bora et al. (2014) review familial and clinical HR studies that have examined cognition.

The work described here combines both longitudinal imaging and cross-sectional imaging with longitudinal clinical data. The latter approach can address predictive capacity, but only the former can probe abnormal neuroprogression *within subjects*. Performing such studies is, however, practically difficult, given factors such as participant dropout, continuity of funding, and technical advances impacting methodological consistency. In addition, illness onset may not coincide with formal diagnosis and features believed to be predictive may be elements of early illness.

In defining risk along genetic lines, we are aligned with the intermediate phenotypic approach (Gottesman and Gould, 2003, Meyer-Lindenberg and Weinberger, 2006). Alternatively, risk groups could be defined according to whether or not participants had previously had a psychotic episode. We consider the latter—tracking progression of an established process rather than its initiation—as qualitatively different from the prospective longitudinal genetic risk–driven design.

Areas for future work

An alternative approach in considering an individual's genetic risk status would be to stratify according to a polygenic risk score, as has been performed successfully in other fields (Belsky et al., 2013).

Another area of expansion is that of combining data across multiple sites. The Enhancing Neuroimaging Genetics through Meta-Analysis initiative (ENIGMA) has allowed for hard-earned longitudinal cohort data to be combined, enhancing statistical power and overall generalizability (Thompson et al., 2014).

In addition to the machine learning method described here and the 5-year risk calculator devised by Hafeman et al. (2017), more innovative methods are being used to predict the onset of psychosis. An interesting example is provided by Fojo et al. (2017), who fit a Bayesian model to weekly symptom severity data, to predict whether or not an individual's newly initiated antipsychotic medication would be likely to fail an 8-week trial. Clinical applications of such techniques are not yet imminent, but this is an area with great promise. Precision medicine is the ultimate goal to evaluate a person's individual characteristics from molecular to societal; determine their likelihood of developing particular illnesses; and identify optimal prevention or management strategies (Ashley, 2016). Genetic HR studies afford a means to get to the underlying mechanisms of psychosis by focusing on people with an established molecular propensity to illness. These genetic clues can then guide therapeutic development, for example, by implicating particular neurochemical cascades and their associated neural receptors. A mechanistic understanding can then be turned toward the people who provided the clues to begin with, perhaps in the form of specific advice for those at genetic risk, e.g., the avoidance of particular substances, especially at certain developmental phases. The other side of the coin concerns genes for resilience. In a population at genetically elevated risk of psychosis, can mechanisms be identified that help contribute to a person remaining well? This too could point out avenues for therapeutics, arguably quite different to any we have considered to date.

Conclusion

In an age of big data, it is perhaps reassuring to see that the connections between people within affected families can make a unique contribution to our understanding of psychosis. Genetic HR studies provide a naturalistic means of examining gene × environment interactions, intrinsically narrowing down the genomic scope of inquiry, while permitting a more tractable look at how the environment then impacts physiology and behavior over time.

In this chapter, we have summarized Edinburgh's longitudinal familial HR studies, which aim to address the onset and progression of psychosis through the use of neuroimaging. One of the key strengths of the longitudinal approach is to establish changes over time and how these relate to clinical developments within the individual. The overarching goal of such research is to develop models capable of predicting the likelihood of illness onset, its severity, and susceptibility to ameliorating factors such as medication and pre-emptive interventions.

The evidence available so far suggests that the road to psychosis is more complex than even theories such as the "two-hit" hypothesis would suggest (Maynard et al., 2001). Genetics lay the foundation for both vulnerability and resilience;

however, early environmental circumstances will dictate whether or not these traits come to fruition. This in turn will build a person's personality and temperament, causing them to engage with subsequent life stressors in varying ways. Adaptive or maladaptive patterns may emerge, inviting further neural and likely epigenetic changes over time. These recurrent and accumulative interactions are likely what give rise to major mental illness in time. Understanding this process is exceptionally challenging, and if any avenues of variance allow themselves to be "nailed down" and held in a relatively controlled or at least quantifiable manner, then those opportunities should be seized. This is why the longitudinal familial HR study has the potential to be such a useful paradigm, offering some means of control over genetic (and some environmental) factors.

Finally, the current generation of HR studies is now establishing which approaches work best, particularly in multisite studies. The wider, societal question is whether people would want to know their risk of developing a psychotic illness, what they would want to do about it, and if anyone is willing to pay for such services (Daws, 2017).

List of acronyms and abbreviations

AHC Amygdala–hippocampal complex
ARMS At Risk Mental State
BD Bipolar disorder
BFS Bipolar Family Study
CON Healthy controls
EHRS Edinburgh High Risk Study
ENIGMA Enhancing Neuroimaging Genetics through Meta-Analysis initiative
FES First-episode schizophrenia
fMRI Functional MRI
GI Gyrification index
HR High-risk
ICD-10 International Classification of Diseases, version 10
MDD Major depressive disorder
MRI Magnetic resonance imaging
PPV Positive predictive value
PSE Present State Examination
ROI Region of interest
sMRI Structural MRI
SNP Single-nucleotide polymorphism
VBM Voxel-based morphometry

Acknowledgments

H.C.W. is supported by a College Fellowship from the University of Edinburgh and John, Margaret, Alfred, and Stewart Sim (JMAS Sim) fellowship from the Royal College of Physicians of Edinburgh. J.E.S. was supported by a Clinical Research Training Fellowship from the Wellcome Trust. A.M.M. was supported by the Health Foundation through a Clinician Scientist Fellowship (reference 2268/4295), by the Brain and Behaviour Research Foundation through a NARSAD Independent Investigator Award, and by a Scottish Funding Council Senior Clinical Fellowship. This work was also supported by a Wellcome Trust Strategic Award (104036/Z/14/Z).

References

Ashley, E.A., 2016. Towards precision medicine. Nat. Rev. Genet. 17 (9), 507–522.

Barker, V., Bois, C., Johnstone, E.C., Owens, D.G.C., Whalley, H.C., McIntosh, A.M., Lawrie, S.M., 2016a. Childhood adversity and cortical thickness and surface area in a population at familial high risk of schizophrenia. Psychol. Med. 46 (4), 891–896.

Barker, V., Bois, C., Neilson, E., Johnstone, E.C., Owens, D.G.C., Whalley, H.C., McIntosh, A.M., Lawrie, S.M., 2016b. Childhood adversity and hippocampal and amygdala volumes in a population at familial high risk of schizophrenia. Schizophr. Res. 175 (1–3), 42–47.

Belsky, D.W., Sears, M.R., Hancox, R.J., Harrington, H., Houts, R., Moffitt, T.E., Sugden, K., Williams, B., Poulton, R., Caspi, A., 2013. Polygenic risk and the development and course of asthma: an analysis of data from a four-decade longitudinal study. Lancet Respir. Med. 1 (6), 453–461.

Bois, C., Levita, L., Ripp, I., Owens, D.C.G., Johnstone, E.C., Whalley, H.C., Lawrie, S.M., 2015a. Hippocampal, amygdala and nucleus accumbens volume in first-episode schizophrenia patients and individuals at high familial risk: a cross-sectional comparison. Schizophr. Res. 165 (1), 45–51.

Bois, C., Ronan, L., Levita, L., Whalley, H.C., Giles, S., McIntosh, A.M., Fletcher, P.C., Owens, D.C., Johnstone, E.C., Lawrie, S.M., 2015b. Cortical surface area differentiates familial high risk individuals who go on to develop schizophrenia. Biol. Psychiatry 78 (6), 413–420.

Bois, C., Whalley, H.C., McIntosh, A.M., Lawrie, S.M., 2015c. Structural magnetic resonance imaging markers of susceptibility and transition to schizophrenia: a review of familial and clinical high risk population studies. J. Psychopharmacol. 29 (2), 144–154.

Bootsman, F., Brouwer, R.M., Kemner, S.M., Schnack, H.G., van der Schot, A.C., Vonk, R., Hillegers, M.H.J., Boomsma, D.I., Hulshoff Pol, H.E., Nolen, W.A., Kahn, R.S., van Haren, N.E.M., 2015a. Contribution of genes and unique environment to cross-sectional and longitudinal measures of subcortical volumes in bipolar disorder. Eur. Neuropsychopharmacol. 25 (12), 2197–2209.

Bootsman, F., Brouwer, R.M., Schnack, H.G., Kemner, S.M., Hillegers, M.H.J., Sarkisyan, G., van der Schot, A.C., Vonk, R., Hulshoff Pol, H.E., Nolen, W.A., Kahn, R.S., van Haren, N.E.M., 2016. A study of genetic and environmental contributions to structural brain changes over time in twins concordant and discordant for bipolar disorder. J. Psychiatr. Res. 79, 116–124.

Bootsman, F., Brouwer, R.M., Schnack, H.G., van Baal, G.C.M., van der Schot, A.C., Vonk, R., Hulshoff Pol, H.E., Nolen, W.A., Kahn, R.S., van Haren, N.E.M., 2015b. Genetic and environmental influences on cortical surface area and cortical thickness in bipolar disorder. Psychol. Med. 45 (1), 193–204.

Bora, E., Fornito, A., Pantelis, C., Yücel, M., 2012. Gray matter abnormalities in major depressive disorder: a meta-analysis of voxel based morphometry studies. J. Affect. Disord. 138 (1–2), 9–18.

Bora, E., Lin, A., Wood, S.J., Yung, A.R., McGorry, P.D., Pantelis, C., 2014. Cognitive deficits in youth with familial and clinical high risk to psychosis: a systematic review and meta-analysis. Acta Psychiatr. Scand. 130 (1), 1–15.

Broome, M.R., He, Z., Iftikhar, M., Eyden, J., Marwaha, S., 2015. Neurobiological and behavioural studies of affective instability in clinical populations: a systematic review. Neurosci. Biobehav. Rev. 51, 243–254.

Cantor-Graae, E., Pedersen, C.B., 2007. Risk of schizophrenia in second-generation immigrants: a Danish population-based cohort study. Psychol. Med. 37 (4), 485–494.

Cash-Padgett, T., Jaaro-Peled, H., 2013. DISC1 mouse models as a tool to decipher gene-environment interactions in psychiatric disorders. Front. Behav. Neurosci. 7, 113.

Cicchetti, D., Rogosch, F.A., 2012. Gene × Environment interaction and resilience: effects of child maltreatment and serotonin, corticotropin releasing hormone, dopamine, and oxytocin genes. Dev. Psychopathol. 24 (02), 411–427.

Dauvermann, M.R., Mukherjee, P., Moorhead, W.T., Stanfield, A.C., Fusar-Poli, P., Lawrie, S.M., Whalley, H.C., 2012. Relationship between gyrification and functional connectivity of the prefrontal cortex in subjects at high genetic risk of schizophrenia. Curr. Pharmaceut. Des. 18 (4), 434–442.

Dauvermann, M.R., Whalley, H.C., Romaniuk, L., Valton, V., Owens, D.G.C., Johnstone, E.C., Lawrie, S.M., Moorhead, T.W.J., 2013. The application of nonlinear Dynamic Causal Modelling for fMRI in subjects at high genetic risk of schizophrenia. Neuroimage 73, 16–29.

Daws, S., 2017. Ethical application of precision medicine to schizophrenia management. New Bioeth. 23 (2), 147–153.

Dempster, E.L., Pidsley, R., Schalkwyk, L.C., Owens, S., Georgiades, A., Kane, F., Kalidindi, S., Picchioni, M., Kravariti, E., Toulopoulou, T., Murray, R.M., Mill, J., 2011. Disease-associated epigenetic changes in monozygotic twins discordant for schizophrenia and bipolar disorder. Hum. Mol. Genet. 20 (24), 4786–4796.

Doucette, S., Levy, A., Flowerdew, G., Horrocks, J., Grof, P., Ellenbogen, M., Duffy, A., 2016. Early parent-child relationships and risk of mood disorder in a Canadian sample of offspring of a parent with bipolar disorder: findings from a 16-year prospective cohort study. Early Interv Psychiatry 10 (5), 381–389.

Drevets, W.C., Price, J.L., Furey, M.L., 2008. Brain structural and functional abnormalities in mood disorders: implications for neurocircuitry models of depression. Brain Struct. Funct. 213 (1–2), 93–118.

Duffy, A., Vandeleur, C., Heffer, N., Preisig, M., 2017. The clinical trajectory of emerging bipolar disorder among the high-risk offspring of bipolar parents: current understanding and future considerations. Int. J. Behav. Dev. 5 (1), 37.

Ellenbogen, M.A., Hodgins, S., Walker, C.-D., Couture, S., Adam, S., 2006. Daytime cortisol and stress reactivity in the offspring of parents with bipolar disorder. Psychoneuroendocrinology 31 (10), 1164–1180.

Elliott, R., Zahn, R., Deakin, J.F.W., Anderson, I.M., 2010. Affective cognition and its disruption in mood disorders. Neuropsychopharmacology 36 (1), 153–182.

Fojo, A.T., Musliner, K.L., Zandi, P.P., Zeger, S.L., 2017. A precision medicine approach for psychiatric disease based on repeated symptom scores. J. Psychiatr. Res. 95, 147–155.

Fusar-Poli, P., Howes, O., Bechdolf, A., Borgwardt, S., 2012. Mapping vulnerability to bipolar disorder: a systematic review and meta-analysis of neuroimaging studies. J. Psychiatry Neurosci. 37 (3), 170–184.

Fusar-Poli, P., McGorry, P.D., Kane, J.M., 2017. Improving outcomes of first-episode psychosis: an overview. World Psychiatry 16 (3), 251–265.

Ganzola, R., Nickson, T., Bastin, M.E., Giles, S., Macdonald, A., Sussmann, J., McIntosh, A.M., Whalley, H.C., Duchesne, S., 2017. Longitudinal differences in white matter integrity in youth at high familial risk for bipolar disorder. Bipolar Disord. 19 (3), 158–167.

Gottesman, I.I., 1991. Schizophrenia Genesis: The Origins of Madness. WH Freeman/Times Books/Henry Holt & Co.

Gottesman, I.I., Gould, T.D., 2003. The endophenotype concept in psychiatry: etymology and strategic intentions. Am. J. Psychiatry 160 (4), 636–645.

Haarman, B.C.M., Riemersma-Van der Lek, R.F., Burger, H., Netkova, M., Drexhage, R.C., Bootsman, F., Mesman, E., Hillegers, M.H., Spijker, A.T., Hoencamp, E., Drexhage, H.A., Nolen, W.A., 2014. Relationship between clinical features and inflammation-related monocyte gene expression in bipolar disorder – towards a better understanding of psychoimmunological interactions. Bipolar Disord. 16 (2), 137–150.

Hafeman, D.M., Merranko, J., Goldstein, T.R., Axelson, D., Goldstein, B.I., Monk, K., Hickey, M.B., Sakolsky, D., Diler, R., Iyengar, S., Brent, D.A., Kupfer, D.J., Kattan, M.W., Birmaher, B., 2017. Assessment of a person-level risk calculator to predict new-onset bipolar spectrum disorder in youth at familial risk. JAMA Psychiatry 74 (8), 841–847.

Hajek, T., Kopecek, M., Kozeny, J., Gunde, E., Alda, M., Höschl, C., 2009. Amygdala volumes in mood disorders—meta-analysis of magnetic resonance volumetry studies. J. Affect. Disord. 115 (3), 395–410.

Harris, J.M., Moorhead, T.W.J., Miller, P., McIntosh, A.M., Bonnici, H.M., Owens, D.G.C., Johnstone, E.C., Lawrie, S.M., 2007. Increased prefrontal gyrification in a large high-risk cohort characterizes those who develop schizophrenia and reflects abnormal prefrontal development. Biol. Psychiatry 62 (7), 722–729.

Harris, J.M., Whalley, H.C., Yates, S., Miller, P., Johnstone, E.C., Lawrie, S.M., 2004a. Abnormal cortical folding in high-risk individuals: a predictor of the development of schizophrenia? Biol. Psychiatry 56 (3), 182–189.

Harris, J.M., Yates, S., Miller, P., Best, J.J.K., Johnstone, E.C., Lawrie, S.M., 2004b. Gyrification in first-episode schizophrenia: a morphometric study. Biol. Psychiatry 55 (2), 141–147.

Henikoff, S., Matzke, M.A., 1997. Exploring and explaining epigenetic effects. Trends Genet. 13 (8), 293–295.

Hoy, K., Barrett, S., Shannon, C., Campbell, C., Watson, D., Rushe, T., Shevlin, M., Bai, F., Cooper, S., Mulholland, C., 2012. Childhood trauma and hippocampal and amygdalar volumes in first-episode psychosis. Schizophr. Bull. 38 (6), 1162–1169.

Iacono, V., Beaulieu, L., Hodgins, S., Ellenbogen, M.A., 2017. Parenting practices in middle childhood mediate the relation between growing up with a parent having bipolar disorder and offspring psychopathology from childhood into early adulthood. Dev. Psychopathol. 128, 1–15.

Job, D.E., Whalley, H.C., Johnstone, E.C., Lawrie, S.M., 2005. Grey matter changes over time in high risk subjects developing schizophrenia. Neuroimage 25 (4), 1023–1030.

Job, D.E., Whalley, H.C., McConnell, S., Glabus, M., Johnstone, E.C., Lawrie, S.M., 2002. Structural gray matter differences between first-episode schizophrenics and normal controls using voxel-based morphometry. Neuroimage 17 (2), 880–889.

Job, D.E., Whalley, H.C., McConnell, S., Glabus, M., Johnstone, E.C., Lawrie, S.M., 2003. Voxel-based morphometry of grey matter densities in subjects at high risk of schizophrenia. Schizophr. Res. 64 (1), 1–13.

Job, D.E., Whalley, H.C., McIntosh, A.M., Owens, D.G.C., Johnstone, E.C., Lawrie, S.M., 2006. Grey matter changes can improve the prediction of schizophrenia in subjects at high risk. BMC Med. 4, 29.

Johnstone, E.C., Cosway, R., Lawrie, S.M., 2002. Distinguishing characteristics of subjects with good and poor early outcome in the Edinburgh High-Risk Study. Br. J. Psychiatr. Suppl. 43, s26–s29.

Johnstone, E.C., Ebmeier, K.P., Miller, P., Owens, D.G.C., Lawrie, S.M., 2005. Predicting schizophrenia: findings from the Edinburgh high-risk study. Br. J. Psychiatry 186 (1), 18–25.

Kebir, O., Chaumette, B., Rivollier, F., Miozzo, F., Lemieux Perreault, L.P., Barhdadi, A., Provost, S., Plaze, M., Bourgin, J., ICAAR team, Gaillard, R., Mezger, V., Dubé, M.-P., Krebs, M.-O., 2017. Methylomic changes during conversion to psychosis. Mol. Psychiatry 22 (4), 512–518.

Lawrie, S., 2007. Distinguishing vulnerability, prediction, and progression in the preschizophrenic brain. Arch. Gen. Psychiatr. 64 (2), 250–251 author reply 252–3.

Lawrie, S.M., Whalley, H.C., Abukmeil, S.S., Kestelman, J.N., Donnelly, L., Miller, P., Best, J.J., Owens, D.G.C., Johnstone, E.C., 2001. Brain structure, genetic liability, and psychotic symptoms in subjects at high risk of developing schizophrenia. Biol. Psychiatry 49 (10), 811–823.

Lawrie, S.M., Whalley, H.C., Abukmeil, S.S., Kestelman, J.N., Miller, P., Best, J.J.K., Owens, D.G.C., Johnstone, E.C., 2002. Temporal lobe volume changes in people at high risk of schizophrenia with psychotic symptoms. Br. J. Psychiatry 181, 138–143.

Lawrie, S.M., Whalley, H.C., Kestelman, J.N., Abukmeil, S.S., Byrne, M., Hodges, A., Rimmington, J.E., Best, J.J., Owens, D.G.C., Johnstone, E.C., 1999. Magnetic resonance imaging of brain in people at high risk of developing schizophrenia. Lancet 353 (9146), 30–33.

LeDoux, J.E., 2000. Emotion circuits in the brain. Annu. Rev. Neurosci. 23, 155–184.

Lymer, G.K.S., Job, D.E., William, T., Moorhead, J., McIntosh, A.M., Owens, D.G.C., Johnstone, E.C., Lawrie, S.M., 2006. Brain-behaviour relationships in people at high genetic risk of schizophrenia. Neuroimage 33 (1), 275–285.

Mannie, Z.N., Filippini, N., Williams, C., Near, J., Mackay, C.E., Cowen, P.J., 2014. Structural and functional imaging of the hippocampus in young people at familial risk of depression. Psychol. Med. 44 (14), 2939–2948.

Marjoram, D., Job, D.E., Whalley, H.C., Gountouna, V.-E., McIntosh, A.M., Simonotto, E., Cunningham-Owens, D., Johnstone, E.C., Lawrie, S., 2006. A visual joke fMRI investigation into Theory of Mind and enhanced risk of schizophrenia. Neuroimage 31 (4), 1850–1858.

Matthews, S.C., Strigo, I.A., Simmons, A.N., Yang, T.T., Paulus, M.P., 2008. Decreased functional coupling of the amygdala and supragenual cingulate is related to increased depression in unmedicated individuals with current major depressive disorder. J. Affect. Disord. 111 (1), 13–20.

Maynard, T.M., Sikich, L., Lieberman, J.A., LaMantia, A.S., 2001. Neural development, cell-cell signaling, and the 'two-hit' hypothesis of schizophrenia. Schizophr. Bull. 27 (3), 457–476.

McGuffin, P., Rijsdijk, F., Andrew, M., Sham, P., Katz, R., Cardno, A., 2003. The heritability of bipolar affective disorder and the genetic relationship to unipolar depression. Arch. Gen. Psychiatr. 60 (5), 497–502.

McIntosh, A.M., Owens, D.C., Moorhead, W.J., Whalley, H.C., Stanfield, A.C., Hall, J., Johnstone, E.C., Lawrie, S.M., 2011. Longitudinal volume reductions in people at high genetic risk of schizophrenia as they develop psychosis. Biol. Psychiatry 69 (10), 953–958.

Meyer-Lindenberg, A., Weinberger, D.R., 2006. Intermediate phenotypes and genetic mechanisms of psychiatric disorders. Nat. Rev. Neurosci. 7 (10), 818–827.

Miskowiak, K.W., Kjærstad, H.L., Meluken, I., Petersen, J.Z., Maciel, B.R., Köhler, C.A., Vinberg, M., Kessing, L.V., Carvalho, A.F., 2017. The search for neuroimaging and cognitive endophenotypes: a critical systematic review of studies involving unaffected first-degree relatives of individuals with bipolar disorder. Neurosci. Biobehav. Rev. 73, 1–22.

Neilson, E., Bois, C., Gibson, J., Duff, B., Watson, A., Roberts, N., Brandon, N.J., Dunlop, J., Hall, J., McIntosh, A.M., Whalley, H.C., Lawrie, S.M., 2017. Effects of environmental risks and polygenic loading for schizophrenia on cortical thickness. Schizophr. Res. 184, 128–136.

Nickson, T., Chan, S.W.Y., Papmeyer, M., Romaniuk, L., Macdonald, A., Stewart, T., Kielty, S., Lawrie, S.M., Hall, J., Sussmann, J.E., McIntosh, A.M., Whalley, H.C., 2016. Prospective longitudinal voxel-based morphometry study of major depressive disorder in young individuals at high familial risk. Psychol. Med. 46 (11), 2351–2361.

Papmeyer, M., Giles, S., Sussmann, J.E., Kielty, S., Stewart, T., Lawrie, S.M., Whalley, H.C., McIntosh, A.M., 2015. Cortical thickness in individuals at high familial risk of mood disorders as they develop major depressive disorder. Biol. Psychiatry 78 (1), 58–66.

Papmeyer, M., Sussmann, J.E., Stewart, T., Giles, S., Centola, J.G., Zannias, V., Lawrie, S.M., Whalley, H.C., McIntosh, A.M., 2016. Prospective longitudinal study of subcortical brain volumes in individuals at high familial risk of mood disorders with or without subsequent onset of depression. Psychiatry Res. 248, 119–125.

Phillips, M.L., Drevets, W.C., Rauch, S.L., Lane, R., 2003. Neurobiology of emotion perception II: implications for major psychiatric disorders. Biol. Psychiatry 54 (5), 515–528.

Piguet, C., Fodoulian, L., Aubry, J.-M., Vuilleumier, P., Houenou, J., 2015. Bipolar disorder: functional neuroimaging markers in relatives. Neurosci. Biobehav. Rev. 57, 284–296.

Sacher, J., Neumann, J., Fünfstück, T., Soliman, A., Villringer, A., Schroeter, M.L., 2012. Mapping the depressed brain: a meta-analysis of structural and functional alterations in major depressive disorder. J. Affect. Disord. 140 (2), 142–148.

Savitz, J.B., Drevets, W.C., 2009. Imaging phenotypes of major depressive disorder: genetic correlates. Neuroscience 164 (1), 300–330.

Schneider, M.R., DelBello, M.P., McNamara, R.K., Strakowski, S.M., Adler, C.M., 2012. Neuroprogression in bipolar disorder. Bipolar Disord. 14 (4), 356–374.

Snitz, B.E., MacDonald, A.W., Carter, C.S., 2006. Cognitive deficits in unaffected first-degree relatives of schizophrenia patients: a meta-analytic review of putative endophenotypes. Schizophr. Bull. 32 (1), 179–194.

Song, J., Bergen, S.E., Kuja-Halkola, R., Larsson, H., Landén, M., Lichtenstein, P., 2015. Bipolar disorder and its relation to major psychiatric disorders: a family-based study in the Swedish population. Bipolar Disord. 17 (2), 184–193.

Sprooten, E., Fleming, K.M., Thomson, P.A., Bastin, M.E., Whalley, H.C., Hall, J., Sussmann, J.E., McKirdy, J., Blackwood, D., Lawrie, S.M., McIntosh, A.M., 2013a. White matter integrity as an intermediate phenotype: exploratory genome-wide association analysis in individuals at high risk of bipolar disorder. Psychiatry Res. 206 (2–3), 223–231.

Sprooten, E., Papmeyer, M., Smyth, A.M., Vincenz, D., Honold, S., Conlon, G.A., Moorhead, T.W.J., Job, D., Whalley, H.C., Hall, J., McIntosh, A.M., Owens, D.C.G., Johnstone, E.C., Lawrie, S.M., 2013b. Cortical thickness in first-episode schizophrenia patients and individuals at high familial risk: a cross-sectional comparison. Schizophr. Res. 151 (1–3), 259–264.

Sprooten, E., Sussmann, J.E., Clugston, A., Peel, A., McKirdy, J., Moorhead, T.W.J., Anderson, S., Shand, A.J., Giles, S., Bastin, M.E., Hall, J., Johnstone, E.C., Lawrie, S.M., McIntosh, A.M., 2011. White matter integrity in individuals at high genetic risk of bipolar disorder. Biol. Psychiatry 70 (4), 350–356.

Thompson, P.M., Stein, J.L., Medland, S.E., Hibar, D.P., Vasquez, A.A., Renteria, M.E., Toro, R., Jahanshad, N., Schumann, G., Franke, B., Wright, M.J., Martin, N.G., Agartz, I., Alda, M., Alhusaini, S., Almasy, L., Almeida, J., Alpert, K., Andreasen, N.C., Andreassen, O.A., Apostolova, L.G., Appel, K., Armstrong, N.J., Aribisala, B., Bastin, M.E., Bauer, M., Bearden, C.E., Bergmann, O., Binder, E.B., Blangero, J., Bockholt, H.J., Bøen, E., Bois, C., Boomsma, D.I., Booth, T., Bowman, I.J., Bralten, J., Brouwer, R.M., Brunner, H.G., Brohawn, D.G., Buckner, R.L., Buitelaar, J., Bulayeva, K., Bustillo, J.R., Calhoun, V.D., Cannon, D.M., Cantor, R.M., Carless, M.A., Caseras, X., Cavalleri, G.L., Chakravarty, M.M., Chang, K.D., Ching, C.R.K., Christoforou, A., Cichon, S., Clark, V.P., Conrod, P., Coppola, G., Crespo-Facorro, B., Curran, J.E., Czisch, M., Deary, I.J., de Geus, E.J.C., Braber, den, A., Delvecchio, G., Depondt, C., de Haan, L., de Zubicaray, G.I., Dima, D., Dimitrova, R., Djurovic, S., Dong, H., Donohoe, G., Duggirala, R., Dyer, T.D., Ehrlich, S., Ekman, C.J., Elvsåshagen, T., Emsell, L., Erk, S., Espeseth, T., Fagerness, J., Fears, S., Fedko, I., Fernández, G., Fisher, S.E., Foroud, T., Fox, P.T., Francks, C., Frangou, S., Frey, E.M., Frodl, T., Frouin, V., Garavan, H., Giddaluru, S., Glahn, D.C., Godlewska, B., Goldstein, R.Z., Gollub, R.L., Grabe, H.J., Grimm, O., Gruber, O., Guadalupe, T., Gur, R.E., Gur, R.C., Göring, H.H.H., Hagenaars, S., Hajek, T., Hall, G.B., Hall, J., Hardy, J., Hartman, C.A., Hass, J., Hatton, S.N., Haukvik, U.K., Hegenscheid, K., Heinz, A., Hickie, I.B., Ho, B.-C., Hoehn, D., Hoekstra, P.J., Hollinshead, M., Holmes, A.J., Homuth, G., Hoogman, M., Hong, L.E., Hosten, N., Hottenga, J.-J., Hulshoff Pol, H.E., Hwang, K.S., Jack, C.R., Jenkinson, M., Johnston, C., Jönsson, E.G., Kahn, R.S., Kasperaviciute, D., Kelly, S., Kim, S., Kochunov, P., Koenders, L., Krämer, B., Kwok, J.B.J., Lagopoulos, J., Laje, G., Landén, M., Landman, B.A., Lauriello, J., Lawrie, S.M., Lee, P.H., Le Hellard, S., Lemaître, H., Leonardo, C.D., Li, C.-S., Liberg, B., Liewald, D.C., Liu, X., Lopez, L.M., Loth, E., Lourdusamy, A., Luciano, M., Macciardi, F., Machielsen, M.W.J., Macqueen, G.M., Malt, U.F., Mandl, R., Manoach, D.S., Martinot, J.-L., Matarin, M., Mather, K.A., Mattheisen, M., Mattingsdal, M., Meyer-Lindenberg, A., McDonald, C., McIntosh, A.M., McMahon, F.J., McMahon, K.L., Meisenzahl, E., Melle, I., Milaneschi, Y., Mohnke, S., Montgomery, G.W., Morris, D.W., Moses, E.K., Mueller, B.A., Muñoz Maniega, S., Mühleisen, T.W., Müller-Myhsok, B., Mwangi, B., Nauck, M., Nho, K., Nichols, T.E., Nilsson, L.-G., Nugent, A.C., Nyberg, L., Olvera, R.L., Oosterlaan, J., Ophoff, R.A., Pandolfo, M., Papalampropoulou-Tsiridou, M., Papmeyer, M., Paus, T., Pausova, Z., Pearlson, G.D., Penninx, B.W., Peterson, C.P., Pfennig, A., Phillips, M., Pike, G.B., Poline, J.-B., Potkin, S.G., Pütz, B., Ramasamy, A., Rasmussen, J., Rietschel, M., Rijpkema, M., Risacher, S.L., Roffman, J.L., Roiz-Santiañez, R., Romanczuk-Seiferth, N., Rose, E.J., Royle, N.A., Rujescu, D., Ryten, M., Sachdev, P.S., Salami, A., Satterthwaite, T.D., Savitz, J., Saykin, A.J., Scanlon, C., Schmaal, L., Schnack, H.G., Schork, A.J., Schulz, S.C., Schür, R., Seidman, L., Shen, L., Shoemaker, J.M., Simmons, A., Sisodiya, S.M., Smith, C., Smoller, J.W., Soares, J.C., Sponheim, S.R., Sprooten, E., Starr, J.M., Steen, V.M., Strakowski, S., Strike, L., Sussmann, J., Sämann, P.G., Teumer, A., Toga, A.W., Tordesillas-Gutierrez, D., Trabzuni, D., Trost, S., Turner, J., Van den Heuvel, M., van der Wee, N.J., van Eijk, K., van Erp, T.G.M., van Haren, N.E.M., van 't Ent, D., van Tol, M.-J., Valdés Hernández, M.C., Veltman, D.J., Versace, A., Völzke, H., Walker, R., Walter, H., Wang, L., Wardlaw, J.M., Weale, M.E., Weiner, M.W., Wen, W., Westlye, L.T., Whalley, H.C., Whelan, C.D., White, T., Winkler, A.M., Wittfeld, K., Woldehawariat, G., Wolf, C., Zilles, D., Zwiers, M.P., Thalamuthu, A., Schofield, P.R., Freimer, N.B., Lawrence, N.S., Drevets, W., Alzheimer's Disease Neuroimaging Initiative, EPIGEN Consortium, IMAGEN Consortium, Saguenay Youth Study (SYS) Group, 2014. The ENIGMA Consortium: large-scale collaborative analyses of neuroimaging and genetic data. Brain Imaging Behav. 8 (2), 153–182.

Tomassi, S., Tosato, S., 2017. Epigenetics and gene expression profile in first-episode psychosis: the role of childhood trauma. Neurosci. Biobehav. Rev. 83, 226–237.

van Winkel, R., van Nierop, M., Myin-Germeys, I., Van Os, J., 2013. Childhood trauma as a cause of psychosis: linking genes, psychology, and biology. Can. J. Psychiatry (Rev. Canad. Psychiatr.) 58 (1), 44–51.

Wahlberg, K.-E., Wynne, L.C., Hakko, H., Läksy, K., Moring, J., Miettunen, J., Tienari, P., 2004. Interaction of genetic risk and adoptive parent communication deviance: longitudinal prediction of adoptee psychiatric disorders. Psychol. Med. 34 (8), 1531–1541.

Welch, K.A., McIntosh, A.M., Job, D.E., Whalley, H.C., Moorhead, T.W., Hall, J., Owens, D.G.C., Lawrie, S.M., Johnstone, E.C., 2011a. The impact of substance use on brain structure in people at high risk of developing schizophrenia. Schizophr. Bull. 37 (5), 1066–1076.

Welch, K.A., Moorhead, T.W., McIntosh, A.M., Owens, D.G.C., Johnstone, E.C., Lawrie, S.M., 2013. Tensor-based morphometry of cannabis use on brain structure in individuals at elevated genetic risk of schizophrenia. Psychol. Med. 43 (10), 2087–2096.

Welch, K.A., Stanfield, A.C., McIntosh, A.M., Whalley, H.C., Job, D.E., Moorhead, T.W., Owens, D.G.C., Lawrie, S.M., Johnstone, E.C., 2011b. Impact of cannabis use on thalamic volume in people at familial high risk of schizophrenia. Br. J. Psychiatry 199 (5), 386–390.

Whalley, H.C., Papmeyer, M., Sprooten, E., Romaniuk, L., Blackwood, D.H., Glahn, D.C., Hall, J., Lawrie, S.M., Sussmann, J., McIntosh, A.M., 2012. The influence of polygenic risk for bipolar disorder on neural activation assessed using fMRI. Transl. Psychiatry 2 (7), e130.

Whalley, H.C., Simonotto, E., Flett, S., Marshall, I., Ebmeier, K.P., Owens, D.G.C., Goddard, N.H., Johnstone, E.C., Lawrie, S.M., 2004. fMRI correlates of state and trait effects in subjects at genetically enhanced risk of schizophrenia. Brain 127 (Pt 3), 478–490.

Whalley, H.C., Simonotto, E., Marshall, I., Owens, D.G.C., Goddard, N.H., Johnstone, E.C., Lawrie, S.M., 2005. Functional disconnectivity in subjects at high genetic risk of schizophrenia. Brain 128 (Pt 9), 2097–2108.

Whalley, H.C., Simonotto, E., Moorhead, W., McIntosh, A., Marshall, I., Ebmeier, K.P., Owens, D.G.C., Goddard, N.H., Johnstone, E.C., Lawrie, S.M., 2006. Functional imaging as a predictor of schizophrenia. Biol. Psychiatry 60 (5), 454–462.

Whalley, H.C., Sussmann, J.E., Romaniuk, L., Stewart, T., Kielty, S., Lawrie, S.M., Hall, J., McIntosh, A.M., 2015. Dysfunction of emotional brain systems in individuals at high risk of mood disorder with depression and predictive features prior to illness. Psychol. Med. 45 (6), 1207–1218.

Whalley, H.C., Sussmann, J.E., Romaniuk, L., Stewart, T., Papmeyer, M., Sprooten, E., Hackett, S., Hall, J., Lawrie, S.M., McIntosh, A.M., 2013. Prediction of depression in individuals at high familial risk of mood disorders using functional magnetic resonance imaging. PLoS One 8 (3), e57357.

Whyte, M.-C., Brett, C., Harrison, L.K., Byrne, M., Miller, P., Lawrie, S.M., Johnstone, E.C., 2006. Neuropsychological performance over time in people at high risk of developing schizophrenia and controls. Biol. Psychiatry 59 (8), 730–739.

Wing, J.K., Cooper, J.E., Sartorius, N., 1974. The description and classification of psychiatric symptoms. In: An Instruction Manual for the PSE and Catego Systems. Cambridge University Press, Cambridge, UK.

Zarogianni, E., Storkey, A.J., Borgwardt, S., Smieskova, R., Studerus, E., Riecher-Rössler, A., Lawrie, S.M., 2017a. Individualized prediction of psychosis in subjects with an at-risk mental state. Schizophr. Res. Pii: S0920–9964(17)30537-6.

Zarogianni, E., Storkey, A.J., Johnstone, E.C., Owens, D.G.C., Lawrie, S.M., 2017b. Improved individualized prediction of schizophrenia in subjects at familial high risk, based on neuroanatomical data, schizotypal and neurocognitive features. Schizophr. Res. 181, 6–12.

CHAPTER

7

Psychotic-like experiences in the general population

Colm Healy[1], *Mary Cannon*[1,2]

[1]Department of Psychiatry, Royal College of Surgeons in Ireland, Dublin, Leinster, Ireland; [2]Trinity College Institute of Neuroscience, Trinity College Dublin, Dublin, Leinster, Ireland

History and definition of psychotic-like experiences

Hallucinations and delusions are two of the most recognizable features of psychosis and capture a range of qualitatively diverse phenomena (Andreasen, 1997; Kelleher and Cannon, 2016).These experiences have been seen as the primary feature distinguishing patients with a psychotic disorder from those with a neurotic disorder (Kelleher and Cannon, 2016). Over the last two decades, there has been a dramatic increase in epidemiological investigations into the prevalence of hallucination and delusions in the general population (Bak et al., 2005; Degenhardt and Hall, 2001; Laurens et al., 2007; Poulton et al., 2000; Spauwen et al., 2003; Kelleher et al., 2012a,b, see Fig. 7.1). These investigations have indicated that "psychotic" experiences are relatively common within the general population.

Classically, hallucinations and delusions have been considered phenomena that are clearly beyond the realm of "normal" experience, a symptom of a disease entity (Johns and van Os, 2001). From a medical perspective, dichotomizing such symptoms made practical sense as it facilitated the creation of operational criteria for medical practitioners to use when patients presented with such complaints/complications (Johns and van Os, 2001) and provided the first step in a treatment plan, i.e., identification. However, epidemiologically, a clear dichotomous division may be inappropriate when so many individuals within the general population also report these symptoms.

Strauss (1969) is identified as one of the first to query the dichotomous nature of psychotic symptoms (Strauss, 1969). He reported that dichotomizing patients "do not reflect accurately the difficulty in determining the presence or absence of the phenomena in the many patients who present with intermediate kinds of expression." He proposed that a more accurate conceptualization was that a psychotic symptom could be represented as a point on a continuum that had multiple dimensions. However, Strauss's observations were primarily considered in the context of patients with a psychotic disorder. In fact, from the 1960's, the concept of a schizotypy

Risk Factors for Psychosis
https://doi.org/10.1016/B978-0-12-813201-2.00007-7

FIGURE 7.1 Publication trends plot of Google Scholar search results frequency by year.

continuum emerged as a dimensional framework for investigating schizophrenia-spectrum psychopathology, and it was proposed that this could be informed through understanding individual variability in schizotypy in the general population (Kwapil and Barrantes-Vidal, 2014; Claridge and Beech, 1995). Others acknowledged that psychotic symptoms were actually relatively common in the general population (Bentall and Slade, 1985; Chapman and Chapman, 1980; Claridge, 1987; Romme and Escher, 1989); however, this concept would not become widely accepted for another 30 years.

One catalyst for the renewed interest in psychotic symptoms within the general population was the seminal work of Poulton and colleagues (Poulton et al., 2000). Using the Dunedin birth cohort, they demonstrated that children who report psychotic symptoms at age 11 were at a 16-fold increase risk of developing a schizophreniform disorder by age 26. Such findings were highly important for two reasons; firstly, for understanding the etiology of psychotic disorders, and secondly, in their use as a marker for the development of such a disorder. The latter was of paramount importance as it presented a real window of opportunity for early intervention services as it is prior to the onset of other severe symptoms of a psychotic disorder and loss of functioning.

The following year Johns and Van Os reviewed the evidence of psychotic symptoms within the general population and proposed the concept of an extended psychosis phenotype (Johns and van Os, 2001). This suggested that psychotic symptoms reported by the general population and symptoms reported by patients with a psychotic disorder lie on the same multidimensional continua. Psychotic disorders represent the narrow tail end of this continuum, while "nonpsychotic" individuals may have similar but less severe experiences, thus encompassing a broader expression of the psychosis phenotype. Opinions differ on the operational definition of this broader phenotype (Lee et al., 2016); however, by its very nature, experiences within the general population should reflect some degree of intact reality testing or at least should not be accompanied by clinical levels of distress or functional impairment. However, as we shall argue, subclinical does not mean that they are not clinically meaningful. There is now a wealth of research indicating their relevance in a broad range of psychiatric conditions (psychotic and nonpsychotic), and when reported by children and adolescents, they can be considered a marker of current psychopathology as well as a marker for vulnerability to subsequent psychiatric conditions. For simplicity, we have referred to these phenomena as psychotic-like experiences (PLEs), and we formally define them as psychotic symptoms or experiences (nondistressing symptoms) that are reported in the absence of a psychotic disorder. They are also referred to in the literature as psychotic experiences, subclinical psychotic experiences/symptoms, or psychotic-like symptoms. Some investigations are now also extending this concept to include "subclinical" negative and disorganized symptoms of psychosis: psychotic spectrum features (Calkins et al., 2014, 2017). However the majority of the literature to date focuses on the positive symptoms or PLEs.

What is the prevalence of psychotic-like experiences?

Current estimates suggest that between 5% and 7% of adults in the general population will experience PLEs in their lifetime (McGrath et al., 2015; van Os et al., 2009; Maijer et al., 2018). Children and adolescents appear to be more susceptible to these experiences with 17% of children and 7.5% of adolescents reporting such phenomena (Kelleher et al., 2012a). For most individuals (~80%), these experiences are transient, and 64% of those who experience PLEs will have less than five such experiences in their lifetime (McGrath et al., 2015) with an annual incidence rate of 2.5% (Linscott and Van Os, 2013).

Persistence of PLEs: For around 20% of individuals, these experiences may persist over time. Individuals who experience multimodal PLEs (for example, auditory and visual hallucinations or auditory hallucinations and persecutory thought content) have an increased rates of persistence. For example, roughly 10% of those with two types of PLEs and 25% of those with three or more types report experiencing over 100 PLEs within their lifetime (McGrath et al., 2015). Unsurprisingly, individuals with persistent PLEs are susceptible to a range of poorer outcomes including increased risk of transitioning to a psychotic disorder, susceptibility to common psychiatric conditions, poorer functioning, and lower rate of employment (Calkins et al., 2017; De Loore et al., 2011; DeVylder et al., 2015b; Dominguez et al., 2011).

It is worth noting that while prevalence estimates are informative, in essence, they pose the same dichotomous classification problem that was pointed out by Strauss (1969) as a person "either has them or he does not" (Strauss, 1969). In truth, even "subclinical" experiences of these symptoms vary across a number of qualities in the general population. Such qualities include (but are not limited to) variability in frequency, degrees of conviction on reality testing, associated distress, preoccupation with the experience, and strength of the experience. One of the major challenges in investigating PLEs, as with any psychological construct, is that our understanding is limited by our methods of investigation, and simply counting does not accurately capture the complexity and variety of experiences.

How can we measure PLEs?

In the general population, PLEs have most often been investigated using clinical measures. Overall, two common approaches have been used: clinical interview and questionnaires. Lee and colleagues reviewed the literature investigating PLEs and reported that there were 41 different assessment types used across 76 studies (Lee et al., 2016).

Prevalence rates in childhood are relatively similar, whether identified by questionnaire or clinical interview (Kelleher et al., 2012a). However, there is some suggestion that within adult populations, self-rated questionnaires may generate more false-positives ratings, particularly in relation to clinically relevant PLEs (van Os et al., 2009; Schultze-Lutter et al., 2014). Few studies have independently validated item endorsement from these questionnaires, which does raise concerns about the use of single-item endorsement as a valid measure. Within young adolescents, it appears that both receiving a score of two or more (no = 0, maybe = 0.5, and yes = 1) on the Adolescent Psychosis Symptoms Screener or endorsing "do you hear voices that no one else can hear" alone have strong predictive value relative to clinical interview using the Kiddie Schedule for Affective Disorders and Schizophrenia (Kelleher et al., 2011). The latter concurs with similar finding suggesting that questions about auditory hallucinations appear to be the most valid single-item measure of PLEs on a questionnaire with child and adolescent populations (Laurens et al., 2012b; Horwood et al., 2008; Granö et al., 2016). Broader validity of the Community Assessment of Psychotic Experiences (CAPE) dimensions (CAPE-42, assessment of positive and negative dimensions) has also been demonstrated within an adult sample (Konings et al., 2006). However, it is possible that a number of characteristics will influence the validity of any item on a questionnaire. These may include but are not limited to demographic variables (age), an individual's attribution of the particular item (what does the question mean to me?), and cultural factors (religious beliefs).

Rather than investigating the individual or summed endorsement of items, others have opted for approaches that attempt to mimic continua. These investigations have included exploratory and confirmatory factor analysis (Capra et al., 2015; Shevlin et al., 2017; Therman and Ziermans, 2016; Yung et al., 2009) or latent constructs analysis and structural equation modeling (Mamah et al., 2013; Laurens et al., 2012a; Sullivan et al., 2015). These approaches may capture the dynamic complexities of a continuum better than dichotomous groupings and also provide a useful theoretic framework for how symptoms may vary in severity and interact with symptoms of other disorders (psychotic and nonpsychotic). Interestingly, these investigations report that only some of these constructs, particularly paranoid ideation and unusual perceptual experiences (hallucinations), are related to poorer outcomes such as suicidal ideation or psychopathology (Capra et al., 2015; Yung et al., 2009, respectively). However, one clear disadvantage to using these approaches is that it can be difficult to translate their utility into something meaningful that can be applied to "real world" setting.

What are the risk factors for PLEs?

There appears to be a relatively large overlap between the risk factors of psychotic disorders and risk factors of PLEs (Kaymaz et al., 2012; Kelleher and Cannon, 2011). This overlap is apparent over a wide range of factors including heritability, early life development, psychosocial outcomes, and environmental stressors.

Heritability and genetics: Heritability has been shown to play a role in PLEs; however, the genetic origins of PLEs are to date unclear (Kelleher and Cannon, 2011). Polanczyk and colleagues reported that genetics accounted for 43% of the variance in PLEs expression (Polanczyk et al., 2010). This appears to vary by gender with genetic variance almost twice as high for men compared to women (Nesvåg et al., 2016). Young adolescents with PLEs are 2.5 times more likely to have a mother with a psychosis spectrum disorder (Polanczyk et al., 2010).

There does not appear to be a relationship between polygenic risk score for schizophrenia or the schizophrenia-associated variants (genome-wide association) and PLEs in adolescent cohorts (Owrutsky et al., 2017; Zammit et al., 2013b). Perhaps, this is unsurprising given that most individuals with PLEs do not go on to develop a psychotic disorder. Interestingly, an association has been shown between negative symptoms of psychosis and depression with the polygenic score for schizophrenia (Jones et al., 2016; Owrutsky et al., 2017).

Early development and health factors: There appears to be an increased risk of pre/perinatal complications and childhood infection in individuals with PLEs. Increased rates of maternal infections, reduced APGAR scores, and maternal smoking have been associated with increased risk of PLEs (Zammit et al., 2009). Higher birth weight has been associated with a reduced risk of PLEs (Thomas et al., 2009).

Early childhood infection has also been associated with PLEs. 25% of adolescent with PLEs had been exposed to Epstein–Barr virus by age 4, reflecting a fivefold increased risk relative to those without PLEs (Khandaker et al., 2014a). Atopic conditions in childhood such as asthma and eczema have been shown to increase risk of adolescent PLEs (Khandaker et al., 2014b). All of these results appear to suggest a role for early life development and the immune system in vulnerability to PLEs, which is consistent with the literature on psychotic disorders (Cannon et al., 2014).

Social environment: Ethnic minority status (particularly for individuals of African-Caribbean origin) has been associated with an increased risk of PLEs (Laurens et al., 2008; Morgan et al., 2009). Individuals living in "urban" areas (generally defined by city status or population density) have a higher risk of PLEs than those not living in urban areas (Krabbendam and Van Os, 2005). This association has been linked with the prevalence of one's own group ethnicity group with a local population (reduction in one's own population density increases risk of PLEs), social adversity, and victimization by violent crime (Das-Munshi et al., 2012; Newbury et al., 2017). There is also evidence that urbanicity moderates genetic family clustering of PLEs (Grech and van Os, 2017).

Lower income status is a commonly reported risk factor for PLEs (Kaymaz et al., 2012). However, McGrath and colleagues reported that in middle- and high-income countries, there is a higher prevalence of adults reporting PLEs when compared with low-income countries (7.2% and 6.8% vs. 3.2%, respectively; McGrath et al., 2015). One possible explanation for this is that the magnitude of economic disparity within a country may be contributing to an individual's likelihood of having PLEs; however, this remains to be explored.

Trauma and victimization: Childhood (or early) trauma is a broad term that encompasses a range of adverse experiences, including sexual, physical, emotional abuse, and neglect (Morgan and Fisher, 2007), and has been extended to include bullying and victimization (Fisher et al., 2013b; Kelleher et al., 2013c; Catone et al., 2015), institutional care (Bentall et al., 2012), and death of a close family member or friend (van Nierop et al., 2014). Recent reviews have highlighted that there is a link between childhood trauma/adversity and psychosis (Coughlan and Cannon, 2017; Varese et al., 2012). There is now also a wealth of evidence linking adverse childhood experiences to PLEs (Arseneault et al., 2011; Bentall et al., 2012; Fisher et al., 2013b; Kelleher et al., 2013c; Peters et al., 2016; Trotta et al., 2015). Some studies have demonstrated a dose–response relationship between traumatic experiences exposure and PLEs (Shevlin et al., 2011). Others have linked specific types of traumas events such as sexual abuse (Bentall et al., 2012), physical abuse (Fisher et al., 2013a,b), and emotional abuse (Daalman et al., 2012) to risk of psychotic experiences.

Recently, a lot of interest has been generated by the dynamic relationship between bullying and PLEs (Fisher et al., 2013b; Kelleher et al., 2013c; Catone et al., 2015; Moffa et al., 2017). Within a year long follow-up study of adolescents, Kelleher et al. (2013c) observed that even after controlling for baseline PLEs, there is a relationship between new incidents of PLEs and bullying/physical assault, suggesting a causal relationship. Another interesting finding was that cessation of the bullying or physical assault was associated with a reduction in PLEs. This finding is important for two reasons. One, it highlights the severity of the impact of bullying has on adolescent psychopathology, and two, intervening in bullying may impact the incidence of PLEs. Additionally, the underlying mechanism of this relationship suggests that bullying has an indirect effect on PLEs (specifically hallucinations) via persecutory ideation and depression (Moffa et al., 2017).

Cannabis use, cigarette use, and other substances: Cannabis use/abuse has been linked to both psychotic disorders (Moore et al., 2007) and PLEs (Linscott and Van Os, 2013; Degenhardt et al., 2018). Kuepper and colleagues neatly demonstrated that incidence of cannabis use was associated with an increased risk of subsequent PLEs and that continued cannabis use was associated with a greater risk of persistent PLEs (Kuepper et al., 2011). The relationship between cannabis use disorder and PLEs has been explained by a shared genetic and environmental factors as well as a causal effect of cannabis use disorder on PLEs (Nesvåg et al., 2016). However, to date, there is conflicting evidence regarding the interaction between potential gene polymorphisms (COMT rs4680) and cannabis use in the development of psychotic experience (Vinkers et al., 2013; Zammit et al., 2011).

Individuals with PLEs also are to be more likely to smoke cigarette (Mackie et al., 2011; Gage et al., 2014; Bhavsar et al., 2018), and this increased risk is independent of, and similar in magnitude to, cannabis use (Van Gastel et al., 2013; Linscott and Van Os, 2013; Degenhardt et al., 2018). Recent epidemiological evidence has shown that PLEs are also associated with a history of amphetamine, crack, ecstasy, and cocaine use (Bhavsar et al., 2018). This concurs with the very high rates of PLEs seen in individuals with amphetamine abuse/dependence (Rognli et al., 2017). Interestingly, nonillicit substance use, such as higher caffeine intake, has also been shown to predict hallucination-like experiences but not persecutory ideation (Jones and Fernyhough, 2009).

Gene–environment interactions (GxE): There has been a surge in interest into GxE as a potential mechanism to explain different outcomes among individuals who are exposed to the same environmental risk or who carry the same genetic vulnerability. Within the context of PLEs, certain single-nucleotide polymorphisms (SNPs) and risk haplotypes (combining "risk-associated" alleles or SNPs found on the same chromosome) have been shown to interact with environmental exposure to traumatic experiences and/or cannabis use resulting in an increased risk of PLEs (Alemany et al., 2011; Ramsey et al., 2013; Alemany et al., 2014; Zammit et al., 2011; Cristobal-Narvaez et al., 2016). The results from these investigations reveal a complex pattern of interactions. For example, Zammit and colleagues found no evidence for an interaction between cannabis use and COMT-158 genotype (Zammit et al., 2011). Similarly, Alemany and colleagues did not observe this interaction but did observe a significant three-way interaction between cannabis use, COMT-158 genotype, and child abuse. This interaction was such that (1) experiencing child abuse and using cannabis was associated with an increase in PLEs in Val genotype carriers (homozygous); (2) experiencing child abuse but not using cannabis was associated with an increased risk of PLEs in Met genotype carriers (homozygous); and (3) none of the other combinations of child abuse genotype and cannabis use were associated with an increased risk of PLEs. Interactions have also been shown to be polymorphism specific. Ramsey and colleagues's results suggest that an interaction between COMT-158 and childhood trauma increases the risk of PLEs, but the interaction between BNDF-66 and childhood trauma does not increase the risk of PLEs (Ramsey et al., 2013). However, research on GxE, in the context of PLEs, is still in its infancy, and much of the interplay between genetics vulnerability and environmental variables is yet to be examined.

Psychopathology: PLEs are linked to an increased vulnerability to subsequent or co-occurring psychiatric conditions (Bartels-Velthuis et al., 2011; Kelleher et al., 2012b). However, there also appears to be a relationship between early psychopathology and subsequent PLEs (McGrath et al., 2016). Emotional and behavioral problems in childhood have been shown to increase risk of subsequent delusional-like experiences (Scott et al., 2009), and this concurs with finding for nonaffective psychosis, particularly in males (Welham et al., 2009). Lancefield and colleagues have demonstrated that there is a relationship between persistent PLEs and concomitant internalizing and externalizing psychopathology but that children with remittent psychopathology were no more likely to develop PLEs during adolescents than those who had never had any psychopathology (Lancefield et al., 2016).

What are the neurodevelopmental correlates of PLEs?

Cognitive and motor dysfunction: Individuals who report PLEs appear to have subtle cognitive dysfunction. The domains identified are consistent with those that appear in psychotic disorders, albeit with smaller effects. These include gross domain such as intellectual quotient (Cannon et al., 2002; Horwood et al., 2008) and focal domain deficits such as speed of processing, psychomotor speed, receptive language, emotional recognition processing, and executive functioning (Blanchard et al., 2010; Daalman et al., 2011; Dickson et al., 2014; Gur et al., 2014; Kelleher et al., 2013a; Roddy et al., 2012). However, it is not yet fully established whether cognitive dysfunction precedes the onset of PLEs or whether they co-occur. Cannon and colleagues did find an association between poorer receptive language abilities, IQ scores, and motor development in early childhood and the occurrence of PLEs in adolescence (Cannon et al., 2002). Also Khandaker and colleagues reported a relationship between adolescent PLEs and a more broad inclusion of any neurodevelopmental disorder, primarily dyslexia (Khandaker et al., 2015). However, more evidence is necessary to fully evaluate early cognitive development and subsequent risk of PLEs. Moreover, given that most individuals will only have transient PLEs, it is unclear whether the cognitive dysfunction observed in these individuals represents a brief deviation from the typical cognitive trajectory (does the cognitive dysfunction resolve with the cessation of the symptoms) or is it stable but atypical trajectory of cognitive development. Finally, it is unclear to date whether certain cognitive domains can be used to distinguish individuals who go on to have persistent PLEs from those whose symptoms are transient. Mapping the cognitive trajectory of individuals with transient, persistent, and emergent PLEs over time will help resolve this question (Calkins et al., 2017).

Early neuroanatomical markers: Similar to cognitive dysfunction, neuroanatomical difference has been observed between those reporting PLEs and those not. These include global differences in gray matter volume and white matter integrity (Satterthwaite et al., 2015), structural and functional differences in frontotemporal regions (Jacobson et al., 2010; O'Hanlon et al., 2015), default network disruption (Jacobson McEwen et al., 2014; Orr et al., 2014), and reductions in global network efficiency and density (indices of regional connectivity) (Drakesmith et al., 2015). Some investigators have also advocated a role for the dysfunction of the metabolic processing of glutamate and glutamine in individuals with PLEs (Modinos et al., 2017). However, similar to the cognitive markers, it is unclear whether the differences in brain structure and functioning precede the onset of PLEs and the same questions regarding typical/atypical trajectories with transient PLEs also apply. Our understanding of the neuroanatomical differences between those who have PLEs and those who do not is still only in its infancy.

What are the risk factors for persistent PLEs?

While much work has been done over the last decade identifying risk factors for PLEs, less has been accomplished regarding risk factors associated with persistent PLEs. This is partially a consequence of the low rates of persistence. We know very little about risk factors and/or markers that distinguish individuals who are likely to have persistent PLEs from those whose symptoms are likely to be transient. Nevertheless, some studies have been able to investigate risk factors in those with persistent PLEs.

Comparison to typically developing individuals: Wigman and colleagues reported that persistent psychotic symptoms were more evident in those from ethnic minority groups, frequent cannabis users and those with early developmental problems, early stressful life events, and early adolescent trauma/victimization (Wigman et al., 2011). Similarly, Mackie et al. (2011) reported that peer victimization and common mental disorders at baseline distinguish persistent symptoms from those with low/no symptoms (Mackie et al., 2011).

Comparison to transient PLEs: Some studies have successfully investigated characteristics that distinguish persistent PLEs from transient PLEs. Baseline worry levels have been shown to be higher in those with persistent paranoid thinking (Freeman et al., 2012). Cougnard and colleagues reported that persistent PLEs were associated with a higher number of cumulative environmental risk factors (cannabis use, trauma, and/or urbanicity) compared to those with transient symptoms and that risk factors and baseline PLEs accounted for ~20–80% of the risk of persistent PLEs (Cougnard et al., 2007). Calkins and colleagues reported that they successfully distinguished 68% of individuals with persistent PLEs from those with transient PLEs using baseline characteristics (Calkins et al., 2017). The baseline characteristics that were independently able to distinguish the two groups were "psychosis" factor scores, global functioning, and psychiatric medication. Thus, it appears that those who have more severe symptoms and poorer functioning early on are those who are likely to have persistent PLEs. In fact, 25% of individuals with persistent PLEs were already being treated with psychiatric medication at baseline (Calkins et al., 2017).

Overview of risk factors and correlates

The current evidence highlights the overlap between risk factors for psychotic disorder and the extended psychosis phenotype. Some of these are stable biological characteristics (such as heritable traits or neural development) or are characteristics that an individual is likely to have little control over (urbanicity or social class). Others present an opportunity for intervention (such as victimization or psychopathology, see Fig. 7.2). Intervening early may abate the risk of transitioning to a psychotic disorder and/or reduce the rates of psychopathology in individuals with PLE's who are known to be vulnerable to a wide variety of disorders. This may be particularly important for individuals with persistent PLE whom we already know are at the more extreme end of the continuum in their early years and pose a greater risk of transitioning to psychotic disorders and have increased rates of psychopathology (De Loore

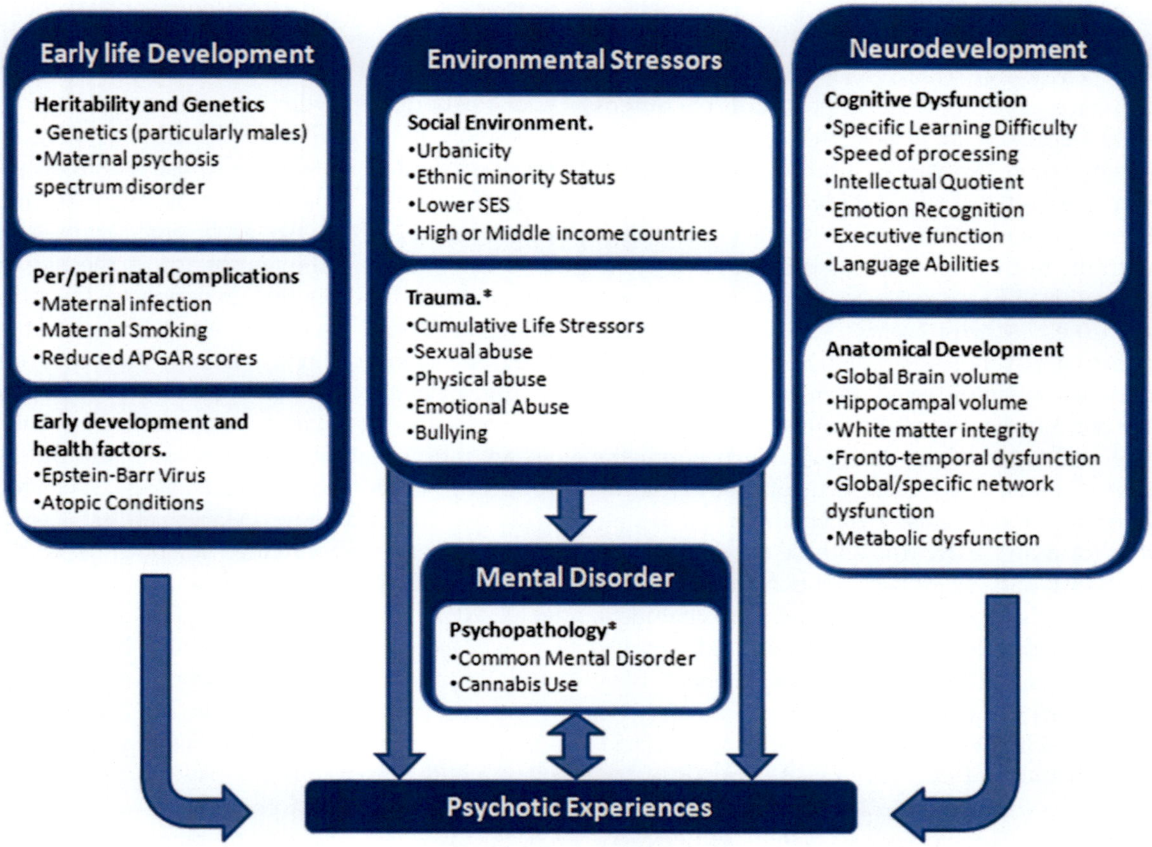

FIGURE 7.2 Risk factors for psychotic experiences. Note: * = Areas where interventions are possible and may have an impact on the prevalence of psychotic experiences.

et al., 2011; Dominguez et al., 2011; Calkin et al., 2017).

Finally, several studies have identified useful risk factors that are associated with an increased likelihood of having PLEs (transient or persistent). In truth, it is likely that a number of identified (and unidentified) factors interact with one another on multiple levels, and the outcome is the dynamic and diverse expression of PLEs. It is possible that more higher-order statistical analysis such as network analysis (Wigman et al., 2016), Direct Acyclic Graphic (Moffa et al., 2017), or machine learning algorithms (Bourque et al., 2017) may mimic this type of interactive environment better than standard statistical techniques. Moving toward these styles of investigation is an important step for psychiatric epidemiology.

What are the outcomes of having PLEs?

The term "subclinical psychotic experiences/symptoms" is somewhat misleading. As stated in the introduction section, "subclinical" does not mean that they are not clinically meaningful. Many individuals who experience these symptoms meet diagnostic criteria for common psychiatric conditions such as anxiety disorders, mood disorders or substance abuse (Kaymaz et al., 2012; McGrath et al., 2016; Healy et al., 2019a), and comorbidities are evident from a very early age (Bartels-Velthuis et al., 2011; Downs et al., 2013; Kelleher et al., 2012b; Laurens et al., 2012). Individuals who experience PLE (even when transient) are known to have increased rates of suicidal ideation and attempts (Honings et al., 2016a). Thus, while the intention with the terminology is to state that the individuals with PLEs do not have a "psychotic" disorder (i.e., schizophrenia spectrum disorder, affective psychosis, or delusional disorder), it may be misinterpreted that these individuals are entirely free from all forms of psychopathology and are nonhelp seeking, which is not the case (Calkins et al., 2017; Murphy et al., 2010). Notwithstanding this, it is also important to state that for many individuals, these experiences are transient and/or benign. The relationship between PLEs and poorer outcomes is probabilistic rather than deterministic (Linscott and Van Os, 2013), and for many, these experiences have limited effects (Johns et al., 2014; Peters et al., 2016). Below, we highlight some of the associations between PLEs and psychiatric disorders as well as discuss some of the psychosocial outcomes that have been associated with experiencing PLEs.

Psychotic disorders: Unsurprisingly, psychotic symptoms have been investigated as a risk factor for psychotic disorders (Poulton et al., 2000; Zammit et al., 2013a). Indeed, it has been reported that 38% of those with psychotic disorders have a history of PLEs (Dominguez et al., 2011), and a systematic review indicated that 7.4% of individuals who report early PLEs go on to develop a psychotic disorder (Linscott and Van Os, 2013). This is a 3.5 fold increase risk of transitioning to a psychotic disorder relative to those who never had PLEs (Kaymaz et al., 2012). This association is evident even after controlling for childhood psychopathology (Fisher et al., 2013a; Healy et al., 2019b) and appears to be particularly strong in males (Welham et al., 2009). Moreover, there is evidence that the severity of PLEs and the persistence of PLEs are associated with an increased risk of psychotic disorders. For example, Dominguez and colleagues reported that those with persistent psychotic experiences were 9.9 times more likely to transition to a psychotic disorder relative to those without PLEs, whereas those with reoccurring PLEs were only 5 times as likely (Dominguez et al., 2011). An additional component that has shown to be important is the age at which an individual first experiences the phenomena. It has been suggested that because PLEs are common in childhood, they may be, in the majority of cases, a marker of transient deviation from typical development. However, as an individual progress toward adulthood, reporting PLEs become increasing associated with psychopathology (Maijer et al., 2018). This has been demonstrated for both psychotic (Calkin et al., 2017) and nonpsychotic (Kelleher et al., 2012b) disorders.

Nonpsychotic outcomes: While PLEs were initially investigated as a potential precursor to psychotic disorder (Poulton et al., 2000), there is now a wealth of studies indicating their association with a myriad of nonpsychotic disorders. This relationship between PLEs and nonpsychotic disorders appears to take one of three possible forms: (1) psychopathology precedes the onset of PLEs; (2) they cooccur; and (3) PLEs precedes the onset of psychopathology.

(a) The first form has been addressed in the risk factor section, and it is relatively clear that psychopathology is related to an increased risk of subsequent PLEs (Lancefield et al., 2016; McGrath et al., 2016; Scott et al., 2009).

(b) The second association indicates that PLEs are associated with an increased risk of co-occurring common psychiatric conditions. Multiple studies have demonstrated this relationship (Armando et al., 2010; Kelleher et al., 2012b; Laurens et al., 2012; Wigman et al., 2012; Yung et al., 2009; Yoshizumi et al., 2004; Healy et al., 2019a). For example, Kelleher and colleagues, using four independent samples of young and middle adolescents (two interview based and two questionnaire based), demonstrated several interesting findings (Kelleher et al., 2012b). Firstly, the majority of those with PLEs met criteria for a clinical (nonpsychotic) disorder, and this association was accentuated in midadolescence. Secondly, those who were at the more severe end of the psychopathology scale were more likely to have PLEs indicating a link between psychopathology severity and PLEs. Finally, those who had PLEs had elevated risk of psychiatric multimorbidity. These results highlight the importance of screening for psychotic experiences in individuals with common psychiatric conditions.

(c) The third possibility is that PLEs, even when transient, are a marker for vulnerability to the subsequent onset of psychiatric conditions. This was initially studied quite extensively in almost an incidental fashion as researchers attempted to bridge a link between PLEs and psychotic disorders. Fisher and colleagues reported that of the participants in the Dunedin study who report PLEs at age 11, only 7.7% did not have a diagnosable psychiatric condition by age 38 (Fisher et al., 2013a). McGrath and colleagues demonstrated, using a the World Health Organization survey of 31,000 individuals, a bidirectional temporal relationship between PLEs and common mental health disorders with PLEs related to the onset of subsequent anxiety disorders, mood disorders, eating disorders, and substance-use disorders (OR range: 1.3–2, McGrath et al., 2016). In a meta-analysis of longitudinal prospective studies, Kaymaz et al. (2012) robustly demonstrated that there was a relationship between nonpsychotic disorders and PLEs; however, this relationship was less strong than their relationship to psychotic disorders (Kaymaz et al., 2012; Healy et al., 2019a). Thus, it appears that PLEs increase vulnerability to psychopathology more generally but still have a strong relationship with psychotic disorders. This is particularly apparent for those with persistent PLEs who have an increased risk of general psychopathology, depression, and delusional ideation (De Loore et al., 2011), as well as for more general internalizing and externalizing behavioral problems (Downs et al., 2013).

Experiencing PLEs has also been associated with a significant increased risk of suicidal ideation, suicidal attempts, and death by suicide (Cederlof et al., 2017; DeVylder et al., 2015b; Fisher et al., 2013a; Kelleher et al., 2013b; Sharifi et al., 2015; Sullivan et al., 2015). A metaanalysis has indicated that those with PLEs were at a two- to threefold increased risk of suicidal behaviors (Honings et al., 2016a). These results were still evident even after adjusting for depression. Additionally, it has been demonstrated that those with persistent PLEs have a greater risk of suicidal ideation, plans, and action relative to those with transient PLEs (Connell et al., 2016). Thus, there appears to be a stable link between PLEs and suicide. However, it is difficult to disentangle the triadic association between severe psychopathology, PLE, and suicidal behavior.

Negative psychosocial outcomes

Given the strong relationship between PLEs and psychiatric conditions, it is not surprising that individuals who have PLEs have poor outcomes across a range of psychosocial measure. For example, it has been reported that those who experience PLEs have poorer global functioning (Kelleher et al., 2015; Yung et al., 2009; Healy et al., 2018), lower quality of life (Svirskis et al., 2007), less social support (Saha et al., 2012), a more externalizing locus of control (Sullivan et al., 2017), and higher rates of perceived social stigma (Lien et al., 2015). Individuals who report PLEs are less likely to marry and more likely to be divorced (Linscott and Van Os, 2013; Scott et al., 2006) and have higher rates of unemployment (Scott et al., 2006; van Os et al., 2009), physical health problems (Moreno et al., 2013), hospitalization for nonpsychotic disorder (Werbeloff et al., 2012), shorter life span (Sharifi et al., 2015), and higher rates of physical violence perpetration and arrest (Honings et al., 2016b).

Poorer outcomes have also been identified in adolescent samples with reports indicating that those with PLEs have lower optimism, low self-esteem, high avoidance coping and school misconduct (Dolphin et al., 2015), poorer functioning (Calkins et al., 2017; Kelleher et al., 2015), and increased risk of sleep disturbances (Taylor et al., 2015). Thus, it is apparent that PLEs (directly or indirectly) influence a range of psychosocial issues that results in a poorer quality of life for those with such experiences.

In terms of distinguishing outcomes, DeVylders and colleagues reported that the only sociodemographic factors that distinguished individuals with persistent PLEs from those with transient PLEs were marital status and higher education (DeVylder et al., 2015a). Those with persistent PLEs were less likely to be married and have lower rates of third-level education than those whose symptoms were transient. However, the temporal direction of the relationship between these factors and PLEs remains to be established.

Other outcomes associated with psychotic-like experiences

Much of the literature on the extended psychosis phenotype has understandably concentrated on linking PLEs to negative outcomes. Limited attention has been paid to the fact that, at least in childhood, a substantial proportion of those who report PLEs do not report negative outcomes and these experiences appear to be relatively benign. Moreover, some have even investigated the potential psychosocial and societal benefits of having such experiences. The persistence of the phenotype indicates that there may be some degree of evolutionary advantage to such experiences. Kelleher, Jenners, and Cannon have previously reported that such experience could be evolutionarily advantageous for threat detection as they manifest as hypervigilance toward neutral stimuli (Kelleher et al., 2010). While this would result in greater number of "false alarm" threats detected, it would more often result in the positive detection of actual threats. This hypothesis, originally proposed by Dodgson and Gordon (2009), coincides with observations that high-positive schizotypy subscale scores correlate with the miscategorization of nonthreatening facial expressions as angry and fearful (van 't Wout et al., 2004). Additionally, hyper-theory-of-mind (low total theory of mind score combined with high mentalizing attributions of the intention of others) is more common in individuals with PLEs, particularly in those with paranoid content PLEs (Clemmensen et al., 2014). This suggests a hypervigilance to the intention of others, in line with the threat detection theory, but difficulty in accurately appraising those intention.

There is a relationship between creativity and schizotypy (Nettle, 2006; O'Reilly et al., 2001). Nettle (2006) reported that poets, artists, and mathematicians have higher schizotypy scores than novices at each profession. It has been suggested that there appears to be a nonlinear relationship between schizotypy and creativity, where the presence of some schizotypy is beneficial to creative outputs but too much will hinder the output (Yung et al., 2009).

Faccio and colleagues examined the content of auditory hallucinations in participants who have persistent PLEs but who never been in contact with mental health services and reported that a certain proportion of individuals did not find hearing voices distressing or worrying, and they developed their own belief about the positive function of the voice (Faccio et al., 2013). It has been suggested that the content of an auditory hallucination may differ between nonpsychiatric patients and patients with psychotic disorders, with patients with psychotic disorders more likely to report negative content (Johns et al., 2002). However, this study reported that both patients and nonpatients found the perceived lack of control over the experience distressing. Finally, positive experiences of hearing voices have also been linked with assigning protective attributions to the voice (Jenner et al., 2008), potentially indicating its use as a coping mechanism.

Overview of outcomes associated with psychotic-like experiences

It is clear that there is a strong link between PLEs and psychopathology. Because this link varies by temporality and across psychiatric condition, it has been suggested that PLEs may be a transdiagnostic marker for general psychopathology. Those with persistent PLEs appear to be susceptible to greater levels of psychopathology and have a more pronounced relationship with psychotic disorder. Specifically, PLEs have been linked to severity of ongoing psychopathology, psychiatric multimorbidity, subsequent psychiatric conditions, suicidal behavior, and a large range of negative psychosocial outcomes. However, not all individuals who experience PLEs have negative experiences, and the likelihood of poorer outcomes appears to be linked with the content of experiences, the perceived lack of control over the experiences, and the individual's attribution of such an experience, particularly in the context of voice hearing.

Current challenges

Difficulty with definition: One of the central challenges in studying PLEs is deciding what constitutes a PLE. The term "psychotic-like" itself evokes a degree of ambiguity, and to date, there is no consensus on what formally characterizes a "PLE." For example, should PLEs be measured as a categorical entity (taxonomic approach that is captured by a clinical interview, for example Poulton et al., 2000; Kelleher et al., 2012a,b; Dominguez et al., 2011; Zammiet et al., 2013) or should they be measured as a continuum (dimensional approach captured by factor/latent construct analysis, for example Capra et al., 2015; Laurens et al., 2012a; Sullivan et al., 2015)? If it is measured as a continuum, at what point along the continuum (if any) should we be concerned about underlying psychopathology? Lee and colleagues' review of the literature on PLEs reported that there is great variability in both definitions of PLEs and assessment tools with 76 studies adopting three different definition of what constitutes PLEs and 41 different assessment types (Lee et al., 2016). The inconsistencies in what constitutes a PLE will obviously pose a problem when deciding which ones have the potential to be pathological and which one are relatively benign.

Heterogeneity of phenomenology: Another difficulty is the heterogeneity of the phenomena that are reported. These experiences vary as widely as those in psychotic disorders. Hallucinations can vary across the sensory modalities (auditory, visual, tactile, and olfactory, etc.) but also within sensory modality. For example, auditory hallucinations may vary in quality (tone, pitch, volume) and content (some may hear whispers or their own voice aloud, while other may hear a third person persecutory commentary). Delusional phenomena also vary widely in content with the phenomenological manifestation of a delusional beliefs being in part determined by the individuals own experiences (Kiran and Chaudhury, 2009). Thus, with the wide variety of possible symptoms, which may stem from different origins, it is naturally difficult to tell which ones are causes for concern within the general population. Moreover, even research using cohort samples tend to group such experiences together for convenience and power requirements. Evidence has started to emerge suggesting that certain subtypes of PLEs may be more pathological than others (Armando et al., 2010). For example, Yung and colleagues reported that in a sample of (nonpsychotic) help-seeking youths, persecutory ideation and bizarre experiences (including hallucinations) were associated with distress, depression, and function, while magical thinking (items from the CAPE relating to telepathy, witchcraft, and grandeur) was not associated with poorer outcomes unless accompanied by distress (Yung et al., 2006). This has led to the conclusion that PLEs should not be treated as homogenous entities (Armando et al., 2010). Another potential hypothesis is that it may not be the type of experience itself that results in poorer outcomes but the distress associated with such an experience. This has been studied within a number of western populations (Kelleher et al., 2015; Murphy et al., 2017; Yung et al., 2006); however, as with much of the PLEs literature to date, there is a dearth of data from nonwestern societies.

What can be done? Given the difficulties regarding definitions, the large degree of heterogeneity, and complexity of the experiences as well as the limited resources any one study can acquire, it is clear that a standardized approach to studying the phenomena might prove more fruitful. One, in-depth and systematic approach to investigating this phenomenon in the general population was proposed by Kelleher and Cannon (Kelleher and Cannon, 2014). Unlike a simple item endorsement on a Likert scale, the SOCRATES method is a structured approach to questioning these symptoms in both children and adults. This approach attempts to maximize the information captured about the nature of PLEs while providing a structure for the direction of the questioning (http://epubs.rcsi.ie/cgi/viewcontent.cgi?article=1019&context=psychart). This method proposes a number of lines of questioning that follows the acronym: Source; Onset, duration, and frequency; Content; Reality testing; Attribution; Timing; Effects on functioning; and Severity of distress (for further information, see Kelleher, 2014). It is possible that merging in-depth questionnaire style with some of the more nuanced research designs (mixed methods approaches) and styles of analysis, such as a machine learning, directed acyclic graphs, and network analysis, may improve our understanding of why these symptoms persist in some individuals but not in other and also may highlight which ones are likely to be pathological.

What do we do about PLE?

The cumulative evidence to date raises interesting questions about PLEs in the general population. Firstly, what should we do about them? Should there be an intervention for individuals within the general population who have PLEs? What would such an intervention look like? And more importantly would an intervention be useful? Oh and colleagues reviewed this topic and confirmed the view that these questions are quite difficult to answer (Oh et al., 2014). As can be seen from above, there is still a need for clarity and consensus from within the research community about the assessment of PLEs and more standardization would facilitate this. Additionally, while PLEs are associated with increased risk of poorer outcomes, such as mental disorder, not everyone who has PLEs has such outcomes, thus a one-size-fits-all approach is not appropriate. Identifying risk factors that separate those who have PLEs with and without mental disorder would be helpful for identifying those who may require intervention. Particularly, identifying early risk factors that are specifically associated with persistent PLEs would be useful as we know these individuals are at an increase risk of mental disorder and thus more likely to require intervention. Alternatively identifying mediating factors that reduce the likelihood of mental disorder in individual who experience PLEs would also be helpful. Specifically, the data available are too limited to tell whether an intervention is necessary.

Nevertheless, there is one important initiative that should be acted upon promptly, namely to demystify "PLEs" in the general population. This point has already been raised by Kelleher (2016). Of all the symptoms of psychiatric illness, psychotic symptoms or PLEs draw some the greatest stigma within the general population (Kelleher, 2016). Increasing public awareness of how common PLEs are and providing psychoeducational information on what they mean may significantly reduce stigma attached to having such experiences and may lead to more help seeking from those reluctant to discuss the phenomena.

Overall conclusion

Over the past two decades, a growing body of literature is challenging the dichotomous conceptualization of psychotic phenomena. This literature has assisted in reshaping the perception of such phenomena and their meaning within psychiatry. This shift in view is particularly owing to epidemiological research that has steered the focus of these symptoms away from phenomena limited to those with severely debilitating psychiatric conditions to something which are now viewed as an extension of normal human experience and is evident with the general population. For most individuals, these experiences are transient and are generally related to difficulties coping with current life adversity. However, for some, these experiences will persist over time and have more of an insidious nature as reflected by a range of poorer outcomes. The origins of these experiences appear to be related to the interaction between a biological vulnerability and certain types of environmental exposures—particularly early development, social environment factors, trauma, types of social isolation, and cannabis use. The experiences themselves seem to be a marker for vulnerability to numerous poorer outcomes including current or a subsequent psychiatric condition, poorer functioning, and suicidal behavior.

While much work has been done regarding the prevalence and relevance of these experiences, particularly in relation to their associations with general and specific psychiatric disorders, a lot remains unanswered. Of particular importance is a consensus on how to measure PLEs, the identification of which types of experiences are likely to be pathological, which types of PLEs are likely to persist, and whether someone with persistent PLEs is likely to benefit from intervention. Finally, given that these experiences are so prevalent within the general population, it is highly important to disseminate the evidence to date about what these symptoms actually mean. Guiding this narrative will hopefully cause shifts in public opinion regarding psychosis.

Glossary

APGAR scores An index used to evaluate the condition of a newborn infant.

Atopy (atopic conditions) A genetic predisposition toward developing allergies such as asthma or eczema.

BNDF-66 Brain-derived neurotropic factor 66 is a single-nucleotide polymorphism that lies on chromosome 11. The nucleotide substitution results in changes in valine and methionine at amino acid residue 66.

COMT-158/rs4680 A single-nucleotide polymorphism in the catechol-O-methyltransferase that lies on chromosome 22. The nucleotide substitution results in changes in valine and methionine at codon 158.

Default network disruption Interference in the resting state connectivity of the "default" brain network that anatomically utilizes the posterior cingulate, medial prefrontal cortex, and the angular gyrus.

Direct Acyclic Graphic A graphical technique for modeling probability, connectivity, and causality between variables such that directionality is implied and without connectivity cycles.

Epstein–Barr virus A herpes family virus (human herpes virus 4) that causes infectious mononucleosis (glandular fever).

Extended psychosis phenotype The subclinical expression of the positive symptoms of psychosis within the general population.

Genetic family clustering The clustering of certain traits or behaviors within a given family.

Global network efficiency A metric of the effectiveness of information processing between distant regions of the brain.

Latent constructs analysis The construction and examination of a variable that is not directly observed but is inferred from other observed variables.

Machine learning algorithms Application of artificial intelligence that provides systems the ability to automatically learn and improve from experience without being explicitly programmed.

Network analysis A statistical technique for measuring the interconnectivity between multiple variables.

Polygenic risk score for schizophrenia A risk index for schizophrenia that is based on the aggregated effect of multiple genetic polymorphisms that have been associated with the disorder.

Pre/perinatal complications Prenatal complications: Complication that occur during pregnancy. Perinatal complications: Complications that occur immediately before and after birth.

Psychosis Spectrum Features The subclinical positive, negative, and disorganized symptoms of psychosis.

Psychotic-like experiences The subclinical positive symptoms of psychosis such as hallucinations or delusions that are experienced in the absence of a psychiatric disorder.

Reality testing The objective evaluation of an emotion or thought against real life, as a faculty.

Risk haplotypes A combined set of "risk" nucleotide polymorphisms that occur on the same chromosome and have been associated with a specific disorder.

Schizophreniform disorder The presentation of schizophrenia but total duration last at least 1 month but no more than 6 months. It is also typically associated with less impairment in social or occupational functioning.

Single-nucleotide polymorphisms A DNA sequence variation that occurs on a single nucleotide (adenine, thymine, cytosine, and guanine).

Structural equation modeling A multivariate statistical analysis that is used to analyze the structural relationship between variables.

References

Alemany, S., Arias, B., Aguilera, M., Villa, H., Moya, J., Ibáņez, M.I., et al., 2011. Childhood abuse, the BDNF-Val66Met polymorphism and adult psychotic-like experiences. Br. J. Psychiatry 199 (1), 38–42.

Alemany, S., Arias, B., Fatjó-Vilas, M., Villa, H., Moya, J., Ibanez, M.I., Ortet, G., Gastó, C., Fananas, L., 2014 Jan. Psychosis-inducing effects of cannabis are related to both childhood abuse and COMT genotypes. Acta Psychiatrica Scandinavica 129 (1), 54–62.

Andreasen, N.C., 1997. The evolving concept of schizophrenia: from Kraepelin to the present and future. Schizophr. Res. 28 (2–3), 105–109. https://doi.org/10.1016/S0920-9964(97)00112-6.

Armando, M., Nelson, B., Yung, A.R., Ross, M., Birchwood, M., Girardi, P., Fiori Nastro, P., 2010. Psychotic-like experiences and correlation with distress and depressive symptoms in a community sample of adolescents and young adults. Schizophr. Res. 119 (1–3), 258–265. https://doi.org/10.1016/j.schres.2010.03.001.

Arseneault, L., Cannon, M., Fisher, H.L., Polanczyk, G., Moffi, T.E., Caspi, A., January 2011. Childhood trauma and children's emerging psychotic symptoms: a genetically sensitive longitudinal cohort study. Am. J. Psychiatry 168, 65–72. https://doi.org/10.1176/appi.ajp.2010.10040567.

Bak, M., Myin-Germeys, I., Delespaul, P., Vollebergh, W., de Graaf, R., van Os, J., 2005. Do different psychotic experiences differentially predict need for care in the general population? Compr. Psychiatr. 46 (3), 192–199. https://doi.org/10.1016/j.comppsych.2004.08.003.

Bartels-Velthuis, A.A., van de Willige, G., Jenner, J.A., van Os, J., Wiersma, D., 2011. Course of auditory vocal hallucinations in childhood: 5-year follow-up study. Br. J. Psychiatry 199 (4), 296–302. https://doi.org/10.1192/bjp.bp.110.086918.

Bentall, R., Slade, P., 1985. Reliability of a scale measuring disposition towards hallucination: a brief report. Personal. Individ. Differ. 6 (4), 527–529.

Bentall, R.P., Wickham, S., Shevlin, M., Varese, F., 2012. Do specific early-life adversities lead to specific symptoms of psychosis? A study from the 2007 the Adult Psychiatric Morbidity Survey. Schizophr. Bull. 38 (4), 734–740. https://doi.org/10.1093/schbul/sbs049.

Bhavsar, V., Jauhar, S., Murray, R.M., Hotopf, M., Hatch, S.L., McNeill, A., et al., 2018. Tobacco smoking is associated with psychotic experiences in the general population of South London. Psychol. Med. 48 (1), 123–131.

Blanchard, M.M., Jacobson, S., Clarke, M.C., Connor, D., Kelleher, I., Garavan, H., et al., 2010. Language, motor and speed of processing deficits in adolescents with subclinical psychotic symptoms. Schizophr. Res. 123 (1), 71–76. https://doi.org/10.1016/j.schres.2010.05.028.

Bourque, J., Spechler, P.A., Potvin, S., Whelan, R., Banaschewski, T., Bokde, A.L.W., et al., 2017. Functional neuroimaging predictors of self-reported psychotic symptoms in adolescents. Am. J. Psychiatry 174 (6), 566–575. https://doi.org/10.1176/appi.ajp.2017.16080897.

Calkins, M.E., Moore, T.M., Merikangas, K.R., Burstein, M., Satterthwaite, T.D., Bilker, W.B., Ruparel, K., Chiavacci, R., Wolf, D.H., Mentch, F., Qiu, H., 2014. The psychosis spectrum in a young U.S. community sample: findings from the Philadelphia Neurodevelopmental Cohort. World Psychiatry 13 (3), 296–305. https://doi.org/10.1002/wps.20152.

Calkins, M.E., Moore, T.M., Satterthwaite, T.D., Wolf, D.H., Turetsky, B.I., Roalf, D.R., et al., 2017. Persistence of psychosis spectrum symptoms in the Philadelphia Neurodevelopmental Cohort: a prospective two-year follow-up. World Psychiatry 16 (1), 62–76. https://doi.org/10.1002/wps.20386.

Cannon, M., Caspi, A., Moffitt, T.E., Harrington, H., Taylor, A., Murray, R.M., Poulton, R., 2002. Evidence for early-childhood, pan-developmental impairment specific to schizophreniform disorder: results from a longitudinal birth cohort. Arch. Gen. Psychiatr. 59 (5), 449–456.

Cannon, M., Clarke, M.C., Cotter, D.R., 2014. Priming the brain for psychosis: maternal inflammation during fetal development and the risk of later psychiatric disorder. Am. J. Psychiatry 171 (9), 901–905.

Capra, C., Kavanagh, D.J., Hides, L., Scott, J.G., 2015. Subtypes of psychotic-like experiences are differentially associated with suicidal ideation, plans and attempts in young adults. Psychiatry Res. 228 (3), 894–898. https://doi.org/10.1016/j.psychres.2015.05.002.

Catone, G., Marwaha, S., Kuipers, E., Lennox, B., Freeman, D., Bebbington, P., Broome, M., 2015. Bullying victimisation and risk of psychotic phenomena: analyses of British national survey data. Lancet Psychiatry 2 (7), 618–624.

Cederlof, M., Kuja-Halkola, R., Larsson, H., Sjolander, A., Ostberg, P., Lundstrom, S., et al., 2017. A longitudinal study of adolescent psychotic experiences and later development of substance use disorder and suicidal behavior. Schizophr. Res. 181, 13–16. https://doi.org/10.1016/j.schres.2016.08.029.

Chapman, L.J., Chapman, J.P., 1980. Scales for rating psychotic and psychotic-like experiences as continua. Schizophr. Bull. 6 (3), 477–489. https://doi.org/10.1093/schbul/6.3.476.

Claridge, G., 1987. "The schizophrenias as nervous types" revisited. Br. J. Psychiatry 151 (6), 735–743.

Claridge, G., Beech, T., 1995. Fully and quasi-dimensional constructions of schizotypy. Schizotypal Pers. 29, 192–216.

Clemmensen, L., van Os, J., Skovgaard, A.M., Vaever, M., Blijd-Hoogewys, E.M.A., Bartels-Velthuis, A.A., Jeppesen, P., 2014. Hyper-theory-of-mind in children with psychotic experiences. PLoS One 9 (11), e113082. https://doi.org/10.1371/journal.pone.0113082.

Connell, M., Betts, K., McGrath, J.J., Alati, R., Najman, J., Clavarino, A., Mamun, A., Williams, G., Scott, J.G., 2016. Hallucinations in adolescents and risk for mental disorders and suicidal behaviour in adulthood: prospective evidence from the MUSP birth cohort study. Schizophr. Res. 176 (2–3), 546–551. https://doi.org/10.1016/j.schres.2016.06.009.

Coughlan, H., Cannon, M., 2017. Does childhood trauma play a role in the aetiology of psychosis? A review of recent evidence. BJPsych Adv. 23 (5), 307–315.

Cougnard, A., Marcelis, M., Myin-Germeys, I., de Graaf, R., Vollebergh, W., Krabbendam, L., et al., 2007. Does normal developmental expression of psychosis combine with environmental risk to cause persistence of psychosis? A psychosis proneness—persistence model. Psychol. Med. 37 (4), 513–527. https://doi.org/10.1017/S0033291706009731.

Cristóbal-Narváez, P., Sheinbaum, T., Rosa, A., Ballespí, S., de Castro-Catala, M., Peña, E., et al., 2016. The interaction between childhood bullying and the FKBP5 gene on psychotic-like experiences and stress reactivity in real life. PLoS One 11 (7), e0158809.

Daalman, K., Diederen, K.M.J., Derks, E.M., van Lutterveld, R., Kahn, R.S., Sommer, I.E.C., 2012. Childhood trauma and auditory verbal hallucinations. Psychol. Med. 42 (12), 2475–2484. https://doi.org/10.1017/S0033291712000761.

Daalman, K., van Zandvoort, M., Bootsman, F., Boks, M., Kahn, R., Sommer, I., 2011. Auditory verbal hallucinations and cognitive functioning in healthy individuals. Schizophr. Res. 132 (2–3), 203–207. https://doi.org/10.1016/j.schres.2011.07.013.

Das-Munshi, J., Bécares, L., Boydell, J.E., Dewey, M.E., Morgan, C., Stansfeld, S.A., Prince, M.J., 2012 Oct. Ethnic density as a buffer for psychotic experiences: findings from a national survey (EMPIRIC). The British Journal of Psychiatry 201 (4), 282–290.

De Loore, E., Gunther, N., Drukker, M., Feron, F., Sabbe, B., Deboutte, D., et al., 2011. Persistence and outcome of auditory hallucinations in adolescence: a longitudinal general population study of 1800 individuals. Schizophr. Res. 127 (1–3), 252–256. https://doi.org/10.1016/j.schres.2011.01.015.

Degenhardt, L., Hall, W., 2001. The association between psychosis and problematical drug use among Australian adults: findings from the National Survey of Mental Health and Well-Being. Psychol. Med. 31 (4), 659–668. https://doi.org/10.1017/S0033291701003865.

Degenhardt, L., Saha, S., Lim, C.C., Aguilar-Gaxiola, S., Al-Hamzawi, A., Alonso, J., Andrade, L.H., Bromet, E.J., Bruffaerts, R., Caldas-de-Almeida, J.M., de Girolamo, G., 2018 May. The associations between psychotic experiences and substance use and substance use disorders: findings from the World Health Organization World Mental Health surveys. Addiction 113 (5), 924–934.

DeVylder, J.E., Lehmann, M., Chen, F.-P., 2015a. Social and clinical correlates of the persistence of psychotic experiences in the general population. Schizophr. Res. 169 (1–3), 286–291. https://doi.org/10.1016/j.schres.2015.08.039.

DeVylder, J.E., Lukens, E.P., Link, B.G., Lieberman, J.A., 2015b. Suicidal ideation and suicide attempts among adults with psychotic experiences: data from the Collaborative Psychiatric Epidemiology Surveys. JAMA Psychiatry 72 (3), 219–225. https://doi.org/10.1001/jamapsychiatry.2014.2663.

Dickson, H., Calkins, M.E., Kohler, C.G., Hodgins, S., Laurens, K.R., 2014. Misperceptions of facial emotions among youth aged 9–14 years who present multiple antecedents of schizophrenia. Schizophr. Bull. 40 (2), 460–468. https://doi.org/10.1093/schbul/sbs193.

Dodgson, G., Gordon, S., 2009. Avoiding false negatives: are some auditory hallucinations an evolved design flaw? Behav. Cognit. Psychother. 37 (3), 325–334.

Dolphin, L., Dooley, B., Fitzgerald, A., 2015. Prevalence and correlates of psychotic like experiences in a nationally representative community sample of adolescents in Ireland. Schizophr. Res. 169 (1–3), 241–247. https://doi.org/10.1016/j.schres.2015.09.005.

Dominguez, M.D.G., Wichers, M., Lieb, R., Wittchen, H.-U., van Os, J., 2011. Evidence that onset of clinical psychosis is an outcome of progressively more persistent subclinical psychotic experiences: an 8-year cohort study. Schizophr. Bull. 37 (1), 84–93. https://doi.org/10.1093/schbul/sbp022.

Downs, J.M., Cullen, A.E., Barragan, M., Laurens, K.R., 2013. Persisting psychotic-like experiences are associated with both externalising and internalising psychopathology in a longitudinal general population child cohort. Schizophr. Res. 144 (1–3), 99–104. https://doi.org/10.1016/j.schres.2012.12.009.

Drakesmith, M., Caeyenberghs, K., Dutt, A., Zammit, S., Evans, C.J., Reichenberg, A., et al., 2015. Schizophrenia-like topological changes in the structural connectome of individuals with subclinical psychotic experiences. Hum. Brain Mapp. 36 (7), 2629–2643.

Faccio, E., Romaioli, D., Dagani, J., Cipolletta, S., 2013. Auditory hallucinations as a personal experience: analysis of non-psychiatric voice hearers' narrations. J. Psychiatr. Ment. Health Nurs. 20 (9), 761–767.

Fisher, H., Caspi, A., Poulton, R., Meier, M., Houts, R., Harrington, H., et al., 2013a. Specificity of childhood psychotic symptoms for predicting schizophrenia by 38 years of age: a birth cohort study. Psychol. Med. 43 (10), 2077. https://doi.org/10.1007/s00787-013-0423-9.

Fisher, H.L., Schreier, A., Zammit, S., Maughan, B., Munafo, M.R., Lewis, G., Wolke, D., 2013b. Pathways between childhood victimization and psychosis-like symptoms in the ALSPAC birth cohort. Schizophr. Bull. 39 (5), 1045–1055. https://doi.org/10.1093/schbul/sbs088.

Freeman, D., Stahl, D., McManus, S., Meltzer, H., Brugha, T., Wiles, N., Bebbington, P., 2012. Insomnia, worry, anxiety and depression as predictors of the occurrence and persistence of paranoid thinking. Soc. Psychiatry Psychiatr. Epidemiol. 47 (8), 1195–1203.

Gage, S.H., Hickman, M., Heron, J., Munafò, M.R., Lewis, G., Macleod, J., Zammit, S., 2014. Associations of cannabis and cigarette use with psychotic experiences at age 18: findings from the Avon Longitudinal Study of Parents and Children. Psychol. Med. 44 (16), 3435–3444.

Granö, N., Kallionpää, S., Karjalainen, M., Roine, M., Ranta, K., Heinimaa, M., 2016. Discrepancy between self-reported and interviewed psychosis risk symptoms: auditory distortions are the most reliably reported symptom by self-report. Early Interv. Psychiatry 10 (2), 129–136.

Grech, A., van Os, J., 2017. Evidence that the urban environment moderates the level of familial clustering of positive psychotic symptoms. Schizophr. Bull. 43 (2), 325–331.

Gur, R.C., Calkins, M.E., Satterthwaite, T.D., Ruparel, K., Bilker, W.B., Moore, T.M., et al., 2014. Neurocognitive growth charting in psychosis spectrum youths. JAMA Psychiatry 71 (4), 366–374. https://doi.org/10.1001/jamapsychiatry.2013.4190.

Healy, C., Brannigan, R., Dooley, N., Coughlan, H., Clarke, M., Kelleher, I., Cannon, M., 2019a. Childhood and adolescent psychotic experiences and risk of mental disorder: a systematic review and meta-analysis. Psychological Medicine 49 (10), 1589–1599.

Healy, C., Campbell, D., Coughlan, H., Clarke, M., Kelleher, I., Cannon, M., 2018. Childhood psychotic experiences are associated with poorer global functioning throughout adolescence and into early adulthood. Acta Psychiatrica Scandinavica 138 (1), 26–34.

Healy, C., Gordon, A.A., Coughlan, H., Clarke, M., Kelleher, I., Cannon, M., 2019b. Do childhood psychotic experiences improve the prediction of adolescent psychopathology? A longitudinal population - based study. Early intervention in psychiatry 13 (5), 1245–1251.

Honings, S., Drukker, M., Groen, R., van Os, J., 2016a. Psychotic experiences and risk of self-injurious behaviour in the general population: a systematic review and meta-analysis. Psychol. Med. 46 (2), 237–251. https://doi.org/10.1017/S0033291715001841.

Honings, S., Drukker, M., Ten Have, M., de Graaf, R., van Dorsselaer, S., van Os, J., 2016b. Psychotic experiences and risk of violence perpetration and arrest in the general population: a prospective study. PLoS One 11 (7), e0159023. https://doi.org/10.1371/journal.pone.0159023.

Horwood, J., Salvi, G., Thomas, K., Duffy, L., Gunnell, D., Hollis, C., et al., 2008. IQ and non-clinical psychotic symptoms in 12-year-olds: results from the ALSPAC birth cohort. Br. J. Psychiatry 193 (3), 185–191. https://doi.org/10.1192/bjp.bp.108.051904.

Jacobson McEwen, S.C., Connolly, C.G., Kelly, A.M.C., Kelleher, I., O'Hanlon, E., Clarke, M., et al., 2014. Resting-state connectivity deficits associated with impaired inhibitory control in non-treatment-seeking adolescents with psychotic symptoms. Acta Psychiatr. Scand. 129 (2), 134–142. https://doi.org/10.1111/acps.12141.

Jacobson, S., Kelleher, I., Harley, M., Murtagh, A., Clarke, M., Blanchard, M., et al., 2010. Structural and functional brain correlates of subclinical psychotic symptoms in 11–13 year old schoolchildren. Neuroimage 49 (2), 1875–1885. https://doi.org/10.1016/j.neuroimage.2009.09.015.

Jenner, J.A., Rutten, S., Beuckens, J., Boonstra, N., Sytema, S., 2008. Positive and useful auditory vocal hallucinations: prevalence, characteristics, attributions, and implications for treatment. Acta Psychiatr. Scand. 118 (3), 238–245.

Johns, L.C., Hemsley, D., Kuipers, E., 2002. A comparison of auditory hallucinations in a psychiatric and non-psychiatric group. Br. J. Clin. Psychol. 41 (1), 81–86. https://doi.org/10.1348/014466502163813.

Johns, L.C., Kompus, K., Connell, M., Humpston, C., Lincoln, T.M., Longden, E., et al., 2014. Auditory verbal hallucinations in persons with and without a need for care. Schizophr. Bull. 40 (Suppl. 4), S255–S264. https://doi.org/10.1093/schbul/sbu005.

Johns, L.C., van Os, J., 2001. The continuity of psychotic experiences in the general population. Clin. Psychol. Rev. 21 (8), 1125–1141.

Jones, S.R., Fernyhough, C., 2009. Caffeine, stress, and proneness to psychosis-like experiences: a preliminary investigation. Personal. Individ. Differ. 46 (4), 562–564.

Jones, H.J., Stergiakouli, E., Tansey, K.E., Hubbard, L., Heron, J., Cannon, M., et al., 2016. Phenotypic manifestation of genetic risk for schizophrenia during adolescence in the general population. JAMA Psychiatry 73 (3), 221. https://doi.org/10.1001/jamapsychiatry.2015.3058.

Kaymaz, N., Drukker, M., Lieb, R., Wittchen, H.-U., Werbeloff, N., Weiser, M., et al., 2012. Do subthreshold psychotic experiences predict clinical outcomes in unselected non-help-seeking population-based samples? A systematic review and meta-analysis, enriched with new results. Psychol. Med. 42 (11), 2239–2253. https://doi.org/10.1017/S0033291711002911.

Kelleher, I., 2014. The Assessment of Psychosis: A Reference Book and Rating Scales for Research and Practice, p. 123.

Kelleher, I., 2016. Auditory hallucinations in the population: what do they mean and what should we do about them? Acta Psychiatr. Scand. 134 (1), 3–5.

Kelleher, I., Cannon, M., 2011 Jan. Psychotic-like experiences in the general population: characterizing a high-risk group for psychosis. Psychological medicine 41 (1), 1–6.

Kelleher, I., Cannon, M., 2014. SOCRATES Assessment of Perceptual Abnormalities and Unusual Thought Content.

Kelleher, I., Cannon, M., 2016. Putting psychosis in its place. Am. J. Psychiatry 173 (10), 951–952. https://doi.org/10.1176/appi.ajp.2016.16070810.

Kelleher, I., Clarke, M.C., Rawdon, C., Murphy, J., Cannon, M., 2013a. Neurocognition in the extended psychosis phenotype: performance of a community sample of adolescents with psychotic symptoms on the MATRICS neurocognitive battery. Schizophr. Bull. 39 (5), 1018–1026. https://doi.org/10.1093/schbul/sbs086.

Kelleher, I., Connor, D., Clarke, M.C., Devlin, N., Harley, M., Cannon, M., 2012a. Prevalence of psychotic symptoms in childhood and adolescence: a systematic review and meta-analysis of population-based studies. Psychol. Med. 42 (9), 1857–1863. https://doi.org/10.1017/S0033291711002960.

Kelleher, I., Corcoran, P., Keeley, H., Wigman, J.T.W., Devlin, N., Ramsay, H., et al., 2013b. Psychotic symptoms and population risk for suicide attempt: a prospective cohort study. JAMA Psychiatry 70 (9), 940–948. https://doi.org/10.1001/jamapsychiatry.2013.140.

Kelleher, I., Harley, M., Murtagh, A., Cannon, M., 2011. Are screening instruments valid for psychotic-like experiences? A validation study of screening questions for psychotic-like experiences using in-depth clinical interview. Schizophr. Bull. 37 (2), 362–369. https://doi.org/10.1093/schbul/sbp057.

Kelleher, I., Jenner, J.A., Cannon, M., 2010 Sep. Psychotic symptoms in the general population—an evolutionary perspective. The British Journal of Psychiatry 197 (3), 167–169.

Kelleher, I., Keeley, H., Corcoran, P., Lynch, F., Fitzpatrick, C., Devlin, N., et al., 2012b. Clinicopathological significance of psychotic experiences in non-psychotic young people: evidence from four population-based studies. Br. J. Psychiatry 201 (1), 26–32. https://doi.org/10.1192/bjp.bp.111.101543.

Kelleher, I., Keeley, H., Corcoran, P., Ramsay, H., Wasserman, C., Carli, V., et al., 2013c. Childhood trauma and psychosis in a prospective cohort study: cause, effect, and directionality. Am. J. Psychiatry 170 (7), 734–741.

Kelleher, I., Wigman, J.T.W., Harley, M., O'Hanlon, E., Coughlan, H., Rawdon, C., et al., 2015. Psychotic experiences in the population: association with functioning and mental distress. Schizophr. Res. 165 (1), 9–14. https://doi.org/10.1016/j.schres.2015.03.020.

Khandaker, G.M., Stochl, J., Zammit, S., Lewis, G., Jones, P.B., 2014a. Childhood Epstein-Barr Virus infection and subsequent risk of psychotic experiences in adolescence: a population-based prospective serological study. Schizophr. Res. 158 (1), 19–24.

Khandaker, G.M., Stochl, J., Zammit, S., Lewis, G., Jones, P.B., 2015. A population-based prospective birth cohort study of childhood neurocognitive and psychological functioning in healthy survivors of early life meningitis. Ann. Epidemiol. 25 (4), 236–242. https://doi.org/10.1016/j.annepidem.2014.11.013.

Khandaker, G.M., Zammit, S., Lewis, G., Jones, P.B., 2014b. A population-based study of atopic disorders and inflammatory markers in childhood before psychotic experiences in adolescence. Schizophr. Res. 152 (1), 139–145. https://doi.org/10.1016/j.schres.2013.09.021.

Kiran, C., Chaudhury, S., 2009 Jan. Understanding delusions. Industrial psychiatry journal 18 (1), 3.

Konings, M., Bak, M., Hanssen, M., Van Os, J., Krabbendam, L., 2006. Validity and reliability of the CAPE: a self-report instrument for the measurement of psychotic experiences in the general population. Acta Psychiatr. Scand. 114 (1), 55–61.

Krabbendam, L., Van Os, J., 2005. Schizophrenia and urbanicity: a major environmental influence—conditional on genetic risk. Schizophr. Bull. 31 (4), 795–799.

Kuepper, R., van Os, J., Lieb, R., Wittchen, H.-U., Hofler, M., Henquet, C., 2011. Continued cannabis use and risk of incidence and persistence of psychotic symptoms: 10 year follow-up cohort study. Br. Med. J. 342, d738.

Kwapil, T.R., Barrantes-Vidal, N., 2014. Schizotypy: looking back and moving forward. Schizophr. Bull. 41 (Suppl. 2), S366–S373.

Lancefield, K.S., Raudino, A., Downs, J.M., Laurens, K.R., 2016. Trajectories of childhood internalizing and externalizing psychopathology and psychotic-like experiences in adolescence: a prospective population-based cohort study. Dev. Psychopathol. 28 (2), 527–536. https://doi.org/10.1017/S0954579415001108.

Laurens, K.R., Downs, J.M., Cullen, A.E., Barragan, M., To, M., 2012a. The relationship of subclinical psychotic experiences to internalising and externalising psychopathology in childhood. Schizophr. Res. 136, S85–S86. Retrieved from: http://www.embase.com/search/results?subaction=viewrecord&from=export&id=L71729600.

Laurens, K.R., Hobbs, M.J., Sunderland, M., Green, M.J., Mould, G.L., 2012b. Psychotic-like experiences in a community sample of 8000 children aged 9 to 11 years: an item response theory analysis. Psychol. Med. 42 (7), 1495–1506. https://doi.org/10.1017/S0033291711002108.

Laurens, K.R., Hodgins, S., Maughan, B., Murray, R.M., Rutter, M.L., Taylor, E.A., 2007. Community screening for psychotic-like experiences and other putative antecedents of schizophrenia in children aged 9-12??years. Schizophr. Res. 90 (1–3), 130–146. https://doi.org/10.1016/j.schres.2006.11.006.

Laurens, K.R., West, S.A., Murray, R.M., Hodgins, S., 2008. Psychotic-like experiences and other antecedents of schizophrenia in children aged 9-12 years: a comparison of ethnic and migrant groups in the United Kingdom. Psychol. Med. 38 (8), 1103–1111. https://doi.org/10.1017/S0033291707001845.

Lee, K.W., Chan, K.W., Chang, W.C., Lee, E.H.M., Hui, C.L.M., Chen, E.Y.H., 2016. A systematic review on definitions and assessments of psychotic-like experiences. Early Interv. Psychiatry 10 (1), 3–16.

Lien, Y.-J., Kao, Y.-C., Liu, Y.-P., Chang, H.-A., Tzeng, N.-S., Lu, C.-W., et al., 2015. Relationships of perceived public stigma of mental illness and psychosis-like experiences in a non-clinical population sample. Soc. Psychiatry Psychiatr. Epidemiol. 50 (2), 289–298. https://doi.org/10.1007/s00127-014-0929-6.

Linscott, R.J., Van Os, J., 2013. An updated and conservative systematic review and meta-analysis of epidemiological evidence on psychotic experiences in children and adults: on the pathway from proneness to persistence to dimensional expression across mental disorders. Psychol. Med. 43 (6), 1133–1149.

Mackie, C.J., Castellanos-Ryan, N., Conrod, P.J., 2011. Developmental trajectories of psychotic-like experiences across adolescence: impact of victimization and substance use. Psychol. Med. 41 (1), 47–58. https://doi.org/10.1017/S0033291710000449.

Maijer, K., Begemann, M.J.H., Palmen, S.J.M.C., Leucht, S., Sommer, I.E.C., 2018. Auditory hallucinations across the lifespan: a systematic review and meta-analysis. Psychol. Med. 48 (6), 879–888.

Mamah, D., Owoso, A., Mbwayo, A.W., Mutiso, V.N., Muriungi, S.K., Khasakhala, L.I., et al., 2013. Classes of psychotic experiences in Kenyan children and adolescents. Child Psychiatr. Hum. Dev. 44 (3), 452–459.

McGrath, J.J., Saha, S., Al-Hamzawi, A., Alonso, J., Bromet, E.J., Bruffaerts, R., et al., 2015. Psychotic experiences in the general population. JAMA Psychiatry 72 (7), 697. https://doi.org/10.1001/jamapsychiatry.2015.0575.

McGrath, J.J., Saha, S., Al-Hamzawi, A., Andrade, L., Benjet, C., Bromet, E.J., et al., 2016. The bidirectional associations between psychotic experiences and DSM-IV mental disorders. Am. J. Psychiatry 173 (10), 997–1006. https://doi.org/10.1176/appi.ajp.2016.15101293.

Modinos, G., McLaughlin, A., Egerton, A., McMullen, K., Kumari, V., Barker, G.J., et al., 2017. Corticolimbic hyper-response to emotion and glutamatergic function in people with high schizotypy: a multimodal fMRI-MRS study. Transl. Psychiatry 7 (4), e1083.

Moffa, G., Catone, G., Kuipers, J., Kuipers, E., Freeman, D., Marwaha, S., et al., 2017. Using directed acyclic graphs in epidemiological research in psychosis: an analysis of the role of bullying in psychosis. Schizophr. Bull. 1–7 https://doi.org/10.1093/schbul/sbx013.

Moore, T.H., Zammit, S., Lingford-Hughes, A., Barnes, T.R., Jones, P.B., Burke, M., Lewis, G., 2007. Cannabis use and risk of psychotic or affective mental health outcomes: a systematic review. Lancet 370 (9584), 319–328.

Moreno, C., Nuevo, R., Chatterji, S., Verdes, E., Arango, C., Ayuso-Mateos, J.L., 2013. Psychotic symptoms are associated with physical health problems independently of a mental disorder diagnosis: results from the WHO World Health Survey. World Psychiatry 12 (3), 251–257. https://doi.org/10.1002/wps.20070.

Morgan, C., Fisher, H., 2007. Environment and schizophrenia: environmental factors in schizophrenia: childhood trauma—a critical review. Schizophr. Bull. 33 (1), 3–10.

Morgan, C., Fisher, H., Hutchinson, G., Kirkbride, J., Craig, T.K., Morgan, K., et al., 2009. Ethnicity, social disadvantage and psychotic-like experiences in a healthy population based sample. Acta Psychiatr. Scand. 119 (3), 226–235. https://doi.org/10.1111/j.1600-0447.2008.01301.x.

Murphy, J., McBride, O., Fried, E., Shevlin, M., 2017. Distress, impairment and the extended psychosis phenotype: a network analysis of psychotic experiences in an US general population sample. Schizophr. Bull. sbx134.

Murphy, J., Shevlin, M., Houston, J., Adamson, G., 2010. A population based analysis of subclinical psychosis and help-seeking behavior. Schizophr. Bull. sbq092.

Nesvåg, R., Reichborn-Kjennerud, T., Gillespie, N.A., Knudsen, G.P., Bramness, J.G., Kendler, K.S., Ystrom, E., 2016. Genetic and environmental contributions to the association between cannabis use and psychotic-like experiences in young adult twins. Schizophr. Bull. sbw101. https://doi.org/10.1093/schbul/sbw101.

Nettle, D., 2006. Schizotypy and mental health amongst poets, visual artists, and mathematicians. J. Res. Personal. 40 (6), 876–890. https://doi.org/10.1016/j.jrp.2005.09.004.

Newbury, J., Arseneault, L., Caspi, A., Moffitt, T.E., Odgers, C.L., Fisher, H.L., 2017. Cumulative effects of neighborhood social adversity and personal crime victimization on adolescent psychotic experiences. Schizophr. Bull. 11–21. https://doi.org/10.1093/schbul/sbx060.

O'Hanlon, E., Leemans, A., Kelleher, I., Clarke, M.C., Roddy, S., Coughlan, H., et al., 2015. White matter differences among adolescents reporting psychotic experiences. JAMA Psychiatry 72 (7), 668. https://doi.org/10.1001/jamapsychiatry.2015.0137.

O'Reilly, T., Dunbar, R., Bentall, R., 2001. Schizotypy and creativity: an evolutionary connection? Personal. Individ. Differ. 31 (7), 1067–1078.

Oh, H., DeVylder, J.E., Chen, F.P., 2014. To treat or not to treat: responding to psychotic experiences. Br. J. Soc. Work 45 (7), 2003–2019.

Orr, J.M., Turner, J.A., Mittal, V.A., 2014. Widespread brain dysconnectivity associated with psychotic-like experiences in the general population. NeuroImage. Clin. 4, 343–351. https://doi.org/10.1016/j.nicl.2014.01.006.

Owrutsky, Z., Ahmed, A., Berman, K., Dickinson, K., 2017. Associations of subclinical psychosis spectrum characteristics, childhood depressive symptoms, and schizophrenia polygenic risk score: analyses from the philadelphia neurodevelopmental cohort. Schizophr. Bull. 43, S116. Retrieved from: http://www.embase.com/search/results?subaction=viewrecord&from=export&id=L616525018.

Peters, E., Ward, T., Jackson, M., Morgan, C., Charalambides, M., McGuire, P., et al., 2016. Clinical, socio-demographic and psychological characteristics in individuals with persistent psychotic experiences with and without a need for care. World Psychiatry 15 (1), 41–52. https://doi.org/10.1002/wps.20301.

Polanczyk, G., Moffitt, T.E., Arseneault, L., Cannon, M., Ambler, A., Keefe, R.S.E., et al., 2010. Etiological and clinical features of childhood psychotic symptoms. Arch. Gen. Psychiatr. 67 (4), 328. https://doi.org/10.1001/archgenpsychiatry.2010.14.

Poulton, R., Caspi, A., Moffitt, T.E., Cannon, M., Murray, R.M., Harrington, H., 2000. Children's self-reported psychotic symptoms and adult schizophreniform disorder: a 15-year longitudinal study. Arch. Gen.

Psychiatr. 57 (11), 1053–1058. https://doi.org/10.1001/archpsyc.57.11.1053.

Ramsay, H., Kelleher, I., Flannery, P., Clarke, M.C., Lynch, F., Harley, M., et al., 2013. Relationship between the COMT-Val158Met and BDNF-Val66Met polymorphisms, childhood trauma and psychotic experiences in an adolescent general population sample. PLoS One 8 (11), e79741.

Roddy, S., Tiedt, L., Kelleher, I., Clarke, M.C., Murphy, J., Rawdon, C., et al., 2012. Facial emotion recognition in adolescents with psychotic-like experiences: a school-based sample from the general population. Psychol. Med. 42 (10), 2157–2166. https://doi.org/10.1017/S0033291712000311.

Rognli, E.B., Bramness, J.G., Skurtveit, S., Bukten, A., 2017. Substance use and sociodemographic background as risk factors for lifetime psychotic experiences in a non-clinical sample. J. Subst. Abus. Treat. 74, 42–47.

Romme, M., Escher, A., 1989. Hearing voices. Schizophr. Bull. 15, 209–216.

Saha, S., Scott, J., Varghese, D., McGrath, J., 2012. Social support and delusional-like experiences: a nationwide population-based study. Epidemiol. Psychiatr. Sci. 21 (2), 203–212. https://doi.org/10.1017/S2045796011000862.

Satterthwaite, T.D., Vandekar, S.N., Wolf, D.H., Bassett, D.S., Ruparel, K., Shehzad, Z., et al., 2015. Connectome-wide network analysis of youth with Psychosis-Spectrum symptoms. Mol. Psychiatry 20 (12), 1508–1515.

Schultze-Lutter, F., Renner, F., Paruch, J., Julkowski, D., Klosterkötter, J., Ruhrmann, S., 2014. Self-reported psychotic-like experiences are a poor estimate of clinician-rated attenuated and frank delusions and hallucinations. Psychopathology 47 (3), 194–201.

Scott, J., Chant, D., Andrews, G., McGrath, J., 2006. Psychotic-like experiences in the general community: the correlates of CIDI psychosis screen items in an Australian sample. Psychol. Med. 36 (2), 231–238. https://doi.org/10.1017/S0033291705006392.

Scott, J., Martin, G., Welham, J., Bor, W., Najman, J., O'Callaghan, M., et al., 2009. Psychopathology during childhood and adolescence predicts delusional-like experiences in adults: a 21-year birth cohort study. Am. J. Psychiatry 166 (5), 567–574. https://doi.org/10.1176/appi.ajp.2008.08081182.

Sharifi, V., Eaton, W.W., Wu, L.T., Roth, K.B., Burchett, B.M., Mojtabai, R., 2015 Jul. Psychotic experiences and risk of death in the general population: 24–27 year follow-up of the Epidemiologic Catchment Area study. The British Journal of Psychiatry 207 (1), 30–36.

Shevlin, M., McElroy, E., Bentall, R.P., Reininghaus, U., Murphy, J., 2017. The psychosis continuum: testing a bifactor model of psychosis in a general population sample. Schizophr. Bull. 43 (1), 133–141. https://doi.org/10.1093/schbul/sbw067.

Shevlin, M., Murphy, J., Read, J., Mallett, J., Adamson, G., Houston, J., 2011. Childhood adversity and hallucinations: a community-based study using the National Comorbidity Survey Replication. Soc. Psychiatry Psychiatr. Epidemiol. 46 (12), 1203–1210.

Spauwen, J., Krabbendam, L., Lieb, R., Wittchen, H.U., van Os, J., 2003. Sex differences in psychosis: normal or pathological? Schizophr. Res. 62 (1–2), 45–49.

Strauss, J.S., 1969. Hallucinations and delusions as points on continua function: rating scale evidence. Arch. Gen. Psychiatr. 21 (5), 581–586.

Sullivan, S.A., Lewis, G., Gunnell, D., Cannon, M., Mars, B., Zammit, S., 2015. The longitudinal association between psychotic experiences, depression and suicidal behaviour in a population sample of adolescents. Soc. Psychiatry Psychiatr. Epidemiol. 50 (12), 1809–1817. https://doi.org/10.1007/s00127-015-1086-2.

Sullivan, S.A., Thompson, A., Kounali, D., Lewis, G., Zammit, S., 2017. The longitudinal association between external locus of control, social cognition and adolescent psychopathology. Soc. Psychiatry Psychiatr. Epidemiol. 0 (0), 1–13. https://doi.org/10.1007/s00127-017-1359-z.

Svirskis, T., Korkeila, J., Heinimaa, M., Huttunen, J., Ilonen, T., Ristkari, T., et al., 2007. Quality of life and functioning ability in subjects vulnerable to psychosis. Compr. Psychiatr. 48 (2), 155–160.

Taylor, M.J., Gregory, A.M., Freeman, D., Ronald, A., 2015. Do sleep disturbances and psychotic-like experiences in adolescence share genetic and environmental influences? J. Abnorm. Psychol. 124 (3), 674–684. https://doi.org/10.1037/abn0000057.

Therman, S., Ziermans, T.B., 2016. Confirmatory factor analysis of psychotic-like experiences in a general population sample. Psychiatry Res. 235, 197–199.

Thomas, K., Harrison, G., Zammit, S., Lewis, G., Horwood, J., Heron, J., et al., 2009. Association of measures of fetal and childhood growth with non-clinical psychotic symptoms in 12-year-olds: the ALSPAC cohort. Br. J. Psychiatry 194 (6), 521–526. https://doi.org/10.1192/bjp.bp.108.051730.

Trotta, A., Murray, R.M., Fisher, H.L., 2015. The impact of childhood adversity on the persistence of psychotic symptoms: a systematic review and meta-analysis. Psychol. Med. 45 (12), 2481–2498. https://doi.org/10.1017/S0033291715000574.

Van Gastel, W.A., Maccabe, J.H., Schubart, C.D., Vreeker, A., Tempelaar, W., Kahn, R.S., Boks, M.P.M., 2013. Cigarette smoking and cannabis use are equally strongly associated with psychotic-like experiences: a cross-sectional study in 1929 young adults. Psychol. Med. 43 (11), 2393–2401.

van Nierop, M., Lataster, T., Smeets, F., Gunther, N., van Zelst, C., de Graaf, R., et al., 2014. Psychopathological mechanisms linking childhood traumatic experiences to risk of psychotic symptoms: analysis of a large, representative population-based sample. Schizophr. Bull. 40 (Suppl. 2), S123–S130.

van Os, J., Linscott, R.J., Myin-Germeys, I., Delespaul, P., Krabbendam, L., 2009. A systematic review and meta-analysis of the psychosis continuum: evidence for a psychosis proneness–persistence–impairment model of psychotic disorder. Psychol. Med. 39 (2), 179. https://doi.org/10.1017/S0033291708003814.

van't Wout, M., Aleman, A., Kessels, R.P., Larøi, F., Kahn, R.S., 2004. Emotional processing in a non-clinical psychosis-prone sample. Schizophr. Res. 68 (2), 271–281.

Varese, F., Smeets, F., Drukker, M., Lieverse, R., Lataster, T., Viechtbauer, W., et al., 2012. Childhood adversities increase the risk of psychosis: a meta-analysis of patient-control, prospective- and cross-sectional cohort studies. Schizophr. Bull. 38 (4), 661–671. https://doi.org/10.1093/schbul/sbs050.

Vinkers, C.H., Van Gastel, W.A., Schubart, C.D., Van Eijk, K.R., Luykx, J.J., Van Winkel, R., et al., 2013. The effect of childhood maltreatment and cannabis use on adult psychotic symptoms is modified by the COMT Val(1)(5)(8)Met polymorphism. Schizophr. Res. 150 (1), 303–311. https://doi.org/10.1016/j.schres.2013.07.020.

Welham, J., Scott, J., Williams, G., Najman, J., Bor, W., O'Callaghan, M., McGrath, J., 2009. Emotional and behavioural antecedents of young adults who screen positive for non-affective psychosis: a 21-year birth cohort study. Psychol. Med. 39 (4), 625. https://doi.org/10.1017/S0033291708003760.

Werbeloff, N., Drukker, M., Dohrenwend, B.P., Levav, I., Yoffe, R., van Os, J., et al., 2012. Self-reported attenuated psychotic symptoms as forerunners of severe mental disorders later in life. Arch. Gen. Psychiatr. 69 (5), 467–475.

Wigman, J.T., de Vos, S., Wichers, M., van Os, J., Bartels-Velthuis, A.A., 2016. A transdiagnostic network approach to psychosis. Schizophr. Bull. sbw095.

Wigman, J.T.W., van Nierop, M., Vollebergh, W.A.M., Lieb, R., Beesdo-Baum, K., Wittchen, H.-U., van Os, J., 2012. Evidence that psychotic symptoms are prevalent in disorders of anxiety and depression, impacting on illness onset, risk, and severity—implications for diagnosis and ultra-high risk research. Schizophr. Bull. 38 (2), 247–257. https://doi.org/10.1093/schbul/sbr196.

Wigman, J.T.W., van Winkel, R., Raaijmakers, Q.A.W., Ormel, J., Verhulst, F.C., Reijneveld, S.A., et al., 2011. Evidence for a persistent, environment-dependent and deteriorating subtype of subclinical psychotic experiences: a 6-year longitudinal general population study. Psychol. Med. 41 (11), 2317–2329. https://doi.org/10.1017/S0033291711000304.

Yoshizumi, T., Murase, S., Honjo, S., Kaneko, H., Murakami, T., 2004. Hallucinatory experiences in a community sample of Japanese children. J. Am. Acad. Child Adolesc. Psychiatry 43 (8), 1030–1036. https://doi.org/10.1097/01.chi.0000126937.44875.6b.

Yung, A.R., Buckby, J.A., Cotton, S.M., Cosgrave, E.M., Killackey, E.J., Stanford, C., et al., 2006. Psychotic-like experiences in nonpsychotic help-seekers: associations with distress, depression, and disability. Schizophr. Bull. 32 (2), 352–359. https://doi.org/10.1093/schbul/sbj018.

Yung, A.R., Nelson, B., Baker, K., Buckby, J.A., Baksheev, G., Cosgrave, E.M., 2009. Psychotic-like experiences in a community sample of adolescents: implications for the continuum model of psychosis and prediction of schizophrenia. Aust. N. Z. J. Psychiatr. 43 (2), 118–128. https://doi.org/10.1080/00048670802607188.

Zammit, S., Hamshere, M., Dwyer, S., Georgiva, L., Timpson, N., Moskvina, V., et al., 2013b. A population-based study of genetic variation and psychotic experiences in adolescents. Schizophr. Bull. 40 (6), 1254–1262.

Zammit, S., Kounali, D., Cannon, M., David, A.S., Gunnell, D., Heron, J., et al., 2013a. Psychotic experiences and psychotic disorders at age 18 in relation to psychotic experiences at age 12 in a longitudinal population-based cohort study. Am. J. Psychiatry 170 (7), 742–750.

Zammit, S., Odd, D., Horwood, J., Thompson, A., Thomas, K., Menezes, P., et al., 2009. Investigating whether adverse prenatal and perinatal events are associated with non-clinical psychotic symptoms at age 12 years in the ALSPAC birth cohort. Psychol. Med. 39 (9), 1457–1467. https://doi.org/10.1017/S0033291708005126.

Zammit, S., Owen, M.J., Evans, J., Heron, J., Lewis, G., 2011. Cannabis, COMT and psychotic experiences. Br. J. Psychiatry 199 (5), 380–385.

CHAPTER

8

22q11.2 deletion syndrome: a neurodevelopmental model of psychosis

Corrado Sandini[1], *Stephan Eliez*[1], *Maude Schneider*[1,2,a], *Marco Armando*[1,a]

[1]**Developmental Imaging and Psychopathology Lab, Department of Psychiatry, School of Medicine, University of Geneva, Geneva, Switzerland;** [2]**Center for Contextual Psychiatry, Department of Neurosciences, Leuven, Belgium**

Introduction

A growing body of literature is converging on the notion that schizophrenia should be considered as a neurodevelopmental disorder (Insel, 2010; Marin, 2016; Murray et al., 1991; Weinberger, 1987). Consequently, clinical manifestations of full-blown psychosis should not be considered as the onset but rather as the end stage of a disease trajectory, starting potentially several years earlier (Insel, 2010). Critically, the neurodevelopmental model predicts that the effectiveness of therapeutic interventions is influenced by disease stage, with earlier interventions having a better hope of altering the path toward psychosis (Marin, 2016; Millan et al., 2016). Therefore, an improved characterization of the early neurodevelopmental and clinical trajectories toward psychosis is going to be critical for informing future therapeutic interventions and for stratifying patients who might benefit from such treatments (Millan et al., 2016; Marin, 2016).

Still, despite growing efforts, several challenges have hindered the comprehension of how genetic predisposition and environmental risk factors lead to early aberrant neurodevelopment and vulnerability to psychosis (Insel, 2010).

For instance, while several epidemiological studies confirmed that cognitive symptoms of schizophrenia, including deficits in social cognition, complex cognition, and receptive vocabulary, are already observable in the early stages of the trajectory (Kremen et al., 2010; Gur et al., 2014), the neurodevelopmental pathophysiology of these early disease stages remains elusive, as patients who are not yet help-seeking are typically not accessible to investigation (Kahn and Keefe, 2013). Moreover, while genetic predisposition is a key determinant in the

[a] These authors are shared senior authors on this work.

https://doi.org/10.1016/B978-0-12-813201-2.00008-9

pathogenesis of psychosis, with over 10,000 risk genes currently described (Schizophrenia Working Group of the Psychiatric Genomics, 2014), each genes contributes only a modest and a specific effect on the phenotype (Gejman et al., 2010). The resulting genetic heterogeneity complicates the assessment of the neurodevelopmental significance of these candidate human mutations (Meechan et al., 2015a). Genetic heterogeneity has furthermore hindered the relevance of reliable translational animal models (Meechan et al., 2015a). Such animal models are indispensable both for understating neurodevelopmental mechanisms leading to psychosis and for developing novel treatments that might interfere with this trajectory (Insel, 2010).

It was suggested that 22q11.2 Deletions Syndrome (22q11DS) could offer a unique opportunity to address some of these challenges and to contribute to the comprehension of the neurodevelopmental pathophysiology of psychosis (Insel, 2010).

Indeed, 22q11DS is the single highest molecular risk factor for psychosis with over 30% of patients developing a schizophrenia spectrum disorder (SSD) by adulthood (Schneider et al., 2014a). Only having either both the parents or a monozygotic twin affected by schizophrenia are associated with a higher genetic risk (McGuffin et al., 1995). Compared to other high-risk populations, in 22q11DS, genetic load for psychosis is relatively homogenous, with patients sharing a common heterozygous genetic deletion of 35–60 genes located on the long arm of chromosome 22 (McDonald-McGinn et al., 2015). Importantly, in 22q11DS, clinical manifestations are largely indistinguishable to psychosis occurring without this specific genetic etiology (Monks et al., 2014; Bassett et al., 2003). Moreover, patients are typically identified at a young age and independently of psychiatric symptoms, due to a complex somatic phenotype that characteristically entails cardiac malformations, dysmorphic features, and cleft palate (McDonald-McGinn et al., 2015). The longitudinal follow-up of affected children offers therefore a rare opportunity of mapping the earliest clinical and neurodevelopmental stages of the trajectory toward psychosis (Insel, 2010). A further unique aspect of 22q11DS is that the affected genetic region is highly conserved throughout mammalian and non mammalian species leading to availability of an animal model carrying a homologous genetic deletion (Meechan et al., 2015a). Importantly, the mice harboring the homologous genetic deletion manifest several behavioral and cognitive alterations common to both 22q11DS and idiopathic schizophrenia, which have been related anomalies in circuit formation and synaptic plasticity (Karayiorgou et al., 2010; Meechan et al., 2015a). Given these considerations, 22q11DS is rapidly emerging as a precious neurodevelopmental and translational model for understating the pathogenesis of psychosis (Karayiorgou et al., 2010).

Here we review the core aspects of the neurocognitive and psychiatric phenotype of 22q11DS throughout different stages of development emphasizing both parallels and divergences with idiopathic psychosis.

We then review the most relevant findings regarding the neurodevelopmental pathogenesis of 22q11DS coming from both the neuroimaging literature and from translational animal work. We argue that an iterative and integrative approach between clinical and neuroimaging human work and mechanistic biology in animal models of 22q11DS could contribute significantly to understating the neurodevelopmental pathogenesis of psychosis. The resulting insights could help inform future treatment strategies that are more effective in altering the trajectory toward psychosis.

Epidemiology and genetic pathophysiology

22q11.2DS is caused by a heterozygous deletion of region 11.2 on the long arm of chromosome 22 and represents the most common genetic microdeletion in humans (McDonald-McGinn et al., 2015). The prevalence of 22q11DS is currently estimated to range from 1 per 3000 to 1 per 6000 live births with approximately 90%−95% of cases arising from de novo microdeletions and only a minority being inherited (McDonald-McGinn et al., 2001; Oskarsdottir, 2004). Both sexes are equally affected, while cases of maternal inheritance are slightly more common than paternal ones (Delio et al., 2013). The prevalence of 22q11DS is higher among populations that show enriched aspects of the syndrome's phenotype, such as congenital cardiac malformations (i.e., 16% of cases of tetralogy of Fallot) (Goldmuntz et al., 1993), developmental disabilities (2%−3%) (Rauch et al., 2006), and schizophrenia (0.5%−1%) (Bassett et al., 2010; Horowitz et al., 2005).

The particular vulnerability of the 22q11.2 region is related to the presence of 4 large blocks of highly homologous duplicated sequences, known as low copy repeats (LCR22A to D from proximal to distal) (Edelmann et al., 1999). The deletion arises from non homologous recombination between such LCRs during meiosis and can vary in size from 3 to 1.5 MB according to the specific LCRs involved (Edelmann et al., 1999; Shaikh et al., 2000). The most common 3 MB deletion, present in 85% of subjects, arises from non homologous recombination of LCR22A and D that bracket the region (Edelmann et al., 1999; Shaikh et al., 2000). Smaller nested deletions affect LCR22B or C, with proximal A-B and A-C deletions accounting for 5%−10% of cases (Shaikh et al., 2000). Distal B-C and B-D deletions are less common and are missed by standard genetic testing based on fluorescent in situ hybridization (FISH) (Rump et al., 2014). Variability in deletion size may account for some of the variability in phenotype observed in 22q11DS, with smaller distal deletions having a somewhat milder phenotype and a more frequent parental origin (Rump et al., 2014).

The typical 3 MB A-D deletion involves 46 protein-coding genes along with 7 microRNAs (Guna et al., 2015).

Haploinsufficiency occurs when both copies of a gene are required for the manifestation of a normal (wild-type) phenotype. Haploinsufficiency of several individual genes in the 22q11 regions has been related to aspects of the somatic and neurocognitive phenotype in animal models.

For instance, the selective heterozygous knockout of the TBX1 gene in mice recapitulates several aspects of the somatic phenotype of the syndrome including cardiovascular, thymic, and parathyroid defects (Lindsay et al., 2001). Moreover, TBX1 has been implicated in the development of cerebral microvasculature (Cioffi et al., 2014) and may contribute to the neuropsychiatric phenotype of the syndrome (Paylor et al., 2006). TBX1 is located in region A-B and is the target of current genetic testing with FISH (Lindsay et al., 2001).

Other genes that have been individually implicated in contributing to the phenotype include ZDHHC8, coding for a palmitoyltransferase and involved in axonal growth and arborization, synaptic stability, and cognitive functioning (Moutin et al., 2016).

The distally located CRKL may be responsible for cardiac anomalies in individuals with nested deletions (Zheng et al., 2015), while DGCR8 may be responsible for neuronal deficits by mediating the biogenesis of microRNAs (Karayiorgou et al., 2010).

The catechol-*O*-methyltransferase enzyme (COMT) is involved in the degradation of catecholamines including dopamine, leading to speculation that its haploinsufficiency might be related to the susceptibility to psychosis. However, while enzymatic activity of COMT can vary across subjects according to a specific Val/Met polymorphism in the remaining allele, such variability has not been consistently related to schizophrenia or to intellectual functioning (Armando et al., 2012b).

Given the complexity of the 22q11.2 region, a variety of mechanisms of signaling pathways are likely to cooperate in the emergence of the phenotype. Indeed, a striking feature of the syndrome is its considerable phenotypic variability even among subjects harboring identical deletions (Vergaelen et al., 2015). The source of such variable expression is a field of ongoing research and is likely to arise from modifier genes located outside the deleted regions (Driscoll, 2006). Moreover, the variability observed even between monozygotic twins points to the role of epigenetic mechanisms along with environmental and stochastic effects (Singh et al., 2002).

Somatic phenotype

The initial clinical description of 22q11DS, which lead to its first eponymous definition by Dr. Angelo DiGeorge, involved the association of immunodeficiency due to thymic aplasia or hypocalcemia due to parathyroid aplasia and congenital cardiac malformations (DiGeorge, 1965). From the embryological perspective, this triad of DiGeorge syndrome has been related to altered development of the third and fourth pharyngeal pouches (Takao et al., 1980).

Congenital cardiac malformations are present in approximately 75% of patients with varying severity and typically include conotruncal defects due to malformations of the cardiac outflow tract such as tetralogy of Fallot, truncus arteriosus, interrupted aortic arch type B, and ventricular septal defects (Marino et al., 1999).

Anomalies of the aortic arch or pulmonary arteries are also frequent and may either be associated to cardiac malformations (60% of cases) or less frequently occur in isolation (40% of cases) (Peyvandi et al., 2013). Overall congestive heart failure represents the main cause (87%) of mortality in children with 22q11.2DS (Repetto et al., 2014).

Some level of immunodeficiency due to thymic aplasia and resulting impaired T cell production affects approximately 75% of patients (Morsheimer et al., 2017). As with other aspects of the phenotype, however, T cell production and immunological status can vary significantly across patients with severe immunodeficiency affecting only a minority of patients and chronic infections involving particularly the upper respiratory tract being the most common manifestation (Morsheimer et al., 2017). The humoral response can also be impaired resulting in poor response to vaccines (Morsheimer et al., 2017). Immunological self-tolerance is likewise affected leading to the overrepresentation of a plethora of autoimmune conditions (Morsheimer et al., 2017).

Hypocalcemia resulting from hypoparathyroidism can affect 50%–65% of patients and manifest with tetany, seizure, feeding difficulties, stridor, and fatigue (Weinzimer, 2001). Transient bouts of hypocalcemia can recur throughout the lifespan in periods of stress such as illness, surgery, or pregnancy (Weinzimer, 2001).

Aside from the triad of immunodeficiency, hypoparathyroidism, and cardiac malformations, the recognition of the genetic origin of DiGeorge Syndrome has led the description of a complex somatic phenotype potentially affecting virtually all organ systems with variable prevalence and severity.

Palatal abnormalities are among the most common manifestations of the syndrome, observed in over 75% of patients and ranging in severity to over cleft palate to milder forms such as occult submucosal cleft palate, bifid uvula, and velopharyngeal dysfunction (Dyce et al., 2002). The clinical manifestations of palate abnormalities vary according to their severity and to developmental stage and may initially become apparent as polyhydramnios in utero or nasal regurgitation in infancy (Dyce et al., 2002). With the development of speech, abnormal nasal resonance and nasal emissions along with compensatory articulation errors become apparent and contribute to worsening speech intelligibility in affected children (McDonald-McGinn et al., 2015).

Slight dysmorphic facial features that can aid in diagnosis include hooded eyelids, upslanted palpebral fissures, auricular abnormalities, small mouth particularly in infancy, and micrognathia (McDonald-McGinn et al., 2015).

Gastrointestinal manifestations particularly involving the upper gastrointestinal tract such as esophageal dysmotility and gastroesophageal and nasopharyngeal reflux are highly prevalent and can lead to feeding and swallowing problems (Giardino et al., 2014). Constipation is also extremely common.

Genitourinary abnormalities are collectively present in approximately one-third of patients and include a vast array of malformations spanning from renal agenesis to cystic kidney to cryptorchidism to hypospadias and absent uterus (Wu et al., 2002).

Neurocognitive profile

22q11DS is associated with a complex profile of cognitive impairments throughout development. Indeed, when the diagnosis is not suggested by facets of the somatic phenotype, the presence of learning difficulties is often the presenting symptom that leads to genetic testing (McDonald-McGinn et al., 2015).

Global cognition

As a population, individuals with 22q11DS present an average reduction in overall intelligence quantifiable in approximately 1.5–2 standard deviations, with a mean IQ of 70–75 (De Smedt et al., 2007). However, variability of cognitive performance in 22q11DS has been reported to follow a typical Gaussian distribution, meaning that 49% of patients present average or borderline cognitive performance (IQ > 70), 42% have mild cognitive impairment (IQ between 50 and 69), 8% have a moderate cognitive impairment (IQ between 35 and 49), and 1% have severe cognitive impairment (IQ < 35) (De Smedt et al., 2007). The source of such variability across subjects, as with other aspects of the syndrome's phenotype, is a source of ongoing investigation.

Cognitive difficulties associated with 22q11DS have been shown to affect different cognitive domains in a heterogeneous manner. For instance, the majority of individuals with 22q11DS present a disproportionate impairment in the nonverbal visuospatial cognition, measured with performance IQ, compared to a relatively preserved verbal reasoning, measured with verbal IQ (Moss et al., 1999). Importantly, language impairments exist independently and must be distinguished from speech difficulties linked to velopharyngeal malformations, which are also highly prevalent in the syndrome. However, the dissociation between verbal and nonverbal reasoning might not be universal to all stages of development (Swillen, 2016). For instance, toddlers with 22q11DS present a significant delay in language onset, with 70% being a verbal or speaking only a few words until 24 months of age or older (Solot et al., 2001). Moreover, the syndrome is associated with atypical trajectories of cognitive development, with a mean progressive decline of approximately 7 IQ points from childhood to early adulthood (Vorstman et al., 2015). Such cognitive decline disproportionately affects verbal reasoning, with a mean decline of 9 points for verbal IQ compared to 5 points for performance IQ (Vorstman et al., 2015). These differential trajectories imply that the preferential impairment in visuospatial cognition might become less prevalent with age (Swillen, 2016). Importantly, the decline in verbal reasoning has been specifically related to the subsequent emergence of psychosis in 22q11DS (Vorstman et al., 2015).

Apart from the dissociation between verbal and nonverbal intelligence, 22q11DS is characterized by several domains of relative cognitive weakness. For instance, mathematical reasoning, involving simple arithmetical operations or numerical comparisons, is disproportionately affected by the syndrome (Eliez et al., 2001). Individuals with the deletions have also been reported to present particular difficulties in temporal perception (Gabriel Mounir et al., 2011; Debbane et al., 2005).

Executive functioning

Executive functions are a group of high-level cognitive processes, encompassing among others working memory, cognitive flexibility, inhibition, and programming, which are required for goal-directed behavior and are particularly solicited during novel tasks and situations (Diamond, 2013). Cognitive difficulties particularly involving executive functions such as working memory are now thought to represent a core and early manifestation of psychosis in the general population (Freedman and Brown, 2011). In 22q11DS, several studies have reported deficits in executive functions particularly involving working memory, cognitive flexibility, and planning in 22q11DS (Campbell et al., 2010b; Lajiness-O'Neill et al., 2006; Dufour et al., 2008). Interestingly, a recent study conducted in a large longitudinal cohort demonstrated that deficits in executive functions arise from atypical developmental trajectories particularly during late childhood and adolescence (Maeder et al., 2016). Given findings relating executive functioning impairments to schizophrenia in the general population (Freedman and Brown, 2011), atypical development of executive functions during childhood and adolescence might be linked to the emergence of psychosis in 22Q11DS (Maeder et al., 2016).

Social cognition

Lastly, social cognition represents an area of significant cognitive weakness in 22q11DS. Social cognition is defined as a set of competences that are necessary to interpret and appropriately navigate a social environment (Green et al., 2015). The ability to recognize emotional states of others through visual processing of faces is a core social competence that is consistently impaired in nonsyndromic schizophrenia (Green et al., 2015). Similarly, impairments in both affective and non affective facial processing have been described in 22q11DS (Campbell et al., 2010a; Glaser et al., 2010; Schneider et al., 2017; Andersson et al., 2008). Impaired facial recognition has been related to an atypical visual exploration pattern, with a reduction of time spent looking at the eyes and a corresponding increase in time spent looking at the mouth region (Campbell et al., 2010a; Glaser et al., 2010). Interestingly, aberrant facial processing has been specifically related to more severe social anxiety and more severe negative symptoms in individuals with 22q11DS (Glaser et al., 2010; Schneider et al., 2017). Neuroimaging studies have also suggested that deficits in face recognition in 22q11DS might be related to insufficient and a specific activation of the fusiform gyrus, the brain regions responsible for facial processing (Andersson et al., 2008). Moreover, altered cortical morphology of the fusiform gyrus has been related to higher negative symptoms and specifically to more severe social anhedonia (Mihailov et al., 2017). Aside from deficits in facial processing, several studies in 22q11DS have demonstrated impairments in higher-level social cognition competences, such as the ability to attribute mental states to others, also known as theory of mind (TOM) (Jalbrzikowski et al., 2012; Campbell et al., 2011). As expected, deficits in TOM were furthermore correlated with daily life social competences in children and adolescents affected by the syndrome (Campbell et al., 2011; Jalbrzikowski et al., 2012).

Taken together, findings suggest that deficits in social cognition could contribute to the emergence of social behavioral deficits, social retreat and negative symptoms, which are both highly prevalent and debilitating in 22q11DS.

Psychiatric phenotype (other than psychosis) throughout development

Aside from increased risk for psychosis, 22q11DS is characterized by a complex psychiatric phenotype with several conditions being highly prevalent at different stages of development. A collaborative study conducted on a large cohort, obtained from an international consortium, contributed to characterize the psychiatric phenotype of 22q11DS (Schneider et al., 2014a). Here we will discuss each main psychiatric diagnosis separately. However, it should be noted that studies that have specifically quantified the co-occurrence of multiple diagnoses report high rates of psychiatric comorbidity in 22q11DS, particularly between attention-deficit/hyperactivity disorder (ADHD) and anxiety disorders and between anxiety, mood disorders, and SSDs (Niarchou et al., 2014; Schneider et al., 2014a). Increased risk for anxiety and mood disorders have also been consistently reported in the general population in patients at different stages of psychosis (Buckley et al., 2009). The origin and nature of psychiatric comorbidity is a source of ongoing debate in field of schizophrenia (Buckley et al., 2009) and psychiatric nosology in general (Lilienfeld and Treadway, 2016; Insel et al., 2010; Caspi et al., 2014). In this sense, 22q11DS could serve as a model to investigate how common genetic load predisposes to multiple domains of psychopathology and how these different domains then interact with one another and evolve of the course of development.

Attention-deficit/hyperactivity disorder

Children with 22q11DS (ages 6–12) presented very high rates (37%) of ADHD. Rates of ADHD were reduced to 23% during adolescence and further to 15% during adulthood. As is the case in the general population, ADHD was significantly more prevalent among males, who accounted for 65% of cases. About 63% of 22q11DS individuals who receive a diagnosis of ADHD manifest in the purely inattentive subtype, followed by 30.5% of cases of mixed inattentive-hyperactive subtype and only a minority of cases (6.5%) classified as predominantly hyperactive (Schneider et al., 2014a).

Anxiety disorders

Among children, anxiety disorders represent the second most common psychiatric manifestation with all anxiety disorder taken together having a prevalence of over 35% (Schneider et al., 2014a). The two most common types of anxiety disorders are represented by specific phobia and social anxiety. While rates of specific phobia may be comparable to those of other populations with comparable intellectual disability, increased rates of social anxiety may be a hallmark of 22q11DS (Schneider et al., 2014a). Rates of anxiety disorders remained high and stable during adolescence, while they were reduced to 24% and 25% during early and mature adulthood, respectively. All anxiety disorders taken together were furthermore more common among females. However, this difference was selectively driven by adult women, while the prevalence of anxiety disorders was equally distributed between sexes among children and adolescents. Importantly, the presence of anxiety disorders has been reported to represent a risk factor for the subsequent development of psychosis (Schneider et al., 2014b; Gothelf et al., 2007).

Autism spectrum disorder

The presence of autism spectrum disorder (ASD) in 22q11DS has been a source of considerable debate with different groups reporting prevalence ranging from 0% to 54% (Schneider et al., 2014a). It has been suggested that differences in ascertainment strategies and diagnostic standards could significantly contribute to this heterogeneity (Angkustsiri et al., 2014). Indeed, groups basing the diagnosis of ASD exclusively on self-report questionnaires report rates as high as 50% (Vorstman et al., 2006), while groups who perform a gold standard evaluation based on both parent reports and standardized observation measure report much lower rates with only a 3% prevalence of autism and 15% prevalence of ASD (Angkustsiri et al., 2014). While deficits in social interaction, characteristically manifesting with social avoidance and difficulty in initiating social interactions, are undoubtedly highly prevalent in 22q11DS, concerns have been expressed regarding the assumption that ASD is responsible for such deficits (Eliez, 2007). Indeed, compared to idiopathic ASD, social difficulties in 22q11DS manifest much later, increasing in prevalence during the course of adolescence, and lack the characteristic sex distribution of ASD with equal prevalence among males and females (Schneider et al., 2014a). Moreover, it has been suggested that such deficits in social behavior might be context-dependent and lack the pervasiveness observed in idiopathic ASD (Angkustsiri et al., 2014; Eliez, 2007). Furthermore, while social communication deficits have been described in 22q11DS (Angkustsiri et al., 2014), language impairment is typically less severe compared to idiopathic ASD (Eliez, 2007), with verbal IQ being on average higher than performance IQ (Swillen, 2016). The atypical developmental course of ASD increasing in prevalence during adolescence (Schneider et al., 2014b) has raised the suspicion that autistic-like deficits in social behavior might represent features of the psychotic prodrome in 22q11DS (Eliez, 2007). However, recent prospective and retrospective studies have failed to demonstrate an association between autistic traits and increased risk for psychosis in 22q11DS, leading the authors to conclude that two should be considered independent psychiatric manifestations of the syndrome (Fiksinski et al., 2017; Vorstman et al., 2013).

Mood disorders

Mood disorders become progressively more prevalent with age in 22q11DS, affecting more than 20% of adults, after the age of 35 years (Schneider et al., 2014a). More than 80% of mood disorders in 22q11DS are represented by major depressive disorders, while bipolar disorder affects only 3% of adult patients and does not appear to be more prevalent than in the general population. Mood disorders show a high comorbidity with anxiety disorders, and as is the case with anxiety disorders, they are also characteristically more frequent among adult females. This increase in prevalence in internalizing disorder among postpubertal females may point to the role of hormonal changes (Hayward and Sanborn, 2002).

Substance abuse

Finally, substance-related disorders are exceedingly rare among children and adolescents with 22q11DS and only affect less than 5% of adults (Schneider et al., 2014b; Tang et al., 2017). It has been proposed that deficits in social behavior, characteristic of 22q11DS, may render these individuals less likely to engage in substance abuse (Schneider et al., 2014b; Tang et al., 2017). An alternative preliminary hypothesis is that alterations in dopamine metabolism, estimated from genetic variability in the non deleted COMT allele, might be associated with reduced substance use in 22q11DS (Vingerhoets et al., 2015). The authors propose that this association might be mediated by dysfunction in reward signal processing that might render this population less vulnerable to addiction and more susceptible to negative symptoms such as motivation and anhedonia (van Duin et al., 2016). Although intriguing, this hypothesis is supported by preliminary evidence and warrants further investigation as it contrasts with the high comorbidity between schizophrenia and substance use disorder, observed in the general population.

Psychosis in 22q11DS

SSDs are among the most frequent psychiatric manifestations of 22q11DS. Indeed, since the original description of the syndrome, increased rates of SSDs have consistently reported in cohorts of patients recruited throughout the world (Gothelf et al., 2007; Monks et al., 2014; Murphy et al., 1999).

Recently, the institution of the International Consortium on Brain and Behavior in 22q11.2DS has allowed to estimate the prevalence of SSDs in a very large cohort of patients affected by the deletion (N = 1400), recruited from a total of 15 research centers, throughout Europe, North America, Australia, and the Middle East (Gur et al., 2017).

According to such collaborative work, the rate of SSDs, including schizophrenia, schizoaffective disorder, schizophreniform disorder, brief psychotic disorder, delusional disorder, and psychotic disorder not otherwise specified, was estimated at 41% among adults older than 25 years (Schneider et al., 2014a). Such prevalence represents a 20–40-fold increase in risk compared to the general population and makes 22q11DS the third highest genetic risk factor for the development of schizophrenia, preceded only by having a monozygotic twin or both parents diagnosed with schizophrenia (Gejman et al., 2010). Indeed, it has been estimated that 22q11DS is responsible for 1%–3% of all cases of schizophrenia (Bassett et al., 2010; Horowitz et al., 2005). The rate of SSDs in 22q11DS also significantly increased with age, being estimated at 2% among children (ages 6–12), at 10% among adolescents (ages 12–18) and at 23% among young adults (ages 18–24), while it remained stable at 41% after the age of 25 years (Schneider et al., 2014a). This trajectory is in accordance with longitudinal studies in 22q11DS, reporting a mean age of transition to psychosis between 17 and 23 years (Vorstman et al., 2015; Bassett et al., 2003; Gothelf et al., 2013) and furthermore is similar to what has been described in idiopathic psychosis (Millan et al., 2016).

Among SSDs, schizophrenia was the most common disorder accounting for 60% of SSDs after the age of 25, followed by psychotic disorder not otherwise specified (27% of SSDs) and schizoaffective disorder (13% of SSDs). However, psychotic disorder not otherwise specified was more frequently diagnosed among children (89% of SSDs) and adolescents (40% of SSDs), while schizoaffective disorders peaked in prevalence between 26 and 35 (16% of SSDs). Brief psychotic disorder was diagnosed almost exclusively among adolescents, when it accounted for 11% of cases of SSDs (Schneider et al., 2014a).

A key requisite for validating 22q11DS as a model for idiopathic schizophrenia is the presence of comparable clinical manifestations of the disorder in the syndrome and in the general population. In this regard, both studies that have compared symptoms of schizophrenia diagnosed in 22q11DS and in the general population did not report significant differences in the respective clinical profiles, including no difference in the severity of positive and negative symptoms (Bassett et al., 2003; Monks et al., 2014).

Moreover, several efforts have recently been devoted to characterize the clinical manifestation preceding full-blown psychosis in 22q11DS, in accordance with the clinical high-risk (CHR) model that has been developed to detect an increased risk of developing psychosis in the general population (for more details see Yung et al., 2005; Fusar-Poli et al., 2013). In this sense, several studies have described a high prevalence of subthreshold psychotic symptoms in 22q11DS. At least one subthreshold symptom of at least moderate intensity was reported in 54% of individuals with 22q11DS, with 10% presenting exclusively positive symptoms, 25% presenting both positive and negative/disorganized symptoms, and 21% presenting exclusively negative/disorganized symptoms (Weisman et al., 2017).

The prevalence of a full ultra high risk (UHR) condition (Fusar-Poli et al., 2013) has been furthermore estimated at 27% (Armando et al., 2017). Importantly, it was demonstrated that the three-solution factorial structure (i.e., positive, negative, and generalized dimensions) of the semistructured interviews developed to detect an increased risk of psychosis in the general population was conserved in 22q11DS (Schneider et al., 2012b). A three-solution factorial structure of subthreshold psychotic symptoms was also reported in an independent cohort of individuals with 22q11DS, in this case separating positive, negative, and disorganized dimension (Tang et al., 2014). Collectively, these reports provide initial evidence that clinical manifestations of prodromal psychosis are broadly comparable in 22q11DS and the general population (Tang et al., 2014; Schneider et al., 2012b). Moreover, the clinical validity of the semistructured interview for prodromal syndromes has been confirmed in 22q11DS, with UHR status predicting conversion to psychosis within 36 months in 27% of cases (Schneider et al., 2016). Such predictive value of UHR diagnosis is comparable to what described in the general population (Fusar-Poli et al., 2013) and would suggest a similar clinical trajectory toward full-blown psychosis. However, the relatively high rate (4.5%) of conversion to psychosis reported among individuals without a UHR diagnosis (false negative diagnosis) points to the importance of identifying additional predictors of vulnerability to psychosis in 22q11DS (Schneider et al., 2016).

To more precisely characterize both similarities and specificities in clinical manifestations of prodromal psychosis, groups have started to compare individuals with 22q11DS to patients at CHR recruited from the general population. A first study with this aim compared deleted to non deleted UHR patients and demonstrated similar overall and positive symptom severity, while 22q11DS subjects were characterized by higher negative symptoms, lower IQ, and more severe impairment of global functioning (Armando et al., 2012a). The authors conclude that increased representation of negative symptom represents a specific feature of the prodromal phenotype of 22q11DS that should be the focus of targeted treatment strategies (Armando et al., 2012a). A second study compared a cohort of individuals with 22q11DS to a non deleted sample that was matched for the presence of psychotic symptoms (Tang et al., 2017). In accordance with previous results, this study demonstrated higher levels of negative and generalized symptoms, such as avolition, ideational richness, and stress tolerance, as well as stronger cognitive defects and more severe impairments in global functioning in the 22q11DS group (Tang et al., 2017).

The overrepresentation and the clinical relevance of negative symptoms in 22q11DS were furthermore demonstrated by a series of studies conducted on independent cohort of subjects. Indeed, negative symptoms affected approximately 85% of adolescents and young adults, being considerably more prevalent than positive symptoms (Schneider et al., 2012b). Moreover, particularly the motivational dimension was strongly correlated with lower levels of daily life functioning (Schneider et al., 2012b). Clustering of patients according to their symptomatic profile revealed that about one-third of the individuals with 22q11DS presented predominantly negative symptoms, while the rest of the sample presented either low or high levels of both positive and negative symptoms

(Schneider et al., 2014c). Patients with predominantly negative symptoms were significantly impaired both in terms of global functioning and several cognitive domains compared to the rest of the cohort (Schneider et al., 2014c). Patients with predominantly negative symptoms were not, however, more likely to have a diagnosis of mood disorder or to be taking neuroleptic medication, suggesting that these factors cannot solely account for negative symptomatology in 22q11DS. Both the association between negative symptoms and neurocognitive deficits (Harvey et al., 2006) and the determining role of negative symptoms in predicting patient's overall outcome are in accordance with findings in the general population (Makinen et al., 2008). However, the observation that positive symptoms manifest almost exclusively in association to negative ones differs from what has been described in the general population and suggests that negative symptoms represent the predominant clinical characteristic of psychotic expression in 22q11DS.

Taken together, these considerations point to 22q11DS as a valuable model to understand particularly the negative dimension of psychotic symptomatology, which represents a major unmet medical need in idiopathic schizophrenia.

Pathophysiology of the neuropsychiatric phenotype

The recognition that 22q11DS could represent a valuable model of schizophrenia has led to growing interest in disentangling neurobiological underpinnings responsible for such increased vulnerability in this population. Indeed, it has been argued that particularly neurodevelopmental mechanisms that are currently implicated in the pathophysiology of schizophrenia might be more accessible in 22q11DS than in the general population, as patients with the syndrome are typically diagnosed at a younger age and independently of psychotic symptoms (Insel, 2010).

Evidence from neuroimaging studies in humans

Neuroimaging tools allow to non invasively investigate both structural and functional features of the developing human brain, and several neuroimaging tools have been employed to probe the pathophysiology of 22q11DS.

Alterations in brain morphology, as described using magnetic resonance imaging (MRI), were among the first features of the brain phenotype to be described in 22q11DS. Overall brain volume was found to be significantly reduced in the syndrome, with reductions affecting disproportionately white matter and parietooccipital and medial gray matter, with a posterior to anterior and medial to lateral gradient of volumetric reductions (Eliez et al., 2000). Moreover, the syndrome is characterized by increased ventricular volume that represents the single most replicated neuroimaging finding in idiopathic schizophrenia (Chow et al., 2002). Importantly, alterations in brain morphology have been related to features of the neurocognitive and psychiatric phenotype. For instance, reduced gray matter volume of the anterior cingulate cortex (ACC) has been related to the presence of psychotic symptoms and to more severe impairments in executive functioning (Dufour et al., 2008).

Given current theories implicating neurodevelopmental processes in the pathogenesis of psychosis, neuroimaging studies are starting to investigate brain maturation in longitudinal cohorts of children and adolescents affected by the deletion. A first landmark study on this topic demonstrated aberrant developmental trajectories of cortical maturation, which were particularly striking at the level of the prefrontal and ACC. Indeed, youth affected by the deletion presented increased cortical thickness during childhood followed by accelerated cortical thinning during adolescence (Schaer et al., 2009). Moreover, reduced cortical thickness was related to the presence of psychosis (Schaer et al., 2009). Acceleration of the normal process of adolescent cortical thinning is also a highly replicated finding in idiopathic psychosis and is thought to arise from a process of excessive synaptic pruning particularly in prefrontal cortices (Cannon et al., 2015). Findings on aberrant cortical thinning in relation to psychosis have since been replicated in independent cohorts of adolescents with 22q11DS (Ramanathan et al., 2016) and provide evidence of a shared neurodevelopmental pathophysiology with idiopathic schizophrenia (Cannon et al., 2015).

Aside from the study of cortical morphology, MRI allows to probe features of structural and functional brain connectivity. Since Bleuler's original definition, it has been proposed that schizophrenia might arise from altered communication between separate brain regions. Several studies have demonstrated widespread microstructural alterations of particularly long-range and midline white matter tracts in 22q11DS, which impacted on the overall architecture of white matter networks (Scariati et al., 2016). Moreover, a recent study demonstrated that structural connectivity alterations of the ACC, as estimated from structural covariance, were specifically related to the presence of prodromal psychotic symptoms (Sandini et al., 2017). Several studies have also investigated functional connectivity between brain regions in 22q11DS using functional MRI. Indeed, two studies conducted on independent cohorts demonstrated reductions in functional connectivity in a brain network connecting anterior and posterior cingulate cortex, known as the default mode network (DMN) (Schreiner et al., 2014; Padula et al., 2015). The DMN is composed of a series of brain regions that are characteristically more active during rest and deactivate when a subject engages in goal-directed behavior (Broyd et al., 2009). The DMN has been implicated in mind wandering and self-referential cognitive processes, and altered functional connectivity of this network has been consistently reported in idiopathic schizophrenia (Broyd et al., 2009).

Among regions classically included in the DMN, dysfunction of the ACC has been specifically related the presence of psychosis in 22q11DS. Indeed, the connectivity pattern of the ACC helped discriminate patients with 22q11DS presenting moderate to severe psychotic symptoms from nonsymptomatic patients (Scariati et al., 2014). Moreover, the activation pattern of the ACC during self-referential cognitive processing was significantly correlated to the severity of prodromal psychotic symptoms (Schneider et al., 2012a). Aberrant activation of the dorsal portion of the ACC was furthermore reported using EEG in 22q11DS and was once again correlated to the severity of prodromal psychotic symptoms (Tomescu et al., 2014; Rihs et al., 2013).

From the perspective of pathophysiology, the dorsal portion of the ACC together with the insula is thought to be critically involved in the attribution of subjective salience to both internally and externally generated stimuli (Kapur, 2003). It has been proposed that disturbances in this process of salience attribution, linked to structural and functional alterations of the ACC, might be the common denominator of several positive psychotic symptoms such as hallucinations and delusions (Kapur, 2003).

Interestingly, recent evidence suggests that functional alterations of the ACC might arise from aberrant developmental trajectories during adolescence in 22q11DS, specifically affecting patients presenting with psychotic symptoms (Zoller et al., 2017). Such findings provide further evidence of the relationship between aberrant neurodevelopment and the pathophysiology of psychosis in 22q11DS. Moreover, they suggest that therapeutic interventions targeted at normalizing structural and functional maturation of the ACC during childhood and adolescents might influence the trajectory toward psychosis. Indeed, several treatment strategies currently proposed to reduce transition to psychosis (Nelson et al., 2017), including cognitive behavioral therapy (Yoshimura et al., 2017), and both antidepressant (Faria et al., 2014) and neuroleptic medication (Sarpal et al., 2015) have been shown to act by modifying functional connectivity of the ACC.

Evidence from animal models

Neuroimaging studies have yielded important insights regarding candidate biomarkers and neurodevelopmental mechanisms contributing to the pathogenesis of psychosis in 22q11DS. Still neuroimaging tools are limited in terms of their ability to discern the precise neurobiological mechanisms underlying such candidate biomarkers, as they offer a relatively coarse and indirect representation of neural functioning and structure. Moreover, the earliest stages of neurodevelopment occurring in utero are to date largely inaccessible to neuroimaging studies. An in-depth understanding of the underlying developmental neurobiology and the ability to experientially manipulate candidate pathophysiological mechanisms are going to be essential for the development of novel treatment interventions (Insel, 2010). Compared to idiopathic psychosis, a unique aspect of 22q11DS is the availability of a well-defined mouse model that carries a homologous genetic deletion, known as a long deletion or LgDel mouse model (Meechan et al., 2015a). Importantly, while psychotic symptoms are not quantifiable in animal models, mice carrying the deletion present several aspects of the neurocognitive phenotype observed in 22q11DS (Meechan et al., 2015a). For instance, it has been shown LgDel mice show characteristic impairments in executive cognition and cognitive flexibility that mirror findings in both 22q11DS and idiopathic schizophrenia (Meechan et al., 2015b). Studies are beginning to disentangle the neurobiological and neurodevelopmental mechanisms that might be accounting for such neurocognitive deficits.

Early neurodevelopmental events appear to be already critically affected in mouse models of 22q11DS, with alterations in the migration of and cortical distribution of parvalbumin positive inhibitory interneurons (Meechan et al., 2012). This particular class of neurons is critical for the synchronization of neuronal activity and has been highly implicated in idiopathic schizophrenia (Nakazawa et al., 2012). Moreover, mice with the deletion present altered neurogenesis of pyramidal neurons destined to layers II–III of the cortex (Meechan et al., 2015b). Interestingly, neuronal deficits, specifically at the level of the mouse cortical region corresponding to the human ACC, have been related to deficits in executive functions and mental flexibility (Meechan et al., 2015b). Such finding replicates correlations between volumetric deficits of the ACC and executive dysfunction observed in 22q11DS and points to a shared neurodevelopmental pathophysiology of executive impairment (Meechan et al., 2015a; Dufour et al., 2008).

Aside from disturbances occurring during early development, studies suggest that trajectories of late maturation are also altered in LgDel mouse models. Indeed typical development is characterized by progressive stabilization of functionally relevant synaptic connections that is critical for mediating learning (Moutin et al., 2016). On the other hand, LgDel mouse models are deficient in this stabilization process leading to an increased prevalence of immature and fragile synapses, particularly at the level of the hippocampus (Moutin et al., 2016). The resulting synaptic immaturity could render such connections particularly vulnerable to a process of aberrant pruning during adolescence that has been highly implicated in the pathophysiology of idiopathic schizophrenia (Cannon et al., 2015). Intriguingly, genetic rescue of a of the zdhhc8 enzyme, involved in the palmitoylation of synaptic proteins, improved stability of synaptic connections in mice, potentially opening avenues for novel treatment strategies (Moutin et al., 2016).

Finally, in accordance with the dysconnectivity hypothesis of schizophrenia, findings in mouse models of 22q11DS suggest that part of the neuropsychiatric phenotype emerges from altered communication between separate brain regions (Sigurdsson et al., 2010). Indeed, mice harboring the 22q11 deletion presented reduced theta-band synchronization between the hippocampus and the prefrontal cortex that was responsible for behavioral deficits in spatial working memory (Sigurdsson et al., 2010). Such findings replicate correlations between long-range dysconnectivity and executive dysfunction observed in both 22q11DS and idiopathic schizophrenia (Sheffield and Barch, 2016; Padula et al., 2017). Research in animal models of 22q11DS is still underway to characterize precise neurobiological mechanisms responsible for such long-range dysconnectivity, as well as treatment strategies that might ameliorate both dysconnectivity and cognitive impairment. Importantly, cognitive symptoms are the most important contributors to functional impairments in idiopathic psychosis and are largely unresponsive to current treatments (Kahn and Keefe, 2013).

Conclusion

Findings presented in this review collectively highlight the value of 22q11.2DS as a model to investigate both clinical and neurodevelopmental mechanisms contributing to the pathophysiology of psychosis. However, several key questions warrant further investigation. Importantly, it should be noted that, given the syndrome's overall prevalence, research in 22q11DS has until recently been constrained by availability of longitudinal samples of sufficient size and with sufficient long-term follow-up. The institution of the International Consortium on Brain and Behavior in 22q11.2DS (ICBB) (Gur et al., 2017), together with the maturation of several single-center longitudinal cohorts throughout the world is likely to open several promising avenues for future research.

An issue that has been largely neglected in the 22q11DS literature to date is the role of environmental risk factors contributing to the emergence of the neuropsychiatric phenotype. Indeed, a recurrent theme with the syndrome's phenotype is the presence of considerable variability across subjects, even among monozygotic twins (Vergaelen et al., 2015; Singh et al., 2002). It is reasonable to assume that at least a proportion of this variability can be accounted for by environmental factors, possibly through the mediation of epigenetic mechanisms. Indeed, life events, such as exposure to stress or cannabis consumption, have been highly implicated in the pathogenesis of idiopathic psychosis (Insel, 2010). Still, disentangling how the environment interacts with genetic predisposition has proved difficult in the general population due to confounding factors such as pleiotropy or reverse causation (Owen et al., 2016). In this sense, investigating a population where the genetic load for schizophrenia is high and largely homogenous across subjects could help dissect the contributing role of environmental risk factors and their mechanism of interaction with genetic predisposition.

It has also been argued that a considerable source of variability in the syndrome's neuropsychiatric phenotype might arise from genetic variance both within and outside the 22q11 locus (Gur et al., 2017). However, until recently, studies have been largely underpowered to detect such effects. Within the framework of the ICBB, whole-genome sequencing is currently available for more than 1600 individuals with 22q11DS who have been furthermore precisely characterized from a clinical and neuropsychiatric perspective (Gur et al., 2017). This rich dataset is already contributing important insights, demonstrating for instance an association between schizophrenia and enriched copy number variations in 22q11DS (Bassett et al., 2017). A more detailed characterization of genetic and epigenetic factors influencing the neuropsychiatric phenotype of 22q11DS is likely to be a key topic of future research.

From a clinical perspective, most research to date has focused on characterizing the syndrome's positive symptom profile. However, recent studies suggest that subthreshold positive symptoms might offer only an incomplete representation of clinical vulnerability to psychosis in 22q11DS, given the relatively high rate of false negative UHR diagnoses (Schneider et al., 2016). Larger longitudinal cohorts will allow identifying additional risk factors, potentially characterizing the earliest clinical stages of the trajectory toward psychosis, which are mostly inaccessible in the general population. For instance, research conducted by the IBBC has revealed that verbal cognitive decline can predate the onset of psychosis by several years (Vorstman et al., 2015). Neurodevelopmental underpinnings of such atypical cognitive maturation, however, still remain unclear. In a more general sense, a shift in focus from psychotic symptoms to a broader representation of clinical factors contributing psychiatric morbidity in 22q11DS is likely to characterize future research in the field. For instance, cross-sectional studies have suggested that negative symptoms are a major predictor of patients' overall functional outcome in 22q11DS (Schneider et al., 2012b), in accordance with what has been reported in idiopathic schizophrenia (Makinen et al., 2008). Future longitudinal work will allow characterizing longitudinal trajectories of negative symptoms in 22q11DS and might furthermore help identify factors that predict their emergence and that might be amenable to treatment. For instance, impairments in social cognition have been implicated in the emergence of several aspects of the clinical profile in 22q11DS (Schneider et al., 2017; Glaser et al., 2010). These impairments might benefit from cognitive remediation strategies that have already been implemented in pediatric populations with developmental disabilities (Glaser et al., 2012).

Finally, a unique aspect of 22q11DS is the availability of a genetically homologous animal model that has already contributed important insights regarding neurobiological mechanisms responsible for aspects of the syndrome's neuropsychiatric phenotype (Meechan et al., 2015a). Important avenues of future research will likely involve the reinforcement of translational paradigms between mice and human subjects. For instance, a recent study combined neuroimaging in mice and humans to demonstrate that a specific polymorphism of COMT gene exerted a common effect on cortical morphology and executive functions in postpubertal females selectively (Sannino et al., 2017). More generally, future translational research should focus on understanding how trajectories of late peripubertal circuit maturation differ in developing mice and humans carrying the 22q11.2 deletion. Indeed, converging evidence has demonstrated that adolescence is a critical period for the emergence of psychotic symptoms and cognitive decline both in 22q11DS (Vorstman et al., 2015) and in the general population (Marin, 2016). A precise characterization of the underlying neurodevelopmental mechanisms through translational research will be crucial for the development of treatments that have a better hope of modifying the trajectory toward psychosis.

Collectively, current research on 22q11DS, reviewed in this chapter, has largely focused on validating the syndrome as a model for the study of psychosis, both through clinical characterization of patients at different disease stages and by describing candidate neurodevelopmental mechanisms that mirror what has been observed in the general population. This first validation stage is a prerequisite for any finding in 22q11DS to be eventually relevant to psychosis in general. As genetic, environmental, and neurodevelopmental mechanisms leading to psychosis in 22q11DS become increasing well-characterized, the next exciting phase of research will likely focus on the development of novel treatment strategies. In this sense work in animal models of 22q11DS is already moving from the description of circuit dysfunction to the implementation of treatment strategies aimed at normalizing behavioral and neural biomarkers (Moutin et al., 2016; Marissal et al., 2018). While such efforts are to date admittedly far from having direct clinical relevance, they do suggest that 22q11DS might be ideally suited for the future implementation of a "bench to bedside" approach to treatment development in the field of psychosis. Moreover, aside from the discovery of novel treatments, it has been argued that development timing might potentially strongly moderate the effectiveness of currently available interventions (Marin, 2016). 22q11DS could serve as a model to test this hypothesis given that patients are both typically diagnosed at a young age and at very high risk for psychosis, justifying the implementation of preventive interventions. For instance, a clinical trial is currently underway to test the effectiveness of dietary supplementation of omega-3 polyunsaturated fatty acids on conversion to psychosis in 22q11DS (Armando et al., 2016).

References

Andersson, F., Glaser, B., Spiridon, M., Debbane, M., Vuilleumier, P., Eliez, S., 2008. Impaired activation of face processing networks revealed by functional magnetic resonance imaging in 22q11.2 deletion syndrome. Biol. Psychiatry 63, 49–57.

Angkustsiri, K., Goodlin-Jones, B., Deprey, L., Brahmbhatt, K., Harris, S., Simon, T.J., 2014. Social impairments in chromosome 22q11.2 deletion syndrome (22q11.2DS): autism spectrum disorder or a different endophenotype? J. Autism Dev. Disord. 44, 739–746.

Armando, M., De Crescenzo, F., Vicari, S., Digilio, M.C., Pontillo, M., Papaleo, F., Amminger, G.P., 2016. Indicated prevention with long-chain polyunsaturated omega-3 fatty acids in patients with 22q11DS genetically at high risk for psychosis. Protocol of a randomized, double-blind, placebo-controlled treatment trial. Early Interv. Psychiatry 10, 390–396.

Armando, M., Girardi, P., Vicari, S., Menghini, D., Digilio, M.C., Pontillo, M., Saba, R., Mazzone, L., Lin, A., Klier, C.M., Schäfer, M.R., Amminger, G.P., 2012a. Adolescents at ultra-high risk for psychosis with and without 22q11 deletion syndrome: a comparison of prodromal psychotic symptoms and general functioning. Schizophr. Res. 139, 151–156.

Armando, M., Papaleo, F., Vicari, S., 2012b. COMT implication in cognitive and psychiatric symptoms in chromosome 22q11 microdeletion syndrome: a selective review. CNS Neurol. Disord. Drug Targets 11, 273–281.

Armando, M., Schneider, M., Pontillo, M., Vicari, S., Debbane, M., Schultze-Lutter, F., Eliez, S., 2017. No age effect in the prevalence and clinical significance of ultra-high risk symptoms and criteria for psychosis in 22q11 deletion syndrome: confirmation of the genetically driven risk for psychosis? PLoS One 12, e0174797.

Bassett, A.S., Chow, E.W., Abdelmalik, P., Gheorghiu, M., Husted, J., Weksberg, R., 2003. The schizophrenia phenotype in 22q11 deletion syndrome. Am. J. Psychiatry 160, 1580–1586.

Bassett, A.S., Costain, G., Fung, W.L., Russell, K.J., Pierce, L., Kapadia, R., Carter, R.F., Chow, E.W., Forsythe, P.J., 2010. Clinically detectable copy number variations in a Canadian catchment population of schizophrenia. J. Psychiatr. Res. 44, 1005–1009.

Bassett, A.S., Lowther, C., Merico, D., Costain, G., Chow, E.W.C., van amelsvoort, T., Mcdonald-Mcginn, D., Gur, R.E., Swillen, A., Van Den Bree, M., Murphy, K., Gothelf, D., Bearden, C.E., Eliez, S., Kates, W., Philip, N., Sashi, V., Campbell, L., Vorstman, J., Cubells, J., Repetto, G.M., Simon, T., Boot, E., Heung, T., Evers, R., Vingerhoets, C., Van Duin, E., Zackai, E., Vergaelen, E., Devriendt, K., Vermeesch, J.R., Owen, M., Murphy, C., Michaelovosky, E., Kushan, L., Schneider, M., Fremont, W., Busa, T., Hooper, S., Mccabe, K., Duijff, S., Isaev, K., Pellecchia, G., Wei, J., Gazzellone, M.J., Scherer, S.W., Emanuel, B.S., Guo, T., Morrow, B.E., Marshall, C.R., International 22q, D.S.B., Behavior, C., 2017. Rare genome-wide copy number variation and expression of schizophrenia in 22q11.2 deletion syndrome. Am. J. Psychiatry 174, 1054–1063.

Broyd, S.J., Demanuele, C., Debener, S., Helps, S.K., James, C.J., Sonuga-Barke, E.J., 2009. Default-mode brain dysfunction in mental disorders: a systematic review. Neurosci. Biobehav. Rev. 33, 279–296.

Buckley, P.F., Miller, B.J., Lehrer, D.S., Castle, D.J., 2009. Psychiatric comorbidities and schizophrenia. Schizophr. Bull. 35, 383–402.

Campbell, L., Mccabe, K., Leadbeater, K., Schall, U., Loughland, C., Rich, D., 2010a. Visual scanning of faces in 22q11.2 deletion syndrome: attention to the mouth or the eyes? Psychiatry Res. 177, 211–215.

Campbell, L.E., Azuma, R., Ambery, F., Stevens, A., Smith, A., Morris, R.G., Murphy, D.G., Murphy, K.C., 2010b. Executive functions and memory abilities in children with 22q11.2 deletion syndrome. Aust. N. Z. J. Psychiatr. 44, 364–371.

Campbell, L.E., Stevens, A.F., Mccabe, K., Cruickshank, L., Morris, R.G., Murphy, D.G., Murphy, K.C., 2011. Is theory of mind related to social dysfunction and emotional problems in 22q11.2 deletion syndrome (velo-cardio-facial syndrome)? J. Neurodev. Disord. 3, 152–161.

Cannon, T.D., Chung, Y., He, G., Sun, D., Jacobson, A., Van Erp, T.G., Mcewen, S., Addington, J., Bearden, C.E., Cadenhead, K., Cornblatt, B., Mathalon, D.H., Mcglashan, T., Perkins, D., Jeffries, C., Seidman, L.J., Tsuang, M., Walker, E., Woods, S.W., Heinssen, R., North American Prodrome Longitudinal Study, C., 2015. Progressive reduction in cortical thickness as psychosis develops: a multisite longitudinal neuroimaging study of youth at elevated clinical risk. Biol. Psychiatry 77, 147–157.

Caspi, A., Houts, R.M., Belsky, D.W., Goldman-Mellor, S.J., Harrington, H., Israel, S., Meier, M.H., Ramrakha, S., Shalev, I., Poulton, R., Moffitt, T.E., 2014. The p factor: one general psychopathology factor in the structure of psychiatric disorders? Clin. Psychol. Sci. 2, 119–137.

Chow, E.W., Zipursky, R.B., Mikulis, D.J., Bassett, A.S., 2002. Structural brain abnormalities in patients with schizophrenia and 22q11 deletion syndrome. Biol. Psychiatry 51, 208–215.

Cioffi, S., Martucciello, S., Fulcoli, F.G., Bilio, M., Ferrentino, R., Nusco, E., Illingworth, E., 2014. Tbx1 regulates brain vascularization. Hum. Mol. Genet. 23, 78–89.

De Smedt, B., Devriendt, K., Fryns, J.P., Vogels, A., Gewillig, M., Swillen, A., 2007. Intellectual abilities in a large sample of children with Velo-Cardio-Facial Syndrome: an update. J. Intellect. Disabil. Res. 51, 666–670.

Debbane, M., Glaser, B., GEX-Fabry, M., Eliez, S., 2005. Temporal perception in velo-cardio-facial syndrome. Neuropsychologia 43, 1754–1762.

Delio, M., Guo, T., Mcdonald-Mcginn, D.M., Zackai, E., Herman, S., Kaminetzky, M., Higgins, A.M., Coleman, K., Chow, C., Jalbrzikowski, M., Bearden, C.E., Bailey, A., Vangkilde, A., Olsen, L., Olesen, C., Skovby, F., Werge, T.M., Templin, L., Busa, T., Philip, N., Swillen, A., Vermeesch, J.R., Devriendt, K., Schneider, M., Dahoun, S., Eliez, S., Schoch, K., Hooper, S.R., Shashi, V., Samanich, J., Marion, R., Van Amelsvoort, T., Boot, E., Klaassen, P., Duijff, S.N., Vorstman, J., Yuen, T., Silversides, C., Chow, E., Bassett, A., Frisch, A., Weizman, A., Gothelf, D., Niarchou, M., Van Den Bree, M., Owen, M.J., Suner, D.H., Andreo, J.R., Armando, M., Vicari, S., Digilio, M.C., Auton, A., Kates, W.R., Wang, T., Shprintzen, R.J., Emanuel, B.S., Morrow, B.E., 2013. Enhanced maternal origin of the 22q11.2 deletion in velocardiofacial and DiGeorge syndromes. Am. J. Hum. Genet. 92, 439–447.

Diamond, A., 2013. Executive functions. Annu. Rev. Psychol. 64, 135–168.

Digeorge, A., 1965. Discussion on a new concept of cellular immunology. J. Pediatr. 65, 907.

Driscoll, D.A., 2006. Molecular and genetic aspects of DiGeorge/velocardiofacial syndrome. Methods Mol. Med. 126, 43–55.

Dufour, F., Schaer, M., Debbane, M., Farhoumand, R., Glaser, B., Eliez, S., 2008. Cingulate gyral reductions are related to low executive functioning and psychotic symptoms in 22q 11.2 deletion syndrome. Neuropsychologia 46, 2986–2992.

Dyce, O., Mcdonald-Mcginn, D., Kirschner, R.E., Zackai, E., Young, K., Jacobs, I.N., 2002. Otolaryngologic manifestations of the 22q11.2 deletion syndrome. Arch. Otolaryngol. Head Neck Surg. 128, 1408–1412.

Edelmann, L., Pandita, R.K., Morrow, B.E., 1999. Low-copy repeats mediate the common 3-Mb deletion in patients with velo-cardio-facial syndrome. Am. J. Hum. Genet. 64, 1076–1086.

Eliez, S., 2007. Autism in children with 22q11.2 deletion syndrome. J. Am. Acad. Child Adolesc. Psychiatry 46, 433–434 author reply 434-4.

Eliez, S., Blasey, C.M., Menon, V., White, C.D., Schmitt, J.E., Reiss, A.L., 2001. Functional brain imaging study of mathematical reasoning abilities in velocardiofacial syndrome (del22q11.2). Genet. Med. 3, 49–55.

Eliez, S., Schmitt, J.E., White, C.D., Reiss, A.L., 2000. Children and adolescents with velocardiofacial syndrome: a volumetric MRI study. Am. J. Psychiatry 157, 409–415.

Faria, V., Ahs, F., Appel, L., Linnman, C., Bani, M., Bettica, P., Pich, E.M., Wahlstedt, K., Fredrikson, M., Furmark, T., 2014. Amygdala-frontal couplings characterizing SSRI and placebo response in social anxiety disorder. Int. J. Neuropsychopharmacol. 17, 1149–1157.

Fiksinski, A.M., Breetvelt, E.J., Duijff, S.N., Bassett, A.S., Kahn, R.S., Vorstman, J.A.S., 2017. Autism Spectrum and psychosis risk in the 22q11.2 deletion syndrome. Findings from a prospective longitudinal study. Schizophr. Res. 188, 59–62.

Freedman, D., Brown, A.S., 2011. The developmental course of executive functioning in schizophrenia. Int. J. Dev. Neurosci. 29, 237–243.

Fusar-Poli, P., Borgwardt, S., Bechdolf, A., Addington, J., Riecher-Rossler, A., Schultze-Lutter, F., Keshavan, M., Wood, S., Ruhrmann, S., Seidman, L.J., Valmaggia, L., Cannon, T., Velthorst, E., De Haan, L., Cornblatt, B.,

Bonoldi, I., Birchwood, M., Mcglashan, T., Carpenter, W., Mcgorry, P., Klosterkotter, J., Mcguire, P., Yung, A., 2013. The psychosis high-risk state: a comprehensive state-of-the-art review. JAMA Psychiatry 70, 107–120.

Gabriel Mounir, D., Debbane, M., Schaer, M., Glaser, B., Eliez, S., 2011. Time processing in the velo-cardio-facial syndrome (22q11) and its link with the caudate nucleus. Encephale 37 (Suppl. 1), S42–S49.

Gejman, P.V., Sanders, A.R., Duan, J., 2010. The role of genetics in the etiology of schizophrenia. Psychiatr. Clin. N. Am. 33, 35–66.

Giardino, G., Cirillo, E., Maio, F., Gallo, V., Esposito, T., Naddei, R., Grasso, F., Pignata, C., 2014. Gastrointestinal involvement in patients affected with 22q11.2 deletion syndrome. Scand. J. Gastroenterol. 49, 274–279.

Glaser, B., Debbane, M., Ottet, M.C., Vuilleumier, P., Zesiger, P., Antonarakis, S.E., Eliez, S., 2010. Eye gaze during face processing in children and adolescents with 22q11.2 deletion syndrome. J. Am. Acad. Child Adolesc. Psychiatry 49, 665–674.

Glaser, B., Lothe, A., Chabloz, M., Dukes, D., Pasca, C., Redoute, J., Eliez, S., 2012. Candidate socioemotional remediation program for individuals with intellectual disability. Am. J. Intellect. Dev. Disabil. 117, 368–383.

Goldmuntz, E., Driscoll, D., Budarf, M.L., Zackai, E.H., Mcdonald-Mcginn, D.M., Biegel, J.A., Emanuel, B.S., 1993. Microdeletions of chromosomal region 22q11 in patients with congenital conotruncal cardiac defects. J. Med. Genet. 30, 807–812.

Gothelf, D., Feinstein, C., Thompson, T., Gu, E., Penniman, L., Van Stone, E., Kwon, H., Eliez, S., Reiss, A.L., 2007. Risk factors for the emergence of psychotic disorders in adolescents with 22q11.2 deletion syndrome. Am. J. Psychiatry 164, 663–669.

Gothelf, D., Schneider, M., Green, T., Debbane, M., Frisch, A., Glaser, B., Zilkha, H., Schaer, M., Weizman, A., Eliez, S., 2013. Risk factors and the evolution of psychosis in 22q11.2 deletion syndrome: a longitudinal 2-site study. J. Am. Acad. Child Adolesc. Psychiatry 52, 1192–1203.e3.

Green, M.F., Horan, W.P., Lee, J., 2015. Social cognition in schizophrenia. Nat. Rev. Neurosci. 16, 620–631.

Guna, A., Butcher, N.J., Bassett, A.S., 2015. Comparative mapping of the 22q11.2 deletion region and the potential of simple model organisms. J. Neurodev. Disord. 7, 18.

Gur, R.C., Calkins, M.E., Satterthwaite, T.D., Ruparel, K., Bilker, W.B., Moore, T.M., Savitt, A.P., Hakonarson, H., Gur, R.E., 2014. Neurocognitive growth charting in psychosis spectrum youths. JAMA Psychiatry 71, 366–374.

Gur, R.E., Bassett, A.S., Mcdonald-Mcginn, D.M., Bearden, C.E., Chow, E., Emanuel, B.S., Owen, M., Swillen, A., Van Den Bree, M., Vermeesch, J., Vorstman, J.A.S., Warren, S., Lehner, T., Morrow, B., 2017. A neurogenetic model for the study of schizophrenia spectrum disorders: the International 22q11.2 Deletion Syndrome Brain Behavior Consortium. Mol. Psychiatry 22, 1664–1672.

Harvey, P.D., Koren, D., Reichenberg, A., Bowie, C.R., 2006. Negative symptoms and cognitive deficits: what is the nature of their relationship? Schizophr. Bull. 32, 250–258.

Hayward, C., Sanborn, K., 2002. Puberty and the emergence of gender differences in psychopathology. J. Adolesc. Health 30, 49–58.

Horowitz, A., Shifman, S., Rivlin, N., Pisante, A., Darvasi, A., 2005. A survey of the 22q11 microdeletion in a large cohort of schizophrenia patients. Schizophr. Res. 73, 263–267.

Insel, T., Cuthbert, B., Garvey, M., Heinssen, R., Pine, D.S., Quinn, K., Sanislow, C., Wang, P., 2010. Research domain criteria (RDoC): toward a new classification framework for research on mental disorders. Am. J. Psychiatry 167, 748–751.

Insel, T.R., 2010. Rethinking schizophrenia. Nature 468, 187–193.

Jalbrzikowski, M., Carter, C., Senturk, D., Chow, C., Hopkins, J.M., Green, M.F., Galvan, A., Cannon, T.D., Bearden, C.E., 2012. Social cognition in 22q11.2 microdeletion syndrome: relevance to psychosis? Schizophr. Res. 142, 99–107.

Kahn, R.S., Keefe, R.S., 2013. Schizophrenia is a cognitive illness: time for a change in focus. JAMA Psychiatry 70, 1107–1112.

Kapur, S., 2003. Psychosis as a state of aberrant salience: a framework linking biology, phenomenology, and pharmacology in schizophrenia. Am. J. Psychiatry 160, 13–23.

Karayiorgou, M., Simon, T.J., Gogos, J.A., 2010. 22q11.2 microdeletions: linking DNA structural variation to brain dysfunction and schizophrenia. Nat. Rev. Neurosci. 11, 402–416.

Kremen, W.S., Vinogradov, S., Poole, J.H., Schaefer, C.A., Deicken, R.F., Factor-Litvak, P., Brown, A.S., 2010. Cognitive decline in schizophrenia from childhood to midlife: a 33-year longitudinal birth cohort study. Schizophr. Res. 118, 1–5.

Lajiness-O'neill, R., Beaulieu, I., Asamoah, A., Titus, J.B., Bawle, E., Ahmad, S., Kirk, J.W., Pollack, R., 2006. The neuropsychological phenotype of velocardiofacial syndrome (VCFS): relationship to psychopathology. Arch. Clin. Neuropsychol. 21, 175–184.

Lilienfeld, S.O., Treadway, M.T., 2016. Clashing diagnostic approaches: DSM-ICD versus RDoC. Annu. Rev. Clin. Psychol. 12, 435–463.

Lindsay, E.A., Vitelli, F., Su, H., Morishima, M., Huynh, T., Pramparo, T., Jurecic, V., Ogunrinu, G., Sutherland, H.F., Scambler, P.J., Bradley, A., Baldini, A., 2001. Tbx1 haploinsufficieny in the DiGeorge syndrome region causes aortic arch defects in mice. Nature 410, 97–101.

Maeder, J., Schneider, M., Bostelmann, M., Debbane, M., Glaser, B., Menghetti, S., Schaer, M., Eliez, S., 2016. Developmental trajectories of executive functions in 22q11.2 deletion syndrome. J. Neurodev. Disord. 8, 10.

Makinen, J., Miettunen, J., Isohanni, M., Koponen, H., 2008. Negative symptoms in schizophrenia: a review. Nord. J. Psychiatry 62, 334–341.

Marin, O., 2016. Developmental timing and critical windows for the treatment of psychiatric disorders. Nat. Med. 22, 1229–1238.

Marino, B., Digilio, M.C., Toscano, A., Giannotti, A., Dallapiccola, B., 1999. Congenital heart defects in patients with DiGeorge/velocardiofacial syndrome and del22q11. Genet. Couns. 10, 25–33.

Marissal, T., Salazar, R.F., Bertollini, C., Mutel, S.D.E., Roo, M., Rodriguez, I., Muller, D., Carleton, A., 2018. Restoring wild-type-like CA1 network dynamics and behavior during adulthood in a mouse model of schizophrenia. Nat Neurosci 21, 1412–1420.

Mcdonald-Mcginn, D.M., Sullivan, K.E., Marino, B., Philip, N., Swillen, A., Vorstman, J.A., Zackai, E.H., Emanuel, B.S., Vermeesch, J.R., Morrow, B.E., Scambler, P.J., Bassett, A.S., 2015. 22q11.2 deletion syndrome. Nat. Rev. Dis. Primers 1, 15071.

Mcdonald-Mcginn, D.M., Tonnesen, M.K., Laufer-Cahana, A., Finucane, B., Driscoll, D.A., Emanuel, B.S., Zackai, E.H., 2001. Phenotype of the 22q11.2 deletion in individuals identified through an affected relative: cast a wide FISHing net! Genet. Med. 3, 23–29.

Mcguffin, P., Owen, M.J., Farmer, A.E., 1995. Genetic basis of schizophrenia. Lancet 346, 678–682.

Meechan, D.W., Maynard, T.M., Tucker, E.S., Fernandez, A., Karpinski, B.A., Rothblat, L.A., Lamantia, A.S., 2015a. Modeling a model: mouse genetics, 22q11.2 Deletion Syndrome, and disorders of cortical circuit development. Prog. Neurobiol. 130, 1–28.

Meechan, D.W., Rutz, H.L., Fralish, M.S., Maynard, T.M., Rothblat, L.A., Lamantia, A.S., 2015b. Cognitive ability is associated with altered medial frontal cortical circuits in the LgDel mouse model of 22q11.2DS. Cerebr. Cortex 25, 1143–1151.

Meechan, D.W., Tucker, E.S., Maynard, T.M., Lamantia, A.S., 2012. Cxcr4 regulation of interneuron migration is disrupted in 22q11.2 deletion syndrome. Proc. Natl. Acad. Sci. USA 109, 18601–18606.

Mihailov, A., Padula, M.C., Scariati, E., Schaer, M., Schneider, M., Eliez, S., 2017. Morphological brain changes associated with negative symptoms in patients with 22q11.2 Deletion Syndrome. Schizophr. Res. 188, 52–58.

Millan, M.J., Andrieux, A., Bartzokis, G., Cadenhead, K., Dazzan, P., Fusar-Poli, P., Gallinat, J., Giedd, J., Grayson, D.R., Heinrichs, M., Kahn, R., Krebs, M.O., Leboyer, M., Lewis, D., Marin, O., Marin, P., Meyer-Lindenberg, A., Mcgorry, P., Mcguire, P., Owen, M.J., Patterson, P., Sawa, A., Spedding, M., Uhlhaas, P., Vaccarino, F., Wahlestedt, C., Weinberger, D., 2016. Altering the course of schizophrenia: progress and perspectives. Nat. Rev. Drug Discov. 15, 485–515.

Monks, S., Niarchou, M., Davies, A.R., Walters, J.T., Williams, N., Owen, M.J., Van Den Bree, M.B., Murphy, K.C., 2014. Further evidence for high rates of schizophrenia in 22q11.2 deletion syndrome. Schizophr. Res. 153, 231–236.

Morsheimer, M., Brown Whitehorn, T.F., Heimall, J., Sullivan, K.E., 2017. The immune deficiency of chromosome 22q11.2 deletion syndrome. Am. J. Med. Genet. 173, 2366–2372.

Moss, E.M., Batshaw, M.L., Solot, C.B., Gerdes, M., Mcdonald-Mcginn, D.M., Driscoll, D.A., Emanuel, B.S., Zackai, E.H., Wang, P.P., 1999. Psychoeducational profile of the 22q11.2 microdeletion: a complex pattern. J. Pediatr. 134, 193–198.

Moutin, E., Nikonenko, I., Stefanelli, T., Wirth, A., Ponimaskin, E., De Roo, M., Muller, D., 2016. Palmitoylation of cdc42 promotes spine stabilization and rescues spine density deficit in a mouse model of 22q11.2 deletion syndrome. Cerebr. Cortex 3618–3629.

Murphy, K.C., Jones, L.A., Owen, M.J., 1999. High rates of schizophrenia in adults with velo-cardio-facial syndrome. Arch. Gen. Psychiatr. 56, 940–945.

Murray, R.M., Jones, P., O'callaghan, E., 1991. Fetal brain development and later schizophrenia. Ciba Found. Symp. 156, 155–163. Discussion 163-70.

Nakazawa, K., Zsiros, V., Jiang, Z., Nakao, K., Kolata, S., Zhang, S., Belforte, J.E., 2012. GABAergic interneuron origin of schizophrenia pathophysiology. Neuropharmacology 62, 1574–1583.

Nelson, B., Amminger, G.P., Yuen, H.P., Wallis, N., J Kerr, M., Dixon, L., Carter, C., Loewy, R., Niendam, T.A., Shumway, M., Morris, S., Blasioli, J., Mcgorry, P.D., 2017. Staged Treatment in Early Psychosis: a sequential multiple assignment randomised trial of interventions for ultra high risk of psychosis patients. Early Interv. Psychiatry.

Niarchou, M., Zammit, S., VAN Goozen, S.H., Thapar, A., Tierling, H.M., Owen, M.J., Van Den Bree, M.B., 2014. Psychopathology and cognition in children with 22q11.2 deletion syndrome. Br. J. Psychiatry 204, 46–54.

Oskarsdottir, S., 2004. Incidence and prevalence of the 22q11 deletion syndrome: a population-based study in Western Sweden. Arch. Dis. Child. 89, 148–151.

Owen, M.J., Sawa, A., Mortensen, P.B., 2016. Schizophrenia. Lancet 388, 86–97.

Padula, M.C., Schaer, M., Scariati, E., Maeder, J., Schneider, M., Eliez, S., 2017. Multimodal investigation of triple network connectivity in patients with 22q11DS and association with executive functions. Hum. Brain Mapp. 38, 2177–2189.

Padula, M.C., Schaer, M., Scariati, E., Schneider, M., Van De Ville, D., Debbane, M., Eliez, S., 2015. Structural and functional connectivity in the default mode network in 22q11.2 deletion syndrome. J. Neurodev. Disord. 7, 23.

Paylor, R., Glaser, B., Mupo, A., Ataliotis, P., Spencer, C., Sobotka, A., Sparks, C., Choi, C.H., Oghalai, J., Curran, S., Murphy, K.C., Monks, S., Williams, N., O'donovan, M.C., Owen, M.J., Scambler, P.J., Lindsay, E., 2006. Tbx1 haploinsufficiency is linked to behavioral disorders in mice and humans: implications for 22q11 deletion syndrome. Proc. Natl. Acad. Sci. USA 103, 7729–7734.

Peyvandi, S., Lupo, P.J., Garbarini, J., Woyciechowski, S., Edman, S., Emanuel, B.S., Mitchell, L.E., Goldmuntz, E., 2013. 22q11.2 deletions in patients with conotruncal defects: data from 1,610 consecutive cases. Pediatr. Cardiol. 34, 1687–1694.

Ramanathan, S., Mattiaccio, L.M., Coman, I.L., Botti, J.C., Fremont, W., Faraone, S.V., Antshel, K.M., Kates, W.R., 2016. Longitudinal trajectories of cortical thickness as a biomarker for psychosis in individuals with 22q11.2 deletion syndrome. Schizophr. Res. 188, 35–41.

Rauch, A., Hoyer, J., Guth, S., Zweier, C., Kraus, C., Becker, C., Zenker, M., Huffmeier, U., Thiel, C., Ruschendorf, F., Nurnberg, P., Reis, A., Trautmann, U., 2006. Diagnostic yield of various genetic approaches in patients with unexplained developmental delay or mental retardation. Am. J. Med. Genet. 140, 2063–2074.

Repetto, G.M., Guzman, M.L., Delgado, I., Loyola, H., Palomares, M., Lay-Son, G., Vial, C., Benavides, F., Espinoza, K., Alvarez, P., 2014. Case fatality rate and associated factors in patients with 22q11 microdeletion syndrome: a retrospective cohort study. BMJ Open 4, e005041.

Rihs, T.A., Tomescu, M.I., Britz, J., Rochas, V., Custo, A., Schneider, M., Debbane, M., Eliez, S., Michel, C.M., 2013. Altered auditory processing in frontal and left temporal cortex in 22q11.2 deletion syndrome: a group at high genetic risk for schizophrenia. Psychiatry Res. 212, 141–149.

Rump, P., DE Leeuw, N., Van Essen, A.J., Verschuuren-Bemelmans, C.C., Veenstra-Knol, H.E., Swinkels, M.E., Oostdijk, W., Ruivenkamp, C., Reardon, W., De Munnik, S., Ruiter, M., Frumkin, A., Lev, D., Evers, C., Sikkema-Raddatz, B., Dijkhuizen, T., Van Ravenswaaij-Arts, C.M., 2014. Central 22q11.2 deletions. Am. J. Med. Genet. 164A, 2707–2723.

Sandini, C., Scariati, E., Padula, M.C., Schneider, M., Schaer, M., Van De Ville, D., Eliez, S., 2017. Cortical dysconnectivity measured by structural covariance is associated with the presence of psychotic symptoms in 22q11.2 deletion syndrome. Biol. Psychiatry Cogn. Neurosci. Neuroimaging 3, 433–442.

Sannino, S., Padula, M.C., Manago, F., Schaer, M., Schneider, M., Armando, M., Scariati, E., Sloan-Bena, F., Mereu, M., Pontillo, M., Vicari, S., Contarini, G., Chiabrera, C., Pagani, M., Gozzi, A., Eliez, S., Papaleo, F., 2017. Adolescence is the starting point of sex-dichotomous COMT genetic effects. Transl. Psychiatry 7, e1141.

Sarpal, D.K., Robinson, D.G., Lencz, T., Argyelan, M., Ikuta, T., Karlsgodt, K., Gallego, J.A., Kane, J.M., Szeszko, P.R., Malhotra, A.K., 2015. Antipsychotic treatment and functional connectivity of the striatum in first-episode schizophrenia. JAMA Psychiatry 72, 5–13.

Scariati, E., Padula, M.C., Schaer, M., Eliez, S., 2016. Long-range dysconnectivity in frontal and midline structures is associated to psychosis in 22q11.2 deletion syndrome. J. Neural Transm. 123, 823–839.

Scariati, E., Schaer, M., Richiardi, J., Schneider, M., Debbane, M., Van De Ville, D., Eliez, S., 2014. Identifying 22q11.2 deletion syndrome and psychosis using resting-state connectivity patterns. Brain Topogr. 27, 808–821.

Schaer, M., Debbane, M., Bach Cuadra, M., Ottet, M.C., Glaser, B., Thiran, J.P., Eliez, S., 2009. Deviant trajectories of cortical maturation in 22q11.2 deletion syndrome (22q11DS): a cross-sectional and longitudinal study. Schizophr. Res. 115, 182–190.

Schizophrenia Working Group Of The Psychiatric Genomics, C, 2014. Biological insights from 108 schizophrenia-associated genetic loci. Nature 511, 421–427.

Schneider, M., Armando, M., Pontillo, M., Vicari, S., Debbane, M., Schultze-Lutter, F., Eliez, S., 2016. Ultra high risk status and transition to psychosis in 22q11.2 deletion syndrome. World Psychiatry 15, 259–265.

Schneider, M., Debbane, M., Bassett, A.S., Chow, E.W., Fung, W.L., Van Den Bree, M., Owen, M., Murphy, K.C., Niarchou, M., Kates, W.R., Antshel, K.M., Fremont, W., Mcdonald-Mcginn, D.M., Gur, R.E., Zackai, E.H., Vorstman, J., Duijff, S.N., Klaassen, P.W., Swillen, A., Gothelf, D., Green, T., Weizman, A., VAN Amelsvoort, T., Evers, L., Boot, E., Shashi, V., Hooper, S.R., Bearden, C.E., Jalbrzikowski, M., Armando, M., Vicari, S., Murphy, D.G., Ousley, O., Campbell, L.E., Simon, T.J., Eliez, S., 2014a. Psychiatric disorders from childhood to adulthood in 22q11.2 deletion syndrome: results from the international consortium on brain and behavior in 22q11.2 deletion syndrome. Am. J. Psychiatry 171, 627–639.

Schneider, M., Debbane, M., Lagioia, A., Salomon, R., D'argembeau, A., Eliez, S., 2012a. Comparing the neural bases of self-referential processing in typically developing and 22q11.2 adolescents. Dev. Cogn. Neurosci. 2, 277−289.

Schneider, M., Schaer, M., Mutlu, A.K., Menghetti, S., Glaser, B., Debbane, M., Eliez, S., 2014b. Clinical and cognitive risk factors for psychotic symptoms in 22q11.2 deletion syndrome: a transversal and longitudinal approach. Eur. Child Adolesc. Psychiatry 23, 425−436.

Schneider, M., Van Der Linden, M., Glaser, B., Rizzi, E., Dahoun, S.P., Hinard, C., Bartoloni, L., Antonarakis, S.E., Debbane, M., Eliez, S., 2012b. Preliminary structure and predictive value of attenuated negative symptoms in 22q11.2 deletion syndrome. Psychiatry Res. 196, 277−284.

Schneider, M., van der linden, M., Menghetti, S., Debbane, M., Eliez, S., 2017. Negative and paranoid symptoms are associated with negative performance beliefs and social cognition in 22q11.2 deletion syndrome. Early Interv. Psychiatry 11, 156−164.

Schneider, M., Van Der Linden, M., Menghetti, S., Glaser, B., Debbane, M., Eliez, S., 2014c. Predominant negative symptoms in 22q11.2 deletion syndrome and their associations with cognitive functioning and functional outcome. J. Psychiatr. Res. 48, 86−93.

Schreiner, M.J., Karlsgodt, K.H., Uddin, L.Q., Chow, C., Congdon, E., Jalbrzikowski, M., Bearden, C.E., 2014. Default mode network connectivity and reciprocal social behavior in 22q11.2 deletion syndrome. Soc. Cogn. Affect. Neurosci. 9, 1261−1267.

Shaikh, T.H., Kurahashi, H., Saitta, S.C., O'hare, A.M., Hu, P., Roe, B.A., Driscoll, D.A., Mcdonald-Mcginn, D.M., Zackai, E.H., Budarf, M.L., Emanuel, B.S., 2000. Chromosome 22-specific low copy repeats and the 22q11.2 deletion syndrome: genomic organization and deletion endpoint analysis. Hum. Mol. Genet. 9, 489−501.

Sheffield, J.M., Barch, D.M., 2016. Cognition and resting-state functional connectivity in schizophrenia. Neurosci. Biobehav. Rev. 61, 108−120.

Sigurdsson, T., Stark, K.L., Karayiorgou, M., Gogos, J.A., Gordon, J.A., 2010. Impaired hippocampal-prefrontal synchrony in a genetic mouse model of schizophrenia. Nature 464, 763−767.

Singh, S.M., Murphy, B., O'reilly, R., 2002. Monozygotic twins with chromosome 22q11 deletion and discordant phenotypes: updates with an epigenetic hypothesis. J. Med. Genet. 39, e71.

Solot, C.B., Gerdes, M., Kirschner, R.E., Mcdonald-Mcginn, D.M., Moss, E., Woodin, M., Aleman, D., Zackai, E.H., Wang, P.P., 2001. Communication issues in 22q11.2 deletion syndrome: children at risk. Genet. Med. 3, 67−71.

Swillen, A., 2016. The importance of understanding cognitive trajectories: the case of 22q11.2 deletion syndrome. Curr. Opin. Psychiatr. 29, 133−137.

Takao, A., Ando, M., Cho, K., Kinouchi, A., Murakami, Y., 1980. Etiology and Morphogenesis of Congenital Heart Disease.

Tang, S.X., Moore, T.M., Calkins, M.E., Yi, J.J., Savitt, A., Kohler, C.G., Souders, M.C., Zackai, E.H., Mcdonald-Mcginn, D.M., Emanuel, B.S., Gur, R.C., Gur, R.E., 2017. The psychosis spectrum in 22q11.2 deletion syndrome is comparable to that of nondeleted youths. Biol. Psychiatry 82, 17−25.

Tang, S.X., Yi, J.J., Moore, T.M., Calkins, M.E., Kohler, C.G., Whinna, D.A., Souders, M.C., Zackai, E.H., Mcdonald-Mcginn, D.M., Emanuel, B.S., Bilker, W.B., Gur, R.C., Gur, R.E., 2014. Subthreshold psychotic symptoms in 22q11.2 deletion syndrome. J. Am. Acad. Child Adolesc. Psychiatry 53, 991−1000.e2.

Tomescu, M.I., Rihs, T.A., Becker, R., Britz, J., Custo, A., Grouiller, F., Schneider, M., Debbane, M., Eliez, S., Michel, C.M., 2014. Deviant dynamics of EEG resting state pattern in 22q11.2 deletion syndrome adolescents: a vulnerability marker of schizophrenia? Schizophr. Res. 157, 175−181.

VAN Duin, E.D.A., Goossens, L., Hernaus, D., Da Silva Alves, F., Schmitz, N., Schruers, K., Van Amelsvoort, T., 2016. Neural correlates of reward processing in adults with 22q11 deletion syndrome. J. Neurodev. Disord. 8, 25.

Vergaelen, E., Swillen, A., Van Esch, H., Claes, S., Van Goethem, G., Devriendt, K., 2015. 3 generation pedigree with paternal transmission of the 22q11.2 deletion syndrome: intrafamilial phenotypic variability. Eur. J. Med. Genet. 58, 244−248.

Vingerhoets, W.A.M., Oudenaren, M.J.F.V., Van Duin, E.D.A., Bloemen, O.J.N., Booij, J., Evers, L.J.M., Boot, E., Vergaelen, E., Vogels, A., Swillen, A., Van Amelsvoort, T.A.M.J., 2015. Prevalence of substance use and the relation with psychosis and catechol-O-methyltransferase in patients with chromosome 22q11 deletion syndrome. Eur. Neuropsychopharmacol 25, S630.

Vorstman, J.A., Breetvelt, E.J., Duijff, S.N., Eliez, S., Schneider, M., Jalbrzikowski, M., Armando, M., Vicari, S., Shashi, V., Hooper, S.R., Chow, E.W., Fung, W.L., Butcher, N.J., Young, D.A., Mcdonald-Mcginn, D.M., Vogels, A., VAN Amelsvoort, T., Gothelf, D., Weinberger, R., Weizman, A., Klaassen, P.W., Koops, S., Kates, W.R., Antshel, K.M., Simon, T.J., Ousley, O.Y., Swillen, A., Gur, R.E., Bearden, C.E., Kahn, R.S., Bassett, A.S., International

Consortium On, B., Behavior In 22q11.2 Deletion, S., 2015. Cognitive decline preceding the onset of psychosis in patients with 22q11.2 deletion syndrome. JAMA Psychiatry 72, 377–385.

Vorstman, J.A., Breetvelt, E.J., Thode, K.I., Chow, E.W., Bassett, A.S., 2013. Expression of autism spectrum and schizophrenia in patients with a 22q11.2 deletion. Schizophr. Res. 143, 55–59.

Vorstman, J.A., Morcus, M.E., Duijff, S.N., Klaassen, P.W., Heineman-De Boer, J.A., Beemer, F.A., Swaab, H., Kahn, R.S., VAN Engeland, H., 2006. The 22q11.2 deletion in children: high rate of autistic disorders and early onset of psychotic symptoms. J. Am. Acad. Child Adolesc. Psychiatry 45, 1104–1113.

Weinberger, D.R., 1987. Implications of normal brain development for the pathogenesis of schizophrenia. Arch. Gen. Psychiatr. 44, 660–669.

Weinzimer, S.A., 2001. Endocrine aspects of the 22q11.2 deletion syndrome. Genet. Med. 3, 19–22.

Weisman, O., Guri, Y., Gur, R.E., Mcdonald-Mcginn, D.M., Calkins, M.E., Tang, S.X., Emanuel, B., Zackai, E.H., Eliez, S., Schneider, M., Schaer, M., Kates, W.R., Antshel, K.M., Fremont, W., Shashi, V., Hooper, S.R., Armando, M., Vicari, S., Pontillo, M., Kushan, L., Jalbrzikowski, M., Bearden, C.E., Cubells, J.F., Ousley, O.Y., Walker, E.F., Simon, T.J., Stoddard, J., Niendam, T.A., Van Den Bree, M.B.M., Gothelf, D., International Consortium On, B., Behavior In 22q11.2 Deletion, S., 2017. Subthreshold psychosis in 22q11.2 deletion syndrome: multisite naturalistic study. Schizophr. Bull. 43, 1079–1089.

Wu, H.Y., Rusnack, S.L., Bellah, R.D., Plachter, N., Mcdonald-Mcginn, D.M., Zackai, E.H., Canning, D.A., 2002. Genitourinary malformations in chromosome 22q11.2 deletion. J. Urol. 168, 2564–2565.

Yoshimura, S., Okamoto, Y., Matsunaga, M., Onoda, K., Okada, G., Kunisato, Y., Yoshino, A., Ueda, K., Suzuki, S.I., Yamawaki, S., 2017. Cognitive behavioral therapy changes functional connectivity between medial prefrontal and anterior cingulate cortices. J. Affect. Disord. 208, 610–614.

Yung, A.R., Yuen, H.P., Mcgorry, P.D., Phillips, L.J., Kelly, D., Dell'olio, M., Francey, S.M., Cosgrave, E.M., Killackey, E., Stanford, C., Godfrey, K., Buckby, J., 2005. Mapping the onset of psychosis: the comprehensive assessment of at-risk mental states. Aust. N. Z. J. Psychiatr. 39, 964–971.

Zheng, P., Noroski, L.M., Hanson, I.C., Chen, Y., Lee, M.E., Huang, Y., Zhu, M.X., Banerjee, P.P., Makedonas, G., Orange, J.S., Shearer, W.T., Liu, D., 2015. Molecular mechanisms of functional natural killer deficiency in patients with partial DiGeorge syndrome. J. Allergy Clin. Immunol. 135, 1293–1302.

Zoller, D., Padula, M.C., Sandini, C., Schneider, M., Scariati, E., Van De Ville, D., Schaer, M., Eliez, S., 2017. Psychotic symptoms influence the development of anterior cingulate BOLD variability in 22q11.2 deletion syndrome. Schizophr. Res. 193, 319–328.

Specific areas and risk

CHAPTER

9

Neuroimaging studies in people at clinical high risk for psychosis

George Gifford[1], Robert McCutcheon[1], Philip McGuire[1,2,3]

[1]Department of Psychosis Studies, Institute of Psychiatry, Psychology & Neuroscience, King's College London, London, United Kingdom; [2]OASIS Service, South London and the Maudsley NHS National Health Service Foundation Trust, London, United Kingdom; [3]National Institute for Health Research (NIHR) Biomedical Research Centre for Mental Health, Institute of Psychiatry, Psychology & Neuroscience, King's College London, London, United Kingdom

Introduction

A key problem in the clinical management of people at clinical high risk (CHR) for psychosis (otherwise described as those with an ultra high risk for psychosis) is that it is difficult to predict whether an individual will go on to develop psychosis, will have persistent symptoms, or will recover on the basis of their clinical features. Neuroimaging has the potential to allow clinicians to differentiate between individuals with distinct clinical outcomes and thus provide more personalized care.

This chapter aims to provide an overview of techniques currently used in studying psychosis risk, highlighting their benefits, limitations, and possible future directions.

Structural magnetic resonance imaging

Structural magnetic resonance imaging (MRI) uses scanning sequences that produce images with high contrast between gray and white matter, allowing for the quantification of gray and white matter volume. A widely used method for processing structural MRI scans is voxel-based morphometry, a process typically used for extracting voxels inferred to be gray matter to make group comparisons of gray matter volume (Ashburner and Friston, 2000). Structural MRI scans can also be used to assess the degree of cortical folding (Zilles et al., 1988) or the pattern and variation of cortical gyrification (Thompson et al., 2004).

Psychosis has been associated with localized changes in gray matter volume and with progressive reductions of gray matter over the course of the disorder (Lawrie and Abukmeil, 1998; Steen et al., 2006; Vita et al., 2006; Wright et al., 2000). Gray matter volume reductions have been shown in CHR versus controls in cross-sectional comparisons (Meisenzahl et al., 2008), in longitudinal studies within the CHR group (Pantelis et al., 2003), and between CHR subgroups showing more marked reductions in

Risk Factors for Psychosis
https://doi.org/10.1016/B978-0-12-813201-2.00009-0

those found to later develop psychosis (Fusar-Poli et al., 2011a; Sun et al., 2009; Takahashi et al., 2009; Ziermans et al., 2010). Several studies have investigated whether cortical gyrification and surface measures in CHR are related to eventual transition to psychosis (Harris et al., 2004; Sun et al., 2009). Within the CHR population, subsequent transition to psychosis has been linked to volume reductions in the hippocampus (Mechelli et al., 2011; Pantelis et al., 2003) and anterior cingulate gyrus (Yücel et al., 2003). A metaanalyses of the literature suggest transition in CHR groups to be associated with volumetric reductions in the right inferior frontal gyrus and the right superior temporal gyrus.

There is, however, heterogeneity in findings across studies (Bois et al., 2015; Velakoulis et al., 2006; Wood et al., 2005), which may be due to the use of small sample sizes and the subsequent challenges of obtaining a sufficient CHR-T sample (Borgwardt and Schmidt, 2016). For instance, a large CHR study of 274 CHR participants had 35 CHR-T participants at 12-month follow-up (Cannon et al., 2015). Multisite studies aiming to recruit sufficient CHR numbers are discussed below. In addition, outcomes other than transition to a full psychotic disorder may be equally or even more meaningful for those at CHR. Successful cortical surface–based pattern recognition has been demonstrated in predicting between CHR participants with good and bad functional outcome (Kambeitz-Ilankovic et al., 2016) and a large multicenter study has used structural MRI to predict 1 year functional outcome in a sample of 116 CHR individuals (Koutsouleris et al., 2018).

Structural MRI scans are a staple of clinical imaging services because of their use in detecting macroscopic abnormalities. As such they can be easily implemented, allowing for their use in nonspecialized research settings. This is of particular benefit for hard-to-recruit populations as it allows for the inclusion of more sites in multicenter studies. It is also a benefit that structural MRI is relatively clear in its interpretation as there are fewer layers of abstraction in postscanning analysis as compared to other imaging techniques. A potential limitation is that structural changes may occur at a relatively late stage of the pathological process that leads to psychosis onset: there is some evidence that they are predated by neurochemical and functional changes (Schobel et al., 2013). In addition, as with all imaging techniques, resolution for structural MRI is limited and it is not possible to separate some structures into component parts. Given these limitations, identifying a structural change does not necessarily provide useful information regarding the underlying mechanism of psychosis development and care should be taken in their interpretation.

Diffusion tensor imaging

Diffusion tensor imaging (DTI) involves the measurement of magnetic resonance signal sensitive to the motion of water molecules to estimate the direction and integrity of white matter tracts (Basser et al., 1994). Common integrity measures include fractional anisotropy and mean diffusivity (Basser and Pierpaoli, 2011), which are thought to relate to features such as white matter myelination, density, and quality (Mori and Zhang, 2006). Using the diffusion tensor, it is also possible to map out a brain's white matter architecture, allowing for whole brain structural connectivity to be modeled.

There is strong evidence for reductions in white matter integrity in patients with schizophrenia (Kanaan et al., 2005, 2009). Cross-sectional DTI studies have suggested that there are white matter abnormalities in CHR groups compared to controls (Clemm von Hohenberg et al., 2013; Karlsgodt et al., 2009; Krakauer et al., 2017) and longitudinal reductions in white matter integrity have been reported in CHR versus control groups (Carletti et al., 2012; Katagiri et al., 2015). White matter integrity has also been related to measures of functioning in CHR populations (Karlsgodt et al., 2009; Katagiri et al., 2018; Krakauer et al., 2017). It has been suggested that abnormal polyunsaturated fatty acid levels in psychosis may mediate the relationships between psychosis risk, functioning, and white matter (Vijayakumar et al., 2016).

DTI provides an index of anatomical connectivity, taking into account the direction and organization of white matter tracts. However, questions remain about what diffusion measures represent, as the direction of movement of white matter molecules can be affected by features other than those reflecting white matter integrity, such as the presence of microglia or when multiple fibers cross within a voxel. There are methods aiming to account for such issues, for example, high angular resolution diffusion imaging uses a greater number of gradient directions compared to typical DTI, allowing for the modeling of multiple fiber directions within a voxel giving a more realistic representation of white matter structure (Tuch et al., 2002).

Head movement during DTI can have a profound impact on the results. A study involving 949 DTI scans found systematic changes in structural connectivity measures from subtle movements, despite the use of standard movement correction procedures (Baum et al., 2017). At present, the effects of head movement on DTI results are not well-characterized. This suggests for psychosis populations even subtle differences in movement between patient and control groups could bias results. Postprocessing methods for dealing with motion such as entering movement as a covariate in multivariate analysis are advisable.

Functional magnetic resonance imaging

Functional MRI (fMRI) makes use of the fact that oxygenated and deoxygenated blood have different magnetic properties; this allows blood flow to be quantified which permits one to make inferences regarding neural activity (Logothetis and Wandell, 2004). fMRI studies typically involve tasks aimed at engaging a particular function of the brain or are conducted in the resting state, when subjects are instructed to simply relax and let their mind wander.

Studies have explored activation in CHR populations versus controls during tasks thought to engage executive functioning (Broome et al., 2009; Morey et al., 2005), salience processing (Schmidt et al., 2017), working memory (Falkenberg et al., 2017; Fusar-Poli et al., 2010), spatial working memory (Goghari et al., 2014), and movement generation (Broome et al., 2010). Overall, task-based fMRI studies suggest that abnormalities in CHR subjects are qualitatively similar to those seen in patients with psychosis, involving the prefrontal cortex, anterior cingulate cortex, basal ganglia, cerebellum, and medial temporal areas (Fusar-Poli et al., 2007).

Functional connectivity studies investigate the degree to which areas of the brain activate and deactivate in synchrony with one another. Typically this approach allows one to comment on whether regions show a functional connection, i.e., whether activity within regions shows a significant correlation, but it does not allow one to draw conclusions regarding the direction of the interaction. Dynamic causal modeling (DCM) is a method of analysis that can be applied to fMRI data to potentially overcome this limitation. This is done in a way that treats interactions as nonlinear and dynamic and that can take into account external inputs such as task demand (Friston et al., 2003). Although complexity limits DCM to a small number of brain areas, it provides a biologically plausible model of effective connectivity (i.e., it allows one to comment on the direction of information flow, as opposed to solely the strength of a connection). It has been applied to CHR groups to explore areas of the brain implicated in psychosis onset, suggesting that reward-induced modulation of connectivity between the basal ganglia (ventral striatum/pallidum) and the midbrain is higher in CHR participants (Winton-Brown et al., 2017).

Resting state fMRI studies are typically used to study functional connectivity on the basis that brain areas working together will exhibit similar activation over time (Van Den Heuvel and Pol, 2010). These studies have suggested that functional connectivity may be perturbed in CHR groups, particularly in the default mode network (Anticevic et al., 2013; Kwon et al., 2010; Shim et al., 2010; Wang et al., 2016; Wotruba et al., 2013) in the connections of the thalamus (Anticevic et al., 2013), and in frontotemporal connectivity (Kwon et al., 2010; Yoon et al., 2015).

There are multiple methodological issues relevant to use of fMRI. First, the choice of atlas used to define regions of interest in fMRI can have a profound influence on the results (Smith et al., 2011). Accurate parcellation of the human brain is an ongoing challenge which initiatives such as the Human Connectome Project hope to address (Glasser et al., 2016). Some functional connectivity studies infer links between areas without direct anatomical connections, which is thought to reflect the involvement of additional areas indirectly mediating connections (Damoiseaux and Greicius, 2009). In defining functional connectivity, it may therefore be appropriate to use methods such as partial correlation or sparse inverse covariance estimation, although this is dependent on there being sufficient data points (Smith et al., 2013). Another limitation is head movement, which can be systematically worse in patient populations and the effects of which are hard to remove from fMRI data. Methodological steps to limit the impact of head motion have been proposed; however, it remains a significant issue (Power et al., 2015).

Most fMRI studies are based on the blood-oxygen-level-dependent signal, which reflects changes in the oxygenation of blood following neuronal activity, although there is a significant time lag. The vascular response to neural activity can also be assessed using arterial spin labeling (ASL), which magnetically labels bloods as it enters tissue from the arteries (perfusion). The signal from labeled blood is 0.5%–1.5% of the total tissue signal from a voxel, meaning that ASL has a low signal to noise ratio (Petcharunpaisan et al., 2010). ASL does, however, provide a more immediate approximation of neural activity, as shown by a more dynamic signal. In studies of CHR subjects, ASL has been used to show increased resting state perfusion in the hippocampus and basal ganglia compared to controls (Allen et al., 2015, 2017).

Electroencephalography

In electroencephalography (EEG), electrodes are placed on the skull to measure electronic signals that reflect neural activity. This is a noninvasive method which is limited in spatial resolution but benefits from a much higher temporal resolution than fMRI, with recorded frequencies giving dynamics in the space of milliseconds rather than seconds. The physical equipment of EEG is less expensive than other imaging modalities, although the data require careful processing to produce meaningful results.

Event-related potentials (ERPs) are a type of EEG signal generated in response to a stimulus. A number of ERPs have been reported to be altered in CHR subjects, including the P300, which is a recorded signal during auditory or visual "oddball" paradigms, where repeated sequences of stimuli are interspersed with an irregular "oddball" stimulus, and mismatch negativity, which is a recorded signal during presentation of a rare event in a repetitive sequence of events (van Tricht et al., 2010, 2011). The P300 signal has also been related to differences in gray matter volume in a CHR group (Fusar-Poli et al., 2011b) and EEG parameters such as frontal theta and delta, and occipital–parietal alpha peak frequencies have been suggested to be useful in the prediction of later transition to psychosis in CHR groups (van Tricht et al., 2014). Another study found a range of EEG bands and SANS scores to be predictive of psychosis transition in comparing

CHR-T and CHR-NT participants (Zimmermann et al., 2010).

An interesting use of EEG is in simultaneous acquisition with an fMRI scan (Horovitz et al., 2008). This potentially exploits both the high temporal resolution of EEG with the relatively high spatial resolution of fMRI. In addition, although EEG is limited in the number of brain areas it can represent, it can be used to define functional connectivity between regions, allowing for complex network analysis (Rubinov et al., 2009). Such studies are yet to be performed in CHR populations.

Proton magnetic resonance spectroscopy

Proton magnetic resonance spectroscopy (1-hydrogen magnetic resonance imaging [1H-MRS]) is similar to MRI in that it involves the analysis of signals generated from hydrogen protons, but instead of indicating water and fat concentrations it infers the abundance of certain metabolites (Urenjak et al., 1993). This has been used to determine the concentration of certain neurotransmitters in CHR individuals in vivo. Limitations of 1H-MRS include the fact that volumes of interest are typically a minimum of $15 \times 15 \times 15$ mm in size, and the method does not allow distinction between intra- and extracellular components. Furthermore, in the absence of information regarding receptor densities, it is not possible to determine whether observed differences in metabolite concentrations are primary deficits or alternatively adaptive responses to pathological alterations in receptor density and function. It is, however, a relatively inexpensive and noninvasive neurochemical imaging technique.

1H-MRS has been widely used to study brain glutamate function, in areas where there is no positron emission tomography (PET) tracer available that is specific for the N-methyl-D-aspartate receptor (Knol et al., 2009). The measurement of glutamate and GABA is of particular interest in CHR subjects, as animal models of psychosis suggest that increased striatal dopaminergic activity is driven by cortical glutamatergic pyramidal neurons that are overactive as a result of a loss of local GABAergic inhibition (Lodge and Grace, 2008). One such study found increased striatal glutamate concentrations in CHR-T participants compared to CHR-NT participants (de la Fuente-Sandoval et al., 2012). The only study in CHR individuals to use 1H-MRS and also obtain PET measures of striatal dopamine function found that reduced hippocampal glutamate concentrations were associated with increased striatal dopamine levels and that this relationship was particularly marked in those individuals who later transitioned to psychosis (Stone et al., 2010). A recent study additionally found a significant difference in the correlation between GABA levels in the medial prefrontal cortex with left hippocampal resting state perfusion between CHR-T and CHR-NT participants, providing preliminary evidence for this relationship to be associated with psychosis risk (Modinos et al., 2018).

While studies reporting metabolite measures in CHR participants potentially provide evidence for the above animal model, a systematic review found that the only consistently observed alteration in CHR individuals was a small elevation in medial frontal cortex glutamate levels (Merritt et al., 2016). In addition, differences in GABA levels have not been consistently demonstrated in CHR individuals (Egerton et al., 2017).

Positron emission tomography

PET uses radioactive ligands to image specific molecular targets. It provides greater neurochemical specificity compared to 1H-MRS and can provide information on a broader range of targets. In CHR individuals, it is the dopamine system that has received the most attention.

In keeping with findings in schizophrenia (Howes et al., 2012), studies examining postsynaptic D2 receptor densities have not indicated any marked differences compared to controls (Mizrahi et al., 2017). Imaging the competition between endogenous dopamine and D2 receptor ligands following a pharmacological (e.g., amphetamine) or psychological (e.g., stress) challenge enables the measurement of dopamine release. A study employing a stress challenge paradigm demonstrated increased dopamine release in CHR individuals in the associative striatum (Mizrahi et al., 2012).

Presynaptic dopamine function can also be quantified using radiolabeled L-dihydroxyphenylalanine (L-DOPA). Measurement of the uptake and conversion of this precursor to radiolabeled dopamine in mesostriatal neurons gives an index of presynaptic dopamine synthesis capacity. Studies of 18F-DOPA in CHR subjects indicate that elevated dopamine synthesis capacity in mesostriatal areas at clinical presentation is associated with subsequent transition to psychosis, and longitudinal scanning suggests that there is a progressive increase in dopamine function as subjects develop psychosis (Howes et al., 2011a,b). Elevated dopamine synthesis capacity in CHR subjects has also been related to abnormal activation in the prefrontal (Fusar-Poli et al., 2011c) and medial temporal cortex (Allen et al., 2012). These findings are consistent with the key role of elevated dopamine activity in models of psychosis (Howes and Kapur, 2009; Modinos et al., 2015).

Other PET targets imaged in CHR individuals include the GABA-A (Kang et al., 2013) and 5HT-2A receptors (Hurlemann et al., 2008), both of which have suggested reduced receptor densities in the patient group within the striatum. A PET study using [^{11}C]PBR28, a marker of microglial activity, suggests that this may be higher in CHR than in control participants (Bloomfield et al., 2015), although other studies using different ligands for microglial activity have failed to find differences between CHR subjects and controls (Di Biase et al., 2017; Hafizi et al., 2017). Such examples show scope for studying neurochemical processes outside of the dopamine system with PET.

PET studies are relatively expensive and time-consuming and PET scanning facilities are not widely accessible. Although PET is able to lend a high degree of molecular specificity, unless it is able to demonstrate marked benefit over other lower burden modalities (Table 9.1), it is not currently ideally suited for use as part of a predictive tool to be used in mainstream clinical settings. Future work in this field may explore additional ligands, as well as ways of reducing participant and logistical burden.

Multicenter studies

The CHR population is heterogeneous. Up to one-third will develop psychosis within 3 years of presenting to services, but the remainder will not. Moreover, within the latter subgroup, some will have persistent CHR symptoms, others will become symptom-free (Fusar-Poli et al., 2012), and some will go on to meet criteria for another psychiatric disorder (Lin et al., 2015). This heterogeneity in clinical profile is likely mirrored in underlying neurobiology which means that in small samples pathoetiologically relevant findings may pass undetected as they may only exist in a subset of the population. For imaging studies, this heterogeneity is particularly problematic because of the high dimensionality of neuroimaging data, with statistical tests often being performed over hundreds of thousands of voxels. This can result in a high rate of false positives and require stringent multiple comparisons procedures.

In recognition of the need for larger samples, increasing efforts have been made toward multicenter studies in the CHR population. Mechelli et al. (2011) combined MRI data from four studies to form a total sample of 182 CHR (48 CHR-T) and 167 controls. This made it possible to conduct one of the largest comparisons of CHR subjects who did or did not develop

TABLE 9.1 Key advantages and disadvantages of common neuroimaging techniques.

Neuroimaging modality	Advantages	Disadvantages
Structural magnetic resonance imaging	Clear interpretation Easily implemented across centers and relatively low cost	Structural changes possibly a secondary phenomenon
Diffusion tensor imaging	Provides unique measurements of white matter tract integrity Allows for structural connectivity modeling	Issues with crossing fibers and presence of microglia Effects of head movement poorly characterized
Functional magnetic resonance imaging	High spatial resolution (compared to EEG and PET) Relatively easily implemented across centers	Low temporal resolution (compared to EEG) Head movement a serious confounder
Electroencephalography (EEG)	High temporal resolution Well-established EEG endophenotypic markers Easily implemented across centers	Low spatial resolution Noisy data which require careful processing
Proton magnetic resonance spectroscopy	Allows for measurement of neurotransmitter concentrations (glutamate/GABA)	Confined to one voxel of interest per scan Interpretation difficult in absence of information on receptor density
Positron emission tomography (PET)	Provides unique insight into neurochemical mechanisms	High participant burden, including exposure to radiation Requires specialist scanning centers High cost

psychosis to date. Subsequent conversion to psychosis was associated with less gray matter volume in the left parahippocampal cortex.

Several other multicenter studies of CHR individuals have recently been completed or are underway, such as EU-GEI (European Network of National Schizophrenia Networks Studying Gene-Environment Interactions), NAPLS (North American Prodrome Study), PRONIA (Personalised Prognostic Tools for Early Psychosis Management), and PSYSCAN. Some results from these consortia have been published. For example, the NAPLS consortia found that CHR participants who developed psychosis had higher longitudinal gray matter loss after 12 months in the right superior frontal, middle frontal, and medial orbitofrontal regions and an enlargement of the third ventricle compared to CHR participants who did not become psychotic. This study included 274 ARMS (35 ARMS-T) and 135 control participants scanned at baseline and 12-month follow-up (Cannon et al., 2015). From the PRONIA study, a support vector machine was used to predict 1 year social/occupational functioning outcomes in a sample of 116 CHR participants using structural MRI data with an accuracy of 76.9% (Koutsouleris et al., 2018).

Multicenter imaging studies do have their own methodological shortcomings. For instance, additional variance can be introduced by between-site differences in the type of scanner/hardware, scanning sequences, and upgrades to software (Han et al., 2006; Jovicich et al., 2009; Kempton et al., 2011), as well as demographic and clinical differences in the samples recruited at different centers. The disadvantage of added variance may be outweighed by the advantage of greater sample size, and any measure that is to be of clinical use will need to be reliable in a multicenter setting. For example, a study of multicenter gray matter volume suggested modest variance from using multiple scanners to assess group differences in high variance brain areas (Suckling et al., 2010). Between-center variance can also be reduced by the use of scanning sequences that are explicitly designed to minimize variance between sites, such as those used in the multicenter Alzheimer's imaging project ADNI (Jack et al., 2008). The harmonization of scanning protocols and methodology in multicenter studies may allow for more accurate results than the metaanalysis of results from multiple unrelated studies with different methods of acquisition and analysis (Thompson et al., 2004) but are much more logistically demanding.

Machine learning

Most imaging studies of psychosis risk have studied group differences. Although this research has significantly advanced our understanding of the mechanisms underlying psychosis onset (Modinos et al., 2015), to be applied in a clinical setting, neuroimaging data need to be used at an individual level. For example, to predict whether a CHR subject is likely to develop psychosis, the clinician needs to be provided with a personalized estimate of risk derived from that particular individual's imaging data. This challenge has been addressed by the application of machine learning methods, specifically those designed for classification. Machine learning classification techniques involve a classifier trained on a large number of examples, which can then be applied to a single case, giving a binary or probabilistic outcome (Gifford et al., 2017).

An early application of this approach used a support vector machine in MRI data from 45 CHR subjects (Koutsouleris et al., 2009). They reported that CHR subjects who did or did not develop psychosis could be classified with accuracies of 86%–93%. This study relied on a nested cross-validation method, cross-validation describing a process in which the data are repeatedly split into training and test data to give an average accuracy and nested describing a process where no model tuning is performed on the testing data.

While nested cross-validation strategies can be used to separate training and test data, high-dimensional data with a small sample size will overfit to the data giving a high accuracy of classification for the study sample only. The most valid way to evaluate a classifier is to test it on a completely independent dataset. However, a recent review of studies of machine learning methods in CHR groups found that none of the 91 studies included (of these, 7 involved MRI and 10 EEG) used an independent sample for validation (Studerus et al., 2017). This partly reflects the lack of availability of appropriate independent CHR datasets and is likely to be less of an obstacle as more and larger studies are conducted in this field (see below).

Overfitting is a particular problem for machine learning methods. The high dimensionality of neuroimaging data makes overfitting likely, and typically a machine learning process needs to involve feature selection, where redundant information is removed from the model. An alternative to this is to use deep learning methods, such as artificial neural networks. These create models using a hierarchy of complexity, higher more abstracted levels being less affected by noise (Vieira et al., 2017). There is also the interesting possibility of using neural networks tailored to sequential data (recurrent neural networks) on fMRI time series data (Dakka et al., 2017). Deep learning methods have existed for decades but have become more accessible with increased availability to computers that can process in parallel, making such analyses plausible for most researchers.

It is likely that large neuroimaging datasets of CHR participants will become available in the future and validation of classifiers could be facilitated by being made available for other researchers to apply to their own datasets

(Young et al., 2016). It is also in the nature of machine learning techniques for their effectiveness to be dependent on large and diverse datasets that encapsulate the intended population. Efforts to share classifiers or datasets are therefore key to the development of accurate classifiers, as are large-scale collaborative efforts such as the HARMONY initiative, which aims to pool data from the ongoing multicenter consortia of CHR populations including PRONIA, PSYSCAN, NAPLS, and the Philadelphia Neurodevelopmental Cohort.

Network analysis

The notion that psychosis may reflect a state of aberrant connectivity, rather than a localized lesion, is as old as current conceptions of the disorder (Wernicke, 1906). Dysconnectivity may be assessed by measuring fractional anisotropy in DTI data or by examining correlations between time courses in fMRI data. Overall, the results from imaging studies of connectivity in CHR subjects have been inconsistent (Pettersson-Yeo et al., 2011). This may reflect the complexity of brain networks as well as the heterogeneity of the CHR population.

Methods for systematically studying dysconnectivity have been greatly aided by developments in a branch of mathematics known as graph theory, in which complex networks are represented as nodes linked by edges. In the case of neuroimaging, data can be broken down into nodes consisting of brain areas and edges consisting of the connections between them. In the case of fMRI data edges tend to represent the magnitude of correlation coefficients (or another relationship metric) between activity at different nodes, while with DTI data edges will often represent the "fiber count" (the number of computed white matter tracts) between nodes. Network measures can then be applied to the resulting "graph" to provide quantification of how the brain is organized into distinct modules (segregation), how efficiently information can be transferred across the brain (integration), and more general measures of how connected the brain is (Bullmore and Bassett, 2011) (Fig. 9.1).

Like other naturally occurring networks, brain graphs also tend to have a minority of highly connected nodes and a majority of sparsely connected nodes (Bullmore and Bassett, 2011). The human brain also appears to organize in a "rich club" of connections, with highly connected hubs having a greater probability of connecting to each other, forming central hubs that likely have an important role in the brain's organization (Van Den Heuvel and Sporns, 2011). Graph analysis based on DTI data has been used to show reduced rich club connectivity in schizophrenia patients and their unaffected siblings (Collin et al., 2013), as well as in a sample of CHR participants (Schmidt et al., 2016).

Graph analysis has been used with fMRI to show lower "centrality" (a measure of network importance) of the dorsal anterior cingulate cortex in CHR-T participants compared to CHR-NT (Lord et al., 2011). Another study used the variance of graph measures across overlapping sliding time windows of correlation matrices across the time courses of resting state scans, suggesting altered "dynamic" functional connectivity in their CHR group to show similar but less severe alterations to FEP participants (Du et al., 2017).

Network analysis is also a potentially useful tool for the elegant integration of multimodal data. For instance, a study in genetic high-risk individuals found that disruption in the anatomical connectome of schizophrenia offspring was associated with increased modularity of their functional connectome (Collin et al., 2017). Techniques developed in fields such as network control theory (Medaglia et al., 2017) can be used to investigate how neurochemical factors measured using PET or 1H-MRS may influence dynamic measures taken from fMRI data, on top of a relatively constant structural backbone inferred from DTI.

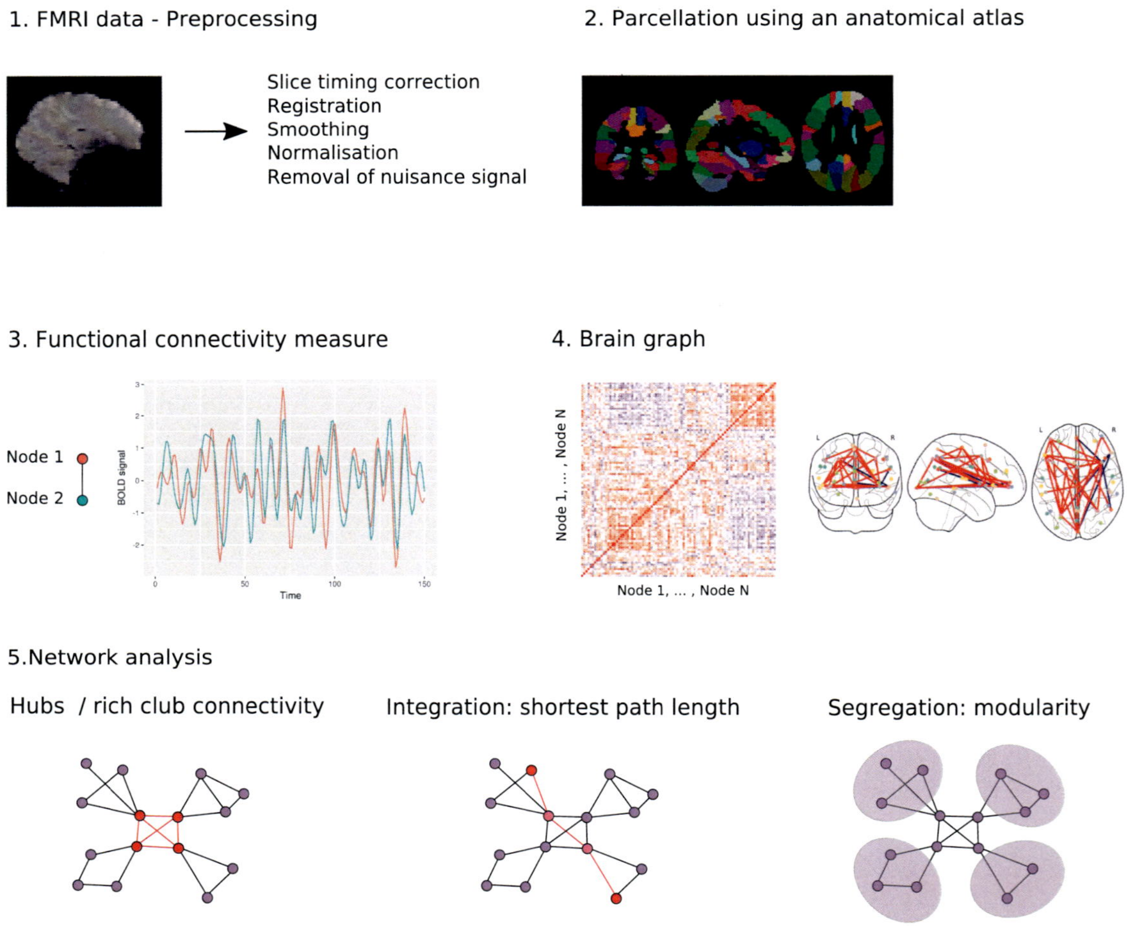

FIGURE 9.1 Process of inferring network analysis results from functional magnetic resonance imaging data. (1) Raw EPI images are collected and preprocessed. (2) Nodes are formed by averaging activation within predefined anatomical areas or by using each voxel as a node (whole brain). (3 and 4) A functional connectivity measure is used to define functional connectivity between nodes forming a graph of the brain. (5) Network analysis is performed on the graph, providing measures such as rich club connectivity/location of hubs, measures of integration such as the characteristic shortest path length, and measures of segregation such as modularity.

Brain graphs can also be used in machine learning classifiers. It is, however, challenging to successfully represent the differences between graphs with machine learning techniques, which rely on some measure of distance or similarity between data points. Representation of the similarity between graphs can be computed using graph kernels, which can allow them to be readily used in techniques such as support vector machines (Foggia et al., 2014).

Conclusion

Further advances in this field will require neuroimaging studies with sufficient scale to stratify CHR subjects according to different clinical outcomes. This requires large samples, which can only be gathered through multicenter collaboration. To date, most neuroimaging studies in this field have been cross-sectional and more longitudinal data that can show how the brain changes as an individual develops psychosis are needed. These issues are widely recognized and a number of large multicenter longitudinal studies are ongoing.

There are also ongoing efforts to harmonize scanning procedures and preprocessing pipelines across sites to reduce the between-center and between-study variability. There are a large number of steps between scan acquisitions to reporting results, so reducing this variability is crucial to minimize methodological confounders.

Developments in the application of machine learning and network analysis, as well as advances in image acquisition have opened up new ways to identify neural mechanisms. The speed at which new methods develop is such that there is a risk that the technical complexity may delay the translation of valid research findings into clinical practice. However, if machine learning classifiers or other image analysis methods allow clinicians to reliably distinguish CHR subjects who will or will not develop psychosis, this would represent a major translational step forward. Making trained machine learning classifiers publicly available would be helpful, as this would allow researchers at other centers to apply them to their own datasets without having to independently develop their own classifiers.

Glossary

- Arterial spin labeling—A magnetic resonance imaging (MRI) technique which magnetically labels blood as it enters tissues, the rate of which gives a measure of surrounding activation.
- Dynamic causal modeling (DCM)—A model for realistically representing functional interactions between brain areas. DCM treats effective relationships as nonlinear and dynamic and readily takes into account external inputs such as task demands.
- Diffusion tensor imaging/diffusion weighted imaging (DWI)—An imaging technique that uses signals from water as it diffuses to infer the shape of surrounding non–water-soluble white matter fibers.
- Electroencephalography (EEG)—Electrophysiological recordings using electrodes placed on the scalp, which can be used to model underlying neural phenomenon at a high temporal resolution.
- Event-related potential (ERP)—Measured neural activity which is influenced by a specific stimulus.
- Functional magnetic resonance imaging (fMRI)—A magnetic resonance imaging technique which uses signal influenced by the oxygenation level of blood, the change of which is inferred to reflect underlying neural activity.
- High angular resolution diffusion imaging—DWI using a high number of gradient directions to model complex wiring of white matter tracts.
- MRI—Imaging techniques which use strong magnetic fields to generate images, which can be used in vivo.
- N-methyl-D-aspartate receptor—A glutamate receptor which is prolific in the brain.
- Positron emission tomography (PET)—An imaging technique which uses radioactively labeled ligands to study neurochemical processes.
- 6-[18F]fluoro-L-DOPA—A radioactively labeled ligand used in PET imaging to measure dopamine synthesis capacity.
- [^{18}F]FEPA—A radioactively labeled ligand used in PET imaging as an index for microglial activity.
- [^{11}C]PBR28—A radioactively labeled ligand used in PET imaging as an index for microglial activity.
- P300—An EEG-based ERP recorded during auditory or visual "oddball" paradigms.
- 1-Hydrogen magnetic resonance spectroscopy—A magnetic resonance

imaging technique used to measure the abundance of metabolites such as glutamate and GABA within a voxel of interest.

List of acronyms and abbreviations

18F-DOPA 6-[18F]fluoro-L-DOPA
1H-MRS 1-Hydrogen magnetic resonance imaging
ADNI Alzheimer's Disease Neuroimaging Initiative
ASL Arterial spin labeling
CHR-NT Clinical high risk-no transition
CHR-T Clinical high risk-transition
DCM Dynamic causal modeling
DTI Diffusion tensor imaging
EEG Electroencephalography
ERP Event-related potential
EU-GEI European Network of National Schizophrenia Networks Studying Gene-Environment Interactions
fMRI Functional magnetic resonance imaging
GABA Gamma-aminobutyric acid
HARDI High angular resolution diffusion imaging
NAPLS North American Prodrome Study
PET Positron emission tomography
PRONIA Personalised Prognostic Tools for Early Psychosis Management

References

Allen, P., Azis, M., Modinos, G., Bossong, M.G., Bonoldi, I., Samson, C., et al., 2017. Increased resting hippocampal and basal ganglia perfusion in people at ultra high risk for psychosis: replication in a second Cohort. Schizophr. Bull. 44 (6), 1323–1331.

Allen, P., Chaddock, C.A., Egerton, A., Howes, O.D., Bonoldi, I., Zelaya, F., et al., 2015. Resting hyperperfusion of the hippocampus, midbrain, and basal ganglia in people at high risk for psychosis. Am. J. Psychiatry 173 (4), 392–399.

Allen, P., Luigjes, J., Howes, O.D., Egerton, A., Hirao, K., Valli, I., et al., 2012. Transition to psychosis associated with prefrontal and subcortical dysfunction in ultra high-risk individuals. Schizophr. Bull. 38 (6), 1268–1276.

Anticevic, A., Tang, Y., Cho, Y.T., Repovs, G., Cole, M.W., Savic, A., et al., 2013. Amygdala connectivity differs among chronic, early course, and individuals at risk for developing schizophrenia. Schizophr. Bull. 40 (5), 1105–1116.

Ashburner, J., Friston, K.J., 2000. Voxel-based morphometry—the methods. Neuroimage 11 (6), 805–821.

Basser, P.J., Mattiello, J., LeBihan, D., 1994. MR diffusion tensor spectroscopy and imaging. Biophys. J. 66 (1), 259–267.

Basser, P.J., Pierpaoli, C., 2011. Microstructural and physiological features of tissues elucidated by quantitative-diffusion-tensor MRI. J. Magn. Reson. 213 (2), 560–570.

Baum, G.L., Roalf, D.R., Cook, P.A., Ciric, R., Rosen, A.F.G., Xia, C., et al., 2017. The impact of in-scanner head motion on structural connectivity derived from diffusion tensor imaging. bioRxiv 185397.

Bloomfield, P.S., Selvaraj, S., Veronese, M., Rizzo, G., Bertoldo, A., Owen, D.R., et al., 2015. Microglial activity in people at ultra high risk of psychosis and in schizophrenia: an [11C] PBR28 PET brain imaging study. Am. J. Psychiatry 173 (1), 44–52.

Bois, C., Levita, L., Ripp, I., Owens, D.C.G., Johnstone, E.C., Whalley, H.C., Lawrie, S.M., 2015. Hippocampal, amygdala and nucleus accumbens volume in first-episode schizophrenia patients and individuals at high familial risk: a cross-sectional comparison. Schizophr. Res. 165 (1), 45–51.

Borgwardt, S., Schmidt, A., 2016. Is neuroimaging clinically useful in subjects at high risk for psychosis? World Psychiatry 15 (2), 178–179.

Broome, M.R., Matthiasson, P., Fusar-Poli, P., Woolley, J.B., Johns, L.C., Tabraham, P., et al., 2009. Neural correlates of executive function and working memory in the "at-risk mental state. Br. J. Psychiatry 194 (1), 25–33.

Broome, M.R., Matthiasson, P., Fusar-Poli, P., Woolley, J.B., Johns, L.C., Tabraham, P., et al., 2010. Neural correlates of movement generation in the 'at-risk mental state'. Acta Psychiatr. Scand. 122 (4), 295–301.

Bullmore, E.T., Bassett, D.S., 2011. Brain graphs: graphical models of the human brain connectome. Annu. Rev. Clin. Psychol. 7, 113–140.

Cannon, T.D., Chung, Y., He, G., Sun, D., Jacobson, A., Van Erp, T.G.M., et al., 2015. Progressive reduction in cortical thickness as psychosis develops: a multisite longitudinal neuroimaging study of youth at elevated clinical risk. Biol. Psychiatry 77 (2), 147–157.

Carletti, F., Woolley, J.B., Bhattacharyya, S., Perez-Iglesias, R., Fusar Poli, P., Valmaggia, L., et al., 2012. Alterations in white matter evident before the onset of psychosis. Schizophr. Bull. 38 (6), 1170–1179.

Clemm von Hohenberg, C., Pasternak, O., Kubicki, M., Ballinger, T., Vu, M.-A., Swisher, T., et al., 2013. White matter microstructure in individuals at clinical high risk of psychosis: a whole-brain diffusion tensor imaging study. Schizophr. Bull. 40 (4), 895–903.

Collin, G., Kahn, R.S., de Reus, M.A., Cahn, W., van den Heuvel, M.P., 2013. Impaired rich club connectivity in unaffected siblings of schizophrenia patients. Schizophr. Bull. 40 (2), 438–448.

Collin, G., Scholtens, L.H., Kahn, R.S., Hillegers, M.H.J., van den Heuvel, M.P., 2017. Affected anatomical rich club and structural–functional coupling in young offspring of schizophrenia and bipolar disorder patients. Biol. Psychiatry 82 (10), 746–755.

Dakka, J., Bashivan, P., Gheiratmand, M., Rish, I., Jha, S., Greiner, R., 2017. Learning Neural Markers of Schizophrenia Disorder Using Recurrent Neural Networks arXiv Preprint arXiv:1712.00512.

Damoiseaux, J.S., Greicius, M.D., 2009. Greater than the sum of its parts: a review of studies combining structural connectivity and resting-state functional connectivity. Brain Struct. Funct. 213 (6), 525–533.

de la Fuente-Sandoval, C., León-Ortiz, P., Azcárraga, M., Favila, R., Stephano, S., Graff-Guerrero, A., 2012. Striatal glutamate and the conversion to psychosis: a prospective 1H-MRS imaging study. Int. J. Neuropsychopharmacol. 16 (2), 471–475.

Di Biase, M.A., Zalesky, A., O'keefe, G., Laskaris, L., Baune, B.T., Weickert, C.S., et al., 2017. PET imaging of putative microglial activation in individuals at ultra-high risk for psychosis, recently diagnosed and chronically ill with schizophrenia. Transl. Psychiatry 7 (8), e1225.

Du, Y., Fryer, S.L., Fu, Z., Lin, D., Sui, J., Chen, J., et al., 2017. Dynamic functional connectivity impairments in early schizophrenia and clinical high-risk for psychosis. Neuroimage 180, 632–645.

Egerton, A., Modinos, G., Ferrera, D., McGuire, P., 2017. Neuroimaging studies of GABA in schizophrenia: a systematic review with meta-analysis. Transl. Psychiatry 7 (6), e1147.

Falkenberg, I., Valli, I., Raffin, M., Broome, M.R., Fusar-Poli, P., Matthiasson, P., et al., 2017. Pattern of activation during delayed matching to sample task predicts functional outcome in people at ultra high risk for psychosis. Schizophr. Res. 181, 86–93.

Foggia, P., Percannella, G., Vento, M., 2014. Graph matching and learning in pattern recognition in the last 10 years. Int. J. Pattern Recognit. Artif. Intell. 28 (1), 1450001.

Friston, K.J., Harrison, L., Penny, W., 2003. Dynamic causal modelling. Neuroimage 19 (4), 1273–1302.

Fusar-Poli, P., Bonoldi, I., Yung, A.R., Borgwardt, S., Kempton, M.J., Valmaggia, L., et al., 2012. Predicting psychosis: meta-analysis of transition outcomes in individuals at high clinical risk. Arch. Gen. Psychiatr. 69 (3), 220–229.

Fusar-Poli, P., Borgwardt, S., Crescini, A., Deste, G., Kempton, M.J., Lawrie, S., et al., 2011a. Neuroanatomy of vulnerability to psychosis: a voxel-based meta-analysis. Neurosci. Biobehav. Rev. 35 (5), 1175–1185.

Fusar-Poli, P., Broome, M.R., Matthiasson, P., Woolley, J.B., Johns, L.C., Tabraham, P., et al., 2010. Spatial working memory in individuals at high risk for psychosis: longitudinal fMRI study. Schizophr. Res. 123 (1), 45–52.

Fusar-Poli, P., Crossley, N., Woolley, J., Carletti, F., Perez-Iglesias, R., Broome, M., et al., 2011b. White matter alterations related to P300 abnormalities in individuals at high risk for psychosis: an MRI–EEG study. J. Psychiatry Neurosci.: JPN 36 (4), 239.

Fusar-Poli, P., Howes, O.D., Allen, P., Broome, M., Valli, I., Asselin, M.C., et al., 2011c. Abnormal prefrontal activation directly related to pre-synaptic striatal dopamine dysfunction in people at clinical high risk for psychosis. Mol. Psychiatry 16 (1), 67–75.

Fusar-Poli, P., Perez, J., Broome, M., Borgwardt, S., Placentino, A., Caverzasi, E., et al., 2007. Neurofunctional correlates of vulnerability to psychosis: a systematic review and meta-analysis. Neurosci. Biobehav. Rev. 31 (4), 465–484.

Gifford, G., Crossley, N., Fusar-Poli, P., Schnack, H.G., Kahn, R.S., Koutsouleris, N., et al., 2017. Using neuroimaging to help predict the onset of psychosis. Neuroimage 145, 209–217.

Glasser, M.F., Coalson, T.S., Robinson, E.C., Hacker, C.D., Harwell, J., Yacoub, E., et al., 2016. A multi-modal parcellation of human cerebral cortex. Nature 536 (7615), 171–178.

Goghari, V.M., Brett, C., Tabraham, P., Johns, L., Valmaggia, L., Broome, M., et al., 2014. Spatial working memory ability in individuals at ultra high risk for psychosis. J. Psychiatr. Res. 50, 100–105.

Hafizi, S., Da Silva, T., Gerritsen, C., Kiang, M., Bagby, R.M., Prce, I., et al., 2017. Imaging microglial activation in individuals at clinical high risk for psychosis: an in vivo PET study with [18 F] FEPPA. Neuropsychopharmacology 42 (13), 2474.

Han, X., Jovicich, J., Salat, D., van der Kouwe, A., Quinn, B., Czanner, S., et al., 2006. Reliability of MRI-derived measurements of human cerebral cortical thickness: the effects of field strength, scanner upgrade and manufacturer. Neuroimage 32 (1), 180–194.

Harris, J.M., Whalley, H., Yates, S., Miller, P., Johnstone, E.C., Lawrie, S.M., 2004. Abnormal cortical folding in high-risk individuals: a predictor of the development of schizophrenia? Biol. Psychiatry 56 (3), 182–189.

Horovitz, S.G., Fukunaga, M., de Zwart, J.A., van Gelderen, P., Fulton, S.C., Balkin, T.J., Duyn, J.H., 2008. Low frequency BOLD fluctuations during resting wakefulness and light sleep: a simultaneous EEG-fMRI study. Hum. Brain Mapp. 29 (6), 671–682.

Howes, O., Bose, S., Turkheimer, F., Valli, I., Egerton, A., Stahl, D., et al., 2011a. Progressive increase in striatal dopamine synthesis capacity as patients develop psychosis: a PET study. Mol. Psychiatry 16 (9), 885–886.

Howes, O.D., Bose, S.K., Turkheimer, F., Valli, I., Egerton, A., Valmaggia, L.R., et al., 2011b. Dopamine synthesis capacity before onset of psychosis: a prospective [18F]-DOPA PET imaging study. Am. J. Psychiatry 168 (12), 1311–1317.

Howes, O.D., Kambeitz, J., Kim, E., Stahl, D., Slifstein, M., Abi-Dargham, A., Kapur, S., 2012. The nature of dopamine dysfunction in schizophrenia and what this means for treatment: meta-analysis of imaging studies. Arch. Gen. Psychiatr. 69 (8), 776–786.

Howes, O.D., Kapur, S., 2009. The dopamine hypothesis of schizophrenia: version III—the final common pathway. Schizophr. Bull. 35 (3), 549–562.

Hurlemann, R., Matusch, A., Kuhn, K.-U., Berning, J., Elmenhorst, D., Winz, O., et al., 2008. 5-HT 2A receptor density is decreased in the at-risk mental state. Psychopharmacology 195 (4), 579–590.

Jack, C.R., Bernstein, M.A., Fox, N.C., Thompson, P., Alexander, G., Harvey, D., et al., 2008. The Alzheimer's disease neuroimaging initiative (ADNI): MRI methods. J. Magn. Reson. Imaging 27 (4), 685–691.

Jovicich, J., Czanner, S., Han, X., Salat, D., van der Kouwe, A., Quinn, B., et al., 2009. MRI-derived measurements of human subcortical, ventricular and intracranial brain volumes: reliability effects of scan sessions, acquisition sequences, data analyses, scanner upgrade, scanner vendors and field strengths. Neuroimage 46 (1), 177–192.

Kambeitz-Ilankovic, L., Meisenzahl, E.M., Cabral, C., von Saldern, S., Kambeitz, J., Falkai, P., et al., 2016. Prediction of outcome in the psychosis prodrome using neuroanatomical pattern classification. Schizophr. Res. 173 (3), 159–165.

Kanaan, R.A.A., Borgwardt, S., McGuire, P.K., Craig, M.C., Murphy, D.G.M., Picchioni, M., et al., 2009. Microstructural organization of cerebellar tracts in schizophrenia. Biol. Psychiatry 66 (11), 1067–1069.

Kanaan, R.A.A., Kim, J.-S., Kaufmann, W.E., Pearlson, G.D., Barker, G.J., McGuire, P.K., 2005. Diffusion tensor imaging in schizophrenia. Biol. Psychiatry 58 (12), 921–929.

Kang, J.I., Park, H.-J., Kim, S.J., Kim, K.R., Lee, S.Y., Lee, E., et al., 2013. Reduced binding potential of GABA-A/benzodiazepine receptors in individuals at ultra-high risk for psychosis: an [18F]-fluoroflumazenil positron emission tomography study. Schizophr. Bull. 40 (3), 548–557.

Karlsgodt, K.H., Niendam, T.A., Bearden, C.E., Cannon, T.D., 2009. White matter integrity and prediction of social and role functioning in subjects at ultra-high risk for psychosis. Biol. Psychiatry 66 (6), 562–569.

Katagiri, N., Pantelis, C., Nemoto, T., Tsujino, N., Saito, J., Hori, M., et al., 2018. Symptom recovery and relationship to structure of corpus callosum in individuals with an "at risk mental state. Psychiatry Res. Neuroimaging 272, 1–6.

Katagiri, N., Pantelis, C., Nemoto, T., Zalesky, A., Hori, M., Shimoji, K., et al., 2015. A longitudinal study investigating sub-threshold symptoms and white matter changes in individuals with an "at risk mental state"(ARMS). Schizophr. Res. 162 (1), 7–13.

Kempton, M.J., Underwood, T.S.A., Brunton, S., Stylios, F., Schmechtig, A., Ettinger, U., et al., 2011. A comprehensive testing protocol for MRI neuroanatomical segmentation techniques: evaluation of a novel lateral ventricle segmentation method. Neuroimage 58 (4), 1051–1059.

Knol, R.J.J., De Bruin, K., Eck-Smit, V., Berthe, L.F., Pimlott, S., Wyper, D.J., Booij, J., 2009. In vivo [123I] CNS-1261 binding to D-serine-activated and MK801-blocked NMDA receptors: a storage phosphor imaging study in rats. Synapse 63 (7), 557–564.

Koutsouleris, N., Kambeitz-Ilankovic, L., Ruhrmann, S., Rosen, M., Ruef, A., Dwyer, D.B., et al., 2018. Prediction models of functional outcomes for individuals in the clinical high-risk state for psychosis or with recent-onset depression: a multimodal, multisite machine learning analysis. JAMA Psychiatry 75 (11), 1156–1172.

Koutsouleris, N., Meisenzahl, E.M., Davatzikos, C., Bottlender, R., Frodl, T., Scheuerecker, J., et al., 2009. Use of neuroanatomical pattern classification to identify subjects in at-risk mental states of psychosis and predict disease transition. Arch. Gen. Psychiatr. 66 (7), 700–712.

Krakauer, K., Ebdrup, B.H., Glenthøj, B.Y., Raghava, J.M., Nordholm, D., Randers, L., et al., 2017. Patterns of white matter microstructure in individuals at ultra-high-risk for psychosis: associations to level of functioning and clinical symptoms. Psychol. Med. 1–19.

Kwon, J.S., Shim, G., Oh, J.S., Jung, W.H., Jang, J.H., Choi, C.-H., et al., 2010. Altered functional connectivity in subjects with Ultra-High Risk for psychosis using resting state fMRI. Schizophr. Res. 117 (2), 348–349.

Lawrie, S.M., Abukmeil, S.S., 1998. Brain abnormality in schizophrenia: a systematic and quantitative review of volumetric magnetic resonance imaging studies. Br. J. Psychiatry 172 (2), 110–120.

Lin, A., Wood, S.J., Nelson, B., Beavan, A., McGorry, P., Yung, A.R., 2015. Outcomes of nontransitioned cases in a sample at ultra-high risk for psychosis. Am. J. Psychiatry 172 (3), 249–258.

Lodge, D.J., Grace, A.A., 2008. Hippocampal dysfunction and disruption of dopamine system regulation in an animal model of schizophrenia. Neurotox. Res. 14 (2), 97–104.

Logothetis, N.K., Wandell, B.A., 2004. Interpreting the BOLD signal. Annu. Rev. Physiol. 66, 735–769.

Lord, L.-D., Allen, P., Expert, P., Howes, O., Lambiotte, R., McGuire, P., et al., 2011. Characterization of the anterior cingulate's role in the at-risk mental state using graph theory. Neuroimage 56 (3), 1531–1539.

Mechelli, A., Riecher-Rössler, A., Meisenzahl, E.M., Tognin, S., Wood, S.J., Borgwardt, S.J., et al., 2011. Neuroanatomical abnormalities that predate the onset of psychosis: a multicenter study. Arch. Gen. Psychiatr. 68 (5), 489–495.

Medaglia, J.D., Pasqualetti, F., Hamilton, R.H., Thompson-Schill, S.L., Bassett, D.S., 2017. Brain and cognitive reserve: translation via network control theory. Neurosci. Biobehav. Rev. 75, 53–64.

Meisenzahl, E.M., Koutsouleris, N., Gaser, C., Bottlender, R., Schmitt, G.J.E., McGuire, P., et al., 2008. Structural brain alterations in subjects at high-risk of psychosis: a voxel-based morphometric study. Schizophr. Res. 102 (1), 150–162.

Merritt, K., Egerton, A., Kempton, M.J., Taylor, M.J., McGuire, P.K., 2016. Nature of glutamate alterations in schizophrenia: a meta-analysis of proton magnetic resonance spectroscopy studies. JAMA Psychiatry 73 (7), 665–674.

Mizrahi, R., Addington, J., Rusjan, P.M., Suridjan, I., Ng, A., Boileau, I., et al., 2012. Increased stress-induced dopamine release in psychosis. Biol. Psychiatry 71 (6), 561–567.

Mizrahi, R., Hafizi, S., Gerritsen, C., Kiang, M., Bargby, M., Houle, S., et al., 2017. 117.3 imaging microglial activation in clinical high risk for psychosis and untreated first-episode psychosis: a PET StudyW [18F] FEPPA. Schizophr. Bull. 43 (Suppl. 1), S64.

Modinos, G., Allen, P., Grace, A.A., McGuire, P., 2015. Translating the MAM model of psychosis to humans. Trends Neurosci. 38 (3), 129–138.

Modinos, G., Şimşek, F., Azis, M., Bossong, M., Bonoldi, I., Samson, C., et al., 2018. Prefrontal GABA levels, hippocampal resting perfusion and the risk of psychosis. Neuropsychopharmacology 1.

Morey, R.A., Inan, S., Mitchell, T.V., Perkins, D.O., Lieberman, J.A., Belger, A., 2005. Imaging frontostriatal function in ultra-high-risk, early, and chronic schizophrenia during executive processing. Arch. Gen. Psychiatr. 62 (3), 254–262.

Mori, S., Zhang, J., 2006. Principles of diffusion tensor imaging and its applications to basic neuroscience research. Neuron 51 (5), 527–539.

Pantelis, C., Velakoulis, D., McGorry, P.D., Wood, S.J., Suckling, J., Phillips, L.J., et al., 2003. Neuroanatomical abnormalities before and after onset of psychosis: a cross-sectional and longitudinal MRI comparison. Lancet 361 (9354), 281–288.

Petcharunpaisan, S., Ramalho, J., Castillo, M., 2010. Arterial spin labeling in neuroimaging. World J. Radiol. 2 (10), 384.

Pettersson-Yeo, W., Allen, P., Benetti, S., McGuire, P., Mechelli, A., 2011. Dysconnectivity in schizophrenia: where are we now? Neurosci. Biobehav. Rev. 35 (5), 1110–1124. https://doi.org/10.1016/j.neubiorev.2010.11.004.

Power, J.D., Schlaggar, B.L., Petersen, S.E., 2015. Recent progress and outstanding issues in motion correction in resting state fMRI. Neuroimage 105, 536–551.

Rubinov, M., Knock, S.A., Stam, C.J., Micheloyannis, S., Harris, A.W.F., Williams, L.M., Breakspear, M., 2009. Small-world properties of nonlinear brain activity in schizophrenia. Hum. Brain Mapp. 30 (2), 403–416.

Schmidt, A., Antoniades, M., Allen, P., Egerton, A., Chaddock, C.A., Borgwardt, S., et al., 2017. Longitudinal alterations in motivational salience processing in ultra-high-risk subjects for psychosis. Psychol. Med. 47 (2), 243–254.

Schmidt, A., Crossley, N.A., Harrisberger, F., Smieskova, R., Lenz, C., Riecher-Rössler, A., et al., 2016. Structural network disorganization in subjects at clinical high risk for psychosis. Schizophr. Bull. 43 (3), 583–591.

Schobel, S.A., Chaudhury, N.H., Khan, U.A., Paniagua, B., Styner, M.A., Asllani, I., et al., 2013. Imaging patients with psychosis and a mouse model establishes a spreading pattern of hippocampal dysfunction and implicates glutamate as a driver. Neuron 78 (1), 81–93.

Shim, G., Oh, J.S., Jung, W.H., Jang, J.H., Choi, C.-H., Kim, E., et al., 2010. Altered resting-state connectivity in subjects at ultra-high risk for psychosis: an fMRI study. Behav. Brain Funct. 6 (1), 58.

Smith, S.M., Miller, K.L., Salimi-Khorshidi, G., Webster, M., Beckmann, C.F., Nichols, T.E., et al., 2011. Network modelling methods for FMRI. Neuroimage 54 (2), 875–891.

Smith, S.M., Vidaurre, D., Beckmann, C.F., Glasser, M.F., Jenkinson, M., Miller, K.L., et al., 2013. Functional connectomics from resting-state fMRI. Trends Cogn. Sci. 17 (12), 666–682.

Steen, R.G., Mull, C., Mcclure, R., Hamer, R.M., Lieberman, J.A., 2006. Brain volume in first-episode schizophrenia. Br. J. Psychiatry 188 (6), 510–518.

Stone, J.M., Howes, O.D., Egerton, A., Kambeitz, J., Allen, P., Lythgoe, D.J., et al., 2010. Altered relationship between hippocampal glutamate levels and striatal dopamine function in subjects at ultra high risk of psychosis. Biol. Psychiatry 68 (7), 599–602.

Studerus, E., Ramyead, A., Riecher-Rössler, A., 2017. Prediction of transition to psychosis in patients with a clinical high risk for psychosis: a systematic review of methodology and reporting. Psychol. Med. 47 (7), 1163–1178.

Suckling, J., Barnes, A., Job, D., Brenan, D., Lymer, K., Dazzan, P., et al., 2010. Power calculations for multicenter imaging studies controlled by the false discovery rate. Hum. Brain Mapp. 31 (8), 1183–1195.

Sun, D., Phillips, L., Velakoulis, D., Yung, A., McGorry, P.D., Wood, S.J., et al., 2009. Progressive brain structural changes mapped as psychosis develops in "at risk"individuals. Schizophr. Res. 108 (1), 85–92.

Takahashi, T., Wood, S.J., Yung, A.R., Soulsby, B., McGorry, P.D., Suzuki, M., et al., 2009. Progressive gray matter reduction of the superior temporal gyrus during transition to psychosis. Arch. Gen. Psychiatr. 66 (4), 366–376.

Thompson, P.M., Hayashi, K.M., Sowell, E.R., Gogtay, N., Giedd, J.N., Rapoport, J.L., et al., 2004. Mapping cortical change in Alzheimer's disease, brain development, and schizophrenia. Neuroimage 23, S2–S18.

Tuch, D.S., Reese, T.G., Wiegell, M.R., Makris, N., Belliveau, J.W., Wedeen, V.J., 2002. High angular resolution diffusion imaging reveals intravoxel white matter fiber heterogeneity. Magn. Reson. Med. 48 (4), 577–582.

Urenjak, J., Williams, S.R., Gadian, D.G., Noble, M., 1993. Proton nuclear magnetic resonance spectroscopy unambiguously identifies different neural cell types. J. Neurosci. 13 (3), 981–989.

Van Den Heuvel, M.P., Pol, H.E.H., 2010. Exploring the brain network: a review on resting-state fMRI functional connectivity. Eur. Neuropsychopharmacol. 20 (8), 519–534.

Van Den Heuvel, M.P., Sporns, O., 2011. Rich-club organization of the human connectome. J. Neurosci. 31 (44), 15775–15786.

van Tricht, M.J., Nieman, D.H., Koelman, J.H.T.M., Bour, L.J., van der Meer, J.N., van Amelsvoort, T.A., et al., 2011. Auditory ERP components before and after transition to a first psychotic episode. Biol. Psychol. 87 (3), 350–357.

van Tricht, M.J., Nieman, D.H., Koelman, J.H.T.M., van der Meer, J.N., Bour, L.J., de Haan, L., Linszen, D.H., 2010. Reduced parietal P300 amplitude is associated with an increased risk for a first psychotic episode. Biol. Psychiatry 68 (7), 642–648.

van Tricht, M.J., Ruhrmann, S., Arns, M., Müller, R., Bodatsch, M., Velthorst, E., et al., 2014. Can quantitative EEG measures predict clinical outcome in subjects at clinical high risk for psychosis? A prospective multicenter study. Schizophr. Res. 153 (1), 42–47.

Velakoulis, D., Wood, S.J., Wong, M.T.H., McGorry, P.D., Yung, A., Phillips, L., et al., 2006. Hippocampal and amygdala volumes according to psychosis stage and diagnosis: a magnetic resonance imaging study of chronic schizophrenia, first-episode psychosis, and ultra–high-risk individuals. Arch. Gen. Psychiatr. 63 (2), 139–149.

Vieira, S., Pinaya, W.H.L., Mechelli, A., 2017. Using deep learning to investigate the neuroimaging correlates of psychiatric and neurological disorders: methods and applications. Neurosci. Biobehav. Rev. 74, 58–75.

Vijayakumar, N., Bartholomeusz, C., Whitford, T., Hermens, D.F., Nelson, B., Rice, S., et al., 2016. White matter integrity in individuals at ultra-high risk for psychosis: a systematic review and discussion of the role of polyunsaturated fatty acids. BMC Psychiatry 16 (1), 287.

Vita, A., De Peri, L., Silenzi, C., Dieci, M., 2006. Brain morphology in first-episode schizophrenia: a meta-analysis of quantitative magnetic resonance imaging studies. Schizophr. Res. 82 (1), 75–88.

Wang, H., Guo, W., Liu, F., Wang, G., Lyu, H., Wu, R., et al., 2016. Patients with first-episode, drug-naive schizophrenia and subjects at ultra-high risk of psychosis shared increased cerebellar-default mode network connectivity at rest. Sci. Rep. 6.

Wernicke, C., 1906. Grundrisse der psychiatrie. Verlag von Georg Thieme, Leipzig.

Winton-Brown, T., Schmidt, A., Roiser, J.P., Howes, O.D., Egerton, A., Fusar-Poli, P., et al., 2017. Altered activation and connectivity in a hippocampal–basal ganglia–midbrain circuit during salience processing in subjects at ultra high risk for psychosis. Transl. Psychiatry 7 (10), e1245.

Wood, S.J., Yücel, M., Velakoulis, D., Phillips, L.J., Yung, A.R., Brewer, W., et al., 2005. Hippocampal and anterior cingulate morphology in subjects at ultra-high-risk for psychosis: the role of family history of psychotic illness. Schizophr. Res. 75 (2), 295–301.

Wotruba, D., Michels, L., Buechler, R., Metzler, S., Theodoridou, A., Gerstenberg, M., et al., 2013. Aberrant coupling within and across the default mode, task-positive, and salience network in subjects at risk for psychosis. Schizophr. Bull. 40 (5), 1095–1104.

Wright, I.C., Rabe-Hesketh, S., Woodruff, P.W.R., David, A.S., Murray, R.M., Bullmore, E.T., 2000. Meta-analysis of regional brain volumes in schizophrenia. Am. J. Psychiatry 157 (1), 16–25.

Yoon, Y.B., Yun, J.-Y., Jung, W.H., Cho, K.I.K., Kim, S.N., Lee, T.Y., et al., 2015. Altered fronto-temporal functional connectivity in individuals at ultra-high-risk of developing psychosis. PLoS One 10 (8), e0135347.

Young, J., Kempton, M.J., McGuire, P., 2016. Using machine learning to predict outcomes in psychosis. Lancet Psychiatry 3 (10), 908.

Yücel, M., Wood, S.J., Phillips, L.J., Stuart, G.W., Smith, D.J., Yung, A., et al., 2003. Morphology of the anterior cingulate cortex in young men at ultra-high risk of developing a psychotic illness. Br. J. Psychiatry 182 (6), 518–524.

Ziermans, T.B., Schothorst, P.F., Schnack, H.G., Koolschijn, P.C.M.P., Kahn, R.S., van Engeland, H., Durston, S., 2010. Progressive structural brain changes during development of psychosis. Schizophr. Bull. 38 (3), 519–530.

Zilles, K., Armstrong, E., Schleicher, A., Kretschmann, H.-J., 1988. The human pattern of gyrification in the cerebral cortex. Anat. Embryol. 179 (2), 173–179.

Zimmermann, R., Gschwandtner, U., Wilhelm, F.H., Pflueger, M.O., Riecher-Rössler, A., Fuhr, P., 2010. EEG spectral power and negative symptoms in at-risk individuals predict transition to psychosis. Schizophr. Res. 123 (2), 208–216.

CHAPTER

10

Genetic studies of psychosis

Hannah J. Jones[1], *Stanley Zammit*[2], *James T.R. Walters*[2]

[1]**Centre for Academic Mental Health, Department of Population Health Sciences, Bristol Medical School, University of Bristol, Bristol, United Kingdom;** [2]**MRC Centre for Neuropsychiatric Genetics and Genomics, Division of Psychological Medicine and Clinical Neurosciences, School of Medicine, Cardiff University, Cardiff, United Kingdom**

Introduction

Genetic epidemiological studies investigating family members have shown that the risk of schizophrenia in siblings of an individual with the disorder is approximately 8% compared to a lifetime risk of 0.8%–1% in the general population (Lichtenstein et al., 2009). Monozygotic (genetically identical) twin pairs have a higher concordance for schizophrenia (41%–65%) than dizygotic (nonidentical) pairs (0%–28%) (Cardno and Gottesman, 2000), indicating that the observed familial effects are, in part, due to genetic influences. Substantial heritability estimates have also been reported in family-based studies, with approximately 60%–80% of the variability in risk of schizophrenia due to genetic factors (Sullivan et al., 2003; Lichtenstein et al., 2009). Similar patterns of familial aggregation and heritability are also evident in bipolar disorder and schizoaffective disorder, with evidence of overlapping genetic influences with schizophrenia (Cardno et al., 2002; Lichtenstein et al., 2009). Therefore, there is clearly merit in investigating genetic factors to better understand the pathogenesis of psychotic conditions.

With major advances in genotyping technologies, computing power, statistical methodology, and international collaboration leading to larger samples, research into the genetic basis of psychiatric disorders is rapidly advancing. In this chapter, we touch upon early efforts in disentangling the role of genetics in psychosis and give a broad overview of recent research developments and results. We note that the majority of published findings so far have focused on the genetic etiology of schizophrenia; however, we also highlight some advances in our understanding of the genetics of bipolar disorder.

Candidate gene studies

Owing to the high heritability of schizophrenia, early investigations sought to uncover specific causal genes or high penetrance alleles

Risk Factors for Psychosis
https://doi.org/10.1016/B978-0-12-813201-2.00010-7

(i.e., alleles that when present at a specific locus carry a high risk of disease manifestation in the carrier). Genotyping in the late 20th century was both expensive and labor-intensive, and therefore investigators tended to genotype a limited number of genetic markers within candidate genes of interest. These genes were selected based on their hypothesized function (for example, genes related to dopaminergic function were strong candidates for schizophrenia given the known importance of dopamine receptor blockade for antipsychotic efficacy) or on results of existing genetic linkage studies of multiply affected families (where large genomic regions containing multiple genes are identified as areas of interest without reference to gene function).

Although thousands of candidate gene studies were reported for schizophrenia and bipolar disorder, none delivered robust or replicable findings. A review of the literature by Farrell et al. (2015) included a meta-analysis and expert evaluations of 25 historical schizophrenia candidate genes (for example, *DISC1*, *CHRNA7*, *DRD2*, and *COMT*). The review found that many studies reported genetic variants with large relative risks (often >1.5) but that these were clearly overestimates, with large genome-wide association studies (GWASs) (see next section for more detail) failing to support effect sizes of this magnitude or provide genome-wide evidence of association for these variants. The meta-analysis results of the 25 genes mostly converged on the null, and thus the authors concluded that historical candidate gene studies did not yield clear insights into the genetic basis of schizophrenia (Farrell et al., 2015).

Many candidate genes studied in schizophrenia have also been studied in bipolar disorder (for example, *BDNF*, *SLC6A3*, *DISC1*, and *DAOA*), although, similarly to schizophrenia, none have been established as bipolar disorder susceptibility genes (Barnett and Smoller, 2009). The most comprehensive meta-analysis of bipolar disorder linkage studies found genome-wide evidence of association with regions on the long arms of chromosomes 6 and 8 (McQueen et al., 2005); however, no specific genes have been robustly implicated (Bigdeli et al., 2013).

There are a number of potential reasons why candidate gene studies were not fruitful. It is clear that limitations in our understanding of the pathophysiology of psychotic disorders meant that biological hypotheses driving the selection of genes of interest were too simplistic and it was unlikely that a small number of genes would account for the vast variation in disorder risk. The low prior probability that the few genetic polymorphisms studied out of all genetic variation across the genome were causally related to psychosis, coupled with, as we know now, the small effect size of common variants, meant that statistically significant results from such studies were highly unlikely to reflect true effects (Button et al., 2013).

Candidate gene studies were vastly underpowered, with studies initially reporting strong genetic associations in novel genes that were due to chance, and that subsequent studies failed to replicate (Ioannidis et al., 2001). Furthermore, this early literature was hampered by inaccurate reports of "replication", based on different alleles or polymorphisms from the original studies, and inadequate adjustment for the multiple variants often tested. The full extent of the challenge to delineate the genetic risk of complex disorders such as schizophrenia has only become apparent in recent years as the polygenic nature of these conditions has emerged, thanks to the advent of genome-wide genotyping technologies and the vast increases in sample sizes enabled by international collaboration.

The era of genome-wide association studies

With the completion of the human genome project in 2003 and the International HapMap project in 2005, along with advances in genomic array technology that enabled the capture of many thousands of genetic variants across the genome, researchers were equipped to investigate genetic contributions to complex diseases in a "hypothesis-free" manner. This means that, in contrast to candidate gene approaches, no prior knowledge of the most likely genes involved in disease pathogenesis is required. GWASs enable hundreds of thousands of common (nonrare) genetic variants, called single-nucleotide polymorphisms (SNPs, often pronounced "snips"), to be assayed in individuals. Common genetic variants are usually defined as occurring in the population at a frequency of >1%. GWASs enable identification of SNPs that are either causally associated with the phenotype of interest or are associated indirectly through being highly correlated (i.e., in linkage disequilibrium [LD]) with the causal SNP. Given the large number of SNP–phenotype association tests in a GWAS and hence the high likelihood for false positives at conventional p-thresholds, adequately strong evidence of association is most commonly defined by a SNP–phenotype association p-value of $\leq 5 \times 10^{-8}$ (a Bonferroni corrected p-value for 1 million tests). SNPs reaching this significance threshold are generally referred to as "genome-wide significant hits."

Early GWASs of schizophrenia tended to use small samples of cases and controls or genetic data from pooled deoxyribonucleic acid (DNA) samples. Although no genome-wide significant polymorphisms were identified in two early GWASs using individual genotype data in samples including 178–738 cases (Lencz et al., 2007; Sullivan et al., 2008), a third GWAS aimed to validate suggestive SNP associations identified in an initial sample consisting of 479 cases and 2937 controls in larger replication cohorts (up to 7308 cases and 12,834 controls). Using this approach, the GWAS found strong evidence of association with a SNP in a gene of unknown function (zinc finger protein 804A [*ZNF804A*]), which reached the genome-wide threshold for significance when the phenotype was broadened to include cases of bipolar disorder (9173 combined cases) (O'Donovan et al., 2008). Early bipolar disorder GWASs also failed to identify genome-wide significant polymorphisms; however, by combining data from three separate studies resulting in a sample of 4387 cases and 6209 controls, a SNP within the ankyrin G gene (*ANK3*) reached genome-wide significance (Ferreira et al., 2008). Results of these GWASs indicated that individual common genetic variants confer small risk effects with odds ratios typically less than 1.5. Therefore, it became clear from these early studies that for GWASs of psychotic conditions to become fruitful, large sample sizes were essential and that this would only be obtained through collaboration between multiple research groups.

Following the parallel publication of three GWASs of schizophrenia by the International Schizophrenia Consortium (ISC) (International Schizophrenia Consortium et al., 2009), SGENE (Stefansson et al., 2009), and molecular genetics of schizophrenia (Shi et al., 2009), the drive to increase collaboration grew. This motivation resulted in the formation of the Psychiatric Genomics Consortium (PGC) and with it major advancement in our understanding of the genetics of psychiatric disorders. The initial aim of the PGC was to conduct large-scale meta-analyses of GWAS data for five major psychiatric disorders: attention-deficit/hyperactivity disorder (ADHD), autism, bipolar disorder, major depressive disorder, and schizophrenia (O'Donovan, 2015). The scope of this has since been expanded to include anorexia nervosa, obsessive–compulsive disorder/Tourette syndrome, posttraumatic stress disorders, and substance use disorder.

The first schizophrenia GWAS published by the Schizophrenia Working Group of the PGC in 2011 identified 7 genome-wide significant genomic regions (loci) using a total sample of up to 17,836 cases and 33,859 controls (Schizophrenia Psychiatric Genome-Wide Association Study Consortium, 2011). This number increased to 22 in a later study using up to 21,246 cases and 38,072 controls (Ripke et al., 2013), and in 2014, in a landmark study using a sample of up to 39,989 cases and 113,075 controls of mainly European ancestry, 128 genome-wide hits were identified spanning 108 independent genomic loci (Schizophrenia Working Group of the Psychiatric Genomics Consortium, 2014). Of these 108 loci, 75% contained protein-coding genes and a further 8% were within 20 kb of a gene. It is an important point that the causal polymorphisms within the associated genomic loci may not directly implicate specific genes within those regions and much present effort is going into using functional annotation data (for instance, from gene expression or chromatin characteristics) as a means of moving from associated loci to implicate causal genes. However, it remains reasonable to highlight genes of interest based on their location within loci. For example, among the 108 loci, interesting genes were highlighted including *DRD2* (encoding the dopamine D2 receptor, the site of action of all effective antipsychotic drugs), genes involved in glutamatergic neurotransmission and synaptic plasticity (including *GRM3*, *GRIN2A*, *SRR*, and *GRIA1*), genes involved in neurodevelopment (including *FXR1* and *SATB2*), and genes encoding voltage-gated calcium channel subunits (including *CACNA1C*, *CACNB2*, and *CACNA1I*), which, as mentioned later, are also implicated in bipolar disorder (Ferreira et al., 2008; Hamshere et al., 2013). The strongest associated region identified was within a region on chromosome 6 spanning the major histocompatibility complex (MHC), a family of surface proteins involved in regulating the adaptive immune system, thus building on previous GWAS findings (Shi et al., 2009; Stefansson et al., 2009; International Schizophrenia Consortium et al., 2009) and indicating a possible contribution of the immune system in the pathology of schizophrenia. The association across the MHC region is thought to arise from three independent signals and one of these has been found to be driven by alleles within complement component 4 (*C4*) gene, thus implicating the complement system in the development of schizophrenia (Sekar et al., 2016).

The most recent GWAS for schizophrenia utilized a large sample of schizophrenia cases from CLOZUK, a cohort consisting of individuals with treatment-resistant schizophrenia. After meta-analyzing the CLOZUK sample GWAS with independent PGC datasets, resulting in a total sample of 40,675 cases and 64,643 controls, 179 independent genome-wide hits were identified spanning 145 genomic loci (Pardiñas et al., 2018). The majority of these loci overlapped with those previously reported by the PGC; however, 50 novel loci were reported. The study also performed gene-based analyses using Multi-marker Analysis of GenoMic Annotation (MAGMA) software and showed that common variants associated with schizophrenia are highly enriched among loss-of-function intolerant genes (protein-coding genes that are intolerant to mutations that disrupt their function and so are under strong selection). The study also investigated 134 central nervous system (CNS)-related gene sets previously shown to be enriched in copy number variants (CNVs) associated with schizophrenia (Pocklington et al., 2015) (see later section for discussion of CNVs). The study identified six CNS-related gene sets that were independently associated with schizophrenia. The most strongly associated set comprised of targets of the fragile X mental retardation protein, previously implicated in schizophrenia, autism, and intellectual disability (Iossifov et al., 2012; Fromer et al., 2014).

The first bipolar disorder PGC GWAS meta-analysis, using up to 11,974 cases and 51,792 controls of European ancestry, identified two genome-wide hits: one within *CACNA1C*, which encodes the α subunit of the L-type voltage-gated calcium channel, and one within the first intron (sections of a gene not translated into protein) of *ODZ4*, a human homolog of a *Drosophila* pair-rule gene (Psychiatric GWAS Consortium Bipolar Disorder Working Group, 2011). By combining bipolar disorder and schizophrenia cases, a multigene region containing *NEK4-ITIH1-ITIH3-ITIH* was also implicated. A later bipolar disorder GWAS identified two additional novel loci, an intron of *ERBB2*, which encodes a receptor tyrosine kinase, and an intergenic region on chromosome 9 (Hou et al., 2016). However, the most striking bipolar disorder GWAS to date was performed by the bipolar disorder working group of the PGC and comprised of a sample of up to 29,764 cases and 169,118 controls of European descent. This GWAS identified 30 bipolar loci, 20 of which were novel, containing genes encoding ion channels and transporters, neurotransmitter transporters and synaptic components (Stahl et al., 2019).

It is difficult to be sure that SNPs identified in a GWAS are themselves causally related to the disease as they might just be correlated to the true causal SNPs (i.e., through LD). Further genotyping (or using specific statistical approaches; see below) is often used to fine-map the locus with the aim of finding the SNPs that are most likely to be causal. Once these high probability SNPs are identified, various downstream functional approaches, including animal and cellular modeling, can be employed to examine the effects of the SNP and understand the causal mechanisms. For example, following the identification of a strong association between a SNP within *ZNF804A* (rs1344706) and psychosis (O'Donovan et al., 2008; Zhang et al., 2011; Steinberg et al., 2011), subsequent studies focused on the potential mechanisms underlying *ZNF804A* and disease risk. Girgenti et al. (2012) found that overexpression of *ZNF804A* in rat forebrain neural progenitor cell cultures affected the expression of several schizophrenia-associated genes including *PRSS16*, *COMT*, *PDE4B*, and *DRD2*. Furthermore, complimentary work using human neural progenitor cells and human-induced pluripotent stem cells found that knockdown of *ZNF804A* resulted in differential expression of genes involved in cell adhesion, neural migration, synaptogenesis, and cytokine signaling (Hill et al., 2012; Chen et al., 2015). A proposed mechanism involves altered expression of *ZNF804A* through differential binding of rs1344706 alleles to nuclear proteins (Hill and Bray, 2011); however, rs1344706 has not been shown to consistently associate with differential expression of *ZNF804A* in adult tissue (Williams et al., 2011; Guella et al., 2014), although some evidence of association has been observed in fetal tissue (Hill and Bray, 2012; Tao et al., 2014). Therefore, it is clear to see that identifying a "genome-wide" association between a SNP and disorder in a GWAS does not immediately lead to a mechanistic understanding or pinpointing a specific therapeutic target.

Approaches are continually being developed with the aim of traversing the gap between identifying loci of interest in a GWAS and implicating potential pathological genes. As previously mentioned, these include fine-mapping methods which attempt to identify specific causal SNPs out of multiple SNPs in LD (Sekar et al., 2016; Pardiñas et al., 2018) as well as transcriptome-wide association studies (TWASs) that integrate information from gene expression and GWAS data to identify causal genes and better understand gene-based mechanisms underlying complex traits (Gusev et al., 2018). These approaches, together with methods using chromatin interaction (Hi-C) data, have been used to identify schizophrenia-related genes with potential regulatory roles in the developing brain (Pardiñas et al., 2018; Gusev et al., 2018). In even more recent work, GWAS results have also been integrated with single-cell RNA sequencing data from different brain cell types to show that common schizophrenia variants display a degree of cellular specificity in expression patterns. This study demonstrated enriched schizophrenia allele expression in pyramidal cells, medium spiny neurons and interneurons, but not other neuronal of glial cell types, thus implicating potential pathological mechanisms at a cellular level (Skene et al., 2018).

The ever-developing methodologies and increases in sample sizes make it an exciting time in the field of psychiatric genetics. However, it is clear that only through triangulation of evidence from multiple sources can GWAS results be translated into practices that will, in the long run, benefit patients.

SNP-based heritability and genetic correlations

Due to the major collaborative success of the PGC, the PGC Cross Disorder Working Group was able to explore SNP-based heritabilities and genetic correlations between the five major psychiatric disorders listed within its original aims. Using univariate and bivariate linear mixed models, implemented through the freely available Genome-wide Complex Trait Analysis (GCTA) software, the proportion of variance in liability explained by SNPs (i.e., SNP-based heritability) was 23% for schizophrenia and 25% for bipolar disorder (assuming a lifetime risk of 0.01 for each disorder) (Cross-Disorder Group of the Psychiatric Genomics Consortium et al., 2013). These SNP-based heritability estimates were markedly lower than heritability estimates based on twin and family studies, which is approximately 80% for schizophrenia (Sullivan et al., 2003) and 75% for bipolar disorder (Smoller and Finn, 2003). This so-called "missing heritability" was initially seen as a problem for GWAS but these lower estimates are to be expected as SNP-based estimates are only based on common (i.e., nonrare) variants correlated with SNPs included on genome-wide genotyping chips within samples of potentially limited power (Wray et al., 2014).

The PGC Cross Disorder Working Group study also showed that SNP-based genetic correlations are moderate to high between psychiatric disorders. For example, the genetic correlation was moderate between major depressive disorder and both schizophrenia and bipolar disorder (ranging from 0.43 to 0.47), whereas the correlation between schizophrenia and bipolar disorder was high (0.68) (Cross-Disorder Group of the Psychiatric Genomics Consortium et al., 2013). These findings have since been replicated by Anttila et al. (2018) who showed that, in contrast to neurological disorders including (but not limited to) Alzheimer's disease, multiple sclerosis, and epilepsy, psychiatric disorders share between them a considerable portion of common genetic variation. The study showed strong evidence of positive genetic correlations between schizophrenia and bipolar disorder, ADHD, anorexia nervosa, autism spectrum disorder, major depressive disorder, and obsessive–compulsive disorder (Anttila et al., 2018).

Genetic correlations between disorders and traits may reflect pleiotropic effects whereby genetic variants independently confer risk to multiple psychiatric disorders. However, genetic correlations may also reflect phenotypic overlap between disorders or possibly sample heterogeneity. For example, Allardyce et al. (2018) demonstrated a gradient of genetic liability across schizophrenia and bipolar disorder subtypes. It was found that, across clinical phenotypes, genetic liability for schizophrenia was most strongly associated with schizophrenia diagnosis, followed by schizoaffective bipolar disorder, bipolar I disorder, and lastly bipolar II disorder. In addition, bipolar disorder cases with psychosis carried a higher burden of schizophrenia risk alleles compared with bipolar disorder without a history of psychosis, and schizophrenia polygenic score was also increased among those with psychosis that was mood-incongruent, the psychotic symptoms that most closely overlap with schizophrenia (Allardyce et al., 2018).

The modest to high genetic correlations between psychiatric disorders, as well as evidence of shared gene expression perturbations (Gandal et al., 2018) and neurogenetic architecture (Lee et al., 2016) across disorders, blur the boundaries between current psychiatric diagnostic classifications. However, the degree of within-disorder common genetic correlation is greater than the degree of between-disorder correlation (Bulik-Sullivan et al., 2015; Cross-Disorder Group of the Psychiatric Genomics Consortium et al., 2013) suggesting that current genetic evidence alone does not provide a foundation for the restructuring of psychiatric classifications. Nevertheless, understanding the shared genetic etiology between psychotic disorders has the potential to inform mapping of shared pathophysiological processes and biologically valid stratification of cases (O'Donovan and Owen, 2016) and, in the long run, may be utilized to improve disorder prediction accuracy (Maier et al., 2015).

Investigating polygenic liability

Polygenic inheritance is thought to explain the continuous distribution of complex traits within the population. Polygenicity implies that phenotypic variation arises from the independent, random inheritance of a large number of genetic variants, with each producing a small effect on the presentation of the phenotype. As for all multifactorial complex diseases, crossing the liability threshold into disorder is based on an individual's genetic susceptibility as well as exposure to environmental risk factors (Gottesman and Shields, 1967). Position on the liability curve will vary from person to person which may explain the variation in the presentation of schizophrenia spectrum disorders or psychopathology.

Before the first PGC schizophrenia study, the ISC published a schizophrenia GWAS using 3322 cases and 3587 controls (International Schizophrenia Consortium et al., 2009). Using results from their GWAS, the ISC study was the first to test the theory of polygenic inheritance by investigating the effect of genetic variants *en masse* on schizophrenia case–control status. To do this, the researchers generated aggregate risk scores for individuals within the sample based on the GWAS SNP information. Using these "polygenic scores," the ISC study not only provided evidence for a substantial polygenic basis of schizophrenia (i.e., risk involves many common genetic variants, each with small effect) but also showed that this polygenicity was markedly shared with bipolar disorder (International Schizophrenia Consortium et al., 2009).

Polygenic scores, also known as polygenic risk scores, genetic risk scores, and allelic scores, have since become a very popular tool in psychiatric genetics and have been predominantly used to investigate genetic overlap between disorders and phenotypic manifestation of genetic liability across development, as well as prediction of disorder, progression, and treatment response in clinical samples (Bogdan et al., 2018).

An individual's polygenic score for a particular trait of interest in a target sample is generally constructed by summing the number of risk alleles an individual has, as determined by the trait's GWAS results in a discovery sample, across multiple SNP loci across the genome (Fig. 10.1). To capture polygenicity, scores are commonly generated using SNPs that meet a liberal GWAS p-value threshold in the discovery study (e.g., p-value ≤ 0.5), although researchers may also generate multiple scores based on a range of GWAS p-value thresholds (e.g., ≤ 0.5 to $\leq 5 \times 10^{-8}$) to investigate the influence of incorporating SNPs with higher or lower likelihood of true effects. Before score construction, SNPs are normally subjected to quality control procedures so that scores are based on common, high-quality, independent (i.e., in linkage equilibrium) variants. Weighted scores can be constructed by multiplying the number of risk

Discovery Sample

1. Obtain SNP-phenotype association information from a well powered GWAS of the phenotype of interest.

2. Select SNPs needed for score construction, e.g. based on a specific GWAS p-value threshold.

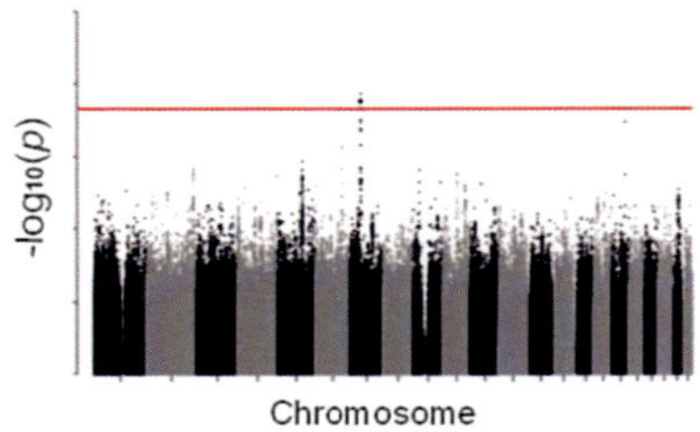

3. For each SNP, note risk allele, frequency, effect size and p-value.

4. Filter SNP list based on pre-determined quality control parameters, e.g. minor allele frequency >1%.

SNP ID	Risk allele	Risk allele frequency Cases	Risk allele frequency Controls	log(OR)	P value
SNP1	G	0.731	0.724	0.032	0.415
SNP2	G	0.799	0.792	0.053	0.219
SNP3	G	0.814	0.810	0.020	0.656
SNP4	A	0.915	0.907	0.108	0.079
SNP5	G	0.886	0.885	0.008	0.876
SNP6	A	0.869	0.871	0.031	0.534
... SNP*n*	...	...	...	...	...

Target Sample

5. Count number of risk alleles per SNP per individual in the target sample.

6. Multiply each allele count by the SNP weight (normally the effect size from the discovery sample GWAS)

7. Add up all the weighted allele counts to generate each individual's polygenic score for the phenotype of interest

		no. risk alleles			no. risk alleles x weight (SNP score)		
SNP ID	SNP weight	Individual 1	Individual 2	Individual 3	Individual 1	Individual 2	Individual 3
SNP1	0.032	GG (2)	TG (1)	TT (0)	0.064	0.032	0
SNP2	0.053	GA (1)	GG (2)	AG (1)	0.053	0.106	0.053
SNP3	0.020	AA (0)	AG (1)	GG (2)	0	0.020	0.040
SNP4	0.108	AT (1)	AT (1)	AA (2)	0.108	0.108	0.216
... SNP*n*	...	...	...	...	...	...	...
Polygenic score (SNP1 score + SNP2 score + ... + SNP*n* score):					0.225	0.266	0.309

FIGURE 10.1 Steps in generating a polygenic score using SNP information from a discovery sample and genotype information from a target sample. *GWAS*, genome-wide association study; *log(OR)*, logarithm of the SNP-phenotype odds ratio; *SNP*, single-nucleotide polymorphism.

alleles at each SNP locus (0, 1, or 2) by the SNP's reported GWAS effect size (e.g., regression beta or log odds ratio). This ensures that SNPs yielding larger effect sizes have a greater weighting in the score than SNPs yielding smaller effect sizes. The discovery and target samples should not have any sample overlap (i.e., they are independent from one another) to avoid introducing biases within analyses using polygenic scores.

As part of the 2014 PGC schizophrenia GWAS, polygenic scores were generated within a subgroup of the samples available, based on a range of GWAS p-value thresholds, and regressed against case–control status in the remaining, independent groups. It was shown that including more SNPs, even those showing extremely weak evidence of association with schizophrenia, explained more variance in case–control status than including fewer SNPs with strong associations (Schizophrenia Working Group of the Psychiatric Genomics Consortium, 2014). Based on SNPs meeting a GWAS p-value threshold of ≤ 0.05, the schizophrenia polygenic score was shown to explain approximately 7% of phenotype variation on the liability scale, with odds ratios for schizophrenia ranging from 7.8 to 20.3 when comparing the highest to lowest score deciles in different samples.

The predictive utility of schizophrenia polygenic scores has since been investigated in clinical samples (Bogdan et al., 2018), including those of first-episode psychosis patients. For example, Vassos et al. (2017) found that schizophrenia polygenic scores were higher in first-episode psychosis subjects who went on to develop schizophrenia compared to unaffected controls and those who went on to develop other psychoses; however, their discriminative accuracy was low.

In an attempt to improve prediction accuracy, more advanced methods have been developed, which are beyond the scope of this chapter to describe, that exploit information from multiple, genetically correlated traits (for further information about these methods, please see the studies cited below). For example, Li et al. (2014) employed a method called bivariate ridge regression and showed that utilizing genetic correlations between schizophrenia and bipolar disorder improved predictive power in a dataset consisting of cases of bipolar and related disorders, cases of schizophrenia, and unaffected controls. Another method developed by Maier et al. (2015), which they called multivariate genomic best linear unbiased predictor (MTGBLUP), was also shown to significantly increase risk prediction accuracy in independent case–control samples of schizophrenia, bipolar disorder, and major depressive disorder, as utilizing data from multiple phenotypes allowed them to better estimate disease-specific SNP effects. This method has also been extended to utilize summary statistics from multiple GWASs of different traits, although with slightly less accuracy in prediction than MTGBLUP (Maier et al., 2018).

Currently, the low predictive power arising from identified common risk alleles places limits on any clinical application, and hence the clinical utility of polygenic risk scores in psychiatry is at present unclear. This situation may change as prediction improves with advances in statistical methods such as the ones discussed above and more powerful GWASs, but at the present time it is not obvious how such scores could be validly applied in a clinical, let alone public health, context. Hence, any promotion of such tests requires caution and careful scrutiny. In contrast, there are perhaps more apparent and valid arguments for the potential clinical utility of detecting rare risk variants given their large effect sizes and, in some case, large exposed attributable risk (Gershon and Alliey-Rodriguez, 2013). We consider this issue and associated ethical considerations in the later section on CNVs.

One major strength of polygenic scores is that they are not limited to use in clinical samples. That is, if there are individual-level genetic data present and a well-powered discovery GWAS, genetic liability for traits of interest can be quantified in any sample of individuals (as long as the target sample was not used in the original discovery GWAS). Therefore, polygenic scores have emerged as valuable research tools in investigating life-time manifestation of genetic liability within non-ill individuals in the general population.

Within childhood and adolescence, weighted polygenic scores for schizophrenia have been shown to associate with negative symptoms and anxiety (Jones et al., 2016), depression (Nivard et al., 2017), internalizing and externalizing scores (Jansen et al., 2018), and cognitive deficits including performance IQ and social and communication difficulties (Hubbard et al., 2016; Riglin et al., 2017; St Pourcain et al., 2018). Schizophrenia polygenic scores have also been shown to associate with potential risk-related factors including cannabis use (Verweij et al., 2017) and greater population density (Colodro-Conde et al., 2018). Interestingly, the schizophrenia scores also strongly associate with lower participation of mothers and children within the Avon Longitudinal Study of Parents and Children (ALSPAC) longitudinal birth cohort (Martin et al., 2016) implying that caution is needed when interpreting genetic analyses in recruited samples which may be subject to selection bias.

In contrast to schizophrenia, fewer studies have found strong evidence of association between bipolar disorder polygenic scores and outcomes of interest (Mistry et al., 2018a), although this might be due to the smaller sample sizes used in the discovery samples, and hence greater measurement error in risk scores. Using published data and simulations, Dudbridge (2013) has shown that large samples are necessary in providing adequate statistical power to identify associations between polygenic scores and traits of interest, with the size of the discovery sample being particularly important in risk prediction.

With the ever-increasing GWAS sample sizes and availability of GWAS summary statistics, there is likely to be a large increase in studies utilizing polygenic scores. Although effect sizes are often only modest, polygenic score studies show their worth in identifying potential early phenotypic manifestations of adult-onset disorders and indicating if observed associations between risk factors and disorder may be partly driven by shared genetic etiology. There is, however, a need for a standardized approach to reporting effects sizes and confidence intervals to allow for meta-analyses across studies (Mistry et al., 2018b). There is also a need for analyses to move beyond simply identifying associations with broad genetic risk, for example, by examining polygenic scores based on SNPs in specific gene pathways or based on expression data of genes identified by TWASs (Gusev et al., 2018) or by implementing more sophisticated mediation-based analyses. Such advances have the potential to help identify targets for prevention or intervention in at-risk populations and improve our understanding of the disorder neuropathogenesis.

Copy number variants and sequencing studies

In parallel to the advances that have been made from GWAS in elucidating the role of common genetic variation in schizophrenia, rare genetic variants have also been identified as important risk factors. The CNVs demonstrated as increasing risk of developing schizophrenia are rare ($<1\%$ population frequency) duplications or deletions of large segments of DNA, typically from 100,000 to several million base pairs (Lee and Scherer, 2010). More recently, very rare (less than 0.1% population frequency) single-base exonic mutations (variants within the protein-coding portion of the genome) have been shown to increase risk of neurodevelopmental disorders including schizophrenia. These classes of variant can be inherited or occur *de novo* (new, not inherited). Taken together with the contribution of common genetic variation, these findings indicate schizophrenia is a complex condition with a genetic architecture arising from a mixed economy of common and rare genetic risk factors. While at present the majority of the attributable genetic risk for schizophrenia arises from common variation, rarer genetic risk factors have proved fruitful from a research perspective not least due to their larger risk compared to common genetic factors.

Large chromosomal deletions or duplications were originally identified using light microscopy and solid staining techniques. Fluorescence in situ hybridization methods then allowed for detection of more subtle chromosomal changes. More recently, the emergence of high-throughput microarray and next-generation whole-genome sequencing technologies has meant that a large number of smaller CNVs could be rapidly measured from across the genome in large samples.

While structural variation is a common phenomenon across the human genome, it is a restricted group of large and rare CNVs that have been associated with neurodevelopmental disorders. The first CNV linked to schizophrenia was a recurrent deletion of approximately 3 million base pairs on chromosomal band 22q11.2 (i.e., the 11.2 band of the long "q" arm of chromosome 22). This CNV causes a congenital disorder formerly known as velocardiofacial syndrome or DiGeorge syndrome and is associated with intellectual disability and a range of dysmorphic features (see Chapter 8 for more detail). It was noted that the syndrome involving 22q11.2 deletion was often accompanied by psychosis (Chow et al., 1994) and it is estimated that 24% of deletion carriers develop schizophrenia (Murphy et al., 1999). More recently, using microarray technology that captured information on rare CNVs in a sample of 21,094 schizophrenia cases and 20,227 controls, Marshall et al. (2017) found that this deletion was present in 0.3% of cases and conveyed an estimated 68-fold increased odds of developing schizophrenia. Interestingly, the incidence of schizophrenia in carriers of large (1.5–3.0 Mb) duplications of 22q11.2 is lower than in the general population implying that duplication of this chromosomal band is protective against the disorder (Rees et al., 2014a).

Following the development of genome-wide array technologies, many additional schizophrenia-associated CNVs have been identified. In 2008, the earliest genotype-led study in a small sample of parent-proband trios found two likely pathogenic CNVs. One was an inherited deletion at 2p16.3 that disrupts *NRXN1* and the other was a *de novo* duplication at 15q13.1 spanning *APBA2* (Kirov et al., 2008). Although these results were only suggestive, disorder-associated disruptions at *NRXN1* were later confirmed using a larger case–control sample (Rujescu et al., 2009). Later in 2008, two larger studies provided robust evidence of the involvement of deletions at 15q13.3, 1q21.11, and 15q11.2 in schizophrenia (International Schizophrenia Consortium, 2008) and a combined sample of schizophrenia and related psychoses, including unspecified functional psychosis, schizoaffective disorder, schizophreniform, and persistent delusional disorder (Stefansson et al., 2008). These deletions are large and implicate multiple genes.

In 2014, a meta-analysis of previously implicated candidate CNVs identified 11 CNV loci robustly associated with schizophrenia (Rees et al., 2014b) (Table 10.1). It was found that approximately 2.5% of schizophrenia cases and 0.9% of the control group carried at least 1 of the candidate CNVs, which were associated with a wide range of odds ratios ranging from approximately 2 to >50. Large effects were also observed in a recent PGC schizophrenia CNV study which found definitive evidence implicating 8 of the 11 previously reported CNVs as risk factors for schizophrenia as well as suggestive evidence for 6 novel loci (Marshall et al.,

TABLE 10.1 List of 11 CNVs implicated as schizophrenia risk loci by Rees et al. (2014b) and the 22q11.2 duplication protective locus (Rees et al., 2014a, 2016).

CNV	Type of CNV	Region	OR (95% CI)[a]	Reported by Marshall et al. (2017)
1q21.1	Deletion	chr1:146,527,987–147,394,444	13.90 (2.02–596.86)	Yes[b]
1q21.1	Duplication	chr1:146,527,987–147,394,444	1.89 (0.45–9.13)	Yes[b]
NRXN1 exonic	Deletion	chr2:50,145,643–51,259,674	1.58 (0.34–7.96)	Yes (2p16.3)[b]
3q29	Deletion	chr3:195,720,167–197,354,826	5.05 (0.50–248.45)	Yes[b]
WBS (7q11.23)	Duplication	chr7:72,744,915–74,142,892	Inf (0.032–Inf)	Yes[b]
15q11.2	Deletion	chr15:22,805,313–23,094,530	2.38 (1.42–4.08)	Yes
PWS/AS	Duplication	chr15:22,805,313–28,390,339	Inf (1.49–Inf)	No
15q13.3	Deletion	chr15:31,080,645–32,462,776	3.79 (0.30–198.64)	Yes[b]
16p13.11	Duplication	chr16:15,511,655–16,293,689	2.38 (1.28–4.58)	No
16p12.1	Deletion	chr16:21,950,135–22,431,889	3.3 (1.61–7.05)	No
16p11.2	Duplication	chr16:29,650,840–30,200,773	8.43 (2.50–44.37)	Yes[b]
22q11.2	Deletion	chr22:19,037,332–21,466,726	Inf (4.19–Inf)	Yes[b]
22q11.2	Duplication	chr22:19,037,332–21,466,726	0.32 (0.0064–3.19)	Yes

chr, chromosome; *CI*, confidence interval; *CNV*, copy number variant; *Inf*, infinity; *OR*, odds ratio; *PWS/AS*, Prader–Willi Syndrome/Angelman Syndrome; *WBS*, Williams–Beuren Syndrome.

[a] *ORs and 95% CIs reported by Rees et al. (2016).*

[b] *Reached genome-wide significance (family-wise error rate < 0.05) in Marshall et al. (2017).*

2017). Carrying one of these eight CNVs was estimated to explain approximately 1% in schizophrenia liability. The study also reported that the global enrichment of CNV burden in cases was enriched for genes associated with synaptic function and neurobehavioral phenotypes (Marshall et al., 2017). In addition, Pocklington et al. (2015) examined genes deleted or duplicated by CNVs in schizophrenia and showed they were enriched for those involved in glutamate and GABAergic signaling. It is worth noting, however, that given the rarity of the CNVs identified so far (population frequency <1%), it is not surprising that only those with large effects have been identified. In fact, power analyses have indicated that even a sample size of 100,000 cases and 100,000 controls may still be too small to confidently implicate very rare CNVs (Rees et al., 2016).

A notable feature of all CNVs implicated in schizophrenia is that they are pleiotropic, in that they also confer risk for other neurodevelopmental disorders, including autism, epilepsy, intellectual disability, and ADHD (Williams et al., 2010). For example, a recent study of 20,403 schizophrenia cases and 26,628 controls found that CNVs implicated in intellectual disability were also enriched in schizophrenia (Rees et al., 2016). In contrast to the overlap between childhood neurodevelopmental disorders and schizophrenia, previously reported schizophrenia CNVs are not so strongly associated with bipolar disorder. A study by Green et al. (2016) using a sample of 2591 bipolar disorder cases, 6882 schizophrenia cases, and 8842 controls reported that out of 15 schizophrenia-related CNV loci, only 1 CNV (16p11.2) was associated with bipolar disorder at genome-wide significance, with 2 additional CNVs being nominally associated.

As pointed out by Rees et al. (2014b), a number of CNVs implicated in neurodevelopmental disorders are also associated with physical disease phenotypes including congenital heart disease (1q21.1 and 22q11.2), microcephaly (1q21.1, 3q29, and 16p11.2), renal cysts and diabetes (17p12), and obesity (distal 16p11.2). Thus, the seemingly large and pleiotropic effects of carrying rare CNVs raise the question of their possible clinical utility for genetic testing or genetic counseling and patient management (International Society of Psychiatric Genetics, 2017). For example, identifying individuals with schizophrenia who have a CNV may not only have implications for physical health monitoring but also raise potentially important questions for them and their family about risk in relatives. The ethical issues raised by clinical CNV testing in psychiatric settings are complex and for many of the implicated CNVs we do not yet have a solid evidence base to accurately convey the risks involved. Research is underway to address these questions and key to this effort will be to engage all stakeholders including patients and their families (see Gershon and Alliey-Rodriguez, 2013 for a review of ethical issues surrounding genetic testing in psychiatry).

In addition to CNVs, high-throughput next-generation sequencing technology has also allowed for the investigation of rare variants within the 1% of the genome that is protein coding (the exome) at single-base resolution (i.e., single-nucleotide variants [SNVs] and insertions/deletions of a small number of bases). The implication is that rare protein-coding variations, particularly those changes associated with loss of function of the protein, are likely to evoke more severe consequences on the phenotype in comparison to variations within the noncoding genome (which are primarily involved in gene regulation).

Studies using exome sequencing data have not reported consistent or substantially raised frequencies of rare (whether inherited or *de novo*) SNVs in schizophrenia cases compared to controls (Xu et al., 2012; Fromer et al., 2014; McCarthy et al., 2017). Fromer et al. (2014) showed that small, nonsynonymous (amino acid altering) *de novo* mutations in schizophrenia were overrepresented in gene sets encoding glutamatergic postsynaptic proteins within the activity-regulated cytoskeleton-associated protein (ARC) and N-methyl-D-aspartate receptor (NMDAR) complexes, which are also enriched in *de novo* CNVs (Kirov et al., 2012). A complimentary case–control study demonstrated a polygenic burden of very rare (allele frequency less than 1 in 10,000) variants in schizophrenia and also found enrichment in the ARC complex, as well as in gene sets related to calcium channels (Purcell et al., 2014). Although encouraging, these earlier studies were too small to implicate individual genes at a genome-wide level. A more recent meta-analysis, however, has proved more promising in this respect by examining larger datasets. Singh et al. (2016) investigated a sample of 4264 schizophrenia cases, 9343 controls, and 1077 parent-proband trios and identified a genome-wide significant association between schizophrenia and rare loss-of-function variants in *SETD1A*, a gene that encodes an enzyme involved in the methylation of lysine residues in histone H3. Rare loss-of-function variants in *SETD1A* were also found to associate with developmental delay (Singh et al., 2016). This is consistent with previous work showing that individuals with schizophrenia or autism who carry rare *de novo* mutations have higher rates of cognitive impairment (Fromer et al., 2014; Iossifov et al., 2014) as has recently been shown to be the case in a schizophrenia case–control study (Singh et al., 2017).

Compared to schizophrenia, there are fewer studies investigating bipolar disorder using exome sequencing. In the largest published exome sequencing study of bipolar disorder to date, involving 8 multiplex families and a case–control meta-analysis of 3541 cases and 4774 controls, rare damaging variants in bipolar disorder were enriched for genes previously linked to schizophrenia and autism (Goes et al., 2016). Larger studies are underway and will provide important insights into the contribution of rare exonic variation to bipolar disorder but at present, findings suggest this role is less than in neurodevelopmental disorders, certainly for the classes of variants studied to date.

Common variant loci and gene sets identified by GWAS show some overlap with genes associated with rare and *de novo* mutations (Schizophrenia Working Group of the Psychiatric Genomics Consortium, 2014; Pardiñas et al., 2018). Exome array studies have also shown that schizophrenia and bipolar disorder rare variants are enriched among genes that map to autosomal loci identified in GWASs (Richards et al., 2016; Lescai et al., 2017). Furthermore, individuals with schizophrenia who have known pathogenic CNVs carry an excess burden of common risk alleles (Tansey et al., 2016). Convergence of common and rare signals therefore supports a liability threshold model of psychotic disorders, where the outcome of rare variants may be determined by the complement of common variants (Owen and O'Donovan, 2017).

Causal inference using Mendelian randomization

Identification of genetic variants that increase risk of specific phenotypes can also aid in determining which environmental exposures have causal effects on psychiatric disorders. Observational findings between environmental risk factors and disease are often limited by unknown confounding factors that may explain associations between an exposure and outcome or reverse causality whereby disease-related factors exert an effect on the exposure. While randomized controlled trials (RCTs) are designed to overcome these limitations, such studies are not always possible because of ethical or practical reasons.

Mendelian randomization (MR) is an approach which has been developed to overcome the limitations of observational analyses by using genetic variants, either identified from candidate gene studies or GWAS, as proxies for modifiable exposures of interest (Davey Smith and Ebrahim, 2003; Davey Smith and Hemani, 2014). Associations between these genetic variants, often referred to in MR as genetic instrumental variables, and an outcome of interest imply a causal relationship if certain assumptions are met. The core assumptions are that (i) the genetic variables are strongly associated with the exposure, i.e., they are strong instruments, (ii) the genetic variants are not associated with any confounding factors between the exposure and outcome, and (iii) the genetic variants are only associated with the outcome through the exposure (Fig. 10.2). As an individual's alleles are randomly allocated at conception, an MR is likened to a "natural" RCT whereby an individual is randomly allocated as to whether they have a certain genetic signal for the exposure or not. This random allocation should also

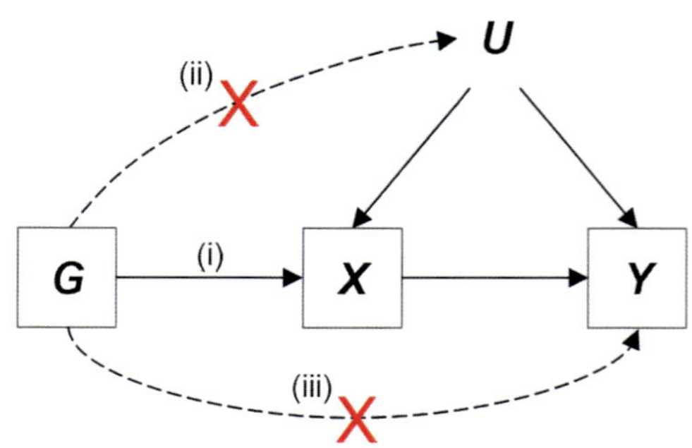

FIGURE 10.2 Diagram of the main Mendelian randomization assumptions. (i) The genetic instrumental variable *G* is strongly associated with the exposure *X*, (ii) the genetic instrumental variable *G* is not associated with any confounding factors *U* between exposure *X* and outcome *Y*, and (iii) the genetic instrumental variable *G* is only associated with outcome *Y* through exposure *X*. Solid arrows represent an effect of one variable on another, dashed arrows with a cross represent no effect of one variable on another.

mean that genetic instruments are not associated with other characteristics that might act as confounders in observational studies.

MR studies can be carried out within population samples that have exposure, outcome, and genetic data available or by using SNP summary statistics from exposure and outcome GWASs. The latter approach is commonly known as two-sample MR (Burgess et al., 2013; Bowden et al., 2017) as SNP information on the exposure is obtained from a different, nonoverlapping sample to the sample used to obtain SNP information on the outcome. Two-sample MR has greatly increased the scope of MR, especially when it is difficult to measure an exposure and outcome in a single sample. Methodological advancements, such as MR Egger regression (Bowden et al., 2015), have also allowed for the assessment of MR assumptions, vastly improving our understanding of the strengths and limitations of MR analyses.

Within psychiatric epidemiology, MR has been used to investigate the causality of a number of behavioral and biological risk factors for psychosis. For example, Wium-Andersen et al. (2015) investigated the potential causal relationship between heaviness of tobacco use and schizophrenia by using a SNP within the *CHRNA3* gene cluster (rs1051730), known to be associated with smoking intensity, as a genetic instrumental variable. They found that the SNP was associated with smoking heaviness in smokers, thus satisfying the first core assumption of MR. They then went on to test whether the SNP was associated with antipsychotic medication use, schizophrenia, antidepressant medication use, and/or depression in the general population within smokers and nonsmokers. Stratifying the sample by smoking status allowed the authors to determine if an association between smoking heaviness and each outcome is indicative of pleiotropy or of a casual association. Pleiotropy occurs when a genetic variant influences more than one independent phenotype and therefore would be evident if the genetic variant for smoking heaviness was associated with an outcome in nonsmokers. This would mean that the genetic instrumental variable is associated with the outcome via an alternative path not through the exposure, thus violating the third core assumption of MR. If the association between the genetic variant and the outcome is only seen in smokers, then evidence of a casual association is much more robust. The study found that the smoking heaviness SNP was associated with antipsychotic medication use in smokers but not in nonsmokers, but there was very weak evidence of an association with the other outcomes of interest (Wium-Andersen et al., 2015). An additional two-sample MR study showed little evidence of a causal association between cigarette smoking initiation and schizophrenia risk (Gage et al., 2017b), while a two-sample MR of cannabis initiation and schizophrenia risk provided evidence for causal pathways operating in both directions (Gage et al., 2017a). However, while the lack of adequate samples and strong genetic instruments for regular cannabis or cigarette use limited previous MR studies of these exposures, advances are now being made through the publications of larger, more robust GWASs of substance use.

MR studies have also investigated causal relationships between schizophrenia and intelligence (Savage et al., 2018), inflammatory biomarkers (Hartwig et al., 2017), and neuroticism (Nagel et al., 2018). However, most that have investigated bidirectional associations (i.e., using schizophrenia liability as an exposure as well as studying schizophrenia as an outcome) have found causal effects in both directions. Bidirectional MR is useful in investigating reverse causation, but when there is little understanding of the biological effects of the genetic instruments, which is true for most SNPs associated with schizophrenia, bidirectional associations can be misleading (Davey Smith and Hemani, 2014). It is only with a good understanding of each variants' biological effects that we can infer causality from MR results with confidence.

Well-powered, well-designed MR studies are invaluable in supplementing information where RCTs are not feasible and in informing whether future RCTs are worthwhile. Despite current limitations, it is likely that as GWAS sample sizes increase, enabling the identification of stronger genetic instruments, and methodologies continue to improve, MR will have an important role in determining whether associations with modifiable environmental exposures are in fact causal.

Epigenetics

Within this chapter we have so far only referred to structural genetic variation; however, epigenetics may also play a role in psychosis risk. Epigenetic processes regulate gene expression through modifications of DNA, histone proteins, and chromatin (Dempster et al., 2013). These processes are essential for cellular development and differentiation and, in contrast to the DNA sequence, are highly dynamic and modifiable. Most epigenetic studies focus on DNA methylation which is a process that decreases or increases gene expression through mechanisms such as physically impeding transcription factor binding or attraction of methyl-binding proteins that initiate formation of regions of compact, inactive chromatin.

Involvement of the epigenome in disease can be investigated through epigenome-wide association studies (EWASs) that use methylation chips to profile DNA methylation across the genome. However, EWASs are inherently harder to design and interpret than GWASs (Dempster et al., 2013).

One factor hindering the interpretation of most EWASs is that the studies tend to investigate peripheral samples (i.e., blood) to the primary target tissue (i.e., the brain). This is because it is generally unfeasible to access the large numbers of high-quality postmortem brain samples needed for an adequately powered study. EWASs are also subject to issues that are faced by conventional epidemiological studies including reverse causality and confounding. For example, while controlling for potential confounding factors including sex, age, and cell type, a recent schizophrenia EWAS identified 160 statistically significant (p-value of $\leq 1 \times 10^{-7}$) schizophrenia-associated differentially methylated positions (DMPs) (Hannon et al., 2016). However, it was noted that the top-ranked DMPs had previously been associated with cigarette smoking. Given that individuals with schizophrenia have elevated rates of smoking, the authors decided to adjust for each individual's cigarette smoking score (based on DNA methylation values at smoking-related sites). Once adjusting for smoking scores, only one of the smoking-related sites were still found to be differentially methylated in schizophrenia (Hannon et al., 2016). After excluding all smoking-related sites and controlling for smoking, age, sex, and cell composition, the study identified 25 DMPs associated with schizophrenia (Hannon et al., 2016).

In spite of the methodological concerns, EWASs provide insight into the functional complexity of the genome and, through integration with genetic approaches, may improve understanding of the molecular mechanisms underlying disease (Hannon et al., 2016).

Conclusion

In the last 10 years, major advances in technology, statistical methodologies, and international collaboration have led to substantial advances in our knowledge of the genetic basis of psychiatric disorders. It is clear that the genetic architecture of psychotic disorders involves a combination of common and rare risk variants. The genetic variants identified to date contribute substantially to increasing the risk of developing psychotic disorders, although do not yet fully explain the heritability of the conditions. Additional areas of interest not covered in detail here include the investigation of nonadditive effects or epigenetic mechanisms that may explain additional variance in disorder risk. Although there has been limited clinical translation of genetic research, methods aimed at fine mapping variants to genes, investigating how genetic liability is manifested, and improving causal inference will inevitably strengthen our understanding of pathological mechanisms. In the long run, triangulation of these methods is needed to produce robust findings that will drive the development of disorder prediction and novel treatments.

Glossary

Allele One of the possible alternative forms of a gene or genetic variant.

Chromatin A complex of DNA and proteins that forms chromosomes within the nucleus of eukaryotic cells. Packaging DNA into chromatin allows for mitosis and meiosis, prevents chromosome breakage, and controls gene expression and DNA replication.

Complex trait A trait resulting from the interplay of multiple factors, each of which has a relatively small effect. It has a genetic component that is not strictly Mendelian.

Copy number variant Structural variations within DNA that generally affect at least 1000 nucleotide base pairs.

Epigenome-wide association study Study of associations between a phenotype and epigenetic variation, such as DNA methylation, across the whole genome.

Exon Coding section of DNA or a ribonucleic acid (RNA) transcript that is translated into a protein.

Genome-wide association study Hypothesis free study of associations between a phenotype and many genetic variants covering the whole genome.

Heritability Proportion of phenotypic variation in a population that is explained by genetic variation.

Hi-C A method based on Chromosome Conformation Capture that detects chromatin interactions in the mammalian nucleus.

Intron Noncoding section of DNA or a ribonucleic acid (RNA) transcript that is spliced out before the RNA molecule is translated into a protein.

Linkage disequilibrium The nonindependence of alleles at different genomic loci, i.e., when combinations of alleles across loci occur more often than expected by chance based on their allele frequencies in the population.

Mendelian randomization A causal inference technique that uses genetic instrumental variables.

Odds ratio The odds of disease in exposed individuals compared with unexposed individuals, as used in case–control studies.

Penetrance The extent to which a genetic variant is expressed in the phenotypes of individuals carrying it, measured by the proportion of carriers showing the characteristic phenotype.

Phenotype Any observable characteristic or trait of an organism such as its development, morphology, biochemical/physiological properties, or behavior. Phenotypes are normally influenced by the combination of the individual's genotype and the environment.

Polygenic scores A single, additive score used to quantify an individual's genetic risk for a particular disorder or trait, created by summing the number of risk alleles present for each single-nucleotide polymorphism. Polygenic scores can be weighted by the SNP effect size for the disorder or trait of interest.

Rare variant Rare genetic variation, commonly defined as having a minor allele frequency $<1\%$.

Single-nucleotide polymorphism A single nucleotide base change leading to an alternative allele within a sequence of DNA.

Transcriptome-wide association studies A method that identifies significant gene expression trait associations by integrating information from gene expression and GWAS data.

List of acronyms and abbreviations

ADHD Attention-deficit/hyperactivity disorder
ALSPAC Avon Longitudinal Study of Parents and Children
ARC Activity-regulated cytoskeleton-associated protein
CNS Central nervous system
CNV Copy number variant
DMP Differentially methylated position
DNA Deoxyribonucleic acid
EWAS Epigenome-wide association study
FISH Fluorescence in situ hybridization
FMRP Fragile X mental retardation protein
GCTA Genome-wide Complex Trait Analysis
GWAS Genome-wide association study
ISC International Schizophrenia Consortium
LD Linkage disequilibrium
MAGMA Multi-marker Analysis of GenoMic Annotation
MHC Major histocompatibility complex
MR Mendelian randomization
MTGBLUP Multivariate genomic best linear unbiased predictor
NMDAR N-methyl-D-aspartate receptor

PGC Psychiatric Genomics Consortium
RCT Randomized controlled trial
SNP Single-nucleotide polymorphism
SNV single-nucleotide variant

References

Allardyce, J., Leonenko, G., Hamshere, M., Pardinas, A.F., Forty, L., Knott, S., Gordon-Smith, K., Porteous, D.J., Haywood, C., DI Florio, A., Jones, L., Mcintosh, A.M., Owen, M.J., Holmans, P., Walters, J.T.R., Craddock, N., Jones, I., O'donovan, M.C., Escott-Price, V., 2018. Association between schizophrenia-related polygenic liability and the occurrence and level of mood-incongruent psychotic symptoms in bipolar disorder. JAMA Psychiatry 75, 28–35.

Anttila, V., Bulik-Sullivan, B., Finucane, H.K., Walters, R.K., Bras, J., Duncan, L., Escott-Price, V., Falcone, G.J., Gormley, P., Malik, R., Patsopoulos, N.A., Ripke, S., Wei, Z., Yu, D.M., Lee, P.H., Turley, P., Grenier-Boley, B., Chouraki, V., Kamatani, Y., Berr, C., Letenneur, L., Hannequin, D., Amouyel, P., Boland, A., Deleuze, J.F., Duron, E., Vardarajan, B.N., Reitz, C., Goate, A.M., Huentelman, M.J., Kamboh, M.I., Larson, E.B., Rogaeva, E., St George-Hyslop, P., Hakonarson, H., Kukull, W.A., Farrer, L.A., Barnes, L.L., Beach, T.G., Demirci, F.Y., Head, E., Hulette, C.M., Jicha, G.A., Kauwe, J.S.K., Kaye, J.A., Leverenz, J.B., Levey, A.I., Lieberman, A.P., Pankratz, V.S., Poon, W.W., Quinn, J.F., Saykin, A.J., Schneider, L.S., Smith, A.G., Sonnen, J.A., Stern, R.A., Van Deerlin, V.M., Van Eldik, L.J., Harold, D., Russo, G., Rubinsztein, D.C., Bayer, A., Tsolaki, M., Proitsi, P., Fox, N.C., Hampel, H., Owen, M.J., Mead, S., Passmore, P., Morgan, K., Nothen, M.M., Rossor, M., Lupton, M.K., Hoffmann, P., Kornhuber, J., Lawlor, B., Mcquillin, A., Al-Chalabi, A., Bis, J.C., Ruiz, A., Boada, M., Seshadri, S., Beiser, A., Rice, K., Van Der Lee, S.J., De Jager, P.L., Geschwind, D.H., Riemenschneider, M., Riedel-Heller, S., Rotter, J.I., Ransmayr, G., Hyman, B.T., Cruchaga, C., Alegret, M., Winsvold, B., Palta, P., Farh, K.H., Cuenca-Leon, E., Furlotte, N., Kurth, T., et al., 2018. Analysis of shared heritability in common disorders of the brain. Science 360.

Barnett, J.H., Smoller, J.W., 2009. The genetics of bipolar disorder. Neuroscience 164, 331–343.

Bigdeli, T.B., Maher, B.S., Zhao, Z.M., Sun, J.C., Medeiros, H., Akula, N., Mcmahon, F.J., Carvalho, C., Ferreira, S.R., Azevedo, M.H., Knowles, J.A., Pato, M.T., Pato, C.N., Fanous, A.H., 2013. Association study of 83 candidate genes for bipolar disorder in chromosome 6q selected using an evidence-based prioritization algorithm. Am. J. Med. Genet. B 162, 898–906.

Bogdan, R., Baranger, D.A.A., Agrawal, A., 2018. Polygenic risk scores in clinical psychology: bridging genomic risk to individual differences. Annu. Rev. Clin. Psychol. 14, 119–157.

Bowden, J., Davey Smith, G., Burgess, S., 2015. Mendelian randomization with invalid instruments: effect estimation and bias detection through Egger regression. Int. J. Epidemiol. 44, 512–525.

Bowden, J., Del Greco, M.F., Minelli, C., Davey Smith, G., Sheehan, N., Thompson, J., 2017. A framework for the investigation of pleiotropy in two-sample summary data Mendelian randomization. Stat. Med. 36, 1783–1802.

Bulik-Sullivan, B., Finucane, H.K., Anttila, V., Gusev, A., Day, F.R., Loh, P.R., Reprogen Consortium, Psychiatric Genomics Consortium, Genetic Consortium For Anorexia Nervosa Of The Wellcome Trust Case Control Consortium, Duncan, L., Perry, J.R., Patterson, N., Robinson, E.B., Daly, M.J., Price, A.L., Neale, B.M., 2015. An atlas of genetic correlations across human diseases and traits. Nat. Genet. 47, 1236–1241.

Burgess, S., Butterworth, A., Thompson, S.G., 2013. Mendelian randomization analysis with multiple genetic variants using summarized data. Genet. Epidemiol. 37, 658–665.

Button, K.S., Ioannidis, J.P., Mokrysz, C., Nosek, B.A., Flint, J., Robinson, E.S., Munafò, M.R., 2013. Power failure: why small sample size undermines the reliability of neuroscience. Nat. Rev. Neurosci. 14, 365–376.

Cardno, A.G., Gottesman, I.I., 2000. Twin studies of schizophrenia: from bow-and-arrow concordances to star wars Mx and functional genomics. Am. J. Med. Genet. 97, 12–17.

Cardno, A.G., Rijsdijk, F.V., Sham, P.C., Murray, R.M., Mcguffin, P., 2002. A twin study of genetic relationships between psychotic symptoms. Am. J. Psychiatry 159, 539–545.

Chen, J., Lin, M.Y., Hrabovsky, A., Pedrosa, E., Dean, J., Jain, S., Zheng, D.Y., Lachman, H.M., 2015. *ZNF804A* transcriptional networks in differentiating neurons derived from induced pluripotent stem cells of human origin. PLoS One 10.

Chow, E.W., Bassett, A.S., Weksberg, R., 1994. Velo-cardio-facial syndrome and psychotic disorders: implications for psychiatric genetics. Am. J. Med. Genet. 54, 107–112.

Colodro-Conde, L., Couvy-Duchesne, B., Whitfield, J.B., Streit, F., Gordon, S., Kemper, K.E., Yengo, L., Zheng, Z., Trzaskowski, M., De Zeeuw, E.L., Nivard, M.G., Das, M., Neale, R.E., Macgregor, S., Olsen, C.M., Whiteman, D.C., Boomsma, D.I., Yang, J., Rietschel, M., Mcgrath, J.J., Medland, S.E., Martin, N.G.,

2018. Association between population density and genetic risk for schizophrenia. JAMA Psychiatry 75, 901–910.

Cross-Disorder Group Of The Psychiatric Genomics Consortium, Lee, S.H., Ripke, S., Neale, B.M., Faraone, S.V., Purcell, S.M., Perlis, R.H., Mowry, B.J., Thapar, A., Goddard, M.E., Witte, J.S., Absher, D., Agartz, I., Akil, H., Amin, F., Andreassen, O.A., Anjorin, A., Anney, R., Anttila, V., Arking, D.E., Asherson, P., Azevedo, M.H., Backlund, L., Badner, J.A., Bailey, A.J., Banaschewski, T., Barchas, J.D., Barnes, M.R., Barrett, T.B., Bass, N., Battaglia, A., Bauer, M., Bayes, M., Bellivier, F., Bergen, S.E., Berrettini, W., Betancur, C., Bettecken, T., Biederman, J., Binder, E.B., Black, D.W., Blackwood, D.H.R., Bloss, C.S., Boehnke, M., Boomsma, D.I., Breen, G., Breuer, R., Bruggeman, R., Cormican, P., Buccola, N.G., Buitelaar, J.K., Bunney, W.E., Buxbaum, J.D., Byerley, W.F., Byrne, E.M., Caesar, S., Cahn, W., Cantor, R.M., Casas, M., Chakravarti, A., Chambert, K., Choudhury, K., Cichon, S., Cloninger, C.R., Collier, D.A., Cook, E.H., Coon, H., Cormand, B., Corvin, A., Coryell, W.H., Craig, D.W., Craig, I.W., Crosbie, J., Cuccaro, M.L., Curtis, D., Czamara, D., Datta, S., Dawson, G., Day, R., DE Geus, E.J., Degenhardt, F., Djurovic, S., Donohoe, G.J., Doyle, A.E., Duan, J.B., Dudbridge, F., Duketis, E., Ebstein, R.P., Edenberg, H.J., Elia, J., Ennis, S., Etain, B., Fanous, A., Farmer, A.E., Ferrier, I.N., Flickinger, M., Fombonne, E., Foroud, T., Frank, J., Franke, B., et al., 2013. Genetic relationship between five psychiatric disorders estimated from genome-wide SNPs. Nat. Genet. 45, 984–994.

Davey Smith, G., Ebrahim, S., 2003. 'Mendelian randomization': can genetic epidemiology contribute to understanding environmental determinants of disease? Int. J. Epidemiol. 32, 1–22.

Davey Smith, G., Hemani, G., 2014. Mendelian randomization: genetic anchors for causal inference in epidemiological studies. Hum. Mol. Genet. 23, R89–R98.

Dempster, E., Viana, J., Pidsley, R., Mill, J., 2013. Epigenetic studies of schizophrenia: progress, predicaments, and promises for the future. Schizophr. Bull. 39, 11–16.

Dudbridge, F., 2013. Power and predictive accuracy of polygenic risk scores. PLoS Genet. 9, e1003348.

Farrell, M.S., Werge, T., Sklar, P., Owen, M.J., Ophoff, R.A., O'donovan, M.C., Corvin, A., Cichon, S., Sullivan, P.F., 2015. Evaluating historical candidate genes for schizophrenia. Mol. Psychiatry 20, 555–562.

Ferreira, M.A., O'donovan, M.C., Meng, Y.A., Jones, I.R., Ruderfer, D.M., Jones, L., Fan, J., Kirov, G., Perlis, R.H., Green, E.K., Smoller, J.W., Grozeva, D., Stone, J., Nikolov, I., Chambert, K., Hamshere, M.L., Nimgaonkar, V.L., Moskvina, V., Thase, M.E., Caesar, S., Sachs, G.S., Franklin, J., GORDON-Smith, K., Ardlie, K.G., Gabriel, S.B., Fraser, C., Blumenstiel, B., Defelice, M., Breen, G., Gill, M., Morris, D.W., Elkin, A., Muir, W.J., Mcghee, K.A., Williamson, R., Macintyre, D.J., Maclean, A.W., St, C.D., Robinson, M., Van Beck, M., Pereira, A.C., Kandaswamy, R., Mcquillin, A., Collier, D.A., Bass, N.J., Young, A.H., Lawrence, J., Ferrier, I.N., Anjorin, A., Farmer, A., Curtis, D., Scolnick, E.M., Mcguffin, P., Daly, M.J., Corvin, A.P., Holmans, P.A., Blackwood, D.H., Gurling, H.M., Owen, M.J., Purcell, S.M., Sklar, P., Craddock, N., Wellcome Trust Case Control, C., 2008. Collaborative genome-wide association analysis supports a role for *ANK3* and *CACNA1C* in bipolar disorder. Nat. Genet. 40, 1056–1058.

Fromer, M., Pocklington, A.J., Kavanagh, D.H., Williams, H.J., Dwyer, S., Gormley, P., Georgieva, L., Rees, E., Palta, P., Ruderfer, D.M., Carrera, N., Humphreys, I., Johnson, J.S., Roussos, P., Barker, D.D., Banks, E., Milanova, V., Grant, S.G., Hannon, E., Rose, S.A., Chambert, K., Mahajan, M., Scolnick, E.M., Moran, J.L., Kirov, G., Palotie, A., Mccarroll, S.A., Holmans, P., Sklar, P., Owen, M.J., Purcell, S.M., O'donovan, M.C., 2014. *De novo* mutations in schizophrenia implicate synaptic networks. Nature 506, 179–184.

Gage, S.H., Jones, H.J., Burgess, S., Bowden, J., Smith, G.D., Zammit, S., Munafo, M.R., 2017a. Assessing causality in associations between cannabis use and schizophrenia risk: a two-sample Mendelian randomization study. Psychol. Med. 47, 971–980.

Gage, S.H., Jones, H.J., Taylor, A.E., Burgess, S., Zammit, S., Munafo, M.R., 2017b. Investigating causality in associations between smoking initiation and schizophrenia using Mendelian randomization. Sci. Rep. 7.

Gandal, M.J., Haney, J.R., Parikshak, N.N., Leppa, V., Ramaswami, G., Hartl, C., Schork, A.J., Appadurai, V., Buil, A., Werge, T.M., Liu, C., White, K.P., Commonmind Consortium, Psychencode Consortium, Ipsych-Broad Working Group, Horvath, S., Geschwind, D.H., 2018. Shared molecular neuropathology across major psychiatric disorders parallels polygenic overlap. Science 359, 693–697.

Gershon, E.S., Alliey-Rodriguez, N., 2013. New ethical issues for genetic counseling in common mental disorders. Am. J. Psychiatry 170, 968–976.

Girgenti, M.J., Loturco, J.J., Maher, B.J., 2012. *ZNF804a* regulates expression of the schizophrenia-associated genes *PRSS16, Comt, Pde4B,* and *DRD2*. PLoS One 7, e32404.

Goes, F.S., Pirooznia, M., Parla, J.S., Kramer, M., Ghiban, E., Mavruk, S., Chen, Y.C., Monson, E.T., Willour, V.L., Karchin, R., Flickinger, M., Locke, A.E., Levy, S.E., Scott, L.J., Boehnke, M., Stahl, E., Moran, J.L., Hultman, C.M., Landen, M., Purcell, S.M., Sklar, P.,

Zandi, P.P., Mccombie, W.R., Potash, J.B., 2016. Exome sequencing of familial bipolar disorder. JAMA Psychiatry 73, 590–597.

Gottesman, I.I., Shields, J., 1967. A polygenic theory of schizophrenia. Proc. Natl. Acad. Sci. USA 58, 199–205.

Green, E.K., Rees, E., Walters, J.T., Smith, K.G., Forty, L., Grozeva, D., Moran, J.L., Sklar, P., Ripke, S., Chambert, K.D., Genovese, G., Mccarroll, S.A., Jones, I., Jones, L., Owen, M.J., O'donovan, M.C., Craddock, N., Kirov, G., 2016. Copy number variation in bipolar disorder. Mol. Psychiatry 21, 89–93.

Guella, I., Sequeira, A., Rollins, B., Morgan, L., Myers, R.M., Watson, S.J., Akil, H., Bunney, W.E., Delisi, L.E., Byerley, W., Vawter, M.P., 2014. Evidence of allelic imbalance in the schizophrenia susceptibility gene *ZNF804A* in human dorsolateral prefrontal cortex. Schizophr. Res. 152, 111–116.

Gusev, A., Mancuso, N., Won, H., Kousi, M., Finucane, H.K., Reshef, Y., Song, L., Safi, A., Schizophrenia Working Group Of The Psychiatric Genomics, C., Mccarroll, S., Neale, B.M., Ophoff, R.A., O'donovan, M.C., Crawford, G.E., Geschwind, D.H., Katsanis, N., Sullivan, P.F., Pasaniuc, B., Price, A.L., 2018. Transcriptome-wide association study of schizophrenia and chromatin activity yields mechanistic disease insights. Nat. Genet. 50, 538–548.

Hamshere, M.L., Walters, J.T., Smith, R., Richards, A.L., Green, E., Grozeva, D., Jones, I., Forty, L., Jones, L., Gordon-Smith, K., Riley, B., O'neill, F.A., Kendler, K.S., Sklar, P., Purcell, S., Kranz, J., Schizophrenia Psychiatric Genome-Wide Association Study, C., Wellcome Trust Case Control, C., Wellcome Trust Case Control, C., Morris, D., Gill, M., Holmans, P., Craddock, N., Corvin, A., Owen, M.J., O'donovan, M.C., 2013. Genome-wide significant associations in schizophrenia to *ITIH3/4, CACNA1C* and *SDCCAG8*, and extensive replication of associations reported by the Schizophrenia PGC. Mol. Psychiatry 18, 708–712.

Hannon, E., Dempster, E., Viana, J., Burrage, J., Smith, A.R., Macdonald, R., ST Clair, D., Mustard, C., Breen, G., Therman, S., Kaprio, J., Toulopoulou, T., Hulshoff Pol, H.E., Bohlken, M.M., Kahn, R.S., Nenadic, I., Hultman, C.M., Murray, R.M., Collier, D.A., Bass, N., Gurling, H., Mcquillin, A., Schalkwyk, L., Mill, J., 2016. An integrated genetic-epigenetic analysis of schizophrenia: evidence for co-localization of genetic associations and differential DNA methylation. Genome Biol. 17, 176.

Hartwig, F.P., Borges, M.C., Horta, B.L., Bowden, J., Davey Smith, G., 2017. Inflammatory biomarkers and risk of schizophrenia: a 2-sample Mendelian randomization study. JAMA Psychiatry 74, 1226–1233.

Hill, M.J., Bray, N.J., 2011. Allelic differences in nuclear protein binding at a genome-wide significant risk variant for schizophrenia in ZNF804A. Mol. Psychiatry 16, 787–789.

Hill, M.J., Bray, N.J., 2012. Evidence that schizophrenia risk variation in the *ZNF804A* gene exerts its effects during fetal brain development. Am. J. Psychiatry 169, 1301–1308.

Hill, M.J., Jeffries, A.R., Dobson, R.J.B., Price, J., Bray, N.J., 2012. Knockdown of the psychosis susceptibility gene *ZNF804A* alters expression of genes involved in cell adhesion. Hum. Mol. Genet. 21, 1018–1024.

Hou, L., Bergen, S.E., Akula, N., Song, J., Hultman, C.M., Landen, M., Adli, M., Alda, M., Ardau, R., Arias, B., Aubry, J.M., Backlund, L., Badner, J.A., Barrett, T.B., Bauer, M., Baune, B.T., Bellivier, F., Benabarre, A., Bengesser, S., Berrettini, W.H., Bhattacharjee, A.K., Biernacka, J.M., Birner, A., Bloss, C.S., Brichant-Petitjean, C., Bui, E.T., Byerley, W., Cervantes, P., Chillotti, C., Cichon, S., Colom, F., Coryell, W., Craig, D.W., Cruceanu, C., Czerski, P.M., Davis, T., Dayer, A., Degenhardt, F., Del Zompo, M., Depaulo, J.R., Edenberg, H.J., Etain, B., Falkai, P., Foroud, T., Forstner, A.J., Frisen, L., Frye, M.A., Fullerton, J.M., Gard, S., Garnham, J.S., Gershon, E.S., Goes, F.S., Greenwood, T.A., Grigoroiu-Serbanescu, M., Hauser, J., Heilbronner, U., Heilmann-Heimbach, S., Herms, S., Hipolito, M., Hitturlingappa, S., Hoffmann, P., Hofmann, A., Jamain, S., Jimenez, E., Kahn, J.P., Kassem, L., Kelsoe, J.R., Kittel-Schneider, S., Kliwicki, S., Koller, D.L., Konig, B., Lackner, N., Laje, G., Lang, M., Lavebratt, C., Lawson, W.B., Leboyer, M., Leckband, S.G., Liu, C., Maaser, A., Mahon, P.B., Maier, W., Maj, M., Manchia, M., Martinsson, L., Mccarthy, M.J., Mcelroy, S.L., Mcinnis, M.G., Mckinney, R., Mitchell, P.B., Mitjans, M., Mondimore, F.M., Monteleone, P., Muhleisen, T.W., Nievergelt, C.M., Nothen, M.M., Novak, T., Nurnberger JR., J.I., Nwulia, E.A., Osby, U., et al., 2016. Genome-wide association study of 40,000 individuals identifies two novel loci associated with bipolar disorder. Hum. Mol. Genet. 25, 3383–3394.

Hubbard, L., Tansey, K.E., Rai, D., Jones, P., Ripke, S., Chambert, K.D., Moran, J.L., Mccarroll, S.A., Linden, D.E.J., Owen, M.J., O'donovan, M.C., Walters, J.T.R., Zammit, S., 2016. Evidence of common genetic overlap between schizophrenia and cognition. Schizophr. Bull. 42, 832–842.

International Schizophrenia Consortium, 2008. Rare chromosomal deletions and duplications increase risk of schizophrenia. Nature 455, 237–241.

International Schizophrenia Consortium, Purcell, S.M., Wray, N.R., Stone, J.L., Visscher, P.M.,

O'donovan, M.C., Sullivan, P.F., Sklar, P., 2009. Common polygenic variation contributes to risk of schizophrenia and bipolar disorder. Nature 460, 748–752.

International Society Of Psychiatric Genetics, 2017. Genetic Testing Taskforce: Genetic Testing Statement [Online]. Available from: https://ispg.net/genetic-testing-statement/.

Ioannidis, J.P., Ntzani, E.E., Trikalinos, T.A., Contopoulos-Ioannidis, D.G., 2001. Replication validity of genetic association studies. Nat. Genet. 29, 306–309.

Iossifov, I., O'roak, B.J., Sanders, S.J., Ronemus, M., Krumm, N., Levy, D., Stessman, H.A., Witherspoon, K.T., Vives, L., Patterson, K.E., Smith, J.D., Paeper, B., Nickerson, D.A., Dea, J., Dong, S., Gonzalez, L.E., Mandell, J.D., Mane, S.M., Murtha, M.T., Sullivan, C.A., Walker, M.F., Waqar, Z., Wei, L., Willsey, A.J., Yamrom, B., Lee, Y.H., Grabowska, E., Dalkic, E., Wang, Z., Marks, S., Andrews, P., Leotta, A., Kendall, J., Hakker, I., Rosenbaum, J., Ma, B., Rodgers, L., Troge, J., Narzisi, G., Yoon, S., Schatz, M.C., Ye, K., Mccombie, W.R., Shendure, J., Eichler, E.E., State, M.W., Wigler, M., 2014. The contribution of de novo coding mutations to autism spectrum disorder. Nature 515, 216–221.

Iossifov, I., Ronemus, M., Levy, D., Wang, Z., Hakker, I., Rosenbaum, J., Yamrom, B., Lee, Y.H., Narzisi, G., Leotta, A., Kendall, J., Grabowska, E., Ma, B., Marks, S., Rodgers, L., Stepansky, A., Troge, J., Andrews, P., Bekritsky, M., Pradhan, K., Ghiban, E., Kramer, M., Parla, J., Demeter, R., Fulton, L.L., Fulton, R.S., Magrini, V.J., Ye, K., Darnell, J.C., Darnell, R.B., Mardis, E.R., Wilson, R.K., Schatz, M.C., Mccombie, W.R., Wigler, M., 2012. *De novo* gene disruptions in children on the autistic spectrum. Neuron 74, 285–299.

Jansen, P.R., Polderman, T.J.C., Bolhuis, K., Van Der Ende, J., Jaddoe, V.W.V., Verhulst, F.C., White, T., Posthuma, D., Tiemeier, H., 2018. Polygenic scores for schizophrenia and educational attainment are associated with behavioural problems in early childhood in the general population. J. Child Psychol. Psychiatry 59, 39–47.

Jones, H.J., Stergiakouli, E., Tansey, K.E., Hubbard, L., Heron, J., Cannon, M., Holmans, P., Lewis, G., Linden, D.E.J., Jones, P.B., Smith, G.D., O'donovan, M.C., Owen, M.J., Walters, J.T., Zammit, S., 2016. Phenotypic manifestation of genetic risk for schizophrenia during adolescence in the general population. JAMA Psychiatry 73, 221–228.

Kirov, G., Gumus, D., Chen, W., Norton, N., Georgieva, L., Sari, M., O'donovan, M.C., Erdogan, F., Owen, M.J., Ropers, H.H., Ullmann, R., 2008. Comparative genome hybridization suggests a role for *NRXN1* and *APBA2* in schizophrenia. Hum. Mol. Genet. 17, 458–465.

Kirov, G., Pocklington, A.J., Holmans, P., Ivanov, D., Ikeda, M., Ruderfer, D., Moran, J., Chambert, K., Toncheva, D., Georgieva, L., Grozeva, D., Fjodorova, M., Wollerton, R., Rees, E., Nikolov, I., Van De Lagemaat, L.N., Bayes, A., Fernandez, E., Olason, P.I., Bottcher, Y., Komiyama, N.H., Collins, M.O., Choudhary, J., Stefansson, K., Stefansson, H., Grant, S.G., Purcell, S., Sklar, P., O'donovan, M.C., Owen, M.J., 2012. *De novo* CNV analysis implicates specific abnormalities of postsynaptic signalling complexes in the pathogenesis of schizophrenia. Mol. Psychiatry 17, 142–153.

Lee, C., Scherer, S.W., 2010. The clinical context of copy number variation in the human genome. Expert Rev. Mol. Med. 12, e8.

Lee, P.H., Baker, J.T., Holmes, A.J., Jahanshad, N., Ge, T., Jung, J.Y., Cruz, Y., Manoach, D.S., Hibar, D.P., Faskowitz, J., Mcmahon, K.L., DE Zubicaray, G.I., Martin, N.H., Wright, M.J., Ongur, D., Buckner, R., Roffman, J., Thompson, P.M., Smoller, J.W., 2016. Partitioning heritability analysis reveals a shared genetic basis of brain anatomy and schizophrenia. Mol. Psychiatry 21, 1680–1689.

Lencz, T., Morgan, T.V., Athanasiou, M., Dain, B., Reed, C.R., Kane, J.M., Kucherlapati, R., Malhotra, A.K., 2007. Converging evidence for a pseudoautosomal cytokine receptor gene locus in schizophrenia. Mol. Psychiatry 12, 572–580.

Lescai, F., Als, T.D., Li, Q., Nyegaard, M., Andorsdottir, G., Biskopstø, M., Hedemand, A., Fiorentino, A., O'brien, N., Jarram, A., Liang, J., Grove, J., Pallesen, J., Eickhardt, E., Mattheisen, M., Bolund, L., Demontis, D., Wang, A.G., Mcquillin, A., Mors, O., Wang, J., Børglum, A.D., 2017. Whole-exome sequencing of individuals from an isolated population implicates rare risk variants in bipolar disorder. Transl. Psychiatry 7, e1034.

Li, C., Yang, C., Gelernter, J., Zhao, H., 2014. Improving genetic risk prediction by leveraging pleiotropy. Hum. Genet. 133, 639–650.

Lichtenstein, P., Yip, B.H., Björk, C., Pawitan, Y., Cannon, T.D., Sullivan, P.F., Hultman, C.M., 2009. Common genetic determinants of schizophrenia and bipolar disorder in Swedish families: a population-based study. Lancet 373, 234–239.

Maier, R., Moser, G., Chen, G.B., Ripke, S., Coryell, W., Potash, J.B., Scheftner, W.A., Shi, J.X., Weissman, M.M., Hultman, C.M., Landen, M., Levinson, D.F., Kendler, K.S., Smoller, J.W., Wray, N.R., Lee, S.H., Cross-Disorder Working Group Of The Psychiatric Genomics Consortium, 2015. Joint analysis of psychiatric disorders increases accuracy of risk Prediction for schizophrenia, bipolar disorder, and major depressive disorder. Am. J. Hum. Genet. 96, 283–294.

Maier, R.M., Zhu, Z.H., Lee, S.H., Trzaskowski, M., Ruderfer, D.M., Stahl, E.A., Ripke, S., Wray, N.R., Yang, J., Visscher, P.M., Robinson, M.R., 2018. Improving genetic prediction by leveraging genetic correlations among human diseases and traits. Nat. Commun. 9.

Marshall, C.R., Howrigan, D.P., Merico, D., Thiruvahindrapuram, B., Wu, W., Greer, D.S., Antaki, D., Shetty, A., Holmans, P.A., Pinto, D., Gujral, M., Brandler, W.M., Malhotra, D., Wang, Z., Fajarado, K.V.F., Maile, M.S., Ripke, S., Agartz, I., Albus, M., Alexander, M., Amin, F., Atkins, J., Bacanu, S.A., Belliveau Jr., R.A., Bergen, S.E., Bertalan, M., Bevilacqua, E., Bigdeli, T.B., Black, D.W., Bruggeman, R., Buccola, N.G., Buckner, R.L., Bulik-Sullivan, B., Byerley, W., Cahn, W., Cai, G., Cairns, M.J., Campion, D., Cantor, R.M., Carr, V.J., Carrera, N., Catts, S.V., Chambert, K.D., Cheng, W., Cloninger, C.R., Cohen, D., Cormican, P., Craddock, N., Crespo-Facorro, B., Crowley, J.J., Curtis, D., Davidson, M., Davis, K.L., Degenhardt, F., Del Favero, J., Delisi, L.E., Dikeos, D., Dinan, T., Djurovic, S., Donohoe, G., Drapeau, E., Duan, J., Dudbridge, F., Eichhammer, P., Eriksson, J., Escott-Price, V., Essioux, L., Fanous, A.H., Farh, K.H., Farrell, M.S., Frank, J., Franke, L., Freedman, R., Freimer, N.B., Friedman, J.I., Forstner, A.J., Fromer, M., Genovese, G., Georgieva, L., Gershon, E.S., Giegling, I., Giusti-Rodriguez, P., Godard, S., Goldstein, J.I., Gratten, J., DE Haan, L., Hamshere, M.L., Hansen, M., Hansen, T., Haroutunian, V., Hartmann, A.M., Henskens, F.A., Herms, S., Hirschhorn, J.N., Hoffmann, P., Hofman, A., Huang, H., Ikeda, M., Joa, I., Kahler, A.K., et al., 2017. Contribution of copy number variants to schizophrenia from a genome-wide study of 41,321 subjects. Nat. Genet. 49, 27–35.

Martin, J., Tilling, K., Hubbard, L., Stergiakouli, E., Thapar, A., Davey Smith, G., O'donovan, M.C., Zammit, S., 2016. Association of genetic risk for schizophrenia with nonparticipation over time in a population-based cohort study. Am. J. Epidemiol. 183, 1149–1158.

McCarthy, N.S., Melton, P.E., Ward, S.V., Allan, S.M., Dragovic, M., Clark, M.L., Morar, B., Rubio, J.P., Blangero, J., Badcock, J.C., Morgan, V.A., Moses, E.K., Jablensky, A., 2017. Exome array analysis suggests an increased variant burden in families with schizophrenia. Schizophr. Res. 185, 9–16.

McQueen, M.B., Devlin, B., Faraone, S.V., Nimgaonkar, V.L., Sklar, P., Smoller, J.W., Abou Jamra, R., Albus, M., Bacanu, S.A., Baron, M., Barrett, T.B., Berrettini, W., Blacker, D., Byerley, W., Cichon, S., Coryell, W., Craddock, N., Daly, M.J., Depaulo, J.R., Edenberg, H.J., Foroud, T., Gill, M., Gilliam, T.C., Hamshere, M., Jones, I., Jones, L., Juo, S.H., Kelsoe, J.R., Lambert, D., Lange, C., Lerer, B., Liu, J., Maier, W., Mackinnon, J.D., Mcinnis, M.G., Mcmahon, F.J., Murphy, D.L., Nothen, M.M., Nurnberger, J.I., Pato, C.N., Pato, M.T., Potash, J.B., Propping, P., Pulver, A.E., Rice, J.P., Rietschel, M., Scheftner, W., Schumacher, J., Segurado, R., Van Steen, K., Xie, W., Zandi, P.P., Laird, N.M., 2005. Combined analysis from eleven linkage studies of bipolar disorder provides strong evidence of susceptibility loci on chromosomes 6q and 8q. Am. J. Hum. Genet. 77, 582–595.

Mistry, S., Harrison, J.R., Smith, D.J., Escott-Price, V., Zammit, S., 2018a. The use of polygenic risk scores to identify phenotypes associated with genetic risk of bipolar disorder and depression: a systematic review. J. Affect. Disord. 234, 148–155.

Mistry, S., Harrison, J.R., Smith, D.J., Escott-Price, V., Zammit, S., 2018b. The use of polygenic risk scores to identify phenotypes associated with genetic risk of schizophrenia: systematic review. Schizophr. Res. 197, 2–8.

Murphy, K.C., Jones, L.A., Owen, M.J., 1999. High rates of schizophrenia in adults with velo-cardio-facial syndrome. Arch. Gen. Psychiatr. 56, 940–945.

Nagel, M., Jansen, P.R., Stringer, S., Watanabe, K., DE Leeuw, C.A., Bryois, J., Savage, J.E., Hammerschlag, A.R., Skene, N.G., Munoz-Manchado, A.B., 23andme Research Team, White, T., Tiemeier, H., Linnarsson, S., Hjerling-Leffler, J., Polderman, T.J.C., Sullivan, P.F., Van Der Sluis, S., Posthuma, D., 2018. Meta-analysis of genome-wide association studies for neuroticism in 449,484 individuals identifies novel genetic loci and pathways. Nat. Genet. 50, 920–927.

Nivard, M.G., Gage, S.H., Hottenga, J.J., Van Beijsterveldt, C.E.M., Abdellaoui, A., Bartels, M., Baselmans, B.M.L., Ligthart, L., ST Pourcain, B., Boomsma, D.I., Munafo, M.R., Middeldorp, C.M., 2017. Genetic overlap between schizophrenia and developmental psychopathology: longitudinal and multivariate polygenic risk prediction of common psychiatric traits during development. Schizophr. Bull. 43, 1197–1207.

O'Donovan, M.C., 2015. What have we learned from the Psychiatric Genomics Consortium. World Psychiatry 14, 291–293.

O'Donovan, M.C., Craddock, N., Norton, N., Williams, H., Peirce, T., Moskvina, V., Nikolov, I., Hamshere, M., Carroll, L., Georgieva, L., Dwyer, S., Holmans, P., Marchini, J.L., Spencer, C.C., Howie, B., Leung, H.T., Hartmann, A.M., Moller, H.J., Morris, D.W., Shi, Y., Feng, G., Hoffmann, P., Propping, P., Vasilescu, C., Maier, W., Rietschel, M., Zammit, S., Schumacher, J., Quinn, E.M., Schulze, T.G., Williams, N.M., Giegling, I.,

Iwata, N., Ikeda, M., Darvasi, A., Shifman, S., He, L., Duan, J., Sanders, A.R., Levinson, D.F., Gejman, P.V., Cichon, S., Nothen, M.M., Gill, M., Corvin, A., Rujescu, D., Kirov, G., Owen, M.J., Buccola, N.G., Mowry, B.J., Freedman, R., Amin, F., Black, D.W., Silverman, J.M., Byerley, W.F., Cloninger, C.R., Molecular Genetics Of Schizophrenia, C., 2008. Identification of loci associated with schizophrenia by genome-wide association and follow-up. Nat. Genet. 40, 1053–1055.

O'donovan, M.C., Owen, M.J., 2016. The implications of the shared genetics of psychiatric disorders. Nat. Med. 22, 1214–1219.

Owen, M.J., O'Donovan, M.C., 2017. Schizophrenia and the neurodevelopmental continuum: evidence from genomics. World Psychiatry 16, 227–235.

Pardiñas, A.F., Holmans, P., Pocklington, A.J., Escott-Price, V., Ripke, S., Carrera, N., Legge, S.E., Bishop, S., Cameron, D., Hamshere, M.L., Han, J., Hubbard, L., Lynham, A., Mantripragada, K., Rees, E., Maccabe, J.H., Mccarroll, S.A., Baune, B.T., Breen, G., Byrne, E.M., Dannlowski, U., Eley, T.C., Hayward, C., Martin, N.G., Mcintosh, A.M., Plomin, R., Porteous, D.J., Wray, N.R., Caballero, A., Geschwind, D.H., Huckins, L.M., Ruderfer, D.M., Santiago, E., Sklar, P., Stahl, E.A., Won, H., Agerbo, E., Als, T.D., Andreassen, O.A., Bækvad-Hansen, M., Mortensen, P.B., Pedersen, C.B., Børglum, A.D., Bybjerg-Grauholm, J., Djurovic, S., Durmishi, N., Pedersen, M.G., Golimbet, V., Grove, J., Hougaard, D.M., Mattheisen, M., Molden, E., Mors, O., Nordentoft, M., Pejovic-Milovancevic, M., Sigurdsson, E., Silagadze, T., Hansen, C.S., Stefansson, K., Stefansson, H., Steinberg, S., Tosato, S., Werge, T., Gerad1 Consortium, Crestar Consortium, Collier, D.A., Rujescu, D., Kirov, G., Owen, M.J., O'donovan, M.C., Walters, J.T.R., 2018. Common schizophrenia alleles are enriched in mutation-intolerant genes and in regions under strong background selection. Nat. Genet. 50, 381–389.

Pocklington, A.J., Rees, E., Walters, J.T., Han, J., Kavanagh, D.H., Chambert, K.D., Holmans, P., Moran, J.L., Mccarroll, S.A., Kirov, G., O'donovan, M.C., Owen, M.J., 2015. Novel findings from CNVs implicate inhibitory and excitatory signaling complexes in schizophrenia. Neuron 86, 1203–1214.

Psychiatric Gwas Consortium Bipolar Disorder Working Group, 2011. Large-scale genome-wide association analysis of bipolar disorder identifies a new susceptibility locus near *ODZ4*. Nat. Genet. 43, 977–983.

Purcell, S.M., Moran, J.L., Fromer, M., Ruderfer, D., Solovieff, N., Roussos, P., O'dushlaine, C., Chambert, K., Bergen, S.E., Kahler, A., Duncan, L., Stahl, E., Genovese, G., Fernandez, E., Collins, M.O., Komiyama, N.H., Choudhary, J.S., Magnusson, P.K., Banks, E., Shakir, K., Garimella, K., Fennell, T., Depristo, M., Grant, S.G., Haggarty, S.J., Gabriel, S., Scolnick, E.M., Lander, E.S., Hultman, C.M., Sullivan, P.F., Mccarroll, S.A., Sklar, P., 2014. A polygenic burden of rare disruptive mutations in schizophrenia. Nature 506, 185–190.

Rees, E., Kendall, K., Pardinas, A.F., Legge, S.E., Pocklington, A., Escott-Price, V., Maccabe, J.H., Collier, D.A., Holmans, P., O'donovan, M.C., Owen, M.J., Walters, J.T.R., Kirov, G., 2016. Analysis of intellectual disability copy number variants for association with schizophrenia. JAMA Psychiatry 73, 963–969.

Rees, E., Kirov, G., Sanders, A., Walters, J.T.R., Chambert, K.D., Shi, J., Szatkiewicz, J., O'dushlaine, C., Richards, A.L., Green, E.K., Jones, I., Davies, G., Legge, S.E., Moran, J.L., Pato, C., Pato, M., Genovese, G., Levinson, D., Duan, J., Moy, W., Goring, H.H.H., Morris, D., Cormican, P., Kendler, K.S., O'neill, F.A., Riley, B., Gill, M., Corvin, A., Wellcome Trust Case Control Consortium, Craddock, N., Sklar, P., Hultman, C., Sullivan, P.F., Gejman, P.V., Mccarroll, S.A., O'donovan, M.C., Owen, M.J., 2014a. Evidence that duplications of 22q11.2 protect against schizophrenia. Mol. Psychiatry 19, 37–40.

Rees, E., Walters, J.T.R., Georgieva, L., Isles, A.R., Chambert, K.D., Richards, A.L., Mahoney-Davies, G., Legge, S.E., Moran, J.L., Mccarroll, S.A., O'donovan, M.C., Owen, M.J., Kirov, G., 2014b. Analysis of copy number variations at 15 schizophrenia-associated loci. Br. J. Psychiatry 204, 108–114.

Richards, A.L., Leonenko, G., Walters, J.T., Kavanagh, D.H., Rees, E.G., Evans, A., Chambert, K.D., Moran, J.L., Goldstein, J., Neale, B.M., Mccarroll, S.A., Pocklington, A.J., Holmans, P.A., Owen, M.J., O'donovan, M.C., 2016. Exome arrays capture polygenic rare variant contributions to schizophrenia. Hum. Mol. Genet. 25, 1001–1007.

Riglin, L., Collishaw, S., Richards, A., Thapar, A.K., Maughan, B., O'donovan, M.C., Thapar, A., 2017. Schizophrenia risk alleles and neurodevelopmental outcomes in childhood: a population-based cohort study. Lancet Psychiatry 4, 57–62.

Ripke, S., O'dushlaine, C., Chambert, K., Moran, J.L., Kahler, A.K., Akterin, S., Bergen, S.E., Collins, A.L., Crowley, J.J., Fromer, M., Kim, Y., Lee, S.H., Magnusson, P.K., Sanchez, N., Stahl, E.A., Williams, S., Wray, N.R., Xia, K., Bettella, F., Borglum, A.D., Bulik-Sullivan, B.K., Cormican, P., Craddock, N., DE Leeuw, C., Durmishi, N., Gill, M., Golimbet, V., Hamshere, M.L., Holmans, P., Hougaard, D.M., Kendler, K.S., Lin, K., Morris, D.W., Mors, O., Mortensen, P.B., Neale, B.M., O'neill, F.A., Owen, M.J., Milovancevic, M.P., Posthuma, D., Powell, J.,

Richards, A.L., Riley, B.P., Ruderfer, D., Rujescu, D., Sigurdsson, E., Silagadze, T., Smit, A.B., Stefansson, H., Steinberg, S., Suvisaari, J., Tosato, S., Verhage, M., Walters, J.T., Multicenter Genetic Studies Of Schizophrenia, C., Levinson, D.F., Gejman, P.V., Kendler, K.S., Laurent, C., Mowry, B.J., O'donovan, M.C., Owen, M.J., Pulver, A.E., Riley, B.P., Schwab, S.G., Wildenauer, D.B., Dudbridge, F., Holmans, P., Shi, J., Albus, M., Alexander, M., Campion, D., Cohen, D., Dikeos, D., Duan, J., Eichhammer, P., Godard, S., Hansen, M., Lerer, F.B., Liang, K.Y., Maier, W., Mallet, J., Nertney, D.A., Nestadt, G., Norton, N., O'neill, F.A., Papadimitriou, G.N., Ribble, R., Sanders, A.R., Silverman, J.M., Walsh, D., Williams, N.M., Wormley, B., Psychosis Endophenotypes International, C., Arranz, M.J., Bakker, S., Bender, S., Bramon, E., Collier, D., Crespo-Facorro, B., et al., 2013. Genome-wide association analysis identifies 13 new risk loci for schizophrenia. Nat. Genet. 45, 1150–1159.

Rujescu, D., Ingason, A., Cichon, S., Pietilainen, O.P.H., Barnes, M.R., Toulopoulou, T., Picchioni, M., Vassos, E., Ettinger, U., Bramon, E., Murray, R., Ruggeri, M., Tosato, S., Bonetto, C., Steinberg, S., Sigurdsson, E., Sigmundsson, T., Petursson, H., Gylfason, A., Olason, P.I., Hardarsson, G., Jonsdottir, G.A., Gustafsson, O., Fossdal, R., Giegling, I., Moller, H.J., Hartmann, A.M., Hoffmann, P., Crombie, C., Fraser, G., Walker, N., Lonnqvist, J., Suvisaari, J., Tuulio-Henriksson, A., Djurovic, S., Melle, I., Andreassen, O.A., Hansen, T., Werge, T., Kiemeney, L.A., Franke, B., Veltman, J., Buizer-Voskamp, J.E., Investigators, G., Sabatti, C., Ophoff, R.A., Rietschel, M., Nothen, M.M., Stefansson, K., Peltonen, L., ST Clair, D., Stefansson, H., Collier, D.A., 2009. Disruption of the neurexin 1 gene is associated with schizophrenia. Hum. Mol. Genet. 18, 988–996.

Savage, J.E., Jansen, P.R., Stringer, S., Watanabe, K., Bryois, J., De Leeuw, C.A., Nagel, M., Awasthi, S., Barr, P.B., Coleman, J.R.I., Grasby, K.L., Hammerschlag, A.R., Kaminski, J.A., Karlsson, R., Krapohl, E., Lam, M., Nygaard, M., Reynolds, C.A., Trampush, J.W., Young, H., Zabaneh, D., Hagg, S., Hansell, N.K., Karlsson, I.K., Linnarsson, S., Montgomery, G.W., Munoz-Manchado, A.B., Quinlan, E.B., Schumann, G., Skene, N.G., Webb, B.T., White, T., Arking, D.E., Avramopoulos, D., Bilder, R.M., Bitsios, P., Burdick, K.E., Cannon, T.D., Chiba-Falek, O., Christoforou, A., Cirulli, E.T., Congdon, E., Corvin, A., Davies, G., Deary, I.J., Derosse, P., Dickinson, D., Djurovic, S., Donohoe, G., Conley, E.D., Eriksson, J.G., Espeseth, T., Freimer, N.A., Giakoumaki, S., Giegling, I., Gill, M., Glahn, D.C., Hariri, A.R., Hatzimanolis, A., Keller, M.C., Knowles, E., Koltai, D., Konte, B., Lahti, J., LE Hellard, S., Lencz, T., Liewald, D.C., London, E., Lundervold, A.J., Malhotra, A.K., Melle, I., Morris, D., Need, A.C., Ollier, W., Palotie, A., Payton, A., Pendleton, N., Poldrack, R.A., Raikkonen, K., Reinvang, I., Roussos, P., Rujescu, D., Sabb, F.W., Scult, M.A., Smeland, O.B., Smyrnis, N., Starr, J.M., Steen, V.M., Stefanis, N.C., Straub, R.E., Sundet, K., Tiemeier, H., Voineskos, A.N., Weinberger, D.R., Widen, E., Yu, J., Abecasis, G., Andreassen, O.A., Breen, G., Christiansen, L., et al., 2018. Genome-wide association meta-analysis in 269,867 individuals identifies new genetic and functional links to intelligence. Nat. Genet. 50, 912–919.

Schizophrenia Psychiatric Genome-Wide Association Study Consortium, 2011. Genome-wide association study identifies five new schizophrenia loci. Nat. Genet. 43, 969–976.

Schizophrenia Working Group Of The Psychiatric Genomics Consortium, 2014. Biological insights from 108 schizophrenia-associated genetic loci. Nature 511, 421–427.

Sekar, A., Bialas, A.R., DE Rivera, H., Davis, A., Hammond, T.R., Kamitaki, N., Tooley, K., Presumey, J., Baum, M., Van Doren, V., Genovese, G., Rose, S.A., Handsaker, R.E., Schizophrenia Working Group Of The Psychiatric Genomics Consortium, Daly, M.J., Carroll, M.C., Stevens, B., Mccarroll, S.A., 2016. Schizophrenia risk from complex variation of complement component 4. Nature 530, 177–183.

Shi, J., Levinson, D.F., Duan, J., Sanders, A.R., Zheng, Y., PE'er, I., Dudbridge, F., Holmans, P.A., Whittemore, A.S., Mowry, B.J., Olincy, A., Amin, F., Cloninger, C.R., Silverman, J.M., Buccola, N.G., Byerley, W.F., Black, D.W., Crowe, R.R., Oksenberg, J.R., Mirel, D.B., Kendler, K.S., Freedman, R., Gejman, P.V., 2009. Common variants on chromosome 6p22.1 are associated with schizophrenia. Nature 460, 753–757.

Singh, T., Kurki, M.I., Curtis, D., Purcell, S.M., Crooks, L., Mcrae, J., Suvisaari, J., Chheda, H., Blackwood, D., Breen, G., Pietilainen, O., Gerety, S.S., Ayub, M., Blyth, M., Cole, T., Collier, D., Coomber, E.L., Craddock, N., Daly, M.J., Danesh, J., Diforti, M., Foster, A., Freimer, N.B., Geschwind, D., Johnstone, M., Joss, S., Kirov, G., Korkko, J., Kuismin, O., Holmans, P., Hultman, C.M., Iyegbe, C., Lonnqvist, J., Mannikko, M., Mccarroll, S.A., Mcguffin, P., Mcintosh, A.M., Mcquillin, A., Moilanen, J.S., Moore, C., Murray, R.M., Newbury-Ecob, R., Ouwehand, W., Paunio, T., Prigmore, E., Rees, E., Roberts, D., Sambrook, J., Sklar, P., ST Clair, D., Veijola, J., Walters, J.T., Williams, H., Swedish Schizophrenia Study, Interval Study, Ddd Study, Uk10 K Consortium, Sullivan, P.F.,

Hurles, M.E., O'donovan, M.C., Palotie, A., Owen, M.J., Barrett, J.C., 2016. Rare loss-of-function variants in *SETD1A* are associated with schizophrenia and developmental disorders. Nat. Neurosci. 19, 571–577.

Singh, T., Walters, J.T.R., Johnstone, M., Curtis, D., Suvisaari, J., Torniainen, M., Rees, E., Iyegbe, C., Blackwood, D., Mcintosh, A.M., Kirov, G., Geschwind, D., Murray, R.M., DI Forti, M., Bramon, E., Gandal, M., Hultman, C.M., Sklar, P., Interval Study, Uk10K Consortium, Palotie, A., Sullivan, P.F., O'donovan, M.C., Owen, M.J., Barrett, J.C., 2017. The contribution of rare variants to risk of schizophrenia in individuals with and without intellectual disability. Nat. Genet. 49, 1167–1173.

Skene, N.G., Bryois, J., Bakken, T.E., Breen, G., Crowley, J.J., Gaspar, H.A., Giusti-Rodriguez, P., Hodge, R.D., Miller, J.A., Muñoz-Manchado, A.B., O'donovan, M.C., Owen, M.J., Pardinas, A.F., Ryge, J., Walters, J.T.R., Linnarsson, S., Lein, E.S., Major Depressive Disorder Working Group Of The Psychiatric Genomics Consortium, Sullivan, P.F., Hjerling-Leffler, J., 2018. Genetic identification of brain cell types underlying schizophrenia. Nat. Genet. 50, 825–833.

Smoller, J.W., Finn, C.T., 2003. Family, twin, and adoption studies of bipolar disorder. Am. J. Med. Genet. C 123c, 48–58.

St Pourcain, B., Robinson, E.B., Anttila, V., Sullivan, B.B., Maller, J., Golding, J., Skuse, D., Ring, S., Evans, D.M., Zammit, S., Fisher, S.E., Neale, B.M., Anney, R.J.L., Ripke, S., Hollegaard, M.V., Werge, T., Ronald, A., Grove, J., Hougaard, D.M., Borglum, A.D., Mortensen, P.B., Daly, M.J., Smith, G.D., Ipsych-Ssi-Broad Autism Group, 2018. ASD and schizophrenia show distinct developmental profiles in common genetic overlap with population-based social communication difficulties. Mol. Psychiatry 23, 263–270.

Stahl, E.A., Breen, G., Forstner, A.J., McQuillin, A., Ripke, S., Trubetskoy, V., Mattheisen, M., Wang, Y., Coleman, J.R.I., Gaspar, H.A., de Leeuw, C.A., Steinberg, S., Pavlides, J.M.W., Trzaskowski, M., Byrne, E.M., Pers, T.H., Holmans, P.A., Richards, A.L., Abbott, L., Agerbo, E., Akil, H., Albani, D., Alliey-Rodriguez, N., Als, T.D., Anjorin, A., Antilla, V., Awasthi, S., Badner, J.A., Bækvad-Hansen, M., Barchas, J.D., Bass, N., Bauer, M., Belliveau, R., Bergen, S.E., Pedersen, C.B., Bøen, E., Boks, M.P., Boocock, J., Budde, M., Bunney, W., Burmeister, M., Bybjerg-Grauholm, J., Byerley, W., Casas, M., Cerrato, F., Cervantes, P., Chambert, K., Charney, A.W., Chen, D., Churchhouse, C., Clarke, T.K., Coryell, W., Craig, D.W., Cruceanu, C., Curtis, D., Czerski, P.M., Dale, A.M., de Jong, S., Degenhardt, F., Del-Favero, J., DePaulo, J.R., Djurovic, S., Dobbyn, A.L., Dumont, A., Elvsåshagen, T., Escott-Price, V., Fan, C.C., Fischer, S.B., Flickinger, M., Foroud, T.M., Forty, L., Frank, J., Fraser, C., Freimer, N.B., Frisén, L., Gade, K., Gage, D., Garnham, J., Giambartolomei, C., Pedersen, M.G., Goldstein, J., Gordon, S.D., Gordon-Smith, K., Green, E.K., Green, M.J., Greenwood, T.A., Grove, J., Guan, W., Guzman-Parra, J., Hamshere, M.L., Hautzinger, M., Heilbronner, U., Herms, S., Hipolito, M., Hoffmann, P., Holland, D., Huckins, L., Jamain, S., Johnson, J.S., Juréus, A., Kandaswamy, R., Karlsson, R., Kennedy, J.L., Kittel-Schneider, S., Knowles, J.A., Kogevinas, M., Koller, A.C., Kupka, R., Lavebratt, C., Lawrence, J., Lawson, W.B., Leber, M., Lee, P.H., Levy, S.E., Li, J.Z., Liu, C., Lucae, S., Maaser, A., MacIntyre, D.J., Mahon, P.B., Maier, W., Martinsson, L., McCarroll, S., McGuffin, P., McInnis, M.G., McKay, J.D., Medeiros, H., Medland, S.E., Meng, F., Milani, L., Montgomery, G.W., Morris, D.W., Mühleisen, T.W., Mullins, N., Nguyen, H., Nievergelt, C.M., Adolfsson, A.N., Nwulia, E.A., O'Donovan, C., Loohuis, L.M.O., Ori, A.P.S., Oruc, L., Ösby, U., Perlis, R.H., Perry, A., Pfennig, A., Potash, J.B., Purcell, S.M., Regeer, E.J., Reif, A., Reinbold, C.S., Rice, J.P., Rivas, F., Rivera, M., Roussos, P., Ruderfer, D.M., Ryu, E., Sánchez-Mora, C., Schatzberg, A.F., Scheftner, W.A., Schork, N.J., Shannon Weickert, C., Shehktman, T., Shilling, P.D., Sigurdsson, E., Slaney, C., Smeland, O.B., Sobell, J.L., Søholm Hansen, C., Spijker, A.T., St Clair, D., Steffens, M., Strauss, J.S., Streit, F., Strohmaier, J., Szelinger, S., Thompson, R.C., Thorgeirsson, T.E., Treutlein, J., Vedder, H., Wang, W., Watson, S.J., Weickert, T.W., Witt, S.H., Xi, S., Xu, W., Young, A.H., Zandi, P., Zhang, P., Zöllner, S., Adolfsson, R., Agartz, I., Alda, M., Backlund, L., Baune, B.T., Bellivier, F., Berrettini, W.H., Biernacka, J.M., Blackwood, D.H.R., Boehnke, M., Børglum, A.D., Corvin, A., Craddock, N., Daly, M.J., Dannlowski, U., Esko, T., Etain, B., Frye, M., Fullerton, J.M., Gershon, E.S., Gill, M., Goes, F., Grigoroiu-Serbanescu, M., Hauser, J., Hougaard, D.M., Hultman, C.M., Jones, I., Jones, L.A., Kahn, R.S., Kirov, G., Landén, M., Leboyer, M., Lewis, C.M., Li, Q.S., Lissowska, J., Martin, N.G., Mayoral, F., McElroy, S.L., McIntosh, A.M., McMahon, F.J., Melle, I., Metspalu, A., Mitchell, P.B., Morken, G., Mors, O., Mortensen, P.B., Müller-Myhsok, B., Myers, R.M., Neale, B.M., Nimgaonkar, V., Nordentoft, M., Nöthen, M.M., O'Donovan, M.C., Oedegaard, K.J., Owen, M.J., Paciga, S.A., Pato, C., Pato, M.T., Posthuma, D., Ramos-Quiroga, J.A., Ribasés, M., Rietschel, M., Rouleau, G.A., Schalling, M., Schofield, P.R., Schulze, T.G., Serretti, A., Smoller, J.W.,

Stefansson, H., Stefansson, K., Stordal, E., Sullivan, P.F., Turecki, G., Vaaler, A.E., Vieta, E., Vincent, J.B., Werge, T., Nurnberger, J.I., Wray, N.R., Di Florio, A., Edenberg, H.J., Cichon, S., Ophoff, R.A., Scott, L.J., Andreassen, O.A., Kelsoe, J., Sklar, P., 2019. The Bipolar Disorder Working Group of the Psychiatric Genomics Consortium, Genome-wide association study identifies 30 loci associated with bipolar disorder. Nat. Genet. 51, 793–803.

Stefansson, H., Ophoff, R.A., Steinberg, S., Andreassen, O.A., Cichon, S., Rujescu, D., Werge, T., Pietilainen, O.P., Mors, O., Mortensen, P.B., Sigurdsson, E., Gustafsson, O., Nyegaard, M., Tuulio-Henriksson, A., Ingason, A., Hansen, T., Suvisaari, J., Lonnqvist, J., Paunio, T., Borglum, A.D., Hartmann, A., FINK-Jensen, A., Nordentoft, M., Hougaard, D., Norgaard-Pedersen, B., Bottcher, Y., Olesen, J., Breuer, R., Moller, H.J., Giegling, I., Rasmussen, H.B., Timm, S., Mattheisen, M., Bitter, I., Rethelyi, J.M., Magnusdottir, B.B., Sigmundsson, T., Olason, P., Masson, G., Gulcher, J.R., Haraldsson, M., Fossdal, R., Thorgeirsson, T.E., Thorsteinsdottir, U., Ruggeri, M., Tosato, S., Franke, B., Strengman, E., Kiemeney, L.A., Genetic, R., Outcome In, P., Melle, I., Djurovic, S., Abramova, L., Kaleda, V., Sanjuan, J., DE Frutos, R., Bramon, E., Vassos, E., Fraser, G., Ettinger, U., Picchioni, M., Walker, N., Toulopoulou, T., Need, A.C., Ge, D., Yoon, J.L., Shianna, K.V., Freimer, N.B., Cantor, R.M., Murray, R., Kong, A., Golimbet, V., Carracedo, A., Arango, C., Costas, J., Jonsson, E.G., Terenius, L., Agartz, I., Petursson, H., Nothen, M.M., Rietschel, M., Matthews, P.M., Muglia, P., Peltonen, L., St Clair, D., Goldstein, D.B., Stefansson, K., Collier, D.A., 2009. Common variants conferring risk of schizophrenia. Nature 460, 744–747.

Stefansson, H., Rujescu, D., Cichon, S., Pietilainen, O.P.H., Ingason, A., Steinberg, S., Fossdal, R., Sigurdsson, E., Sigmundsson, T., Buizer-Voskamp, J.E., Hansen, T., Jakobsen, K.D., Muglia, P., Francks, C., Matthews, P.M., Gylfason, A., Halldorsson, B.V., Gudbjartsson, D., Thorgeirsson, T.E., Sigurdsson, A., Jonasdottir, A., Jonasdottir, A., Bjornsson, A., Mattiasdottir, S., Blondal, T., Haraldsson, M., Magnusdottir, B.B., Giegling, I., Moller, H.J., Hartmann, A., Shianna, K.V., Ge, D.L., Need, A.C., Crombie, C., Fraser, G., Walker, N., Lonnqvist, J., Suvisaari, J., Tuulio-Henriksson, A., Paunio, T., Toulopoulou, T., Bramon, E., DI Forti, M., Murray, R., Ruggeri, M., Vassos, E., Tosato, S., Walshe, M., Li, T., Vasilescu, C., Muhleisen, T.W., Wang, A.G., Ullum, H., Djurovic, S., Melle, I., Olesen, J., Kiemeney, L.A., Franke, B., Group, Sabatti, C., Freimer, N.B., Gulcher, J.R., Thorsteinsdottir, U., Kong, A., Andreassen, O.A., Ophoff, R.A., Georgi, A., Rietschel, M., Werge, T., Petursson, H., Goldstein, D.B., Nothen, M.M., Peltonen, L., Collier, D.A., ST Clair, D., Stefansson, K., 2008. Large recurrent microdeletions associated with schizophrenia. Nature 455. 232-U61.

Steinberg, S., Mors, O., Borglum, A.D., Gustafsson, O., Werge, T., Mortensen, P.B., Andreassen, O.A., Sigurdsson, E., Thorgeirsson, T.E., Bottcher, Y., Olason, P., Ophoff, R.A., Cichon, S., Gudjonsdottir, I.H., Pietilainen, O.P., Nyegaard, M., Tuulio-Henriksson, A., Ingason, A., Hansen, T., Athanasiu, L., Suvisaari, J., Lonnqvist, J., Paunio, T., Hartmann, A., Jurgens, G., Nordentoft, M., Hougaard, D., Norgaard-Pedersen, B., Breuer, R., Moller, H.J., Giegling, I., Glenthoj, B., Rasmussen, H.B., Mattheisen, M., Bitter, I., Rethelyi, J.M., Sigmundsson, T., Fossdal, R., Thorsteinsdottir, U., Ruggeri, M., Tosato, S., Strengman, E., Genetic, R., Outcome In, P., Kiemeney, L.A., Melle, I., Djurovic, S., Abramova, L., Kaleda, V., Walshe, M., Bramon, E., Vassos, E., Li, T., Fraser, G., Walker, N., Toulopoulou, T., Yoon, J., Freimer, N.B., Cantor, R.M., Murray, R., Kong, A., Golimbet, V., Jonsson, E.G., Terenius, L., Agartz, I., Petursson, H., Nothen, M.M., Rietschel, M., Peltonen, L., Rujescu, D., Collier, D.A., Stefansson, H., St Clair, D., Stefansson, K., 2011. Expanding the range of *ZNF804A* variants conferring risk of psychosis. Mol. Psychiatry 16, 59–66.

Sullivan, P.F., Kendler, K.S., Neale, M.C., 2003. Schizophrenia as a complex trait: evidence from a meta-analysis of twin studies. Arch. Gen. Psychiatr. 60, 1187–1192.

Sullivan, P.F., Lin, D., Tzeng, J.Y., Van Den Oord, E., Perkins, D., Stroup, T.S., Wagner, M., Lee, S., Wright, F.A., Zou, F., Liu, W., Downing, A.M., Lieberman, J., Close, S.L., 2008. Genomewide association for schizophrenia in the CATIE study: results of stage 1. Mol. Psychiatry 13, 570–584.

Tansey, K.E., Rees, E., Linden, D.E., Ripke, S., Chambert, K.D., Moran, J.L., Mccarroll, S.A., Holmans, P., Kirov, G., Walters, J., Owen, M.J., O'donovan, M.C., 2016. Common alleles contribute to schizophrenia in CNV carriers. Mol. Psychiatry 21, 1085–1089.

Tao, R., Cousijn, H., Jaffe, A.E., Burnet, P.W., Edwards, F., Eastwood, S.L., Shin, J.H., Lane, T.A., Walker, M.A., Maher, B.J., Weinberger, D.R., Harrison, P.J., Hyde, T.M., Kleinman, J.E., 2014. Expression of *ZNF804A* in human brain and alterations in schizophrenia, bipolar disorder, and major depressive disorder: a novel transcript fetally regulated by the psychosis risk variant rs1344706. JAMA Psychiatry 71, 1112–1120.

Vassos, E., Di Forti, M., Coleman, J., Iyegbe, C., Prata, D., Euesden, J., O'reilly, P., Curtis, C., Kolliakou, A.,

Patel, H., Newhouse, S., Traylor, M., Ajnakina, O., Mondelli, V., Marques, T.R., Gardner-Sood, P., Aitchison, K.J., Powell, J., Atakan, Z., Greenwood, K.E., Smith, S., Ismail, K., Pariante, C., Gaughran, F., Dazzan, P., Markus, H.S., David, A.S., Lewis, C.M., Murray, R.M., Breen, G., 2017. An examination of polygenic score risk prediction in individuals with first-episode psychosis. Biol. Psychiatry 81, 470–477.

Verweij, K.J.H., Abdellaoui, A., Nivard, M.G., Cort, A.S., Ligthart, L., Draisma, H.H.M., Minica, C.C., International Cannabis Consortium, Gillespie, N.A., Willemsen, G., Hottenga, J.J., Boomsma, D.I., Vink, J.M., 2017. Short communication: genetic association between schizophrenia and cannabis use. Drug Alcohol Depend. 171, 117–121.

Williams, H.J., Norton, N., Dwyer, S., Moskvina, V., Nikolov, I., Carroll, L., Georgieva, L., Williams, N.M., Morris, D.W., Quinn, E.M., Giegling, I., Ikeda, M., Wood, J., Lencz, T., Hultman, C., Lichtenstein, P., Thiselton, D., Maher, B.S., Molecular Genetics Of Schizophrenia Collaboration (Mgs), International Schizophrenia Consortium (Isc), Sgene-Plus, Group, Malhotra, A.K., Riley, B., Kendler, K.S., Gill, M., Sullivan, P., Sklar, P., Purcell, S., Nimgaonkar, V.L., Kirov, G., Holmans, P., Corvin, A., Rujescu, D., Craddock, N., Owen, M.J., O'donovan, M.C., 2011. Fine mapping of ZNF804A and genome-wide significant evidence for its involvement in schizophrenia and bipolar disorder. Mol. Psychiatry 16, 429–441.

Williams, N.M., Zaharieva, I., Martin, A., Langley, K., Mantripragada, K., Fossdal, R., Stefansson, H., Stefansson, K., Magnusson, P., Gudmundsson, O.O., Gustafsson, O., Holmans, P., Owen, M.J., O'donovan, M., Thapar, A., 2010. Rare chromosomal deletions and duplications in attention-deficit hyperactivity disorder: a genome-wide analysis. Lancet 376, 1401–1408.

Wium-Andersen, M.K., Ørsted, D.D., Nordestgaard, B.G., 2015. Tobacco smoking is causally associated with antipsychotic medication use and schizophrenia, but not with antidepressant medication use or depression. Int. J. Epidemiol. 44, 566–577.

Wray, N.R., Lee, S.H., Mehta, D., Vinkhuyzen, A.A.E., Dudbridge, F., Middeldorp, C.M., 2014. Research review: polygenic methods and their application to psychiatric traits. J. Child Psychol. Psychiatry 55, 1068–1087.

Xu, B., Ionita-Laza, I., Roos, J.L., Boone, B., Woodrick, S., Sun, Y., Levy, S., Gogos, J.A., Karayiorgou, M., 2012. *De novo* gene mutations highlight patterns of genetic and neural complexity in schizophrenia. Nat. Genet. 44, 1365–1369.

Zhang, R., Lu, S.M., Qiu, C., Liu, X.G., Gao, C.G., Guo, T.W., Valenzuela, R.K., Deng, H.W., Ma, J., 2011. Population-based and family-based association studies of *ZNF804A* locus and schizophrenia. Mol. Psychiatry 16, 360–361.

CHAPTER

11

Immune processes and risk of psychosis

Adam Al-Diwani[1], *Thomas A. Pollak*[2]

[1]Wellcome Research Training Fellow, Department of Psychiatry, University of Oxford, Oxford, United Kingdom; [2]Department of Psychosis Studies, Institute of Psychiatry, Psychology and Neuroscience, King's College London, London, United Kingdom

Introduction

Based on observations including those from tissue allograft experiments and an apparent lack of draining lymphatic vessels, it was thought for many years that the brain was an "immune privileged" organ, in effect hidden from the immune system. Therefore, despite occasional interest and epidemiological hints such as a well-established association between schizophrenia and maternal infection, immunology has been until recently a largely peripheral player in psychosis risk research. However, several lines of evidence including the elucidation of meningeal lymphatic vessels (Louveau et al., 2015b), the complex and dynamic properties of the blood–brain barrier (Pollak et al., 2018), and a more nuanced understanding of microglia have modified the concept of brain immune privilege (Louveau et al., 2015a). This has helped to foment a renewed interest in the role of immunity in the regulation of brain function as applied to psychosis, complementary to neuroscience and psychology.

Reciprocal immune interactions between the brain and the periphery have been described in both health and perturbed brain states operating during neurodevelopment in utero, postnatally, and into adolescent and adult life (Kipnis, 2018). Not only classical notions of immunity such as inflammatory mediators and neuronal autoantibodies have been studied but also an increasingly sophisticated understanding of the central role of glial biology in brain homeostasis has blurred traditional concepts of inflammation providing novel perspectives (Masgrau et al., 2017). Fundamental cellular communication mechanisms such as synapses, secreted chemical messengers, and receptors are shared between neural and immunological systems. Indeed, "memory," emerging from interactions with the environment and physically encoded in such substrates, is both a neural and an immune phenomenon, fundamental to the function of both systems. This may well be more than coincidence: a cohesive argument has been made that during the coevolution of both neural and immunological systems, fundamental aspects of human cognition and behavior have emerged, driven by their interaction, which in turn has been subject to natural selection. This may eventually provide a unifying framework to align

Risk Factors for Psychosis
https://doi.org/10.1016/B978-0-12-813201-2.00011-9

basic neuroimmunology and clinical observations relevant to psychotic disorders (Nataf, 2017).

Here, we review immune processes and psychosis risk aiming to focus on mechanisms by which immunity mediates long-established risk factors in early priming, illness onset, and perpetuation, with potential clues for therapeutic interventions.

Genetics

As discussed in detail in Chapters 6, 8, and 10, there is now an established genetic contribution to schizophrenia with at least 108 independent loci of genome-wide significance and probably more to be discovered as sample sizes further increase (Schizophrenia Working Group of the Psychiatric Genomics, 2014). Importantly, this line of investigation provides direct unbiased evidence to guide the search for pathogenic mechanisms. Such genomic data have contributed to enthusiasm for immune pathways because of signals from regions such as the major histocompatibility complex (MHC), which includes human leukocyte antigen (HLA) genes involved in peptide antigen presentation, and a region containing B lymphocyte surface markers CD19 and CD20. Nonetheless, moving from potential insights to modeling these in vitro and in vivo is challenging. In addition, inferring immune relevance from the MHC association in particular is problematic as it is intrinsically highly polymorphic and in linkage disequilibrium (Dendrou et al., 2018; Kennedy et al., 2017).

One approach to contextualize these genomic signals has come from pathway analyses, which attempt to identify whether genes cluster together in certain tissues or physiological pathways. One such analysis found an enrichment of enhancer and regulatory elements in immune processes including TGF-β signaling, B cell activation, and T cell activation (Network and Pathway Analysis Subgroup of Psychiatric Genomics, 2015). Indeed, this was shared across risk for schizophrenia, bipolar disorder, and major depression providing a reminder that biological pathways contributing to psychosis risk do not necessarily respect clinically defined categorical boundaries (Network and Pathway Analysis Subgroup of Psychiatric Genomics, 2015).

Genetic signals provide a guide to help focus real-world experimental neuroimmunology. Nonetheless, when looked at from both brain and immune system perspectives, investigators are faced with the counterintuitive yet scientifically exciting situation that immune signals may have more to do with brain development and function and vice versa. So while the study of neuroimmune interactions has already shed much light on potential mechanisms in psychotic disorders (Khandaker and Dantzer, 2016), it increasingly appears that psychiatric research in the coming years will be shaped by a reconceptualization of what we understand as "neuro" and what as "immune." For example, it has recently been shown that antigen-specific immune responses involving signaling between T follicular helper cells and B cells at the "immunological synapse" are amplified by dopamine via dopamine D1 receptors, a process that is usually considered in the context of brain activity (Papa et al., 2017). Conversely, complement genes such as complement component 4 (C4), which have traditionally been considered to be involved in antibody-dependent inflammatory reactions involved in host defense against microorganisms, now have an established role in microglia-mediated synaptic pruning and hence brain development (Sekar et al., 2016). This direct relationship between C4 alleles, pattern of brain expression, and synaptic abnormalities consistent with schizophrenia was experimentally demonstrated in a mouse model. Here, it was shown that variation at the C4 locus, located within the MHC region, was associated with schizophrenia risk in proportion to the allele's tendency to cause brain expression of C4a, clearly demonstrating that MHC associations with schizophrenia need not always reflect functional variation of the "HLA system" as traditionally conceived. There are many other highly variable "immune" loci in the population, meaning that there are likely similar as yet unexplored pathways. Moreover, the theme of genetic risk via apparent immune pathways manifesting through neurodevelopmental processes adds richness to long-held paradigms in psychosis risk research and could in the future offer translational opportunities (Borglum et al., 2014; Sekar et al., 2016).

Genomics focusing on common variation in many genes of small effect is complemented by studies of rarer, higher penetrance microchromosomal abnormalities such as copy number variants. One of the best described is 22q11.2 deletion syndrome, which has a striking association with psychosis: up to 40% affected can develop a psychotic disorder by their third decade (Schneider et al., 2014), discussed in detail in Chapter 8. From an immune perspective, this is intriguing because these patients tend to have a hypoplastic or sometimes absent thymus, the organ where T cells mature and undergo central tolerance against self-antigens. Accordingly, patients tend to have prominent immune dysregulation, not only being both immune-deficient and therefore more vulnerable to infection but also experiencing higher rates of autoimmunity, possibly owing to persistent autoreactive T cell that escaped central tolerance and underdeveloped T regulatory cell populations unable to provide adequate peripheral tolerance. Beyond T cell–dependent mechanisms, these patients also can have B cell deficiency and dysregulation giving rise to a combined immunodeficiency syndrome (Morsheimer et al., 2017). Despite the likelihood that in at least a subgroup of patients their immune and psychiatric risks are etiologically linked (Vann Jones et al., 2017), until recently little work has been done to explore this further. Vergaelen et al. found enrichment of Th17 cells in peripheral blood compared to healthy volunteers which appeared to correlate with PANSS score (Vergaelen et al., 2018). However, the relationship was modest and further work with larger samples is now needed to validate this pilot study and importantly probe whether they are causally related, for example, by interacting with neurodevelopment, or merely parallel aspects of a broader syndrome.

Epidemiology and infection

One substantial thread of evidence for an immune etiology of psychotic disorders comes from studies showing associations between psychosis and either immune and inflammatory (chiefly autoimmune) disorders or infection. In the 1950s, it was recognized that rheumatoid arthritis is negatively associated with schizophrenia (Pilkington, 1956), with recent estimates suggesting it confers a risk of less than 29% of that in the general population (Oken and Schulzer, 1999). It remains the only autoimmune disorder with a consistently reported negative association with psychosis: reasons are not fully understood but may reflect a small but robust negative SNP–genetic correlation between the two disorders (Lee et al., 2015).

Subsequent studies have assessed the relationship between autoimmune disorders, considered individually or as a whole, with psychotic disorders: significant associations have been described for celiac disease, Grave's disease, pemphigoid, psoriasis, autoimmune anemia, type 1 diabetes, Crohn's disease, and SLE (systemic lupus erythematosus), although negative associations have also been reported (Benros et al., 2012, 2014a; Chen et al., 2011). Recent population-level studies have reported the incident rate ratio of any (ICD-8 or ICD-10 listed) autoimmune disorder in people with psychosis as 1.53 (Benros et al., 2014b), with the risk of psychotic disorder in people with an autoimmune disorder similarly increased (relative risk 1.4 for any autoimmune disorder, between 0.7 and 5.4 for specific disorders (Benros et al., 2012)). Associations have also been noted with family history of autoimmune disorder (Benros et al., 2012, 2014a; Eaton et al., 2010). There is some evidence that the risk associated with autoimmune disorder is synergistic with that associated with infection, insofar that risk of subsequent diagnosis of psychotic disorder in people with autoimmune disorders was further increased when those individuals had been hospitalized for a serious infection (Benros et al., 2012). Interestingly, risk of a new diagnosis of psychotic disorder was increased in a time-dependent manner with exposure to antibiotic medication in a population-wide Danish cohort (Kohler et al., 2017); whether the pathogenically relevant exposure here is the initial infection for which the antibiotic medication was prescribed or the direct effects of the treatment (e.g., on the microbiome—see below) remains unexplored.

While infectious and postinfectious cases of psychosis have been reported for well over a century—largely in association with syphilis (Kaplan, 2010a,b) or influenza (Honigsbaum, 2013; Yudofsky, 2009) infection—renewed interest in infection *as a risk factor* began with the recognition that people born in winter are at increased risk of developing schizophrenia. Many hundreds of papers from both world hemispheres have reported, with a remarkable degree of convergence, a winter–spring excess of between 5% and 8% for schizophrenia and mania/bipolar disorder (Torrey et al., 1997). Infection, thought by many to be the mechanism underlying the seasonality effect, took central stage after Mednick et al., in 1988, demonstrated an excess of schizophrenia in children born to mothers who were pregnant during the 1957 Asian influenza epidemic (Mednick et al., 1988).

These findings were soon replicated in ecological studies (Wright et al., 1995; Takei et al., 1995, 1996; Sham et al., 1992; Adams et al., 1993) and in serological studies of stored serum (Brown et al., 2004). However, negative studies, often by the same authors, have also been reported (Cannon et al., 1996; Grech et al., 1997; Kendell and Kemp, 1989) (see Khandaker et al. (2013) and Brown and Derkits (2010) for reviews). A recent metaanalysis has cast doubt on the serological evidence for the association with influenza (Selten and Termorshuizen, 2017) but has been criticized on methodological grounds (Brown and Meyer, 2017).

The psychiatric specificity of the influenza-psychosis association is unclear: increased rates of major depression (Machon et al., 1997) and, inconsistently, autism spectrum disorders (Zerbo et al., 2013; Atladottir et al., 2010; Deykin and Macmahon, 1979) have both been reported in children born to mothers who were pregnant during influenza epidemics. Whether the effect is specific to influenza is unclear too, with associations similarly reported in ecological or epidemiological designs for maternal infection with rubella (Brown et al., 2000, 2001) and serological studies reporting associations with maternal HSV-2 (Buka et al., 2001, 2008) and toxoplasma (Xiao et al., 2009) infection. Maternal HSV-1 and CMV (cytomegalovirus) infection do not appear to be associated with increased risk of adult schizophrenia (Khandaker et al., 2013). Trimester differences have been observed inconsistently, with one review suggesting the greatest effect of prenatal infection occurring for influenza in the first trimester (Khandaker et al., 2013).

Importantly, as maternal infection is often associated with increased drug (including analgesic) exposure, low birth weight, and pregnancy complications, both of which have independently been associated with schizophrenia risk (Wright et al., 1995), these may have a confounding role.

Childhood infection may also increase psychosis risk. A 2012 metaanalysis found that childhood CNS (central nervous system) infection, particularly viral, conferred a nearly twofold risk of the development of adult schizophrenia (Khandaker et al., 2012). A UK population-based prospective serological study found that evidence of Epstein–Barr virus infection at age 4 predicted the occurrence of psychotic experiences, itself a risk factor for the later development of psychotic disorder, in adolescence (Khandaker et al., 2014b). Interestingly, schizophrenia polygenic risk score does not appear to account for the association between schizophrenia and infections, suggesting that genetic predisposition to psychosis does not also appreciably predispose to infection (Benros et al., 2016).

The human microbiome has, with the advent of 16S RNA (metagenomic) sequencing, become a focus of interest in psychiatry as in most branches of medicine. Group-level differences in microbial diversity between patients with psychotic disorders and healthy controls have been described (Schwarz et al., 2017; Shen et al., 2018; Castro-Nallar et al., 2015; Yolken et al., 2015), but because these studies are cross-sectional, the question of whether the observed changes are a cause or consequence of the disease remains unanswered.

Adult infections have been the object of considerable attention in relation to psychotic disorders, with most studies employing cross-sectional designs and measuring the presence of IgG antibodies to various pathogens (indexing putative lifetime exposure): as such, inferring causality from these studies is difficult. Nonetheless, some pathogens have been consistently implicated, with toxoplasma emerging most frequently (Yolken and Torrey, 2008; Gutierrez-Fernandez et al., 2015; Radua et al., 2018). One metaanalysis concluded the evidence for toxoplasma was "overwhelming," with an associated odds ratio of 1.43 after correction for publication bias (Sutterland et al., 2015). Furthermore, phase-specific analysis found a strongest signal for recent onset psychosis and increased serointensity in schizophrenia cases suggesting either active infection or a stronger immune response (Sutterland et al., 2015). Indeed, a recent vast umbrella review of 55 metaanalyses and systematic reviews, corresponding to 170 protective and risk factors, identified toxoplasma IgG as the *only* blood-based measure that associated with schizophrenia risk (Radua et al., 2018). Relatedly, as cat ownership is thought to be a risk factor for toxoplasma infection, there has been fiercely fought debate regarding whether childhood cat ownership increases schizophrenia risk (Solmi et al., 2017; Scharko, 2015; Fuller Torrey et al., 2015).

Human immunodeficiency virus (HIV) infection is associated with an increased risk of a subsequent diagnosis of both schizophrenia (incidence rate ratio (IRR) 4.09) and acute psychosis (IRR 7.15); the underlying reasons are likely to be multifactorial, taking in immune deficiency, lifestyle factors, increased medical surveillance, and increased susceptibility to infection, but intriguingly the risk of diagnosis of schizophrenia following HIV infection falls after antiretroviral treatment is initiated (Helleberg et al., 2015).

Beyond the risk of caseness, emerging evidence suggests that within the psychosis construct, infections may affect risk of particular aspects of the phenotype, including cognitive dysfunction (Dickerson et al., 2003, 2006, 2014a, 2016; Hamdani et al., 2017; Prasad et al., 2012; Houenou et al., 2014), suicidality (Dickerson et al., 2017; Okusaga et al., 2011), and death by natural causes (Dickerson et al., 2007, 2014b).

Given that risk appears to be present in relation to multiple autoimmune disorders and with multiple infectious agents across viruses, bacteria, and protozoa, a reasonable strategy would be to ask whether inflammation per se is the common denominator which underlies risk and which, mechanistically, is involved in the pathogenesis of psychotic disorders. To this end, attention has turned to studies of psychosis risk in relation to markers of inflammation, such as cytokines, and to animal models seeking to illuminate how inflammatory challenges can result in the kind of brain and behavioral pathology observed in psychotic disorders.

Cytokines, microglial phenotype, and neurodevelopment

Cytokines are small proteins whose classical role is to signal over short and long range between innate and adaptive immune cells. They work in concert with chemokines, which form concentration gradients along which immune cells traffic, for example, to a germinal center or to a site of infection. Alongside these prototypical functions, cytokines are increasingly recognized to influence the brain both during neurodevelopment and homeostasis, via multiple distinct pathways (Savitz and Harrison, 2018; Mondelli et al., 2017). These molecules appear to be as central to brain function as canonical neurotransmitters and can modulate neuronal networks leading to stereotyped behavioral phenotypes. This has been most clearly described in association with systemic inflammation in the context of infection typically leading to hypersomnia, fatigue, and anhedonia (Capuron and Miller, 2011) and recently molecular and neural network details have begun to be experimentally modeled. For example, in an immunodeficient mouse which manifests abnormal social behavior associated with frontal hyperconnectivity, adoptive transfer of T cells lead to an interferon-γ (IFN-γ)–dependent restoration of behavior and functional magnetic resonance imaging pattern associated with increased inhibitory GABAergic neuron signaling (Filiano et al., 2017). Intriguingly, the IFN-γ was secreted from T cells circulating within meningeal lymphatics, providing a mechanism by which circulating immune cells may modulate brain activity via cytokines without necessarily directly infiltrating the brain.

In addition, the maternal immune activation (MIA) paradigm provides evidence implicating cytokines in shaping psychosis risk via neurodevelopmental processes. Here, various infection-mimicking stimuli such as poly(I:C) or lipopolysaccharides are given to pregnant mice to model the neurodevelopmental effects of physiological perturbations associated with psychosis risk, e.g., maternal infection (Estes and Mcallister, 2016; Meyer, 2014), resulting in behavioral, neurophysiological, and neuropathological phenotypes of relevance to psychosis and other putatively neurodevelopmental disorders. Central to MIA models are the activation state of microglia which via synaptic pruning is thought to influence neural circuits, connectivity, and ultimately thoughts and behaviors (Salter and Stevens, 2017). Increasingly, intermediate steps in this pathway are becoming elucidated, with clear relevance for pathogenesis of psychotic disorders. For example, it was recently demonstrated that this process is dependent on Th17 cell–derived IL-17α, which could be selectively blocked, rescuing the phenotype (Salter and Stevens, 2017). Another MIA model found that maternal inflammation can delay the switch of GABAergic signaling from excitatory to inhibitory (Corradini et al., 2017). Of the multiplicity of cytokines of relevance to these paradigms, IL-6 has emerged as potentially the most important immunological mediator of the association between maternal infection and abnormal neurodevelopment (Samuelsson et al., 2006; Smith et al., 2007; Rudolph et al., 2018), perhaps because of the relative ease with which it crosses the placental barrier. If validated, such mechanisms may represent a potential therapeutic target to offset risk associated with maternal immune state.

Beyond this promising work linking cytokines with psychosis risk via neurodevelopment in model systems, most studies in people have measured cytokine panels in peripheral blood and occasionally cerebrospinal fluid (CSF) at intervals ranging from the perinatal and neonatal periods, to childhood and clinical high-risk states, and through to first episode and established illness. Despite including prodromal states and medication-naïve participants in acute stages, these studies are still prone to confounding by multiple other factors known to influence cytokine profile including, nonexhaustively, age, sex, smoking, alcohol, and adiposity (O'Connor et al., 2009). Nonetheless, there have been sufficient numbers of these studies to justify metaanalyses, which have attempted to adjust for confounding factors, yielding reasonably consistent patterns of cytokine fluctuations operating as trait and state factors in psychotic disorders. For example, IL-1, IL-6, and TGF-β appear to rise and fall in line with symptom fluctuation resembling state factors, whereas IL-12, IFN-γ, and TNF-α can be elevated independent of clinical state, therefore resembling trait factors (Miller et al., 2011; Goldsmith et al., 2016).

Cohort studies have identified longitudinal relationships between cytokines and psychosis risk, potentially interacting with other associations. For example, childhood trauma has been associated with altered cytokine profiles in adults, suggesting a plausible mechanism by which an environmental risk factor might affect neurodevelopment and predispose to the development of psychotic disorder (Baumeister et al., 2016). At age 9, IL-6 levels predicted, in a dose-dependent manner, the occurrence of psychotic experiences at age 18 (Khandaker et al., 2014a). Studies focusing on prodromal psychosis have identified IL-6 (Zeni-Graiff et al., 2016), along with IL-12p40, as a marker of transition to psychosis (Focking et al., 2016), although panels of inflammatory markers (rather than individual candidates) may emerge as having the greatest predictive utility (Perkins et al., 2015). Nonetheless, there have been few mechanistic studies directly assessing causality of individual cytokines; therefore, whether this work is aiming for biomarkers or for therapeutic targets, as is emerging in depression (Kohler et al., 2014), will depend on future studies. Notably, a trial of an anti-IL-6 receptor monoclonal antibody, tocilizumab, failed to improve residual symptoms or cognitive deficits over placebo in participants with established, moderately stable schizophrenia (Girgis et al., 2017).

Anchored in inflammatory paradigms, most investigators have interpreted their findings as indicative of "proinflammatory" or "antiinflammatory" states, implying that interventions which shift the balance of putative inflammation may be therapeutic. It is tempting to suggest that cardiometabolic comorbidities that often emerge with psychotic disorder may be manifestations of a common "inflammatory" process, perhaps linked with psychosocial stress (Howes and Mccutcheon, 2017). For example, there is evidence from positron emission tomography brain imaging using ligands that bind translocator-binding protein (TSPO), a marker of microglial activation, suggesting that microglial state correlates with the transition from high risk to first-episode psychosis in clinical populations (Bloomfield et al., 2016). However, not only is the concept of a binary microglial "activation—inactivation" increasingly questioned, but also it has been shown that translocator-binding proteins are expressed on multiple CNS cell types, reducing the specificity of this approach (Notter et al., 2017). In addition to these caveats, few other groups have replicated the observation of increased TSPO signal in psychosis (Holmes et al., 2016; Hafizi et al., 2017; Collste et al., 2017).

Therefore, as understanding of brain—immune interactions evolves, the extent to which a traditional "inflammation" model is appropriate will continue to be evaluated and the concept is likely to become more nuanced. This is hinted at by recent evidence using two-sample Mendelian randomization to offset the effects of confounding factors to evaluate the relationship between peripheral inflammatory markers and schizophrenia risk (Hartwig et al., 2017). This study found that contrary to the more common finding of raised inflammatory signals associating with raised risk, elevated CRP (C-reactive protein) appears to be associated with a reduced schizophrenia risk. The authors hypothesize that this may be the result of better clearance of early life infections. Clearly, further studies are needed to validate the main observation, but linear accounts of inflammation driving psychosis will probably require revision.

Lymphocytes and antigen-specific autoimmunity

Lymphocyte biology has not been explored extensively in psychosis risk research. However, as drivers of cytokine networks and mediators of antigen-specific reactivity underlying autoimmunity, this area of immunology is likely to receive increased attention. Studies of peripheral blood have suggested abnormal lymphocyte populations during transition to psychosis. Compared to healthy volunteers, in patients with acute psychosis, elevated B cells ($CD19^+$), reduced T cells ($CD3^+$), and an elevated T helper to T effector cell ($CD4^+$:$CD8^+$) ratio have been described (Steiner et al., 2010; Maino et al., 2007; Miller et al., 2013). However, little functional work has been done to model a causative role in psychosis.

Lumbar punctures are infrequently performed in the evaluation of psychosis; however, in centers that routinely sample CSF, a perhaps surprisingly large number of patients had abnormalities consistent with the "mild encephalitis" hypothesis of schizophrenia (Bechter, 2013). For example, in one series of 180 patients sampled from 2006 to 2013 (34 nonconsecutive and 146 consecutive), 42.2% had raised protein and 3.4% had pleocytosis (Endres et al., 2015). Therefore, although some of this sample was enriched by testing patients with a higher probability of an organic component to their illness, most of it was not. This has been further demonstrated by Oviedo-Salcedo et al. in a series of 124 patients with a clinical diagnosis of a schizophrenia spectrum disorder (excluding drug-induced psychoses and those with signs suspicious for autoimmune encephalitis) who underwent lumbar puncture as part of a routine clinical work-up. They found that 6 of the 124 (4.8%) patients had a mild pleocytosis (range 6–14 cells/μL) and 18 of the 124 (14.6%) patients showed unmatched intrathecal oligoclonal bands suggestive of antibody production within the central nervous system (Oviedo-Salcedo et al., 2018). Given that the exclusion criteria reduced the prior likelihood of an autoimmune/inflammatory brain condition, these findings provide impetus to prospectively investigate the immunophenotype of CSF in patients with psychotic disorders, potentially yielding pathophysiologically relevant data.

One key reason investigators have become more interested in sampling CSF routinely in patients with psychotic disorders is to exclude autoimmune encephalopathies. Autoimmune encephalopathies are brain diseases mediated by antibodies against neuronal surface targets including the NMDA, GABA, and AMPA receptors and synaptic proteins such as LGI1 and CASPR2. This disease process has provided a model for considering how an immune process can directly lead to acquired mental illness and has emerged as an important differential diagnosis of acute-onset psychotic illness with recent estimates suggesting it may be at least as common as viral encephalitis (Dubey et al., 2018).

NMDAR-antibody encephalitis has a distinctive clinical picture: patients are typically young females presenting with an abrupt onset of multiple psychiatric symptoms including psychosis, mood disturbance, bizarre behaviors, and catatonia (Dalmau et al., 2017; Al-Diwani et al., 2019). Here, B cells directed against the GluN1 subunit of the NMDA receptor mature into NMDAR-antibody secreting cells. In most patients, the triggers of this event and timing relative to symptom onset are not entirely clear. Antibody is detectable in serum and CSF of most patients, but the disease is defined by CSF antibody, whereas serum antibody can be found in healthy individuals (Gresa-Arribas et al., 2014; Graus et al., 2016). About 1/4 of patients have an ovarian teratoma, usually unilateral, mature, and benign, although bilateral and/or immature cases have been reported. Tumor removal usually expedites recovery, which is consistent with lymphocytic infiltrates and abnormal neurons observed in teratoma tissue in associated cases, which may be sufficient to generate the pool of pathogenic cells (Dabner et al., 2012; Day et al., 2014). Additionally, there is a subgroup of patients who develop NMDAR-antibody encephalitis following recovery from herpes simplex virus encephalitis (Armangue et al., 2014); however, the mechanistic basis of this is not yet well understood. The divalent IgG antibodies cross-link epitopes on spatially clustered NMDA receptors, triggering their internalization and leading to widespread network disruption probably because of reduction of GABAergic interneuron activity secondary to NMDAR hypofunction/reduced availability (Hughes et al., 2010).

Autoimmune encephalitis provides an important paradigm linking lymphocyte autoreactivity with disturbed neural circuit function and transition from health to psychotic symptoms mediated by pathogenic antibodies. Since the first publications on these diseases, awareness from psychiatrists has been highlighted as crucial (Dalmau et al., 2008; Lejuste et al., 2016). Indeed, Scott et al. reported serum antibody data collected in 113 unwell inpatients with psychotic illness and found 3 cases that would satisfy criteria for NMDAR-antibody encephalitis cases and had been admitted to psychiatric wards and initially given primary psychiatric diagnoses. These patients responded to immunotherapeutic approaches as would be expected from established knowledge of the disease (Titulaer et al., 2013), but they also found a further patient with serum NMDAR reactivity and noted an immunotherapy response (Scott et al., 2018). This adds to the growing literature looking at the prevalence of serum autoreactivity in clinically undifferentiated psychoses that do not satisfy criteria for autoimmune encephalitis and whether immunotherapies may be beneficial (Al-Diwani et al., 2017). Several studies have reported a positive association with psychosis spectrum presentations and noted responses to immunotherapy (Pollak et al., 2014; Pearlman and Najjar, 2014; Lennox et al., 2017; Scott et al., 2018); however, most work has been done using single assay approaches without paired CSF samples or neurophysiological parameters of relevance, making it hard to determine the significance of these findings. Furthermore, given placebo effects, regression to the mean and inconsistent long-term follow-up, in the absence of double-blind randomized trial data, treatment responses should be considered cautiously as evidence of antibody pathogenicity. Those studies that included CSF have not found neuronal surface antibodies in patients with psychotic disorders at clinical low risk for autoimmune encephalitis (Oviedo-Salcedo et al., 2018), but further work comparing assay platforms and sampling within inpatient populations remains necessary.

Nonetheless, innovative imaging techniques capable of tracking the effect of antibodies on single receptors on live neurons in vitro imply that NMDAR antibodies in patients with psychosis without autoimmune encephalitis have the capacity to bind neurons and disrupt the clustering of the receptor potentially affecting synaptic function and that this does not occur in NMDAR-antibody positive samples from healthy volunteers (Jezequel et al., 2017a,b). These intriguing findings do not yet equate to a demonstration of pathogenicity in the patients from whom the samples were sourced, but does mean that serum NMDAR autoreactivity in patients with psychosis remains an important research topic. Investigators in this area are now moving beyond prevalence estimates alone and increasingly considering functional approaches at both molecular and system levels, as well as randomized clinical trials to determine antibody relevance outside of classical phenotypes.

While comparison with autoimmune encephalitis provides evidence of acquired autoimmunity, neuronal surface antibodies have also been found to be of relevance to psychosis risk via vertical (maternofetal) transfer and perturbation of neurodevelopment (Jagota et al., 2014; Hilderink et al., 2015; Lamale-Smith et al., 2015; Chourasia et al., 2017). As demonstrated with MIA models, early physiological challenges in utero appear to be able to mediate significant neurodevelopmental effects that ultimately manifest in abnormal brain architecture at cellular and system levels and propensity to psychiatric syndromes. For example, several case reports report neurological disability in infants born to clinically remitted but persistently NMDAR-antibody seropositive mothers (Jagota et al., 2014; Hilderink et al., 2015; Lamale-Smith et al., 2015; Chourasia et al., 2017). This fits with observations from Coutinho and colleagues who found in a Scandinavian cohort an association between CASPR2-antibody seropositive mothers and neurodevelopmental disorder in offspring (Coutinho et al., 2017a). They investigated this further with a passive transfer mouse model of CASPR2 antibodies to pregnant dams, noting abnormal social behaviors that correlated with histological evidence of microglial activation and abnormal synaptic pruning demonstrating profound long-term effects of a brief exposure to neuronal surface antibodies during development in utero (Coutinho et al., 2017b).

Pathogenic autoantibodies therefore can affect disease-relevant pathology in neurodevelopment and in adult life. We have seen that cytokines and infectious pathogens have been evaluated thoroughly in regard to psychosis risk; it is a logical next step that autoantibodies, as potential determinants of disturbed brain development and function, are assessed similarly from the perinatal period through to the development of psychotic disorder in adults.

Conclusion

Immunology and neuroscience are experiencing a productive period of cross-fertilization, which has challenged many long-held preconceptions, immune privilege, glia, and lymphatic drainage, long since considered settled matters are particularly hot topics. But does this have practical implications for psychosis risk? These findings have largely not yet yielded findings of practical relevance for everyday psychiatric practice. However, for research, they have reinvigorated interest in classic findings, e.g., maternal infection, through the lens of mechanisms not known about by previous generations of investigators, e.g., microglial activation states and synaptic pruning. Indeed, even relatively recent work such as the various studies of cytokine levels should be considered carefully. For example, complement C4, formerly thought relevant to inflammation, is now considered a crucial molecule in brain development not requiring inflammation to be invoked at all to understand its relevance to psychosis.

As outlined elsewhere in other chapters, psychosis is a complex phenomenon where unitary causes are rare. Immunology and neuroimmunology need to be considered as a further covariate alongside existing factors operating across biopsychosocial domains throughout pre- and postnatal development to modify likelihood of psychosis. Similar to psychiatric genetics, where signal and noise have required large sample sizes to tease out robust results, investigators assessing immune mechanisms will likely also need to collaborate to achieve large sample sizes and consistent assay approaches to increase clinical relevance. Despite this important caveat, the field has produced some examples where causality is linear: autoimmune encephalitis being the most striking example. Although relatively rare, NMDAR-antibody encephalitis is principally a disease of young women, and in abrupt onset polymorphic psychosis the prior probability of this disease is enhanced. In addition to providing a clear clinical example where everyday psychiatry is increasingly engaging with immune paradigms, this condition provides an important disease model where psychosis is unequivocally linked to immune dysfunction via a neuronal surface receptor–specific pathogenic antibody. However, the field is evolving at such pace that perhaps by the next edition of this book further robustly demonstrated immune-mediated subtypes of psychosis will emerge.

References

Adams, W., Kendell, R.E., Hare, E.H., Munk-Jorgensen, P., 1993. Epidemiological evidence that maternal influenza contributes to the aetiology of schizophrenia. An analysis of Scottish, English, and Danish data. Br. J. Psychiatry 163, 522–534.

Al-Diwani, A., Pollak, T.A., Langford, A.E., Lennox, B.R., 2017. Synaptic and neuronal autoantibody-associated psychiatric syndromes: controversies and hypotheses. Front. Psychiatry 8, 13.

Al-Diwani, A., Handel, A., Townsend, L., Pollak, T., Leite, M.I., Harrison, P.J., Lennox, B.R., Okai, D., Manohar, S.G., Irani, S.R., 2019. The psychopathology of NMDAR-antibody encephalitis in adults: a systematic review and phenotypic analysis of individual patient data. Lancet Psychiatry 6, 235–246.

Armangue, T., Leypoldt, F., Malaga, I., Raspall-Chaure, M., Marti, I., Nichter, C., Pugh, J., Vicente-Rasoamalala, M., Lafuente-Hidalgo, M., Macaya, A., Ke, M., Titulaer, M.J., Hoftberger, R., Sheriff, H., Glaser, C., Dalmau, J., 2014. Herpes simplex virus encephalitis is a trigger of brain autoimmunity. Ann. Neurol. 75, 317–323.

Atladottir, H.O., Thorsen, P., Ostergaard, L., Schendel, D.E., Lemcke, S., Abdallah, M., Parner, E.T., 2010. Maternal infection requiring hospitalization during pregnancy and autism spectrum disorders. J. Autism Dev. Disord. 40, 1423–1430.

Baumeister, D., Akhtar, R., Ciufolini, S., Pariante, C.M., Mondelli, V., 2016. Childhood trauma and adulthood inflammation: a meta-analysis of peripheral C-reactive protein, interleukin-6 and tumour necrosis factor-alpha. Mol. Psychiatry 21, 642–649.

Bechter, K., 2013. Updating the mild encephalitis hypothesis of schizophrenia. Prog. Neuropsychopharmacol. Biol. Psychiatry 42, 71–91.

Benros, M.E., Eaton, W.W., Mortensen, P.B., 2014a. The epidemiologic evidence linking autoimmune diseases and psychosis. Biol. Psychiatry 75, 300–306.

Benros, M.E., Mortensen, P.B., Eaton, W.W., 2012. Autoimmune diseases and infections as risk factors for schizophrenia. Ann. N. Y. Acad. Sci. 1262, 56–66.

Benros, M.E., Pedersen, M.G., Rasmussen, H., Eaton, W.W., Nordentoft, M., Mortensen, P.B., 2014b. A nationwide study on the risk of autoimmune diseases in individuals with a personal or a family history of schizophrenia and related psychosis. Am. J. Psychiatry 171, 218–226.

Benros, M.E., Trabjerg, B.B., Meier, S., Mattheisen, M., Mortensen, P.B., Mors, O., Borglum, A.D., Hougaard, D.M., Norgaard-Pedersen, B., Nordentoft, M., Agerbo, E., 2016. Influence of polygenic risk scores on the association between infections and schizophrenia. Biol. Psychiatry 80, 609–616.

Bloomfield, P.S., Selvaraj, S., Veronese, M., Rizzo, G., Bertoldo, A., Owen, D.R., Bloomfield, M.A., Bonoldi, I., Kalk, N., Turkheimer, F., Mcguire, P., De Paola, V., Howes, O.D., 2016. Microglial activity in people at ultra high risk of psychosis and in schizophrenia: an [(11)C] PBR28 PET brain imaging study. Am. J. Psychiatry 173, 44–52.

Borglum, A.D., Demontis, D., Grove, J., Pallesen, J., Hollegaard, M.V., Pedersen, C.B., Hedemand, A., Mattheisen, M., Investigators, G., Uitterlinden, A., Nyegaard, M., Orntoft, T., Wiuf, C., Didriksen, M., Nordentoft, M., Nothen, M.M., Rietschel, M., Ophoff, R.A., Cichon, S., Yolken, R.H., Hougaard, D.M., Mortensen, P.B., Mors, O., 2014. Genome-wide study of association and interaction with maternal cytomegalovirus infection suggests new schizophrenia loci. Mol. Psychiatry 19, 325–333.

Brown, A.S., Begg, M.D., Gravenstein, S., Schaefer, C.A., Wyatt, R.J., Bresnahan, M., Babulas, V.P., Susser, E.S., 2004. Serologic evidence of prenatal influenza in the etiology of schizophrenia. Arch. Gen. Psychiatr. 61, 774–780.

Brown, A.S., Cohen, P., Greenwald, S., Susser, E., 2000. Nonaffective psychosis after prenatal exposure to rubella. Am. J. Psychiatry 157, 438–443.

Brown, A.S., Cohen, P., Harkavy-Friedman, J., Babulas, V., Malaspina, D., Gorman, J.M., Susser, E.S., 2001. A.E. Bennett Research Award. Prenatal rubella, premorbid abnormalities, and adult schizophrenia. Biol. Psychiatry 49, 473–486.

Brown, A.S., Derkits, E.J., 2010. Prenatal infection and schizophrenia: a review of epidemiologic and translational studies. Am. J. Psychiatry 167, 261–280.

Brown, A.S., Meyer, U., 2017. Comment on: "The serological evidence for maternal influenza as risk factor for psychosis in offspring is insufficient: critical review and meta-analysis". Schizophr. Res. 189, 223–224.

Buka, S.L., Cannon, T.D., Torrey, E.F., Yolken, R.H., Collaborative Study, Group On The Perinatal Origins Of Severe Psychiatric, D., 2008. Maternal exposure to herpes simplex virus and risk of psychosis among adult offspring. Biol. Psychiatry 63, 809–815.

Buka, S.L., Tsuang, M.T., Torrey, E.F., Klebanoff, M.A., Bernstein, D., Yolken, R.H., 2001. Maternal infections and subsequent psychosis among offspring. Arch. Gen. Psychiatr. 58, 1032–1037.

Cannon, M., Cotter, D., Coffey, V.P., Sham, P.C., Takei, N., Larkin, C., Murray, R.M., O'callaghan, E., 1996. Prenatal exposure to the 1957 influenza epidemic and adult schizophrenia: a follow-up study. Br. J. Psychiatry 168, 368–371.

Capuron, L., Miller, A.H., 2011. Immune system to brain signaling: neuropsychopharmacological implications. Pharmacol. Ther. 130, 226–238.

Castro-Nallar, E., Bendall, M.L., Perez-Losada, M., Sabuncyan, S., Severance, E.G., Dickerson, F.B., Schroeder, J.R., Yolken, R.H., Crandall, K.A., 2015. Composition, taxonomy and functional diversity of the oropharynx microbiome in individuals with schizophrenia and controls. PeerJ 3, e1140.

Chen, Y.J., Wu, C.Y., Lin, M.W., Chen, T.J., Liao, K.K., Chen, Y.C., Hwang, C.Y., Chu, S.Y., Chen, C.C., Lee, D.D., Chang, Y.T., Wang, W.J., Liu, H.N., 2011. Comorbidity profiles among patients with bullous pemphigoid: a nationwide population-based study. Br. J. Dermatol. 165, 593–599.

Chourasia, N., Watkins, M.W., Lankford, J.E., Kass, J.S., Kamdar, A., 2017. An infant born to a mother with anti-N-Methyl-d-Aspartate receptor encephalitis. Pediatr. Neurol. 79, 65–68.

Collste, K., Plaven-Sigray, P., Fatouros-Bergman, H., Victorsson, P., Schain, M., Forsberg, A., Amini, N., Aeinehband, S., Karolinska Schizophrenia Project, C., Erhardt, S., Halldin, C., Flyckt, L., Farde, L., Cervenka, S., 2017. Lower levels of the glial cell marker TSPO in drug-naive first-episode psychosis patients as measured using PET and [(11)C]PBR28. Mol. Psychiatry 22, 850–856.

Corradini, I., Focchi, E., Rasile, M., Morini, R., Desiato, G., Tomasoni, R., Lizier, M., Ghirardini, E., Fesce, R., Morone, D., Barajon, I., Antonucci, F., Pozzi, D., Matteoli, M., 2017. Maternal immune activation delays excitatory-to-inhibitory gamma-aminobutyric acid switch in offspring. Biol. Psychiatry 83, 680–691.

Coutinho, E., Jacobson, L., Pedersen, M.G., Benros, M.E., Norgaard-Pedersen, B., Mortensen, P.B., Harrison, P.J., Vincent, A., 2017a. CASPR2 autoantibodies are raised during pregnancy in mothers of children with mental retardation and disorders of psychological development but not autism. J. Neurol. Neurosurg. Psychiatry 88, 718–721.

Coutinho, E., Menassa, D.A., Jacobson, L., West, S.J., Domingos, J., Moloney, T.C., Lang, B., Harrison, P.J., Bennett, D.L.H., Bannerman, D., Vincent, A., 2017b. Persistent microglial activation and synaptic loss with behavioral abnormalities in mouse offspring exposed to CASPR2-antibodies in utero. Acta Neuropathol. 134, 567–583.

Dabner, M., Mccluggage, W.G., Bundell, C., Carr, A., Leung, Y., Sharma, R., Stewart, C.J., 2012. Ovarian teratoma associated with anti-N-methyl D-aspartate receptor encephalitis: a report of 5 cases documenting prominent intratumoral lymphoid infiltrates. Int. J. Gynecol. Pathol. 31, 429–437.

Dalmau, J., Geis, C., Graus, F., 2017. Autoantibodies to synaptic receptors and neuronal cell surface proteins in autoimmune diseases of the central nervous system. Physiol. Rev. 97, 839–887.

Dalmau, J., Gleichman, A.J., Hughes, E.G., Rossi, J.E., Peng, X., Lai, M., Dessain, S.K., Rosenfeld, M.R., Balice-Gordon, R., Lynch, D.R., 2008. Anti-NMDA-receptor encephalitis: case series and analysis of the effects of antibodies. Lancet Neurol. 7, 1091–1098.

Day, G.S., Laiq, S., Tang-Wai, D.F., Munoz, D.G., 2014. Abnormal neurons in teratomas in NMDAR encephalitis. JAMA Neurol. 71, 717–724.

Dendrou, C.A., Petersen, J., Rossjohn, J., Fugger, L., 2018. HLA variation and disease. Nat. Rev. Immunol. 18, 325–339.

Deykin, E.Y., Macmahon, B., 1979. Viral exposure and autism. Am. J. Epidemiol. 109, 628–638.

Dickerson, F., Adamos, M.B., Katsafanas, E., Khushalani, S., Origoni, A., Savage, C.L.G., Schroeder, J., Schweinfurth, L.A.B., Stallings, C., Sweeney, K., Yolken, R., 2016. The association among smoking, HSV-1 exposure, and cognitive functioning in schizophrenia, bipolar disorder, and non-psychiatric controls. Schizophr. Res. 176, 566–571.

Dickerson, F., Boronow, J., Stallings, C., Origoni, A., Yolken, R., 2007. Toxoplasma gondii in individuals with schizophrenia: association with clinical and demographic factors and with mortality. Schizophr. Bull. 33, 737–740.

Dickerson, F., Kirkpatrick, B., Boronow, J., Stallings, C., Origoni, A., Yolken, R., 2006. Deficit schizophrenia: association with serum antibodies to cytomegalovirus. Schizophr. Bull. 32, 396–400.

Dickerson, F., Stallings, C., Origoni, A., Katsafanas, E., Schweinfurth, L., Savage, C., Khushalani, S., Yolken, R., 2014a. Antibodies to Toxoplasma gondii and cognitive functioning in schizophrenia, bipolar disorder, and nonpsychiatric controls. J. Nerv. Ment. Dis. 202, 589–593.

Dickerson, F., Stallings, C., Origoni, A., Schroeder, J., Khushalani, S., Yolken, R., 2014b. Mortality in schizophrenia: clinical and serological predictors. Schizophr. Bull. 40, 796–803.

Dickerson, F., Wilcox, H.C., Adamos, M., Katsafanas, E., Khushalani, S., Origoni, A., Savage, C., Schweinfurth, L., Stallings, C., Sweeney, K., Yolken, R., 2017. Suicide attempts and markers of immune response in individuals with serious mental illness. J. Psychiatr. Res. 87, 37–43.

Dickerson, F.B., Boronow, J.J., Stallings, C., Origoni, A.E., Ruslanova, I., Yolken, R.H., 2003. Association of serum antibodies to herpes simplex virus 1 with cognitive deficits in individuals with schizophrenia. Arch. Gen. Psychiatr. 60, 466–472.

Dubey, D., Pittock, S.J., Kelly, C.R., Mckeon, A., Lopez-Chiriboga, A.S., Lennon, V.A., Gadoth, A., Smith, C.Y., Bryant, S.C., Klein, C.J., Aksamit, A.J., Toledano, M., Boeve, B.F., Tillema, J.M., Flanagan, E.P., 2018. Autoimmune encephalitis epidemiology and a comparison to infectious encephalitis. Ann. Neurol. 83, 166–177.

Eaton, W.W., Pedersen, M.G., Nielsen, P.R., Mortensen, P.B., 2010. Autoimmune diseases, bipolar disorder, and non-affective psychosis. Bipolar Disord. 12, 638–646.

Endres, D., Perlov, E., Baumgartner, A., Hottenrott, T., Dersch, R., Stich, O., Tebartz Van Elst, L., 2015. Immunological findings in psychotic syndromes: a tertiary care hospital's CSF sample of 180 patients. Front. Hum. Neurosci. 9, 476.

Estes, M.L., Mcallister, A.K., 2016. Maternal immune activation: implications for neuropsychiatric disorders. Science 353, 772–777.

Filiano, A.J., Gadani, S.P., Kipnis, J., 2017. How and why do T cells and their derived cytokines affect the injured and healthy brain? Nat. Rev. Neurosci. 18, 375–384.

Focking, M., Dicker, P., Lopez, L.M., Cannon, M., Schafer, M.R., Mcgorry, P.D., Smesny, S., Cotter, D.R., Amminger, G.P., 2016. Differential expression of the inflammation marker IL12p40 in the at-risk mental state for psychosis: a predictor of transition to psychotic disorder? BMC Psychiatry 16, 326.

Fuller Torrey, E., Simmons, W., Yolken, R.H., 2015. Is childhood cat ownership a risk factor for schizophrenia later in life? Schizophr. Res. 165, 1–2.

Girgis, R.R., Ciarleglio, A., Choo, T., Haynes, G., Bathon, J.M., Cremers, S., Kantrowitz, J.T., Lieberman, J.A., Brown, A.S., 2017. A randomized, double-blind, placebo-controlled clinical trial of tocilizumab, an interleukin-6 receptor antibody, for residual symptoms in schizophrenia. Neuropsychopharmacology 43, 1317–1323.

Goldsmith, D.R., Rapaport, M.H., Miller, B.J., 2016. A meta-analysis of blood cytokine network alterations in psychiatric patients: comparisons between schizophrenia, bipolar disorder and depression. Mol. Psychiatry 21, 1696–1709.

Graus, F., Titulaer, M.J., Balu, R., Benseler, S., Bien, C.G., Cellucci, T., Cortese, I., Dale, R.C., Gelfand, J.M., Geschwind, M., Glaser, C.A., Honnorat, J., Hoftberger, R., Iizuka, T., Irani, S.R., Lancaster, E., Leypoldt, F., Pruss, H., RAE-Grant, A., Reindl, M., Rosenfeld, M.R., Rostasy, K., Saiz, A., Venkatesan, A., Vincent, A., Wandinger, K.P., Waters, P., Dalmau, J., 2016. A clinical approach to diagnosis of autoimmune encephalitis. Lancet Neurol. 15, 391–404.

Grech, A., Takei, N., Murray, R.M., 1997. Maternal exposure to influenza and paranoid schizophrenia. Schizophr. Res. 26, 121–125.

Gresa-Arribas, N., Titulaer, M.J., Torrents, A., Aguilar, E., Mccracken, L., Leypoldt, F., Gleichman, A.J., Balice-Gordon, R., Rosenfeld, M.R., Lynch, D., Graus, F., Dalmau, J., 2014. Antibody titres at diagnosis and during follow-up of anti-NMDA receptor encephalitis: a retrospective study. Lancet Neurol. 13, 167–177.

Gutierrez-Fernandez, J., Luna Del Castillo Jde, D., Mananes-Gonzalez, S., Carrillo-Avila, J.A., Gutierrez, B., Cervilla, J.A., Sorlozano-Puerto, A., 2015. Different presence of *Chlamydia pneumoniae*, herpes simplex virus type 1, human herpes virus 6, and *Toxoplasma gondii* in schizophrenia: meta-analysis and analytical study. Neuropsychiatric Dis. Treat. 11, 843–852.

Hafizi, S., Tseng, H.H., Rao, N., Selvanathan, T., Kenk, M., Bazinet, R.P., Suridjan, I., Wilson, A.A., Meyer, J.H., Remington, G., Houle, S., Rusjan, P.M., Mizrahi, R., 2017. Imaging microglial activation in untreated first-episode psychosis: a PET study with [(18)F]FEPPA. Am. J. Psychiatry 174, 118–124.

Hamdani, N., Daban-Huard, C., Godin, O., Laouamri, H., Jamain, S., Attiba, D., Delavest, M., Lepine, J.P., Le Corvoisier, P., Houenou, J., Richard, J.R., Yolken, R.H., Krishnamoorthy, R., Tamouza, R., Leboyer, M., Dickerson, F.B., 2017. Effects of cumulative herpesviridae and *Toxoplasma gondii* infections on cognitive function in healthy, bipolar, and schizophrenia subjects. J. Clin. Psychiatry 78, e18–e27.

Hartwig, F.P., Borges, M.C., Horta, B.L., Bowden, J., Davey Smith, G., 2017. Inflammatory biomarkers and risk of schizophrenia: a 2-sample Mendelian randomization study. JAMA Psychiatry 74, 1226–1233.

Helleberg, M., Pedersen, M.G., Pedersen, C.B., Mortensen, P.B., Obel, N., 2015. Associations between HIV and schizophrenia and their effect on HIV treatment outcomes: a nationwide population-based cohort study in Denmark. Lancet HIV 2, e344–e350.

Hilderink, M., Titulaer, M.J., Schreurs, M.W., Keizer, K., Bunt, J.E., 2015. Transient anti-NMDAR encephalitis in a newborn infant due to transplacental transmission. Neurol. Neuroimmunol. Neuroinflamm. 2, e126.

Holmes, S.E., Hinz, R., Drake, R.J., Gregory, C.J., Conen, S., Matthews, J.C., Anton-Rodriguez, J.M., Gerhard, A., Talbot, P.S., 2016. In vivo imaging of brain microglial activity in antipsychotic-free and medicated schizophrenia: a [(11)C](R)-PK11195 positron emission tomography study. Mol. Psychiatry 21, 1672–1679.

Honigsbaum, M., 2013. "An inexpressible dread": psychoses of influenza at fin-de-siecle. Lancet 381, 988–989.

Houenou, J., D'albis, M.A., Daban, C., Hamdani, N., Delavest, M., Lepine, J.P., Vederine, F.E., Carde, S., Lajnef, M., Cabon, C., Dickerson, F., Yolken, R.H., Tamouza, R., Poupon, C., Leboyer, M., 2014. Cytomegalovirus seropositivity and serointensity are associated with hippocampal volume and verbal memory in schizophrenia and bipolar disorder. Prog. Neuropsychopharmacol. Biol. Psychiatry 48, 142–148.

Howes, O.D., Mccutcheon, R., 2017. Inflammation and the neural diathesis-stress hypothesis of schizophrenia: a reconceptualization. Transl. Psychiatry 7, e1024.

Hughes, E.G., Peng, X., Gleichman, A.J., Lai, M., Zhou, L., Tsou, R., Parsons, T.D., Lynch, D.R., Dalmau, J., Balice-Gordon, R.J., 2010. Cellular and synaptic mechanisms of anti-NMDA receptor encephalitis. J. Neurosci. 30, 5866–5875.

Jagota, P., Vincent, A., Bhidayasiri, R., 2014. Transplacental transfer of NMDA receptor antibodies in an infant with cortical dysplasia. Neurology 82, 1662–1663.

Jezequel, J., Johansson, E.M., Dupuis, J.P., Rogemond, V., Grea, H., Kellermayer, B., Hamdani, N., Le Guen, E., Rabu, C., Lepleux, M., Spatola, M., Mathias, E., Bouchet, D., Ramsey, A.J., Yolken, R.H., Tamouza, R., Dalmau, J., Honnorat, J., Leboyer, M., Groc, L., 2017a. Dynamic disorganization of synaptic NMDA receptors triggered by autoantibodies from psychotic patients. Nat. Commun. 8, 1791.

Jezequel, J., Rogemond, V., Pollak, T., Lepleux, M., Jacobson, L., Grea, H., Iyegbe, C., Kahn, R., Mcguire, P., Vincent, A., Honnorat, J., Leboyer, M., Groc, L., 2017b. Cell- and single molecule-based methods to detect anti-N-Methyl-D-Aspartate receptor autoantibodies in patients with first-episode psychosis from the OPTiMiSE project. Biol. Psychiatry 82, 766–772.

Kaplan, R.M., 2010a. Syphilis, sex and psychiatry, 1789-1925: Part 1. Australas. Psychiatr. 18, 17–21.

Kaplan, R.M., 2010b. Syphilis, sex and psychiatry, 1789-1925: Part 2. Australas. Psychiatr. 18, 22–27.

Kendell, R.E., Kemp, I.W., 1989. Maternal influenza in the etiology of schizophrenia. Arch. Gen. Psychiatr. 46, 878–882.

Kennedy, A.E., Ozbek, U., Dorak, M.T., 2017. What has GWAS done for HLA and disease associations? Int. J. Immunogenet. 44, 195–211.

Khandaker, G.M., Dantzer, R., 2016. Is there a role for immune-to-brain communication in schizophrenia? Psychopharmacology (Berl) 233, 1559–1573.

Khandaker, G.M., Pearson, R.M., Zammit, S., Lewis, G., Jones, P.B., 2014a. Association of serum interleukin 6 and C-reactive protein in childhood with depression and psychosis in young adult life: a population-based longitudinal study. JAMA Psychiatry 71, 1121–1128.

Khandaker, G.M., Stochl, J., Zammit, S., Lewis, G., Jones, P.B., 2014b. Childhood Epstein-Barr Virus infection and subsequent risk of psychotic experiences in adolescence: a population-based prospective serological study. Schizophr. Res. 158, 19–24.

Khandaker, G.M., Zimbron, J., Dalman, C., Lewis, G., Jones, P.B., 2012. Childhood infection and adult schizophrenia: a meta-analysis of population-based studies. Schizophr. Res. 139, 161–168.

Khandaker, G.M., Zimbron, J., Lewis, G., Jones, P.B., 2013. Prenatal maternal infection, neurodevelopment and adult schizophrenia: a systematic review of population-based studies. Psychol. Med. 43, 239–257.

Kipnis, J., 2018. Immune system: the "seventh sense". J. Exp. Med. 215, 397–398.

Kohler, O., Benros, M.E., Nordentoft, M., Farkouh, M.E., Iyengar, R.L., Mors, O., Krogh, J., 2014. Effect of anti-inflammatory treatment on depression, depressive symptoms, and adverse effects: a systematic review and meta-analysis of randomized clinical trials. JAMA Psychiatry 71, 1381–1391.

Kohler, O., Petersen, L., Mors, O., Mortensen, P.B., Yolken, R.H., Gasse, C., Benros, M.E., 2017. Infections and exposure to anti-infective agents and the risk of severe mental disorders: a nationwide study. Acta Psychiatr. Scand. 135, 97–105.

Lamale-Smith, L.M., Moore, G.S., Guntupalli, S.R., Scott, J.B., 2015. Maternal-fetal transfer of anti-N-methyl-D-aspartate receptor antibodies. Obstet. Gynecol. 125, 1056–1058.

Lee, S.H., Byrne, E.M., Hultman, C.M., Kahler, A., Vinkhuyzen, A.A., Ripke, S., Andreassen, O.A., Frisell, T., Gusev, A., Hu, X., Karlsson, R., Mantzioris, V.X., Mcgrath, J.J., Mehta, D., Stahl, E.A., Zhao, Q., Kendler, K.S., Sullivan, P.F., Price, A.L., O'donovan, M., Okada, Y., Mowry, B.J., Raychaudhuri, S., Wray, N.R., Schizophrenia Working Group Of The Psychiatric Genomics, C, Rheumatoid Arthritis Consortium, I., Schizophrenia Working Group Of The Psychiatric Genomics Consortium, A., Byerley, W., Cahn, W., Cantor, R.M., Cichon, S., Cormican, P., Curtis, D., Djurovic, S., Escott-Price, V., Gejman, P.V., Georgieva, L., Giegling, I., Hansen, T.F., Ingason, A., Kim, Y., Konte, B., Lee, P.H., Mcintosh, A., Mcquillin, A., Morris, D.W., Nothen, M.M., O'dushlaine, C., Olincy, A., Olsen, L., Pato, C.N., Pato, M.T., Pickard, B.S., Posthuma, D., Rasmussen, H.B., Rietschel, M., Rujescu, D., Schulze, T.G., Silverman, J.M., Thirumalai, S., Werge, T., Schizophrenia Working Group Of The Psychiatric

Genomics Consortium, C., Agartz, I., Amin, F., Azevedo, M.H., Bass, N., Black, D.W., Blackwood, D.H., Bruggeman, R., Buccola, N.G., Choudhury, K., Cloninger, R.C., Corvin, A., Craddock, N., Daly, M.J., Datta, S., Donohoe, G.J., Duan, J., Dudbridge, F., Fanous, A., Freedman, R., Freimer, N.B., Friedl, M., Gill, M., Gurling, H., De Haan, L., Hamshere, M.L., Hartmann, A.M., Holmans, P.A., Kahn, R.S., Keller, M.C., Kenny, E., Kirov, G.K., Krabbendam, L., Krasucki, R., Lawrence, J., Lencz, T., Levinson, D.F., Lieberman, J.A., Lin, D.Y., et al., 2015. New data and an old puzzle: the negative association between schizophrenia and rheumatoid arthritis. Int. J. Epidemiol. 44, 1706–1721.

Lejuste, F., Thomas, L., Picard, G., Desestret, V., Ducray, F., Rogemond, V., Psimaras, D., Antoine, J.C., Delattre, J.Y., Groc, L., Leboyer, M., Honnorat, J., 2016. Neuroleptic intolerance in patients with anti-NMDAR encephalitis. Neurol. Neuroimmunol. Neuroinflamm. 3, e280.

Lennox, B.R., Palmer-Cooper, E.C., Pollak, T., Hainsworth, J., Marks, J., Jacobson, L., Lang, B., Fox, H., Ferry, B., Scoriels, L., Crowley, H., Jones, P.B., Harrison, P.J., Vincent, A., Team, P.P.S., 2017. Prevalence and clinical characteristics of serum neuronal cell surface antibodies in first-episode psychosis: a case-control study. Lancet Psychiatry 4, 42–48.

Louveau, A., Harris, T.H., Kipnis, J., 2015a. Revisiting the mechanisms of CNS immune privilege. Trends Immunol. 36, 569–577.

Louveau, A., Smirnov, I., Keyes, T.J., Eccles, J.D., Rouhani, S.J., Peske, J.D., Derecki, N.C., Castle, D., Mandell, J.W., Lee, K.S., Harris, T.H., Kipnis, J., 2015b. Structural and functional features of central nervous system lymphatic vessels. Nature 523, 337–341.

Machon, R.A., Mednick, S.A., Huttunen, M.O., 1997. Adult major affective disorder after prenatal exposure to an influenza epidemic. Arch. Gen. Psychiatr. 54, 322–328.

Maino, K., Gruber, R., Riedel, M., Seitz, N., Schwarz, M., Muller, N., 2007. T- and B-lymphocytes in patients with schizophrenia in acute psychotic episode and the course of the treatment. Psychiatry Res. 152, 173–180.

Masgrau, R., Guaza, C., Ransohoff, R.M., Galea, E., 2017. Should we stop saying 'glia' and 'neuroinflammation'? Trends Mol. Med. 23, 486–500.

Mednick, S.A., Machon, R.A., Huttunen, M.O., Bonett, D., 1988. Adult schizophrenia following prenatal exposure to an influenza epidemic. Arch. Gen. Psychiatr. 45, 189–192.

Meyer, U., 2014. Prenatal poly(i:C) exposure and other developmental immune activation models in rodent systems. Biol. Psychiatry 75, 307–315.

Miller, B.J., Buckley, P., Seabolt, W., Mellor, A., Kirkpatrick, B., 2011. Meta-analysis of cytokine alterations in schizophrenia: clinical status and antipsychotic effects. Biol. Psychiatry 70, 663–671.

Miller, B.J., Gassama, B., Sebastian, D., Buckley, P., Mellor, A., 2013. Meta-analysis of lymphocytes in schizophrenia: clinical status and antipsychotic effects. Biol. Psychiatry 73, 993–999.

Mondelli, V., Vernon, A.C., Turkheimer, F., Dazzan, P., Pariante, C.M., 2017. Brain microglia in psychiatric disorders. Lancet Psychiatry 4, 563–572.

Morsheimer, M., Brown Whitehorn, T.F., Heimall, J., Sullivan, K.E., 2017. The immune deficiency of chromosome 22q11.2 deletion syndrome. Am. J. Med. Genet. 173, 2366–2372.

Nataf, S., 2017. Evolution, immunity and the emergence of brain superautoantigens. F1000Res. 6, 171.

Network & Pathway Analysis Subgroup Of Psychiatric Genomics, C, 2015. Psychiatric genome-wide association study analyses implicate neuronal, immune and histone pathways. Nat. Neurosci. 18, 199–209.

Notter, T., Coughlin, J.M., Gschwind, T., Weber-Stadlbauer, U., Wang, Y., Kassiou, M., Vernon, A.C., Benke, D., Pomper, M.G., Sawa, A., Meyer, U., 2017. Translational evaluation of translocator protein as a marker of neuroinflammation in schizophrenia. Mol. Psychiatry 23, 323–334.

O'connor, M.F., Bower, J.E., Cho, H.J., Creswell, J.D., Dimitrov, S., Hamby, M.E., Hoyt, M.A., Martin, J.L., Robles, T.F., Sloan, E.K., Thomas, K.S., Irwin, M.R., 2009. To assess, to control, to exclude: effects of biobehavioral factors on circulating inflammatory markers. Brain Behav. Immun. 23, 887–897.

Oken, R.J., Schulzer, M., 1999. At issue: schizophrenia and rheumatoid arthritis: the negative association revisited. Schizophr. Bull. 25, 625–638.

Okusaga, O., Langenberg, P., Sleemi, A., Vaswani, D., Giegling, I., Hartmann, A.M., Konte, B., Friedl, M., Groer, M.W., Yolken, R.H., Rujescu, D., Postolache, T.T., 2011. *Toxoplasma gondii* antibody titers and history of suicide attempts in patients with schizophrenia. Schizophr. Res. 133, 150–155.

Oviedo-Salcedo, T., De Witte, L., Kumpfel, T., Kahn, R.S., Falkai, P., Eichhorn, P., Luykx, J., Hasan, A., 2018. Absence of cerebrospinal fluid antineuronal antibodies in schizophrenia spectrum disorders. Br. J. Psychiatry 212, 318–320.

Papa, I., Saliba, D., Ponzoni, M., Bustamante, S., Canete, P.F., Gonzalez-Figueroa, P., Mcnamara, H.A., Valvo, S., Grimbaldeston, M., Sweet, R.A., Vohra, H., Cockburn, I.A., Meyer-Hermann, M., Dustin, M.L., Doglioni, C., Vinuesa, C.G., 2017. TFH-derived dopamine accelerates productive synapses in germinal centres. Nature 547, 318–323.

Pearlman, D.M., Najjar, S., 2014. Meta-analysis of the association between N-methyl-d-aspartate receptor antibodies and schizophrenia, schizoaffective disorder, bipolar disorder, and major depressive disorder. Schizophr. Res. 157, 249–258.

Perkins, D.O., Jeffries, C.D., Addington, J., Bearden, C.E., Cadenhead, K.S., Cannon, T.D., Cornblatt, B.A., Mathalon, D.H., Mcglashan, T.H., Seidman, L.J., Tsuang, M.T., Walker, E.F., Woods, S.W., Heinssen, R., 2015. Towards a psychosis risk blood diagnostic for persons experiencing high-risk symptoms: preliminary results from the NAPLS project. Schizophr. Bull. 41, 419–428.

Pilkington, T.L., 1956. The coincidence of rheumatoid arthritis and schizophrenia. J. Nerv. Ment. Dis. 124, 604–606.

Pollak, T.A., Drndarski, S., Stone, J.M., David, A.S., Mcguire, P., Abbott, N.J., 2018. The blood-brain barrier in psychosis. Lancet Psychiatry 5, 79–92.

Pollak, T.A., Mccormack, R., Peakman, M., Nicholson, T.R., David, A.S., 2014. Prevalence of anti-N-methyl-D-aspartate (NMDA) receptor [corrected] antibodies in patients with schizophrenia and related psychoses: a systematic review and meta-analysis. Psychol. Med. 44, 2475–2487.

Prasad, K.M., Watson, A.M., Dickerson, F.B., Yolken, R.H., Nimgaonkar, V.L., 2012. Exposure to herpes simplex virus type 1 and cognitive impairments in individuals with schizophrenia. Schizophr. Bull. 38, 1137–1148.

Radua, J., Ramella-Cravaro, V., Ioannidis, J.P.A., Reichenberg, A., Phiphopthatsanee, N., Amir, T., Yenn Thoo, H., Oliver, D., Davies, C., Morgan, C., Mcguire, P., Murray, R.M., FUSAR-Poli, P., 2018. What causes psychosis? An umbrella review of risk and protective factors. World Psychiatry 17, 49–66.

Rudolph, M.D., Graham, A.M., Feczko, E., Miranda-Dominguez, O., Rasmussen, J.M., Nardos, R., Entringer, S., Wadhwa, P.D., Buss, C., Fair, D.A., 2018. Maternal IL-6 during pregnancy can be estimated from newborn brain connectivity and predicts future working memory in offspring. Nat. Neurosci. 21, 765–772.

Salter, M.W., Stevens, B., 2017. Microglia emerge as central players in brain disease. Nat. Med. 23, 1018–1027.

Samuelsson, A.M., Jennische, E., Hansson, H.A., Holmang, A., 2006. Prenatal exposure to interleukin-6 results in inflammatory neurodegeneration in hippocampus with NMDA/GABA(A) dysregulation and impaired spatial learning. Am. J. Physiol. Regul. Integr. Comp. Physiol. 290, R1345–R1356.

Savitz, J., Harrison, N.A., 2018. Interoception and inflammation in psychiatric disorders. Biol. Psychiatry Cogn. Neurosci. Neuroimaging 3, 514–524.

Scharko, A.M., 2015. Cat ownership and schizophrenia. Schizophr. Res. 169, 504.

Schizophrenia Working Group Of The Psychiatric Genomics, C, 2014. Biological insights from 108 schizophrenia-associated genetic loci. Nature 511, 421–427.

Schneider, M., Debbane, M., Bassett, A.S., Chow, E.W., Fung, W.L., Van Den Bree, M., Owen, M., Murphy, K.C., Niarchou, M., Kates, W.R., Antshel, K.M., Fremont, W., Mcdonald-Mcginn, D.M., Gur, R.E., Zackai, E.H., Vorstman, J., Duijff, S.N., Klaassen, P.W., Swillen, A., Gothelf, D., Green, T., Weizman, A., Van Amelsvoort, T., Evers, L., Boot, E., Shashi, V., Hooper, S.R., Bearden, C.E., Jalbrzikowski, M., Armando, M., Vicari, S., Murphy, D.G., Ousley, O., Campbell, L.E., Simon, T.J., Eliez, S., International Consortium On, B., Behavior In 22q11.2 Deletion, S, 2014. Psychiatric disorders from childhood to adulthood in 22q11.2 deletion syndrome: results from the international consortium on brain and behavior in 22q11.2 deletion syndrome. Am. J. Psychiatry 171, 627–639.

Schwarz, E., Maukonen, J., Hyytiainen, T., Kieseppa, T., Oresic, M., Sabunciyan, S., Mantere, O., Saarela, M., Yolken, R., Suvisaari, J., 2017. Analysis of microbiota in first episode psychosis identifies preliminary associations with symptom severity and treatment response. Schizophr. Res. 192, 398–403.

Scott, J.G., Gillis, D., Ryan, A.E., Hargovan, H., Gundarpi, N., Mckeon, G., Hatherill, S., Newman, M.P., Parry, P., Prain, K., Patterson, S., Wong, R.C.W., Wilson, R.J., Blum, S., 2018. The prevalence and treatment outcomes of antineuronal antibody-positive patients admitted with first episode of psychosis. BJPsych Open 4, 69–74.

Sekar, A., Bialas, A.R., De Rivera, H., Davis, A., Hammond, T.R., Kamitaki, N., Tooley, K., Presumey, J., Baum, M., Van Doren, V., Genovese, G., Rose, S.A., Handsaker, R.E., Schizophrenia Working Group Of The Psychiatric Genomics, C., Daly, M.J., Carroll, M.C., Stevens, B., Mccarroll, S.A., 2016. Schizophrenia risk from complex variation of complement component 4. Nature 530, 177–183.

Selten, J.P., Termorshuizen, F., 2017. The serological evidence for maternal influenza as risk factor for psychosis in offspring is insufficient: critical review and meta-analysis. Schizophr. Res. 183, 2–9.

Sham, P.C., O'callaghan, E., Takei, N., Murray, G.K., Hare, E.H., Murray, R.M., 1992. Schizophrenia following pre-natal exposure to influenza epidemics between 1939 and 1960. Br. J. Psychiatry 160, 461–466.

Shen, Y., Xu, J., Li, Z., Huang, Y., Yuan, Y., Wang, J., Zhang, M., Hu, S., Liang, Y., 2018. Analysis of gut microbiota diversity and auxiliary diagnosis as a biomarker in patients with schizophrenia: a cross-sectional study. Schizophr. Res. 197, 470–477.

Smith, S.E., Li, J., Garbett, K., Mirnics, K., Patterson, P.H., 2007. Maternal immune activation alters fetal brain development through interleukin-6. J. Neurosci. 27, 10695–10702.

Solmi, F., Hayes, J.F., Lewis, G., Kirkbride, J.B., 2017. Curiosity killed the cat: no evidence of an association between cat ownership and psychotic symptoms at ages 13 and 18 years in a UK general population cohort. Psychol. Med. 47, 1659–1667.

Steiner, J., Jacobs, R., Panteli, B., Brauner, M., Schiltz, K., Bahn, S., Herberth, M., Westphal, S., Gos, T., Walter, M., Bernstein, H.G., Myint, A.M., Bogerts, B., 2010. Acute schizophrenia is accompanied by reduced T cell and increased B cell immunity. Eur. Arch. Psychiatry Clin. Neurosci. 260, 509–518.

Sutterland, A.L., Fond, G., Kuin, A., Koeter, M.W., Lutter, R., Van Gool, T., Yolken, R., Szoke, A., Leboyer, M., DE Haan, L., 2015. Beyond the association. Toxoplasma gondii in schizophrenia, bipolar disorder, and addiction: systematic review and meta-analysis. Acta Psychiatr. Scand. 132, 161–179.

Takei, N., Mortensen, P.B., Klaening, U., Murray, R.M., Sham, P.C., O'callaghan, E., Munk-Jorgensen, P., 1996. Relationship between in utero exposure to influenza epidemics and risk of schizophrenia in Denmark. Biol. Psychiatry 40, 817–824.

Takei, N., Van Os, J., Murray, R.M., 1995. Maternal exposure to influenza and risk of schizophrenia: a 22 year study from The Netherlands. J. Psychiatr. Res. 29, 435–445.

Titulaer, M.J., Mccracken, L., Gabilondo, I., Armangue, T., Glaser, C., Iizuka, T., Honig, L.S., Benseler, S.M., Kawachi, I., Martinez-Hernandez, E., Aguilar, E., Gresa-Arribas, N., Ryan-Florance, N., Torrents, A., Saiz, A., Rosenfeld, M.R., Balice-Gordon, R., Graus, F., Dalmau, J., 2013. Treatment and prognostic factors for long-term outcome in patients with anti-NMDA receptor encephalitis: an observational cohort study. Lancet Neurol. 12, 157–165.

Torrey, E.F., Miller, J., Rawlings, R., Yolken, R.H., 1997. Seasonality of births in schizophrenia and bipolar disorder: a review of the literature. Schizophr. Res. 28, 1–38.

Vann Jones, S., Banerjee, S., Smith, A.D., Refsum, H., Lennox, B., 2017. Elevated homocysteine and N-methyl-d-aspartate-receptor antibodies as a cause of behavioural and cognitive decline in 22q11.2 deletion syndrome. Oxf. Med. Case Rep. 2017, omx076.

Vergaelen, E., Schiweck, C., Van Steeland, K., Counotte, J., Veling, W., Swillen, A., Drexhage, H., Claes, S., 2018. A pilot study on immuno-psychiatry in the 22q11.2 deletion syndrome: a role for Th17 cells in psychosis? Brain Behav. Immun. 70, 88–95.

Wright, P., Takei, N., Rifkin, L., Murray, R.M., 1995. Maternal influenza, obstetric complications, and schizophrenia. Am. J. Psychiatry 152, 1714–1720.

Xiao, J., Buka, S.L., Cannon, T.D., Suzuki, Y., Viscidi, R.P., Torrey, E.F., Yolken, R.H., 2009. Serological pattern consistent with infection with type I *Toxoplasma gondii* in mothers and risk of psychosis among adult offspring. Microb. Infect. 11, 1011–1018.

Yolken, R.H., Severance, E.G., Sabunciyan, S., Gressitt, K.L., Chen, O., Stallings, C., Origoni, A., Katsafanas, E., Schweinfurth, L.A., Savage, C.L., Banis, M., Khushalani, S., Dickerson, F.B., 2015. Metagenomic sequencing indicates that the oropharyngeal phageome of individuals with schizophrenia differs from that of controls. Schizophr. Bull. 41, 1153–1161.

Yolken, R.H., Torrey, E.F., 2008. Are some cases of psychosis caused by microbial agents? A review of the evidence. Mol. Psychiatry 13, 470–479.

Yudofsky, S.C., 2009. Contracting schizophrenia: lessons from the influenza epidemic of 1918–1919. J. Am. Med. Assoc. 301, 324–326.

Zeni-Graiff, M., Rizzo, L.B., Mansur, R.B., Maurya, P.K., Sethi, S., Cunha, G.R., Asevedo, E., Pan, P., Zugman, A., Yamagata, A.S., Higuchi, C., Bressan, R.A., Gadelha, A., Brietzke, E., 2016. Peripheral immuno-inflammatory abnormalities in ultra-high risk of developing psychosis. Schizophr. Res. 176, 191–195.

Zerbo, O., Iosif, A.M., Walker, C., Ozonoff, S., Hansen, R.L., Hertz-Picciotto, I., 2013. Is maternal influenza or fever during pregnancy associated with autism or developmental delays? Results from the CHARGE (CHildhood Autism Risks from Genetics and Environment) study. J. Autism Dev. Disord. 43, 25–33.

CHAPTER

12

Neurochemical models of psychosis risk and onset

Dominic Oliver[1], Gemma Modinos[1,2], Philip McGuire[1,3,4]

[1]Department of Psychosis Studies, Institute of Psychiatry, Psychology & Neuroscience, King's College London, London, United Kingdom; [2]Department of Neuroimaging, Institute of Psychiatry, Psychology & Neuroscience, King's College London, London, United Kingdom; [3]OASIS Service, South London and the Maudsley NHS National Health Service Foundation Trust, London, United Kingdom; [4]National Institute for Health Research (NIHR) Biomedical Research Centre for Mental Health, Institute of Psychiatry, Psychology & Neuroscience, King's College London, London, United Kingdom

Introduction

Individuals at clinical high risk for psychosis (CHR, CHR-P, or UHR) typically present with attenuated psychotic symptoms that are often associated with a recent decline in overall functioning. Clinical intervention at this stage is of great interest, as it has the potential to reduce or delay the onset of frank psychosis (McGorry, 2013). However, although a series of different pharmacological interventions have been evaluated in CHR subjects since 2002, there is still no licensed pharmacological treatment for the prevention of psychosis in this group (Davies et al., 2018). The development of novel compounds that may be effective preventative treatments can be informed by preclinical modeling of the mechanisms that underlie vulnerability to psychosis and the onset of a psychotic disorder. The age of CHR subjects varies, but in centers where they present to clinical services, the mean age is around 24 years (Fusar-Poli et al., 2013). Up to a third of CHR individuals will subsequently transition to a full blown psychosis within 3 years (Fusar-Poli et al., 2012). However, after this relatively brief "window of risk," the likelihood of developing psychosis reverts to a relatively lower level (Nelson et al., 2013). It is thought that most patients who develop psychosis have done so following a CHR phase (Meares, 1959; Häfner et al., 2003; Schultze-Lutter et al., 2010). The onset of psychosis thus appears to occur in young adults who are already vulnerable to developing the disorder and seems to have a neurodevelopmental component. Ideally, preclinical models of psychosis should reflect these features.

Preclinical research utilizes animal models to inform our understanding of psychosis.

Risk Factors for Psychosis
https://doi.org/10.1016/B978-0-12-813201-2.00012-0

For these models to accurately reflect the disorder, ideally, they should be valid in a number of different capacities, including construct validity, predictive validity, internal validity, external validity, and face validity. Construct validity refers to how accurate a model is in representing what it is designed to model. Predictive validity relates to a model's capacity to predict the outcome in question. Internal validity reflects both the reliability and replicability of a model. External validity refers to the extent that the findings garnered from a model can be generalized to the general population. Face validity is particularly important in this context and reflects the comparability of the model to the symptoms and behaviors observed in humans.

Human neuroimaging studies permit the measurement of the structure, function, and neurochemistry of the brain in vivo. However, neurobiological measurements across human and animal studies may not be directly comparable. Neuroimaging methods often provide indirect measures of neuronal activity, neurotransmitter function, or connectivity, whereas these can be measured more directly in preclinical experiments. For example, functional magnetic resonance imaging assesses the vascular response to neural activity via the blood oxygen level–dependent signal (Logothetis, 2002), but neural activity can be measured directly in animal studies by implanting electrodes into brain tissue. Microdialysis can be used in freely moving animals to sample and quantify the synaptic concentrations of neurotransmitters dynamically (Chefer et al., 2009). In humans, molecular imaging techniques can estimate neurotransmitter release by measuring the displacement of a receptor ligand using positron emission tomography (Heurling et al., 2017). The concentration of neurotransmitters can also be quantified in humans using magnetic resonance spectroscopy (MRS), but this provides an average measure across white matter, cerebrospinal fluid, vascular tissue, and gray matter. In animals, in vivo MRS can be subsequently verified using biochemical assays ex vivo (Waschkies et al., 2014).

The use of animal models is thus paramount to increasing our understanding of the neurobiological basis of psychosis onset and for the development of novel drugs with preventative efficacy for psychosis. This chapter will explore four key rodent neurochemical models of psychosis risk and onset, assessing the strengths and weaknesses of each model and the characteristics of psychosis development they seek to replicate (see Table 12.1). The following models will be explored: prenatal immune activation, neonatal hippocampal lesion, chronic PCP (phencyclidine), and methylazoxymethanol acetate (MAM) rodent model.

Neonatal hippocampal lesion

In the CHR state, later transition to psychosis has been associated with reduced hippocampal volume (Pantelis et al., 2007), increased resting hippocampal metabolism and perfusion (Allen et al., 2018; Schobel et al., 2013), altered hippocampal activation (Allen et al., 2012), and increased hippocampal glutamate levels (Bossong et al., 2018).

There is thus a clear rationale for a model of psychosis based on lesions of the hippocampus, especially if these are made at the appropriate stage of development. In rodent models, hippocampal lesions have been induced using ibotenic acid, typically in the ventral hippocampus and subiculum, which correspond to the anterior hippocampus in humans (Lipska and Weinberger, 2000). This lesion is thought to disrupt the development of cortical and subcortical networks that involve the hippocampus (Tseng et al., 2007; Lipska and Weinberger, 2002; Mattei et al., 2015).

TABLE 12.1 Summary of model effects on behavior, pharmacology, neurophysiology, and histology and anatomy.

Model	Behavior	Pharmacology	Neurophysiology	Histology and anatomy
Neonatal hippocampal lesion	Sensorimotor gating deficits Reduced social interaction Impaired working memory Impaired spatial and associative learning Impaired attentional processing	Increased responsivity to psychostimulants	Abnormal corticosubcortical synaptic transmission Prefrontal cortical glutamatergic hyperactivity	Decreased synaptic density, dendritic length, and branching in PFC
Poly(I:C)	Sensorimotor gating deficits Latent inhibition reversal learning Reduced social interaction Reduced exploration Impaired working memory	Increased responsivity to psychostimulants	Reduced GABAergic transmission in hippocampus	Decreased RELN expression in hippocampus
Chronic PCP	Sensorimotor gating deficits Latent inhibition reversal learning Reduced social interaction Impaired set-shifting Anticipatory deficits Flat affect	Increased responsivity to psychostimulants	Reduced glutamate and dopamine levels in PFC Increased activity in mesolimbic tract	Activation of astroglia in cortex and hippocampus
MAM	Cognitive dysfunction Sensorimotor gating deficits Latent inhibition reversal learning Reduced social interaction Perseverative responding	Increased responsivity to psychostimulants Rapid antipsychotic drug effects on dopaminergic neurons	Abnormal corticocortical synaptic transmission Striatal hyperdopaminergia Altered glutamate levels in hippocampus Disrupted evoked gamma rhythms	Decreased cortical thickness in hippocampus, parahippocampal cortex, medial PFC No differences in neocortical neuron number Decreased parvalbumin expression in ventral hippocampus, medial and orbital PFC

Anatomical and neurophysiological effects

It has been proposed that hippocampal lesions model the cognitive deficits of schizophrenia through disrupting the development of connections to the prefrontal cortex (PFC) (O'Donnell et al., 2002). The mesocorticolimbic dopaminergic signaling is thought to be impaired due to the effects of antipsychotics on reversing behavioral and physiological changes (Le Pen and Moreau, 2002; Lipska and Weinberger, 1994; Goto and O'Donnell, 2002). Moreover, PFC neurons in rats with neonatal hippocampal lesions have been shown to have abnormal responses to inputs from mesocortical neurons (O'Donnell et al., 2002). In addition, removing PFC neurons in adult rats reverses the behavioral and electrophysiological changes seen as a result of a neonatal hippocampal lesion (Lipska and Weinberger, 1998; Goto and O'Donnell, 2004). Together, this suggests that this model results in impaired connections to the PFC, and this aberrant connectivity results in the observable effects. Furthermore, anatomical studies in lesioned rodents have suggested that this impaired hippocampal–PFC connection may result in reduced dendritic length and spine density of PFC neurons (Alquicer et al., 2008; Flores et al., 2005). Neonatal hippocampal lesions also lead to PFC hyperexcitability, which is potentially the result of attenuated cortical γ-aminobutyric acid (GABA) neurotransmission (Manoach, 2003). Early impairment of the hippocampus could lead to altered maturation of PFC inhibitory interneurons and thereby reduced basal inhibitory tone (Tseng and O'Donnell, 2007). Therefore, the changes to PFC function in this model could be associated with impaired GABAergic function in conjunction with disrupted mesocortical dopaminergic modulation of PFC glutamate neurotransmission (Tseng et al., 2009).

Behavioral effects

Lesioning the ventral hippocampus in rodents results in a number of behavioral changes that may correspond to psychotic symptoms. These include amphetamine-induced hyperactivity, deficits in prepulse inhibition (PPI) of the startle reflex, impaired social interaction, and cognitive deficits. Some of these effects can be normalized through the administration of antipsychotic drugs (Le Pen and Moreau, 2002; Lipska, 2004). However, the effects of the lesion are only evident when it is induced in the neonatal period (around postnatal day 7; equivalent to 36–40 weeks' gestation in humans (Lipska et al., 1993; Semple et al., 2013)); when induced later in life, these changes do not occur. This suggests that the timing of the lesioning is critical and that its effects depend on the neurodevelopmental context (for further detail on translating rodent development to human stages, see Semple et al. (2013)).

Amphetamine-induced hyperactivity is a behavioral measurement in animal models that is thought to correspond to positive psychotic symptoms in humans (Marona-Lewicka et al., 2011) and is thought to reflect a hyperdopaminergic state. This aspect is reproduced in the hippocampal lesion rodent model with lesioned animals showing hypersensitivity to dopaminergic drugs such as amphetamine (Flores et al., 1996; Drouin-Ouellet et al., 2011; Naert et al., 2013; Vázquez-Roque et al., 2012), apomorphine (François et al., 2009), and cocaine (Macedo et al., 2010). Additionally, the neonatal hippocampal model also displays locomotor hyperactivity in absence of drug that emerges upon reaching adulthood, reflecting overall total hyperdopaminergia (Lipska et al., 1993; Wan et al., 1996; Ahnaou et al., 2007).

The PPI paradigm is a well-established method of evaluating filtering of sensory information. The PPI refers to the inhibition of a startle reflex by presentation of a prepulse immediately before the startle stimulus. Disruption of PPI is a robust finding in patients with schizophrenia (Braff et al. 1978, 1992; Bolino et al., 1994; Perry and Braff, 1994). PPI is an inherent behavior and shows remarkable similarity among rats and humans (Swerdlow et al., 1994). However, similar to amphetamine-induced locomotion, there is a lack of anatomical specificity, with both limbic and associative dopaminergic projections implicated. This highlights a clear weakness of using this behavior as an analogue for psychotic symptoms (Koch, 1999; Baldan Ramsey et al., 2011; Swerdlow et al., 1992).

Impairments in social interaction can also be seen in the neonatal ventral hippocampal lesion model. When rats are lesioned in this area at day 7 after birth, they display reduced social behavior. In this study, two rats were placed into an open arena, and social behavior was tested for 7 minutes (Becker et al., 1999). This was split into aggressive (kicking, boxing, wrestling, aggressive grooming, and biting) and nonaggressive (sniffing, following, grooming, and social play) behavior. Lesioned rats showed reduced time spent in social interaction. In particular, there was a reduction in time spent in nonaggressive social behavior and an increase in aggressive social behavior.

The rodent neonatal hippocampal lesion model is also associated with cognitive deficits, such as reduced performance in memory (Lipska et al., 2002), spatial and associative learning (Le Pen et al., 2000), and attentional processing (Le Pen et al., 2003). This is consistent with meta-analytical results of cognitive function in CHR individuals, showing impairments in multiple domains of memory, visual memory, and attention (Bora et al., 2014). While the complete spectrum of cognitive deficits seen in CHR individuals and patients with schizophrenia is not evident, this is a generic feature of preclinical models of psychosis. Indeed, this may reflect the inherent differences in the types of cognitive processes found in humans and rodents. For example, in animals that lack spoken language, it would be difficult to reproduce deficits in verbal working memory or fluency or symptoms like thought disorder, auditory verbal hallucinations, or poverty of speech.

Limitations

One weakness of the neonatal hippocampal lesion model is its limited construct validity, as there is little evidence that psychotic disorders result from a macroscopic neurodevelopmental lesion in the hippocampus. Nevertheless, both psychosis and the high risk state have been linked to alterations in the structure, function, and neurochemistry of the hippocampus (Heckers and Konradi, 2015), and there is more evidence implicating the hippocampus in the onset of psychosis than for any other brain region (Nakahara et al., 2018). There is less evidence that the connectivity of the hippocampus is altered in CHR subjects, although to date only a few studies have examined this directly, and the sample sizes were small. Benetti et al. (2009) described altered functional connectivity between hippocampus and PFC in both CHR and first episode psychosis (FEP) subjects. Bernard et al. (2015) reported differences in thalamohippocampal white matter integrity between CHR individuals and controls and linked this to prodromal psychotic symptom levels. Moreover, there is evidence of decreased developmental hippocampal growth in familial high-risk groups compared to controls, and this is not linked to clinical outcome (Bois et al., 2016).

Because human neuroimaging studies in CHR subjects have mainly been limited to people who were already in late adolescence or early adulthood, it is not known when these hippocampal abnormalities detected first emerge. Thus, the most appropriate time in development to induce a hippocampal lesion is unclear. Given that the hippocampal alterations in CHR subjects do not involve gross macroscopic changes, the model might be improved by inducing a more neurotransmitter-specific type of lesion, for example, one that targeted the GABA or glutamate system in this area, and involved longitudinal assessment of the associated behavioral, anatomical, and neurophysiological changes in order to clarify the timing of their development. While these issues do occur, the changes seen in the hippocampal lesion model recapitulate the observations in humans of increased hippocampal activity/perfusion in CHR individuals (Allen et al. 2016, 2018; Schobel et al., 2013), reduced hippocampal volume in CHR individuals who transition to a psychotic disorder (Mechelli et al., 2011), and increased glutamate levels (Bossong et al., 2018).

Prenatal immune activation

Prenatal infection is a neurodevelopmental risk factor for several mental health conditions, including psychosis. There was initially evidence for the influence of neuroimmune mechanisms on psychosis in 1919 when an association was found between influenza exposure and subsequent psychotic disorders following an influenza pandemic (Menninger, 1919). More recent interest has focused on the potential impact of infections during the prenatal period, when there is greater sensitivity to environmental insults like infections (Meyer et al., 2007). Infection of the mother during pregnancy can lead to changes in the fetal environment, which can influence the course of fetal brain development (Meyer et al., 2007). Exposure of expectant mothers to a number of infectious agents has been associated with an increased risk of schizophrenia, including rubella (Brown et al., 2001), measles (Torrey et al., 1988), polio (Suvisaari et al., 1999), herpes simplex (Buka et al., 2001), bacterial infections causing sinusitis, tonsillitis and pneumonia (Sørensen et al., 2009), genital and/or reproductive infections (Babulas et al., 2006) and toxoplasma gondii (Mortensen et al., 2007).

A variety of methods can be used to reproduce this immune activation in preclinical models, including maternal gestational exposure to the human influenza virus, and administration of the viral mimic polyriboinosinic-polyribocytidilic acid (Poly[I:C]), the bacterial endotoxin lipopolysaccharide, the locally acting inflammatory agent turpentine, or selected inflammatory cytokines (Nyffeler et al., 2006; Jaaro-Peled et al., 2010). This type of model is based on the human influenza virus model, and the most prominent version in use today is the poly(I:C) model. Hence, this section will focus on the human influenza virus and poly(I:C) models.

Human influenza virus

The notion of the immune system being involved in the development of schizophrenia began through links to the human influenza virus (Martin and Dombrowski, 2008). Early models of prenatal immune activation primarily used human influenza virus in mice. Pregnant mice received an intranasal infusion of a sublethal dose of a mouse-adapted strain of human influenza on gestational day 9 (~36–40 weeks' gestation in humans). This leads to a number of neuropathological effects in the offspring, including aberrant cortical development, leading to impaired corticogenesis, and brain atrophy (Fatemi et al., 2008). More specifically, there is evidence of reduced corpus callosum and hippocampal volume, decreased expression of GABA markers, such as reelin (RELN), and reduced serotonin levels, but no change in dopamine levels. Affected mice also show impaired PPI, reduced exploration in open-field and novel object tests, and reduced social interaction (Shi et al., 2003).

Polyriboinosinic-polyribocytidilic acid

Other models use agents that initiate cytokine-associated immune responses in the mother without direct exposure to pathogens. These models were initially developed to investigate whether changes in maternal/fetal cytokine expression play a key role in the effect of maternal infection during pregnancy and altered neurodevelopment in the offspring. Administering poly(I:C) is one of the most common methods of inducing a cytokine-driven immune response in animal models. Poly(I:C) is a synthetic analogue of double-stranded RNA, which is generated during a viral infection. Viruses produce this both as an intermediate for single-stranded RNA and also as a by-product of symmetrical transcription in DNA viruses (Akira and Takeda, 2004). This is recognized as foreign through transmembrane protein toll-like receptor 3 (Akira and Takeda, 2004). When poly(I:C) binds to the toll-like receptor 3, this results in the expression of a variety of innate immune response genes and proteins. Proinflammatory cytokines are produced and released, including interleukin (IL)-1β, IL-6, and tumor necrosis factor-α (Cunningham et al., 2007), in addition to type I interferons (IFNs): IFN-α and IFN-β (Kimura et al., 1994). Administration of poly(I:C) thus leads to a response that mimics the acute-phase response to viral infection, and this initiates inflammatory processes in the fetal brain (Meyer et al., 2006).

The poly(I:C) has a significant impact on the GABAergic system, particularly reducing RELN positive GABAergic cells in the hippocampus. RELN is a glycoprotein that is a key regulator of cortical development in mammals (Sekine et al., 2014). The absence of RELN in humans leads to lissencephaly (smooth brain) and cerebellar hypoplasia (Hong et al., 2000). In addition to regulating neural positioning in early brain development, RELN is also crucial for synaptic plasticity as well as the acquisition, consolidation, and expression of memory. In humans, postmortem studies have found GABAergic impairments in the PFC and hippocampus of individuals with schizophrenia (Volk and Lewis, 2013; Gonzalez-Burgos and Lewis, 2012). However, in vivo neurochemical studies in patients have been inconclusive (Egerton et al., 2017). In CHR individuals specifically compared to healthy controls, studies have reported increases (de la Fuente-Sandoval et al., 2015), decreases (Menschikov et al., 2016), and no differences (Wang et al., 2016; Modinos et al., 2018a) between these groups. However, there is evidence that GABA concentrations in the MPFC are associated with the severity of positive symptoms in CHR individuals (Modinos et al., 2018a) and directly related to hippocampal perfusion in CHR individuals who subsequently transitioned to psychosis (Modinos et al., 2018b).

The poly(I:C) model also shows a delay in myelination and reduced axonal development in the offspring (Makinodan et al., 2008). Aberrant myelination, particularly relating to oligodendrocytes, has been similarly shown in schizophrenia through microarrays (Tkachev et al., 2003; Hakak et al., 2001; Sugai et al., 2004), diffusion tensor imaging (Foong et al. 2000, 2002; Lim et al., 1999), and ultrastructural studies (Uranova et al., 2001).

Poly(I:C) model rodents display alterations in PPI, latent inhibition, social behavior, exploration, and working memory (see Section 2.2. for further details on these behaviors) (Meyer et al., 2009) and have increased sensitivity to dopamine receptor and NMDA receptor agonists (Meyer et al., 2009).

Amphetamine increases dopamine release in humans. In rodents, amphetamines increase locomotion, and this is thought to largely due to limbic dopamine release. Activating limbic dopaminergic projections with chemogenetic tools increases locomotion, with no effect when associative dopaminergic projections are activated (Boekhoudt et al., 2016). In contrast, some neuroimaging studies implicate the associative, as opposed to the limbic striatum in the dopamine dysfunction associated with the CHR state and psychosis (Kegeles et al., 2010), although the extent to which the findings in humans are specific to the associative subdivision is unclear.

The poly(I:C) model has a number of advantages over the human influenza virus model. Firstly, there is greater scope for controlling the intensity of the cytokine-driven immune response. This has been shown in numerous dose-response studies (Cunningham et al., 2007; Shi et al., 2003; Meyer et al., 2005). This is especially important for investigating gene–environment and environment–environment interactions. For example, studies can investigate two-hit approaches, such as prenatal immune activation and peripubertal stress (Giovanoli et al., 2014). Secondly, the immune response is far more time-limited, lasting between 24 and 48 h (Shi et al., 2003; Meyer et al., 2005; Cunningham et al., 2007), which facilitates investigation of the effects of different timings of maternal immune activation on neurodevelopment.

Following prenatal immune activation in rodents, there are a multitude of behavioral, cognitive, and neurochemical changes. The severity of these alterations appears to be dependent on the intensity of the cytokine-driven immune response (Meyer et al., 2005). This suggests that the more severe the prenatal immune response, the more distinct the structural and functional abnormalities (Ellman et al., 2010). Moreover, these changes only emerge after the offspring enter late adolescence or early adulthood (Zuckerman et al., 2003; Vuillermot et al., 2010; Ozawa et al., 2006; Piontkewitz et al., 2011), in line with the natural history of psychotic disorders (Tandon and Fleischhacker, 2005).

However, the poly(I:C) model is limited by the fact that it does not precisely correspond to the immunological insults that occur in humans. Instead, it mimics the acute-phase response to viruses triggered by cytokine activation in both the mother and the fetal environment. Moreover, recent studies have suggested that there is insufficient evidence to link the maternal immune response and psychosis (Selten and Termorshuizen, 2017; Cai et al., 2015; Radua et al., 2018). A recent umbrella review summarizing evidence from 14 studies on maternal influenza found no significant effect on psychosis risk despite a large sample of over 7500 patients (Radua et al., 2018). Evidence from such a large sample sheds serious doubt on the construct validity of maternal infection as a model for psychosis risk. This needs to be taken in consideration when using this model but, pragmatically, does not erase its potential to recapitulate the behavioral, functional, and anatomical features of psychosis. Even if a model does not entirely replicate the entirety of the pathophysiology of a disorder, it can still be useful if it is able to capture some key characteristics of pathology (such as impaired GABAergic system function (Menschikov et al., 2016)) and behavior.

Chronic phencyclidine

Phencyclidine and ketamine in humans

Single doses of noncompetitive NMDA antagonists, such as PCP and ketamine, can exacerbate psychotic symptoms in patients with schizophrenia and can induce acute psychotic symptoms in healthy controls (Lahti et al., 1995b; Krystal, 1994). Their acute effects in volunteers are distinct from those seen with illicit psychoactive substances that stimulate dopamine activity, like cocaine or amphetamine, in that they can include negative symptoms and cognitive dysfunction in addition to positive symptoms and thus more closely resemble the features of schizophrenia. This adds to the model's face validity.

Chronic use of PCP can lead to more persistent psychotic symptoms, cognitive dysfunction, and social withdrawal (Allen and Young, 1978;

Rainey and Crowder, 1975). While acute exposure can be associated with visual hallucinations, euphoria, and catatonia (Krystal, 1994), chronic use has been linked to auditory hallucinations, paranoia, and anxiety (Allen and Young, 1978).

The use of PCP in experimental human research is prohibited due to its neurotoxic effects (Olney et al. 1989, 1991), but this does not apply to ketamine, which can be safely administered in research studies. Acute administration of ketamine is associated with an increase in cerebral blood flow in anterior cingulate and frontomedial regions (Lahti et al., 1995a; Vollenweider et al., 1997; Breier et al., 1997). In contrast, long-term use of PCP has been linked to reduced frontal lobe blood flow (Hertzmann et al., 1990; Wu et al., 1991).

Results from MRS studies have more recently given us scope for understanding the neurochemical effects of ketamine on humans. In particular, these have shown that there is an increase in glutamate in the anterior cingulate gyrus following ketamine administration, which is related to increases in PANSS positive scores (Stone et al., 2012). Moreover, pretreating healthy controls with lamotrigine (an anticonvulsant that impairs glutamate release) attenuates the increase in positive symptom scores as well as preventing cognitive impairments in memory and learning domains (Anand et al., 2000; Deakin et al., 2008). This suggests that the increased glutamate release is key in ketamine's psychotomimetic effects.

Phencyclidine in rodents

Due to the similarities between the effects of chronic PCP use in humans and psychosis, the PCP model was applied in rodents. The effects in rodents can be divided into anatomical, neurophysiological, and behavioral effects.

Anatomical effects

Acute doses of PCP increase forebrain dopamine activity (Jentsch et al., 1997; Verma and Moghaddam, 1996; Hertel et al., 1995; Deutch et al., 1987; Doherty et al., 1980; Bowers and Hoffman, 1984). In contrast, chronic administration of PCP to rodents is associated with reduced frontal dopamine transmission (Jentsch et al., 1998a,b). Chronic PCP administration in rodents can induce activation of astroglia both in the cortex and hippocampus (Zhu et al., 2014). This results in the release of numerous immune mediators with a combination of neuroprotective and neurotoxic effects (Farina et al., 2007). Activated astroglia can also form a glial scar, which can obstruct regeneration of axons and outgrowth of neurites (Sofroniew, 2005). These changes are not specific to those seen in schizophrenia. However, there is some evidence from neuroimaging studies that both CHR subjects and patients with psychosis show evidence of glial activation (Bloomfield et al., 2016; Barron et al., 2017).

Neurophysiological effects

Chronic PCP administration in rodents leads to reduced innervation of the PFC via the mesocortical tract, thus reducing glutamate release and reducing activation of GABAergic interneurons (Kabanova et al., 2015). This leads to disinhibition of the ventral VTA, increasing activity of the mesolimbic pathway (French, 1988; Kabanova et al., 2015). Chronic PCP administration also dysregulates firing of dopaminergic neurons in the mesocortical pathway, leading to reduced dopamine release into the PFC (Kabanova et al., 2015).

As elevated subcortical dopaminergic function is evident in both CHR subjects and patients with psychosis, these effects suggest that the chronic PCP model has face validity.

Behavioral effects

In rodents, the PCP model is associated with impaired PPI (Pietraszek and Ossowska, 1998), as well as deficits in reversal learning. Reversal learning involves the animal associating an outcome (e.g., a reward) with an action (e.g., pulling a lever) after repeated pairing of action and outcome. After the animal reaches certain criteria for accuracy, the contingencies for reward are reversed (Izquierdo and Jentsch, 2012). The animal is thus required to inhibit the initially trained response and instead respond using cues it previously learned to be useless. Reversal learning is thought to reflect cognitive flexibility (Fellini et al., 2014). Lesioning orbital PFC impairs performance on reversal learning tasks (Bohn et al., 2003; McAlonan and Brown, 2003; Boulougouris et al., 2007). The PCP model is also associated with anticipatory deficits and negative affect, which may correspond to negative symptoms (Sahin et al., 2016).

Chronic PCP-treated rats display impaired set-shifting, particularly when switching to a previously irrelevant stimulus (extradimensional shift) (Rodefer et al., 2005). Moreover, chronic PCP administration in rodents also leads to reduced social interaction (Neill et al., 2010).

One limitation of the PCP model is that it does not involve a neurodevelopmental component because due to its chronicity, the model is restricted to adult rodents. In addition, it lacks construct validity as it does not incorporate genetic or environmental risk factors for schizophrenia.

Methylazoxymethanol acetate

The MAM model is particularly useful for understanding the neurochemical mechanisms underlying psychosis onset, as opposed to the established disorder. Administration of MAM to pregnant rats can selectively disrupt brain development (Haddad et al., 1972). While the exact mechanism is unknown, it is thought that MAM alters cell division and gene expression through DNA methylation (Moore et al., 2006). Administration of MAM before gestational day 15 leads to microencephaly, but administration at gestational day 17 (Moore et al., 2006) leads to a reduced density of parvalbumin-positive GABAergic interneurons in mediodorsal thalamus, hippocampus, parahippocampal, and PFC (Lodge et al., 2009; Gastambide et al., 2012). A reduction in parvalbumin-positive GABAergic interneurons has also been shown in patients with schizophrenia (Akbarian et al., 1995), particularly in the anterior cingulate cortex and hippocampus (reviewed in Benes (2010)), as well as in the DLPFC (reviewed in Lewis (2012)), and may result in the disinhibition of glutamatergic pyramidal cells (Marín, 2012). Abnormally increased glutamatergic outputs from the hippocampus are found to lead to the higher levels of striatal dopaminergic activity observed in MAM-treated rats (Lodge and Grace, 2007), as for example inactivating the ventral subiculum (the output region of the medial temporal region) in this model normalizes subcortical hyperdopaminergia (Lodge and Grace, 2007). Furthermore, electrical stimulation of the subiculum in healthy rats results in increased dopaminergic activity (Hammad and Wagner, 2006). It has been proposed that a loss of parvalbumin-positive GABA interneurons in the ventral hippocampus leads to increased pyramidal cell activity, resulting in greater glutamate release in the striatum through direct projections (Gonzalez-Burgos and Lewis, 2012). This increased glutamatergic tone has complex repercussions due to excitotoxicity. It has been shown in animal models that extremely high levels of glutamate can induce hippocampal hypermetabolism and atrophy (Schobel et al., 2013; Lieberman et al., 2018), similar to what is seen in CHR individuals who transition to psychosis (Schobel et al., 2013; Allen et al., 2016). As mentioned above, a functional loss of parvalbumin-positive interneurons in hippocampus leads to an overdrive of glutamatergic pyramidal neurons innervating the nucleus accumbens, which inhibits the ventral pallidum. This thereby increases activity of dopamine neurons in the midbrain that project back to the associative striatum. This increased level of striatal dopamine is proposed to underlie the dopamine-dependent positive symptoms of schizophrenia (Fig. 12.1) (Grace, 2016; Krystal and Anticevic, 2015; Krystal et al., 2017; Lieberman et al., 2018; Modinos et al., 2015; Lisman et al., 2008). Because the ventral subiculum is also involved in several other circuits, its hyperactivity may also disrupt (1) prefrontal function, leading to cognitive deficits, and (2) basolateral amygdala function, leading to a lack of control over emotional reactivity, possibly associated with negative symptoms (Grace, 2016). As such, ventral hippocampal disruption could potentially contribute to the three main symptom dimensions of schizophrenia.

These effects show some correspondence with neuroimaging data from CHR subjects. Thus, CHR individuals show increased resting activity in the medial temporal lobe (Allen et al., 2016) and increased glutamate levels in the MTL and striatum (Kindler et al., 2018; de la Fuente-Sandoval et al., 2013; Valli et al., 2011). GABA levels in the MTL in CHR subjects have yet to be examined, as this region is difficult to assess using MRS, but recent data suggest an

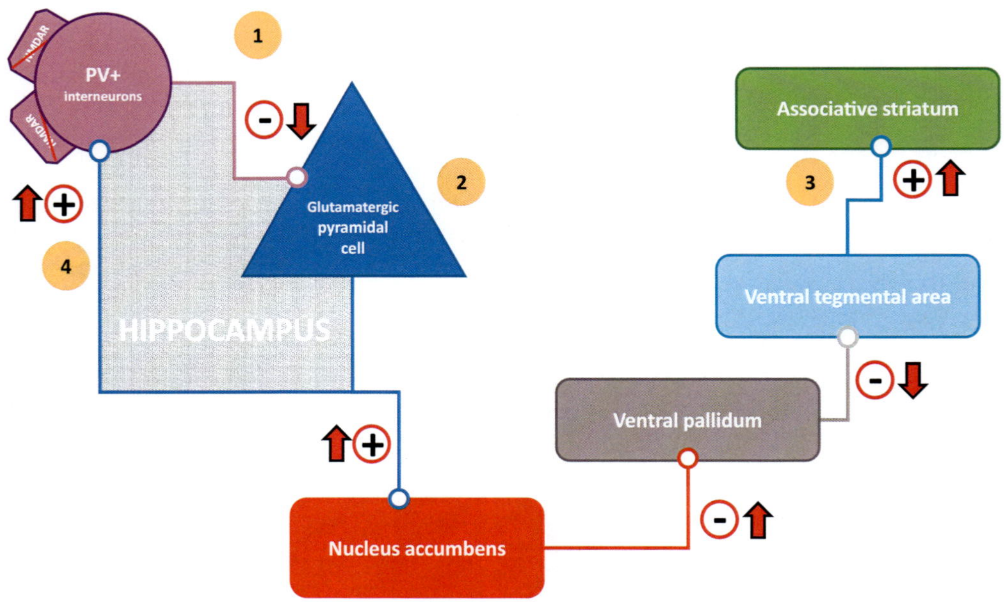

FIGURE 12.1 Simplified schematic of proposed neural circuit mechanisms leading to psychosis *pathophysiology*. In (1), NMDARs hypofunction results in decreased GABAergic interneurons function and thus a reduction of inhibition (disinhibition) of glutamatergic pyramidal cells, thereby increasing excitation. However, by disinhibiting pyramidal cells (and thus increasing glutamate signaling) in this dysfunctional neural environment, the potential homeostatic adaptation becomes allostatic (2). This imbalance drives increased nucleus accumbens activity, enhancing inhibitory projections to the ventral pallidum. In (3), disinhibition of the VTA by the ventral pallidum neurons leads to an overdrive in the responsivity of VTA neurons that project to the associative striatum. Completing the (simplified) circuit, local glutamatergic tone is increased in (4) but is not detected as such by hypofunctioning NMDARs on GABAergic interneurons. For original diagrams and discussion of evidence for this proposed circuit, see Krystal et al., 2017; Krystal and Anticevic, 2015; Lieberman et al., 2018; Lisman et al., 2008; Modinos et al., 2015; Davies et al., 2019. *GABA*, γ-aminobutyric acid; *NMDAR*, N-methyl-D-aspartate receptor; and *VTA*, ventral tegmental area.

association between prefrontal GABA levels and hippocampal resting perfusion in CHR subjects, particularly in those who later develop psychosis (Modinos et al., 2018b). Furthermore, because the hypersensitivity to amphetamine in MAM only develops in the adult rat, but is not present prepubertally (Moore et al., 2006), the model has high face validity as it is consistent with the late adolescence/early adulthood onset of psychosis in humans. This comparable time frame of abnormalities is particularly interesting as a window of opportunity for prevention. Restoring hippocampal activity by modulation of the GABAergic system is shown to reverse dopaminergic hyperactivity in the MAM model of schizophrenia. This is achieved either with a GABA-A α5-selective positive allosteric modulator applied systemically or directly into the hippocampus (Gill et al., 2011), as well as with diazepam applied systemically (Du and Grace, 2013). Such interventions during the peripubertal period in MAM-treated rats normalized amygdala hyperactivity and hyperanxiety (Du and Grace, 2016a) and reduced loss of PV+ interneurons in the hippocampus (Du and Grace, 2016b). These effects were only evident when treatment was administered in the premorbid phase and in animals that had not been previously exposed to antipsychotics (Gill et al., 2014). Thus, this evidence suggests that enhancing GABAergic function premorbidly may delay or prevent the onset of psychosis, with enormous translational potential in the CHR state.

Dopamine

A key prediction of the MAM model is that there is an increase in dopamine function in the striatum. MAM rats show higher activity in the ventral tegmental area (VTA), with preferentially increased firing of dopaminergic neurons in the lateral VTA, which projects to the equivalent of the associative striatum (Lodge and Grace, 2012). The model also predicts there is increased activity in the ventral striatum, which is the target of projections from the hippocampus.

Data from neuroimaging studies confirm that striatal dopamine function is elevated in CHR individuals and in patients with schizophrenia (Howes et al. 2009, 2011; Egerton et al., 2013), with the findings mainly in the associative subdivision.

Glutamate

The MAM model suggests that there is an increase in glutamate levels in cortical areas due to reduced GABAergic inhibition of pyramidal neurons (Lisman and Grace, 2005) (Fig. 12.1). MRS studies of glutamate metabolites in CHR subjects and patients with schizophrenia indicate that these are indeed altered but that the findings vary depending on the cortical region and on the stage of the disorder with differential findings in CHR, FEP, and chronic schizophrenia (Merritt et al., 2016). A key cortical region implicated in the MAM model is the MTL, and MRS studies in this region suggest that glutamate levels are elevated in CHR subjects who later develop psychosis (Bossong et al., 2018), and that MTL glutamate levels are related to subcortical dopamine function (Stone et al., 2010).

The MAM model also predicts that in the long term, increased levels of glutamate in the MTL would lead to progressive degeneration of neurons in the hippocampus, in particular of parvalbumin interneurons (Modinos et al., 2015; Deutsch et al., 2001; Grace and Gomes, 2019). This is consistent with evidence of longitudinal reductions in MTL volume as CHR subjects make the transition to psychosis (Pantelis et al., 2003; Schobel et al., 2013).

γ-Aminobutyric acid

The MAM model is associated with reduced numbers of cortical GABAergic interneurons that would lead to increased hippocampal activity. However, the results from MR spectroscopy and PET studies of GABA levels or receptors in CHR subjects and psychosis have been inconsistent (Egerton et al., 2017). This could be due to heterogeneity between studies, which have examined different types of patient, used different regions of interest, and used different neuroimaging protocols. A recent circuit-based study, combining multiple MRI-based techniques (MRS and ASL), suggests an association between prefrontal GABA levels and hippocampal resting perfusion in CHR subjects, particularly in those who later develop psychosis (Modinos et al., 2018b).

Limitations of the model

The original MAM model did not incorporate genetic and environmental risk factors for psychosis. However, recent work (Grace and Gomes, 2019) suggests that psychosocial stress at critical developmental periods can have similar effects to MAM and that genetic risk factors might influence the model by moderating the response to psychosocial stress.

Conclusion

Preclinical models have improved our understanding of the putative neurochemical mechanisms underlying the onset of psychosis, which is fundamental to the development of preventative treatments. In particular, animal models have allowed for a greater understanding of the effects of glutamatergic and GABAergic dysfunction on striatal dopamine as well as the neurodevelopmental staging of psychosis and its onset. These models also provide insights into strategies to circumvent the development of psychosis by targeting hippocampal GABA—glutamate balance during the premorbid stage. Ongoing efforts to translate these findings to humans may significantly advance the identification of new therapeutic targets to delay or prevent the onset of psychosis.

The neonatal hippocampal lesion model is limited by its construct validity: although psychosis is associated with alterations in the hippocampus, there is not a gross macroscopic lesion. The poly(I:C) model does not accurately reflect the acute phase of maternal infection and thus has limited external validity. Moreover, the link between prenatal infection and psychosis remains controversial. The chronic PCP model lacks a neurodevelopmental component and therefore lacks construct validity. Finally, the MAM model reproduces many of the neurobiological features of psychosis, although the roles of environmental and genetic factors in this model are less clearly defined and are the subject of ongoing research.

References

Ahnaou, A., et al., 2007. Sleep and EEG profile in neonatal hippocampal lesion model of schizophrenia. Physiol. Behav. 92 (3), 461—467.

Akbarian, S., et al., 1995. Gene expression for glutamic acid decarboxylase is reduced without loss of neurons in prefrontal cortex of schizophrenics. Arch. Gen. Psychiatr. 52 (4), 258—266.

Akira, S., Takeda, K., 2004. Toll-like receptor signalling. Nat. Rev. Immunol. 4 (7), 499—511.

Allen, R.M., Young, S.J., 1978. Phencyclidine-induced psychosis. Am. J. Psychiatry 135 (9), 1081—1084.

Allen, P., et al., 2012. Transition to psychosis associated with prefrontal and subcortical dysfunction in ultra high-risk individuals. Schizophr. Bull. 38 (6), 1268—1276.

Allen, P., et al., 2016. Resting hyperperfusion of the hippocampus, midbrain, and basal ganglia in people at high risk for psychosis. Am. J. Psychiatry 173 (4), 392—399.

Allen, P., et al., 2018. Increased resting hippocampal and basal ganglia perfusion in people at ultra high risk for psychosis: replication in a second cohort. Schizophr. Bull. 44 (6), 1323—1331.

Alquicer, G., et al., 2008. Postweaning social isolation enhances morphological changes in the neonatal ventral hippocampal lesion rat model of psychosis. J. Chem. Neuroanat. 35 (2), 179—187.

Anand, A., et al., 2000. Attenuation of the neuropsychiatric effects of ketamine with lamotrigine: support for hyperglutamatergic effects of N-methyl-D-aspartate receptor antagonists. Arch. Gen. Psychiatr. 57 (3), 270–276.

Babulas, V., et al., 2006. Prenatal exposure to maternal genital and reproductive infections and adult schizophrenia. Am. J. Psychiatry 163 (5), 927–929.

Baldan Ramsey, L.C., et al., 2011. Lesions of the dorsomedial striatum disrupt prepulse inhibition. Neuroscience 180, 222–228.

Barron, H., et al., 2017. Neuroinflammation and oxidative stress in psychosis and psychosis risk. Int. J. Mol. Sci. 18 (3).

Becker, A., et al., 1999. Social behaviour in rats lesioned with ibotenic acid in the hippocampus: quantitative and qualitative analysis. Psychopharmacology 144 (4), 333–338.

Benes, F.M., 2010. Amygdalocortical circuitry in schizophrenia: from circuits to molecules. Neuropsychopharmacology 35 (1), 239–257.

Benetti, S., et al., 2009. Functional integration between the posterior hippocampus and prefrontal cortex is impaired in both first episode schizophrenia and the at risk mental state. Brain 132 (Pt 9), 2426–2436.

Bernard, J.A., Orr, J.M., Mittal, V.A., 2015. Abnormal hippocampal-thalamic white matter tract development and positive symptom course in individuals at ultra-high risk for psychosis. NPJ Schizophr. 1.

Bloomfield, P.S., et al., 2016. Microglial activity in people at ultra high risk of psychosis and in schizophrenia: an [(11)C]PBR28 PET brain imaging study. Am. J. Psychiatry 173 (1), 44–52.

Boekhoudt, L., et al., 2016. Chemogenetic activation of dopamine neurons in the ventral tegmental area, but not substantia nigra, induces hyperactivity in rats. Eur. Neuropsychopharmacol. 26 (11), 1784–1793.

Bohn, I., Giertler, C., Hauber, W., 2003. Orbital prefrontal cortex and guidance of instrumental behaviour in rats under reversal conditions. Behav. Brain Res. 143 (1), 49–56.

Bois, C., et al., 2016. Longitudinal changes in hippocampal volume in the Edinburgh high risk study of schizophrenia. Schizophr. Res. 173 (3), 146–151.

Bolino, F., et al., 1994. Sensorimotor gating and habituation evoked by electro-cutaneous stimulation in schizophrenia. Biol. Psychiatry 36 (10), 670–679.

Bora, E., et al., 2014. Cognitive deficits in youth with familial and clinical high risk to psychosis: a systematic review and meta-analysis. Acta Psychiatr. Scand. 130 (1), 1–15.

Bossong, M.G., et al., 2018. Human striatal response to reward anticipation linked to hippocampal glutamate levels. Int. J. Neuropsychopharmacol. 21 (7), 623–630.

Boulougouris, V., Dalley, J.W., Robbins, T.W., 2007. Effects of orbitofrontal, infralimbic and prelimbic cortical lesions on serial spatial reversal learning in the rat. Behav. Brain Res. 179 (2), 219–228.

Bowers, M.B., Hoffman, F.J., 1984. Homovanillic acid in rat caudate and prefrontal cortex following phencyclidine and amphetamine. Psychopharmacology 84 (1), 136–137.

Braff, D., et al., 1978. Prestimulus effects on human startle reflex in normals and schizophrenics. Psychophysiology 15 (4), 339–343.

Braff, D.L., Grillon, C., Geyer, M.A., 1992. Gating and habituation of the startle reflex in schizophrenic patients. Arch. Gen. Psychiatr. 49 (3), 206–215.

Breier, A., et al., 1997. Association of ketamine-induced psychosis with focal activation of the prefrontal cortex in healthy volunteers. Am. J. Psychiatry 154 (6), 805–811.

Brown, A.S., et al., 2001. Prenatal rubella, premorbid abnormalities, and adult schizophrenia. Biol. Psychiatry 49 (6), 473–486.

Buka, S.L., et al., 2001. Maternal infections and subsequent psychosis among offspring. Arch. Gen. Psychiatr. 58 (11), 1032–1037.

Cai, L., et al., 2015. Gestational influenza increases the risk of psychosis in adults. Med. Chem. 11 (7), 676–682.

Chefer, V.I., et al., 2009. Overview of brain microdialysis. Curr. Protoc. Neurosci. 47 (1), 7.1.1–7.1.28. Chapter 7, p.Unit7.1.

Cunningham, C., et al., 2007. The sickness behaviour and CNS inflammatory mediator profile induced by systemic challenge of mice with synthetic double-stranded RNA (poly I:C). Brain Behav. Immun. 21 (4), 490–502.

Davies, C., et al., 2018. Lack of evidence to favor specific preventive interventions in psychosis: a network meta-analysis. World Psychiatry.

Davies, C., et al., 2019. Oxytocin modulates hippocampal perfusion in people at clinical high risk for psychosis. Neuropsychopharmacology 44 (7), 1300–1309. https://doi.org/10.1038/s41386-018-0311-6.

de la Fuente-Sandoval, C., et al., 2013. Glutamate levels in the associative striatum before and after 4 weeks of antipsychotic treatment in first-episode psychosis: a longitudinal proton magnetic resonance spectroscopy study. JAMA Psychiatry 70 (10), 1057–1066.

de la Fuente-Sandoval, C., et al., 2015. Cortico-striatal GABAergic and glutamatergic dysregulations in subjects at ultra-high risk for psychosis investigated with proton magnetic resonance spectroscopy. Int. J. Neuropsychopharmacol. 19 (3), pyv105.

Deakin, J.F.W., et al., 2008. Glutamate and the neural basis of the subjective effects of ketamine: a pharmaco-magnetic resonance imaging study. Arch. Gen. Psychiatr. 65 (2), 154–164.

Deutch, A.Y., et al., 1987. Mesolimbic and mesocortical dopamine activation induced by phencyclidine: contrasting pattern to striatal response. Eur. J. Pharmacol. 134 (3), 257–264.

Deutsch, S.I., et al., 2001. A revised excitotoxic hypothesis of schizophrenia: therapeutic implications. Clin. Neuropharmacol. 24 (1), 43–49.

Doherty, J.D., et al., 1980. The effect of phencyclidine on dopamine synthesis and metabolic in rat striatum. Eur. J. Pharmacol. 65 (2–3), 139–149.

Drouin-Ouellet, J., et al., 2011. Neuroinflammation is associated with changes in glial mGluR5 expression and the development of neonatal excitotoxic lesions. Glia 59 (2), 188–199.

Du, Y., Grace, A.A., 2013. Peripubertal diazepam administration prevents the emergence of dopamine system hyperresponsivity in the MAM developmental disruption model of schizophrenia. Neuropsychopharmacology 38 (10), 1881–1888.

Du, Y., Grace, A.A., 2016a. Amygdala hyperactivity in MAM model of schizophrenia is normalized by peripubertal diazepam administration. Neuropsychopharmacology 41 (10), 2455–2462.

Du, Y., Grace, A.A., 2016b. Loss of parvalbumin in the hippocampus of MAM schizophrenia model rats is attenuated by peripubertal diazepam. Int. J. Neuropsychopharmacol. 19 (11).

Egerton, A., et al., 2013. Presynaptic striatal dopamine dysfunction in people at ultra-high risk for psychosis: findings in a second cohort. Biol. Psychiatry 74 (2), 106–112.

Egerton, A., et al., 2017. Neuroimaging studies of GABA in schizophrenia: a systematic review with meta-analysis. Transl. Psychiatry 7 (6), e1147.

Ellman, L.M., et al., 2010. Structural brain alterations in schizophrenia following fetal exposure to the inflammatory cytokine interleukin-8. Schizophr. Res. 121 (1–3), 46–54.

Farina, C., Aloisi, F., Meinl, E., 2007. Astrocytes are active players in cerebral innate immunity. Trends Immunol. 28 (3), 138–145.

Fatemi, S.H., et al., 2008. Maternal infection leads to abnormal gene regulation and brain atrophy in mouse offspring: implications for genesis of neurodevelopmental disorders. Schizophr. Res. 99 (1–3), 56–70.

Fellini, L., et al., 2014. Re-evaluating the PCP challenge as a pre-clinical model of impaired cognitive flexibility in schizophrenia. Eur. Neuropsychopharmacol. 24 (11), 1836–1849.

Flores, G., et al., 1996. Enhanced amphetamine sensitivity and increased expression of dopamine D2 receptors in postpubertal rats after neonatal excitotoxic lesions of the medial prefrontal cortex. J. Neurosci. 16 (22), 7366–7375.

Flores, G., et al., 2005. Alterations in dendritic morphology of prefrontal cortical and nucleus accumbens neurons in post-pubertal rats after neonatal excitotoxic lesions of the ventral hippocampus. Neuroscience 133 (2), 463–470.

Foong, J., et al., 2000. Neuropathological abnormalities of the corpus callosum in schizophrenia: a diffusion tensor imaging study. J. Neurol. Neurosurg. Psychiatry 68 (2), 242–244.

Foong, J., et al., 2002. Investigating regional white matter in schizophrenia using diffusion tensor imaging. Neuroreport 13 (3), 333–336.

François, J., et al., 2009. Selective reorganization of GABAergic transmission in neonatal ventral hippocampal-lesioned rats. Int. J. Neuropsychopharmacol. 12 (8), 1097–1110.

French, E.D., 1988. Effects of acute and chronic administration of phencyclidine on the A10 dopaminergic mesolimbic system: electrophysiological and behavioral correlates. Neuropharmacology 27 (8), 791–798.

Fusar-Poli, P., et al., 2012. Predicting psychosis: meta-analysis of transition outcomes in individuals at high clinical risk. Arch. Gen. Psychiatr. 69 (3), 220–229.

Fusar-Poli, P., et al., 2013. Outreach and support in south London (OASIS), 2001–2011: ten years of early diagnosis and treatment for young individuals at high clinical risk for psychosis. Eur. Psychiatry 28 (5), 315–326.

Gastambide, F., et al., 2012. Selective remediation of reversal learning deficits in the neurodevelopmental MAM model of schizophrenia by a novel mGlu5 positive allosteric modulator. Neuropsychopharmacology 37 (4), 1057–1066.

Gill, K.M., et al., 2011. A novel α5GABA(A)R-positive allosteric modulator reverses hyperactivation of the dopamine system in the MAM model of schizophrenia. Neuropsychopharmacology 36 (9), 1903–1911.

Gill, K.M., et al., 2014. Prior antipsychotic drug treatment prevents response to novel antipsychotic agent in the methylazoxymethanol acetate model of schizophrenia. Schizophr. Bull. 40 (2), 341–350.

Giovanoli, S., Weber, L., Meyer, U., 2014. Single and combined effects of prenatal immune activation and peripubertal stress on parvalbumin and reelin expression in the hippocampal formation. Brain Behav. Immun. 40, 48–54.

Gonzalez-Burgos, G., Lewis, D.A., 2012. NMDA receptor hypofunction, parvalbumin-positive neurons, and cortical gamma oscillations in schizophrenia. Schizophr. Bull. 38 (5), 950–957.

Goto, Y., O'Donnell, P., 2002. Delayed mesolimbic system alteration in a developmental animal model of schizophrenia. J. Neurosci. 22 (20), 9070–9077.

Goto, Y., O'Donnell, P., 2004. Prefrontal lesion reverses abnormal mesoaccumbens response in an animal model of schizophrenia. Biol. Psychiatry 55 (2), 172–176.

Grace, A.A., 2016. Dysregulation of the dopamine system in the pathophysiology of schizophrenia and depression. Nat. Rev. Neurosci. 17 (8), 524–532.

Grace, A.A., Gomes, F.V., 2019. The circuitry of dopamine system regulation and its disruption in schizophrenia: insights into treatment and prevention. Schizophr. Bull. 45 (1), 148–157.

Haddad, R.K., Rabe, A., Dumas, R., 1972. Comparison of effects of methylazoxymethanol acetate on brain development in different species. Fed. Proc. 31 (5), 1520–1523.

Häfner, H., et al., 2003. Modeling the early course of schizophrenia. Schizophr. Bull. 29 (2), 325–340.

Hakak, Y., et al., 2001. Genome-wide expression analysis reveals dysregulation of myelination-related genes in chronic schizophrenia. Proc. Natl. Acad. Sci. USA 98 (8), 4746–4751.

Hammad, H., Wagner, J.J., 2006. Dopamine-mediated disinhibition in the CA1 region of rat hippocampus via D3 receptor activation. J. Pharmacol. Exp. Ther. 316 (1), 113–120.

Heckers, S., Konradi, C., 2015. GABAergic mechanisms of hippocampal hyperactivity in schizophrenia. Schizophr. Res. 167 (1–3), 4–11.

Hertel, P., et al., 1995. Effects of D-amphetamine and phencyclidine on behavior and extracellular concentrations of neurotensin and dopamine in the ventral striatum and the medial prefrontal cortex of the rat. Behav. Brain Res. 72 (1–2), 103–114.

Hertzmann, M., Reba, R.C., Kotlyarov, E.V., 1990. Single photon emission computed tomography in phencyclidine and related drug abuse. Am. J. Psychiatry 147 (2), 255–256.

Heurling, K., et al., 2017. Quantitative positron emission tomography in brain research. Brain Res. 1670, 220–234.

Hong, S.E., et al., 2000. Autosomal recessive lissencephaly with cerebellar hypoplasia is associated with human RELN mutations. Nat. Genet. 26 (1), 93–96.

Howes, O.D., et al., 2009. Elevated striatal dopamine function linked to prodromal signs of schizophrenia. Arch. Gen. Psychiatr. 66 (1), 13–20.

Howes, O.D., et al., 2011. Dopamine synthesis capacity before onset of psychosis: a prospective [18F]-DOPA PET imaging study. Am. J. Psychiatry 168 (12), 1311–1317.

Izquierdo, A., Jentsch, J.D., 2012. Reversal learning as a measure of impulsive and compulsive behavior in addictions. Psychopharmacology 219 (2), 607–620.

Jaaro-Peled, H., et al., 2010. Review of pathological hallmarks of schizophrenia: comparison of genetic models with patients and nongenetic models. Schizophr. Bull. 36 (2), 301–313.

Jentsch, J.D., et al., 1997. Phencyclidine increases forebrain monoamine metabolism in rats and monkeys: modulation by the isomers of HA966. J. Neurosci. 17 (5), 1769–1775.

Jentsch, J.D., Taylor, J.R., Roth, R.H., 1998a. Subchronic phencyclidine administration increases mesolimbic dopaminergic system responsivity and augments stress- and psychostimulant-induced hyperlocomotion. Neuropsychopharmacology 19 (2), 105–113.

Jentsch, J.D., Wise, A., et al., 1998b. Alpha-noradrenergic receptor modulation of the phencyclidine- and delta9-tetrahydrocannabinol-induced increases in dopamine utilization in rat prefrontal cortex. Synapse 28 (1), 21–26.

Kabanova, A., et al., 2015. Function and developmental origin of a mesocortical inhibitory circuit. Nat. Neurosci. 18 (6), 872–882.

Kegeles, L.S., et al., 2010. Increased synaptic dopamine function in associative regions of the striatum in schizophrenia. Arch. Gen. Psychiatr. 67 (3), 231–239.

Kimura, M., et al., 1994. Comparison of acute phase responses induced in rabbits by lipopolysaccharide and double-stranded RNA. Am. J. Physiol. 267 (6 Pt 2), R1596–R1605.

Kindler, J., et al., 2018. Increased striatal and reduced prefrontal cerebral blood flow in clinical high risk for psychosis. Schizophr. Bull. 44 (1), 182–192.

Koch, M., 1999. The neurobiology of startle. Prog. Neurobiol. 59 (2), 107–128.

Krystal, J.H., 1994. Subanesthetic effects of the noncompetitive NMDA antagonist, ketamine, in humans. Arch. Gen. Psychiatr. 51 (3), 199.

Krystal, J.H., Anticevic, A., 2015. Toward illness phase-specific pharmacotherapy for schizophrenia. Biol. Psychiatry 78 (11), 738–740.

Krystal, J.H., et al., 2017. Impaired tuning of neural ensembles and the pathophysiology of schizophrenia: a translational and computational neuroscience perspective. Biol. Psychiatry 81 (10), 874–885.

Lahti, A.C., Holcomb, H.H., et al., 1995a. Ketamine activates psychosis and alters limbic blood flow in schizophrenia. Neuroreport 6 (6), 869–872.

Lahti, A.C., Koffel, B., et al., 1995b. Subanesthetic doses of ketamine stimulate psychosis in schizophrenia. Neuropsychopharmacology 13 (1), 9–19.

Le Pen, G., Moreau, J.-L., 2002. Disruption of prepulse inhibition of startle reflex in a neurodevelopmental model of schizophrenia: reversal by clozapine, olanzapine and risperidone but not by haloperidol. Neuropsychopharmacology 27 (1), 1–11.

Le Pen, G., et al., 2000. Spatial and associative learning deficits induced by neonatal excitotoxic hippocampal damage in rats: further evaluation of an animal model of schizophrenia. Behav. Pharmacol. 11 (3–4), 257–268.

Le Pen, G., et al., 2003. Phencyclidine exacerbates attentional deficits in a neurodevelopmental rat model of schizophrenia. Neuropsychopharmacology 28 (10), 1799–1809.

Lewis, D.A., 2012. Cortical circuit dysfunction and cognitive deficits in schizophrenia—implications for preemptive interventions. Eur. J. Neurosci. 35 (12), 1871–1878.

Lieberman, J.A., et al., 2018. Hippocampal dysfunction in the pathophysiology of schizophrenia: a selective review and hypothesis for early detection and intervention. Mol. Psychiatry 23 (8), 1764–1772.

Lim, K.O., et al., 1999. Compromised white matter tract integrity in schizophrenia inferred from diffusion tensor imaging. Arch. Gen. Psychiatr. 56 (4), 367–374.

Lipska, B.K., 2004. Using animal models to test a neurodevelopmental hypothesis of schizophrenia. J. Psychiatry Neurosci. 29 (4), 282–286.

Lipska, B.K., Weinberger, D.R., 1994. Subchronic treatment with haloperidol and clozapine in rats with neonatal excitotoxic hippocampal damage. Neuropsychopharmacology 10 (3), 199–205.

Lipska, B.K., Weinberger, D.R., 1998. Prefrontal cortical and hippocampal modulation of dopamine-mediated effects. Adv. Pharmacol. 42, 806–809.

Lipska, B.K., Weinberger, D.R., 2000. To model a psychiatric disorder in animals: schizophrenia as a reality test. Neuropsychopharmacology 23 (3), 223–239.

Lipska, B.K., Weinberger, D.R., 2002. A neurodevelopmental model of schizophrenia: neonatal disconnection of the hippocampus. Neurotox. Res. 4 (5–6), 469–475.

Lipska, B.K., Jaskiw, G.E., Weinberger, D.R., 1993. Postpubertal emergence of hyperresponsiveness to stress and to amphetamine after neonatal excitotoxic hippocampal damage: a potential animal model of schizophrenia. Neuropsychopharmacology 9 (1), 67–75.

Lipska, B.K., et al., 2002. Neonatal damage of the ventral hippocampus impairs working memory in the rat. Neuropsychopharmacology 27 (1), 47–54.

Lisman, J.E., Grace, A.A., 2005. The hippocampal-VTA loop: controlling the entry of information into long-term memory. Neuron 46 (5), 703–713.

Lisman, J.E., et al., 2008. Circuit-based framework for understanding neurotransmitter and risk gene interactions in schizophrenia. Trends Neurosci. 31 (5), 234–242.

Lodge, D.J., Grace, A.A., 2007. Aberrant hippocampal activity underlies the dopamine dysregulation in an animal model of schizophrenia. J. Neurosci. 27 (42), 11424–11430.

Lodge, D.J., Grace, A.A., 2012. Divergent activation of ventromedial and ventrolateral dopamine systems in animal models of amphetamine sensitization and schizophrenia. Int. J. Neuropsychopharmacol. 15 (1), 69–76.

Lodge, D.J., Behrens, M.M., Grace, A.A., 2009. A loss of parvalbumin-containing interneurons is associated with diminished oscillatory activity in an animal model of schizophrenia. J. Neurosci. 29 (8), 2344–2354.

Logothetis, N.K., 2002. The neural basis of the blood-oxygen-level-dependent functional magnetic resonance imaging signal. Philos. Trans. R. Soc. Lond. Ser. B Biol. Sci. 357 (1424), 1003–1037.

Macedo, C.E., et al., 2010. Acoustic hypersensitivity in adult rats after neonatal ventral hippocampus lesions. Behav. Brain Res. 207 (1), 161–168.

Makinodan, M., et al., 2008. Maternal immune activation in mice delays myelination and axonal development in the hippocampus of the offspring. J. Neurosci. Res. 86 (10), 2190–2200.

Manoach, D.S., 2003. Prefrontal cortex dysfunction during working memory performance in schizophrenia: reconciling discrepant findings. Schizophr. Res. 60 (2–3), 285–298.

Marín, O., 2012. Interneuron dysfunction in psychiatric disorders. Nat. Rev. Neurosci. 13 (2), 107–120.

Marona-Lewicka, D., Nichols, C.D., Nichols, D.E., 2011. An animal model of schizophrenia based on chronic LSD administration: old idea, new results. Neuropharmacology 61 (3), 503–512.

Martin, R.P., Dombrowski, S.C., 2008. Prenatal Exposures: Psychological and Educational Consequences for Children. Springer Science & Business Media.

Mattei, D., Schweibold, R., Wolf, S.A., 2015. Brain in flames – animal models of psychosis: utility and limitations. Neuropsychiatric Dis. Treat. 11, 1313–1329.

McAlonan, K., Brown, V.J., 2003. Orbital prefrontal cortex mediates reversal learning and not attentional set shifting in the rat. Behav. Brain Res. 146 (1–2), 97–103.

McGorry, P.D., 2013. Early clinical phenotypes, clinical staging, and strategic biomarker research: building blocks for personalized psychiatry. Biol. Psychiatry 74 (6), 394–395.

Meares, A., 1959. The diagnosis of prepsychotic schizophrenia. Lancet 273 (7063), 55–58.

Mechelli, A., et al., 2011. Neuroanatomical abnormalities that predate the onset of psychosis: a multicenter study. Arch. Gen. Psychiatr. 68 (5), 489–495.

Menninger, K.A., 1919. Psychoses associated with influenza. J. Am. Med. Assoc. 72 (4), 235.

Menschikov, P.E., et al., 2016. (1)H-MRS and MEGA-PRESS pulse sequence in the study of balance of inhibitory and excitatory neurotransmitters in the human brain of ultra-high risk of schizophrenia patients. Dokl. Biochem. Biophys. 468 (1), 168–172.

Merritt, K., et al., 2016. Nature of glutamate alterations in schizophrenia: a meta-analysis of proton magnetic resonance spectroscopy studies. JAMA Psychiatry 73 (7), 665–674.

Meyer, U., et al., 2005. Towards an immuno-precipitated neurodevelopmental animal model of schizophrenia. Neurosci. Biobehav. Rev. 29 (6), 913–947.

Meyer, U., et al., 2006. The time of prenatal immune challenge determines the specificity of inflammation-mediated brain and behavioral pathology. J. Neurosci. 26 (18), 4752–4762.

Meyer, U., Yee, B.K., Feldon, J., 2007. The neurodevelopmental impact of prenatal infections at different times of pregnancy: the earlier the worse? Neuroscientist 13 (3), 241–256.

Meyer, U., Feldon, J., Fatemi, S.H., 2009. In-vivo rodent models for the experimental investigation of prenatal immune activation effects in neurodevelopmental brain disorders. Neurosci. Biobehav. Rev. 33 (7), 1061–1079.

Modinos, G., et al., 2015. Translating the MAM model of psychosis to humans. Trends Neurosci. 38 (3), 129–138.

Modinos, G., Simsek, F., et al., 2018a. Cortical GABA in subjects at ultra-high risk of psychosis: relationship to negative prodromal symptoms. Int. J. Neuropsychopharmacol. 21 (2), 114–119.

Modinos, G., Şimşek, F., et al., 2018b. Prefrontal GABA levels, hippocampal resting perfusion and the risk of psychosis. Neuropsychopharmacology 43 (13), 2652–2659.

Moore, H., et al., 2006. A neurobehavioral systems analysis of adult rats exposed to methylazoxymethanol acetate on E17: implications for the neuropathology of schizophrenia. Biol. Psychiatry 60 (3), 253–264.

Mortensen, P.B., et al., 2007. Early infections of Toxoplasma gondii and the later development of schizophrenia. Schizophr. Bull. 33 (3), 741–744.

Naert, A., et al., 2013. Behavioural alterations relevant to developmental brain disorders in mice with neonatally induced ventral hippocampal lesions. Brain Res. Bull. 94, 71–81.

Nakahara, S., Matsumoto, M., van Erp, T.G.M., 2018. Hippocampal subregion abnormalities in schizophrenia: a systematic review of structural and physiological imaging studies. Neuropsychopharmacology Rep. 38 (4), 156–166.

Neill, J.C., et al., 2010. Animal models of cognitive dysfunction and negative symptoms of schizophrenia: focus on NMDA receptor antagonism. Pharmacol. Ther. 128 (3), 419–432.

Nelson, B., et al., 2013. Long-term follow-up of a group at ultra high risk ("prodromal") for psychosis: the PACE 400 study. JAMA Psychiatry 70 (8), 793–802.

Nyffeler, M., et al., 2006. Maternal immune activation during pregnancy increases limbic GABAA receptor immunoreactivity in the adult offspring: implications for schizophrenia. Neuroscience 143 (1), 51–62.

Olney, J.W., Labruyere, J., Price, M.T., 1989. Pathological changes induced in cerebrocortical neurons by phencyclidine and related drugs. Science 244 (4910), 1360–1362.

Olney, J.W., et al., 1991. NMDA antagonist neurotoxicity: mechanism and prevention. Science 254 (5037), 1515–1518.

Ozawa, K., et al., 2006. Immune activation during pregnancy in mice leads to dopaminergic hyperfunction and cognitive impairment in the offspring: a neurodevelopmental animal model of schizophrenia. Biol. Psychiatry 59 (6), 546–554.

O'Donnell, P., et al., 2002. Neonatal hippocampal damage alters electrophysiological properties of prefrontal cortical neurons in adult rats. Cerebr. Cortex 12 (9), 975–982.

Pantelis, C., et al., 2003. Neuroanatomical abnormalities before and after onset of psychosis: a cross-sectional and longitudinal MRI comparison. Lancet 361 (9354), 281–288.

Pantelis, C., et al., 2007. Neuroimaging and emerging psychotic disorders: the Melbourne ultra-high risk studies. Int. Rev. Psychiatry 19 (4), 371–381.

Perry, W., Braff, D.L., 1994. Information-processing deficits and thought disorder in schizophrenia. Am. J. Psychiatry 151 (3), 363–367.

Pietraszek, M., Ossowska, K., 1998. Chronic treatment with haloperidol diminishes the phencyclidine-induced sensorimotor gating deficit in rats. Naunyn-Schmiedeberg's Arch. Pharmacol. 357 (4), 466–471.

Piontkewitz, Y., Arad, M., Weiner, I., 2011. Abnormal trajectories of neurodevelopment and behavior following in utero insult in the rat. Biol. Psychiatry 70 (9), 842–851.

Radua, J., et al., 2018. What causes psychosis? An umbrella review of risk and protective factors. World Psychiatry 17 (1), 49–66.

Rainey, J.M., Crowder, M.K., 1975. Prolonged psychosis attributed to phencyclidine: report of three cases. Am. J. Psychiatry 132 (10), 1076–1078.

Rodefer, J.S., Murphy, E.R., Baxter, M.G., 2005. PDE10A inhibition reverses subchronic PCP-induced deficits in attentional set-shifting in rats. Eur. J. Neurosci. 21 (4), 1070–1076.

Sahin, C., Doostdar, N., Neill, J.C., 2016. Towards the development of improved tests for negative symptoms of schizophrenia in a validated animal model. Behav. Brain Res. 312, 93–101.

Schobel, S.A., et al., 2013. Imaging patients with psychosis and a mouse model establishes a spreading pattern of hippocampal dysfunction and implicates glutamate as a driver. Neuron 78 (1), 81–93.

Schultze-Lutter, F., et al., 2010. Basic symptoms and ultrahigh risk criteria: symptom development in the initial prodromal state. Schizophr. Bull. 36 (1), 182–191.

Sekine, K., Kubo, K., Nakajima, K., 2014. How does Reelin control neuronal migration and layer formation in the developing mammalian neocortex? Neurosci. Res. 86, 50–58.

Selten, J.-P., Termorshuizen, F., 2017. The serological evidence for maternal influenza as risk factor for psychosis in offspring is insufficient: critical review and meta-analysis. Schizophr. Res. 183, 2–9.

Semple, B.D., et al., 2013. Brain development in rodents and humans: identifying benchmarks of maturation and vulnerability to injury across species. Prog. Neurobiol. 106 (107), 1–16.

Shi, L., et al., 2003. Maternal influenza infection causes marked behavioral and pharmacological changes in the offspring. J. Neurosci. 23 (1), 297–302.

Sofroniew, M.V., 2005. Reactive astrocytes in neural repair and protection. Neuroscientist 11 (5), 400–407.

Sørensen, H.J., et al., 2009. Association between prenatal exposure to bacterial infection and risk of schizophrenia. Schizophr. Bull. 35 (3), 631–637.

Stone, J.M., et al., 2010. Altered relationship between hippocampal glutamate levels and striatal dopamine function in subjects at ultra high risk of psychosis. Biol. Psychiatry 68 (7), 599–602.

Stone, J.M., et al., 2012. Ketamine effects on brain GABA and glutamate levels with 1H-MRS: relationship to ketamine-induced psychopathology. Mol. Psychiatry 17 (7), 664–665.

Sugai, T., et al., 2004. Prefrontal abnormality of schizophrenia revealed by DNA microarray: impact on glial and neurotrophic gene expression. Ann. N. Y. Acad. Sci. 1025, 84–91.

Suvisaari, J., et al., 1999. Association between prenatal exposure to poliovirus infection and adult schizophrenia. Am. J. Psychiatry 156 (7), 1100–1102.

Swerdlow, N.R., Caine, S.B., Geyer, M.A., 1992. Regionally selective effects of intracerebral dopamine infusion on sensorimotor gating of the startle reflex in rats. Psychopharmacology 108 (1–2), 189–195.

Swerdlow, N.R., et al., 1994. Assessing the validity of an animal model of deficient sensorimotor gating in schizophrenic patients. Arch. Gen. Psychiatr. 51 (2), 139–154.

Tandon, R., Fleischhacker, W.W., 2005. Comparative efficacy of antipsychotics in the treatment of schizophrenia: a critical assessment. Schizophr. Res. 79 (2–3), 145–155.

Tkachev, D., et al., 2003. Oligodendrocyte dysfunction in schizophrenia and bipolar disorder. Lancet 362 (9386), 798–805.

Torrey, E.F., Rawlings, R., Waldman, I.N., 1988. Schizophrenic births and viral diseases in two states. Schizophr. Res. 1 (1), 73–77.

Tseng, K.Y., O'Donnell, P., 2007. D2 dopamine receptors recruit a GABA component for their attenuation of excitatory synaptic transmission in the adult rat prefrontal cortex. Synapse 61 (10), 843–850.

Tseng, K.-Y., et al., 2007. Post-pubertal disruption of medial prefrontal cortical dopamine-glutamate interactions in a developmental animal model of schizophrenia. Biol. Psychiatry 62 (7), 730–738.

Tseng, K.Y., Chambers, R.A., Lipska, B.K., 2009. The neonatal ventral hippocampal lesion as a heuristic neurodevelopmental model of schizophrenia. Behav. Brain Res. 204 (2), 295–305.

Uranova, N., et al., 2001. Electron microscopy of oligodendroglia in severe mental illness. Brain Res. Bull. 55 (5), 597–610.

Valli, I., et al., 2011. Altered medial temporal activation related to local glutamate levels in subjects with prodromal signs of psychosis. Biol. Psychiatry 69 (1), 97–99.

Vázquez-Roque, R.A., et al., 2012. Chronic administration of the neurotrophic agent cerebrolysin ameliorates the behavioral and morphological changes induced by neonatal ventral hippocampus lesion in a rat model of schizophrenia. J. Neurosci. Res. 90 (1), 288–306.

Verma, A., Moghaddam, B., 1996. NMDA receptor antagonists impair prefrontal cortex function as assessed via spatial delayed alternation performance in rats: modulation by dopamine. J. Neurosci. 16 (1), 373–379.

Volk, D.W., Lewis, D.A., 2013. Prenatal ontogeny as a susceptibility period for cortical GABA neuron disturbances in schizophrenia. Neuroscience 248, 154–164.

Vollenweider, F.X., et al., 1997. Metabolic hyperfrontality and psychopathology in the ketamine model of psychosis using positron emission tomography (PET) and [18F]fluorodeoxyglucose (FDG). Eur. Neuropsychopharmacol. 7 (1), 9–24.

Vuillermot, S., et al., 2010. A longitudinal examination of the neurodevelopmental impact of prenatal immune activation in mice reveals primary defects in dopaminergic development relevant to schizophrenia. J. Neurosci. 30 (4), 1270–1287.

Wan, R.Q., et al., 1996. Neonatal hippocampal lesions induced hyperresponsiveness to amphetamine: behavioral and in vivo microdialysis studies. Behav. Brain Res. 78 (2), 211–223.

Wang, J., et al., 2016. Reduced γ-aminobutyric acid and Glutamate+Glutamine levels in drug-Naïve patients with first-episode schizophrenia but not in those at ultrahigh risk. Neural Plast. 2016, 3915703.

Waschkies, C.F., et al., 2014. Neuropharmacological and neurobiological relevance of in vivo ^{1}H-MRS of GABA and glutamate for preclinical drug discovery in mental disorders. Neuropsychopharmacology 39 (10), 2331–2339.

Wu, J.C., Buchsbaum, M.S., Bunney, W.E., 1991. Positron emission tomography study of phencyclidine users as a possible drug model of schizophrenia. Yakubutsu seishin kodo (Japn. J Psychopharmacol.) 11 (1), 47–48.

Zhu, S., et al., 2014. Chronic phencyclidine induces inflammatory responses and activates GSK3β in mice. Neurochem. Res. 39 (12), 2385–2393.

Zuckerman, L., et al., 2003. Immune activation during pregnancy in rats leads to a postpubertal emergence of disrupted latent inhibition, dopaminergic hyperfunction, and altered limbic morphology in the offspring: a novel neurodevelopmental model of schizophrenia. Neuropsychopharmacology 28 (10), 1778–1789.

CHAPTER

13

Clinical risk factors for psychosis

Aswin Ratheesh[1,2], *Jessica A. Hartmann*[1,2], *Barnaby Nelson*[1,2]

[1]**Orygen, Parkville, VIC, Australia;** [2]**Centre for Youth Mental Health, The University of Melbourne, Parkville, VIC, Australia**

Introduction

The genesis of disorders such as schizophrenia and related psychotic disorders has been described across multiple levels including those of heritable factors, environmental factors, and symptomatic risk markers. Given the high heritability (47%–79%) of schizophrenia (Hilker et al., 2017; Chou et al., 2017), genetic risk factors are prominent in its onset. Attempting to understand observable or symptomatic mental phenomena (phenotypes) such as psychoses from the perspective of heritable factors (genotypes) has also led to the description of a range of "endophenotypes" (Gottesman and Gould, 2003), considered to be *intermediary* physiological or behavioral markers between phenotype and underlying genotype. These have been described in terms of brain functioning, structure, or chemistry (Allen et al., 2009). However, nonheritable factors are also associated with the onset of psychosis. These environmental factors have been studied from the perspective of either *specific* risk factors (e.g., maternal rubella infections) or *general risk factors* associated with multiple underlying processes that could affect neurodevelopment (e.g., seasonality of birth or migrant status). Thus, clinicians and researchers often grapple with a range of *risk factors* that are studied with diverse methodologies with their inherent complexities. While some of these details have been or will be described in other chapters in this section, this chapter will attempt to describe the clinical risk factors associated with later development of psychosis. For the purposes of this review, clinical risk factors will cover those that are assessed by clinicians in everyday practice or can be assessed in a clinical manner ("bedside assessment").

Symptomatic factors are one of the most common ways of assessing risk for psychosis, given that the manner in which we define psychosis itself is phenomenological. The phenomenological factors (or states) that predate the onset of psychosis have been described as "microphenotypes" that lie on the continuum toward "macrophenotypes" such as schizophrenia (van Os and Linscott, 2012; McGorry et al., 2018). Such a putative progression of phenotypic states is the basis for "staging" models that have been described for psychosis (McGorry et al., 2006; Cosci and Fava, 2013). Heuristic staging models attempt to separate those at various steps in the continuum from being at risk (Stages 0, 1) and early psychosis

Risk Factors for Psychosis
https://doi.org/10.1016/B978-0-12-813201-2.00013-2

(Stage 2) to persistent or recurrent disorder (Stages 3, 4). The at-risk stages are of particular interest to this chapter, as these represent the states before the onset of threshold psychotic disorders. Within these, Stage 0 refers to the asymptomatic at-risk stage, currently best exemplified by familial high-risk (FHR) individuals. Stage 1 refers to a symptomatic at-risk stage and can be divided into an early stage (1a) where the symptoms are nonspecific (e.g., depressive or anxiety symptoms) and a later "ultra high risk" (UHR) stage (1b) where more specific symptoms of psychosis have emerged. Such conceptualizations are discussed further in Chapter 21.

In this staging model, the at-risk stages may manifest clinical microphenotypes that are discernible to clinicians. For example, subthreshold psychotic symptoms may be a microphenotype for later development of frank psychosis (macrophenotype). These subthreshold symptoms may be considered to be clinical risk factors for psychosis. However, from an epidemiological perspective, a risk factor is defined as "a factor that is *causally* related to a change in the risk of a relevant health process, outcome or condition" (Porta, 2014, page 251). It is not clear at this stage if subthreshold psychotic symptoms play a causal role in the onset of threshold-level psychosis and therefore qualify, from the perspective of epidemiological terminology, as "risk factors." Therefore, "risk marker" may be a more accurate term, at least for the purpose of referring to clinical phenotypes, as this term does not assume a causal role for the variable being referred to. Alternatively, these clinical presentations (e.g., subthreshold psychotic symptoms) may be considered to be "risk states or stages" in which additional discrete risk factors (e.g., cannabis abuse) could increase the risk of later disorder in a causal manner. Thus, the clinical microphenotypes associated with increased risk of evolution to later stages of disorder (psychosis in this instance) can be conceptualized as risk markers or risk states where additional "causal risk factors" (genetic, endophenotypic, or environmental) may exert their influence. In subsequent sections, we will describe these phenotypic states and certain risk factors (defined broadly, rather than in the stricter epidemiological sense mentioned above) that have been identified to be associated with later psychosis in prospective studies in clinical samples.

Method

We conducted a literature search using existing key literature reviews, updated with information from recently published articles. The reviews were identified using the following search terms on PubMed in December 2017: high risk, famil*, clinic*, predict*, psychos*, and schizophreni* (* indicating truncated terms). To update these literature reviews, we conducted a search of all new publications in the 2 years preceding December 2017 (updated to December 2018) using the terms clinical, prodrom*, at-risk, psychos*, and schizophreni* in the same database. To limit the description of these studies to clinical risk factors for psychosis, we included only those studies that referred to an assessment that could be conducted by a clinician with or without a clinically implementable assessment instrument. While some environmental risk factors could be assessed by a clinician (e.g., obstetric complications, seasonality of birth, urbanicity, migration, substance use), these will be covered in one or more other chapters (e.g., Chapter 15). In addition, the original studies and metaanalyses that have concluded that these risk factors are associated with psychosis have utilized measures of these risk factors in a manner that is outside the scope of the average clinician (e.g., tracing and linking hospital records). Studies with a control group where outcomes were examined (controlled studies) as well as those which included a single cohort of UHR individuals which did and did not report clinical outcomes (uncontrolled studies) were included. The most common outcome examined in psychosis prediction studies was "transition" to psychosis, which has generally been operationalized as at least a week of daily positive psychotic symptoms. We included studies with outcomes of a range of psychotic disorders, i.e., schizophrenia, schizophreniform disorder, first-episode psychosis, affective psychosis (if specifically reported), and unspecified or other psychoses. In addition, we only included studies which described baseline risk factors in prospective studies given the risk of biases from retrospective studies. General population studies or birth cohort studies with no enrichment have not been included for detailed description as these do not correspond to a risk stage (0 or 1). However, they have meaningfully contributed to the identification of clinical risk markers for psychosis. When prospective data are available from such studies, the identified risk markers are described in the subsequent sections.

Results

In this section, we will first describe the types of cohorts and studies (risk stages) that have reported on the clinical risk factors before summarizing the identified risk markers from these studies.

Cohorts that have identified clinical risk markers

Familial and "prodromal" approaches have contributed substantially to the identification of clinical predictors of later psychosis. FHR approaches were the result of longitudinal investigations among offspring of parents with schizophrenia, given the high heritability associated with this disorder. While these studies led to identification of a broad range of early and late-life risk factors, the relatively low transition rates and the long duration of follow-up necessary were significant drawbacks. Prodromal or clinical high risk (CHR) approaches resulted from several decades of observation that those with schizophrenia and other psychotic disorders experienced attenuated psychotic symptoms, nonspecific symptoms, and psychosocial impairment well before the onset of frank psychosis. The FHR studies will be described briefly before considering the findings from CHR or UHR studies.

Familial high-risk studies

At-risk research in psychosis started with FHR studies. These studies generally included first-degree relatives of affected participants, generally offspring of one or both parents diagnosed with schizophrenia. The prospective longitudinal FHR studies can be grouped into early or *first-generation* and later or *second-generation* studies. The major *first-generation* longitudinal FHR studies were the Copenhagen High Risk Study (Mednick and McNeil, 1968), the Finnish Adoptive Family Study (Tienari et al., 1987), the Helsinki HR Study (Wrede et al., 1980), the Israeli HR Study (Ingraham et al., 1995; Marcus et al., 1987), New York Infant HR Study (Fish et al., 1992), and the New York HR Study (Erlenmeyer-Kimling et al., 1995). In general, these studies suffered from modest sample sizes and less rigorous methods. To address these concerns, several *second-generation* FHR studies such as the Edinburgh HR study (Johnstone et al., 2000) and the Pittsburgh Risk Evaluation Program (Keshavan et al., 2004) were conducted. In summary, the morbid risk for offspring developing schizophrenia (within the duration of follow-up) from these studies was estimated to be approximately 13% if one parent is affected and 46% if both parents are diagnosed with the illness (Faraone et al., 2002).

Ultra high risk, clinical high risk, and basic symptom studies

As outlined in previous chapters, a greater awareness of the psychosis prodrome and its operationalization using the UHR criteria led to cohort studies of help-seeking youth at clinical risk of incipient psychosis (Cornblatt, 2002; Yung et al., 1996). While the UHR approach was spearheaded by Australasian researchers, a conceptually similar approach was popularized by US researchers, termed the Clinical High Risk (CHR) approach. Pragmatically for the purposes of this chapter, the studies that utilized the Comprehensive Assessment of At-Risk Mental States (CAARMS) (Yung et al., 2005) is referred to as UHR studies, and those that utilized the Structured Interview for Prodromal Symptoms (Miller et al., 2003) is referred to as CHR studies. Briefly, these UHR/CHR states for psychosis include the age group at the highest risk for onset of psychosis (15–25 years), presence of positive psychotic symptoms at subthreshold intensity or frequency, and markers of genetic risk, functional impairment, or schizotypal personality disorder. While the FHR studies had a focus on earlier (or distal) identification of premorbid predictors, the UHR approach focused on identifying youth in the (proximal) prodromal period, i.e., those with risk markers of a high positive predictive value (PPV) for developing psychosis in the short term (typically 1–3 years). In addition, such approaches have also focused on the possibility of indicated prevention, i.e., the possibility of delaying, ameliorating, or even preventing the onset of psychosis in symptomatic, help-seeking individuals (McGorry et al., 2009; Fusar-Poli et al., 2013). The basic symptom (BS) approach evolved out of observations from several decades earlier (Huber and Gross, 1989) that subtle, self-experienced changes in thinking, speech, affect, perception, and motor function predated the onset of schizophrenia. This approach aims to identify those at an earlier stage than the UHR criteria. Several follow-up studies of clinical participants with BS identified that they predicted the later development of schizophrenia (Klosterkötter et al., 2001; Ruhrmann et al., 2010).

UHR/CHR and BS studies have generally followed participants for 6–24 months, with several studies performing further follow-ups over the medium (2–3 years) (van der Gaag et al., 2013; Bechdolf et al., 2012) and longer term (up to 15 years) (Nelson et al., 2013). The main studies that have utilized this approach have included participants from Melbourne, Australia, in the PACE cohort (Yung et al., 2003; Nelson et al., 2013), the NIMH funded Cognitive Assessment and Risk Evaluation (CARE) program (Haroun et al., 2006), the Dutch Prediction of Psychosis Study (DUPS) (Velthorst et al., 2009), the Spanish P3 prevention cohort (Lemos-Giraldez et al., 2009), the German prodromal study (Klosterkötter et al., 2001), the European Prediction of Psychosis Study (EPOS) (Ruhrmann et al., 2010), the Outreach and Support in South London (OASIS) cohort (Fusar-Poli et al., 2010), and the North American Longitudinal Studies (NAPLS and NAPLS-2) (Addington et al., 2007, 2012). In UHR samples, the cumulative transition rate to first-episode psychosis was reported to be 36% at 3 years in a meta-analysis of studies published until January 2011 (Fusar-Poli et al., 2012).

Other high-risk/Stage 1 cohorts

The Wisconsin Psychosis Proneness Project in university students (Chapman and Chapman, 1987; Chapman et al., 1994) adopted an approach similar to that of assessing BS. Chapman developed several interviews (Chapman and Chapman, 1980) and scales to assess individuals in the high-risk age group with psychotic-like experiences. Longer-term follow-up (10 years) of this population identified that a significant subproportion (21.6%) developed schizophrenia spectrum disorders (Kwapil, 1998).

Identified predictors in high-risk familial and clinical cohorts

Demographic risk markers

Within the at-risk age range (15–25 years), older age was associated with a greater risk of transition to psychosis in two studies (Haroun et al., 2006; Lemos-Giraldez et al., 2009), while those below the age of 18 were identified to be at a greater risk of transition to non-affective psychosis in another study (Amminger et al., 2006). Another recent study (Cornblatt et al., 2015) of help-seeking youth with CHR criteria identified that those aged 12–14 years had a very low transition rate (3%) over 3 years.

While gender differences for psychosis risk have not been marked in at-risk studies, one study identified that when psychosis does develop, males were noted to have a faster onset with more prolonged symptoms than females (Lemos-Giraldez et al., 2009). Lower educational level was associated with later psychosis in the EPOS cohort (Ruhrmann et al., 2010).

Psychotic-like symptoms, attenuated psychotic symptoms

Thought disorder/disorganization: One of the earliest studies (Parnas et al., 1982) to identify formal thought disorder as a high-risk marker was a FHR study where offspring of parents with psychosis were prospectively followed. Those with higher levels of formal thought disorder had a higher risk of developing psychosis. Thought blocking was associated with the later development of schizophrenia in an early large controlled study (Klosterkötter et al., 2001) of participants with BS. The early findings from the Recovery and Prevention Program (RAP) also indicated that conceptual disorganization was predictive of psychosis onset (Lencz et al., 2003) in CHR individuals. In a similar sample, the higher risk of disorganization symptoms contributing to transition was confirmed in the CARE program, in a smaller uncontrolled study (Haroun et al., 2006). Scores on the disorganization factor of the CAARMS (Yung et al., 2005) were also identified to predict transition in the OASIS UHR cohort (Demjaha et al., 2012). Thought disorder when present at baseline and when persistent at follow-up was highly predictive of transition to psychosis (DeVylder et al., 2014) in a New York CHR sample. In the large North American Prodrome Longitudinal Study-2 (NAPLS-2) (Addington et al., 2015), which recruited 764 CHR participants, disorganized thought was predictive of transition to psychosis in post hoc analyses. In a latent-class reanalysis of the NAPLS cohorts, those with disorganization symptoms were significantly more likely to transition to psychosis than those with perceptual abnormalities (Ryan et al., 2017). In the longest follow-up study of CHR participants to date (Nelson et al., 2013), baseline thought disorganization was predictive of later psychosis. Thought disorder was one of the strongest predictors of transition in two recent large uncontrolled prospective CHR studies (Brucato et al., 2017; Cornblatt et al., 2015). Disorganization also predicted poor functioning at follow-up, which may be a more significant and stable outcome than transition to psychosis (Ziermans et al., 2014).

Unusual thought content, paranoia, and bizarre ideation: Several CHR studies confirmed this finding. In the North American Prodrome Longitudinal Study (NAPLS) (Cannon et al., 2008), unusual thought content, as well as suspiciousness/paranoia, uniquely contributed to a greater level of psychosis transition risk. The EPOS study that utilized both UHR and BS approaches also reported higher transition rates for those with bizarre thinking (e.g., attenuated passivity experiences or delusions of control) and paranoid ideation at baseline (Ruhrmann et al., 2010). Similarly, in the DUPS that included UHR and healthy control participants, those with higher levels of bizarre ideation had a greater risk of transition (Velthorst et al., 2009).

Severity of attenuated psychotic symptoms: In another longer-term controlled study of

UHR participants over 6 years, the severity of attenuated positive experiences was associated with higher risk of transition to psychosis (Ziermans et al., 2014). This was also noted in several uncontrolled studies (Brucato et al., 2017; Haroun et al., 2006; Lemos-Giraldez et al., 2009; Lencz et al., 2003; Yung et al., 2003). A reanalysis of NAPLS predictors using PACE data also indicated that scores on the unusual thought content scale of the CAARMS were highly predictive of later psychosis (Thompson et al., 2011).

Negative symptoms: In FHR studies, anhedonia was predictive of later psychosis among women in the New York HR study (Erlenmeyer-Kimling et al., 1993), but not in the overall sample. In CHR studies, several studies have noted an association with severity of negative symptoms and onset of psychosis. In large prospective UHR/CHR studies, negative symptoms were associated with later psychosis in the PACE (Nelson et al., 2013), OASIS (Demjaha et al., 2012), Spanish P3 Prevention (Lemos-Giraldez et al., 2009), and NAPLS cohorts (Piskulic et al., 2012). With respect to specific negative symptoms, a latent-class analysis combining two UHR cohorts noted that alogia, avolition/apathy, anhedonia, and social isolation predicted the class with the highest risk of transition to psychosis (Valmaggia et al., 2013). Mason et al. (2004) found that blunted affect, anhedonia, and asociality predicted transition to psychosis in an uncontrolled CHR study (Mason et al., 2004). Social withdrawal also predicted transition to psychosis in four other UHR/CHR studies (Brucato et al., 2017; Mason et al., 2004; Piskulic et al., 2012; Velthorst et al., 2009). Consistent with these findings, lower negative symptoms were associated with symptomatic and functional recovery in a study including CHR participants and healthy controls (Schlosser et al., 2012).

Perceptual abnormalities: One early uncontrolled CHR study identified that those with auditory hallucinations had a greater risk of transition to psychosis (Mason et al., 2004). However, this finding has not been replicated in other studies. In addition, within a UHR cohort, perceptual abnormalities were cross-sectionally associated with depression, borderline personality disorder, and history of childhood trauma (O'Connor et al., 2017).

Duration of psychotic symptoms: Short (Yung et al., 2004) and longer-term follow-up (Nelson et al., 2013; Hartmann et al., 2016) of the PACE cohort has suggested that those with a longer duration of untreated symptoms have a greater risk of developing psychosis.

Combination of attenuated psychotic symptoms: A relatively new approach to combining risk factors was utilized in an FHR youth study from Pittsburgh, USA (Tandon et al., 2012). In this study, Magical Ideation and Perceptual Abnormalities scales from Chapman's Psychoses Proneness scales were combined with scores on Positive Symptoms and Disorganization domains from the Scale of Prodromal Symptoms to create a new measure of psychosis proneness. In this study, those who had a higher score on this new measure had a 59% probability (PPV) of developing psychosis. Another multivariate prediction model identified that a combination of disorganized thought, suspiciousness, verbal memory deficits, and decline in social functioning had a PPV of 82% with respect to transition to psychosis (Cornblatt et al., 2015).

Social difficulties (including schizotypy, social withdrawal, and skill difficulties)

Several UHR/CHR and FHR studies identified that social difficulties may indicate a greater risk of future psychosis. The Copenhagen FHR study (Parnas et al., 1982) suggested that "defective emotional contact" indicated those at greater risk, likely indicating social cognition difficulties, parental relational difficulties, social anxiety, or social withdrawal. This study did, however, include those with psychotic schizophrenia spectrum disorders and those with schizotypal personality disorders as an outcome. A similar finding emerged from the Edinburgh High Risk Study (EHRS) where four components of the Structured Interview for Schizotypy predicted a small sample (n = 7) who later developed schizophrenia (Miller et al., 2002a). A longer follow-up where 20 participants developed the same disorder identified social anxiety and withdrawal as being highly predictive (Johnstone et al., 2005). Schizotypal disorder or traits were also predictive of transition to psychosis in the EPOS cohort (Ruhrmann et al., 2010) and in another uncontrolled study (Mason et al., 2004). Social anhedonia as measured by Chapman's Psychosis Proneness scales was also associated with later psychosis in the recent Pittsburgh FHR study (Shah et al., 2012). These social difficulties have been noted from an early age in some studies. Preschool social adjustment problems were associated with the later development of schizophrenia spectrum psychoses in the Helsinki FHR study (Niemi et al., 2005). Later social adjustment difficulties when combined with lower parietal P300 evoked potential amplitudes were highly predictive of later psychosis in the DUPS FHR cohort (Nieman et al., 2014). In the subgroup with the lowest social adjustment and the least P300 amplitude, 74% developed psychosis. Similarly, difficulties in social adjustment in adolescence when combined with suspiciousness at baseline were associated with a high PPV (59%) for later psychosis (Tarbox et al., 2013) in the NAPLS CHR cohort. The finding that social decline was associated with later psychosis was confirmed in the NAPLS-2 cohort (Carrion et al., 2018).

Functional impairment

Poor functioning was identified to be a predictor of later psychosis in three clinical at-risk cohorts from Melbourne, Australia. This finding was notable for both affective (Amminger et al., 2006) and non-affective psychoses (Yung et al., 2004). Unsurprisingly, the long-term follow-up of this UHR cohort (Nelson et al., 2013) also noted that baseline low functioning was associated with the later development of psychotic disorders in this population. Low functioning was also predictive of later psychosis in the EPOS cohort (Ruhrmann et al., 2010), the Spanish P3 prevention project (Lemos-Giraldez et al., 2009), and the Hong Kong high risk cohort (Lam et al., 2006). Those with low or deteriorating functioning, particularly the latter group, were at a greater risk of developing psychosis (Velthorst et al., 2010, 2013) in the DUPS. In the NAPLS CHR cohort, the specific domain of social functioning was predictive of later psychosis (Cannon et al., 2008). Among those who transitioned to psychosis, those who did so in the longer term (between 7 and 30 months, n = 24) had poorer baseline role functioning compared to those who developed psychosis in the short term (0–6 months follow-up) (Cornblatt et al., 2012). However, this difference was no longer significant after accounting for negative symptoms at baseline. The authors posited that those with negative symptoms may take a longer time to convert to psychosis (i.e., have a slower mode of onset). Moreover, both groups demonstrated baseline social and role functioning impairments compared to nonconverters at baseline. More severe measures of poor functioning such as unemployment were also predictive of later psychosis in a UHR sample (Fusar-Poli et al., 2010). Impaired role functioning as indicated by unemployment and lower global functioning were predictive of later psychosis in a pooled UHR sample from across the United Kingdom and Australia (Valmaggia et al., 2013). Conversely, those who functioned closer to healthy controls at baseline had a significantly greater chance of symptomatic remission at follow-up (Schlosser et al., 2012).

Childhood behavioral antecedents

While childhood behavioral difficulties are often related to the social difficulties described previously or environmental factors such as parenting, these may represent markers of later risk in their own right. Children of parents with psychosis were noted to have greater disruptive school behavior in the Copenhagen and Israeli HR studies (Parnas et al., 1982; Carter et al., 2002; Olin et al., 1995; Marcus et al., 1987). Contrary to these studies, the Edinburgh HR study did not find an association between childhood behavioral problems as rated on the Child Behavioral Checklist (CBCL) and later psychosis. However, ratings on "delinquent" and withdrawn behaviors after age 13 were predictive of later psychosis (Miller et al., 2002b). Childhood behavioral problems were associated with the development of psychoses, particularly among the subgroup where substance abuse was not a significant risk factor (Amminger et al., 1999) in the New York HR study. However, in the same study, behavioral problems were noted to be susceptible to being modified by parenting approaches unlike the more neurodevelopmental difficulties related to attention (Cornblatt et al., 1999). Notably, early-life risk factors such as family instability and relational difficulties with parents have been identified as being associated with psychosis in HR offspring (Carter et al., 2002; Cannon et al., 1990; Schiffman et al., 2002; Tienari et al., 1987, 2004).

Family history of psychosis

Aside from the higher risk of psychosis in several FHR studies, family history of psychosis has also been associated with greater transition risk to psychosis in CHR samples (Thompson et al., 2011; Cannon et al., 2008). As the inclusion criteria for UHR studies require a decrease in functioning or chronic poor functioning associated with a family genetic risk, it should be borne in mind that this finding does not relate to familial genetic risk alone. There is also evidence that familial risk (in combination with a change in functioning) is significantly less predictive than experiencing brief self-limiting psychotic symptoms (the BLIPS group; in UHR samples) (Nelson et al., 2011; Matsumoto et al., 2018).

Depressive, anxiety, or related symptoms

Self-reported and interviewer-rated measures of anxiety were higher in participants who developed schizophrenia spectrum diagnoses by age 26 in the Israeli HR Study (Kugelmass et al., 1995). Depressive symptoms were also predictive of later psychosis in two CHR studies from the PACE clinic (Yung et al., 2003, 2004). Sleep disturbances as measured by the SIPS sleep item were also predictive of later psychosis in the EPOS cohort (Ruhrmann et al., 2010).

Cannabis and substance use

History of substance use was independently associated with a greater risk of transition to psychosis in the NAPLS CHR cohort (Cannon et al., 2008). While earlier reports from the PACE clinic (Phillips et al., 2002) suggested that cannabis abuse or dependence was not associated with later development of psychosis, recent longer-term follow-up data have suggested contrary findings. In the later follow-up, severity of cannabis abuse was indeed associated with transition to psychosis, but this was mostly mediated by the occurrence of attenuated psychotic symptoms while using cannabis among those in the cannabis-using subgroup (McHugh et al., 2017). Data from the CARE CHR program in San Diego also suggested that cannabis use (Kristensen and Cadenhead, 2007) or previous history of substance use (Haroun et al., 2006) was associated with the later development of psychosis. The association between cannabis use and later development of psychosis was confirmed in a meta-analysis of five prospective cohort studies (Marconi et al., 2016) and a recent birth cohort study (Mustonen et al., 2018). In the latter study, cannabis use predicted later psychosis even after controlling for prodromal symptoms at baseline.

Trauma, life events, and bullying

In the PACE UHR cohort, a history of childhood sexual trauma was significantly associated with later psychosis in both longer (Thompson et al., 2014) and short-term follow-up (Bechdolf et al., 2010). A meta-analysis of prospective cohort studies (nonenriched community samples) also identified an association between childhood adverse experiences and later psychosis (Varese et al., 2012). The relationship between adult life events and psychosis onset is less established with no prospective studies establishing an association (Beards et al., 2013). Although not conclusively related to psychotic disorders, childhood bullying and victimization have been associated with psychotic symptoms in several non enriched community samples (Cunningham et al., 2016).

Ethnicity and migration

Migration and ethnicity have long been identified as risk factors for psychosis in epidemiological studies and can help clinicians identify subgroups at greater risk. First and second-generation migrants have a higher risk of psychosis compared with nonmigrants, as confirmed by meta-analytic evidence (Bourque et al., 2011; Kinney et al., 2009). Migrants living in areas of higher and lower ethnic density have a greater risk of psychosis (Bosqui et al., 2014) indicating that postmigration social factors alone may not account for the greater risk. Possibly due to related and independent factors, persons from certain ethnicities have a greater risk of psychosis, for example, those from an Afro-Caribbean ethnicity in England (Tortelli et al., 2015). Environmental risk factor for psychosis is discussed further in Chapter 15.

Gestational and early-life risk factors

Several factors associated with gestational or neonatal neurodevelopment have been associated with psychosis in metaanalyses. These include winter birth in the northern hemisphere (Davies et al., 2003), diabetes in pregnancy (Cannon et al., 2002), antepartum hemorrhage (Geddes et al., 1999; Cannon et al., 2002), obstetric complications in general (Cannon et al., 2002), the need for incubator or resuscitation (Geddes et al., 1999), the presence of congenital malformations (Geddes et al., 1999; Cannon et al., 2002), and low birth weight (Geddes et al., 1999; Cannon et al., 2002).

Markers of aberrant neurodevelopment

Several clinical markers associated with neurodevelopmental difficulties have been associated with later psychosis. The presence of olfactory identification abilities (Moberg et al., 2014), neurological soft signs (Hirnstein and Hugdahl, 2014), and non−right handedness (Neelam et al., 2011) have been associated with psychosis in several meta-analyses. Although the presence of minor physical anomalies was associated with psychosis in some studies, pooled evidence was not indicative of greater risk (Xu et al., 2011). Developmental delays across several milestones have also been associated with later psychosis (Filatova et al., 2017).

Discussion

Clinical risk factors or markers have been described in at-risk stages such as those with a family history of psychosis or UHR status for psychosis. Within these risk stages (stages 0 or 1), certain subclasses or markers are prominent in published literature including (i) conceptual disorganization, (ii) unusual thought content or bizarre ideation, (iii) negative symptoms, (iv) childhood or early social and behavioral difficulties, (v) premorbid decline in general and social functioning, (vi) depressive or anxiety symptoms, (vii) bullying, childhood trauma, and (ix) combinations of the above features. Cannabis use, and more specifically cannabis-associated attenuated psychotic symptoms, is associated with a greater risk of developing psychosis (Box 13.1). Migration and early life or gestational factors are significant risk factors for psychosis as are markers of aberrant neurodevelopment.

BOX 13.1

CLINICAL RISK MARKERS FOR FIRST-EPISODE PSYCHOSIS IN FAMILIAL AND CLINICAL RISK POPULATIONS

1. Demographic characteristics: Age group between 15 and 25 years
2. Psychotic symptoms (presence/absence or higher severity)
   a. Thought disorganization
   b. Unusual thought content
   c. Bizarre ideation
   d. Negative symptoms
   e. Longer duration of psychotic symptoms
   f. Combination of psychotic symptoms
3. Developmental characteristics
   a. Social difficulties including schizotypal disorder and premorbid adjustment difficulties
   b. Functional impairment
   c. Childhood behavioral, attentional, and family relational difficulties
   d. Trauma, adverse life events, and bullying
   e. Cannabis abuse/cannabis-associated psychotic symptoms
   f. Markers of aberrant neurodevelopment
4. Family history of psychosis
5. Ethnicity and migration
6. Gestational and early-life risk factors

Clinical implications of the identified predictors

Several symptomatic markers are particularly associated with psychosis transition. Conceptual disorganization and disturbances of thought content are strong markers of transition, while perceptual abnormalities have few controlled studies that support an association with later psychosis. This differential impact on transition risk has important clinical implications for those with single attenuated psychotic symptoms, particularly those in the hallucinatory domain. While these clients should not be dismissed as having "low risk," further research is needed into the nature, onset, and development of these symptoms. Indeed, depressive disorders, borderline personality disorders, childhood trauma, and bullying were associated with perceptual abnormalities in a UHR population (O'Connor et al., 2017). Further delineation of these symptoms in their clinical assessments may be important. Recent approaches have questioned the validity of the descriptive label of hallucinations (Waters and Fernyhough, 2017; Raballo, 2017), highlighting the multiplicity of subjective phenomena captured by this category, and argued that the process of transformation of mental states associated with hallucinations and thus the trajectory of development of mental phenomena is more significant than this broad psychopathological category in itself (Raballo, 2017).

Severity of attenuated positive psychotic symptoms and longer duration of such symptoms were also associated with transition, indicating that a "watch and wait" approach without any interventions may not be appropriate from a clinical perspective. However, it should be borne in mind that the duration of symptoms identified to be predictive in these studies refers to the duration of any symptom leading to clinical referral, not exclusively duration of attenuated psychotic symptoms.

Depressive symptoms were associated with later psychosis in cohorts from one center, with small effect sizes; these symptoms may not indicate a high-level risk. This is consistent with their consideration as markers of Stage 1a (McGorry et al., 2006), where the risk of later psychosis may be lower than those with more specific symptoms (Stage 1b) of psychosis. The greater risk associated with a combination of psychotic symptoms also indicates the need for further enrichment from a clinical perspective. Clinicians may need to consider the addition of several risk factors (e.g., demographic, clinical, and developmental factors) or a risk calculator (Fusar-Poli et al., 2017; Cannon et al., 2016) in their assessment of youth with clinical or familial risk for psychosis.

Among the demographic characteristics associated with greater risk of transition, there is no clear relationship with age or gender. Those aged below 15 years in UHR cohorts may have a lower risk of transition (Cornblatt et al., 2015). Lower educational attainment may be consistent with the premorbid functioning markers of greater risk. Specifically, lower social functioning and behavioral and social difficulties in childhood may indicate a neurodevelopmental vulnerability (Murray et al., 2017). This is exemplified by the combination of low social adjustment and information processing difficulties predicting a greater risk of transition in a UHR sample (Nieman et al., 2014). Similarly, a decline in functioning and poor functioning at intake were some of the most common predictors of later psychosis in UHR samples, with several different measures of functioning and different analytical methods. In addition, the specific role of social functional impairments was highlighted in the NAPLS cohort (Cannon et al., 2008). Constructs closely related to social functioning are those of social anhedonia (Shah et al., 2012), social withdrawal, or anhedonia in general, which were all associated with transition to psychosis in several studies, as previously described. These then overlap with the other negative symptoms such as alogia, avolition/apathy, and affective blunting, which also contribute to social impairments. Thus, several domains of social impairments and negative symptoms which may have conceptual and temporal overlap in their development play a central role in the risk of transition to psychosis.

The functional impairments are also closely aligned with neurocognitive impairments which have also been identified to be associated with later psychosis in enriched and non enriched samples (Ziermans et al., 2014; Khandaker et al., 2011). The primary clinical implication from the findings related to functional markers of psychosis risk is that these may need to be considered to be as relevant as symptomatic markers in predicting symptomatic transitions as well as poor functional outcomes (Cotter et al., 2014). Functional outcomes, as well as non psychotic diagnostic outcomes (Lin et al., 2012), need to be studied further in UHR samples to avoid the problems associated with focusing on transition to threshold psychoses alone (van Os and Guloksuz, 2017; Yung et al., 2010).

In addition to these proximal markers, early-life and static risk factors merit consideration. While a family history of psychosis alone may not be a sufficiently enriched risk state for incipient risk prediction, the presence of several clinical risk factors such as gestational complications, markers of early neurodevelopment, childhood trauma, and migration with or without a family history of psychosis could indicate the need for additional monitoring. Similarly, those with a family history of psychosis among clinically help-seeking at-risk mental state (ARMS) participants may need a closer focus. Adverse life events, in particular sexual trauma, are another set of early-life risk factors that confer greater risk in enriched and non enriched samples. Hence, those who have experienced such severe adverse life events may need targeted care and specific interventions in clinical settings. Those with other static risk factors such as obstetric complications and spring/winter birth may have greater risk, but the very low predictive power associated with these variables makes them less meaningful in a clinical setting.

An important dynamic risk factor for transition may be cannabis use and more specifically an interaction between cannabis use and attenuated psychotic symptoms (McHugh et al., 2017). The clinical implications of this risk factor are particularly important given its modifiable nature. Factors such as childhood trauma may also be a modifiable risk factor that warrants intervention, not just to decrease the risk of psychosis but for various health outcomes that are associated with such events. The impact and potential implications of this factor will be considered in more detail in a later chapter. In addition to considering that a certain risk factor is associated with a greater risk of transition, a more nuanced understanding of the magnitude of risk within the specific population is important.

Issues to consider in interpreting the nature of risk

Risk for psychosis when determined in prospective or retrospective studies is described using several measures of the magnitude of risk. This "effect size" of risk is typically reported as odds ratios, relative risk, or hazard ratios (HR), in relation to the prevalence or incidence of psychosis in a comparison group (also see Chapter 2 for a discussion regarding interpreting risk). This may not be particularly easy to interpret for clinicians in their everyday clinical work. Predictive values (PPV or negative predictive value), on the other hand, describe the likelihood of an event based on known data, which are more easily interpreted. For example, a cluster of psychotic symptoms and risks were associated with HR of 15.59 for later psychosis (hospitalization) in a primary care sample (Livny et al., 2017). However, the PPV of this cluster was only 2% in the same time period in the same sample, indicating the need to understand risk in absolute as well as relative terms. Such Bayesian inference considers the probability of the disorder *before* a test being conducted (or a risk factor examined)—termed the pretest probability (i.e., the "known" data) and the probability of the disorder after, which is referred to as the posttest probability. In a UHR sample from London, the pretest probability for later psychosis could be enhanced from 14% for the overall sample to 55.5% in a subgroup (5% of the overall sample) that included participants with Caribbean ethnicity and those referred from inpatient mental health services (Fusar-Poli et al., 2016). Here, even before a UHR assessment was performed, the predictive value of the positive test was demonstrated to be substantially different between different groups based on their demographic and referral characteristics. This indicates that clinicians have to consider such probabilities in discussing the meaning of a positive UHR assessment. Such pretest probabilities, as well as the cumulative additional risk posed by diagnoses such as At Risk Mental State (ARMS), bipolar disorder, and brief psychotic disorder have been incorporated into the development of two risk calculators currently available for help-seeking people identified as having an ARMS (Fusar-Poli et al., 2017; Cannon et al., 2016). Several UHR studies have also reported on predictive values based on models that have utilized more than one risk factor or marker (Nieman et al., 2014; Clark et al., 2016). However, none of these models with multiple predictors ("multimodal" prediction) have been validated yet in independent samples or with a sufficient sample size to justify dissemination

(Studerus et al., 2017). Please see Chapter 21 for further details on multimodal prediction approaches.

As demonstrated by the London study on pretest risk probabilities, referral characteristics may have a significant role in an individual's risk of transition (Rice et al., 2018). While this study (Fusar-Poli et al., 2016) indicated that those referred from a mental health service as having an at-risk state had a higher rate of transition, another study from the Netherlands suggested that clients referred for this reason had a lower rate of transition compared to those identified from screening strategies (Rietdijk et al., 2012). This may at least partly be due to strong demographic differences between the two groups in these studies, affected by referral biases. The possible impact of referral characteristics and pathways to care is also evident in the impact of year of recruitment on transition rates in the PACE long-term follow-up study (Nelson et al., 2013). The change from a psychosis-focused service to a broader youth mental health service with a triage and screening process at intake may have changed the probability of psychosis risk among those who were assessed as having an ARMS status. The changes in the pattern of attenuated psychotic symptoms among the included cohorts over time (e.g., lower number of attenuated psychotic symptoms in later cohorts) may reflect changes in referral characteristics in the region (Hartmann et al., 2016). One factor that may be related to referral of those with specific attenuated symptoms might be the ease of identification of specific symptoms. For example, young people may be able to report perceptual abnormalities substantially more easily than thought disorder (Parnas and Urfer-Parnas, 2017). The latter set of signs and symptoms are often not reported and require skilled clinical assessment to be elicited and identified, particularly when in attenuated form. Thus, reliance on self-report and screening could lead to greater identification of those with perceptual abnormalities, a set of symptoms which have less often been associated with transition than thought disorder (Ryan et al., 2017). Thus, biases in the client's observation/report and within-referral pathways could lead to differences in pretest probabilities and thereby impact transition rates among UHR samples.

Other important considerations relating to risk factor interpretation relate to the biases associated with longitudinal research. Some such factors include attrition biases, lead-time biases, length-time bias, and ascertainment bias, both in sample selection and outcome determination (Porta, 2014). Lead-time bias refers to the estimation of increased survival rates among screened populations compared to those who present as help-seeking with clinical symptoms. In the case of the UHR stage, the likelihood that those who screen positive earlier (e.g., UHR positive due to milder symptoms) may take longer to transition than those who have UHR symptoms closer to threshold (Fusar-Poli et al., 2013). In research settings where all participants have similar length of observation, those identified as UHR earlier in their course may have a lower transition rate (Nelson et al., 2016). A related concept is the length-time bias, which is the bias related to the difference in groups due to them having different rates of progression of the underlying disorder (Porta, 2014). For example, a sample that includes more participants who have underlying genetic or environmental factors that lead to slower psychosis progression will have a lower transition rate. Thus, different samples may have differing transition rates due to their UHR symptom severity or other factors that delay progression, despite meeting the same notional criterion threshold. Thus, comparing cohorts that have been followed for the same duration would give the false impression of some cohorts (e.g., the more recent UHR cohorts) having lower transition rates. This implies that extending this window of observation by following patients from recent cohorts for a longer period of time may reveal comparable transition rates to earlier cohorts (Nelson et al., 2016). Several studies have attempted to examine these biases in UHR samples. In an examination of different transition rates across multiple cohorts from the PACE clinic recruited at different time points, shorter duration of symptoms was associated with a lower transition rate, more so than shorter follow-up time (Nelson et al., 2016). This suggested that lead-time bias (i.e., period of follow-up) may not be the primary contributor to the observed decline in transition rates, at least in this sample.

Several limitations of the UHR approaches that have been described (van Os and Guloksuz, 2017) may also be rectified if concerns about referral bias, base-rate problems, or enrichment as well as concerns about stability of diagnosis in psychosis transitioned cases are addressed in future studies. In addition, assessment of not only psychopathology as being predictive of psychotic disorders but also other psychiatric disorders and functioning is an important next step in prediction of mental disorders. Prospective examination of a broad, heterogeneous group of participants with a putative pluripotent ARMS may help understand the development of such diverse outcomes (Hartmann et al., 2017; McGorry et al., 2018) (please see Chapter 21 for further discussion). In addition, moving from cross-sectional assessments of mental states to dynamic assessments of symptom or network systems will help develop a nuanced understanding of progressive changes in persons at clinical risk of psychosis (Nelson et al., 2017).

References

Addington, J., Cadenhead, K.S., Cannon, T.D., Cornblatt, B., Mcglashan, T.H., Perkins, D.O., Seidman, L.J., Tsuang, M., Walker, E.F., Woods, S.W., Heinssen, R., North American Prodrome Longitudinal, S., 2007. North American Prodrome Longitudinal Study: a collaborative multisite approach to prodromal schizophrenia research. Schizophr. Bull. 33, 665–672.

Addington, J., Cadenhead, K.S., Cornblatt, B.A., Mathalon, D.H., Mcglashan, T.H., Perkins, D.O., Seidman, L.J., Tsuang, M.T., Walker, E.F., Woods, S.W., Addington, J.A., Cannon, T.D., 2012. North American prodrome longitudinal study (NAPLS 2): overview and recruitment. Schizophr. Res. 142, 77–82.

Addington, J., Liu, L., Buchy, L., Cadenhead, K.S., Cannon, T.D., Cornblatt, B.A., Perkins, D.O., Seidman, L.J., Tsuang, M.T., Walker, E.F., Woods, S.W., Bearden, C.E., Mathalon, D.H., Mcglashan, T.H., 2015. North American prodrome longitudinal study (NAPLS 2): the prodromal symptoms. J. Nerv. Ment. Dis. 203, 328–335.

Allen, A.J., Griss, M.E., Folley, B.S., Hawkins, K.A., Pearlson, G.D., 2009. Endophenotypes in schizophrenia: a selective review. Schizophr. Res. 109, 24–37.

Amminger, G.P., Leicester, S., Yung, A.R., Phillips, L.J., Berger, G.E., Francey, S.M., Yuen, H.P., Mcgorry, P.D., 2006. Early-onset of symptoms predicts conversion to non-affective psychosis in ultra-high risk individuals. Schizophr. Res. 84, 67–76.

Amminger, G.P., Pape, S., Rock, D., Roberts, S.A., Ott, S.L., Squires-Wheeler, E., Kestenbaum, C., Erlenmeyer-Kimling, L., 1999. Relationship between childhood behavioral disturbance and later schizophrenia in the New York High-Risk Project. Am. J. Psychiatry 156, 525–530.

Beards, S., Gayer-Anderson, C., Borges, S., Dewey, M.E., Fisher, H.L., Morgan, C., 2013. Life events and psychosis: a review and meta-analysis. Schizophr. Bull. 39, 740–747.

Bechdolf, A., Thompson, A., Nelson, B., Cotton, S., Simmons, M.B., Amminger, G.P., Leicester, S., Francey, S.M., Mcnab, C., Krstev, H., Sidis, A., Mcgorry, P.D., Yung, A.R., 2010. Experience of trauma and conversion to psychosis in an ultra-high-risk (prodromal) group. Acta Psychiatr. Scand. 121, 377–384.

Bechdolf, A., Wagner, M., Ruhrmann, S., Harrigan, S., Putzfeld, V., Pukrop, R., Brockhaus-Dumke, A., Berning, J., Janssen, B., Decker, P., Bottlender, R., Maurer, K., Moller, H.J., Gaebel, W., Hafner, H., Maier, W., Klosterkotter, J., 2012. Preventing progression to first-episode psychosis in early initial prodromal states. Br. J. Psychiatry 200, 22–29.

Bosqui, T.J., Hoy, K., Shannon, C., 2014. A systematic review and meta-analysis of the ethnic density effect in psychotic disorders. Soc. Psychiatry Psychiatr. Epidemiol. 49, 519–529.

Bourque, F., van der Ven, E., Malla, A., 2011. A meta-analysis of the risk for psychotic disorders among first- and second-generation immigrants. Psychol. Med. 41, 897–910.

Brucato, G., Masucci, M.D., Arndt, L.Y., Ben-David, S., Colibazzi, T., Corcoran, C.M., Crumbley, A.H., Crump, F.M., Gill, K.E., Kimhy, D., Lister, A., Schobel, S.A., Yang, L.H., Lieberman, J.A., Girgis, R.R., 2017. Baseline demographics, clinical features and predictors of conversion among 200 individuals in a longitudinal prospective psychosis-risk cohort. Psychol. Med. 1–13.

Cannon, M., Jones, P.B., Murray, R.M., 2002. Obstetric complications and schizophrenia: historical and meta-analytic review. Am. J. Psychiatry 159, 1080–1092.

Cannon, T.D., Cadenhead, K., Cornblatt, B., Woods, S.W., Addington, J., Walker, E., Seidman, L.J., Perkins, D., Tsuang, M., Mcglashan, T., Heinssen, R., 2008. Prediction of psychosis in youth at high clinical risk: a multisite longitudinal study in North America. Arch. Gen. Psychiatr. 65, 28–37.

Cannon, T.D., Mednick, S.A., Parnas, J., 1990. Antecedents of predominantly negative- and predominantly positive-symptom schizophrenia in a high-risk population. Arch. Gen. Psychiatr. 47, 622–632.

Cannon, T.D., Yu, C., Addington, J., Bearden, C.E., Cadenhead, K.S., Cornblatt, B.A., Heinssen, R., Jeffries, C.D., Mathalon, D.H., Mcglashan, T.H., Perkins, D.O., Seidman, L.J., Tsuang, M.T., Walker, E.F., Woods, S.W., Kattan, M.W., 2016. An individualized risk calculator for research in prodromal psychosis. Am. J. Psychiatry 173, 980–988.

Carrion, R.E., Auther, A.M., Mclaughlin, D., Olsen, R., Addington, J., Bearden, C.E., Cadenhead, K.S., Cannon, T.D., Mathalon, D.H., Mcglashan, T.H., Perkins, D.O., Seidman, L.J., Tsuang, M.T., Walker, E.F., Woods, S.W., Cornblatt, B.A., 2018. The global functioning: social and role scales-further validation in a large sample of adolescents and young adults at clinical high risk for psychosis. Schizophr. Bull. 45 (4), 763–772. https://doi.org/10.1093/schbul/sby126.

Carter, J.W., Schulsinger, F., Parnas, J., Cannon, T., Mednick, S.A., 2002. A multivariate prediction model of schizophrenia. Schizophr. Bull. 28, 649–682.

Chapman, L.J., Chapman, J.P., 1980. Scales for rating psychotic and psychotic-like experiences as continua. Schizophr. Bull. 6, 477–489.

Chapman, L.J., Chapman, J.P., 1987. The search for symptoms predictive of schizophrenia. Schizophr. Bull. 13, 497–503.

Chapman, L.J., Chapman, J.P., Kwapil, T.R., Eckblad, M., Zinser, M.C., 1994. Putatively psychosis-prone subjects 10 years later. J. Abnorm. Psychol. 103, 171–183.

Chou, I.J., Kuo, C.F., Huang, Y.S., Grainge, M.J., Valdes, A.M., See, L.C., Yu, K.H., Luo, S.F., Huang, L.S., Tseng, W.Y., Zhang, W., Doherty, M., 2017. Familial aggregation and heritability of schizophrenia and Co-aggregation of psychiatric illnesses in affected families. Schizophr. Bull. 43, 1070–1078.

Clark, S.R., Baune, B.T., Schubert, K.O., Lavoie, S., Smesny, S., Rice, S.M., Schafer, M.R., Benninger, F., Feucht, M., Klier, C.M., Mcgorry, P.D., Amminger, G.P., 2016. Prediction of transition from ultra-high risk to first-episode psychosis using a probabilistic model combining history, clinical assessment and fatty-acid biomarkers. Transl. Psychiatry 6, e897.

Cornblatt, B., Obuchowski, M., Roberts, S., Pollack, S., Erlenmeyer-Kimling, L., 1999. Cognitive and behavioral precursors of schizophrenia. Dev. Psychopathol. 11, 487–508.

Cornblatt, B.A., 2002. The New York high risk project to the Hillside recognition and prevention (RAP) program. Am. J. Med. Genet. 114, 956–966.

Cornblatt, B.A., Carrion, R.E., Addington, J., Seidman, L., Walker, E.F., Cannon, T.D., Cadenhead, K.S., Mcglashan, T.H., Perkins, D.O., Tsuang, M.T., Woods, S.W., Heinssen, R., Lencz, T., 2012. Risk factors for psychosis: impaired social and role functioning. Schizophr. Bull. 38, 1247–1257.

Cornblatt, B.A., Carrion, R.E., Auther, A., Mclaughlin, D., Olsen, R.H., John, M., Correll, C.U., 2015. Psychosis prevention: a modified clinical high risk perspective from the recognition and prevention (RAP) program. Am. J. Psychiatry 172, 986–994.

Cosci, F., Fava, G.A., 2013. Staging of mental disorders: systematic review. Psychother. Psychosom. 82, 20–34.

Cotter, J., Drake, R.J., Bucci, S., Firth, J., Edge, D., Yung, A.R., 2014. What drives poor functioning in the at-risk mental state? A systematic review. Schizophr. Res. 159, 267–277.

Cunningham, T., Hoy, K., Shannon, C., 2016. Does childhood bullying lead to the development of psychotic symptoms? A meta-analysis and review of prospective studies. Psychosis 8, 48–59.

Davies, G., Welham, J., Chant, D., Torrey, E.F., Mcgrath, J., 2003. A systematic review and meta-analysis of Northern Hemisphere season of birth studies in schizophrenia. Schizophr. Bull. 29, 587–593.

Demjaha, A., Valmaggia, L., Stahl, D., Byrne, M., Mcguire, P., 2012. Disorganization/cognitive and negative symptom dimensions in the at-risk mental state predict subsequent transition to psychosis. Schizophr. Bull. 38, 351–359.

DeVylder, J.E., Muchomba, F.M., Gill, K.E., BEN-David, S., Walder, D.J., Malaspina, D., Corcoran, C.M., 2014. Symptom trajectories and psychosis onset in a clinical high-risk cohort: the relevance of subthreshold thought disorder. Schizophr. Res. 159, 278–283.

Erlenmeyer-Kimling, L., Cornblatt, B.A., Rock, D., Roberts, S., Bell, M., West, A., 1993. The New York High-Risk Project: anhedonia, attentional deviance, and psychopathology. Schizophr. Bull. 19, 141–153.

Erlenmeyer-Kimling, L., Squires-Wheeler, E., Adamo, U.H., Bassett, A.S., Cornblatt, B.A., Kestenbaum, C.J., Rock, D., Roberts, S.A., Gottesman, I., 1995. The New York High-Risk Project. Psychoses and cluster A personality disorders in offspring of schizophrenic parents at 23 years of follow-up. Arch. Gen. Psychiatr. 52, 857–865.

Faraone, S.V., Taylor, L., Tsuang, M.T., 2002. The molecular genetics of schizophrenia: an emerging consensus. Expert Rev. Mol. Med. 4, 1–13.

Filatova, S., Koivumaa-Honkanen, H., Hirvonen, N., Freeman, A., Ivandic, I., Hurtig, T., Khandaker, G.M., Jones, P.B., Moilanen, K., Miettunen, J., 2017. Early motor developmental milestones and schizophrenia: a systematic review and meta-analysis. Schizophr. Res. 188, 13–20.

Fish, B., Marcus, J., Hans, S.L., Auerbach, J.G., Perdue, S., 1992. Infants at risk for schizophrenia: sequelae of a genetic neurointegrative defect. A review and replication analysis of pandysmaturation in the Jerusalem Infant Development Study. Arch. Gen. Psychiatr. 49, 221–235.

Fusar-Poli, P., Bonoldi, I., Yung, A.R., Borgwardt, S., Kempton, M.J., Valmaggia, L., Barale, F., Caverzasi, E., Mcguire, P., 2012. Predicting psychosis: meta-analysis of transition outcomes in individuals at high clinical risk. Arch. Gen. Psychiatr. 69, 220–229.

Fusar-Poli, P., Borgwardt, S., Bechdolf, A., Addington, J., Riecher-Rossler, A., Schultze-Lutter, F., Keshavan, M., Wood, S., Ruhrmann, S., Seidman, L.J., Valmaggia, L., Cannon, T., Velthorst, E., DE Haan, L., Cornblatt, B., Bonoldi, I., Birchwood, M., Mcglashan, T., Carpenter, W., Mcgorry, P., Klosterkotter, J., Mcguire, P., Yung, A., 2013. The psychosis high-risk state: a comprehensive state-of-the-art review. JAMA Psychiatry 70, 107–120.

Fusar-Poli, P., Byrne, M., Valmaggia, L., Day, F., Tabraham, P., Johns, L., Mcguire, P., Team, O., 2010. Social dysfunction predicts two years clinical outcome in people at ultra high risk for psychosis. J. Psychiatr. Res. 44, 294–301.

Fusar-Poli, P., Rutigliano, G., Stahl, D., Davies, C., Bonoldi, I., Reilly, T., Mcguire, P., 2017. Development and validation of a clinically based risk calculator for the transdiagnostic prediction of psychosis. JAMA Psychiatry 74, 493–500.

Fusar-Poli, P., Rutigliano, G., Stahl, D., Schmidt, A., Ramella-Cravaro, V., Hitesh, S., Mcguire, P., 2016. Deconstructing pretest risk enrichment to optimize prediction of psychosis in individuals at clinical high risk. JAMA Psychiatry 73, 1260–1267.

Geddes, J.R., Verdoux, H., Takei, N., Lawrie, S.M., Bovet, P., Eagles, J.M., Heun, R., Mccreadie, R.G., Mcneil, T.F., O'callaghan, E., Stober, G., Willinger, U., Murray, R.M., 1999. Schizophrenia and complications of pregnancy and labor: an individual patient data meta-analysis. Schizophr. Bull. 25, 413–423.

Gottesman, I., Gould, T.D., 2003. The endophenotype concept in psychiatry: etymology and strategic intentions. Am. J. Psychiatry 160, 636–645.

Haroun, N., Dunn, L., Haroun, A., Cadenhead, K.S., 2006. Risk and protection in prodromal schizophrenia: ethical implications for clinical practice and future research. Schizophr. Bull. 32, 166–178.

Hartmann, J.A., Nelson, B., Spooner, R., Paul Amminger, G., Chanen, A., Davey, C.G., Mchugh, M., Ratheesh, A., Treen, D., Yuen, H.P., Mcgorry, P.D., 2017. Broad clinical high-risk mental state (CHARMS): methodology of a cohort study validating criteria for pluripotent risk. Early Interv. Psychiatry 13 (3), 379–386. https://doi.org/10.1111/eip.12483.

Hartmann, J.A., Yuen, H.P., Mcgorry, P.D., Yung, A.R., Lin, A., Wood, S.J., Lavoie, S., Nelson, B., 2016. Declining transition rates to psychotic disorder in "ultra-high risk" clients: investigation of a dilution effect. Schizophr. Res. 170, 130–136.

Hilker, R., Helenius, D., Fagerlund, B., Skytthe, A., Christensen, K., Werge, T.M., Nordentoft, M., Glenthøj, B., 2017. Heritability of schizophrenia and schizophrenia spectrum based on the Nationwide Danish Twin Register. Biol. Psychiatry 83, 492–498.

Hirnstein, M., Hugdahl, K., 2014. Excess of non-right-handedness in schizophrenia: meta-analysis of gender effects and potential biases in handedness assessment. Br. J. Psychiatry 205, 260–267.

Huber, G., Gross, G., 1989. The concept of basic symptoms in schizophrenic and schizoaffective psychoses. Recent. Prog. Med. 80, 646–652.

Ingraham, L.J., Kugelmass, S., Frenkel, E., Nathan, M., Mirsky, A.F., 1995. Twenty-five-year followup of the Israeli High-Risk Study: current and lifetime psychopathology. Schizophr. Bull. 21, 183–192.

Johnstone, E.C., Abukmeil, S.S., Byrne, M., Clafferty, R., Grant, E., Hodges, A., Lawrie, S.M., Owens, D.G., 2000. Edinburgh high risk study—findings after four years: demographic, attainment and psychopathological issues. Schizophr. Res. 46, 1–15.

Johnstone, E.C., Ebmeier, K.P., Miller, P., Owens, D.G., Lawrie, S.M., 2005. Predicting schizophrenia: findings from the edinburgh high-risk study. Br. J. Psychiatry 186, 18–25.

Keshavan, M.S., Diwadkar, V.A., Montrose, D.M., Stanley, J.A., Pettegrew, J.W., 2004. Premorbid characterization in schizophrenia: the Pittsburgh high risk study. World Psychiatry 3, 163–168.

Khandaker, G.M., Barnett, J.H., White, I.R., Jones, P.B., 2011. A quantitative meta-analysis of population-based studies of premorbid intelligence and schizophrenia. Schizophr. Res. 132, 220–227.

Kinney, D.K., Teixeira, P., Hsu, D., Napoleon, S.C., Crowley, D.J., Miller, A., Hyman, W., Huang, E., 2009. Relation of schizophrenia prevalence to latitude, climate, fish consumption, infant mortality, and skin color: a role for prenatal vitamin d deficiency and infections? Schizophr. Bull. 35, 582–595.

Klosterkötter, J., Hellmich, M., Steinmeyer, E.M., Schultze-Lutter, F., 2001. Diagnosing schizophrenia in the initial prodromal phase. Arch. Gen. Psychiatr. 58, 158–164.

Kristensen, K., Cadenhead, K.S., 2007. Cannabis abuse and risk for psychosis in a prodromal sample. Psychiatry Res. 151, 151–154.

Kugelmass, S., Faber, N., Ingraham, L.J., Frenkel, E., Nathan, M., Mirsky, A.F., Ben Shakhar, G., 1995. Reanalysis of SCOR and anxiety measures in the Israeli high-risk study. Schizophr. Bull. 21, 205–217.

Kwapil, T.R., 1998. Social anhedonia as a predictor of the development of schizophrenia-spectrum disorders. J. Abnorm. Psychol. 107, 558–565.

Lam, M.M., Hung, S.F., Chen, E.Y., 2006. Transition to psychosis: 6-month follow-up of a Chinese high-risk group in Hong Kong. Aust. N.Z. J. Psychiatr. 40, 414–420.

Lemos-Giraldez, S., Vallina-Fernandez, O., Fernandez-Iglesias, P., Vallejo-Seco, G., Fonseca-Pedrero, E., Paino-Pineiro, M., Sierra-Baigrie, S., Garcia-Pelayo, P., Pedrejon-Molino, C., Alonso-Bada, S., Gutierrez-Perez, A., Ortega-Ferrandez, J.A., 2009. Symptomatic and functional outcome in youth at ultra-high risk for psychosis: a longitudinal study. Schizophr. Res. 115, 121–129.

Lencz, T., Smith, C.W., Auther, A.M., Correll, C.U., Cornblatt, B.A., 2003. The assessment of "prodromal schizophrenia": unresolved issues and future directions. Schizophr. Bull. 29, 717–728.

Lin, A., Nelson, B., Wood, S., Beavan, A., Mcgorry, P., Yung, A., 2012. Outcomes of non-transitioned cases in a sample at ultra-high risk for psychosis: a medium to long-term follow-up study. Early Interv. Psychiatry 6, 3.

Livny, A., Reichenberg, A., Fruchter, E., Yoffe, R., Goldberg, S., Fenchel, D., Burshtein, S., Bachar, E., Davidson, M., Weiser, M., 2017. A population-based longitudinal study of symptoms and signs before the onset of psychosis. Am. J. Psychiatry 175 (4), 351–358. https://doi.org/10.1176/appi.ajp.2017.16121384.

Marconi, A., DI Forti, M., Lewis, C.M., Murray, R.M., Vassos, E., 2016. Meta-analysis of the association between the level of cannabis use and risk of psychosis. Schizophr. Bull. 42, 1262–1269.

Marcus, J., Hans, S.L., Nagler, S., Auerbach, J.G., Mirsky, A.F., Aubrey, A., 1987. Review of the NIMH Israeli Kibbutz-City study and the Jerusalem infant development study. Schizophr. Bull. 13, 425–438.

Mason, O., Startup, M., Halpin, S., Schall, U., Conrad, A., Carr, V., 2004. Risk factors for transition to first episode psychosis among individuals with 'at-risk mental states'. Schizophr. Res. 71, 227–237.

Matsumoto, K., Katsura, M., Tsujino, N., Nishiyama, S., Nemoto, T., Katagiri, N., Takahashi, T., Higuchi, Y., Ohmuro, N., Matsuoka, H., Suzuki, M., Mizuno, M., 2018. Federated multi-site longitudinal study of at-risk mental state for psychosis in Japan. Schizophr. Res. 204, 343–352. https://doi.org/10.1016/j.schres.2018.09.001.

McGorry, P.D., Hartmann, J.A., Spooner, R., Nelson, B., 2018. Beyond the "at risk mental state" concept: transitioning to transdiagnostic psychiatry. World Psychiatry 17, 133–142.

Mcgorry, P.D., Hickie, I.B., Yung, A.R., Pantelis, C., Jackson, H.J., 2006. Clinical staging of psychiatric disorders: a heuristic framework for choosing earlier, safer and more effective interventions. Aust. N.Z. J. Psychiatr. 40, 616–622.

Mcgorry, P.D., Nelson, B., Amminger, G.P., Bechdolf, A., Francey, S.M., Berger, G., Riecher-Rossler, A., Klosterkotter, J., Ruhrmann, S., Schultze-Lutter, F., Nordentoft, M., Hickie, I., Mcguire, P., Berk, M., Chen, E.Y., Keshavan, M.S., Yung, A.R., 2009. Intervention in individuals at ultra-high risk for psychosis: a review and future directions. J. Clin. Psychiatry 70, 1206–1212.

McHugh, M.J., Mcgorry, P.D., Yung, A.R., Lin, A., Wood, S.J., Hartmann, J.A., Nelson, B., 2017. Cannabis-induced attenuated psychotic symptoms: implications for prognosis in young people at ultra-high risk for psychosis. Psychol. Med. 47, 616–626.

Mednick, S.A., McNeil, T.F., 1968. Current methodology in research on the etiology of schizophrenia: serious difficulties which suggest the use of the high-risk-group method. Psychol. Bull. 70, 681–693.

Miller, P., Byrne, M., Hodges, A., Lawrie, S.M., Owens, D.G., Johnstone, E.C., 2002a. Schizotypal components in people at high risk of developing schizophrenia: early findings from the Edinburgh High-Risk Study. Br. J. Psychiatry 180, 179–184.

Miller, P.M., Byrne, M., Hodges, A., Lawrie, S.M., Johnstone, E.C., 2002b. Childhood behaviour, psychotic symptoms and psychosis onset in young people at high risk of schizophrenia: early findings from the edinburgh high risk study. Psychol. Med. 32, 173–179.

Miller, T.J., Mcglashan, T.H., Rosen, J.L., Cadenhead, K., Cannon, T., Ventura, J., Mcfarlane, W., Perkins, D.O., Pearlson, G.D., Woods, S.W., 2003. Prodromal assessment with the structured interview for prodromal syndromes and the scale of prodromal symptoms: predictive validity, interrater reliability, and training to reliability. Schizophr. Bull. 29, 703–715.

Moberg, P.J., Kamath, V., Marchetto, D.M., Calkins, M.E., Doty, R.L., Hahn, C.G., Borgmann-Winter, K.E., Kohler, C.G., Gur, R.E., Turetsky, B.I., 2014. Meta-analysis of olfactory function in schizophrenia, first-degree family members, and youths at-risk for psychosis. Schizophr. Bull. 40, 50–59.

Murray, R.M., Bhavsar, V., Tripoli, G., Howes, O., 2017. 30 Years on: how the neurodevelopmental hypothesis of schizophrenia morphed into the developmental risk factor model of psychosis. Schizophr. Bull. 43, 1190–1196.

Mustonen, A., Niemela, S., Nordstrom, T., Murray, G.K., Maki, P., Jaaskelainen, E., Miettunen, J., 2018. Adolescent cannabis use, baseline prodromal symptoms and the risk of psychosis. Br. J. Psychiatry 212, 227–233.

Neelam, K., Garg, D., Marshall, M., 2011. A systematic review and meta-analysis of neurological soft signs in relatives of people with schizophrenia. BMC Psychiatry 11, 139.

Nelson, B., Mcgorry, P.D., Wichers, M., Wigman, J.T.W., Hartmann, J.A., 2017. Moving from static to dynamic models of the onset of mental disorder: a review. JAMA Psychiatry 74, 528–534.

Nelson, B., Yuen, H.P., Lin, A., Wood, S.J., Mcgorry, P.D., Hartmann, J.A., Yung, A.R., 2016. Further examination of the reducing transition rate in ultra high risk for psychosis samples: the possible role of earlier intervention. Schizophr. Res. 174, 43–49.

Nelson, B., Yuen, H.P., Wood, S.J., Lin, A., Spiliotacopoulos, D., Bruxner, A., Broussard, C., Simmons, M., Foley, D.L., Brewer, W.J., Francey, S.M., Amminger, G.P., Thompson, A., Mcgorry, P.D., Yung, A.R., 2013. Long-term follow-up of a group at ultra high risk ("prodromal") for psychosis: the PACE 400 study. JAMA Psychiatry 70, 793–802.

Nelson, B., Yuen, K., Yung, A.R., 2011. Ultra high risk (UHR) for psychosis criteria: are there different levels of risk for transition to psychosis? Schizophr. Res. 125, 62–68.

Nieman, D.H., Ruhrmann, S., Dragt, S., Soen, F., Van Tricht, M.J., Koelman, J.H., Bour, L.J., Velthorst, E., Becker, H.E., Weiser, M., Linszen, D.H., De Haan, L., 2014. Psychosis prediction: stratification of risk estimation with information-processing and premorbid functioning variables. Schizophr. Bull. 40, 1482–1490.

Niemi, L.T., Suvisaari, J.M., Haukka, J.K., Lonnqvist, J.K., 2005. Childhood predictors of future psychiatric morbidity in offspring of mothers with psychotic disorder: results from the Helsinki High-Risk Study. Br. J. Psychiatry 186, 108–114.

O'Connor, K., Nelson, B., Cannon, M., Yung, A., Thompson, A., 2017. Perceptual abnormalities in an ultra-high risk for psychosis population relationship to trauma and co-morbid disorder. Early Interv. Psychiatry 13 (2), 231–240. https://doi.org/10.1111/eip.1246.

Olin, S.S., John, R.S., Mednick, S.A., 1995. Assessing the predictive value of teacher reports in a high risk sample for schizophrenia: a ROC analysis. Schizophr. Res. 16, 53–66.

Parnas, J., Schulsinger, F., Schulsinger, H., Mednick, S.A., Teasdale, T.W., 1982. Behavioral precursors of schizophrenia spectrum. A prospective study. Arch. Gen. Psychiatr. 39, 658–664.

Parnas, J., Urfer-Parnas, A., 2017. The ontology and epistemology of symptoms: the case of auditory verbal hallucinations in schizophrenia. In: Kendler, K., Parnas, J. (Eds.), Philosophical Issues in Psychiatry IV: Classification of Psychiatric Illness. Oxford University Press.

Phillips, L.J., Curry, C., Yung, A.R., Yuen, H.P., Adlard, S., Mcgorry, P.D., 2002. Cannabis use is not associated with the development of psychosis in an 'ultra' high-risk group. Aust. N.Z. J. Psychiatr. 36, 800–806.

Piskulic, D., Addington, J., Cadenhead, K.S., Cannon, T.D., Cornblatt, B.A., Heinssen, R., Perkins, D.O., Seidman, L.J., Tsuang, M.T., Walker, E.F., Woods, S.W., Mcglashan, T.H., 2012. Negative symptoms in individuals at clinical high risk of psychosis. Psychiatry Res. 196, 220–224.

Porta, M., 2014. A Dictionary of Epidemiology. Oxford University Press, New York.

Raballo, A., 2017. From perception to thought: a phenomenological approach to hallucinatory experience. Schizophr. Bull. 43, 18–20.

Rice, S., Polari, A., Thompson, A., Hartmann, J.A., Mcgorry, P.D., Nelson, B., 2018. Does reason for referral to an ultra-high risk clinic predict transition to psychosis? Early Interv. Psychiatry 13 (2), 318–321. https://doi.org/10.1111/eip.12679.

Rietdijk, J., Klaassen, R., Ising, H., Dragt, S., Nieman, D.H., Van De Kamp, J., Cuijpers, P., Linszen, D., Van Der Gaag, M., 2012. Detection of people at risk of developing a first psychosis: comparison of two recruitment strategies. Acta Psychiatr. Scand. 126, 21–30.

Ruhrmann, S., Schultze-Lutter, F., Salokangas, R.K., Heinimaa, M., Linszen, D., Dingemans, P., Birchwood, M., Patterson, P., Juckel, G., Heinz, A., Morrison, A., Lewis, S., VON Reventlow, H.G., Klosterkotter, J., 2010. Prediction of psychosis in adolescents and young adults at high risk: results from the prospective European prediction of psychosis study. Arch. Gen. Psychiatr. 67, 241–251.

Ryan, A.T., Addington, J., Bearden, C.E., Cadenhead, K.S., Cornblatt, B.A., Mathalon, D.H., Mcglashan, T.H., Perkins, D.O., Seidman, L.J., Tsuang, M.T., Woods, S.W., Cannon, T.D., Walker, E.F., 2017. Latent class cluster analysis of symptom ratings identifies distinct subgroups within the clinical high risk for psychosis syndrome. Schizophr. Res. 197, 522–530. https://10.1016/j.schres.2017.12.001.

Schiffman, J., Labrie, J., Carter, J., Cannon, T., Schulsinger, F., Parnas, J., Mednick, S., 2002. Perception of parent-child relationships in high-risk families, and adult schizophrenia outcome of offspring. J. Psychiatr. Res. 36, 41–47.

Schlosser, D.A., Jacobson, S., Chen, Q., Sugar, C.A., Niendam, T.A., Li, G., Bearden, C.E., Cannon, T.D., 2012. Recovery from an at-risk state: clinical and functional outcomes of putatively prodromal youth who do not develop psychosis. Schizophr. Bull. 38, 1225–1233.

Shah, J., Eack, S.M., Montrose, D.M., Tandon, N., Miewald, J.M., Prasad, K.M., Keshavan, M.S., 2012. Multivariate prediction of emerging psychosis in adolescents at high risk for schizophrenia. Schizophr. Res. 141, 189–196.

Studerus, E., Ramyead, A., Riecher-Rössler, A., 2017. Prediction of transition to psychosis in patients with a clinical high risk for psychosis: a systematic review of methodology and reporting. Psychol. Med. 47, 1163–1178.

Tandon, N., Montrose, D., Shah, J., Rajarethinam, R.P., Diwadkar, V.A., Keshavan, M.S., 2012. Early prodromal symptoms can predict future psychosis in familial high-risk youth. J. Psychiatr. Res. 46, 105–110.

Tarbox, S.I., Addington, J., Cadenhead, K.S., Cannon, T.D., Cornblatt, B.A., Perkins, D.O., Seidman, L.J., Tsuang, M.T., Walker, E.F., Heinssen, R., Mcglashan, T.H., Woods, S.W., 2013. Premorbid functional development and conversion to psychosis in clinical high-risk youths. Dev. Psychopathol. 25, 1171–1186.

Thompson, A., Nelson, B., Yung, A., 2011. Predictive validity of clinical variables in the "at risk" for psychosis population: international comparison with results from the North American Prodrome Longitudinal Study. Schizophr. Res. 126, 51–57.

Thompson, A.D., Nelson, B., Yuen, H.P., Lin, A., Amminger, G.P., Mcgorry, P.D., Wood, S.J., Yung, A.R., 2014. Sexual trauma increases the risk of developing psychosis in an ultra high-risk "prodromal" population. Schizophr. Bull. 40, 697–706.

Tienari, P., Sorri, A., Lahti, I., Naarala, M., Wahlberg, K.E., Moring, J., Pohjola, J., Wynne, L.C., 1987. Genetic and psychosocial factors in schizophrenia: the Finnish adoptive family study. Schizophr. Bull. 13, 477–484.

Tienari, P., Wynne, L.C., Sorri, A., Lahti, I., Laksy, K., Moring, J., Naarala, M., Nieminen, P., Wahlberg, K.E., 2004. Genotype-environment interaction in schizophrenia-spectrum disorder. Long-term follow-up study of Finnish adoptees. Br. J. Psychiatry 184, 216–222.

Tortelli, A., Errazuriz, A., Croudace, T., Morgan, C., Murray, R.M., Jones, P.B., Szoke, A., Kirkbride, J.B., 2015. Schizophrenia and other psychotic disorders in Caribbean-born migrants and their descendants in England: systematic review and meta-analysis of incidence rates, 1950–2013. Soc. Psychiatry Psychiatr. Epidemiol. 50, 1039–1055.

Valmaggia, L.R., Stahl, D., Yung, A.R., Nelson, B., Fusar-Poli, P., Mcgorry, P.D., Mcguire, P.K., 2013. Negative psychotic symptoms and impaired role functioning predict transition outcomes in the at-risk mental state: a latent class cluster analysis study. Psychol. Med. 43, 2311–2325.

van der Gaag, M., Smit, F., Bechdolf, A., French, P., Linszen, D.H., Yung, A.R., Mcgorry, P., Cuijpers, P., 2013. Preventing a first episode of psychosis: meta-analysis of randomized controlled prevention trials of 12 month and longer-term follow-ups. Schizophr. Res. 149, 56–62.

van Os, J., Guloksuz, S., 2017. A critique of the "ultra-high risk" and "transition" paradigm. World Psychiatry 16, 200–206.

Van Os, J., Linscott, R.J., 2012. Introduction: the extended psychosis phenotype—relationship with schizophrenia and with ultrahigh risk status for psychosis. Schizophr. Bull. 38, 227–230.

Varese, F., Smeets, F., Drukker, M., Lieverse, R., Lataster, T., Viechtbauer, W., Read, J., Van Os, J., Bentall, R.P., 2012. Childhood adversities increase the risk of psychosis: a meta-analysis of patient-control, prospective- and cross-sectional cohort studies. Schizophr. Bull. 38, 661–671.

Velthorst, E., Nelson, B., Wiltink, S., De Haan, L., Wood, S.J., Lin, A., Yung, A.R., 2013. Transition to first episode psychosis in ultra high risk populations: does baseline functioning hold the key? Schizophr. Res. 143, 132–137.

Velthorst, E., Nieman, D.H., Becker, H.E., Van De Fliert, R., Dingemans, P.M., Klaassen, R., De Haan, L., Van Amelsvoort, T., Linszen, D.H., 2009. Baseline differences in clinical symptomatology between ultra high risk subjects with and without a transition to psychosis. Schizophr. Res. 109, 60–65.

Velthorst, E., Nieman, D.H., Linszen, D., Becker, H., De Haan, L., Dingemans, P.M., Birchwood, M., Patterson, P., Salokangas, R.K., Heinimaa, M., Heinz, A., Juckel, G., VON Reventlow, H.G., French, P., Stevens, H., Schultze-Lutter, F., Klosterkotter, J., Ruhrmann, S., 2010. Disability in people clinically at high risk of psychosis. Br. J. Psychiatry 197, 278–284.

Waters, F., Fernyhough, C., 2017. Hallucinations: a systematic review of points of similarity and difference across diagnostic classes. Schizophr. Bull. 43, 32–43.

Wrede, G., Mednick, S.A., Huttunen, M.O., Nilsson, C.G., 1980. Pregnancy and delivery complications in the births of an unselected series of Finnish children with schizophrenic mothers. Acta Psychiatr. Scand. 62, 369–381.

Xu, T., Chan, R.C., Compton, M.T., 2011. Minor physical anomalies in patients with schizophrenia, unaffected first-degree relatives, and healthy controls: a meta-analysis. PLoS One 6, e24129.

Yung, A.R., Mcgorry, P.D., Mcfarlane, C.A., Jackson, H.J., Patton, G.C., Rakkar, A., 1996. Monitoring and care of young people at incipient risk of psychosis. Schizophr. Bull. 22, 283–303.

Yung, A.R., Nelson, B., Thompson, A., Wood, S.J., 2010. The psychosis threshold in Ultra High Risk (prodromal) research: is it valid? Schizophr. Res. 120, 1–6.

Yung, A.R., Phillips, L.J., Yuen, H.P., Francey, S.M., Mcfarlane, C.A., Hallgren, M., Mcgorry, P.D., 2003. Psychosis prediction: 12-month follow up of a high-risk ("prodromal") group. Schizophr. Res. 60, 21–32.

Yung, A.R., Phillips, L.J., Yuen, H.P., Mcgorry, P.D., 2004. Risk factors for psychosis in an ultra high-risk group: psychopathology and clinical features. Schizophr. Res. 67, 131–142.

Yung, A.R., Yuen, H.P., Mcgorry, P.D., Phillips, L.J., Kelly, D., Dell'olio, M., Francey, S.M., Cosgrave, E.M., Killackey, E., Stanford, C., Godfrey, K., Buckby, J., 2005. Mapping the onset of psychosis: the comprehensive assessment of at-risk mental states. Aust. N.Z. J. Psychiatr. 39, 964–971.

Ziermans, T., De Wit, S., Schothorst, P., Sprong, M., Van Engeland, H., Kahn, R., Durston, S., 2014. Neurocognitive and clinical predictors of long-term outcome in adolescents at ultra-high risk for psychosis: a 6-year follow-up. PLoS One 9, e93994.

CHAPTER

14

Cognitive risk factors for psychosis

Kelly Allott[1,2], *Ashleigh Lin*[3]

[1]Orygen, Parkville, VIC, Australia; [2]Centre for Youth Mental Health, The University of Melbourne, Parkville, VIC, Australia; [3]Telethon Kids Institute, The University of Western Australia, Perth, WA, Australia

Introduction

Cognitive impairment is considered a core clinical feature of established psychotic disorder (Kahn and Keefe, 2013), with moderate to large widespread impairments evident at the onset of psychosis (Mesholam-Gately et al., 2009) that do not appear to deteriorate significantly thereafter (Szoke et al., 2008; Bozikas and Andreou, 2011; Bora and Murray, 2014). Accordingly, cognition has received considerable investigation as a potentially important early risk marker for the development of psychosis, with a large body of evidence suggesting that cognitive difficulties are present early in people who later develop psychotic disorder. Cognitive difficulties are also evident in relatives of people with psychosis (Snitz et al., 2006), with research investigating whether cognition is an endophenotype a large area of inquiry. Apart from being clearly associated with the illness and present during remission (state-independent), the criteria for an endophenotype include that it is heritable, more frequently present in unaffected relatives than in the general population, while still being more frequently present in the affected than unaffected family members, and co-segregates with the illness within families (Gottesman and Gould, 2003). The uniqueness or strength of specificity of cognition as a risk marker for the development of psychosis also needs to be considered if it is to be a useful prognostic marker of risk for the illness. An advantage of focusing research on cognition is that cognitive function can be relatively easily assessed within the clinical setting, so translation of research findings can occur readily. In addition, cognition is potentially modifiable, as shown by the effectiveness of interventions such as cognitive remediation therapy in established psychotic disorder (Wykes et al., 2011).

In this chapter, various lines of evidence for cognition as a risk factor for psychosis will be examined. Research on cognitive functioning in longitudinal cohort studies, individuals who have not experienced symptoms of psychosis, but may be at increased risk due to familial liability or developmental conditions, and individuals deemed at clinical high risk (CHR) because of attenuated symptoms will be reviewed. We will also review studies that have examined cognition as a specific predictor

Risk Factors for Psychosis
https://doi.org/10.1016/B978-0-12-813201-2.00014-4

of transition to psychosis from the CHR state, as well as research into the prediction of other clinically important outcomes. Finally, the trajectory of cognition in individuals identified as being at risk will be examined.

Definition of cognition and cognitive deficit

Cognitive functions are the interrelated unconscious and conscious mental activities of a person that are not directly observed, but inferred from behavior (Lezak et al., 2004; Millan et al., 2012). They encompass receptive functions, memory and learning, thinking skills, and expressive functions. For the current chapter, the term "cognition" includes both neurocognition and social cognition, which have been examined via objective measures rather than self-report (with the exception of some attributional measures). Metacognition (awareness or knowledge of one's thinking or cognition) can also be included under the rubric of cognition, but it is beyond the scope of this chapter. Cognition research in psychosis (including in at-risk individuals) has been dominated by a focus on neurocognition, which can be defined as the mental operations underlying goal-directed behavior, including attention and vigilance, working memory, processing speed, learning and memory, language and visual functions, executive functions, and intelligence (Lezak et al., 2004). Social cognition, a related but independent construct to neurocognition, has received increasing investigation in the field. Social cognition involves the perception, processing, and interpretation of people's intentions, feelings, and thoughts (Fiske and Taylor, 2013; Adolphs, 2009; Pinkham et al., 2014). The domains that have received the most investigation in psychosis include emotion processing, theory of mind or mentalizing, social perception/knowledge, and attributional style. Table 14.1 lists and defines these key cognitive domains.

Broadly speaking, cognitive dysfunction may be conceptualized as deficits or biases. Deficits are evident when an individual or group performance is significantly below a demographically matched normative sample (e.g., greater than 1−1.5 standard deviations below the mean of the general population), or at an individual level, performance in a particular domain is below what is expected based on the individual's estimated premorbid intellectual functioning (e.g., there is a significant discrepancy between cognitive abilities). Biases are not associated with accuracy in performance, but are the tendency to process information in a way that favors certain types of emotional valence or meaning, for example, the tendency to attend to negative information in depression or threatening information in psychotic or anxiety disorders (Mathews and MacLeod, 2005). Given the vast literature in these two areas and space limitations, the current chapter will primarily focus on cognitive deficits or weaknesses, with some attention to attributional style as this is commonly conceptualized under social cognition. It is important to note that poorer cognition may be a risk marker for psychosis without meeting the definition of "deficit" described above.

Individuals without psychotic symptoms

Three lines of investigation have provided evidence that poorer cognition may be an early risk marker for psychosis in individuals who are asymptomatic for psychosis: (1) individuals with neurodevelopmental disorders, (2) large longitudinal birth or conscript cohort studies, and (3) individuals with a family history of psychosis.

Neurodevelopmental disorders

People with intellectual disability (i.e., IQ $<$ 70) are at 3−5 times greater risk of

TABLE 14.1 Definitions of cognitive domains.

Cognitive domains	Definition
Neurocognition	
Intelligence/IQ	"General" intellectual ability subserved by a multiplicity of cognitive functions and examined by a range of tests that provide a composite IQ (intelligence quotient) score. Mean IQ in the general population is 100 with a standard deviation of 15. Premorbid IQ can be estimated based on a graded word reading task.
Language abilities	Ability to perceive and understand/comprehend the meaning of words and sentences/prose and use or express oneself with language. Also includes acquired general knowledge.
Visuospatial/nonverbal abilities	Ability to perceive, integrate, and analyze visual information. Includes spatial and constructional abilities. For example, copying a picture, putting together a puzzle, finding what is missing from a picture, or completing a visual puzzle (e.g., Rubik's cube).
Attention	Several capacities relating to the reception and processing of information (in lay terms referred to as "*concentration*"), including the following: - Attention *span* refers to how much information can be held in mind at once - *Focused* or *selective* attention is the capacity to focus on important information while ignoring or suppressing awareness of competing information (nonrelevant information/distractions) - *Sustained* attention or *vigilance* is the capacity to maintain attention or concentration over a period of time - *Divided* attention is the ability to respond to more than one stimuli or task at a time - Attention *shifting* is the capacity to alternate focus
Working memory	Capacity to mentally hold and manipulate verbal or visual information (e.g., working out how much change you are owed).
Processing speed	Speed with which one can "take in" and process information. May also involve the speed with which one can complete a motor task, referred to as *psychomotor speed*.
Verbal learning and memory	Ability to learn and remember information that is presented verbally or in words (e.g., a conversation, a telephone number). Includes immediate and delayed recall and recognition.
Visual learning and memory	Ability to learn and remember information that is presented visual or in pictures/figures (e.g., route on a map, the scene of a crime). Includes immediate and delayed recall and recognition.
Verbal fluency	Ease, speed, and quantity of verbal production.
Executive functions	Executive functions involve the capacity for engagement in independent, purposive, and self-serving behavior. They relate to *how* an individual goes about performing goal-directed behaviors and *how* they engage with *novel* tasks and situations. Executive functions is the rubric for a number of "higher-level" cognitive abilities including *flexibility/set-shifting*, *inhibition*, *planning*, *organization*, *reasoning*, and *problem solving*.
Social cognition	
Theory of mind	Involves the ability to infer the mental states of others, including their intentions, dispositions, and beliefs. Also referred to as *mentalizing* or *mental state attribution*.
Emotion processing	Ability to perceive, interpret, and use emotions—conveyed via face, body, and speech. Includes the ability to discriminate between and recognize specific emotions, as well as understand and manage emotions.

Continued

TABLE 14.1 Definitions of cognitive domains.—cont'd

Social perception and knowledge	Social perception is the ability to perceive and interpret social cues from social contextual information and communicative (verbal, nonverbal) gestures. Social knowledge is the awareness of the roles, rules, and goals that characterize social situations and guide social interactions.
Attributional style	How an individual tends to explain the causes for positive and negative events or interactions. Includes external personal attributions (i.e., event is caused by others), external situational attributions (i.e., event is caused by situational factors), and internal attributions (i.e., event is caused by oneself). As opposed to other domains, this domain is a measure of *bias* or *tendency*, rather than a deficit (or not) in performance.

Based on Lezak, M.D., Howieson, D.B., Loring, D.W., 2004. Neuropsychological Assessment. Oxford University Press, New York; Pinkham, A.E., Penn, D.L., Green, M.F., Buck, B., Healey, K., Harvey, P.D., 2014. The social cognition psychometric evaluation study: results of the expert survey and RAND panel. Schizophr. Bull. 40, 813–823; Snyder, H.R., Miyake, A., Hankin, B.L., 2015. Advancing understanding of executive function impairments and psychopathology: bridging the gap between clinical and cognitive approaches. Front. Psychol. 6, 24.

experiencing a psychotic disorder than the general population (Morgan et al., 2008) and those with an IQ within the borderline range (~70–80) have a mildly increased risk of psychosis (Morgan et al., 2008). One population-based study showed that compared to individuals with an average IQ, borderline IQ was associated with a 2.5 increase in the odds of being diagnosed with a schizophrenia spectrum disorder (Gigi et al., 2014). Other population-based longitudinal research shows a significant association between neurodevelopmental disorders diagnosed in childhood (such as autism spectrum disorder, dyslexia, and dyspraxia) and a nearly twofold risk of psychotic experiences in adolescence, an association that was partially mediated by IQ in late childhood (Khandaker et al., 2014). Together, these findings suggest that early aberrant neurodevelopmental processes play a role in psychosis (Owen, 2012). Owen and O'Donovan (2017) further suggest that neurodevelopmental disorders and schizophrenia lie along a continuum, upon which a gradient of neurodevelopmental impairment exists, with the gradient being characterized by age of onset of illness, degree of cognitive impairment, and persistence of functional impairment. Support for this model comes from evidence of shared genetic risk across these disorders (Owen and O'Donovan, 2017).

Risk for schizophrenia is particularly interesting in the context of 22q11 deletion syndrome (also known as velocardiofacial or DiGeorge syndrome; see Chapter 8 for further details). Almost one-third of individuals with the syndrome develop schizophrenia or affective psychosis (Bassett et al., 2000; Green et al., 2009), making it one of the largest known early risk factors for psychotic illness. Longitudinal evidence indicates that a decline in verbal IQ (evident as early as age 11) is associated with the later onset of psychotic illness in this population (Vorstman et al., 2015; Gothelf et al., 2013).

Birth or conscript cohort studies

The second line of evidence comes from large birth or conscript cohort studies that have examined the cognitive functioning of individuals before they experience first onset of psychotic disorder (Davidson et al., 1999; Jones et al., 1994; Kremen et al., 1998; Osler et al., 2007; Rabinowitz et al., 2000; Cannon et al., 2002; Reichenberg et al., 2002). Again, most studies have examined IQ, with a lower premorbid IQ being associated with a subsequent diagnosis of psychotic disorder (David et al., 2008; Zammit et al., 2004; Reichenberg et al., 2002; Koenen et al., 2009; Batty et al., 2005). It is important to note that lower IQ is also associated with an increased risk of other psychiatric disorders, including major depression, personality disorders, substance abuse, and anxiety disorders, although the size of the effect is generally smaller than for psychotic disorders (David et al., 2008; Zammit et al., 2004; Koenen et al., 2009). An exception to these findings appears to be in the case of bipolar disorder, where findings suggest no association between premorbid IQ and bipolar disorder (both psychotic and nonpsychotic) (Zammit et al., 2004; Reichenberg et al., 2002) or *higher* IQ being a risk for the onset of nonpsychotic bipolar disorder (Koenen et al., 2009; Gale et al., 2013).

Premorbid IQ (assessed as early as age 3) of approximately 0.5 and 0.33 standard deviations below healthy controls has been found in studies of people who later develop schizophrenia (Khandaker et al., 2011; Woodberry et al., 2008; Reichenberg et al., 2010) and schizophreniform disorder (Cannon et al., 2002), respectively. In their meta-analysis of longitudinal population-based studies, Khandaker et al. (2011) identified a linear relationship between IQ deficit in the premorbid period and future risk of schizophrenia, whereby risk increased by 3.7% for every point decrease in IQ (also see Kendler et al., 2015). Furthermore, greater IQ decrement was associated with an earlier age of psychosis onset (Khandaker et al., 2011), in line with the model proposed by Owen and O'Donovan (2017). Importantly, lower IQ does not appear to be a by-product of insidious prodromal symptoms, as studies have confirmed the association between lower IQ assessed early and the onset of schizophrenia many years later (Kendler et al., 2015; Tiihonen et al., 2005; Reichenberg et al., 2010). There is also evidence showing that a decline (or lack of development) in IQ or specifically verbal ability during adolescence is an independent risk factor for later development of schizophrenia and other psychotic disorders, but not bipolar, depression, or anxiety disorder (Reichenberg et al., 2005; MacCabe et al., 2013). Similarly, evidence from a large cohort of men in Denmark demonstrated that a decline in general neurocognitive performance between ages 12 and 18 was associated with the subsequent development of schizophrenia spectrum disorder, but not bipolar disorder (Osler et al., 2007). In the Dunedin population cohort study, Meier et al. (2014) prospectively examined cognition at 7, 9, 11, 13, and 38 years of age and compared individuals who had been diagnosed with schizophrenia ($n = 31$), persistent depression ($n = 185$), low childhood IQ (80–89; $n = 120$), and healthy controls ($n = 518$). A significant decline in IQ was only observed in the schizophrenia group, with the largest deterioration occurring between age 13 and 38 years of age. More recently, Mollon et al. (2018) mapped IQ change from 18 months, 4, 8, 15, and 20 years of age as part of the Avon Longitudinal Study of Parents and Children (ALSPAC) study. They compared individuals who met criteria for psychotic disorder at 18 years of age ($n = 19$), psychosis with depression ($n = 13$), psychotic experiences rated from the psychosis-like symptom interview ($n = 76$), and depression ($n = 35$) with healthy controls ($n = 114$). The group with psychotic disorder showed increasing significant deficits in Full-Scale and Performance IQ from 18 months to 20 years of age, whereas Verbal IQ declined early in life remaining statically impaired from age 8–20 years. Based on raw scores on identical tests, increasing lag (but not decline) in processing speed, working memory, and attention were observed from age 8–20 years in the psychotic disorder group. The psychotic experience group showed minimal impairment in IQ or specific cognitive functions.

Several studies have also examined the timing of early developmental milestones as risk markers for the later development of schizophrenia. The findings of a recent metaanalysis show a similar pattern to those of IQ, with evidence of delayed development of motor milestones, specifically walking unsupported ($g = 0.46$), standing unsupported ($g = 0.28$), and sitting unsupported ($g = 0.18$) being significantly associated with adult-onset schizophrenia (Filatova et al., 2017). As with lower IQ, delayed motor milestones are also associated with the onset of other mental disorders, albeit to a lesser degree than schizophrenia (Sorensen et al., 2010).

Together, findings of these studies suggest a static continuity of developmental risk (with likely shared genetic etiology) in association with early developmental milestones and premorbid IQ, possibly followed by further relative or actual neurocognitive deterioration in specific domains in people who are diagnosed with full-threshold psychotic disorder. However, the precise timing and period of neurocognitive decline still remains unclear.

Familial high risk

The third line of evidence comes from studies of individuals considered at familial high risk of psychotic illness. It is important to highlight that, as with the studies showing an association with lower IQ and psychiatric disorders generally, milder cognitive difficulties in unaffected first-degree relatives of people with various psychiatric disorders are observed (Weiser et al., 2008). Evidence that cognitive impairments are greater in multiplex (more than one affected sibling) versus simplex (one affected sibling) sibships supports the premise that there is some degree of heritability in cognitive impairment that cuts across psychiatric disorders (Weiser et al., 2008; Faraone et al., 2000).

Relative to other psychiatric disorders, the evidence appears strongest, and cognitive difficulties are largest, in relatives of people with schizophrenia (Sitskoorn et al., 2004; Snitz et al., 2006; Weiser et al., 2008; Bora et al., 2014), including unaffected co-twins from twin pairs discordant for schizophrenia (Cannon et al., 2000). Offspring of parents with schizophrenia also perform more poorly than healthy controls from early in childhood and across a range of neurocognitive domains (Byrne et al., 2003; Seidman et al., 2006). In addition, individuals at familial risk who later develop schizophrenia have been differentiated from those who do not by performances on tasks of attention (Erlenmeyer-Kimling et al., 2000; Mirsky et al., 1995) and verbal memory (Erlenmeyer-Kimling et al., 2000; Johnstone et al., 2005).

According to four meta-analyses of individuals at familial high risk for psychosis (but asymptomatic at the time of assessment), multiple neurocognitive domains are implicated, with small to medium effect sizes evident in attention (including vigilance) (ES = 0.28–0.66), processing speed (ES = 0.35–0.62), executive functions (ES = 0.24–0.51), verbal fluency (ES = 0.39), verbal learning/memory (ES = 0.45–0.54), and visual learning/memory (ES = 0.51) (Sitskoorn et al., 2004; Snitz et al., 2006; Dickinson et al., 2007; Bora et al., 2014). Social cognition has received less attention, but two meta-analyses have reported moderate impairments in first-degree relatives of people with schizophrenia (Lavoie et al., 2013; Bora and Pantelis, 2013). Specifically, poorer functioning in the domains of theory of mind or mentalizing (ES = 0.37–0.48), emotion recognition/processing (ES = 0.41), and social perception (ES = 0.42) have been found (Lavoie et al., 2013; Bora and Pantelis, 2013). These areas of poorer cognitive performance correspond to the pattern of impairment seen in individuals with schizophrenia. Impairments are, however, less severe, and the performance of those with familial risk for psychosis is generally intermediate to that of individuals with the illness and healthy controls. This suggests that while impairments in cognitive function are likely to partly represent familial vulnerability, a proportion of variance in performance must be accounted for by factors associated with the illness itself.

The meta-analysis by Bora et al. (2014) found that the size of neurocognitive impairment was relatively comparable in individuals at familial high risk for psychosis versus those deemed at CHR (reviewed in further detail in the following section), with two exceptions. The first is IQ, which was significantly lower on average in familial high risk compared to clinical high-risk cohorts, both premorbidly (ES = 0.81 vs. ES = 0.40) and currently (ES = 0.64 vs. ES = 0.30) (Bora et al., 2014). In contrast, more impaired visuospatial working memory was observed in CHR than familial high-risk individuals (ES = 0.71 vs. ES = 0.35) (Bora et al., 2014). There is also evidence that social cognitive difficulties in first-degree relatives of people with schizophrenia are significantly correlated with IQ (Andric et al., 2016; Lavoie et al., 2013). Bora and colleagues have speculated that general intelligence might carry more genetic "weight" than individual neurocognitive domains in schizophrenia. This hypothesis would be consistent with the neurodevelopmental gradient model proposed by Owen and O'Donovan (2017), which implicates global intelligence as an important marker of degree of neurodevelopmental impairment and genetic risk (i.e., with intellectual disability at the severe end).

The results of familial high-risk studies showing cognition as being more impaired in nonaffected family members than the general population, but less impaired than in the affected family member, provide some support for cognition as an endophenotype for schizophrenia. However, to meet criteria as an endophenotype, further work is needed to determine whether there are specific genetic abnormalities associated with social and neurocognitive deficits in psychosis and whether there is co-inheritance of cognitive impairment with schizophrenia within families (Andric et al., 2016). Furthermore, future studies should not exclude people with IQ below 70 to allow examination of the full spectrum of intellectual functioning within families and those at risk for psychosis.

The clinical high-risk state

Significant neurocognitive and social cognitive deficits are found in individuals who are experiencing subthreshold or attenuated psychotic symptoms, often described as being at CHR or ultra high risk (UHR) for psychosis (henceforth referred to as CHR) (refer to Chapter 3 for further details on nomenclature). Similar to those at familial high risk, individuals at CHR for psychosis tend to show cognitive performance which is intermediate to the performance of healthy controls and individuals with first-episode psychosis (FEP) (Hauser et al., 2017; Fusar-Poli et al., 2012; Giuliano et al., 2012; van Donkersgoed et al., 2015; Lee et al., 2015).

Five meta-analyses have examined neurocognitive functioning in studies of CHR cohorts (Giuliano et al., 2012; Bora et al., 2014; Hauser et al., 2017; Fusar-Poli et al., 2012; De Herdt et al., 2013). Starting with IQ, small to medium impairments in current IQ (ES = 0.21–0.53) (Giuliano et al., 2012; Fusar-Poli et al., 2012; Bora et al., 2014; Hauser et al., 2017) and small impairments in premorbid IQ (ES = 0.25–0.30) (Bora et al., 2014; Hauser et al., 2017) have been reported in CHR versus healthy control samples. Furthermore, those who transition to full threshold psychosis within 2.5 years of cognitive assessment have a significantly lower IQ than those who do not (Giuliano et al., 2012; Fusar-Poli et al., 2012; Bora et al., 2014; Hauser et al., 2017).

These meta-analyses have shown that multiple neurocognitive domains are also significantly lower in CHR compared to healthy control samples, with small to medium effect sizes evident in olfactory identification (ES = 0.67), language (ES = 0.51), visuospatial skills (ES = 0.42), attention (including vigilance) (ES = 0.17–0.66), processing speed (ES = 0.35–0.62), verbal or visuospatial working memory (ES = 0.24–0.71), executive functions

(ES = 0.22–0.51), verbal fluency (ES = 0.31–0.52), verbal learning/memory (ES = 0.26–0.54), and visual learning/memory (ES = 0.27–0.51) (Giuliano et al., 2012; Fusar-Poli et al., 2012; Bora et al., 2014; Hauser et al., 2017). Hauser et al. (2017) were the only authors to include direct comparison of cognitive performance of CHR with that of FEP samples and found that CHR participants performed significantly better than FEP participants in the domains of attention/vigilance, processing speed, verbal learning/memory, and working memory. However, CHR and FEP samples had a similar degree of impairment in the domain of reasoning/problem solving, suggesting that impairment in this domain may not be specific to full threshold psychosis (there were insufficient data in the domains of visual learning/memory, social cognition, and premorbid IQ) (Hauser et al., 2017).

Several meta-analyses have also examined social cognition in CHR samples (Lee et al., 2015; Fusar-Poli et al., 2012; Hauser et al., 2017; van Donkersgoed et al., 2015; Bora and Pantelis, 2013). These have found significant medium-sized decrements relative to healthy controls in global social cognition (ES = 0.43–0.55) (Fusar-Poli et al., 2012; Lee et al., 2015; van Donkersgoed et al., 2015; Hauser et al., 2017), similar to the effect sizes seen for neurocognitive function. Significant medium decrements in the specific domains of emotion recognition/processing (ES = 0.45–0.46), theory of mind (ES = 0.43–0.45), and social perception (ES = 0.38) have also been reported (Lee et al., 2015; van Donkersgoed et al., 2015; Bora and Pantelis, 2013). Attributional biases are also observed in CHR cohorts, and attributional styles may differentiate those who transition to psychosis from those who do not, but more studies investigating this domain are needed (Lee et al., 2015; van Donkersgoed et al., 2015; Bora and Pantelis, 2013; Thompson et al., 2013).

It is interesting to note that studies comparing the cognitive performance of CHR individuals to a non-CHR help-seeking control group (rather than healthy control group) show far fewer group differences (Magaud et al., 2014; Ilonen et al., 2010; Lindgren et al., 2010). There is evidence to suggest that impaired visual manipulation might have some specificity for vulnerability for psychosis (Lindgren et al., 2010; Ilonen et al., 2010), while other cognitive difficulties are likely to be associated with general psychopathology in young people. There is a notable lack of differences on tasks of processing speed between CHR and non-CHR help-seeking groups (Ilonen et al., 2010; Magaud et al., 2014; Lindgren et al., 2010). These findings speak to the continuum of cognitive difficulties evident across psychiatric disorders, with more investigation necessary to better understand the specificity of cognitive function for risk for psychosis.

Basic symptoms

A further paradigm for investigating risk is that of basic symptoms (see Chapter 4), which are subtle changes in perception, language, cognition, motor function, and motivation present before psychosis onset (Huber, 1966). Basic symptoms are thought to be indicative of the "early prodromal phase" (Klosterkotter et al., 2001), differentiated from the "late prodromal phase" in which attenuated psychotic symptoms are present and the individual meets criteria for CHR. Findings suggest that the neurocognitive performance of individuals meeting basic symptoms criteria is superior to that of the CHR group (Pukrop et al., 2006; Simon et al., 2007; Frommann et al., 2011; Koutsouleris et al., 2012). Simon et al. (2007) found that the basic symptoms group performed below normative levels on tasks of executive function and processing speed, but did not differ in neurocognitive performance from a help-seeking control group (who did not meet criteria for basic symptoms or CHR). Verbal fluency has been demonstrated to differentiate basic symptoms groups from healthy controls (Pukrop et al., 2006), while Frommann et al. (2011) similarly demonstrated that the basic symptoms group showed greater impairment in executive control (including verbal fluency) and processing speed, both relative to healthy controls and their performance on other neurocognitive domains. Together, these findings suggest that processing speed and executive function may be markers of vulnerability before the development of attenuated positive symptoms, although their specificity for risk for psychosis remains unclear.

Cognition in transition versus non transition clinical high-risk individuals

CHR individuals who later transition to full threshold psychosis (within up to 2.5 years follow-up) have been shown to display larger impairments than those who did not transition in several neurocognitive domains, with medium to large effect sizes observed in those who transition relative to healthy controls (Giuliano et al., 2012; Fusar-Poli et al., 2012; Bora et al., 2014; Hauser et al., 2017; De Herdt et al., 2013), although this meta-analytic literature is somewhat inconsistent. Most neurocognitive domains were implicated in the reviews by Giuliano et al. (2012) and Bora et al. (2014). Only verbal fluency, visual and verbal learning/memory, and working memory were found to be significantly poorer in individuals who transitioned to psychosis compared to those who did not in the review by Fusar-Poli et al. (2012), whereas working memory and visual learning/memory were the only two domains that differentiated transitioned and non transitioned cases in the review by De Herdt et al. (2013). In the most recent and comprehensive review of 21 studies by Hauser et al. (2017), CHR samples who transitioned to psychosis performed significantly more poorly than those who did not in the domains of attention, processing speed, and verbal and visual learning/memory. Effect sizes between transition and non transition cases were in the small to medium range (Hauser et al., 2017). Unfortunately, Hauser et al. (2017) did not compare those who transitioned to psychosis with FEP individuals. However, the cross-sectional evidence suggests that the magnitude of baseline impairment of the CHR participants who transition to psychosis (large effect sizes relative to healthy controls) may be comparable to first-episode populations (Mesholam-Gately et al., 2009), particularly in general intelligence (IQ), visual and verbal learning/memory, and processing speed (Hauser et al., 2017). This is confirmed by meta-analytic evidence from longitudinal studies of CHR individuals showing relatively similar levels of impairment to baseline (Hauser et al., 2017), suggesting stability of neurocognitive function in CHR cohorts and that maximal "impairment" may have occurred before the onset of full threshold disorder. Research comparing social cognitive functioning in CHR individuals who transition to psychosis versus those who do not remains very limited at present, although one recent study found no differences in theory of mind, social perception, and emotion recognition between the groups (Piskulic et al., 2016). In summary, current research implicates both general intellectual functioning (IQ) and specific neurocognitive domains, particularly verbal and visual learning/memory and processing speed, as most affected in CHR samples who transition to psychosis.

Cognition as a significant independent predictor of transition to psychotic disorder

Poorer cognitive functioning in at-risk individuals may implicate cognition as being a useful marker for predicting later transition to psychosis within the CHR population (Cadenhead, 2002). However, having lower cognitive functioning in one or more domains does not automatically equate to a psychosis transition outcome. Previous reviews have shown there is considerable overlap in cognitive performance (up to 67%) between those who do and do not transition to psychosis (Bora et al., 2014; Hauser et al., 2017) and many people with full threshold psychosis have cognitive function that is similar to healthy controls (Uren et al., 2017). Furthermore, it is possible that the cognitive domains that are useful predictors of transition may differ from the domains that are most impaired, particularly if some domains are stably impaired from early in life. Thus, cognitive *risk factors* and *predictors* of psychotic disorder may differ. The present research on cognition as a predictor of transition to psychosis has been inconsistent.

The largest study to date (689 CHR individuals and 264 healthy controls) to examine neurocognitive predictors of transition to psychosis showed that those who transitioned to psychosis (within 2 years) had significantly larger impairments in attention, working memory, and verbal and visual memory, compared to the group that did not transition (Seidman et al., 2016). However, after controlling for age and positive symptoms, analyses showed that the period to transition to psychosis was significantly predicted by poorer visual and verbal memory and *higher* verbal abilities, demonstrating the complexity of this area of research and that specific combinations of neurocognitive functioning may be especially pernicious (Seidman et al., 2016). Another large study (325 CHR individuals and 66 healthy controls) with the longest follow-up period to date (up to 14 years, median of 6 years) demonstrated that transition to psychosis was predicted by poorer performance on tasks of visual memory (visual reproduction) and visual reasoning (matrix reasoning) (Lin et al., 2013). Novel multivariate methods, such as machine learning techniques, have shown that individualized prediction of psychosis in CHR individuals can be determined with 91% accuracy for individuals who transition and 89% for non transition cases, with impaired executive functioning (verbal fluency, cognitive flexibility) and verbal learning the most predictive of psychosis transition over 4 years (Koutsouleris et al., 2012).

There is emerging evidence that social cognition (especially emotion recognition and theory of mind) may also be a useful predictor of psychosis transition, but only a few studies with small samples have examined this (Allott et al., 2014; Kim et al., 2011; Corcoran et al., 2015). Interestingly, Corcoran et al. (2015) found that face emotion processing and negative symptoms provided the best model for predicting a schizophrenia diagnosis within 2 years, with accuracy for prediction of 96% and area under the receiver operating characteristic curve of 0.99. Neurocognitive variables were not predictive in the model (Corcoran et al., 2015). Further studies that simultaneously examine neurocognition and social cognition will be useful for refining cognitive prediction models of psychosis transition.

Many predictive studies have examined cognition either in isolation or have only included a small number of variables (often due to low power). As the chapters of this book imply, and based on what is known about psychosis and mental disorder more broadly, many biological, environmental, psychological, and social risk factors interact in complex ways to lead to mental disorder. Thus, the field has moved to combined models of risk factors in an effort to improve prognostic accuracy and clinical utility. Whether cognition is a necessary factor in these models remains open to debate. Symptoms are consistently stronger independent predictors than neurocognition in studies assessing the relative contribution of these factors in predicting psychosis transition in CHR cohorts (Metzler et al., 2016; Ziermans et al., 2014; Riecher-Rossler et al., 2009; Lin et al., 2013; Cornblatt et al., 2015; Cannon et al.,

2016). Nevertheless, some of this evidence suggests that combining cognitive vulnerability markers with clinical symptoms improves the accuracy of psychosis prediction compared with these variables in isolation (Riecher-Rossler et al., 2009; Cornblatt et al., 2015; Cannon et al., 2016). This has led to the development of clinical risk calculators such as the one from the NAPLS group (http://riskcalc.org/napls/) (Cannon et al., 2016). Across multiple studies, the neurocognitive domains of verbal and visual learning/memory and processing speed have received the most investigation in these models and contribute independently to transition to psychotic disorder.

A recent review that applied simulated modeling found that compared to baseline clinical assessment, a three-stage sequential multivariable assessment (including clinical, EEG, MRI, and blood biomarkers) significantly improved prediction of psychosis and achieved a positive predictive value of 98% (Schmidt et al., 2017). Notably, neurocognition was not included in this simulation. However, given the relative ease, cost-effectiveness, and feasibility of assessing cognition in a clinical setting, it is worthwhile exploring whether it could be a useful contributor to such a model. Indeed, recent probabilistic modeling work by Clark et al. (2015) has shown that multiple domains of assessment (including clinical criteria, neurocognitive testing, neuroimaging, and electrophysiology) in CHR individuals are not only necessary but also lead to cumulatively better prediction of psychosis outcome.

Cognitive change in at risk for psychosis: what is the evidence for progression?

A recent meta-analysis of longitudinal studies concluded that there is currently limited evidence for progression of neurocognitive impairment from pre- to post-psychosis onset (Bora and Murray, 2014), although caution with this conclusion is necessitated by the methodology of studies included in the review. For example, many lacked a healthy comparison group, there was variable timing and number of prospective follow-up assessments, and importantly, few studies examined the same individuals from CHR to FEP. The only way to reliably detect whether change in cognition occurs before or after the onset of psychosis is to investigate the same individuals longitudinally, multiple times before and after the development of psychotic disorder. These investigations are rare. In the Edinburgh High Risk study, high-risk participants performed significantly worse than age and sex-matched controls on tests of verbal memory and executive function over an 18-month to 2-year follow-up period. Furthermore, the high-risk participants who experienced psychotic symptoms showed a decline in IQ and poorer performance on tests of verbal memory and executive function than high-risk participants without symptoms. Increasing psychotic symptoms over time were associated with a decline in IQ and improved immediate memory compared to those with no change in symptoms (Cosway et al., 2000). In the Dunedin population cohort study mentioned earlier (Meier et al., 2014), a significant decline in IQ was observed in the schizophrenia group, with the main drop occurring between age 13 and 38 years. Furthermore, IQ was stable over time in an exploratory analysis of "at-risk" individuals (unconventionally defined as low IQ, low socioeconomic status, family history of schizophrenia) who did not transition to psychosis over the follow-up period (Meier et al., 2014). Analysis of specific neurocognitive tests showed that in the schizophrenia group, significant decline was observed on tests of processing speed, verbal learning, mental flexibility, and motor function. These findings provide evidence for "decline" in neurocognition between early adolescence (age 13) and age 38 in people with schizophrenia; however, the 25-year gap between assessments limits understanding of the precise course and timing of decline (i.e., before, during, or after schizophrenia onset). In the earlier mentioned ALSPAC study (Mollon et al., 2018), the group with psychotic disorder showed increasing significant deficits in Full-Scale and Performance IQ from 18 months to 20 years of age, whereas Verbal IQ declined early in life remaining statically impaired from age 8–20 years. Based on raw scores on identical tests, increasing lag (but not decline) in processing speed, working memory, and attention were observed from age 8 to 20 years in the psychotic disorder group. The findings of these studies are extremely important in characterizing cognition throughout childhood and adolescence and lend support to both the neurodevelopmental model of psychosis and progressive processes being part of the psychosis spectrum.

Longitudinal studies of CHR individuals have captured the course of cognitive functioning closer to illness onset. Meta-analytic findings suggest that neurocognition either remains stable or improves over time, although improvement in verbal fluency was significantly less pronounced in CHR individuals compared to healthy controls (Bora and Murray, 2014). However, it is worth noting that existing studies, including those published since the above mentioned meta-analysis, are based on small samples (<60 participants), usually have only two time points, and relatively have short follow-up periods (6–24 months) (Wood et al., 2007; Jahshan et al., 2010; Hawkins et al., 2008; Becker et al., 2010; Woodberry et al., 2013; Carrion et al., 2015). One consistent finding in these studies is that decline in IQ was not observed (Woodberry et al., 2013; Wood et al., 2007; Jahshan et al., 2010; Carrion et al., 2015), which is in contrast to the previous population cohort studies (Meier et al., 2014; Mollon et al., 2018). This may be because the follow-up periods of the CHR studies were too short to show change in global intelligence or that decline in IQ is specific to a diagnosis of schizophrenia. Finally, one study involving the NAPLS-2 cohort has examined social cognition over time (baseline, 1 year, and 2 years) in CHR individuals and found relative stability in the domains of theory of mind and social perception, although non transitioned cases were more likely to improve than transitioned cases (Piskulic et al., 2016).

Cognition as a risk marker for nonpsychosis outcomes

Until recently, the progression to full-blown psychosis has been the primary outcome of interest in studies investigating the CHR population. However, this approach ignores the heterogeneous composition of CHR cohorts, including those who transition, and the fact that the threshold at which frank psychosis is judged is arbitrary (Yung et al., 2010). Evidence suggests that vulnerability within the CHR state is not specific to psychosis alone, with studies implicating significant psychopathology, such as anxiety and major depressive disorders and poor functioning, as highly prevalent and persistent outcomes of the group (Addington et al., 2017; Lin et al., 2015; Rutigliano et al., 2016; Cornblatt et al., 2012; Carrion et al., 2013). Recent research has turned to examining cognitive predictors of clinically relevant nonpsychosis outcomes in CHR, primarily functional outcomes (Yung et al., 2010; Lin et al., 2012), although research is in its infancy.

Processing speed and composite neurocognition have been associated with social and role functioning at identification of risk (Carrion et al., 2011; Meyer et al., 2014). Several studies have examined baseline neurocognition as a predictor of later functional outcomes in CHR individuals. In a small sample of non transition cases, baseline attention and processing speed were associated with global functioning at 12-month follow-up (Sawada et al., 2017). In a larger study with 12-month follow-up, Meyer et al. (2014) found that composite neurocognition and negative symptoms made unique significant contributions to role functioning at follow-up. Negative symptoms were also found to mediate the relationship between neurocognition and functioning (Meyer et al., 2014). Carrion et al. (2013) followed CHR individuals for a mean of 3 years and found that poorer processing speed, social functioning, and disorganized symptoms at baseline predicted poor social outcome at follow-up, although the contribution of processing speed was not significant after transition to psychosis status was accounted for. In contrast, transition to psychosis did not predict poor role outcome at follow-up, which was most strongly predicted by poorer verbal memory, motor disturbances, and role functioning at baseline (Carrion et al., 2013). In the longest follow-up study to date to examine functional outcomes of CHR individuals, poorer verbal memory, processing speed, attention, and verbal fluency at baseline were found to be predictive of poor functional outcome over a mean of 7 years follow-up, regardless of transition status (Lin et al., 2011). In this study, verbal memory in combination with negative symptoms at baseline was the best predictor of functional outcome (Lin et al., 2011).

Two studies have examined whether change in cognition is associated with change in functioning in CHR individuals. Niendam et al. (2007) showed that improved performance in verbal learning and memory and processing speed were associated with improved functioning over an 8-month period. Recently, Shin et al. (2016) found that change in semantic fluency was significantly associated with changes in negative symptoms and functioning over 2 years in a small CHR cohort. Taken together, there is some consistency in the neurocognitive domains associated with functional outcomes in CHR individuals, with processing speed, verbal memory, and possibly verbal fluency being implicated. These findings point to neurocognition as being important to assess, monitor, and consider as part of treatment approaches to improve functional outcomes for people at CHR for psychosis, regardless of transition to psychosis. To our knowledge there is no research on the relationship between cognition and non-psychotic disorder outcomes (e.g., major depressive disorder) in CHR cohorts, which is an important line of future investigation given their prevalence.

Conclusion

Decades of research has examined the role that cognition plays as a risk marker of psychosis, with hopes for identifying a potentially malleable target for prevention of psychosis onset and better clinical and functional outcomes. The neurodevelopmental hypothesis has dominated the field as the most likely explanation for cognitive impairment in psychosis, with clear evidence of cognitive impairments being present early in life and before onset of clinical symptoms. Strong evidence for genetic contributions to cognitive impairments exists from family studies, but less is understood about the co-inheritance of cognitive difficulties with psychosis. Group studies of CHR individuals suggest relative stability of cognition over several years. Those who transition to psychosis have poorer cognitive function relative to those who do not transition, with evidence for impairments that are comparable to FEP samples. Longitudinal cohort studies suggest progressive impairments in cognition, but the timing of progression in relation to psychosis transition remains to be understood. Whether cognition may serve as a useful or necessary predictive marker remains an open question with evidence both for and against the inclusion of cognition in multivariate predictor models. Because cognition can be measured relatively inexpensively and easily in a clinical setting (as opposed to other variables such as neuroimaging or electrophysiology), further investigation into its value as a predictor for psychosis is warranted.

From this overview it is evident that there has been a paucity of longitudinal studies examining cognition premorbidly and after the onset of frank psychosis. This research is essential to improve understanding of the predictive power of cognition in the development of psychosis, as well as the timing of any decrements that may occur. For advancement of our understanding of cognition as a risk for psychosis, particularly in the CHR group, several limitations should be addressed. The inclusion of a psychiatric help-seeking comparison group who are not at CHR, in addition to a healthy control group, is necessary to determine the specificity of cognitive difficulties to psychosis. In addition, the heterogeneity of cognitive measures and batteries used makes comparison across studies difficult and is likely to be a major reason for inconsistent findings. Neurocognitive and social cognitive measurement harmonization initiatives, such as the MATRICS Consensus Cognitive Battery (MCCB) (Nuechterlein et al., 2004), Brief Assessment of Cognition in Schizophrenia (BACS) (Keefe et al., 2004), and Social Cognition Psychometric Evaluation (SCOPE) (Pinkham et al., 2018), have been developed to address this in established psychotic disorder, but are yet to be consistently applied to CHR populations. We would also argue that across research on cognitive risk for psychosis, the inclusion of individuals with IQ < 70 is necessary to better understand the full neurodevelopmental continuum in relation to psychosis. Understanding cognition as it relates to psychosis risk may allow for better prediction of psychotic disorder and other poor outcomes, providing the potential for early intervention and ultimately prevention.

References

Addington, J., Piskulic, D., Liu, L., Lockwood, J., Cadenhead, K.S., Cannon, T.D., Cornblatt, B.A., Mcglashan, T.H., Perkins, D.O., Seidman, L.J., Tsuang, M.T., Walker, E.F., Bearden, C.E., Mathalon, D.H., Woods, S.W., 2017. Comorbid diagnoses for youth at clinical high risk of psychosis. Schizophr. Res. 190, 90–95.

Adolphs, R., 2009. The social brain: neural basis of social knowledge. Annu. Rev. Psychol. 60, 693–716.

Allott, K., Schafer, M.R., Thompson, A., Nelson, B., Bendall, S., Bartholomeusz, C.F., Yuen, H.P., Mcgorry, P.D., Schlogelhofer, M., Bechdolf, A., Amminger, G.P., 2014. Emotion recognition as a predictor of transition to a psychotic disorder in ultra-high risk participants. Schizophr. Res. 153, 25–31.

Andric, S., Maric, N.P., Mihaljevic, M., Mirjanic, T., Van Os, J., 2016. Familial covariation of facial emotion recognition and IQ in schizophrenia. Psychiatry Res. 246, 52–57.

Bassett, A.S., Chow, E.W., Weksberg, R., 2000. Chromosomal abnormalities and schizophrenia. Am. J. Med. Genet. 97, 45–51.

Batty, G.D., Mortensen, E.L., Osler, M., 2005. Childhood IQ in relation to later psychiatric disorder: evidence from a Danish birth cohort study. Br. J. Psychiatry 187, 180–181.

Becker, H.E., Nieman, D.H., Wiltink, S., Dingemans, P.M., De Fliert, J.R.V., Velthorst, E., DE Haan, L., Van Amelsvoort, T.A., Linszen, D.H., 2010. Neurocognitive functioning before and after the first psychotic episode: does psychosis result in cognitive deterioration? Psychol. Med. 40, 1599–1606.

Bora, E., Lin, A., Wood, S.J., Yung, A.R., Mcgorry, P.D., Pantelis, C., 2014. Cognitive deficits in youth with familial and clinical high risk to psychosis: a systematic review and meta-analysis. Acta Psychiatr. Scand. 130, 1–15.

Bora, E., Murray, R.M., 2014. Meta-analysis of cognitive deficits in ultra-high risk to psychosis and first-episode psychosis: do the cognitive deficits progress over, or after, the onset of psychosis? Schizophr. Bull. 40, 744–755.

Bora, E., Pantelis, C., 2013. Theory of mind impairments in first-episode psychosis, individuals at ultra-high risk for psychosis and in first-degree relatives of schizophrenia: systematic review and meta-analysis. Schizophr. Res. 144, 31–36.

Bozikas, V.P., Andreou, C., 2011. Longitudinal studies of cognition in first episode psychosis: a systematic review of the literature. Aust. N. Z. J. Psychiatr. 45, 93–108.

Byrne, M., Clafferty, B.A., Cosway, R., Grant, E., Hodges, A., Whalley, H.C., Lawrie, S.M., Owens, D.G.C., Johnstone, E.C., 2003. Neuropsychology, genetic liability, and psychotic symptoms in those at high risk of schizophrenia. J. Abnorm. Psychol. 112, 38–48.

Cadenhead, K.S., 2002. Vulnerability markers in the schizophrenia spectrum: implications for phenomenology, genetics, and the identification of the schizophrenia prodrome. Psychiatr. Clin. N. Am. 25, 837-+.

Cannon, M., Caspi, A., Moffitt, T.E., Harrington, H., Taylor, A., Murray, R.M., Poulton, R., 2002. Evidence for early-childhood, pan-developmental impairment specific to schizophreniform disorder – results from a longitudinal birth cohort. Arch. Gen. Psychiatr. 59, 449–456.

Cannon, T.D., Huttunen, M.O., Lonnqvist, J., Tuulio-Henriksson, A., Pirkola, T., Glahn, D., Finkelstein, J., Hietanen, M., Kaprio, J., Koskenvuo, M., 2000. The inheritance of neuropsychological dysfunction in twins discordant for schizophrenia. Am. J. Hum. Genet. 67, 369–382.

Cannon, T.D., Yu, C., Addington, J., Bearden, C.E., Cadenhead, K.S., Cornblatt, B.A., Heinssen, R., Jeffries, C.D., Mathalon, D.H., Mcglashan, T.H., Perkins, D.O., Seidman, L.J., Tsuang, M.T., Walker, E.F., Woods, S.W., Kattan, M.W., 2016. An individualized risk calculator for research in prodromal psychosis. Am. J. Psychiatry 173, 980–988.

Carrion, R.E., Goldberg, T.E., Mclaughlin, D., Auther, A.M., Correll, C.U., Cornblatt, B.A., 2011. Impact of neurocognition on social and role functioning in individuals at clinical high risk for psychosis. Am. J. Psychiatry 168, 806–813.

Carrion, R.E., Mclaughlin, D., Auther, A.M., Olsen, R., Correll, C.U., Cornblatt, B.A., 2015. The impact of psychosis on the course of cognition: a prospective, nested case-control study in individuals at clinical high-risk for psychosis. Psychol. Med. 45, 3341–3354.

Carrion, R.E., Mclaughlin, D., Goldberg, T.E., Auther, A.M., Olsen, R.H., Olvet, D.M., Correll, C.U., Cornblatt, B.A., 2013. Prediction of functional outcome in individuals at clinical high risk for psychosis. JAMA Psychiatry 70, 1133–1142.

Clark, S.R., Schubert, K.O., Baune, B.T., 2015. Towards indicated prevention of psychosis: using probabilistic assessments of transition risk in psychosis prodrome. J. Neurotransm. 122, 155–169.

Corcoran, C.M., Keilp, J.G., Kayser, J., Klim, C., Butler, P.D., Bruder, G.E., Gur, R.C., Javitt, D.C., 2015. Emotion recognition deficits as predictors of transition in individuals at clinical high risk for schizophrenia: a neurodevelopmental perspective. Psychol. Med. 45, 2959–2973.

Cornblatt, B.A., Carrion, R.E., Addington, J., Seidman, L., Walker, E.F., Cannon, T.D., Cadenhead, K.S., Mcglashan, T.H., Perkins, D.O., Tsuang, M.T., Woods, S.W., Heinssen, R., Lencz, T., 2012. Risk factors for psychosis: impaired social and role functioning. Schizophr. Bull. 38, 1247–1257.

Cornblatt, B.A., Carrion, R.E., Auther, A., Mclaughlin, D., Olsen, R.H., John, M., Correll, C.U., 2015. Psychosis prevention: a modified clinical high risk perspective from the recognition and prevention (RAP) program. Am. J. Psychiatry 172, 986–994.

Cosway, R., Byrne, M., Clafferty, R., Hodges, A., Grant, E., Abukmeil, S.S., Lawrie, S.M., Miller, P., Johnstone, E.C., 2000. Neuropsychological change in young people at high risk for schizophrenia: results from the first two neuropsychological assessments of the Edinburgh High Risk Study. Psychol. Med. 30, 1111–1121.

David, A.S., Zammit, S., Lewis, G., Dalman, C., Allebeck, P., 2008. Impairments in cognition across the spectrum of psychiatric disorders: evidence from a Swedish conscript cohort. Schizophr. Bull. 34, 1035–1041.

Davidson, M., Reichenberg, A., Rabinowitz, J., Weiser, M., Kaplan, Z., Mark, M., 1999. Behavioral and intellectual markers for schizophrenia in apparently healthy male adolescents. Am. J. Psychiatry 156, 1328–1335.

DE Herdt, A., Wampers, M., Vancampfort, D., De Hert, M., Vanhees, L., Demunter, H., VAN Bouwel, L., Brunner, E., Probst, M., 2013. Neurocognition in clinical high risk young adults who did or did not convert to a first schizophrenic psychosis: a meta-analysis. Schizophr. Res. 149, 48–55.

Dickinson, D., Ramsey, M.E., Gold, J.M., 2007. Overlooking the obvious. A meta-analytic comparison of digit symbol coding tasks and other cognitive measures in schizophrenia. Arch. Gen. Psychiatr. 64, 532–542.

Erlenmeyer-Kimling, L., Rock, D., Roberts, S.A., Janal, M., Kestenbaum, C., Cornblatt, B., Adamo, U.H., Gottesman, I.I., 2000. Attention, memory, and motor skills as childhood predictors of schizophrenia-related psychoses: the New York high-risk project. Am. J. Psychiatry 157, 1416–1422.

Faraone, S.V., Seidman, L.J., Kremen, W.S., Toomey, R., Pepple, J.R., Tsuang, M.T., 2000. Neuropsychologic functioning among the nonpsychotic relatives of schizophrenic patients: the effect of genetic loading. Biol. Psychiatry 48, 120–126.

Filatova, S., Koivumaa-Honkanen, H., Hirvonen, N., Freeman, A., Ivandic, I., Hurtig, T., Khandaker, G.M., Jones, P.B., Moilanen, K., Miettunen, J., 2017. Early motor developmental milestones and schizophrenia: a systematic review and meta-analysis. Schizophr. Res. 188, 13–20.

Fiske, S.T., Taylor, S.E., 2013. Social Cognition: From Brains to Culture. Sage Publications Ltd, London.

Frommann, I., Pukrop, R., Brinkmeyer, J., Bechdolf, A., Ruhrmann, S., Berning, J., Decker, P., Riedel, M., Moller, H.J., Wolwer, W., Gaebel, W., Klosterkotter, J., Maier, W., Wagner, M., 2011. Neuropsychological profiles in different at-risk states of psychosis: executive control impairment in the early—and additional memory dysfunction in the late—prodromal state. Schizophr. Bull. 37, 861–873.

Fusar-Poli, P., Deste, G., Smieskova, R., Barlati, S., Yung, A.R., Howes, O., Stieglitz, R.-D., Vita, A., Mcguire, P., Borgwardt, S., 2012. Cogntive functioning in prodromal psychosis: a meta-analysis. Arch. Gen. Psychiatr. 69, 562–571.

Gale, C.R., Batty, G.D., Mcintosh, A.M., Porteous, D.J., Deary, I.J., Rasmussen, F., 2013. Is bipolar disorder more common in highly intelligent people? A cohort study of a million men. Mol. Psychiatry 18, 190–194.

Gigi, K., Werbeloff, N., Goldberg, S., Portuguese, S., Reichenberg, A., Fruchter, E., Weiser, M., 2014. Borderline intellectual functioning is associated with poor social functioning, increased rates of psychiatric diagnosis and drug use—a cross sectional population based study. Eur. Neuropsychopharmacol. 24, 1793–1797.

Giuliano, A.J., Li, H.J., Mesholam-Gately, R.I., Sorenson, S.M., Woodberry, K.A., Seidman, L.J., 2012. Neurocognition in the psychosis risk syndrome: a quantitative and qualitative review. Curr. Pharmaceut. Des. 18, 399–415.

Gothelf, D., Schneider, M., Green, T., Debbane, M., Frisch, A., Glaser, B., Zilkha, H., Schaer, M., Weizman, A., Eliez, S., 2013. Risk factors and the evolution of psychosis in 22q11.2 deletion syndrome: a longitudinal 2-site study. J. Am. Acad. Child Adolesc. Psychiatry 52, 1192–1203 e3.

Gottesman, I.I., Gould, T.D., 2003. The endophenotype concept in psychiatry: etymology and strategic intentions. Am. J. Psychiatry 160, 636–645.

Green, T., Gothelf, D., Glaser, B., Debbane, M., Frisch, A., Kotler, M., Weizman, A., Eliez, S., 2009. Psychiatric disorders and intellectual functioning throughout development in velocardiofacial (22q11.2 deletion) syndrome. J.Am. Acad. Child Adolesc. Psychiatry 48, 1060–1068.

Hauser, M., Zhang, J.P., Sheridan, E.M., Burdick, K.E., Mogil, R., Kane, J.M., Auther, A., Carrion, R.E., Cornblatt, B.A., Correll, C.U., 2017. Neuropsychological test performance to enhance identification of subjects at clinical high risk for psychosis and to Be most promising for predictive algorithms for conversion to psychosis: a meta-analysis. J. Clin. Psychiatry 78, E28–E40.

Hawkins, K.A., Keefe, R.S.E., Christensen, B.K., Addington, J., Woods, S.W., Callahan, J., Zipursky, R.B., Perkins, D.O., Tohen, M., Breier, A., Mcglashan, T.H., 2008. Neuropsychological course in the prodrome and first episode of psychosis: findings from the PRIME north America double blind treatment study. Schizophr. Res. 105, 1–9.

Huber, G., 1966. Reine Defektsyndrome und Basisstadien endogener Psychosen. Fortschr. Neurol. Psychiatr. 34, 409–426.

Ilonen, T., Heinimaa, M., Korkeila, J., Svirskis, T., Salokangas, R.K., 2010. Differentiating adolescents at clinical high risk for psychosis from psychotic and non-psychotic patients with the Rorschach. Psychiatry Res. 179, 151–156.

Jahshan, C., Heaton, R.K., Golshan, S., Cadenhead, K.S., 2010. Course of neurocognitive deficits in the prodrorne and first episode of schizophrenia. Neuropsychology 24, 109–120.

Johnstone, E.C., Ebmeier, K.P., Miller, P., Owens, D.G.C., Lawrie, S.M., 2005. Predicting schizophrenia: findings from the Edinburgh high-risk study. Br. J. Psychiatry 186, 18.

Jones, P., Rodgers, B., Murray, R., Marmot, M., 1994. Child developmental risk factors for adult schizophrenia in the British 1946 birth cohort. Lancet 344, 1398–1402.

Kahn, R.S., Keefe, R.S.E., 2013. Schizophrenia is a cognitive illness: time for a change in focus. JAMA Psychiatry 70, 1107–1112.

Keefe, R.S.E., Goldberg, T.E., Harvey, P.D., Gold, J.M., Poe, M.P., Coughenour, L., 2004. The Brief Assessment of Cognition in Schizophrenia: reliability, sensitivity, and comparison with a standard neurocognitive battery. Schizophr. Res. 68, 283–297.

Kendler, K.S., Ohlsson, H., Sundquist, J., Sundquist, K., 2015. IQ and schizophrenia in a Swedish national sample: their causal relationship and the interaction of IQ with genetic risk. Am. J. Psychiatry 172, 259–265.

Khandaker, G.M., Barnett, J.H., White, I.R., Jones, P.B., 2011. A quantitative meta-analysis of population-based studies of premorbid intelligence and schizophrenia. Schizophr. Res. 132, 220–227.

Khandaker, G.M., Stochl, J., Zammit, S., Lewis, G., Jones, P.B., 2014. A population-based longitudinal study of childhood neurodevelopmental disorders, IQ and subsequent risk of psychotic experiences in adolescence. Psychol. Med. 44, 3229–3238.

Kim, H.S., Shin, N.Y., Jang, J.H., Kim, E., Shim, G., Park, H.Y., Hong, K.S., Kwon, J.S., 2011. Social cognition and neurocognition as predictors of conversion to psychosis in individuals at ultra-high risk. Schizophr. Res. 130, 170–175.

Klosterkotter, J., Hellmich, M., Steinmeyer, E.M., Schultze-Lutter, F., 2001. Diagnosing schizophrenia in the initial prodromal phase. Arch. Gen. Psychiatr. 58, 158–164.

Koenen, K.C., Moffitt, T.E., Roberts, A.L., Martin, L.T., Kubzansky, L., Harrington, H., Poulton, R., Caspi, A., 2009. Childhood IQ and adult mental disorders: a test of the cognitive reserve hypothesis. Am. J. Psychiatry 166, 50–57.

Koutsouleris, N., Davatzikos, C., Bottlender, R., Patschurek-Kliche, K., Scheuerecker, J., Decker, P., Gaser, C., Moller, H.J., Meisenzahl, E.M., 2012. Early recognition and disease prediction in the at-risk mental states for psychosis using neurocognitive pattern classification. Schizophr. Bull. 38, 1200–1215.

Kremen, W.S., Buka, S.L., Seidman, L.J., Goldstein, J.M., Koren, D., Tsuang, M.T., 1998. IQ decline during childhood and adult psychotic symptoms in a community sample: a 19-Year longitudinal study. Am. J. Psychiatry 155, 672–677.

Lavoie, M.A., Plana, I., Lacroix, J.B., Godmaire-Duhaime, F., Jackson, P.L., Achim, A.M., 2013. Social cognition in first-degree relatives of people with schizophrenia: a meta-analysis. Psychiatry Res. 209, 129–135.

Lee, T.Y., Hong, S.B., Shin, N.Y., Kwon, J.S., 2015. Social cognitive functioning in prodromal psychosis: a meta-analysis. Schizophr. Res. 164, 28–34.

Lezak, M.D., Howieson, D.B., Loring, D.W., 2004. Neuropsychological Assessment. Oxford University Press, New York.

Lin, A., Nelson, B., Yung, A.R., 2012. 'At-risk' for psychosis research: where are we heading? Epidemiol. Psychiatr. Sci. https://doi.org/10.1017/S2045796012000388.

Lin, A., Wood, S.J., Nelson, B., Beavan, A., Mcgorry, P., Yung, A.R., 2015. Outcomes of nontransitioned cases in a sample at ultra-high risk for psychosis. Am. J. Psychiatry 172, 249–258.

Lin, A., Wood, S.J., Nelson, B., Brewer, W.J., Spiliotacopoulos, D., Bruxner, A., Broussard, C., Pantelis, C., Yung, A.R., 2011. Neurocognitive predictors of functional outcome two to 13 years after identification as ultra-high risk for psychosis. Schizophr. Res. 132, 1–7.

Lin, A., Yung, A.R., Nelson, B., Brewer, W.J., Riley, R., Simmons, M., Pantelis, C., Wood, S.J., 2013. Neurocognitive predictors of transition to psychosis: medium- to long-term findings from a sample at ultra-high risk for psychosis. Psychol. Med. 43, 2349–2360.

Lindgren, M., Manninen, M., Laajasalo, T., Mustonen, U., Kalska, H., Suvisaari, J., Moilanen, K., Cannon, T.D., Huttunen, M., Therman, S., 2010. The relationship between psychotic-like symptoms and neurocognitive performance in a general adolescent psychiatric sample. Schizophr. Res. 123, 77–85.

Maccabe, J.H., Wicks, S., Lofving, S., David, A.S., Berndtsson, A., Gustafsson, J.E., Allebeck, P., Dalman, C., 2013. Decline in cognitive performance between ages 13 and 18 years and the risk for psychosis in adulthood: a Swedish longitudinal cohort study in males. JAMA Psychiatry 70, 261–270.

Magaud, E., Morvan, Y., Rampazzo, A., Alexandre, C., Willard, D., Gaillard, R., Kazes, M., Krebs, M.O., 2014. Subjects at Ultra High Risk for psychosis have 'heterogeneous' intellectual functioning profile: a multiple-case study. Schizophr. Res. 152, 415–420.

Mathews, A., Macleod, C., 2005. Cognitive vulnerability to emotional disorders. Annu. Rev. Clin. Psychol. 1, 167–195.

Meier, M.H., Caspi, A., Reichenberg, A., Keefe, R.S.E., Fisher, H.L., Harrington, N., Houts, R., Poulton, R., Moffitt, T.E., 2014. Neuropsychological decline in schizophrenia from the premorbid to the postonset period: evidence from a population-representative longitudinal study. Am. J. Psychiatry 171, 91–101.

Mesholam-Gately, R., Giuliano, A.J., Faraone, S.V., Goff, K.P., Seidman, L.J., 2009. Neurocognition in first-episode schizophrenia: a meta-analytic review. Neuropsychology 23, 315–336.

Metzler, S., Dvorsky, D., Wyss, C., Nordt, C., Walitza, S., Heekeren, K., Rossler, W., Theodoridou, A., 2016. Neurocognition in help-seeking individuals at risk for psychosis: prediction of outcome after 24 months. Psychiatry Res. 246, 188–194.

Meyer, E.C., Carrion, R.E., Cornblatt, B.A., Addington, J., Cadenhead, K.S., Cannon, T.D., Mcglashan, T.H., Perkins, D.O., Tsuang, M.T., Walker, E.F., Woods, S.W., Heinssen, R., Seidman, L.J., 2014. The relationship of neurocognition and negative symptoms to social and role functioning over time in individuals at clinical high risk in the first phase of the north American prodrome longitudinal study. Schizophr. Bull. 40, 1452–1461.

Millan, M.J., Agid, Y., Brüne, M., Bullmore, E.T., Carter, C.S., Clayton, N.S., Connor, R., Davis, S., Deakin, B., Derubeis, R.J., Dubois, B., Geyer, M.A., Goodwin, G.M., Gorwood, P., Jay, T.M., Joëls, M., Mansuy, I.M., MEYER-Lindenberg, A., Murphy, D., Rolls, E., Saletu, B., Spedding, M., Sweeney, J., Whittington, M., Young, L.J., 2012. Cognitive dysfunction in psychiatric disorders: characteristics, causes and the quest for improved therapy. Nat. Rev. Drug Discov. 11, 141–168.

Mirsky, A.F., Ingraham, L.J., Kugelmass, S., 1995. Neuropsychological assessment of attention and its pathology in the Israeli cohort. Schizophr. Bull. 21, 193–204.

Mollon, J., David, A.S., Zammit, S., Lewis, G., Reichenberg, A., 2018. Course of cognitive development from infancy to early adulthood in the psychosis spectrum. JAMA Psychiatry 75, 270–279.

Morgan, V.A., Leonard, H., Bourke, J., Jablensky, A., 2008. Intellectual disability co-occurring with schizophrenia and other psychiatric illness: population-based study. Br. J. Psychiatry 193, 364–372.

Niendam, T.A., Bearden, C.E., Zinberg, J., Johnson, J.K., O'brien, M., Cannon, T.D., 2007. The course of neurocognition and social functioning in individuals at ultra high risk for psychosis. Schizophr. Bull. 33, 772–781.

Nuechterlein, K.H., Barch, D.M., Gold, J.M., Goldberg, T.E., Green, M.F., Heaton, R.K., 2004. Identification of separable cognitive factors in schizophrenia. Schizophr. Res. 72, 29–39.

Osler, M., Lawlor, D.A., Nordentoft, M., 2007. Cognitive function in childhood and early adulthood and hospital admission for schizophrenia and bipolar disorders in Danish men born in 1953. Schizophr. Res. 92, 132–141.

Owen, M.J., 2012. Intellectual disability and major psychiatric disorders: a continuum of neurodevelopmental causality. Br. J. Psychiatry 200, 268–269.

Owen, M.J., O'donovan, M.C., 2017. Schizophrenia and the neurodevelopmental continuum:evidence from genomics. World Psychiatry 16, 227–235.

Pinkham, A.E., Harvey, P.D., Penn, D.L., 2018. Social cognition psychometric evaluation: results of the final validation study. Schizophr. Bull. 44, 737–748.

Pinkham, A.E., Penn, D.L., Green, M.F., Buck, B., Healey, K., Harvey, P.D., 2014. The social cognition psychometric evaluation study: results of the expert survey and RAND panel. Schizophr. Bull. 40, 813–823.

Piskulic, D., Liu, L., Cadenhead, K.S., Cannon, T.D., Cornblatt, B.A., Mcglashan, T.H., Perkins, D.O., Seidman, L.J., Tsuang, M.T., Walker, E.F., Woods, S.W., Bearden, C.E., Mathalon, D.H., Addington, J., 2016. Social cognition over time in individuals at clinical high risk for psychosis: findings from the NAPLS-2 cohort. Schizophr. Res. 171, 176–181.

Pukrop, R., Schultze-Lutter, F., Ruhrmann, S., Brockhaus-Dumke, A., Tendolkar, I., Bechdolf, A., Matuschek, E., Klosterkotter, J., 2006. Neurocognitive functioning in subjects at risk for a first episode of psychosis compared with first- and multiple-episode schizophrenia. J. Clin. Exp. Neuropsychol. 28, 1388–1407.

Rabinowitz, J., Reichenberg, A., Weiser, M., Mark, M., Kaplan, Z., Davidson, M., 2000. Cognitive and behavioural functioning in men with schizophrenia both before and shortly after first admission to hospital: cross-sectional analysis. Br. J. Psychiatry 177, 26–32.

Reichenberg, A., Caspi, A., Harrington, H., Houts, R., Keefe, R.S.E., Murray, R.M., Poulton, R., Moffitt, T.E., 2010. Static and dynamic cognitive deficits in childhood preceding adult schizophrenia: a 30-year study. Am. J. Psychiatry 167, 160–169.

Reichenberg, A., Weiser, M., Rabinowitz, J., Caspi, A., Schmeidler, J., Mark, M., Kaplan, Z., Davidson, M., 2002. A population-based cohort study of premorbid intellectual, language, and behavioral functioning in patients with schizophrenia, schizoaffective disorder, and nonpsychotic bipolar disorder. Am. J. Psychiatry 159, 2027–2035.

Reichenberg, A., Weiser, M., Rapp, M.A., Rabinowitz, J., Caspi, A., Schmeidler, J., Knobler, H.Y., Lubin, G., Nahon, D., Harvey, P.D., Davidson, M., 2005. Elaboration on premorbid intellectual performance in schizophrenia: premorbid intellectual decline and risk for schizophrenia. Arch. Gen. Psychiatr. 62, 1297–1304.

Riecher-Rossler, A., Pflueger, M.O., Aston, J., Borgwardt, S.J., Brewer, W.J., Gschwandtner, U., Stieglitz, R.D., 2009. Efficacy of using cognitive status in predicting psychosis: a 7-year follow-up. Biol. Psychiatry 66, 1023–1030.

Rutigliano, G., Valmaggia, L., Landi, P., Frascarelli, M., Cappucciati, M., Sear, V., Rocchetti, M., DE Micheli, A., Jones, C., Palombini, E., Mcguire, P., Fusar-Poli, P., 2016. Persistence or recurrence of non-psychotic comorbid mental disorders associated with 6-year poor functional outcomes in patients at ultra high risk for psychosis. J. Affect. Disord. 203, 101–110.

Sawada, K., Kanehara, A., Sakakibara, E., Eguchi, S., Tada, M., Satomura, Y., Suga, M., Koike, S., Kasai, K., 2017. Identifying neurocognitive markers for outcome prediction of global functioning in individuals with first-episode and ultra-high-risk for psychosis. Psychiatry Clin. Neurosci. 71, 318–327.

Schmidt, A., Cappucciati, M., Radua, J., Rutigliano, G., Rocchetti, M., Dell'osso, L., Politi, P., Borgwardt, S., Reilly, T., Valmaggia, L., Mcguire, P., Fusar-Poli, P., 2017. Improving prognostic accuracy in subjects at clinical high risk for psychosis: systematic review of predictive models and meta-analytical sequential testing simulation. Schizophr. Bull. 43, 375–388.

Seidman, L.J., Giuliano, A.J., Smith, C.W., Stone, W.S., Glatt, S.J., Meyer, E., Faraone, S.V., Tsuang, M.T., Cornblatt, B., 2006. Neuropsychological functioning in adolescents and young adults at genetic risk for schizophrenia and affective psychoses: results from the Harvard and Hillside Adolescent High Risk Studies. Schizophr. Bull. 32, 507.

Seidman, L.J., Shapiro, D.I., Stone, W.S., Woodberry, K.A., Ronzio, A., Cornblatt, B.A., Addington, J., Bearden, C.E., Cadenhead, K.S., Cannon, T.D., Mathalon, D.H., Mcglashan, T.H., Perkins, D.O., Tsuang, M.T., Walker, E.F., Woods, S.W., 2016. Association of neurocognition with transition to psychosis: baseline functioning in the second phase of the north American prodrome longitudinal study. JAMA Psychiatry 73, 1239–1248.

Shin, Y.S., Kim, S.Y., Lee, T.Y., Hur, J.W., Shin, N.Y., Kim, S.N., Shin, M.S., Kwon, J.S., 2016. Longitudinal change in neurocognition and its relation to symptomatic and functional changes over 2 years in individuals at clinical high-risk for psychosis. Schizophr. Res. 174, 50–57.

Simon, A.E., Cattapan-Ludewig, K., Zmilacher, S., Arbach, D., Gruber, K., Dvorsky, D.N., Roth, B., Isler, E., Zimmer, A., Umbricht, D., 2007. Cognitive functioning in the schizophrenia prodrome. Schizophr. Bull. 33, 761–771.

Sitskoorn, M.M., Aleman, A., Ebisch, S.J.H., Appels, M.C.M., Kahn, R.S., 2004. Cognitive deficits in relatives of patients with schizophrenia: a meta-analysis. Schizophr. Res. 71, 285–295.

Snitz, B.E., Macdonald 3rd, A.W., Carter, C.S., 2006. Cognitive deficits in unaffected first-degree relatives of schizophrenia patients: a meta-analytic review of putative endophenotypes. Schizophr. Bull. 32, 179–194.

Snyder, H.R., Miyake, A., Hankin, B.L., 2015. Advancing understanding of executive function impairments and psychopathology: bridging the gap between clinical and cognitive approaches. Front. Psychol. 6, 24.

Sorensen, H.J., Mortensen, E.L., Schiffman, J., Reinisch, J.M., Maeda, J., Mednick, S.A., 2010. Early developmental milestones and risk of schizophrenia: a 45-year follow-up of the Copenhagen Perinatal Cohort. Schizophr. Res. 118, 41–47.

Szoke, A., Trandafir, A., Dupont, M.-E., Meary, A., Schurhoff, F., Leboyer, M., 2008. Longitudinal studies of cognition in schizophrenia: meta analysis. Br. J. Psychiatry 192, 248–257.

Thompson, A., Papas, A., Bartholomeusz, C., Nelson, B., Yung, A., 2013. Externalized attributional bias in the Ultra High Risk (UHR) for psychosis population. Psychiatry Res. 206, 200–205.

Tiihonen, J., Haukka, J., Henriksson, M., Cannon, M., Kieseppa, T., Laaksonen, I., Sinivuo, J., Lonnqvist, J., 2005. Premorbid intellectual functioning in bipolar disorder and schizophrenia: results from a cohort study of male conscripts. Am. J. Psychiatry 162, 1904–1910.

Uren, J., Cotton, S., Killackey, E., Saling, M., Allott, K., 2017. Cognitive clusters in first-episode psychosis: overlap with healthy controls and relationship to concurrent and prospective symptoms and functioning. Neuropsychology 31, 787–797.

Van Donkersgoed, R.J., Wunderink, L., Nieboer, R., Aleman, A., Pijnenborg, G.H., 2015. Social cognition in individuals at ultra-high risk for psychosis: a meta-analysis. PLoS One 10, e0141075.

Vorstman, J.A., Breetvelt, E.J., Duijff, S.N., Eliez, S., Schneider, M., Jalbrzikowski, M., Armando, M., Vicari, S., Shashi, V., Hooper, S.R., Chow, E.W., Fung, W.L., Butcher, N.J., Young, D.A., Mcdonald-Mcginn, D.M., Vogels, A., VAN Amelsvoort, T., Gothelf, D., Weinberger, R., Weizman, A., Klaassen, P.W., Koops, S., Kates, W.R., Antshel, K.M., Simon, T.J., Ousley, O.Y., Swillen, A., Gur, R.E., Bearden, C.E., Kahn, R.S., Bassett, A.S., 2015. Cognitive decline preceding the onset of psychosis in patients with 22q11.2 deletion syndrome. JAMA Psychiatry 72, 377–385.

Weiser, M., Reichenberg, A., Kravitz, E., Lubin, G., Shmushkevich, M., Glahn, D.C., Gross, R., Rabinowitz, J., Noy, S., Davidson, M., 2008. Subtle cognitive dysfunction in nonaffected siblings of individuals affected by nonpsychotic disorders. Biol. Psychiatry 63, 602–608.

Wood, S.J., Brewer, W.J., Koutsouradis, P., Phillips, L.J., Francey, S.M., Proffitt, T.M., Yung, A.R., Jackson, H.J., Mcgorry, P.D., Pantelis, C., 2007. Cognitive decline following psychosis onset. Data from the PACE clinic. Br. J. Psychiatry 191 (Suppl. 51), s52–s57.

Woodberry, K.A., Giuliano, A.J., Seidman, L.J., 2008. Premorbid IQ in schizophrenia: a meta-analytic review. Am. J. Psychiatry 165, 579–587.

Woodberry, K.A., Mcfarlane, W.R., Giuliano, A.J., Verdi, M.B., Cook, W.L., Faraone, S.V., Seidman, L.J., 2013. Change in neuropsychological functioning over one year in youth at clinical high risk for psychosis. Schizophr. Res. 146, 87–94.

Wykes, T., Huddy, V., Cellard, C., Mcgurk, S.R., Czobor, P., 2011. A meta-analysis of cognitive remediation for schizophrenia: methodology and effect sizes. Am. J. Psychiatry 168, 472–485.

Yung, A.R., Nelson, B., Thompson, A., Wood, S.J., 2010. The psychosis threshold in Ultra High Risk (prodromal) research: is it valid? Schizophr. Res. 120, 1–6.

Zammit, S., Allebeck, P., David, A.S., Dalman, C., Hemmingsson, T., Lundberg, I., Lewis, G., 2004. A longitudinal study of premorbid IQ score and risk of developing schizophrenia, bipolar disorder, severe depression, and other nonaffective psychoses. Arch. Gen. Psychiatr. 61, 354–360.

Ziermans, T., DE Wit, S., Schothorst, P., Sprong, M., VAN Engeland, H., Kahn, R., Durston, S., 2014. Neurocognitive and clinical predictors of long-term outcome in adolescents at ultra-high risk for psychosis: a 6-year follow-up. PLoS One 9, e93994.

CHAPTER

15

Society and risk of psychosis

Craig Morgan[1], *Tessa Roberts*[1], *Brian O. Donoghue*[2], *Andrew D. Thompson*[2,3]

[1]ESRC Centre for Society and Mental Health and Social Epidemiology Research Group, King's College London, London, United Kingdom; [2]Orygen, the National Centre of Excellence in Youth Mental Health, VIC, Australia; [3]Mental Health and Wellbeing Division, Warwick Medical School, University of Warwick, Coventry, United Kingdom

Our understanding of the nature and etiological architecture of psychotic mental disorders has advanced notably in the past 20 years. It is now widely considered that psychotic disorders occur as a consequence of a complex range of factors, spanning multiple domains and levels, that act on several psychological and biological mechanisms (Radua et al., 2018). It is clear from this that no single factor is sufficient or necessary to cause psychoses, which means that multiple interacting causal partners must be involved in determining onset. Indeed, this heterogeneity of risks may underlie variations in the manifestation and course and outcome of psychotic disorders. In recent years, evidence has accumulated that exposure to, broadly, adverse social conditions and experiences over the course of development constitutes part of this complex array of factors that influence onset (Howes and Murray, 2014; Murray et al., 2017). In this chapter, we will provide a critical overview of this evidence, including consideration of outstanding conceptual and methodological issues that limit the causal inferences that can be drawn from existing findings. From this, we will propose and sketch a sociodevelopmental pathway—as one of the several pathways—to psychoses, in which exposure to multiple social risks over time, particularly those involving threat, hostility, and violence, are the primary drivers of onset.

Recent developments

As a basis for considering social risks and psychoses, it is useful to briefly sketch several developments in our understandings of psychoses in general that have occurred over the past 20 years.

Variations in incidence by place and social group

Twenty years ago, it was still widely believed that the incidence of psychoses was broadly similar around the world. This was based on findings from a multinational study conducted in the 1980s by the World Health Organization

Risk Factors for Psychosis
https://doi.org/10.1016/B978-0-12-813201-2.00015-6

of the incidence and presentation of psychoses in 10 countries (Jablensky et al., 1992). The relatively narrow—and nonstatistically significant at p 0.05—range of incidence rates for narrow schizophrenia reported in this study (from 7 per 100,000 in Aarhus, Denmark, to 14 per 100,000 in Nottingham, UK) was widely interpreted as indicating a uniform worldwide incidence of schizophrenia. In retrospect, this interpretation was always flawed, relying arbitrarily on the absence of statistical significance and ignoring the actual evidence of variation (i.e., 2.0-fold variation in narrow schizophrenia; 2.6-fold variation in broad schizophrenia). Further, substantial evidence has accrued since then that the incidence of psychotic disorders does vary widely across geographical areas and social groups. For example, in an early review of 100 incidence studies, McGrath et al. found that rates varied fivefold across place, even after removing the bottom and top 10% reported rates (McGrath et al., 2004). They further found evidence that rates are higher among men, in urban areas, and among minority ethnic groups. These findings have been replicated in several studies and further reviews since, but with important nuances and caveats (Kirkbride et al., 2012; Castillejos et al., 2018). For instance, rates may not be elevated in all urban environments and all migrant and minority groups, and intriguing epidemiological differences have been reported between sites in the global north and south (Jongsma et al., 2018; Fearon et al., 2006; DeVylder et al., 2018). Variations by place and social group are important because understanding them can provide potential clues to etiology.

Heterogeneity

Further, various recent trends in research have converged to highlight the marked heterogeneity of psychoses, both in terms of their manifestations and symptomatology and in their course and outcome. For example, there are now several reports that distinctive symptom profiles are associated with exposure to specific environmental risk factors (Quattrone et al., 2019; Bentall et al., 2014; Oher et al., 2014; Berg et al., 2014; Demjaha et al., 2009; van der Ven et al., 2012; Ajnakina et al., 2016), and recent long-term outcome studies confirm the highly variable course and outcome of psychoses (Volavka and Vevera, 2018; Menezes et al., 2006; Cohen et al., 2007). This heterogeneity—along with evidence that many risk factors are common across diagnoses—has further and increasingly challenged current diagnostic categories and boundaries. This underpins a trend to study psychoses broadly and, more radically, to dispense with diagnoses and focus on particular symptoms and processes (e.g., the RDoC initiative that aimed to investigate dimensions of neurobiology and behavior that cut across diagnostic categories (Insel et al., 2010)).

A continuum of psychosis

At the same time, substantial evidence has emerged that low-level, psychotic—or unusual and anomalous—experiences, such as fleeting hallucinations, suspiciousness, and magical thinking, are common in the general population (McGrath et al., 2015; Johns and Van Os, 2001; Kelleher et al., 2010). It seems, moreover, that these experiences are associated with an increased risk of psychotic disorder and with similar risk factors to those for psychotic disorders (Kelleher and Cannon, 2011). For example, there is evidence that these experiences are also more common in some urban areas and some of the minority ethnic groups in which high rates of psychotic disorder have been reported (Van Os et al., 2009). These developments have stimulated much research on the basis that investigating low-level psychotic experiences may tell us something about psychotic disorder (see Chapter 7). However, this is complicated by more recent findings that show low-level psychotic experiences are also associated with depression, anxiety, PTSD, and suicide (Kaymaz et al., 2012; Honings et al., 2016; Kelleher et al., 2014). This is important for our purposes because low-level perceptual disturbances and unusual beliefs may be general markers of distress and not necessarily indicators of an underlying vulnerability for psychotic disorder. This noted that the important point for the purposes of this chapter is that many studies of social conditions and experiences have, in fact, been of psychotic experiences in general population samples.

Further, this line of research has contributed to renewed interest in continuum models of psychosis, in which psychotic disorder lies at the extreme end of a continuum of psychotic experiences that vary in their frequency, intensity, and severity (Van Os et al., 2009). In this model, the stage before the development of psychotic disorder—when symptoms are distressing, but not sufficiently severe to meet criteria for disorder—has been characterized as an ultra high risk (UHR) mental state (see Chapter 3 for criteria). Researching the UHR stage provides opportunities to investigate the effects of social and other environmental factors on risk in at least two respects. First, it is possible to study predictors of transition to disorder. However, findings have not always been in line with what might be expected. For example, a review of relevant evidence found that a lifetime history of cannabis use was not associated with transition (Kraan et al., 2016), despite being an established risk factor for psychotic disorder. The review did, however, find that current cannabis abuse was associated with transition. Second, it is possible to study factors associated with the development of a UHR state. Again, using the example of cannabis, studies have found that the prevalence of cannabis use in UHR cohorts is higher than in the general population. It is possible, then, that environmental factors exert their effects early, in the progression to a UHR state. These factors may or may not be associated with subsequent risk of transition, which could explain how some environmental risk factors could be overrepresented in first-episode cohorts but not associated with the risk of transition to psychotic disorder in the UHR state.

The etiological architecture of psychoses

Finally, over the past 20 years, our understanding of the broad etiological architecture of psychotic disorders has increased substantially. It is now clear that an array of factors, that are neither necessary nor sufficient in themselves to cause disorder, are associated with an increased risk of psychoses, spanning genetic, neurobiological, psychological, and social domains (Murray et al., 2017). That none are sufficient or necessary means that multiple factors must coparticipate over time, in various combinations, to push individuals along a developmental pathway to psychosis. Further, evidence is converging on interrelated psychological and biological mechanisms through which these factors increase risk, notably via effects on affective and cognitive processes and on physiological stress response and dopamine synthesis (Howes and Murray, 2014). Still, these broad contours of risk need much further elaboration, particularly social risks.

Social risks

These developments provide the general context for considering the potential role of social risks. For example, evidence that the incidence of psychotic disorders varies by social context and social groups has partly driven renewed interest in the possible role of social conditions and experiences in the development of psychoses. Substantial evidence has now accumulated implicating a range of social factors at area, family, and individual levels in the etiology of psychoses. Here, we will provide a selective critical review of this evidence, highlighting ongoing conceptual and methodological issues that limit inferences about the causal role of social risks, focusing in particular on the most extensively studied area: childhood adversity.

Contexts: ecologies of risk

In general, place matters for health. There is, for example, an extensive literature documenting the deleterious consequences for health and development of living in disadvantaged, unequal, and fragmented neighborhoods (March et al., 2008; Kirkbride et al., 2007a,b). Some of the mechanisms that link these exposures with particular health problems may vary, but in general, such conditions likely impact on individuals by constraining access to resources—both financial and social—that buffer against other vulnerabilities and adversities and that promote resilience and well-being (e.g., affordable healthy food, social supports, green spaces, leisure facilities, etc.). It seems that psychotic disorders are more common in these areas.

Interest in social contexts and psychoses has a long history. In a seminal and now highly referenced study in the 1930s, Faris and Dunham reported that hospital admissions for schizophrenia in Chicago, USA, were higher among those who lived in inner-city areas characterized by high levels of social fragmentation (Faris and Dunham, 1939), i.e., a high proportion of single person households and migrants, high crime rates, etc. More recent work has found broadly similar patterns. For example, using data from the AESOP study, Kirkbride et al. found that rates of psychoses were highest in the most disadvantaged and fragmented areas of south London, UK (Kirkbride et al., 2007a,b). Similarly, in the Cavan–Monaghan First Episode Psychosis Study in rural Ireland, Omer et al. found higher rates of disorder in more deprived areas (Omer et al., 2014). It is possible that greater deprivation and fragmentation in part account for the high rates of disorder found in some densely populated urban areas. Allardyce et al. sought to directly test this using data on first admissions to hospital in Scotland (Allardyce et al., 2005). When simultaneously including urbanicity (population density) and indices of area-level disadvantage and fragmentation in regression models, they found that both disadvantage and fragmentation, but not population density, were still associated with an increased risk of psychoses. Zammit et al. found that the association between urbanicity and rates of psychosis in Sweden was primarily explained by school-level social fragmentation—although most of the variation in rates was accounted for by individual-level factors (Zammit et al., 2010). Finally, in an extension to the AESOP study, Kirkbride et al. found that areas with lower neighborhood social capital—measured using voter turnout as a proxy indicator—and greater ethnic segregation had the highest rates of psychosis after controlling for individual-level factors, rather than the most socioeconomically deprived neighborhoods per se, suggesting that local social capital and cohesion may provide a buffer against the social exposures that increase risk for psychosis (Kirkbride et al., 2007a,b).

There are, however, limitations to this work, the most salient being that it is difficult to disentangle cause and effect. It is possible, as many commentators have noted, that those predisposed to developing schizophrenia either more often move into or more often remain in fragmented, inner-city areas. There is some evidence to suggest this does not fully explain observed associations. For example, Pedersen and Mortensen found that living in more densely populated areas during childhood increased risk of later psychoses and that movement out of urban areas decreased risk (Pedersen and Mortensen, 2001). Still, the possibility that associations reflect, in part at least, effects that go from psychosis to urban living means we should remain cautious in interpreting these findings.

Extending this, a body of research has considered interactions between area-level and individual-level characteristics, with some intriguing findings. The most studied example is ethnic density × ethnicity interactions. Ethnic density refers to the proportion of individuals from minority ethnic groups living in an area, with higher proportions indicating higher levels of ethnic density. Several studies have reported that the relative incidence of psychoses is highest among minority ethnic groups when they form a smaller proportion of local populations (i.e., in areas of low ethnic density). This finding is remarkably well-replicated (Bosqui et al., 2014). However, it is not easy to explain in terms of reverse causation, given that areas of low minority ethnic group density tend to be wealthier, or in terms of known biological risk factors. Consequently, speculation about what underlies this finding has focused on social aspects of living in areas of high or low ethnic density. For instance, those living in areas of low ethic density may face more discrimination and have access to fewer social supports. Conversely, those living in areas of high ethnic density may be less isolated and have more access to supportive networks. Das Munshi et al., in a study using data from the Ethnic Minorities Psychiatric Illness Rates in the Community survey, did find some evidence to support these possibilities, i.e., in low ethnic density areas, more social adversity such as discrimination, and reduced social support (Das-Munshi et al., 2012), which were in turn associated with increased risk of subthreshold psychotic experiences. This type of area - × individual interaction may extend to other characteristics. For example, van Os et al. found that being single was associated with greater risk of psychotic experiences in areas where there was a low proportion of single people (Van Os et al., 2000). Thinking more broadly and speculatively, these findings together tentatively suggest risk of psychoses is increased when individuals live in neighborhoods where they feel like outsiders or disadvantaged relative to others. This certainly merits further research.

Migrant and minority ethnic populations

Related to the findings on ethnic density, it is well established that the incidence of psychotic disorders—and the prevalence of low-level psychotic experiences—is higher in several migrant and minority ethnic populations (Selten et al., 2019; Cantor-Graae and Selten, 2005). The current evidence suggests that greater exposure to adverse social contexts and experiences over the life course in many migrant and minority populations underlie these higher rates. This is supported by findings that an earlier age at migration is associated with a greater risk (Kirkbride et al., 2017). By contrast, in the limited research on UHR samples, there is no evidence that migrant and minority groups are overrepresented and no evidence of an increased risk of transition (O'Donoghue et al., 2015; Nelson et al., 2018; Geros et al., 2020) (Byrne et al., 2019). Indeed, several studies in Australia and the United Kingdom suggest migrant and minority groups are underrepresented in UHR cohorts (Kirkbride et al., 2015) or at least underrepresented relative to first-episode cohorts (Byrne et al., 2019). However, this may be an artifact of the UHR model that requires individuals to be help-seeking. We know that those from migrant and minority groups are less likely, for several reasons, to actively seek help from mental health services.

Social position, disadvantage, and experience: vulnerability and resilience

Refining our focus more, there is evidence that several family- and individual-level social indicators and exposures across the life course are associated with increased risk or odds of psychoses. These range from crude indicators of household and individual social position (e.g., social class), disadvantage (e.g., unemployment), and isolation (e.g., living alone; limited social networks) to direct measures of adverse experiences, particularly during childhood (e.g., discrimination, abuse, etc.) (Morgan et al., 2008), and daily hassles (Oorschot et al., 2009). Here, we focus on childhood adversity because this is the most extensively studied social exposure and most clearly illustrates the current state of our knowledge about the role of social risks in the etiology of psychoses and the remaining gaps, limitations, and methodological challenges.

The expansion of research on a wide range of childhood adversities and both psychotic experiences and disorders over the past 20 years is remarkable. There have been at least 13 systematic reviews of related literatures in the past 15 years alone (Morgan and Gayer-Anderson, 2016). On the face of it, the accumulated evidence is clear: all forms of adversity in childhood that have been studied are associated with an increased risk or odds of psychotic disorder in adulthood. This is most clearly illustrated in Varese et al.'s review and metaanalysis of 36 studies, which found that any adversity was associated with 2.78 increased odds of psychosis and that each form of adversity—i.e., parental death, neglect, abuse, and bullying—was associated with a twofold to fourfold increased odds of psychosis (Varese et al., 2012). They further found that these associations held regardless of study type, that is whether cross-sectional, case–control, or longitudinal, and that, when considered, most studies found evidence that the more adversities experienced, the greater the odds of psychosis, suggesting cumulative effects of exposure to multiple adversities (see below). Since this review, further studies have continued to find similar associations. For example, a metaanalysis of research on UHR samples reported a strikingly high mean prevalence of childhood trauma of 87% (Kraan et al., 2015), which included a history of physical, psychological, sexual, and emotional abuse and neglect. Related to this, in a longitudinal study of UHR individuals, in which there was a 5-year transition rate of 23%, a history of sexual abuse was associated with transition, but physical abuse, physical neglect, emotional abuse, and emotional neglect were not (Thompson et al., 2014). This replicated the findings of an earlier, albeit smaller study, which found that the young people at UHR who had experienced sexual trauma had nearly a threefold increased risk of transition to a psychotic disorder (Bechdolf et al., 2010). It appears, then, that childhood adversities, in particular abuse, are associated with an increased risk across the spectrum of psychotic experiences.

However, there are several outstanding issues that limit what can be inferred from this body of research about the nature of the relationship between childhood adversities and psychotic disorders. To begin, much of this research—or at least the most robust, including most prospective studies—are of psychotic experiences in general populations, which, as noted, bear an indeterminate relationship with psychotic disorder and are associated with several other forms of mental distress, including depression, anxiety, and suicidality, all of which frequently occur against a background of childhood adversity. To be sure, when studied, researchers have tended to report high levels of adversities among those with a psychotic disorder (Matheson et al., 2013). But the research has been less consistent, possibly because of methodological differences, and overall is less impressive. There are, moreover, more methodological challenges and therefore limitations to research on psychotic disorder (Morgan and Gayer-Anderson, 2016). The potential for selection and recall biases, given that study designs are generally case–control, has been well rehearsed, and the possibility that these exaggerate differences between cases and controls in reports of childhood adversity cannot be entirely ruled out. It is noteworthy that a recent study found limited overlap in findings when using contemporary and retrospective methods of ascertaining exposure to childhood adversities (Baldwin et al., 2019), a finding that should temper all inferences drawn from the existing literature.

There are further considerations. Most research to date has tended to analyze adversities as dichotomous variables, grouping individuals into exposed versus unexposed to a specific adversity (e.g., separation from a parent, abuse, bullying, etc.) at any time during childhood, without considering the timing, duration, frequency, or severity of exposure. But these dimensions of experience are likely to be profoundly important. The meaning and impact of a relatively brief period of name-calling at school—which would meet most definitions of bullying—is very different from frequent verbal abuse, physical threat, and prolonged ostracism by peers through several years of childhood. Yet, we have tended to group these very different experiences together. Along similar lines, the timing of exposure may matter. Adolescence, for example, is a critical period of development, both biologically and socially. It is a period during which the brain is especially plastic and shaped by experience and during which peer influence can have profound effects on an individual's developing sense of self.

Timing, severity, specificity

The potential importance of these dimensions is apparent in our analyses of data from the Childhood Adversity and Psychosis (CAPsy) study, which was designed specifically to address the limitations of previous research in ignoring, for example, timing and severity. For example, our findings did suggest that exposure to bullying and sexual abuse during adolescence may have stronger effects on odds of later psychosis than exposure during early childhood. More intriguing still, when we analyzed various forms of adversity (i.e., household discord, several forms of abuse, bullying) by severity, we found that the strongest effects were for the most severe forms. In particular, household discord was only associated with increased odds of psychosis when it involved domestic violence (i.e., odds ratios for marked household discord with no violence, 1.22 vs. with violence, 4.40) and bullying was only associated with an increased odds of psychosis when it involved physical assault (i.e., odds ratios for bullying with no assault, 1.27 vs. bullying with assault, 1.91) (submitted).

These findings are particularly striking and bring to mind previous studies that similarly implicate experiences involving threat, hostility, and violence. It is notable, for example, that hostility is one of the elements that constitute negative expressed emotion in family interactions, which is strongly associated with an increased risk of relapse (Alvarez-Jimenez et al., 2012). In more recent work, Arseneault et al., in a study of over 2000 twin pairs assessed at ages 8 and 12, found that experiences of maltreatment and bullying, but not accidents, at age 8 were associated with an increased risk of psychotic experiences at age 12 (Arseneault et al., 2011). These findings tentatively suggest that there may be some degree of specificity for type and severity of adversity and psychoses, i.e., for experiences involving threat, hostility, and violence. Finally, this ties in with hypotheses formulated by Tirril Harris following from her work with George Brown on life events and depression. In finding that experiences involving loss, entrapment, and humiliation were particularly important in the onset of depression, Harris speculated that intrusive events and difficulties, i.e., those that directly intrude into and threaten the integrity of individuals' daily lives, may be especially important for psychoses (Harris, 1987). There is some evidence to support this. In the CAPsy study, intrusive life events were associated with sixfold increased odds of psychosis, compared with threefold increase for nonintrusive events (submitted). Finally, it is notable that one characteristic of socially fragmented areas, in which rates of psychosis are high, is high levels of crime.

It is possible, following from this, that social adversities over the life course have both general and specific effects on risk of mental health problems. All negative and difficult experiences may have deleterious effects on biological processes involved in stress response and on psychological processes relating to self-esteem and sense of self that increase distress. At the same time, the differing nature and meaning of contexts and experiences—and the timing of exposure—may specifically increase risk for particularly types of problems, i.e., low self-esteem and depression following humiliation, etc., and suspiciousness and ideas of reference following prolonged threat.

Clusters of adversity: cumulative and synergistic effects

Adversities tend to cluster; they rarely occur in isolation. Consequently, a full understanding of the impact of adversities on risk of psychoses—and indeed of mental health problems in general—must involve consideration of how social risks combine and interact over time.

To date, approaches to this have tended to be relatively crude, simply counting the number of adversities experienced and analyzing whether there is evidence of cumulative (dose-response like) effects. As noted above, for childhood adversities, there is some evidence that the more the adversities experienced, the greater the risk of psychoses. A more sophisticated approach is to weight each exposure by its established effect size. We have attempted to do this in proposing an environmental risk score, to capture the total risk from a range of environmental exposures, for use in future research and, with further development, possibly in clinical practice (Vassos et al., 2018). These approaches, however, assume the effects of each adversity add up (i.e., are independent) and that the cumulative effect is therefore linear. This may not always be true.

The impact of exposure to multiple adversities over the life course may be such that they compound and amplify each other over time, i.e., they interact to increase risk over and above the effect of each alone (Mansueto and Faravelli, 2017; Stilo et al., 2017). There is some evidence for these synergistic effects in relation to adversity in childhood and adulthood. For example, in analyses of data on psychotic experiences in an adult general population sample, we found some evidence that the combined effect of exposure to both abuse in childhood and threatening events in adulthood on odds of psychotic experiences was greater than the effect of each alone (Morgan et al., 2014a,b; Bhavsar et al., 2017). That is, experiences of both childhood abuse and adult life events was associated with 3.37-fold increased odds of psychotic experiences, compared with 2.16 for abuse alone and 1.65 for events in the past year alone. Others have reported similar findings (Newbury et al., 2017; Morgan et al., 2014a,b), and initial analyses of CAPsy data suggest this holds in relation to psychotic disorder. There are caveats. The number of studies examining combined effects of multiple environmental exposures is small, and detecting all but large synergistic effects is difficult, given the limited sample sizes of most currently available data sets. However, they do tentatively suggest repeated exposure to adversities over time, especially those involving threat, compounds and amplifies risk, pushing some along what might be termed a predominantly sociodevelopmental pathway to psychotic disorder.

Further, social risks may similarly combine synergistically with other risk factors. For example, there is some tentative evidence of synergistic effects of childhood adversities and cannabis use (Morgan et al., 2014a,b). There has also been much speculation about gene - + environment interaction, i.e., the impact of environmental exposures such as social adversities is most pronounced among those with an underlying genetic vulnerability to psychoses. There are some scattered studies, using various measures of genetic risk and environmental exposures, that have found evidence of interaction (van Os et al., 2004; Caspi et al., 2005). For example, Van Os et al. found evidence of interaction between urbanicity and family history of psychosis, with the combined effect of both being greater than the effect of each alone. However, several studies have failed to find any strong evidence of interaction (Zwicker et al., 2018). Indeed, the most recent studies using polygenic risk scores for schizophrenia—considered the current best measure of overall genetic risk—have suggested effects are largely independent (Trotta et al., 2016), albeit with the caveat that most studies of gene–environment interaction in relation to psychosis are underpowered to detect all but large synergistic effects (Modinos et al., 2013).

Mechanisms

In the past, skepticism about the potential role of social risks in the etiology of psychoses stemmed in part from a lack of plausible mechanisms that might explain how common experiences of adversity could impact on the brain and cognitive development to generate hallucinations, delusions, and disturbances of thought. This has changed. While direct evidence is still limited, there are now several interrelated candidate psychological and biological mechanisms that may link social risks and psychotic disorder.

For example, experiences of threat, hostility, and violence may increase risk of psychoses, particularly paranoid delusions, via the development of cognitive biases (e.g., tendencies to ruminate, to jump to conclusions, etc.) and affective processes (e.g., increased anxiety). That is, prolonged exposure to threat, especially during childhood and adolescence, may contribute to a heightened tendency to ruminate and to perceive directed interpersonal threat in everyday interactions, fostering beliefs of persecution that—in the presence of additional risks—develop and crystallize into fixed beliefs about the negative, conspiratorial intentions of others (Bentall et al., 2009). Further, there is evidence that various forms of adversity impact on biological systems implicated in the underlying pathophysiology of psychoses, including dysregulation of the HPA axis and dopamine system (Howes and Murray, 2014). More specifically, it is notable that the dopamine system is involved in the regulation of emotion responses, including fear, and the attachment of salience to external stimuli (Heinz and Schlagenhauf, 2010). It is possible, then, that prolonged exposure to severe threat, hostility, and violence contribute to long-term changes in biological systems that underpin the cognitive and affective processes noted above, i.e., fear in response to and attachment of aberrant levels of salience to relatively neutral contexts and experiences, leading to the development of excessive suspiciousness and, ultimately, delusions of paranoia and reference and other psychotic symptoms.

Variations in population rates of psychoses

Returning briefly to variations in the incidence of psychoses across populations, it is possible that these variations are underpinned by differences in the population distribution of social risks (and indeed others, such as cannabis). Put simply, the more common risks are in a population, the greater the incidence is likely to be. In particular, we have previously speculated that this may explain the high rates of psychoses observed in some migrant and minority ethnic populations (Morgan et al., 2019), i.e., greater exposure over the life course to the cumulative risks of social disadvantage, marginalization, and adversities involving direct threat and hostility, including discrimination.

A sociodevelopmental pathway to psychosis

Putting all of this together, it is possible—as we have suggested before—that there is a distinct sociodevelopmental pathway in which prolonged exposure to threat at critical developmental junctures—possibly against a background of poverty and family breakdown—has lasting effects on cognitive schema, affective processes, and related biological systems that predispose some to developing a psychotic disorder. This vulnerability may be realized in the event of exposure in adulthood to further risks. As noted, in relation to psychotic experiences at least, we have previously found that adult life events and cannabis use combine synergistically with childhood adversity to increase risk, as have others (Morgan et al., 2014a,b). This may also explain variation in the timing of onset. That is, the vulnerability embodied as a consequence of early adversity may manifest only in the presence of subsequent risks, which may occur many years later. More provocatively, this pathway may, in many instances, be independent of genetic and neurodevelopmental risks. Again, as noted, despite much speculation, most studies have failed to find any strong evidence that childhood adversity and genes combine synergistically to increase risk. They appear independent. Of course, this may reflect methodological limitations (e.g., small samples, crude measures of adversity, failure to fully capture and measure genetic risk). But it is also possible that there are multiple, only partly overlapping, pathways to psychosis. This, as others have suggested, may explain the marked heterogeneity in the incidence, development, phenomenology, and course and outcome of psychotic disorders.

Future research

This chapter has provided an overview of the existing evidence on social risks and psychoses and, from this, proposed that some develop psychosis primarily as a consequence of exposure over time to adverse social conditions and experiences, that is, via a sociodevelopmental pathway to disorder. However, there are many methodological limitations that characterize the current evidence base, and this is therefore necessarily a tentative hypothesis. Much more research is needed to test this idea.

To do this, future research needs to extend beyond its current frameworks and methods. For example, we need a stronger organizing conceptual framework for how health develops over time in specific contexts to guide research. Mental health is a dynamic, developmental phenomenon, shaped by complex interactions between individuals and environments. That is, states of mental health emerge and fluctuate over time as a consequence of causes that operate at multiple levels—from societal to molecular. From this perspective, mental health development is time sensitive. There are periods when biological systems are more plastic and adaptable and when socially structured pathways produce linked exposures that create patterns of risks (e.g., multiple, clustered social risks, including poverty, discrimination, violence, etc.) that can have lasting effects on development of mental health. This is particularly so during the developmental transitions through childhood, adolescence, and early adulthood—critical periods for brain maturation and the formation of social relationships and identities—when the impact of exposures is amplified, with long-term health consequences. Research on the influence of social contexts and experiences on the emergence of psychotic phenomena and ultimately disorder consequently needs to incorporate more fully the element of time, the concept of critical periods, and the importance of specific ecologies of risk. To do this, research also needs to embrace—alongside traditional approaches—more innovative methods to enable dynamic social processes and their impacts to be more fully captured. This includes the use of novel technologies to facilitate real-time data collection to capture experience and its impacts in the flow of everyday life, including experience sampling, virtual reality, GPS mapping, social media, sociometry, and wearable technologies. This further extends to the use of more sophisticated statistical techniques and simulations, including agent-based models, microsimulation, and machine learning.

Further, to fully understand the role of the social contexts and experiences in the etiology of psychotic disorders, research needs to extend to a far wider range of settings than have been studied to date. The vast majority of research on psychotic disorders has been conducted in a small number of settings in high-income countries in Europe, North America, and Australasia, which limits the diversity of populations and social exposures studied. There are moves in this direction. For example, the INTREPID program is designed to conduct directly comparable studies of the phenomenology, etiology, and outcome of psychoses in settings in Asia, Africa, and the Caribbean (Morgan et al., 2016).

Implications

The need for more research notwithstanding, there are several important implications that follow from what we already know about social risks and psychosis. We conclude with two examples. First, a population perspective on risk of psychosis that highlights the social conditions and experiences that underpin onset in many cases points toward prevention as the optimum social and public health strategy to reduce incidence and to minimize burden. For example, there is some evidence that reducing bullying decreases risk of psychosis, making this a promising target for prevention (Kelleher et al., 2008). Indeed, if causal, estimates of population impacts for social risks suggest that upward of a third of cases of psychosis could be prevented if these risks—or their effects—were removed (Kirkbride and Jones, 2010; Kirkbride et al., 2010). Potential strategies for prevention include current approaches that target those at UHR of disorder, but also need to extend further into populations to target the underlying, fundamental social causes, including marginalization and discrimination, especially in minority populations, and exposure to contexts and experiences of threat and violence. The detail of which strategies and community-level interventions can mitigate the effects of these exposures and build, at a community level, resilience remains to be worked out and is a priority for research. Second, irrespective of whether the social conditions and contexts reviewed in this chapter are in fact causal, it remains that many of those who present to mental health services with a first-episode psychosis will do so against a background of lifetime exposure to poverty, disadvantage, discrimination, hostility, and violence. For many patients, these facets of their social lives are in fact the major sources of distress. This implies a reorientation of mental health services, particularly early intervention services, to incorporate more thorough assessments of and responses to patients' social needs. To note this is not to suggest social needs are currently ignored. But the extent to which they are addressed—across countries—is, at best, patchy and ad hoc. Ensuring services systematically address the social needs of those with a psychotic disorder, as a primary and mandated part of early intervention, is one change that society can make to mitigate the distress of psychosis without the need for further research on social risks.

References

Ajnakina, O., Trotta, A., Oakley-Hannibal, E., Di Forti, M., Stilo, S., Kolliakou, A., et al., 2016. Impact of childhood adversities on specific symptom dimensions in first-episode psychosis. Psychol. Med. 46 (2), 317–326.

Allardyce, J., Gilmour, H., Atkinson, J., Rapson, T., Bishop, J., McCreadie, R., 2005. Social fragmentation, deprivation and urbanicity: relation to first-admission rates for psychoses. Br. J. Psychiatry 187 (5), 401–406.

Alvarez-Jimenez, M., Priede, A., Hetrick, S., Bendall, S., Killackey, E., Parker, A., et al., 2012. Risk factors for relapse following treatment for first episode psychosis: a systematic review and meta-analysis of longitudinal studies. Schizophr. Res. 139 (1–3), 116–128.

Arseneault, L., Cannon, M., Fisher, H.L., Polanczyk, G., Moffitt, T.E., Caspi, A., 2011. Childhood trauma and children's emerging psychotic symptoms: a genetically sensitive longitudinal cohort study. Am. J. Psychiatry 168 (1), 65–72.

Baldwin, J.R., Reuben, A., Newbury, J.B., Danese, A., 2019. Agreement between prospective and retrospective measures of childhood maltreatment: a systematic review and meta-analysis. JAMA Psychiatry 76 (6), 584–593.

Bechdolf, A., Thompson, A., Nelson, B., Cotton, S., Simmons, M.B., Amminger, G.P., et al., 2010. Experience of trauma and conversion to psychosis in an ultra-high-risk (prodromal) group. Acta Psychiatr. Scand. 121 (5), 377–384.

Bentall, R.P., Rowse, G., Shryane, N., Kinderman, P., Howard, R., Blackwood, N., et al., 2009. The cognitive and affective structure of paranoid delusions: a transdiagnostic investigation of patients with schizophrenia spectrum disorders and depression. Arch. Gen. Psychiatr. 66 (3), 236–247.

Bentall, R.P., de Sousa, P., Varese, F., Wickham, S., Sitko, K., Haarmans, M., et al., 2014. From adversity to psychosis: pathways and mechanisms from specific adversities to specific symptoms. Soc. Psychiatry Psychiatr. Epidemiol. 49 (7), 1011–1022.

Berg, A.O., Andreassen, O.A., Aminoff, S.R., Romm, K.L., Hauff, E., Melle, I., 2014. The impact of immigration and visible minority status on psychosis symptom profile. Soc. Psychiatry Psychiatr. Epidemiol. 49 (11), 1747–1757.

Bhavsar, V., Boydell, J., McGuire, P., Harris, V., Hotopf, M., Hatch, S., et al., 2017. Childhood abuse and psychotic experiences—evidence for mediation by adulthood adverse life events. Epidemiol. Psychiatr. Sci. 1–10.

Bosqui, T.J., Hoy, K., Shannon, C., 2014. A systematic review and meta-analysis of the ethnic density effect in psychotic disorders. Soc. Psychiatry Psychiatr. Epidemiol. 49 (4), 519–529.

Byrne, M., Codjoe, L., Morgan, C., Stahl, D., Day, F., Fearon, P., et al., 2019. The relationship between ethnicity and service access, treatment uptake and the incidence of psychosis among people at ultra high risk for psychosis. Psychiatry Res. 272, 618–627.

Cantor-Graae, E., Selten, J.P., 2005. Schizophrenia and migration: a meta-analysis and review. Am. J. Psychiatry 162 (1), 12–24.

Caspi, A., Moffitt, T.E., Cannon, M., McClay, J., Murray, R., Harrington, H., et al., 2005. Moderation of the effect of adolescent-onset cannabis use on adult psychosis by a functional polymorphism in the catechol-O-methyltransferase gene: longitudinal evidence of a gene X environment interaction. Biol. Psychiatry 57 (10), 1117–1127.

Castillejos, M., Martín-Pérez, C., Moreno-Küstner, B., 2018. A systematic review and meta-analysis of the incidence of psychotic disorders: the distribution of rates and the influence of gender, urbanicity, immigration and socio-economic level. Psychol. Med. 48 (13), 2101–2115.

Cohen, A., Patel, V., Thara, R., Gureje, O., 2007. Questioning an axiom: better prognosis for schizophrenia in the developing world? Schizophr. Bull. 34 (2), 229–244.

Das-Munshi, J., Bécares, L., Boydell, J.E., Dewey, M.E., Morgan, C., Stansfeld, S.A., et al., 2012. Ethnic density as a buffer for psychotic experiences: findings from a national survey (EMPIRIC). Br. J. Psychiatry 201 (4), 282–290.

Demjaha, A., Morgan, K., Morgan, C., Landau, S., Dean, K., Reichenberg, A., et al., 2009. Combining dimensional and categorical representation of psychosis: the way forward for DSM-V and ICD-11? Psychol. Med. 39 (12), 1943–1955.

DeVylder, J.E., Kelleher, I., Lalane, M., Oh, H., Link, B.G., Koyanagi, A., 2018. Association of urbanicity with psychosis in low-and middle-income countries. JAMA Psychiatry 75 (7), 678–686.

Faris, R.E.L., Dunham, H.W., 1939. Mental Disorders in Urban Areas: An Ecological Study of Schizophrenia and Other Psychoses.

Fearon, P., Kirkbride, J.B., Morgan, C., Dazzan, P., Morgan, K., Lloyd, T., et al., 2006. Incidence of schizophrenia and other psychoses in ethnic minority groups: results from the MRC AESOP study. Psychol. Med. 36 (11), 1541–1550.

Geros, H., Sizer, H., Mifsud, N., Reynolds, S., Kim, D.J., Eaton, S., et al., 2020. Migrant status and identification as ultra-high risk for psychosis and transitioning to a psychotic disorder. Acta Psychiatr. Scand. 141 (1), 52–59.

Harris, T., 1987. Recent developments in the study of life events in relation to psychiatric and physical disorders. Psychiatr. Epidemiol. 81–103.

Heinz, A., Schlagenhauf, F., 2010. Dopaminergic dysfunction in schizophrenia: salience attribution revisited. Schizophr. Bull. 36 (3), 472–485.

Honings, S., Drukker, M., Groen, R., van Os, J., 2016. Psychotic experiences and risk of self-injurious behaviour in the general population: a systematic review and meta-analysis. Psychol. Med. 46 (2), 237–251.

Howes, O.D., Murray, R.M., 2014. Schizophrenia: an integrated sociodevelopmental-cognitive model. Lancet 383 (9929), 1677–1687.

Insel, T., Cuthbert, B., Garvey, M., Heinssen, R., Pine, D.S., Quinn, K., et al., 2010. Research domain criteria (RDoC): toward a new classification framework for research on mental disorders. Am. Psychiatr. Assoc.

Jablensky, A., Sartorius, N., Ernberg, G., Anker, M., Korten, A., Cooper, J.E., et al., 1992. Schizophrenia: manifestations, incidence and course in different cultures A world health organization ten-country study. Psychol. Med. Monogr. Suppl. 20, 1–97.

Johns, L.C., Van Os, J., 2001. The continuity of psychotic experiences in the general population. Clin. Psychol. Rev. 21 (8), 1125–1141.

Jongsma, H.E., Gayer-Anderson, C., Lasalvia, A., Quattrone, D., Mulè, A., Szöke, A., et al., 2018. Treated incidence of psychotic disorders in the multinational EU-GEI study. JAMA Psychiatry 75 (1), 36–46.

Kaymaz, N., Drukker, M., Lieb, R., Wittchen, H.-U., Werbeloff, N., Weiser, M., et al., 2012. Do subthreshold psychotic experiences predict clinical outcomes in unselected non-help-seeking population-based samples? A systematic review and meta-analysis, enriched with new results. Psychol. Med. 42 (11), 2239–2253.

Kelleher, I., Cannon, M., 2011. Psychotic-like experiences in the general population: characterizing a high-risk group for psychosis. Psychol. Med. 41 (1), 1–6.

Kelleher, I., Harley, M., Lynch, F., Arseneault, L., Fitzpatrick, C., Cannon, M., 2008. Associations between childhood trauma, bullying and psychotic symptoms among a school-based adolescent sample. Br. J. Psychiatry 193 (5), 378–382.

Kelleher, I., Jenner, J.A., Cannon, M., 2010. Psychotic symptoms in the general population—an evolutionary perspective. Br. J. Psychiatry 197 (3), 167–169.

Kelleher, I., Cederlöf, M., Lichtenstein, P., 2014. Psychotic experiences as a predictor of the natural course of suicidal ideation: a Swedish cohort study. World Psychiatry 13 (2), 184–188.

Kirkbride, J.B., Jones, P.B., 2010. The prevention of schizophrenia—what can we learn from eco-epidemiology? Schizophr. Bull. 37 (2), 262–271.

Kirkbride, J.B., Morgan, C., Fearon, P., Dazzan, P., Murray, R.M., Jones, P.B., 2007a. Neighbourhood-level effects on psychoses: re-examining the role of context. Psychol. Med. 37 (10), 1413–1425.

Kirkbride, J.B., Fearon, P., Morgan, C., Dazzan, P., Morgan, K., Murray, R.M., et al., 2007b. Neighbourhood variation in the incidence of psychotic disorders in Southeast London. Soc. Psychiatry Psychiatr. Epidemiol. 42 (6), 438–445.

Kirkbride, J., Coid, J., Morgan, C., Fearon, P., Dazzan, P., Yang, M., et al., 2010. Translating the epidemiology of psychosis into public mental health: evidence, challenges and future prospects. J. Public Ment. Health 9 (2), 4–14.

Kirkbride, J.B., Errazuriz, A., Croudace, T.J., Morgan, C., Jackson, D., Boydell, J., et al., 2012. Incidence of schizophrenia and other psychoses in England, 1950–2009: a systematic review and meta-analyses. PLoS One 7 (3), e31660.

Kirkbride, J.B., Stochl, J., Zimbron, J., Crane, C.M., Metastasio, A., Aguilar, E., et al., 2015. Social and spatial heterogeneity in psychosis proneness in a multilevel case-prodrome-control study. Acta Psychiatr. Scand. 32 (4), 283–292.

Kirkbride, J.B., Hameed, Y., Ioannidis, K., Ankireddypalli, G., Crane, C.M., Nasir, M., et al., 2017. Ethnic minority status, age-at-immigration and psychosis risk in rural environments: evidence from the SEPEA study. Schizophr. Bull. 43 (6), 1251–1261.

Kraan, T., Velthorst, E., Smit, F., de Haan, L., van der Gaag, M., 2015. Trauma and recent life events in individuals at ultra high risk for psychosis: review and meta-analysis. Schizophr. Res. 161 (2–3), 143–149.

Kraan, T., Velthorst, E., Koenders, L., Zwaart, K., Ising, H.K., van den Berg, D., et al., 2016. Cannabis use and transition to psychosis in individuals at ultra-high risk: review and meta-analysis. Psychol. Med. 46 (4), 673–681.

Mansueto, G., Faravelli, C., 2017. Recent life events and psychosis: the role of childhood adversities. Psychiatry Res. 256, 111–117.

March, D., Hatch, S.L., Morgan, C., Kirkbride, J.B., Bresnahan, M., Fearon, P., et al., 2008. Psychosis and place. Epidemiol. Rev. 30 (1), 84–100.

Matheson, S., Shepherd, A.M., Pinchbeck, R., Laurens, K., Carr, V.J., 2013. Childhood adversity in schizophrenia: a systematic meta-analysis. Psychol. Med. 43 (2), 225–238.

McGrath, J., Saha, S., Welham, J., El Saadi, O., MacCauley, C., Chant, D., 2004. A systematic review of the incidence of schizophrenia: the distribution of rates and the influence of sex, urbanicity, migrant status and methodology. BMC Med. 2 (1), 13.

McGrath, J.J., Saha, S., Al-Hamzawi, A., Alonso, J., Bromet, E.J., Bruffaerts, R., et al., 2015. Psychotic experiences in the general population: a cross-national analysis based on 31 261 respondents from 18 countries. JAMA Psychiatry 72 (7), 697–705.

Menezes, N., Arenovich, T., Zipursky, R., 2006. A systematic review of longitudinal outcome studies of first-episode psychosis. Psychol. Med. 36 (10), 1349–1362.

Modinos, G., Iyegbe, C., Prata, D., Rivera, M., Kempton, M.J., Valmaggia, L.R., et al., 2013. Molecular genetic gene–environment studies using candidate genes in schizophrenia: a systematic review. Schizophr. Res. 150 (2–3), 356–365.

Morgan, C., Gayer-Anderson, C., 2016. Childhood adversities and psychosis: evidence, challenges, implications. World Psychiatry 15 (2), 93–102.

Morgan, C., McKenzie, K., Fearon, P., 2008. Society and Psychosis. Cambridge University Press.

Morgan, C., Reininghaus, U., Reichenberg, A., Frissa, S., Hotopf, M., Hatch, S.L., et al., 2014a. Adversity, cannabis use and psychotic experiences: evidence of cumulative and synergistic effects. Br. J. Psychiatry 204 (5), 346–353.

Morgan, C., Reininghaus, U., Fearon, P., Hutchinson, G., Morgan, K., Dazzan, P., et al., 2014b. Modelling the interplay between childhood and adult adversity in pathways to psychosis: initial evidence from the AESOP study. Psychol. Med. 44 (2), 407–419.

Morgan, C., John, S., Esan, O., Hibben, M., Patel, V., Weiss, H., et al., 2016. The incidence of psychoses in diverse settings, INTREPID (2): a feasibility study in India, Nigeria, and Trinidad. Psychol. Med. 46 (9), 1923–1933.

Morgan, C., Knowles, G., Hutchinson, G., 2019. Migration, ethnicity and psychoses: evidence, models and future directions. World Psychiatry 18 (3), 247–258.

Murray, R.M., Bhavsar, V., Tripoli, G., Howes, O., 2017. 30 years on: how the neurodevelopmental hypothesis of schizophrenia morphed into the developmental risk factor model of psychosis. Schizophr. Bull. 43 (6), 1190–1196.

Nelson, B., Amminger, G.P., Yuen, H.P., Markulev, C., Lavoie, S., Schafer, M.R., et al., 2018. NEURAPRO: a multi-centre RCT of omega-3 polyunsaturated fatty acids versus placebo in young people at ultra-high risk of psychotic disorders-medium-term follow-up and clinical course. NPJ Schizophr. 4 (1), 11.

Newbury, J., Arseneault, L., Caspi, A., Moffitt, T.E., Odgers, C.L., Fisher, H.L., 2017. Cumulative effects of neighborhood social adversity and personal crime victimization on adolescent psychotic experiences. Schizophr. Bull. 44 (2), 348–358.

O'Donoghue, B., Nelson, B., Yuen, H.P., Lane, A., Wood, S., Thompson, A., et al., 2015. Social environmental risk factors for transition to psychosis in an Ultra-High Risk population. Schizophr. Res. 161 (2–3), 150–155.

Oher, F., Demjaha, A., Jackson, D., Morgan, C., Dazzan, P., Morgan, K., et al., 2014. The effect of the environment on symptom dimensions in the first episode of psychosis: a multilevel study. Psychol. Med. 44 (11), 2419–2430.

Omer, S., Kirkbride, J.B., Pringle, D.G., Russell, V., O'Callaghan, E., Waddington, J.L., 2014. Neighbourhood-level socio-environmental factors and incidence of first episode psychosis by place at onset in rural Ireland: the Cavan–Monaghan First Episode Psychosis Study [CAMFEPS]. Schizophr. Res. 152 (1), 152–157.

Oorschot, M., Kwapil, T., Delespaul, P., Myin-Germeys, I., 2009. Momentary assessment research in psychosis. Psychol. Assess. 21 (4), 498.

Pedersen, C.B., Mortensen, P.B., 2001. Evidence of a dose-response relationship between urbanicity during upbringing and schizophrenia risk. Arch. Gen. Psychiatr. 58 (11), 1039–1046.

Quattrone, D., Di Forti, M., Gayer-Anderson, C., Ferraro, L., Jongsma, H.E., Tripoli, G., et al., 2019. Transdiagnostic dimensions of psychopathology at first episode psychosis: findings from the multinational EU-GEI study. Psychol. Med. 49 (8), 1378–1391.

Radua, J., Ramella-Cravaro, V., Ioannidis, J.P., Reichenberg, A., Phiphopthatsanee, N., Amir, T., et al., 2018. What causes psychosis? An umbrella review of risk and protective factors. World Psychiatry 17 (1), 49–66.

Selten, J.P., van der Ven, E., Termorshuizen, F., 2019. Migration and psychosis: a meta-analysis of incidence studies. Psychol. Med. 1–11.

Stilo, S., Gayer-Anderson, C., Beards, S., Hubbard, K., Onyejiaka, A., Keraite, A., et al., 2017. Further evidence of a cumulative effect of social disadvantage on risk of psychosis. Psychol. Med. 47 (5), 913–924.

Thompson, A.D., Nelson, B., Yuen, H.P., Lin, A., Amminger, G.P., McGorry, P.D., et al., 2014. Sexual trauma increases the risk of developing psychosis in an ultra high-risk "prodromal" population. Schizophr. Bull. 40 (3), 697–706.

Trotta, A., Iyegbe, C., Di Forti, M., Sham, P.C., Campbell, D.D., Cherny, S.S., et al., 2016. Interplay between schizophrenia polygenic risk score and childhood adversity in first-presentation psychotic disorder: a pilot study. PLoS One 11 (9), e0163319.

van der Ven, E., Bourque, F., Joober, R., Selten, J.-P., Malla, A.K., 2012. Comparing the clinical presentation of first-episode psychosis across different migrant and ethnic minority groups in Montreal, Quebec. Can. J. Psychiatr. 57 (5), 300–308.

Van Os, J., Driessen, G., Gunther, N., Delespaul, P., 2000. Neighbourhood variation in incidence of schizophrenia: evidence for person-environment interaction. Br. J. Psychiatry 176 (3), 243–248.

van Os, J., Pedersen, C.B., Mortensen, P.B., 2004. Confirmation of synergy between urbanicity and familial liability in the causation of psychosis. Am. J. Psychiatry 161 (12), 2312–2314.

Van Os, J., Linscott, R.J., Myin-Germeys, I., Delespaul, P., Krabbendam, L., 2009. A systematic review and meta-analysis of the psychosis continuum: evidence for a psychosis proneness–persistence–impairment model of psychotic disorder. Psychol. Med. 39 (2), 179–195.

Varese, F., Smeets, F., Drukker, M., Lieverse, R., Lataster, T., Viechtbauer, W., et al., 2012. Childhood adversities increase the risk of psychosis: a meta-analysis of patient-control, prospective-and cross-sectional cohort studies. Schizophr. Bull. 38 (4), 661–671.

Vassos, E., Sham, P., Kempton, M., Trotta, A., Stilo, S.A., Gayer-Anderson, C., et al., 2018. The maudsley environmental risk score for psychosis. bioRxiv 453936.

Volavka, J., Vevera, J., 2018. Very long-term outcome of schizophrenia. Int. J. Clin. Pract. 72 (7), e13094.

Zammit, S., Lewis, G., Rasbash, J., Dalman, C., Gustafsson, J.-E., Allebeck, P., 2010. Individuals, schools, and neighborhood: a multilevel longitudinal study of variation in incidence of psychotic disorders. Arch. Gen. Psychiatr. 67 (9), 914–922.

Zwicker, A., Denovan-Wright, E.M., Uher, R., 2018. Gene–environment interplay in the etiology of psychosis. Psychol. Med. 48 (12), 1925–1936.

CHAPTER

16

Is there sufficient evidence that cannabis use is a risk factor for psychosis?

Marco Colizzi[1,2], *Sagnik Bhattacharyya*[1]

[1]Department of Psychosis Studies, Institute of Psychiatry, Psychology and Neuroscience, King's College London, London, United Kingdom; [2]Section of Psychiatry, Department of Neurosciences, Biomedicine and Movement Sciences, University of Verona, Verona, Italy

Historical background

Suggested in oral traditions, possibly dating back to 2700 BC, the first documented evidence of an effect of cannabis use on mental health dates back to the first to second century BC, when the psychotropic properties of parts of the cannabis plant, possibly the flowers of the seedless female plant, were described in Chinese medicine texts (Brand and Zhao, 2017). Historical evidence suggests that the *Cannabis sativa* plant gained its psychotropic properties when it was transposed to India from Central Asia, where it had been used as a fiber plant. In response to the new environment, the plant developed narcotic-yielding properties and, with its new properties, traveled back to Central Asia and Europe (Dhunjibhoy, 1930). Iban Beitar (1197−1248 BC) was the first physician to highlight the detrimental psychological effects that the use of hemp, a variety of the *C. sativa* plant, had in man (Dhunjibhoy, 1930).

The first reference to the association between cannabis use and psychosis in modern medicine is ascribed to the French psychiatrist Jacque-Joseph Moreau who, in 1845, described the psychoactive effects of hashish, the resin obtained from the cannabis plant (Moreau, 1845). He reported that hashish could precipitate "acute psychotic reactions, generally lasting but a few hours, but occasionally as long as a week; the reaction seemed dose-related and its main features included paranoid ideation, illusions, hallucinations, delusions, depersonalization, confusion, restlessness, and excitement. There can be delirium, disorientation, and marked clouding of consciousness." In light of similarities between the hashish-induced psychosis-like manifestations and the symptoms experienced by psychosis patients, he suggested a potential role of cannabis in the pathophysiology of "aliénation mentale" (translated as mental illness), laying the groundwork for the investigation of the association between cannabis and psychosis in the subsequent century (Abel, 2005). In 1930, long before the identification of delta-9-tetrahydrocannabinol (Δ-9-THC), the main psychoactive ingredient of

Risk Factors for Psychosis
https://doi.org/10.1016/B978-0-12-813201-2.00016-8

cannabis (Gaoni and Mechoulam, 1971) and the discovery of the endocannabinoid system and its cannabinoid receptors (Martin et al., 1999), drug-induced psychosis (DIP), with hemp as the main drug involved, was reported as the most frequent form of psychosis after the affective and nonaffective psychoses (Dhunjibhoy, 1930). In 2007, a review of available evidence suggested an increased risk of any psychotic outcome in individuals who had ever used cannabis (pooled adjusted odds ratio = 1.41, 95% CI 1.20–1.65), with a higher risk in frequent users (2.09, 1.54–2.84) (Moore et al., 2007). The association did not appear to be explained by confounding factors or transient intoxication effects (Moore et al., 2007).

Recent population data indicate that recreational cannabis use has increased over time, with approximately 200 million users worldwide (National Academies of Sciences, 2017), of whom about 13 million have cannabis dependence (Degenhardt et al., 2013b), with an expected increase over time linked to ongoing decriminalization and legalization policies (Hall and Lynskey, 2016; Hasin et al., 2017). Cannabis potency, indexed as Δ-9-THC concentration, has also increased over the last three decades (Cascini et al., 2012). This is of particular concern as the detrimental neurocognitive and psychological effects of cannabis appear to be driven mainly by Δ-9-THC (Colizzi and Bhattacharyya, 2017), with high-potency forms of cannabis carrying a greater risk of psychosis and adverse outcomes of the illness than the traditional low-potency cannabis (Di Forti et al., 2015; Schoeler et al., 2016c).

Finally, since 2008, the recreational use of synthetic cannabinoids, known by a number of brand names including K2, Spice, and many more, has increased dramatically (Law et al., 2015). While THC is a partial agonist with weak affinity for the cannabinoid receptor type 1 (CB1) (Pertwee, 2008), synthetic cannabinoids are full agonists and have higher affinity for the CB1 receptor (Cohen and Weinstein, 2018). Consistent with their effects at the CB1 receptor, synthetic cannabinoids have been associated with a greater health risk compared to cannabinoids in the extract of the cannabis plant (Murray et al., 2017), both in terms of psychotic reactions (Castaneto et al., 2014; Papanti et al., 2013) and long-lasting psychotic disorders (Fattore, 2016).

Cannabis use and subsequent development of persistent psychosis

To date, results of 13 studies from prospective general population longitudinal samples have been published examining whether cannabis use could be a causal component in the development of psychosis (Andreasson et al., 1987; Tien and Anthony, 1990; Arseneault et al., 2002; van Os et al., 2002; Weiser et al., 2002; Zammit et al., 2002; Fergusson et al., 2003; Ferdinand et al., 2005; Henquet et al., 2005; Wiles et al., 2006; Manrique-Garcia et al., 2012; Rossler et al., 2012; Gage et al., 2014; Rognli et al., 2015; Bechtold et al., 2016). The first study was carried out among 50,053 conscripts to the Swedish Military, aged 18–20 years (Andreasson et al., 1987). The study indicated a six-time greater risk of developing psychosis in conscripts who had used cannabis at least 50 times in their life compared to those who had never used cannabis (Andreasson et al., 1987). Subsequent analyses performed on the same cohort confirmed and extended these findings, indicating a dose-response association, with frequent use being associated with a greater risk (Zammit et al., 2002; Manrique-Garcia et al., 2012). In the context of frequent cannabis use, the risk of subsequent development of schizophrenia was also higher compared with other forms of nonaffective psychosis (Manrique-Garcia et al., 2012). Subsequent studies following the initial study with Swedish conscripts have tried to replicate those findings and also address its limitations. In 1990, Tien and Anthony analyzed prospective data from a survey conducted among 4994 adult household residents in the United States, indicating that the risk for onset of positive psychotic symptoms such as delusions or hallucinations was more than two times higher for daily cannabis users compared to nonusers, after controlling for sociodemographic characteristics, preexisting psychiatric conditions, and other substance use (Tien and Anthony, 1990). Later, a study conducted in the Netherlands among a general population of 4045 individuals without psychosis and 59 patients with a diagnosis of psychotic disorder assessed the presence at follow-up of psychosis as a function of cannabis use (van Os et al., 2002). Taking into account confounding factors such as ethnicity, urbanicity, and level of education, results indicated an almost three-time greater risk of presenting with psychotic symptoms at follow-up, which was especially high in those with a baseline diagnosis of psychotic disorder (van Os et al., 2002). Two population cohort studies conducted in New Zealand with more than 1000 participants indicated that those who use cannabis in adolescence (Arseneault et al., 2002) or develop cannabis dependence (Fergusson et al., 2003) are at a greater risk of experiencing psychotic symptoms (Fergusson et al., 2003) or being diagnosed with schizophreniform disorder in adulthood, even after controlling for preexisting psychotic symptoms and other background factors (Fergusson et al., 2003; Arseneault et al., 2002). A study conducted in an Israeli population of 9724 adolescents indicated that the prevalence of self-reported drug abuse was higher in adolescents later hospitalized for schizophrenia compared to the prevalence of drug abuse in adolescents who did not need later hospitalization (Weiser et al., 2002). A further study conducted among 1580 individuals randomly drawn from the Dutch general population indicated that, compared to nonusers, cannabis using healthy individuals have an almost three-time higher risk of presenting with psychotic symptoms in the future (Ferdinand et al., 2005). The authors also found that preexisting psychosis among nonusers was associated with a greater risk of subsequent cannabis use, suggesting a shared vulnerability with varying order of onset or a bidirectional causal relationship between cannabis use and psychosis (Ferdinand et al., 2005). In 2005, Henquet et al. found that cannabis use was associated with an almost two-time higher risk of subsequent development of psychotic symptoms in a population sample of 2437 individuals, after controlling for demographic factors, socioeconomic status, urbanicity, childhood trauma, and preexisting psychosis (Henquet et al., 2005). This study also replicated the evidence of a much greater risk in cannabis users with preexisting psychosis (van Os et al., 2002) while not confirming the bidirectional causal relationship between cannabis use and psychosis (Ferdinand et al., 2005), as predisposition for psychosis at baseline did not significantly predict future cannabis use (Henquet et al., 2005). In a cohort of 6217 Swedish discharged prisoners, Rognli and colleagues found that, among different substances, only cannabis predicted a higher risk of primary psychosis, which was almost three-time higher than in nonusers (Rognli et al., 2015). The most recent study, conducted in 2016 among 1009 adolescents, indicated a greater risk of subsequent subclinical psychotic symptoms, such as subclinical paranoia or hallucinations, for those engaged in regular cannabis use, after controlling for a number of confounders including recent cannabis use (Bechtold et al., 2016). Those results also did not support the bidirectional causal relationship between cannabis use and psychosis (Bechtold et al., 2016).

In summary, these 10 studies found that cannabis users are at significant increased risk of subsequently developing psychotic symptoms or a psychotic disorder. The remaining three studies showed a trend in the same direction but did not reach statistical significance (Wiles et al., 2006; Rossler et al., 2012; Gage et al., 2014). It is worth mentioning that of those, two had a short follow-up period (Gage et al., 2014; Wiles et al., 2006) while the third had limited sample power (Rossler et al., 2012). Details on the strength of the association between cannabis use and psychosis for all of the studies are reported in Table 16.1. Though attenuated in some cases, the greater risk of psychosis associated with cannabis use persisted even after controlling for the confounding effect of age, sex, socioeconomic status, ethnicity, urbanicity, use of other substances, and preexisting psychosis. A recent large metaanalysis conducted among 66,816 study participants reported that the risk of developing psychotic symptoms or a psychotic disorder in individuals who had used cannabis over nonusers is about two-time higher in any cannabis users and almost four-time higher for the group reporting the heaviest use (Marconi et al., 2016). As this metaanalysis specifically focused on the association between the level of cannabis use and risk of psychosis, it included only 5 of the 13 studies mentioned above (Tien and Anthony, 1990; Arseneault et al., 2002; Zammit et al., 2002; Henquet et al., 2005; Wiles et al., 2006).

Alternative explanations for the association between cannabis and psychosis

Although consistent and replicated findings converge on the evidence that cannabis users are at a higher risk of developing a psychotic disorder, some concerns have been raised regarding the nature of this association (Haney and Evins, 2016; Hill, 2015; Ksir and Hart, 2016a; Nutt, 2009). Most common arguments fall within the following categories: (i) the association between cannabis and psychosis would be driven by other factors, especially use of other substance; (ii) preexisting differences between cannabis users and nonusers in terms of psychosis-related symptomatology could explain why cannabis users are more prone to develop psychosis; (iii) preexisting psychosis-related symptomology may lead to a higher cannabis use in an attempt to self-medicate, implying a reverse causation between cannabis use and psychosis; (iv) a common genetic background could predispose to psychotic disorders as well as cannabis use.

Also, the severity of the harmful effects of cannabis is still debated. According to a nine-category matrix of harm developed by Nutt and colleagues to assess the physical and social harms of a range of illicit and legal drugs of potential misuse in an evidence-based fashion, cannabis did not score in the top 10. Interestingly, alcohol and tobacco were both ranked more harmful despite being publicly available and not subject to the UK Misuse of Drugs Act classification of harm severity (Nutt et al., 2007).

The confounding effect of other substance use

Along with evidence that cannabis use can increase the risk of psychosis (Moore et al., 2007; Colizzi and Murray, 2018), a number of studies suggest an association between use of other substances and psychosis. That patients with psychosis are more likely to use and abuse substances than the general population is well-known (Regier et al., 1990). Other than cannabis, most common substances used by patients with psychosis include alcohol, nicotine, and stimulants (Colizzi et al., 2016a). Apart from population-based cohort studies indicating that individuals with substance use disorders, especially methamphetamine, have a higher risk of schizophrenia (Callaghan et al., 2012), DIP animal models are well documented and support a biological plausibility for such associations. Other illicit drugs that can cause psychosis include, but are not limited to, amphetamine, scopolamine, ketamine, phencyclidine (PCP), and lysergic acid diethylamide (LSD) (Ham et al., 2017). Most animal models of DIP

TABLE 16.1 Prospective studies on the effect of cannabis use on risk of psychosis.

Authors, year (country)	Study type	Sample size	Years of FU	OR* (95% CI)	Significance
Tien and Anthony (1990) (US)	Population based	4.494	1	2.4 (1.2–7.1)	✓
Zammit et al. (2002) (Sweden)	Conscript cohort	50.053	27	3.1 (1.7–5.5)	✓
Manrique-Garcia et al. (2012)			35	1.8 (1.3–2.3)	
van Os et al. (2002) (the Netherlands)	Population based	4.045	3	2.8 (1.2–6.5)	✓
Weiser et al. (2002) (Israel)	Population based	9.724	4–15	2.0 (1.3–3.1)	✓
Arseneault et al. (2002) (New Zealand)	Birth cohort	1.034	15	4.5 (1.1–18.2)	✓
Fergusson et al. (2003) (New Zealand)	Birth cohort	1.265	3	1.8 (1.2–2.6)	✓
Ferdinand et al. (2005) (the Netherlands)	Population based	1.580	14	2.8 (1.79–4.43)	✓
Henquet et al. (2005) (Germany)	Population based	2.437	4	1.7 (1.1–1.5)	✓
Rognli et al. (2015) (Sweden)	Cohort of discharged prisoners	6.217	5	2.6 (1.40–5.0)	✓
Bechtold et al. (2016) (US)	Adolescent boys	1.009	5	1.51 (1.08–2.11)	✓
Wiles et al. (2006) (UK)	Population based	8.580	1.5	1.5 (0.55–3.94)	×
Rössler et al. (2012) (Switzerland)	Community survey	591	30	1.8 (0.96–3.2)	×
Gage et al. (2014) (UK)	Birth cohort	1.756	2	1.1 (0.76–1.65)	×
				Twofold increased risk of psychosis	

*, adjusted risk; *FU*, follow-up; *UK*, United Kingdom; *US*, United States.

have explained psychotic symptoms based on the ability of these drugs to alter neurotransmitter systems (Ham et al., 2017). However, while psychoses developed in the absence of substance use are frequently associated with positive, negative, and cognitive symptoms, DIP models of psychosis do not necessarily explain the complexity of the psychotic spectrum symptomatology (Ham et al., 2017). Briefly, amphetamine seems to alter dopaminergic neurotransmission (Hsieh et al., 2014), inducing the manifestation of positive and cognitive symptoms, but not negative symptoms (Ham et al., 2017). A scopolamine administration regimen can modulate both cholinergic and dopaminergic neurotransmission (Barak, 2009), leading to the manifestation of positive and cognitive symptoms as well as a restricted range of negative symptoms (Ham et al., 2017). PCP and ketamine regulate glutamatergic neurotransmission, possibly explaining their ability to induce a broader range of psychotic symptoms, including negative symptoms (Frohlich and Van Horn, 2014). Finally, LSD affects serotoninergic and glutamatergic neurotransmission (Martin et al., 2014), inducing positive symptoms upon acute administration as well as negative symptoms upon chronic exposure (Ham et al., 2017).

Special reference also needs to be made to the confounding role of tobacco. Cannabis and tobacco are often consumed together in the same joint or cigarette, and there is growing evidence of a shared underlying biological vulnerability to both substances (Agrawal et al., 2012). Importantly, heavy consumption of nicotine has been reported to increase the risk of psychosis (Gurillo et al., 2015) and to negatively impact of the global outcome of the disorder (Krishnadas et al., 2012). Although some evidence indicates that chronic use of nicotine might affect dopaminergic neurotransmission (Brody et al., 2010), evidence supporting the biological plausibility of this association still remains limited (Gurillo et al., 2015; Krishnadas et al., 2012).

Following the initial evidence of an association between cannabis use and psychosis in the Swedish conscript cohort (Andreasson et al., 1987), further studies have addressed this issue, controlling for the effect of other drugs (Arseneault et al., 2002) as well as tobacco (Di Forti et al., 2015). They have indicated that the association between cannabis use and psychosis is not explained by use of other drugs. Interestingly, some studies have also suggested that the risk of psychosis is specific to cannabis use as opposed to other drugs. Specifically, when entered jointly with cannabis in the model, the effect of other substances was greatly reduced or negligible, while the effect of cannabis remained (Arseneault et al., 2002; van Os et al., 2002). Similarly, other studies showed that although patients with first-episode psychosis reported having smoked tobacco more than healthy controls, the association between cannabis use and psychosis (Di Forti et al., 2015) or its relapse (Schoeler et al., 2016c,d) did not appear to be explained by a comorbid use of tobacco.

Higher levels of preexisting psychopathology in cannabis users

As a potential explanation for the association between cannabis use and psychosis, it has been suggested that cannabis users may have more psychological problems and deviant behavior compared to those who do not use cannabis. A limited number of studies have attempted to investigate whether temperamental and personality indicators of poor behavioral control, such as aggressiveness, risk-taking, and sensation seeking, are more commonly present in the context of adolescent cannabis use, reporting mixed and inconclusive findings (Creemers et al., 2009). A review of evidence suggested that only a severe deviant behavioral style, including both high approach (increased novelty and experience seeking) and low avoidance (poor harm avoidance and behavioral inhibition) attitudes, was weakly to moderately associated with an increased risk of using cannabis and developing a substance use disorder during adolescence (Creemers et al., 2009).

The hypothesis of preexisting psychopathology is also in contrast with the evidence that, compared to nonusing patients with psychosis, those who use cannabis have a higher intelligence quotient (IQ) (Arnold et al., 2015; Loberg et al., 2014), higher premorbid IQ, and psychosocial functioning (Ferraro et al., 2013), and are less likely to present with neurological soft signs of psychosis (Ruiz-Veguilla et al., 2012). Thus, preexisting psychopathology, such as poor cognitive and psychosocial function, would be higher in cannabis-naïve patients than cannabis-using patients and not the reverse, in line with the neurodevelopmental hypothesis of schizophrenia that psychotic disorders in adulthood have their origins in disturbed development of the nervous system (Owen et al., 2011). Instead, among cannabis-using patients, it would be the use of the substance that accelerates the cascade of events leading to psychosis onset, even in the absence of poor premorbid adjustment.

To address this issue, data from the Swedish conscript cohort were reanalyzed controlling for "disturbed behavior" (truancy, contact with police/childcare authority, running away from home), which still confirmed an increased risk of psychosis among cannabis users (Manrique-Garcia et al., 2012; Zammit et al., 2002). Subsequent studies have also found that controlling for the presence of preexisting psychopathology did not explain the higher risk of presenting with psychotic symptoms in future as a function of cannabis use (Arseneault et al., 2002; Tien and Anthony, 1990; Fergusson et al., 2003). However, some studies have indicated that the risk of developing psychotic symptoms at follow-up was higher if cannabis users had preexisting psychosis (Henquet et al., 2005; van Os et al., 2002). Thus, future studies need to investigate the complex relationship between premorbid psychopathology, cannabis use, and subsequent onset of psychosis.

Self-medication and the reverse causality hypothesis

The self-medication hypothesis posits that patients with psychosis would use cannabis to ameliorate symptoms of the disorder, especially negative ones (Gill et al., 2015; Archie et al., 2013), or in order to attenuate side effects induced by the antipsychotic treatment (Khantzian, 1997). While plausible, there is very limited available evidence in support of this hypothesis. Most studies do not support the relationship between the manifestation of psychotic symptoms and the subsequent increase in substance use including cannabis (Khokhar et al., 2018). In a study encompassing all diagnostic categories, apart from state anxiety, no other clinical dimension including both major psychiatric disorders and personality disorders predicted an increase in cannabis use (Spalletta et al., 2007). Also, the occurrence of medication-related side effects did not seem to be associated with a higher use of the substance (Dequardo et al., 1994). Instead, research evidence seems to indicate that after the onset of psychosis, patients tend to use less cannabis (Fergusson et al., 2003, 2015), probably because of the cannabis-induced exacerbation of the psychotic symptomatology, as experimentally demonstrated (D'Souza et al., 2005) and independently shown in a case–control study (Bianconi et al., 2016). Therefore, in contrast with the self-medication hypothesis, cannabis use can worsen rather than relieve symptoms of psychosis. However, psychosis patients have been shown to also experience more enjoyment than healthy controls when using cannabis, potentially accounting for the maintenance of cannabis use among some patients despite its detrimental effects (Bianconi et al., 2016).

To test the self-medication hypothesis and thus the possibility of a reverse causality, some of the more recent prospective studies have also evaluated whether preexisting psychosis among nonusers would be associated with a greater risk of subsequent cannabis use. While a bidirectional causal relationship between cannabis use and psychosis has been reported (Ferdinand et al., 2005), this evidence has not been replicated in other studies (Henquet et al., 2005; Bechtold et al., 2016; Schoeler et al., 2016d). Thus, research evidence seems to suggest that, if there is a bidirectional relationship between cannabis use and psychosis, it is unlikely that there is full reverse causality, such that the negative effect of psychosis on cannabis use may account for the entire impact of cannabis use on psychosis. Furthermore, it is worth mentioning that in case of a bidirectional relationship between cannabis use and psychosis, with the presence of both a direct (negative effect of cannabis use on psychosis) and reverse (negative effect of psychosis on cannabis use) causality, the actual direct effect of cannabis use on psychosis would be less pronounced than the accumulated evidence seems to suggest.

Shared genetic vulnerability and the unifying hypothesis

An alternative explanation is that, due to a common genetic susceptibility, some individuals are at a greater risk of both psychosis and cannabis use (Hill, 2015; Ksir and Hart, 2016a,b). Over the last few years, imputations from genome-wide single-nucleotide polymorphisms (SNPs) have been used to develop a psychosis-related polygenic risk score from 108 loci identified in a very large genome-wide case–control study (Schizophrenia Working Group of the Psychiatric Genomics, 2014). This risk score was able to increasingly predict case–control status as a function of increasing number of schizophrenia risk alleles (Schizophrenia Working Group of the Psychiatric Genomics, 2014). In very recent years, a limited number of studies have investigated whether this risk score would predict cannabis use. Results seem to indicate that genetic vulnerability to psychosis predicts only a small amount of the variance in cannabis use (Power et al., 2014; Verweij et al., 2017). In another study, compared to cannabis, the risk score explained a similar or slightly higher proportion of the variance in use of other substances such as alcohol, amphetamines, cocaine, and opioids (Reginsson et al., 2018), for which there is less or no evidence of association with increased risk of psychosis (Murray et al., 2017). Reversing the genetic approach, another study indicated that genetic markers associated with ever use of cannabis predicted an increased risk of schizophrenia (Vaucher et al., 2018). A further study confirmed the validity of both genetic models, suggesting direct causality, with genetic risk for cannabis use increasing the likelihood of developing a psychotic disorder (odds ratio [OR]: 1.04, 95% confidence interval [CI] 1.01–1.07), as well as reverse causality, with genetic risk for schizophrenia predicting odds of initiating cannabis use (OR: 1.10, 95% CI 1.05–1.14) (Gage et al., 2017). Thus, to a certain extent, the association between cannabis use and psychosis could be due to a shared genetic vulnerability. However, a common genetic background does not exclude the possibility that cannabis had a causal role in the development of psychosis.

Gene by environment (GxE) interaction studies have tried to clarify whether some individuals carry a genetic vulnerability to the psychosis-inducing effects of cannabis use. So far, studies have suggested that genetic variations in dopamine signaling, including the catechol-methyltransferase (COMT) (Caspi et al., 2005), RAC-alpha serine/threonine-protein kinase (AKT1) (Van Winkel et al., 2011a; Di Forti et al., 2012; Bhattacharyya et al., 2012a), dopamine transporter (DAT1) (Bhattacharyya et al., 2012a), and dopamine receptor D2 (DRD2) (Colizzi et al., 2015c) genes, modulate the risk of psychosis in the context of cannabis use (Di Forti et al., 2012; Van Winkel et al., 2011a; Colizzi et al., 2015c) or sensitivity to transient cannabis-induced psychotic symptoms (Bhattacharyya et al., 2012a). Also, the risk of psychosis among cannabis users increased further if people carried both the risk variants in the DRD2 and AKT1 (Colizzi et al., 2015b) or the AKT1 and DAT1 (Bhattacharyya et al., 2012a) genes. Intriguingly, variations in these genes have been associated with psychosis independently of cannabis use (Colizzi et al., 2015b). However, it is worth mentioning that the interaction between the COMT gene and cannabis use on risk of psychosis has not been consistently replicated (Zammit et al., 2011).

A complimentary possibility is that both psychosis and cannabis use share a common pathophysiology in overlapping neural circuits (Green et al., 1999), which would be dysfunctional in both the conditions. Genetic or environmental risks factors would lead to a dysregulated mesocorticolimbic system involving different neurotransmitter pathways such as dopamine-only, gamma-aminobutyric acid (GABA)-only, glutamate-only, and glutamate—dopamine pathways (Khokhar et al., 2018). On one hand, long before the manifestation of clear psychotic symptoms, such a dysregulated system would lead adolescents to use cannabis at a higher rate than adolescents who do not have this underlying neurobiological abnormality. Cannabis use itself would then lead to the expression of clinical symptoms and potentially persist also after the psychosis onset. On the other hand, even if cannabis use did not predate the onset of psychosis, the dysfunctional mesocorticolimbic system would lead to cannabis use and its maintenance (Khokhar et al., 2018). In order to confirm this unifying hypothesis, future prospective studies need to examine the potential occurrence of neurobiological alterations—such as hypoconnectivity or dopamine hypersensitivity—and related dysfunction of cognition and behavior, prior to adolescents developing symptoms of psychosis or starting use of cannabis.

Evidence for a cause—effect relationship between cannabis use and psychosis

Even though alternative explanations for the association between cannabis use and psychosis have not been conclusive or have been shown to explain only a modest part of the relationship (Murray et al., 2017), skepticism persists today because of two additional arguments: (v) the documented rise in cannabis use worldwide does not seem to have led to an increase in the incidence of psychosis; and (vi) association studies do not necessarily imply a causal relationship. While it is arguably difficult to find definitive evidence in favor or against these arguments, some lines of research can help quantify the strength of the causal hypothesis versus the relational hypothesis.

Should increase in cannabis use be followed by an increase in psychosis incidence?

Recent reports indicate that the use of cannabis for recreational purposes has increased over the last decades, with the drug rating as the most common illicit drug consumed over the world (National Academies of Sciences, 2017). Thus, if cannabis had a causal role in the manifestation of psychotic symptoms and the subsequent development of a full-blown disorder, one would expect a related increase in the incidence of schizophrenia. However, the evidence in this regard remains equivocal. A study exploring the incidence of schizophrenia in South London reported that it doubled between the sixties and the nineties, and that over the same period, there was also a significant increase in the rate of cannabis use in the year before presentation in those presenting with schizophrenia (Boydell et al., 2006). This is of interest, in light of cannabis, especially in its high-potency form skunk, being reported to explain up to 24% of the new episodes of psychosis in this geographical area (Di Forti et al., 2015). This is also in line with other evidence reporting a rise in the rates of first admission for psychotic disorders in younger people in the 1990s in Zurich that paralleled an increase in the consumption and availability of cannabis in the same geographical area (Ajdacic-Gross et al., 2007). In contrast, other studies from the United Kingdom (Frisher et al., 2009) and Australia (Degenhardt et al., 2003) have failed to find evidence of an increase in the incidence of psychotic disorders following from an increase in cannabis consumption at the population level.

However, studies of this nature are particularly challenging because it is difficult to exclude the possibility that population-level changes in other risk factors for psychosis may have occurred over time, thereby affecting the relative population-level influence of various risk factors in a nonintuitive manner, which in turn may have masked the effects of increase in incidence of cannabis use over time on the incidence of psychosis (Macleod and Hickman, 2006). It is also possible that a risk factor may become relatively more important as a cause of disease over time, with a progressive increase in their population attributable risk, while explaining less of the population variation in the disease (Pearce, 2011). Therefore, it would be even more difficult to evaluate variations in the overall incidence of schizophrenia overtime and to imply a causal role of cannabis in the development of psychosis from such variation. Indeed, comprehensive high-quality data collection of even the most well-known risk factors of psychosis in longitudinal cohort samples over a sufficient duration to allow systematic examination of such questions remains challenging. Therefore, existing studies have relied on databases that were not primarily designed to address such nuanced questions.

Another pertinent issue is that until about 25 years ago, only a few studies had provided solid evidence regarding the incidence of schizophrenia. Studies conducted in the last part of the 20th century have been able, at least in part, to provide some answers on this issue. A large systematic review conducted in 2004 examined 158 studies drawn from 33 countries and indicated substantial variations in the incidence of schizophrenia as a function of several risk factors (Mcgrath et al., 2004). In particular, while the median incidence rate was 15.2 per 100.000, the 10%–90% quantiles covered over a fivefold range of rates, from 7.7 to 43.0 per 100.000. Incidence rates tended to be higher in males compared to females, in urban catchment areas compared to rural areas, and in migrants compared to native-born. Interestingly, study quality and methodology as well as diagnostic tools and criteria did not appear to significantly affect incidence rates across the studies reviewed (Mcgrath et al., 2004).

In conclusion, due to the paucity of high-quality, comprehensive studies looking at temporal trends in the incidence of psychosis, especially as a function of cannabis use, there is currently little evidence to either strongly support or disregard the possibility that a population-level increase in cannabis use has led to an increase in the incidence of psychosis worldwide.

Another issue that has generally not been considered thus far but may be pertinent to this debate is worth noting here. This refers to the possibility that a proportion of those cannabis-using individuals who may be at the highest risk of developing a psychotic disorder may self-select themselves out of continued use of cannabis and therefore protecting themselves from the risk of developing a psychotic disorder such as schizophrenia, thereby offsetting the potential effect of any population-level increase in cannabis use on the incidence rates of psychotic disorders. While there exists no definitive evidence in this regard, emerging evidence is consistent with such a possibility. For example, quantitative as well as qualitative analyses of evidence from a large internet-based convenient sample of participants indicate that transient psychotic-like experiences while using cannabis are strongly associated with both cessation of cannabis use subsequently and intention to quit in the future (Sami et al., 2019). Consistent with this, evidence from other studies also suggest that subthreshold psychotic experiences, distinct from cannabis-induced psychotic experiences, whether in general population samples (Van Gastel et al., 2014) or in those presenting to specialist clinical services (Valmaggia et al., 2014) are associated with discontinuation of cannabis use.

What degree of proof do we need to have to infer causality?

A common methodological issue across research evidence is the difficulty to prove causation for the association between a given risk factor and the outcome of investigation. The same issue has been raised for the association between cannabis use and psychosis (Haney and Evins, 2016). While this may have a bearing on the debate on legalizing or decriminalizing cannabis use for recreational purposes, caution is warranted in avoiding conflating the debate about the nature of the relationship between cannabis use and schizophrenia with that around the legal status of cannabis and the most appropriate response from a public health point of view, which needs to consider more widely the harms associated with cannabis use. The judgment as to whether an observed statistical association represents a cause−effect relationship between exposure and disease requires inferences far beyond the data from a single study (Hennekens and Buring, 1987). Aspects to consider include (i) magnitude of the association, (ii) consistency of findings from other studies, (iii) specificity of the association, (iv) temporal sequence of association, (v) biological gradient, (vi) coherence and experimental proof, and (vii) biological plausibility (Hill, 1965). These criteria, known as the Bradford−Hill criteria (Hill, 1965), are not intended as a checklist for evaluating whether a reported association might be interpreted as causal. However, they are widely used in epidemiology as a framework to assess the potential cause−effect relationship of the observed association and can be applied to the extensive available evidence on the association between cannabis use and psychosis.

Magnitude of the association

Though the association between cannabis use and risk of psychosis appears to be consistent, its magnitude seems to be relatively modest (Moore et al., 2007). However, it is worth mentioning that the risk of developing psychosis might be higher in frequent users, especially if using high-potency cannabis (Di Forti et al., 2015) or carrying specific risk alleles of genes modulating dopamine signaling (Bhattacharyya et al., 2012a; Colizzi et al., 2015b). Consistently, even in the context of a well-documented association such as that between tobacco and lung cancer (Mattson and Kessler, 1987), there is considerable variation in the risk of developing the disease depending on patterns of use. For instance, while current cigarette smokers who consume more than 20 cigarettes per day seem to have a 10.6 times greater risk to develop lung cancer, the risk drops to 3 times higher for those smoking less than 20 cigarettes per day (Pavlovska et al., 2012).

Consistency of findings from other studies

Cannabis use and outcome of psychosis: In recent years, several studies have indicated that cannabis use is a major clinical problem not only in provoking the onset of psychosis but also in precipitating relapse of the disorder. A metaanalysis of available studies has suggested that continued use of cannabis after the onset of psychosis is associated with a poorer outcome of the disorder (Schoeler et al., 2016b). In particular, this effect seems to be attributable to the ability of cannabis to exacerbate psychotic symptomatology (Schoeler et al., 2016b; Ouellet-Plamondon et al., 2017; Seddon et al., 2016), increasing the risk of both nonremission (Colizzi et al., 2016a) and relapse (Patel et al., 2016; Schoeler et al., 2016c), even when accounting for a range of important confounders such as sociodemographic and clinical information as well as other confounders previously not generally considered such as medication adherence and other illicit drug use (Schoeler et al., 2016c). Also, in those using cannabis after their first episode of psychosis, the risk of relapse increased in a dose-dependent manner (Schoeler et al., 2016d), implying that this relationship could be of a causal nature. Research evidence from inpatient cohorts followed up for up to 10 years came to the same conclusions (San et al., 2013; Foti et al., 2010), even when the confounding effect of most common substances such as alcohol and tobacco were taken into account (Colizzi et al., 2018a).

The effect of continued cannabis use on poor outcome of psychosis could be due to a number of factors. Studies conducted in incidental cohorts suggest that cannabis use increased the risk of poor medication adherence (Colizzi et al., 2016a; Schoeler et al., 2017a; Foglia et al., 2017), the latter thus lying on the causal pathway between cannabis use and nonremission (Colizzi et al., 2016a) as well as risk of relapse (Schoeler et al., 2017b). Another possibility is that cannabis use in the context of psychosis increased the rates of antipsychotic treatment failure (Patel et al., 2016; Wilson and Bhattacharyya, 2016) and dropout from treatment (Miller et al., 2009). Research evidence suggests that chronic use of cannabis results in blunted dopamine synthesis and release in the striatum (Bloomfield et al., 2014; Murray et al., 2014), which in turn could be further reduced by the dopamine antagonism of most antipsychotics. Therefore, among cannabis-using patients, antipsychotic medication, especially potent DRD2 blockers, could be less effective or less tolerated and be discontinued, precipitating relapse (Brzozowska et al., 2017).

It is worth mentioning that while continued use of cannabis after the onset of psychosis is associated with adverse outcome, including higher relapse rates, longer hospital admissions, and a more severe symptomatology (Schoeler et al., 2016b), its discontinuation seems to mitigate the increased liability of poor outcome (Schoeler et al., 2016b; Colizzi et al., 2016a). Thus, investing in services specialized in the treatment of substance use comorbidity in the context of psychosis might improve patients' clinical outcome and reduce burden on psychiatric services. Consistently, recent research evidence has indicated that, compared to cannabis-using patients, those who do not use cannabis after their onset of psychosis spend on average 23 fewer days in hospital per year (Colizzi et al., 2018a).

Cannabis use and cognitive processing: Many studies have investigated the potential for cannabis use to affect cognition, suggesting similarities with the cognitive impairment observed in schizophrenia (Solowij and Michie, 2007). Studies of association tend to support a detrimental effect of cannabis use on cognitive performance in otherwise healthy people (Bolla et al., 2002; Solowij et al., 2002; Jockers-Scherubl et al., 2007). However, negative findings have also been reported (Pope et al., 2001). Even more unclear is the contribution of cannabis use in patients with schizophrenia, with studies suggesting negative (Mata et al., 2008), positive (Jockers-Scherubl et al., 2007; Schnell et al., 2009), and neutral effects (Sevy et al., 2007).

The association between cannabis use and cognition seems to be of a complex nature. First, cannabis use is unlikely to induce homogenous effects across cognitive domains. Metaanalytic work suggested selective effects of recreational cannabis use, with memory function being the most impaired cognitive domain, particularly prospective and retrospective verbal and visual memory (Grant et al., 2003; Schoeler et al., 2016a). Instead, cannabis use seems to have modest effects on other cognitive domains. However, while effects of cannabis use on learning and memory appear to recover after cessation of use, alterations in attentional processing and psychomotor dysfunction seem to be more persistent (Curran et al., 2016; Broyd et al., 2016).

Second, as highlighted for the association between cannabis and psychosis, some individuals may carry genetic vulnerability to the cognitive effects of cannabis use. GxE interaction studies suggest that genetic variations in genes modulating the cannabinoid (Ruiz-Contreras et al., 2013, 2014; Taurisano et al., 2016;

Colizzi et al., 2015a) and dopamine (Van Winkel et al., 2011b; Bhattacharyya et al., 2014) signaling may increase the likelihood of long-lasting cognitive dysfunction. In line with previous systematic reviews (Broyd et al., 2016), memory was the cognitive domain showing the highest proportion of impairment also across GxE interaction studies.

Third, differences across studies investigating the effects of cannabis use on cognition may also be due to the age of onset, severity, and duration of cannabis use. Severe forms of cannabis use including cannabis dependence have been associated with a loss of IQ from childhood to adulthood, which does not recover after cannabis discontinuation (Meier et al., 2012). These results suggest that using cannabis when the brain is still in an active state of development during adolescence may result in permanent cognitive impairment (Meier et al., 2012). Also, chronic cannabis use per se may result in a more marked and persistent disruption of cognitive processing (Meier et al., 2012). Similarly, impairment in cognitive function appears to be greater in individuals using high-potency cannabis, indicating a cumulative effect of Δ9-THC (Colizzi and Bhattacharyya, 2017).

Moreover, the period of abstinence prior to cognitive testing (Schoeler et al., 2016a) may be relevant to the manifestation of cognitive impairments, in light of the complex pharmacodynamics of cannabinoids (D'Souza et al., 2016a) and their implications for the tolerance phenomenon occurring upon heavy cannabis exposure (Colizzi and Bhattacharyya, 2018a).

Imaging studies: Several studies have investigated the neural impact of cannabis use using different magnetic resonance imaging (MRI) methodologies. Evidence from structural MRI studies tend to support an association between sustained cannabis use and reduced gray matter volumes in brain regions relevant to the pathophysiology of psychosis, including the prefrontal cortex, striatum, hippocampus, and amygdala (Batalla et al., 2013). However, some inconsistencies have also been reported, partially explained by the age of onset (Orr et al., 2016), frequency (Lorenzetti et al., 2015; Yucel et al., 2008; Orr et al., 2016), duration (Orr et al., 2016; Lorenzetti et al., 2015), and severity of cannabis use (Koenders et al., 2016; Cousijn et al., 2012). Limited research evidence suggests that additional risk factors may modulate the detrimental effects of cannabis use on the brain structure (Batalla et al., 2014, 2018). In particular, a large study conducted in 1577 adolescents indicated a global reduction in cortical thickness in the context of cannabis use and only in male adolescents who had started using cannabis before the age of 16 and who also carried a high genetic risk measured with the polygenic risk score (French et al., 2015). However, independent evidence from a sibling and twin dataset indicates that brain structural alterations and cannabis use may have shared genetic or environmental liability, highlighting the challenges in attribution of causality to observed associations between cannabis use and brain structural alterations (Paul and Bhattacharyya, 2018; Pagliaccio et al., 2015).

Another MRI technique, diffusion tensor imaging, was used to mainly assess the integrity of fractional anisotropy (FA) in white matter tracts as a function of cannabis use. Evidence indicated a substantial overlap in terms of brain regions between the cannabis- and psychosis-associated FA reduction (Murray et al., 2017). Moreover, the cannabis-induced aberrance in the diffusivity appeared to be more pronounced in individuals using high-potency cannabis compared to those using less potent forms of cannabis (Rigucci et al., 2016).

Functional MRI studies have been used to provide an indication of neurophysiological activity in the context of cannabis use (reviewed here (Batalla et al., 2013; Blest-Hopley et al., 2018)). This line of research indicates that chronic cannabis use may disrupt brain activity in regions involved in the processing of various cognitive tasks (Jager et al., 2007, 2010) as well as in the key components of a core set of brain functional networks in both adult and adolescent cannabis users (Blest-Hopley et al., 2018). Interestingly, some studies have suggested that cannabis users' brain function may be disrupted even in the absence of a poorer behavioral performance, reflecting a greater neural effort or a change in strategy to sustain performance (Bossong et al., 2014a), and recent evidence indicate that some of these neurophysiological alterations may persist even after cannabis after a prolonged period of abstinence (Blest-Hopley et al., 2019). This could account, at least in part, for some of the discrepancies reported in studies assessing exclusively the behavioral component of cognitive processing (Pope et al., 2001). In line with evidence that cannabis-induced cognitive impairments tend to map on some of the domains affected in psychosis patients, similarities have been reported between the neurophysiological alterations underlying cognitive dysfunction in cannabis users and those underlying the cognitive deficits observed in psychosis (Bossong et al., 2014b; Blest-Hopley et al., 2018).

A number of studies have also investigated the neurochemical underpinnings of cannabis use, focusing mainly on dopamine measures, in light of abnormalities in dopamine neurotransmission being considered the final common pathway to psychosis (reviewed here (Sami et al., 2015; Sami and Bhattacharyya, 2018; Colizzi et al., 2016b)). However, a systematic review of positron emission tomography and single-photon emission computerized tomography studies indicated that there is little evidence to support an alteration of dopamine receptor status in otherwise healthy cannabis users (Sami et al., 2015). Conversely, chronic cannabis use could lead to a blunted dopamine synthesis and dopamine release capacity (Sami et al., 2015). In recent years, some studies have also investigated the effect of cannabis use on brain levels of glutamate, n-acetylaspartic acid, and myo-inositol. Reviews of the available literature tend to suggest a decrease of these metabolites as a consequence of chronic exposure to cannabis (Colizzi et al., 2016b; Sneider et al., 2013).

Specificity of the association

In order to establish whether there is a specificity of the association between cannabis use and psychosis, studies have investigated (i) the effects of other substance use on psychosis and its confounding effect in the association between cannabis use and psychosis and (ii) the effect of cannabis use on other neuropsychiatric conditions. The direct and indirect contribution of other substance use in causing psychosis has been extensively discussed as a potential alternative explanation for the association between cannabis and psychosis. In summary, there seem to be higher rates of conversion to schizophrenia with cannabis compared to other substances (Niemi-Pynttari et al., 2013). However, as a conservative estimate, we can assume a certain role of other substances in the manifestation of psychosis, both with and without comorbid cannabis use.

With reference to the association between cannabis use and development of other neuropsychiatric conditions, research evidence is not as extensive as for the association between cannabis use and psychosis. Some of the prospective studies presented here suggested that cannabis use does not increase the risk of subsequent depression (Arseneault et al., 2002; Manrique-Garcia et al., 2012). However, recent evidence implicates frequent cannabis use during adolescence as a risk factor for later life depression (Schoeler et al., 2018) with the effects being particularly prominent in high-frequency users (OR of 8.83) and also present in low-frequency users (OR of 2.41). A metaanalysis of longitudinal studies suggested a weak association between cannabis use and depression, with an OR of 1.17 (Lev-Ran et al., 2014). A more recent metaanalysis has confirmed these findings, with a slightly higher OR of 1.37 (Gobbi et al., 2019). While metaanalytic evidence may suggest that the strength of the association between cannabis use and depression may be weaker compared to that between cannabis and psychosis (Moore et al., 2007), the limited nature of available evidence and absence of studies directly comparing the association between cannabis use and psychosis with that between cannabis use and depression preclude any firm conclusions. In terms of anxiety disorders, while an earlier study suggested an association between cannabis use and anxiety disorders (Degenhardt et al., 2013a), this was neither confirmed in a subsequent study (Blanco et al., 2016) nor in a recent metaanalysis (Gobbi et al., 2019).

Temporal sequence of association

The issue of direction of any association and temporality has been extensively reviewed when presenting the available prospective studies on the association between cannabis use and psychosis as well as alternative explanations for it, such as the reverse causality and unifying hypotheses. In summary, even though the direct, reverse, and unifying hypotheses may coexist, the available literature weights the balance in favor of a direct causality between cannabis use and psychosis (Murray et al., 2017) as well as poor outcome of the disorder (Schoeler et al., 2016d). Among the majority of psychosis patients with comorbid cannabis use, onset of cannabis use normally precedes the development of psychosis by a number of years. Also, compared to other substances, cannabis use has been shown to drive an earlier onset of the disorder (Large et al., 2011). Moreover, the younger the age of initiation of cannabis use, the higher is the risk to develop psychosis, possibly because of the greater impact of recreational cannabis use on brain maturation during early adolescence (Volkow et al., 2016).

Biological gradient

As also suggested by recent metaanalytic work (Marconi et al., 2016), most studies consistently indicate a dose-response effect for the association between cannabis use and subsequent development of persistent psychosis (Andreasson et al., 1987; Manrique-Garcia et al., 2012; van Os et al., 2002; Fergusson et al., 2005; Davis et al., 2013). Moreover, a retrospective assessment of premorbid cannabis use since the age of 12 in patients with their first episode of psychosis has indicated that escalation of cannabis use in the 5 years prior to the psychosis onset was highly predictive of an increased risk of onset (Kelley et al., 2016). Similarly, research evidence has indicated a dose-response effect for the association between cannabis use after the onset of psychosis and poor outcome (Schoeler et al., 2016d).

Independent of the frequency of cannabis use, another measure contributing to exposure dose is the potency of the cannabis used. Since the initial discovery of the chemical basis of cannabis activity (Mechoulam and Gaoni, 1967), over 100 different cannabinoids have been identified, with Δ9-THC and cannabidiol (CBD) being the most prominent among them (Pertwee, 2008). If Δ9-THC is responsible for most of the psychoactive properties of cannabis (Mechoulam et al., 1970), CBD seems to prevent or reverse the Δ9-THC–induced psychotomimetic effects (Bhattacharyya et al., 2010), potentially having antipsychotic properties on its own (Mcguire et al., 2018; Bhattacharyya et al., 2018b). These opposite clinical effects of the two compounds seem to be due to their almost opposite chemical activity, with Δ9-THC being a partial agonist while CBD an antagonist/reverse agonist (Pertwee, 2008). The Δ9-THC:CBD ratio appears to be of crucial importance with reference to the detrimental effects of cannabis in man (Colizzi and Bhattacharyya, 2017). Consistently, case–control studies suggest that the risk of development (Di Forti et al., 2015) and relapse (Schoeler et al., 2016c) of psychosis in cannabis users depends on both frequency of use and cannabis potency, with the risk being the highest in individuals exposed on a daily basis to cannabis with a high Δ9-THC concentration and almost no CBD (Di Forti et al., 2015; Schoeler et al., 2016c,d, 2017b). Even though products with a more balanced or low Δ9-THC:CBD ratio may be less harmful, it is unclear whether at specific concentrations CBD might outweigh any harmful effects of Δ9-THC on risk of psychosis (Colizzi and Bhattacharyya, 2017).

Coherence and experimental proof

The use of cannabis' key psychoactive ingredient Δ9-THC as an experimental human model for psychosis is well documented in the literature (Colizzi et al., 2019b). There is solid and consistent evidence that Δ9-THC can acutely induce psychosis-like symptoms as well as cognitive impairments in healthy individuals, which are temporally related to its administration (D'Souza et al., 2004; Bhattacharyya et al., 2009, 2012b; Colizzi and Bhattacharyya, 2018b). While they are transient in nature, symptoms are similar to those observed in the context of chronic cannabis use and may vary in terms of temporal onset, intensity, and duration as a function of dose, route of administration, and previous exposure to cannabis (D'Souza et al., 2008; Colizzi and Bhattacharyya, 2018a). Also, genetic predisposition to dysregulated dopamine signaling (Bhattacharyya et al., 2012a, 2014), altered target receptor (cannabinoid type 1 receptor) availability (Bhattacharyya et al., 2017) or baseline neurophysiological function (Bhattacharyya et al., 2018a), and may modulate the psychotomimetic and neurocognitive effects of acute Δ9-THC administration. It is worth mentioning that acute administration of Δ9-THC has been shown to worsen symptoms of psychosis in patient with schizophrenia (D'Souza et al., 2005) as well as induce persistent psychotic symptoms in a small proportion of otherwise healthy individuals requiring clinical intervention (D'Souza et al., 2016b).

In terms of neurobiological underpinnings of the acute effects of Δ9-THC, there is some evidence that acute exposure to cannabis or Δ9-THC increased striatal glutamate levels (Colizzi et al., 2019a) as well as dopamine release in striatal and prefrontal areas, especially in those at a genetic or clinical high risk of psychosis (Sami et al., 2015; Colizzi et al., 2019a). Also, acute Δ9-THC administration is able to disrupt brain function as investigated in electrophysiological (Cortes-Briones et al., 2015; D'Souza et al., 2012) as well as neurophysiological (Bhattacharyya et al., 2009, 2012b, 2015a,b) studies, and this alteration is greater in those individuals presenting with the most prominent impairment in their cognitive performance (Colizzi et al., 2018b,c). Finally, acute challenge studies involving administration of Δ9-THC, CBD, and their combination compliment epidemiological evidence of the contribution of these cannabinoids on the risk of psychosis as a function of cannabis potency. The well-established detrimental effects of Δ9-THC on neuropsychological functioning may be prevented, reversed, or attenuated if Δ9-THC is administered in combination or following CBD treatment (Colizzi and Bhattacharyya, 2017), reflecting opposite effects of Δ9-THC and CBD at a behavioral and neurophysiological level (Bhattacharyya et al., 2010, 2012b, 2015b).

Biological plausibility

Research evidence has indicated that sustained cannabis use would lead to an altered dopamine function of the associative striatum (Van De Giessen et al., 2017). The same neurobiological dysfunction has been robustly associated with the manifestation of positive symptoms of psychosis (Howes et al., 2009), thus indicating biological plausibility for the association between cannabis use and psychosis. However, the association between cannabis use and altered dopamine signaling is unlikely to be a direct effect. Instead, as suggested by preclinical evidence, acute Δ9-THC administration activates CB1 receptors (Pertwee, 2008) at both cortical and subcortical glutamatergic presynaptic terminals (Galanopoulos et al., 2011), inhibiting the glutamatergic neurotransmission (Colizzi et al., 2016b), which in turn may result in a dysfunctional firing of dopaminergic neurons (Pistis et al., 2002). Consistently, glutamate signaling has been reported to be disrupted in otherwise healthy individuals with a history of sustained cannabis use (Colizzi et al., 2016b) or acutely exposed to Δ9-THC (Colizzi et al., 2019a) as well as in individuals at risk of psychosis (De La Fuente-Sandoval et al., 2015) or who have already developed a psychotic disorder (Davison et al., 2018).

Other preclinical evidence also suggests that the Δ9-THC—induced acute activation of CB1 receptors at GABAergic interneurons in the medial temporal cortex (Matsuda et al., 1993; Pertwee and Ross, 2002) may explain the ability of this cannabinoid to induce cognitive impairments (Clarke et al., 2008; Wise et al., 2009). Less clear is whether the long-lasting cognitive deficits observed in the context of sustained cannabis use are a consequence of a persistent alteration in the GABA signaling. However, there is substantial evidence for an alteration of the GABA system in psychosis, also with reference to its complex symptom manifestation including cognitive deficits (Lisman et al., 2008).

In summary, both the acute and persistent neuropsychological effects of cannabis seem to be driven by the extensive disruption operated by Δ9-THC on multiple neurotransmitter systems that are also affected in psychosis. It is worth mentioning that the endocannabinoid system has been reported to be altered in psychosis, even in the absence of a comorbid cannabis use (Appiah-Kusi et al., 2016). A debate has originated whether cannabis-induced psychosis differed from psychotic disorders that have not developed in the context of cannabis use. While early reports did not suggest a substantial difference, a recent review indicates a different clinical profile, underlying neurobiological features, and prognosis in those who developed psychosis following cannabis use compared to those who did not (Sami and Bhattacharyya, 2018). Using imaging techniques, future longitudinal studies may be able to track the trajectory of illness in cannabis-using and nonusing patients, in order to establish whether cannabis use is associated with specific neurobiological alterations and symptom profiles of psychosis. This is compelling for a clear understanding of the pathophysiology of cannabis-associated psychosis and to move toward a more specific and effective treatment of this potential subtype of psychotic disorder.

Conclusion

Collectively, as summarized above, there is robust evidence for an association between cannabis use and psychosis. This association appears to be of a modest strength, though the risk of psychosis may increase significantly as a function of cannabis frequency and potency as well as in the context of specific genetic or neurophysiological vulnerability. From prospective to experimental studies, the available literature points in the direction of a causal relationship (Fig. 16.1). This is further corroborated by evidence of similarities between the neurobiological correlates of cannabis use and psychosis and relapse and poorer outcome of psychosis in those using cannabis after the onset of the disorder. However, the neurobiological mechanisms that may underlie this association remain unclear. Interestingly, some evidence suggests a greater sensitivity to the effects of cannabis use among people who have suffered their first psychotic episode, with both "bad" and "enjoyable" experiences reported more frequently

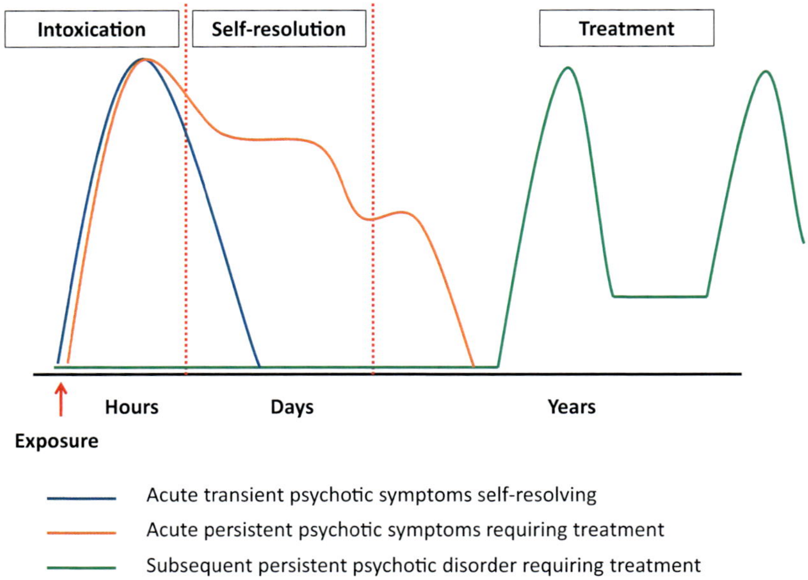

FIGURE 16.1 Temporal sequence of association between cannabis use and psychosis.

compared to a control group. While hypersensitivity to the enjoyable effects of cannabis may explain, at least in part, why some patients continue to use it after their psychosis onset, despite experiencing an exacerbation of their psychotic symptoms (Bianconi et al., 2016), this is another area that remains to be fully understood. Finally, even though it has become increasingly clear that for some individuals there is an increased risk of onset and relapse of psychosis from daily use of high-potency cannabis, the combined effects of Δ9-THC and CBD on mental health are still not completely understood. Future work therefore needs to focus on understanding the mechanisms that operate at the neurobiological as well as at the behavioral level to result in the association between cannabis use and onset and relapse of psychosis and continued use despite clear evidence of harm to many individual users and how the different ingredients of cannabis moderate some of these effects.

References

Abel, E.L., 2005. Jacques Joseph Moreau (1804–1884). Am. J. Psychiatry 162, 458.

Agrawal, A., Budney, A.J., Lynskey, M.T., 2012. The co-occurring use and misuse of cannabis and tobacco: a review. Addiction 107, 1221–1233.

Ajdacic-Gross, V., Lauber, C., Warnke, I., Haker, H., Murray, R.M., Rossler, W., 2007. Changing incidence of psychotic disorders among the young in Zurich. Schizophr. Res. 95, 9–18.

Andreasson, S., Allebeck, P., Engstrom, A., Rydberg, U., 1987. Cannabis and schizophrenia. A longitudinal study of Swedish conscripts. Lancet 2, 1483–1486.

Appiah-Kusi, E., Leyden, E., Parmar, S., Mondelli, V., Mcguire, P., Bhattacharyya, S., 2016. Abnormalities in neuroendocrine stress response in psychosis: the role of endocannabinoids. Psychol. Med. 46, 27–45.

Archie, S., Boydell, K.M., Stasiulis, E., Volpe, T., Gladstone, B.M., 2013. Reflections of young people who have had a first episode of psychosis: what attracted them to use alcohol and illicit drugs? Early Interv. Psychiatry 7, 193–199.

Arnold, C., Allott, K., Farhall, J., Killackey, E., Cotton, S., 2015. Neurocognitive and social cognitive predictors of cannabis use in first-episode psychosis. Schizophr. Res. 168, 231–237.

Arseneault, L., Cannon, M., Poulton, R., Murray, R., Caspi, A., Moffitt, T.E., 2002. Cannabis use in adolescence and risk for adult psychosis: longitudinal prospective study. BMJ 325, 1212–1213.

Barak, S., 2009. Modeling cholinergic aspects of schizophrenia: focus on the antimuscarinic syndrome. Behav. Brain Res. 204, 335–351.

Batalla, A., Bhattacharyya, S., Yucel, M., Fusar-Poli, P., Crippa, J.A., Nogue, S., Torrens, M., Pujol, J., Farre, M., Martin-Santos, R., 2013. Structural and functional imaging studies in chronic cannabis users: a systematic review of adolescent and adult findings. PLoS One 8, e55821.

Batalla, A., Lorenzetti, V., Chye, Y., Yucel, M., Soriano-Mas, C., Bhattacharyya, S., Torrens, M., Crippa, J.A.S., Martin-Santos, R., 2018. The influence of DAT1, COMT, and BDNF genetic polymorphisms on total and subregional hippocampal volumes in early onset heavy cannabis users. Cannabis Cannabinoid Res. 3, 1–10.

Batalla, A., Soriano-Mas, C., Lopez-Sola, M., Torrens, M., Crippa, J.A., Bhattacharyya, S., Blanco-Hinojo, L., Fagundo, A.B., Harrison, B.J., Nogue, S., De La Torre, R., Farre, M., Pujol, J., Martin-Santos, R., 2014. Modulation of brain structure by catechol-O-methyltransferase Val(158) Met polymorphism in chronic cannabis users. Addict. Biol. 19, 722–732.

Bechtold, J., Hipwell, A., Lewis, D.A., Loeber, R., Pardini, D., 2016. Concurrent and sustained cumulative effects of adolescent marijuana use on subclinical psychotic symptoms. Am. J. Psychiatry 173, 781–789.

Bhattacharyya, S., Atakan, Z., Martin-Santos, R., Crippa, J.A., Kambeitz, J., Malhi, S., Giampietro, V., Williams, S., Brammer, M., Rubia, K., Collier, D.A., Mcguire, P.K., 2015a. Impairment of inhibitory control processing related to acute psychotomimetic effects of cannabis. Eur. Neuropsychopharmacol. 25, 26–37.

Bhattacharyya, S., Atakan, Z., Martin-Santos, R., Crippa, J.A., Kambeitz, J., Prata, D., Williams, S., Brammer, M., Collier, D.A., Mcguire, P.K., 2012a. Preliminary report of biological basis of sensitivity to the effects of cannabis on psychosis: AKT1 and DAT1 genotype modulates the effects of delta-9-tetrahydrocannabinol on midbrain and striatal function. Mol. Psychiatry 17, 1152–1155.

Bhattacharyya, S., Crippa, J.A., Allen, P., Martin-Santos, R., Borgwardt, S., Fusar-Poli, P., Rubia, K., Kambeitz, J., O'carroll, C., Seal, M.L., Giampietro, V., Brammer, M., Zuardi, A.W., Atakan, Z., Mcguire, P.K., 2012b. Induction of psychosis by Delta9-tetrahydrocannabinol reflects modulation of prefrontal and striatal function during attentional salience processing. Arch. Gen. Psychiatr. 69, 27–36.

Bhattacharyya, S., Egerton, A., Kim, E., Rosso, L., Riano Barros, D., Hammers, A., Brammer, M., Turkheimer, F.E., Howes, O.D., Mcguire, P., 2017. Acute induction of anxiety in humans by delta-9-tetrahydrocannabinol related to amygdalar cannabinoid-1 (CB1) receptors. Sci. Rep. 7, 15025.

Bhattacharyya, S., Falkenberg, I., Martin-Santos, R., Atakan, Z., Crippa, J.A., Giampietro, V., Brammer, M., Mcguire, P., 2015b. Cannabinoid modulation of functional connectivity within regions processing attentional salience. Neuropsychopharmacology 40, 1343–1352.

Bhattacharyya, S., Fusar-Poli, P., Borgwardt, S., Martin-Santos, R., Nosarti, C., O'carroll, C., Allen, P., Seal, M.L., Fletcher, P.C., Crippa, J.A., Giampietro, V., Mechelli, A., Atakan, Z., Mcguire, P., 2009. Modulation of mediotemporal and ventrostriatal function in humans by Delta9-tetrahydrocannabinol: a neural basis for the effects of Cannabis sativa on learning and psychosis. Arch. Gen. Psychiatr. 66, 442–451.

Bhattacharyya, S., Iyegbe, C., Atakan, Z., Martin-Santos, R., Crippa, J.A., Xu, X., Williams, S., Brammer, M., Rubia, K., Prata, D., Collier, D.A., Mcguire, P.K., 2014. Protein kinase B (AKT1) genotype mediates sensitivity to cannabis-induced impairments in psychomotor control. Psychol. Med. 44, 3315–3328.

Bhattacharyya, S., Morrison, P.D., Fusar-Poli, P., Martin-Santos, R., Borgwardt, S., Winton-Brown, T., Nosarti, C., Cm, O.C., Seal, M., Allen, P., Mehta, M.A., Stone, J.M., Tunstall, N., Giampietro, V., Kapur, S., Murray, R.M., Zuardi, A.W., Crippa, J.A., Atakan, Z., Mcguire, P.K., 2010. Opposite effects of delta-9-tetrahydrocannabinol and cannabidiol on human brain function and psychopathology. Neuropsychopharmacology 35, 764–774.

Bhattacharyya, S., Sainsbury, T., Allen, P., Nosarti, C., Atakan, Z., Giampietro, V., Brammer, M., Mcguire, P.K., 2018a. Increased hippocampal engagement during learning as a marker of sensitivity to psychotomimetic effects of delta-9-THC. Psychol. Med. 48, 2748–2756.

Bhattacharyya, S., Wilson, R., Appiah-Kusi, E., O'neill, A., Brammer, M., Perez, J., Murray, R., Allen, P., Bossong, M.G., Mcguire, P., 2018b. Effect of cannabidiol on medial temporal, midbrain, and striatal dysfunction in people at clinical high risk of psychosis: a randomized clinical trial. JAMA Psychiatry 75, 1107–1117.

Bianconi, F., Bonomo, M., Marconi, A., Kolliakou, A., Stilo, S.A., Iyegbe, C., Gurillo Munoz, P., Homayoun, S., Mondelli, V., Luzi, S., Dazzan, P., Prata, D., LA Cascia, C., O'connor, J., David, A., Morgan, C., Murray, R.M., Lynskey, M., Di Forti, M., 2016. Differences in cannabis-related experiences between patients with a first episode of psychosis and controls. Psychol. Med. 46, 995–1003.

Blanco, C., Hasin, D.S., Wall, M.M., Florez-Salamanca, L., Hoertel, N., Wang, S., Kerridge, B.T., Olfson, M., 2016. Cannabis use and risk of psychiatric disorders: prospective evidence from a US national longitudinal study. JAMA Psychiatry 73, 388–395.

Blest-Hopley, G., Giampietro, V., Bhattacharyya, S., 2018. Residual effects of cannabis use in adolescent and adult brains – a meta-analysis of fMRI studies. Neurosci. Biobehav. Rev. 88, 26–41.

Blest-Hopley, G., Giampietro, V., Bhattacharyya, S., 2019. Regular cannabis use is associated with altered activation of central executive and default mode networks even after prolonged abstinence in adolescent users: results from a complementary meta-analysis. Neurosci. Biobehav. Rev. 96, 45–55.

Bloomfield, M.A., Morgan, C.J., Egerton, A., Kapur, S., Curran, H.V., Howes, O.D., 2014. Dopaminergic function in cannabis users and its relationship to cannabis-induced psychotic symptoms. Biol. Psychiatry 75, 470–478.

Bolla, K.I., Brown, K., Eldreth, D., Tate, K., Cadet, J.L., 2002. Dose-related neurocognitive effects of marijuana use. Neurology 59, 1337–1343.

Bossong, M.G., Jager, G., Bhattacharyya, S., Allen, P., 2014a. Acute and non-acute effects of cannabis on human memory function: a critical review of neuroimaging studies. Curr. Pharmaceut. Des. 20, 2114–2125.

Bossong, M.G., Jansma, J.M., Bhattacharyya, S., Ramsey, N.F., 2014b. Role of the endocannabinoid system in brain functions relevant for schizophrenia: an overview of human challenge studies with cannabis or 9-tetrahydrocannabinol (THC). Prog. Neuropsychopharmacol. Biol. Psychiatry 52, 53–69.

Boydell, J., Van Os, J., Caspi, A., Kennedy, N., Giouroukou, E., Fearon, P., Farrell, M., Murray, R.M., 2006. Trends in cannabis use prior to first presentation with schizophrenia, in South-East London between 1965 and 1999. Psychol. Med. 36, 1441–1446.

Brand, E.J., Zhao, Z., 2017. Cannabis in Chinese medicine: are some traditional indications referenced in ancient literature related to cannabinoids? Front. Pharmacol. 8, 108.

Brody, A.L., London, E.D., Olmstead, R.E., Allen-Martinez, Z., Shulenberger, S., Costello, M.R., Abrams, A.L., Scheibal, D., Farahi, J., Shoptaw, S., Mandelkern, M.A., 2010. Smoking-induced change in intrasynaptic dopamine concentration: effect of treatment for Tobacco Dependence. Psychiatry Res. 183, 218–224.

Broyd, S.J., Greenwood, L.M., Van Hell, H.H., Croft, R.J., Coyle, H., Lee-Bates, B., Todd, J., Johnstone, S.J., Michie, P.T., Solowij, N., 2016. Mismatch negativity and P50 sensory gating in abstinent former cannabis users. Neural Plast. 6526437.

Brzozowska, N.I., De Tonnerre, E.J., Li, K.M., Wang, X.S., Boucher, A.A., Callaghan, P.D., Kuligowski, M., Wong, A., Arnold, J.C., 2017. The differential binding of antipsychotic drugs to the ABC transporter P-Glycoprotein predicts cannabinoid-antipsychotic drug interactions. Neuropsychopharmacology 42, 2222–2231.

Callaghan, P., Phillips, P., Khalil, E., Carter, T., 2012. Meeting the physical health-care needs of people with substance misuse problems: evaluation of a nurse-led blood-borne virus programme. Int. J. Ment. Health Nurs. 21, 248–258.

Cascini, F., Aiello, C., DI Tanna, G., 2012. Increasing delta-9-tetrahydrocannabinol (Delta-9-THC) content in herbal cannabis over time: systematic review and meta-analysis. Curr. Drug Abuse Rev. 5, 32–40.

Caspi, A., Moffitt, T.E., Cannon, M., Mcclay, J., Murray, R., Harrington, H., Taylor, A., Arseneault, L., Williams, B., Braithwaite, A., Poulton, R., Craig, I.W., 2005. Moderation of the effect of adolescent-onset cannabis use on adult psychosis by a functional polymorphism in the catechol-O-methyltransferase gene: longitudinal evidence of a gene X environment interaction. Biol. Psychiatry 57, 1117–1127.

Castaneto, M.S., Gorelick, D.A., Desrosiers, N.A., Hartman, R.L., Pirard, S., Huestis, M.A., 2014. Synthetic cannabinoids: epidemiology, pharmacodynamics, and clinical implications. Drug Alcohol Depend. 144, 12–41.

Clarke, J.R., Rossato, J.I., Monteiro, S., Bevilaqua, L.R.M., Izquierdo, I., Cammarota, M., 2008. Posttraining activation of CB1 cannabinoid receptors in the CA1 region of the dorsal hippocampus impairs object recognition long-term memory. Neurobiol. Learn. Mem. 90, 374–381.

Cohen, K., Weinstein, A.M., 2018. Synthetic and non-synthetic cannabinoid drugs and their adverse effects-A review from public health prospective. Front Public Health 6, 162.

Colizzi, M., Bhattacharyya, S., 2017. Does cannabis composition matter? Differential effects of delta-9-tetrahydrocannabinol and cannabidiol on human cognition. Curr. Addict. Rep. 4, 62–74.

Colizzi, M., Bhattacharyya, S., 2018a. Cannabis use and the development of tolerance: a systematic review of human evidence. Neurosci. Biobehav. Rev. 93, 1–25.

Colizzi, M., Bhattacharyya, S., 2018b. Neurocognitive effects of cannabis: lessons learned from human experimental studies. Prog. Brain Res. 242, 179–216.

Colizzi, M., Burnett, N., Costa, R., De Agostini, M., Griffin, J., Bhattacharyya, S., 2018a. Longitudinal assessment of the effect of cannabis use on hospital readmission rates in early psychosis: a 6-year follow-up in an inpatient cohort. Psychiatry Res. 268, 381–387.

Colizzi, M., Carra, E., Fraietta, S., Lally, J., Quattrone, D., Bonaccorso, S., Mondelli, V., Ajnakina, O., Dazzan, P., Trotta, A., Sideli, L., Kolliakou, A., Gaughran, F., Khondoker, M., David, A.S., Murray, R.M., Maccabe, J.H., Di Forti, M., 2016a. Substance use, medication adherence and outcome one year following a first episode of psychosis. Schizophr. Res. 170, 311–317.

Colizzi, M., Fazio, L., Ferranti, L., Porcelli, A., Masellis, R., Marvulli, D., Bonvino, A., Ursini, G., Blasi, G., Bertolino, A., 2015a. Functional genetic variation of the cannabinoid receptor I and cannabis use interact on prefrontal connectivity and related working memory behavior. Neuropsychopharmacology 40, 640–649.

Colizzi, M., Iyegbe, C., Powell, J., Blasi, G., Bertolino, A., Murray, R.M., Di Forti, M., 2015b. Interaction between DRD2 and AKT1 genetic variations on risk of psychosis in cannabis users: a case-control study. NPJ Schizophr. 1.

Colizzi, M., Iyegbe, C., Powell, J., Ursini, G., Porcelli, A., Bonvino, A., Taurisano, P., Romano, R., Masellis, R., Blasi, G., Morgan, C., Aitchison, K., Mondelli, V., Luzi, S., Kolliakou, A., David, A., Murray, R.M., Bertolino, A., Di Forti, M., 2015c. Interaction between functional genetic variation of DRD2 and cannabis use on risk of psychosis. Schizophr. Bull. 41, 1171–1182.

Colizzi, M., Mcguire, P., Giampietro, V., Williams, S., Brammer, M., Bhattacharyya, S., 2018b. Modulation of acute effects of delta-9-tetrahydrocannabinol on psychotomimetic effects, cognition and brain function by previous cannabis exposure. Eur. Neuropsychopharmacol. 28, 850–862.

Colizzi, M., Mcguire, P., Giampietro, V., Williams, S., Brammer, M., Bhattacharyya, S., 2018c. Previous cannabis exposure modulates the acute effects of delta-9-tetrahydrocannabinol on attentional salience and fear processing. Experimental and clinical psychopharmacology 26, 582–598.

Colizzi, M., Mcguire, P., Pertwee, R.G., Bhattacharyya, S., 2016b. Effect of cannabis on glutamate signalling in the brain: a systematic review of human and animal evidence. Neurosci. Biobehav. Rev. 64, 359–381.

Colizzi, M., Murray, R., 2018. Cannabis and psychosis: what do we know and what should we do? Br. J. Psychiatry 212, 195–196.

Colizzi, M., Weltens, N., Mcguire, P., Lythgoe, D., Williams, S., Van Oudenhove, L., Bhattacharyya, S., 2019a. Delta-9-tetrahydrocannabinol increases striatal glutamate levels in healthy individuals: implications for psychosis. Mol. Psychiatry.

Colizzi, M., Weltens, N., Mcguire, P., Van Oudenhove, L., Bhattacharyya, S., 2019b. Descriptive psychopathology of the acute effects of intravenous delta-9-tetrahydrocannabinol administration in humans. Brain Sci. 9.

Cortes-Briones, J., Skosnik, P.D., Mathalon, D., Cahill, J., Pittman, B., Williams, A., Sewell, R.A., Ranganathan, M., Roach, B., Ford, J., D'souza, D.C., 2015. Delta9-THC disrupts gamma (gamma)-Band neural oscillations in humans. Neuropsychopharmacology 40, 2124–2134.

Cousijn, J., Wiers, R.W., Ridderinkhof, K.R., Van Den Brink, W., Veltman, D.J., Goudriaan, A.E., 2012. Grey matter alterations associated with cannabis use: results of a VBM study in heavy cannabis users and healthy controls. Neuroimage 59, 3845–3851.

Creemers, H.E., Verhulst, F.C., Huizink, A.C., 2009. Temperamental risk factors for adolescent cannabis use: a systematic review of prospective general population studies. Subst. Use Misuse 44, 1833–1854.

Curran, H.V., Freeman, T.P., Mokrysz, C., Lewis, D.A., Morgan, C.J., Parsons, L.H., 2016. Keep off the grass? Cannabis, cognition and addiction. Nat. Rev. Neurosci. 17, 293–306.

D'souza, D.C., Abi-Saab, W.M., Madonick, S., Forselius-Bielen, K., Doersch, A., Braley, G., Gueorguieva, R., Cooper, T.B., Krystal, J.H., 2005. Delta-9-tetrahydrocannabinol effects in schizophrenia: implications for cognition, psychosis, and addiction. Biol. Psychiatry 57, 594–608.

D'souza, D.C., Cortes-Briones, J.A., Ranganathan, M., Thurnauer, H., Creatura, G., Surti, T., Planeta, B., Neumeister, A., Pittman, B., Normandin, M.D., Kapinos, M., Ropchan, J., Huang, Y., Carson, R.E., Skosnik, P.D., 2016a. Rapid changes in cannabinoid 1 receptor availability in cannabis-dependent male subjects after abstinence from cannabis. Biol. Psychiatry Cogn. Neurosci. Neuroimaging 1, 60–67.

D'souza, D.C., Fridberg, D.J., Skosnik, P.D., Williams, A., Roach, B., Singh, N., Carbuto, M., Elander, J., Schnakenberg, A., Pittman, B., Sewell, R.A., Ranganathan, M., Mathalon, D., 2012. Dose-related modulation of event-related potentials to novel and target stimuli by intravenous Delta(9)-THC in humans. Neuropsychopharmacology 37, 1632–1646.

D'souza, D.C., Perry, E., Macdougall, L., Ammerman, Y., Cooper, T., Wu, Y.T., Braley, G., Gueorguieva, R., Krystal, J.H., 2004. The psychotomimetic effects of intravenous delta-9-tetrahydrocannabinol in healthy individuals: implications for psychosis. Neuropsychopharmacology 29, 1558–1572.

D'souza, D.C., Radhakrishnan, R., Sherif, M., Cortes-Briones, J., Cahill, J., Gupta, S., Skosnik, P.D., Ranganathan, M., 2016b. Cannabinoids and psychosis. Curr. Pharmaceut. Des. 22, 6380–6391.

D'souza, D.C., Ranganathan, M., Braley, G., Gueorguieva, R., Zimolo, Z., Cooper, T., Perry, E., Krystal, J., 2008. Blunted psychotomimetic and amnestic effects of delta-9-tetrahydrocannabinol in frequent users of cannabis. Neuropsychopharmacology 33, 2505–2516.

Davis, G.P., Compton, M.T., Wang, S., Levin, F.R., Blanco, C., 2013. Association between cannabis use, psychosis, and schizotypal personality disorder: findings from the National Epidemiologic Survey on Alcohol and Related Conditions. Schizophr. Res. 151, 197–202.

Davison, J., O'gorman, A., Brennan, L., Cotter, D.R., 2018. A systematic review of metabolite biomarkers of schizophrenia. Schizophr. Res. 195, 32–50.

De La Fuente-Sandoval, C., Reyes-Madrigal, F., Mao, X., Leon-Ortiz, P., Rodriguez-Mayoral, O., Solis-Vivanco, R., Favila, R., Graff-Guerrero, A., Shungu, D.C., 2015. Cortico-striatal GABAergic and glutamatergic dysregulations in subjects at ultra-high risk for psychosis investigated with proton magnetic resonance spectroscopy. Int. J. Neuropsychopharmacol. 19, pyv105.

Degenhardt, L., Coffey, C., Romaniuk, H., Swift, W., Carlin, J.B., Hall, W.D., Patton, G.C., 2013a. The persistence of the association between adolescent cannabis use and common mental disorders into young adulthood. Addiction 108, 124–133.

Degenhardt, L., Ferrari, A.J., Calabria, B., Hall, W.D., Norman, R.E., Mcgrath, J., Flaxman, A.D., Engell, R.E., Freedman, G.D., Whiteford, H.A., Vos, T., 2013b. The global epidemiology and contribution of cannabis use and dependence to the global burden of disease: results from the GBD 2010 study. PLoS One 8, e76635.

Degenhardt, L., Hall, W., Lynskey, M., 2003. Testing hypotheses about the relationship between cannabis use and psychosis. Drug Alcohol Depend. 71, 37–48.

Dequardo, J.R., Carpenter, C.F., Tandon, R., 1994. Patterns of substance abuse in schizophrenia: nature and significance. J. Psychiatr. Res. 28, 267–275.

Dhunjibhoy, J.E., 1930. A brief resume of the types of insanity commonly met with in India with a full description of "Indian hemp insanity" peculiar to the country. J. Ment. Sci. 254–264.

Di Forti, M., Iyegbe, C., Sallis, H., Kolliakou, A., Falcone, M.A., Paparelli, A., Sirianni, M., LA Cascia, C., Stilo, S.A., Marques, T.R., Handley, R., Mondelli, V., Dazzan, P., Pariante, C., David, A.S., Morgan, C., Powell, J., Murray, R.M., 2012. Confirmation that the AKT1 (rs2494732) genotype influences the risk of psychosis in cannabis users. Biol. Psychiatry 72, 811–816.

Di Forti, M., Marconi, A., Carra, E., Fraietta, S., Trotta, A., Bonomo, M., Bianconi, F., Gardner-Sood, P., O'connor, J., Russo, M., Stilo, S.A., Marques, T.R., Mondelli, V., Dazzan, P., Pariante, C., David, A.S., Gaughran, F., Atakan, Z., Iyegbe, C., Powell, J., Morgan, C., Lynskey, M., Murray, R.M., 2015. Proportion of patients in south London with first-episode psychosis attributable to use of high potency cannabis: a case-control study. Lancet Psychiatry 2, 233–238.

Fattore, L., 2016. Synthetic cannabinoids-further evidence supporting the relationship between cannabinoids and psychosis. Biol. Psychiatry 79, 539–548.

Ferdinand, R.F., Sondeijker, F., Van Der Ende, J., Selten, J.P., Huizink, A., Verhulst, F.C., 2005. Cannabis use predicts future psychotic symptoms, and vice versa. Addiction 100, 612–618.

Fergusson, D.M., Boden, J.M., Horwood, L.J., 2015. Psychosocial sequelae of cannabis use and implications for policy: findings from the Christchurch Health and Development Study. Soc. Psychiatry Psychiatr. Epidemiol. 50, 1317–1326.

Fergusson, D.M., Horwood, L.J., Ridder, E.M., 2005. Tests of causal linkages between cannabis use and psychotic symptoms. Addiction 100, 354–366.

Fergusson, D.M., Horwood, L.J., Swain-Campbell, N.R., 2003. Cannabis dependence and psychotic symptoms in young people. Psychol. Med. 33, 15–21.

Ferraro, L., Russo, M., O'connor, J., Wiffen, B.D., Falcone, M.A., Sideli, L., Gardner-Sood, P., Stilo, S., Trotta, A., Dazzan, P., Mondelli, V., Taylor, H., Friedman, B., Sallis, H., LA Cascia, C., LA Barbera, D., David, A.S., Reichenberg, A., Murray, R.M., Di Forti, M., 2013. Cannabis users have higher premorbid IQ than other patients with first onset psychosis. Schizophr. Res. 150, 129–135.

Foglia, E., Schoeler, T., Klamerus, E., Morgan, K., Bhattacharyya, S., 2017. Cannabis use and adherence to antipsychotic medication: a systematic review and meta-analysis. Psychol. Med. 47, 1691–1705.

Foti, D.J., Kotov, R., Guey, L.T., Bromet, E.J., 2010. Cannabis use and the course of schizophrenia: 10-year follow-up after first hospitalization. Am. J. Psychiatry 167, 987–993.

French, L., Gray, C., Leonard, G., Perron, M., Pike, G.B., Richer, L., Seguin, J.R., Veillette, S., Evans, C.J., Artiges, E., Banaschewski, T., Bokde, A.W., Bromberg, U., Bruehl, R., Buchel, C., Cattrell, A., Conrod, P.J., Flor, H., Frouin, V., Gallinat, J., Garavan, H., Gowland, P., Heinz, A., Lemaitre, H., Martinot, J.L., Nees, F., Orfanos, D.P., Pangelinan, M.M., Poustka, L., Rietschel, M., Smolka, M.N., Walter, H., Whelan, R., Timpson, N.J., Schumann, G., Smith, G.D., Pausova, Z., Paus, T., 2015. Early cannabis use, polygenic risk score for schizophrenia and brain maturation in adolescence. JAMA Psychiatry 72, 1002–1011.

Frisher, M., Crome, I., Martino, O., Croft, P., 2009. Assessing the impact of cannabis use on trends in diagnosed schizophrenia in the United Kingdom from 1996 to 2005. Schizophr. Res. 113, 123–128.

Frohlich, J., Van Horn, J.D., 2014. Reviewing the ketamine model for schizophrenia. J. Psychopharmacol. 28, 287–302.

Gage, S.H., Hickman, M., Heron, J., Munafo, M.R., Lewis, G., Macleod, J., Zammit, S., 2014. Associations of cannabis and cigarette use with psychotic experiences at age 18: findings from the Avon Longitudinal Study of Parents and Children. Psychol. Med. 44, 3435–3444.

Gage, S.H., Jones, H.J., Burgess, S., Bowden, J., Davey Smith, G., Zammit, S., Munafo, M.R., 2017. Assessing causality in associations between cannabis use and schizophrenia risk: a two-sample Mendelian randomization study. Psychol. Med. 47, 971–980.

Galanopoulos, A., Polissidis, A., Papadopoulou-Daifoti, Z., Nomikos, G.G., Antoniou, K., 2011. Delta(9)-THC and WIN55,212-2 affect brain tissue levels of excitatory amino acids in a phenotype-, compound-, dose-, and region-specific manner. Behav. Brain Res. 224, 65–72.

Gaoni, Y., Mechoulam, R., 1971. The isolation and structure of delta-1-tetrahydrocannabinol and other neutral cannabinoids from hashish. J. Am. Chem. Soc. 93, 217–224.

Gill, K.E., Poe, L., Azimov, N., Ben-David, S., Vadhan, N.P., Girgis, R., Moore, H., Cressman, V., Corcoran, C.M., 2015. Reasons for cannabis use among youths at ultra high risk for psychosis. Early Interv. Psychiatry 9, 207–210.

Gobbi, G., Atkin, T., Zytynski, T., Wang, S., Askari, S., Boruff, J., Ware, M., Marmorstein, N., Cipriani, A., Dendukuri, N., Mayo, N., 2019. Association of Cannabis Use in Adolescence and Risk of Depression, Anxiety, and Suicidality in Young Adulthood: A Systematic Review and Meta-analysis. JAMA Psychiatry 76, 426–434.

Grant, I., Gonzalez, R., Carey, C.L., Natarajan, L., Wolfson, T., 2003. Non-acute (residual) neurocognitive effects of cannabis use: a meta-analytic study. J. Int. Neuropsychol. Soc. 9, 679–689.

Green, A.I., Zimmet, S.V., Strous, R.D., Schildkraut, J.J., 1999. Clozapine for comorbid substance use disorder and schizophrenia: do patients with schizophrenia have a reward-deficiency syndrome that can be ameliorated by clozapine? Harv. Rev. Psychiatry 6, 287–296.

Gurillo, P., Jauhar, S., Murray, R.M., Maccabe, J.H., 2015. Does tobacco use cause psychosis? Systematic review and meta-analysis. Lancet Psychiatry 2, 718–725.

Hall, W., Lynskey, M., 2016. Evaluating the public health impacts of legalizing recreational cannabis use in the United States. Addiction 111, 1764–1773.

Ham, S., Kim, T.K., Chung, S., Im, H.I., 2017. Drug abuse and psychosis: new insights into drug-induced psychosis. Exp. Neurobiol. 26, 11–24.

Haney, M., Evins, A.E., 2016. Does cannabis cause, exacerbate or ameliorate psychiatric disorders? An oversimplified debate discussed. Neuropsychopharmacology 41, 393–401.

Hasin, D.S., Sarvet, A.L., Cerdá, M., Keyes, K.M., Stohl, M., Galea, S., Wall, M.M., 2017. US adult illicit cannabis use, cannabis use disorder, and medical marijuana laws: 1991–1992 to 2012–2013. JAMA Psychiatry 74, 579–588.

Hennekens, C., Buring, J.E., 1987. Need for large sample sizes in randomized trials. Pediatrics 79, 569–571.

Henquet, C., Krabbendam, L., Spauwen, J., Kaplan, C., Lieb, R., Wittchen, H.U., Van Os, J., 2005. Prospective cohort study of cannabis use, predisposition for psychosis, and psychotic symptoms in young people. BMJ 330, 11.

Hill, A.B., 1965. The environment and disease: association or causation? Proc. R. Soc. Med. 58, 295–300.

Hill, M., 2015. Perspective: be clear about the real risks. Nature 525, S14.

Howes, O.D., Montgomery, A.J., Asselin, M.C., Murray, R.M., Valli, I., Tabraham, P., Bramon-Bosch, E., Valmaggia, L., Johns, L., Broome, M., Mcguire, P.K., Grasby, P.M., 2009. Elevated striatal dopamine function linked to prodromal signs of schizophrenia. Arch. Gen. Psychiatr. 66, 13–20.

Hsieh, J.H., Stein, D.J., Howells, F.M., 2014. The neurobiology of methamphetamine induced psychosis. Front. Hum. Neurosci. 8, 537.

Jager, G., Block, R.I., Luijten, M., Ramsey, N.F., 2010. Cannabis use and memory brain function in adolescent boys: a cross-sectional multicenter functional magnetic resonance imaging study. J. Am. Acad. Child Adolesc. Psychiatry 49 (561–72), 572 e1-3.

Jager, G., Van Hell, H.H., DE Win, M.M., Kahn, R.S., Van Den Brink, W., Van Ree, J.M., Ramsey, N.F., 2007. Effects of frequent cannabis use on hippocampal activity during an associative memory task. Eur. Neuropsychopharmacol. 17, 289–297.

Jockers-Scherubl, M.C., Wolf, T., Radzei, N., Schlattmann, P., Rentzsch, J., Gomez-Carrillo De Castro, A., Kuhl, K.P., 2007. Cannabis induces different cognitive changes in schizophrenic patients and in healthy controls. Prog. Neuropsychopharmacol. Biol. Psychiatry 31, 1054–1063.

Kelley, M.E., Wan, C.R., Broussard, B., Crisafio, A., Cristofaro, S., Johnson, S., Reed, T.A., Amar, P., Kaslow, N.J., Walker, E.F., Compton, M.T., 2016. Marijuana use in the immediate 5-year premorbid period is associated with increased risk of onset of schizophrenia and related psychotic disorders. Schizophr. Res. 171, 62–67.

Khantzian, E.J., 1997. The self-medication hypothesis of substance use disorders: a reconsideration and recent applications. Harv. Rev. Psychiatry 4, 231–244.

Khokhar, J.Y., Dwiel, L.L., Henricks, A.M., Doucette, W.T., Green, A.I., 2018. The link between schizophrenia and substance use disorder: a unifying hypothesis. Schizophr. Res. 194, 78–85.

Koenders, L., Cousijn, J., Vingerhoets, W.A., Van Den Brink, W., Wiers, R.W., Meijer, C.J., Machielsen, M.W., Veltman, D.J., Goudriaan, A.E., DE Haan, L., 2016. Grey matter changes associated with heavy cannabis use: a longitudinal sMRI study. PLoS One 11, e0152482.

Krishnadas, R., Jauhar, S., Telfer, S., Shivashankar, S., Mccreadie, R.G., 2012. Nicotine dependence and illness severity in schizophrenia. Br. J. Psychiatry 201, 306–312.

Ksir, C., Hart, C.L., 2016a. Cannabis and psychosis: a critical overview of the relationship. Curr. Psychiatr. Rep. 18, 12.

Ksir, C., Hart, C.L., 2016b. Correlation still does not imply causation. Lancet Psychiatry 3, 401-401.

Large, M., Sharma, S., Compton, M.T., Slade, T., Nielssen, O., 2011. Cannabis use and earlier onset of psychosis: a systematic meta-analysis. Arch. Gen. Psychiatr. 68, 555–561.

Law, R., Schier, J., Martin, C., Chang, A., Wolkin, A., Centers For Disease, C., 2015. Notes from the field: increase in reported adverse health effects related to synthetic cannabinoid use – United States, January–May 2015. MMWR Morb. Mortal. Wkly. Rep. 64, 618–619.

Lev-Ran, S., Roerecke, M., LE Foll, B., George, T.P., Mckenzie, K., Rehm, J., 2014. The association between cannabis use and depression: a systematic review and meta-analysis of longitudinal studies. Psychol. Med. 44, 797–810.

Lisman, J.E., Coyle, J.T., Green, R.W., Javitt, D.C., Benes, F.M., Heckers, S., Grace, A.A., 2008. Circuit-based framework for understanding neurotransmitter and risk gene interactions in schizophrenia. Trends Neurosci. 31, 234–242.

Loberg, E.M., Helle, S., Nygard, M., Berle, J.O., Kroken, R.A., Johnsen, E., 2014. The cannabis pathway to non-affective psychosis may reflect less neurobiological vulnerability. Front. Psychiatry 5, 159.

Lorenzetti, V., Solowij, N., Whittle, S., Fornito, A., Lubman, D.I., Pantelis, C., Yucel, M., 2015. Gross morphological brain changes with chronic, heavy cannabis use. Br. J. Psychiatry 206, 77–78.

Macleod, J., Hickman, M., 2006. Commentary: understanding pathways to cannabis use and from use to harm. Int. J. Epidemiol. 35, 680–682.

Manrique-Garcia, E., Zammit, S., Dalman, C., Hemmingsson, T., Andreasson, S., Allebeck, P., 2012. Cannabis, schizophrenia and other non-affective psychoses: 35 years of follow-up of a population-based cohort. Psychol. Med. 42, 1321–1328.

Marconi, A., Di Forti, M., Lewis, C.M., Murray, R.M., Vassos, E., 2016. Meta-analysis of the association between the level of cannabis use and risk of psychosis. Schizophr. Bull. 42, 1262–1269.

Martin, B.R., Mechoulam, R., Razdan, R.K., 1999. Discovery and characterization of endogenous cannabinoids. Life Sci. 65, 573–595.

Martin, D.A., Marona-Lewicka, D., Nichols, D.E., Nichols, C.D., 2014. Chronic LSD alters gene expression profiles in the mPFC relevant to schizophrenia. Neuropharmacology 83, 1–8.

Mata, I., Rodriguez-Sanchez, J.M., Pelayo-Teran, J.M., Perez-Iglesias, R., Gonzalez-Blanch, C., Ramirez-Bonilla, M., Martinez-Garcia, O., Vazquez-Barquero, J.L., Crespo-Facorro, B., 2008. Cannabis abuse is associated with decision-making impairment among first-episode patients with schizophrenia-spectrum psychosis. Psychol. Med. 38, 1257–1266.

Matsuda, L.A., Bonner, T.I., Lolait, S.J., 1993. Localization of cannabinoid receptor mRNA in rat brain. J. Comp. Neurol. 327, 535–550.

Mattson, M.E., Kessler, L.G., 1987. The use of time-related epidemiologic data on smoking for planning cancer control programs. J. Chronic Dis. 40 (Suppl. 2), 25S–37S.

Mcgrath, J., Saha, S., Welham, J., EL Saadi, O., Maccauley, C., Chant, D., 2004. A systematic review of the incidence of schizophrenia: the distribution of rates and the influence of sex, urbanicity, migrant status and methodology. BMC Med. 2, 13.

Mcguire, P., Robson, P., Cubala, W.J., Vasile, D., Morrison, P.D., Barron, R., Taylor, A., Wright, S., 2018. Cannabidiol (CBD) as an adjunctive therapy in schizophrenia: a multicenter randomized controlled trial. Am. J. Psychiatry 175, 225–231.

Mechoulam, R., Gaoni, Y., 1967. Recent advances in the chemistry of hashish. Fortschr. Chem. Org. Naturstoffe 25, 175–213.

Mechoulam, R., Shani, A., Edery, H., Grunfeld, Y., 1970. Chemical basis of hashish activity. Science 169, 611–612.

Meier, M.H., Caspi, A., Ambler, A., Harrington, H., Houts, R., Keefe, R.S., Mcdonald, K., Ward, A., Poulton, R., Moffitt, T.E., 2012. Persistent cannabis users show neuropsychological decline from childhood to midlife. Proc. Natl. Acad. Sci. USA 109, E2657–E2664.

Miller, R., Ream, G., Mccormack, J., Gunduz-Bruce, H., Sevy, S., Robinson, D., 2009. A prospective study of cannabis use as a risk factor for non-adherence and treatment dropout in first-episode schizophrenia. Schizophr. Res. 113, 138–144.

Moore, T.H., Zammit, S., Lingford-Hughes, A., Barnes, T.R., Jones, P.B., Burke, M., Lewis, G., 2007. Cannabis use and risk of psychotic or affective mental health outcomes: a systematic review. Lancet 370, 319–328.

Moreau, J.J., 1845. Du hachisch et de l'aliénation mentale: études psychologiques. Fortin Masson, Paris.

Murray, R.M., Englund, A., Abi-Dargham, A., Lewis, D.A., Di Forti, M., Davies, C., Sherif, M., Mcguire, P., D'souza, D.C., 2017. Cannabis-associated psychosis: neural substrate and clinical impact. Neuropharmacology 124, 89–104.

Murray, R.M., Mehta, M., Di Forti, M., 2014. Different dopaminergic abnormalities underlie cannabis dependence and cannabis-induced psychosis. Biol. Psychiatry 75, 430–431.

National Academies of Sciences, E., Medicine, 2017. The Health Effects of Cannabis and Cannabinoids: The Current State of Evidence and Recommendations for Research. The National Academies Press, Washington, DC.

Niemi-Pynttari, J.A., Sund, R., Putkonen, H., Vorma, H., Wahlbeck, K., Pirkola, S.P., 2013. Substance-induced psychoses converting into schizophrenia: a register-based study of 18,478 Finnish inpatient cases. J. Clin. Psychiatry 74, e94–e99.

Nutt, D., 2009. Government vs science over drug and alcohol policy. Lancet 374, 1731–1733.

Nutt, D., King, L.A., Saulsbury, W., Blakemore, C., 2007. Development of a rational scale to assess the harm of drugs of potential misuse. Lancet 369, 1047–1053.

Orr, J.M., Paschall, C.J., Banich, M.T., 2016. Recreational marijuana use impacts white matter integrity and subcortical (but not cortical) morphometry. Neuroimage Clin. 12, 47–56.

Ouellet-Plamondon, C., Abdel-Baki, A., Salvat, E., Potvin, S., 2017. Specific impact of stimulant, alcohol and cannabis use disorders on first-episode psychosis: 2-year functional and symptomatic outcomes. Psychol. Med. 47, 2461–2471.

Owen, M.J., O'donovan, M.C., Thapar, A., Craddock, N., 2011. Neurodevelopmental hypothesis of schizophrenia. Br. J. Psychiatry 198, 173–175.

Pagliaccio, D., Barch, D.M., Bogdan, R., Wood, P.K., Lynskey, M.T., Heath, A.C., Agrawal, A., 2015. Shared predisposition in the association between cannabis use and subcortical brain structure. JAMA Psychiatry 72, 994–1001.

Papanti, D., Schifano, F., Botteon, G., Bertossi, F., Mannix, J., Vidoni, D., Impagnatiello, M., Pascolo-Fabrici, E., Bonavigo, T., 2013. "Spiceophrenia": a systematic overview of "spice"-related psychopathological issues and a case report. Hum. Psychopharmacol. 28, 379–389.

Patel, R., Wilson, R., Jackson, R., Ball, M., Shetty, H., Broadbent, M., Stewart, R., Mcguire, P., Bhattacharyya, S., 2016. Association of cannabis use with hospital admission and antipsychotic treatment failure in first episode psychosis: an observational study. BMJ Open 6, e009888.

Paul, S., Bhattacharyya, S., 2018. Does thinner right entorhinal cortex underlie genetic liability to cannabis use? Psychol. Med. 48, 2766–2775.

Pavlovska, I., Orovcanec, N., Tausanova, B., Zafirova, B., 2012. Risk of lung cancer in current cigarette smokers: case-control study results. Eur. Respir. J. P4890.

Pearce, N., 2011. Epidemiology in a changing world: variation, causation and ubiquitous risk factors. Int. J. Epidemiol. 40, 503–512.

Pertwee, R.G., 2008. The diverse CB1 and CB2 receptor pharmacology of three plant cannabinoids: delta9-tetrahydrocannabinol, cannabidiol and delta9-tetrahydrocannabivarin. Br. J. Pharmacol. 153, 199–215.

Pertwee, R.G., Ross, R.A., 2002. Cannabinoid receptors and their ligands. Prostaglandins Leukot. Essent. Fatty Acids 66, 101–121.

Pistis, M., Ferraro, L., Pira, L., Flore, G., Tanganelli, S., Gessa, G.L., Devoto, P., 2002. Delta(9)-tetrahydrocannabinol decreases extracellular GABA and increases extracellular glutamate and dopamine levels in the rat prefrontal cortex: an in vivo microdialysis study. Brain Res. 948, 155–158.

Pope JR., H.G., Gruber, A.J., Yurgelun-Todd, D., 2001. Residual neuropsychologic effects of cannabis. Curr. Psychiatr. Rep. 3, 507–512.

Power, R.A., Verweij, K.J., Zuhair, M., Montgomery, G.W., Henders, A.K., Heath, A.C., Madden, P.A., Medland, S.E., Wray, N.R., Martin, N.G., 2014. Genetic predisposition to schizophrenia associated with increased use of cannabis. Mol. Psychiatry 19, 1201–1204.

Regier, D.A., Farmer, M.E., Rae, D.S., Locke, B.Z., Keith, S.J., Judd, L.L., Goodwin, F.K., 1990. Comorbidity of mental disorders with alcohol and other drug abuse. Results from the Epidemiologic Catchment Area (ECA) Study. JAMA 264, 2511–2518.

Reginsson, G.W., Ingason, A., Euesden, J., Bjornsdottir, G., Olafsson, S., Sigurdsson, E., Oskarsson, H., Tyrfingsson, T., Runarsdottir, V., Hansdottir, I., Steinberg, S., Stefansson, H., Gudbjartsson, D.F., Thorgeirsson, T.E., Stefansson, K., 2018. Polygenic risk scores for schizophrenia and bipolar disorder associate with addiction. Addict. Biol. 23, 485–492.

Rigucci, S., Marques, T.R., Di Forti, M., Taylor, H., Dell'acqua, F., Mondelli, V., Bonaccorso, S., Simmons, A., David, A.S., Girardi, P., Pariante, C.M., Murray, R.M., Dazzan, P., 2016. Effect of high-potency cannabis on corpus callosum microstructure. Psychol. Med. 46, 841–854.

Rognli, E.B., Berge, J., Hakansson, A., Bramness, J.G., 2015. Long-term risk factors for substance-induced and primary psychosis after release from prison. A longitudinal study of substance users. Schizophr. Res. 168, 185–190.

Rossler, W., Hengartner, M.P., Angst, J., Ajdacic-Gross, V., 2012. Linking substance use with symptoms of subclinical psychosis in a community cohort over 30 years. Addiction 107, 1174–1184.

Ruiz-Contreras, A.E., Carrillo-Sanchez, K., Gomez-Lopez, N., Vadillo-Ortega, F., Hernandez-Morales, S., Carnevale-Cantoni, A., Espejel-Nunez, A., Mendez-Diaz, M., Prospero-Garcia, O., 2013. Working memory performance in young adults is associated to the AATn polymorphism of the CNR1 gene. Behav. Brain Res. 236, 62–66.

Ruiz-Contreras, A.E., Carrillo-Sanchez, K., Ortega-Mora, I., Barrera-Tlapa, M.A., Roman-Lopez, T.V., Rosas-Escobar, C.B., Flores-Barrera, L., Caballero-Sanchez, U., Munoz-Torres, Z., Romero-Hidalgo, S., Hernandez-Morales, S., Gonzalez-Barrios, J.A., Vadillo-Ortega, F., Mendez-Diaz, M., Aguilar-Roblero, R., Prospero-Garcia, O., 2014. Performance in working memory and attentional control is associated with the rs2180619 SNP in the CNR1 gene. Genes Brain Behav. 13, 173–178.

Ruiz-Veguilla, M., Callado, L.F., Ferrin, M., 2012. Neurological soft signs in patients with psychosis and cannabis abuse: a systematic review and meta-analysis of paradox. Curr. Pharmaceut. Des. 18, 5156–5164.

Sami, M., Notley, C., Kouimtsidis, C., Lynskey, M., Bhattacharyya, S., 2019. Psychotic-like experiences with cannabis use predict cannabis cessation and desire to quit: a cannabis discontinuation hypothesis. Psychol. Med. 49, 103–112.

Sami, M.B., Bhattacharyya, S., 2018. Are cannabis and Non-cannabis using patients different groups? Towards understanding the neurobiology of cannabis use in psychotic disorders. Journal of Psychopharmacology 32, 825–849.

Sami, M.B., Rabiner, E.A., Bhattacharyya, S., 2015. Does cannabis affect dopaminergic signaling in the human brain? A systematic review of evidence to date. Eur. Neuropsychopharmacol. 25, 1201–1224.

San, L., Bernardo, M., Gomez, A., Pena, M., 2013. Factors associated with relapse in patients with schizophrenia. Int. J. Psychiatry Clin. Pract. 17, 2–9.

Schizophrenia Working Group Of The Psychiatric Genomics, C, 2014. Biological insights from 108 schizophrenia-associated genetic loci. Nature 511, 421–427.

Schnell, T., Koethe, D., Daumann, J., Gouzoulis-Mayfrank, E., 2009. The role of cannabis in cognitive functioning of patients with schizophrenia. Psychopharmacology (Berl) 205, 45–52.

Schoeler, T., Kambeitz, J., Behlke, I., Murray, R., Bhattacharyya, S., 2016a. The effects of cannabis on memory function in users with and without a psychotic disorder: findings from a combined meta-analysis. Psychol. Med. 46, 177–188.

Schoeler, T., Monk, A., Sami, M.B., Klamerus, E., Foglia, E., Brown, R., Camuri, G., Altamura, A.C., Murray, R., Bhattacharyya, S., 2016b. Continued versus discontinued cannabis use in patients with psychosis: a systematic review and meta-analysis. Lancet Psychiatry 3, 215–225.

Schoeler, T., Petros, N., Di Forti, M., Klamerus, E., Foglia, E., Ajnakina, O., Gayer-Anderson, C., Colizzi, M., Quattrone, D., Behlke, I., Shetty, S., Mcguire, P., David, A.S., Murray, R., Bhattacharyya, S., 2016c. Effects of continuation, frequency, and type of cannabis use on relapse in the first 2 years after onset of psychosis: an observational study. Lancet Psychiatry 3, 947–953.

Schoeler, T., Petros, N., Di Forti, M., Klamerus, E., Foglia, E., Murray, R., Bhattacharyya, S., 2017a. Effect of continued cannabis use on medication adherence in the first two years following onset of psychosis. Psychiatry Res. 255, 36–41.

Schoeler, T., Petros, N., Di Forti, M., Klamerus, E., Foglia, E., Murray, R., Bhattacharyya, S., 2017b. Poor medication adherence and risk of relapse associated with continued cannabis use in patients with first-episode psychosis: a prospective analysis. Lancet Psychiatry 4, 627–633.

Schoeler, T., Petros, N., Di Forti, M., Pingault, J.B., Klamerus, E., Foglia, E., Small, A., Murray, R., Bhattacharyya, S., 2016d. Association between continued cannabis use and risk of relapse in first-episode psychosis: a quasi-experimental investigation within an observational study. JAMA Psychiatry 73, 1173–1179.

Schoeler, T., Theobald, D., Pingault, J.B., Farrington, D.P., Coid, J.W., Bhattacharyya, S., 2018. Developmental sensitivity to cannabis use patterns and risk for major depressive disorder in mid-life: findings from 40 years of follow-up. Psychol. Med. 1–8.

Seddon, J.L., Birchwood, M., Copello, A., Everard, L., Jones, P.B., Fowler, D., Amos, T., Freemantle, N., Sharma, V., Marshall, M., Singh, S.P., 2016. Cannabis use is associated with increased psychotic symptoms and poorer psychosocial functioning in first-episode psychosis: a report from the UK national EDEN study. Schizophr. Bull. 42, 619–625.

Sevy, S., Burdick, K.E., Visweswaraiah, H., Abdelmessih, S., Lukin, M., Yechiam, E., Bechara, A., 2007. Iowa gambling task in schizophrenia: a review and new data in patients with schizophrenia and co-occurring cannabis use disorders. Schizophr. Res. 92, 74–84.

Sneider, J.T., Mashhoon, Y., Silveri, M.M., 2013. A review of magnetic resonance spectroscopy studies in marijuana using adolescents and adults. J. Addict. Res. Ther. Suppl 4.

Solowij, N., Michie, P.T., 2007. Cannabis and cognitive dysfunction: parallels with endophenotypes of schizophrenia? J. Psychiatry Neurosci. 32, 30–52.

Solowij, N., Stephens, R.S., Roffman, R.A., Babor, T., Kadden, R., Miller, M., Christiansen, K., Mcree, B., Vendetti, J., Marijuana Treatment Project Research, G., 2002. Cognitive functioning of long-term heavy cannabis users seeking treatment. J. Am. Med. Assoc. 287, 1123–1131.

Spalletta, G., Bria, P., Caltagirone, C., 2007. Differences in temperament, character and psychopathology among subjects with different patterns of cannabis use. Psychopathology 40, 29–34.

Taurisano, P., Antonucci, L.A., Fazio, L., Rampino, A., Romano, R., Porcelli, A., Masellis, R., Colizzi, M., Quarto, T., Torretta, S., DI Giorgio, A., Pergola, G., Bertolino, A., Blasi, G., 2016. Prefrontal activity during working memory is modulated by the interaction of variation in CB1 and COX2 coding genes and correlates with frequency of cannabis use. Cortex 81, 231–238.

Tien, A.Y., Anthony, J.C., 1990. Epidemiological analysis of alcohol and drug use as risk factors for psychotic experiences. J. Nerv. Ment. Dis. 178, 473–480.

Valmaggia, L.R., Day, F.L., Jones, C., Bissoli, S., Pugh, C., Hall, D., Bhattacharyya, S., Howes, O., Stone, J., Fusar-Poli, P., Byrne, M., Mcguire, P.K., 2014. Cannabis use and transition to psychosis in people at ultra-high risk. Psychol. Med. 44, 2503–2512.

Van De Giessen, E., Weinstein, J.J., Cassidy, C.M., Haney, M., Dong, Z., Ghazzaoui, R., Ojeil, N., Kegeles, L.S., Xu, X., Vadhan, N.P., Volkow, N.D., Slifstein, M., Abi-Dargham, A., 2017. Deficits in striatal dopamine release in cannabis dependence. Mol. Psychiatry 22, 68–75.

Van Gastel, W.A., Vreeker, A., Schubart, C.D., Maccabe, J.H., Kahn, R.S., Boks, M.P., 2014. Change in cannabis use in the general population: a longitudinal study on the impact on psychotic experiences. Schizophr. Res. 157, 266–270.

Van Os, J., Bak, M., Hanssen, M., Bijl, R.V., De Graaf, R., Verdoux, H., 2002. Cannabis use and psychosis: a longitudinal population-based study. Am. J. Epidemiol. 156, 319–327.

Van Winkel, R., Genetic, R., Outcome of Psychosis, I, 2011a. Family-based analysis of genetic variation underlying psychosis-inducing effects of cannabis: sibling analysis and proband follow-up. Arch. Gen. Psychiatr. 68, 148–157.

Van Winkel, R., Van Beveren, N.J., Simons, C., Genetic, R., Outcome of Psychosis, I., 2011b. AKT1 moderation of cannabis-induced cognitive alterations in psychotic disorder. Neuropsychopharmacology 36, 2529–2537.

Vaucher, J., Keating, B.J., Lasserre, A.M., Gan, W., Lyall, D.M., Ward, J., Smith, D.J., Pell, J.P., Sattar, N., Pare, G., Holmes, M.V., 2018. Cannabis use and risk of schizophrenia: a Mendelian randomization study. Mol. Psychiatry 23, 1287–1292.

Verweij, K.J., Abdellaoui, A., Nivard, M.G., Sainz Cort, A., Ligthart, L., Draisma, H.H., Minica, C.C., International Cannabis, C., Gillespie, N.A., Willemsen, G., Hottenga, J.J., Boomsma, D.I., Vink, J.M., 2017. Short communication: genetic association between schizophrenia and cannabis use. Drug Alcohol Depend. 171, 117–121.

Volkow, N.D., Swanson, J.M., Evins, A.E., Delisi, L.E., Meier, M.H., Gonzalez, R., Bloomfield, M.A., Curran, H.V., Baler, R., 2016. Effects of cannabis use on human behavior, including cognition, motivation, and psychosis: a review. JAMA Psychiatry 73, 292–297.

Weiser, M., Knobler, H.Y., Noy, S., Kaplan, Z., 2002. Clinical characteristics of adolescents later hospitalized for schizophrenia. Am. J. Med. Genet. 114, 949–955.

Wiles, N.J., Zammit, S., Bebbington, P., Singleton, N., Meltzer, H., Lewis, G., 2006. Self-reported psychotic symptoms in the general population: results from the longitudinal study of the British National Psychiatric Morbidity Survey. Br. J. Psychiatry 188, 519–526.

Wilson, R.P., Bhattacharyya, S., 2016. Antipsychotic efficacy in psychosis with co-morbid cannabis misuse: a systematic review. J. Psychopharmacol. 30, 99–111.

Wise, L.E., Thorpe, A.J., Lichtman, A.H., 2009. Hippocampal CB1 receptors mediate the memory impairing effects of delta(9)-tetrahydrocannabinol. Neuropsychopharmacology 34, 2072–2080.

Yucel, M., Solowij, N., Respondek, C., Whittle, S., Fornito, A., Pantelis, C., Lubman, D.I., 2008. Regional brain abnormalities associated with long-term heavy cannabis use. Arch. Gen. Psychiatr. 65, 694–701.

Zammit, S., Allebeck, P., Andreasson, S., Lundberg, I., Lewis, G., 2002. Self reported cannabis use as a risk factor for schizophrenia in Swedish conscripts of 1969: historical cohort study. BMJ 325, 1199.

Zammit, S., Owen, M.J., Evans, J., Heron, J., Lewis, G., 2011. Cannabis, COMT and psychotic experiences. Br. J. Psychiatry 199, 380–385.

Interventions

CHAPTER

17

The ethics of identifying and treating psychosis risk

Paolo Corsico[1,2], *Ilina Singh*[3]

[1]Centre for Social Ethics and Policy, Department of Law, School of Social Sciences, the University of Manchester, Manchester, United Kingdom; [2]Department of Psychiatry, University of Oxford, Oxford, United Kingdom; [3]Department of Psychiatry & Wellcome Centre for Ethics and Humanities, University of Oxford, Oxford, United Kingdom

Introduction

The attempt to identify and treat psychosis risk ought not to be understood as a purely clinical endeavor. Rather, the attempt to identify individuals at the very early stages of psychotic illness, *before* they suffer from a psychotic episode, has clinical, ethical, and social implications. In recent years, the interest for at-risk mental state (ARMS) and high-risk populations has increased alongside the implementation of early intervention (EI) for psychosis (Fusar-Poli et al., 2013b; Marwaha et al., 2016). Efforts to tackle psychosis risk and the expansion of EI services are historically interconnected. In this chapter, we try to grasp the ethical dimensions of research and clinical efforts in identifying and treating psychosis risk, and we provide some arguments as to why these efforts, as well as the expansion of EI into psychosis *prediction*, require careful ethical scrutiny.

Patrick McGorry has described the need to intervene early in order to tackle psychotic disorders as EI *imperative*. The imperative described by McGorry consists of the fact that the need for care, in the case of psychotic illness, predates the point where a clear diagnosis can be made (McGorry, 2010). The development of EI for psychosis in the past 25 years has to a certain extent followed this imperative, by recognizing the need to engage young individuals in the early phases of psychotic illness with youth-friendly, community-based services, in order to promote social recovery and improve clinical outcomes (Yung, 2012). Delivering interventions to young individuals who have experienced a first episode of psychosis (FEP) situates EI within the realm of secondary prevention of psychotic disorders. Yet, the clinical benefits gained by implementing the EI imperative come with risks and costs. As clinical benefits potentially increase, while we move upstream in the clinical staging model toward the management of psychosis risk and into the realm of primary prevention, so do ethical risks and costs.

Risk Factors for Psychosis
https://doi.org/10.1016/B978-0-12-813201-2.00017-X

In this chapter, we first describe why the involvement of at-risk populations in EI for psychosis generates ethical concerns. Then, we review and summarize the ethical issues that arise from the attempt to identify and treat psychosis risk. Even though there are some overlaps with the ethical debate regarding EI for individuals who have already experienced an FEP, the reflections presented in this chapter shall refer to individuals at risk of psychosis, who have not (yet) transitioned to an FEP. We outline four reasons as to why the clinical management of psychosis risk requires careful ethical scrutiny. More precisely, we argue that there are, at least, four ethical questions that we ought to ask when we try to identify the risk of developing psychotic illness and offer treatment based on such risk. First, the idea of prediction: What are the ethical implications of the attempt to *predict* psychosis, even before an individual has suffered from a psychotic episode? Second, what ethical issues arise in the *assessment* of psychosis risk? Third, how should psychosis risk be *communicated* to healthy or help-seeking individuals and their relevant others? Fourth, what kind of *interventions* ought to be offered based on psychosis risk? We do not pretend to provide a definitive and exhaustive answer to these four questions. Rather, we argue that these questions should remain present to stakeholders involved in the attempt to identify and treat psychosis risk, with the aim to promote critical reflection on the ethical issues arising from this endeavor.

Lastly, we outline a further area of ethicolegal concern. We argue that there is one additional question that stakeholders should address, when evaluating the ethical implications of the attempt to identify and treat psychosis risk: How does the logic of *risk* in the clinical management of psychotic illness interact with the logic of *risk* in mental health legislation? In England and Wales, as in other legal jurisdictions, risk of harm to self or others is currently one of the criteria for involuntary admission to hospital (Department of Health of the UK Government, 2015a), provided further conditions are satisfied. Scholars have highlighted a recent shift toward a "logic of risk" in mental health legislation and expressed concerns over how this logic is implemented in clinical practice (Glover-Thomas, 2011; Szmukler and Rose, 2013). Risk of harm to self or others and risk to develop psychotic illness are, obviously, different types of risk. Nonetheless, we deem it necessary to highlight some overlaps in the two domains. As we shall argue later in this chapter, researchers and clinicians should be wary of the potentially harmful consequences, for individuals who are defined as being at risk of psychosis, of an uncritical reception of the differences between "psychosis risk" and the concept of "risk of harm" in mental health legislation.

The early intervention imperative and at-risk populations

Since the initiation of the EI strategy, it has been clear that EI aimed at challenging previous approaches to psychotic disorders, mainly by contrasting the conception of schizophrenia as a chronic and incurable mental disorder (Yung et al., 1996). This was to be achieved by focusing on the patient rather than the illness, reducing the duration of untreated psychosis (DUP) in order to improve clinical outcomes, and by promoting social recovery (Marwaha et al., 2016; Yung, 2012). In this regard, EI for psychosis aimed to ameliorate previous clinical approaches to psychosis and promote social recovery and normalization of symptoms, as opposed to medicalization of mental illness. Narratives from EI service users, who successfully engaged with the services, are essential to substantiate this point:

> At first I was terrified that I was going to forever be an incapacitated person and I was never going to be able to achieve the things that I was potentially going to achieve before this happened. I had that feeling in the hospital because of the way that the doctors kept talking about it, that I'd be on medication for life. Really, that so countered everything that I believe

> was going to happen and it really terrified me, and it was really good not to be in there [hospital] and to be in the community program, where I quickly got from that, I'm not the problem, the problem is the problem.
> ***Participant from an early psychosis program quoted in Stewart (2013, p. 342).***

Likewise, some of the ethical challenges of delivering EI in the context of psychosis risk, such as the problem of false positives in ARMS populations or what type of intervention is appropriate to offer to high-risk individuals, were apparent since the initiation of the EI strategy (Yung et al., 1996). In this respect, Broome and Fusar-Poli (2012) have highlighted three main areas of ethical concern. First, early detection and intervention in individuals with ARMS are more controversial than EI for individuals who have already experienced an FEP, given the problem of false positives and the variability in transition rates (Fusar-Poli et al., 2012a). Second, there are potential harmful consequences of overtreatment and stigmatization for at-risk individuals who may, but also may not, transition to full-blown psychosis. Third, there are ethical concerns regarding how—or *whether*—to disclose psychosis risk to individuals who have not yet and may never transition to psychotic illness (Broome and Fusar-Poli, 2012; Broome et al., 2013).

A tension between the clinical benefits gained by intervening early and the potential hazards that come from the inclusion of at-risk individuals into clinical services is intrinsic to EI for psychosis. The earlier the intervention, the greater the potential clinical benefits and also the greater the potential hazards. In this respect, it is not primarily relevant which definition is being used to identify individuals at risk of transitioning to psychosis—whether it is ARMS, prodromal, high risk (HR), or ultra high risk (UHR) (Fusar-Poli et al., 2013a). What is relevant, from the standpoint of ethics, is the fact that those individuals are targeted by clinical services and research efforts in the *absence* of full-blown psychosis. We believe this to be the central moral dilemma of EI for psychosis. However, we agree that the hazards of EI might be greater when EI is implemented very early in the putative developmental course. The evidence indicates that a majority of at-risk patients will never transition to psychotic illness, and presumably, the earlier one intervenes in the putative developmental course, the less accurately one is able to predict risk of transition, due to the fact of less developmental evidence. Therefore, the moral dilemma of EI may be exacerbated by the attempt to move the clinical staging model upstream, in order to expand the clinical benefits of EI to those who are *just* at risk of developing psychosis.

By referring to two concepts traditionally used in medical ethics, we can describe the dilemma of EI for psychosis as a conflict between the principle of *beneficence* and the principle of *nonmaleficence* (Beauchamp and Childress, 2013). The principle of beneficence states that clinicians have a moral duty to act in a way that benefits their patients, to promote their best interest. On the other hand, the principle of nonmaleficence states that clinicians have a moral duty to *avoid* causing harm to their patients and that, should harm occur, this should be outweighed by the benefits of the intervention. A conflict of ethical principles is different from a mere cost–benefit analysis, in that it directly involves a discussion regarding moral *duties* of the actors involved and not only an evaluation of benefits and costs. Benefits of involving at-risk individuals in EI include (i) potential improvements in clinical outcomes (Bechdolf et al., 2012; van der Gaag et al., 2013); (ii) promotion of social recovery (Hegelstad et al., 2012); (iii) prevention of comorbidity and disability associated with illness progression (Fusar-Poli et al., 2014); and (iv) impact on societal costs of severe mental illness (McCrone et al., 2010). Potential hazards include (i) risks of overdiagnosis and overtreatment (Raven et al., 2012); (ii) risk of stigmatization (Yang et al., 2012); (iii) risk of harm from excessive or inappropriate use of medication (Jorm, 2012;

Warner, 2005); and (iv) impact on the developing sense of identity and agency of young individuals (Brew et al., 2017).

Despite the fact that some scholars have expressed relevant concerns (Warner, 2005), it appears to be generally accepted that the clinical benefits of EI for FEP populations outweigh the risks of the intervention (Broome and Fusar-Poli, 2012; Marshall and Rathbone, 2011), provided that appropriate ethical safeguards are put in place. There is agreement that it is ethically justifiable to try to engage individuals whose experiences have reached a threshold of clinical significance in youth-friendly, low-stigmatizing, community-based clinical services (Marshall and Rathbone, 2011). However, this may not be the case with respect to at risk populations, specifically those with ARMS status. The moral dilemma posed by the implementation of EI for ARMS individuals is far from being resolved, and we do not pretend to solve it in this chapter. Rather, as we shall argue, clinicians and researchers may—and probably ought to—try to shed light on this dilemma by addressing four ethical questions in the process of identifying risk of developing psychosis.

Predicting psychosis and the ethics of prediction

Question (1): What are the ethical implications of the attempt to predict psychosis, before an individual has suffered from a psychotic episode?

Here lies the main ethical issue that we ought to address when attempting to identify and treat psychosis risk. The attempt to prevent psychotic illness by identifying vulnerability to psychosis (Fusar-Poli et al., 2012b) and predicting psychosis transition is grounded in the principle of beneficence, as it aims to extend the benefits gained by using the clinical staging model to preclinical populations (Yung and McGorry, 2007). More precisely, prediction of psychosis transition would serve as a *medium* for prevention of psychotic disorders. As Fusar-Poli et al. (2013a, p. 107) point out, "This [HR] category was introduced with the goal of developing treatments for prevention of psychotic disorders." This fact situates psychosis prediction in the realm of preventive medicine, specifically primary prevention, notwithstanding undeniable differences between preventive capabilities in physical and mental health. The important consideration is as follows: individuals at risk of psychosis are, by definition, in a preclinical stage, in a health context where conversion rates are low. So, can the principle of beneficence, which establishes a moral duty to intervene, outweigh the principle of nonmaleficence?

(Young) individuals who meet UHR criteria may progress to develop a psychotic disorder; they may develop a form of nonpsychotic illness; or they may just be experiencing transient mental and behavioral difficulties, which will not result in a mental disorder (McGorry, 2010). Two questions arise from these points. First, what are we predicting when we use UHR criteria, given that a proportion of UHR individuals will develop a nonpsychotic illness? A partial answer to this question may lie in the alleged poor clinical utility of standard diagnostic categories in early detection (McGorry, 2010). Second, who should be offered treatment, given that a proportion of UHR individuals may in fact never get ill? Again, McGorry (2010, p. 50) has provided a possible answer to this question, in stating that "the boundary of the initial offer of care should be set where there is a capacity to benefit from it and the potential benefits outweigh any risks of such care." Not only does establishing where to set such boundary require careful clinical evaluation but it also requires ethical reflection.

The preventive strategy of psychosis prediction shows some similarities—at least in terms of ethics appraisal—with the debate over predictive genetic testing (PGT) in young people for late-onset disorders and conditions. In their review of ethical arguments, Mand et al. (2012) show that most clinical guidelines for PGT, at least up to 2006, recommend *against* predictive testing in minors, where there is no potential medical benefit in the near future. As the authors point out, "[o]pponents of testing highlight potential harms, while proponents look to testing as an opportunity to promote benefit" (Mand et al., 2012, p. 519). Arguments in favor or against PGT in young people include considerations regarding psychological harm, autonomy, social discrimination, and impact on the family. The parallel with psychosis prediction is strong. Prevention of psychotic disorders via psychosis prediction is prone to concerns about clinical utility, concerns that may be exacerbated by the young age of at-risk individuals.

Here, we argue that the clinical utility of the clinical staging model must be weighed against the risks of prediction, in order to safeguard vulnerable young individuals. First, just as EI for psychosis aims at normalizing psychotic experiences and at providing appropriate treatment in the early phases of psychotic illness (Lester et al., 2012), there is a risk that the attempt to predict psychosis, before an individual has suffered from a first psychotic episode, might have the opposite effect to *medicalize* behavioral difficulties and adverse life experiences. The attempt to predict psychosis in the absence of an FEP could also increase anxiety and depression. Second, there is a risk of *paternalism*. In medical ethics, paternalism can be defined as a conflict between the principle of beneficence and the duty to respect autonomy (Beauchamp and Childress, 2013). Given that at-risk individuals do not yet suffer from psychotic illness, the attempt to predict a psychosis transition, which *may not* occur, could result in imposing unnecessary clinical burdens on young individuals. Third, the young age of at risk populations mandates that additional ethical safeguards be put in place when using a preventive clinical strategy, especially in mental health. In this regard, the criterion offered by McGorry, "the boundary of the initial offer of care," would have to be carefully discussed and negotiated with the young recipients of such care. Careful attention to developmentally appropriate communication tools and educational resources will be needed to address the challenge of the capacity of young people to assent to preventive strategies.

Ethics and psychosis risk assessment

Question (2): What ethical issues arise in the assessment of psychosis risk?

Ethical issues arise from the attempt to *identify* psychosis risk and from the practice of risk assessment. In their review, Addington et al. (2015) found that the two screening tools most widely used to identify clinical HR for psychosis are the Structured Interview for Psychosis-Risk Syndrome (SIPS) and the Comprehensive Assessment of At-Risk Mental State (CAARMS). Both tools are "structured clinical interviews typically used to identify individuals who seek help and present at clinical settings" (Addington et al., 2015, p. 345). Further, the authors show that transition to psychosis, in individuals identified as being at risk for psychosis through a clinical interview, occurs in approximately 36% of cases within 36 months (Fusar-Poli et al., 2012a). Transition rates of less than 40% based on clinical interview assessment have generally been confirmed by other studies and metaanalyses (Fusar-Poli et al., 2012a). Despite the great operational validity of interview-based risk assessment in the clinical setting, the recognition of (relatively) low transition rates has directed research efforts toward the identification of neurocognitive and neurobiological measures of psychosis risk (Fusar-Poli et al., 2012b; Gifford et al., 2017) to enhance or increase prediction. To date, psychosis risk in help-seeking individuals is established by means of a clinical interview, whereas in the future, it *could* be established via neurocognitive and neurobiological measures. These two models of psychosis risk assessment display different—though often related—ethical and social implications.

Clinical measures

The CAARMS is a "semi structured interview schedule designed for use by [trained] mental health professionals", which operationalizes the UHR criteria in order to assess psychopathology, and "determine if an individual meets criteria for being at ultra high risk (UHR) for onset of first psychotic disorder" (Yung et al., 2005, pp. 964–7). It has good predictive validity, meaning that high CAARMS scores are significantly associated with development of a psychotic disorder and good to excellent interrater reliability, meaning that it can reliably be administered by different providers (Yung et al., 2005). Nonetheless, the transition rate of UHR CAARMS-positive individuals reported by Yung et al. (2005) was 42% over 12 months, and transition rates were smaller in other clinical samples. In their metaanalysis of transition outcomes of HR individuals, Fusar-Poli et al. (2012a) found that, independently from the instrument used, transition risk was 18% after 6 months, 22% after 1 year, 29% after 2 years, and 36% after 3 years. In addition, these figures refer to help-seeking individuals and are not transferable to the general population. As Yung et al. (2005) point out

> [C]aution is needed in using the CAARMS or similar instruments in the general population as the incidence of psychotic disorders in unselected populations is so low that the false-positive rate would be extremely high and make screening inappropriate (p. 971).

Low transition rates and the problem of false positives have been recognized as major ethical concerns in at-risk designation of young individuals (Appelbaum, 2015; Cassetta and Goghari, 2015; Raven et al., 2012). Appelbaum (2015), for instance, poses the issues of informed consent procedures in at-risk research and of what type of intervention should follow at-risk designation. In their comprehensive review of the ethical implications of screening for at risk status, Cassetta and Goghari (2015) remind us of the potential implications that at-risk designation could have on individuals who *will not* transition to psychosis, given that they represent the majority of at-risk individuals. The authors conclude that, in order to respect the self-determination, liberty, and dignity of those individuals, healthcare professionals ought to ensure that effective communication on potential hazards and benefits of screening takes place, and that information acquired through risk assessment should remain private and confidential (with regard to the ethics of risk communication, see also Shanmugaratnam and Edwards, 2016).

Here, we argue that clinicians should be aware of the ethical constraints that low transition rates pose to their clinical agenda. As we specify later, engagement in clinical services should be considered carefully and should directly involve the young at-risk person and her family.

Neurofunctional and neurobiological measures

In recent years, the heterogeneity of UHR populations and the (relatively) low transition rates observed with interview-based risk assessment have sparked great interest for neurocognitive and neurobiological measures of psychosis risk (Gifford et al., 2017; McGorry et al., 2014). Fusar-Poli et al. (2012b) have provided an engaging account of the evidence concerning cognitive, functional, and neurobiological abnormalities that characterize vulnerability to psychosis and predate psychopathology; this book provides an update and further evidence on the matter.

Some recent developments are worth special mention. First, there are ongoing efforts to identify psychosis risk and to predict psychosis transition using neuroanatomical or neurofunctional markers (Koutsouleris et al., 2015;

Mechelli et al., 2011). Recently, for instance, Koutsouleris et al. (2015) were able to predict transition outcomes in 80% of cases, by stratifying individuals into different risk groups and then using magnetic resonance imaging (MRI)–based prediction scores. Second, there is a growing interest for psychosis prediction through integration of different data sets via *machine learning* methods (Pettersson-Yeo et al., 2013; Valli et al., 2016). The integration of genetic, cognitive, structural, and functional MRI data could support group classification and help to predict psychosis transition.

Assessing risk of psychosis by using neuroimaging data is still, and perhaps will remain, just a possibility. Yet, the introduction of neuroimaging in psychiatry poses some significant ethical challenges (Bluhm et al., 2015). Here, we shall mention three. First, there is a risk that using neuroimaging to predict psychosis could reinforce *essentialist* thinking in service users and families. Neuroessentialism may be defined as a tendency to understand biological explanations of mental disorders as complete explanations of mental illness (Haslam, 2011). This could in turn have relevant implications on issues of stigma and labeling (Yang et al., 2012), in so far as labeling a mental illness a problem of biological causation can increase stigma (Illes et al., 2008). Second, using neuroimaging to predict psychosis will likely generate incidental findings (IFs), which could reveal sensitive information regarding the health status of service users (Racine and Illes, 2007). Appropriate guidelines on how to manage IFs in psychosis prediction will have to be put in place, and consideration should be given to how (and whether) to communicate IFs to service users and families. Third, there is an issue of feasibility and cost-effectiveness. Under what criteria will (scarce) resources be allocated, should psychosis prediction via neuroimaging become a reality, given that it will likely involve much higher costs than interview-based risk assessment?

Psychosis prediction via machine learning methods poses even greater ethical and legal challenges. Issues of data privacy and data protection will likely be added to the challenges previously identified (Sejnowski et al., 2014). What impact will machine learning–based technologies have on (young) mental health service users' lives, identities, and social relationships? How will the enormous amount of data gathered with these methods be managed or fed back to service users and families? How should it be (Mittelstadt et al., 2016)?

Ethics and risk communication

Question (3): How should psychosis risk be communicated to healthy or help-seeking individuals and their relevant others?

Psychosis risk assessment may take place in a clinical setting upon engagement of help-seeking individuals or in the context of clinical research. Often, the two settings overlap (Fusar-Poli et al., 2013b). Disclosing an at-risk state or an attenuated psychosis syndrome (APS) (American Psychiatric Association, 2013b), as codified by the *Diagnostic and Statistical Manual of Mental Disorders, 5th Edition* (DSM-5), may have profound implications for healthy or help-seeking (young) individuals, as well as their families (Cassetta and Goghari, 2015).

Communicating psychosis risk poses significant ethical challenges. Do clinicians working in psychosis prevention have a duty to disclose such information to help-seeking individuals and their families? Would clinicians have a duty to respect the right *not to know* of prospective patients/service users, should they express such desire? What about non–help-seeking (healthy?) individuals who agree to take part in clinical research? Most importantly, *how* should clinicians and researchers communicate risk, so that such information may positively influence the clinical course and positively affect identity formation of individuals at a relatively early developmental stage? A moral dilemma is embedded in the practice of communicating psychosis risk. On the one hand, there is a need to communicate risk to individuals who may progress to develop psychotic illness. On the other hand, there is a need to normalize experiences of mental distress over the developmental course, to tailor communication to the age, capacity, and clinical status of the individual and to avoid *harm* that could derive from uncritical disclosure (Mittal et al., 2015). A subset of these harms could have long-term implications for a young person, e.g., identification of psychosis risk in employment searches or armed forces applications. Such identification processes may not make a nuanced distinction between different risk categories as discussed in this chapter, such as ARMS and UHR and FEP or schizophrenia.

With regard to APS, Mittal et al. (2015) have carefully outlined the ethical implications of full disclosure, nondisclosure, and partial disclosure in the clinical setting. First, the authors describe full disclosure as grounded in the principle of *autonomy*. Full disclosure would also be supported by the clinical benefits that can derive by the offer of intervention. At same time, the authors warn against the potential hazards of stigma, distress, and age-related/cultural differences in interpreting the information provided. Second, the authors describe nondisclosure as grounded in the principle of *nonmaleficence*, as this would avoid the potentially negative impacts of stigma and labeling. However, nondisclosure of APS would undermine individuals' autonomy. Eventually, Mittal et al. (2015) argue in favor of a hybrid approach. According to their proposal, disclosure of APS should be tailored based upon factors specific to each case. Parents/legal guardians and at-risk adults should be given full disclosure, provided that culturally sensitive communication strategies are put in place. On the contrary, minors, young adolescents, or individuals with "poorer insight" into their condition (p. 13) should not be given full disclosure—but only information that focuses on symptoms—unless full disclosure is necessary to avoid exposing the individual to additional risk factors, such as drug abuse. Ultimately, as we argue, the decision-making process will rely on the sensibility of the clinician. Awareness of the ethical implications of risk communication will be essential in order to shape this sensibility and to avoid the pitfalls of paternalism or the hazards of uncritical disclosure.

Communication of psychosis risk can also occur in clinical research (Morris and Heinssen, 2014). However, clinical research on psychosis risk may well involve individuals who *are not* currently seeking help. Therefore, Morris and Heinssen (2014) have listed three concepts that informed consent in psychosis-risk research should highlight: (i) risk assessment tools cannot provide a *definitive* measure of who will develop a psychotic disorder, and who will not; (ii) transition to psychosis *is statistically unlikely*; and (iii) at present, treatment options for individuals at risk of psychosis are being debated. Treatment options will thus have to be carefully negotiated with each individual and/or her family. In addition, Morris and Heinssen (2014) remind us of the risk of "therapeutic misconception" in clinical research. Researchers will have to make sure that young individuals and their families do not misinterpret the goals of research and that they are capable to provide fully informed consent to their participation.

Once again, we can merely scratch the surface of such stratified ethical and legal concerns. In this respect, it will be vital to promote ethical *awareness* among clinicians and researchers working in psychosis prevention, so that risk communication and informed consent procedures may fully contribute to empowering prospective patients, service users, and research participants.

Interventions based on psychosis risk

Question (4): What kind of interventions ought to be offered based on psychosis risk?

Should at-risk individuals be offered treatment, and, if so, what kind of treatment should they be offered? These questions have long characterized EI for psychosis. They have sparked great attention primarily with reference to the (proposed) inclusion of a "psychosis risk syndrome" in the latest revision of the DSM (Yung et al., 2010), which was published in 2013 (American Psychiatric Association, 2013a).

Philosophical and conceptual issues are embedded in the definition of psychosis risk (Broome et al., 2013). The very concept of *prodromal* was developed retrospectively, by investigating early signs of psychological distress, which could predate psychopathology in individuals who later developed a psychotic disorder. Further, it has been demonstrated that psychotic experiences are much more frequent in the general population than what the epidemiology of schizophrenia would reveal and what was previously known (Broome and Fusar-Poli, 2012, Healy and Cannon, this book, Chapter 7). The vast majority of people who experience psychotic "symptoms" in their life will not develop a psychotic disorder (Read and Dillon, 2013). These and other ethical considerations played a major role in shaping the debate over the inclusion of a "psychosis risk syndrome" in the latest revision of the DSM-5 (Goncalves et al., 2016). It is worth noticing that the "founders" of EI for psychosis were critical of the proposal to include a psychosis risk syndrome in the DSM-5. Yung et al. (2010), for instance, argued that it was premature and potentially dangerous to include a risk syndrome as a formal diagnosis in the DSM-5. Patrick McGorry supported a substantially modified version of the risk syndrome. By referring to the poor utility of standard diagnostic categories in early detection, which leads to a considerable number of at-risk individuals developing a form of nonpsychotic illness, McGorry supported the idea that a "pluripotential" or "common" risk syndrome could have better clinical and operational validity (Johannessen and McGorry, 2010; McGorry, 2010). Eventually, the latest version of the DSM-5 did include APS (American Psychiatric Association, 2013b). However, even though criteria for APS resemble UHR criteria, the "risk" label was removed from the name of the new syndrome. Moreover, although APS is listed as a diagnosable condition in Section II of the DSM-5 (p. 122), it is also placed in Section III among "Conditions for Further Study." It may be that the "reification" of psychosis risk has been partially avoided, at least for now.

The tension between beneficence and nonmaleficence in EI for at-risk individuals, as we have outlined it in this chapter, would speak *against* the reification of psychosis risk into a new syndrome. However, now that APS is a diagnosable condition, the urgent ethical question is this: Which treatments should be offered to individuals at risk of developing a psychotic disorder? How do we avoid the potential harm that could derive from such treatments?

It is outside the remit of ethics to establish which treatments are more appropriate in a clinical setting. Yet, it is within the duties of ethics to reflect on the boundaries of practice and to evaluate potential harms to prospective patients and service users. In this regard, several scholars have expressed strong concerns over the use of antipsychotic medication with at-risk individuals (Jorm, 2012; Raven et al., 2012; Weiser, 2011). It seems to be an accepted principle—within the scientific community—that antipsychotic medication *should not* be offered to individuals in the prepsychotic phase (NICE, 2014). Recent studies suggest that antipsychotic medication may not be necessary even for FEP individuals. Alternative treatments, such as comprehensive psychological intervention may in fact constitute an equally good or even better option (Francey et al., 2010). However, the principle of nonmaleficence supports the claim that the harms that could derive from prescribing antipsychotic medications to HR individuals outweigh potential benefits. It remains open the question of whether the potential harms that could derive from providing psychological interventions to HR individuals might outweigh potential benefits.

As we saw earlier, the varying thresholds of psychosis risk status require a continuous balancing of the principles of beneficence and nonmaleficence, and this is certainly the case when it comes to consideration of appropriate treatments. Antipsychotic medications have significant side effects, and the risk of harm in cases of psychotic "risk" is very real in all cases, given low conversion rates across the risk continuum. At the same time, it should not be assumed that psychological treatments come without risk, and more evidence is needed to assess the balance of harms and benefits of medications versus psychological treatments across the risk continuum.

One way to implement ethical and clinical recommendations is the development of clinical guidelines. In recent years, a considerable number of guidelines have been produced, in order to address the issue of what type of treatment should be offered based on psychosis risk. The "International clinical practice guidelines for early psychosis," published by the International Early Psychosis Association writing group paved the way for further national and international developments (International Early Psychosis Association Writing Group, 2005). In the United Kingdom, for instance, clinical guidance for EI for psychosis is provided by the National Institute of Health and Care Excellence (NICE). UK guidelines are relatively consistent in suggesting treatment options for individuals who are at risk for psychosis (NICE, 2013; NICE, 2014). According to NICE guidelines, individuals should be referred for assessment to EI or alternatively, in the case of children and young people, Child and Adolescent Mental Health Services. Treatment options include cognitive behavioral therapy, family intervention, psychosocial support, and vocational support. NICE and NHS England guidelines explicitly recommend *against* offering antipsychotic medications in the prepsychotic phase or "with the aim of decreasing the risk of psychosis" (NICE, 2013; NHS England et al., 2016).

Guidelines will change as clinical evidence evolves. However, two facts will remain essential, regardless of what type of intervention may be offered to at-risk individuals. First, engagement with clinical services as well as treatment options ought to be *offered* and not imposed to individuals at risk of psychosis. This must be upheld particularly when the individual with at-risk status is a minor. In the case of minors, the principle of autonomy is still relevant, albeit in a relational framing. Given that psychosis treatment is likely to involve discussions with a young person and his/her family, there is a particular danger that paternalistic inclinations on the part of EI service providers override both the young person's and parental instincts. Any engagement with clinical services should be carefully discussed with each individual and with family members where appropriate (McGorry, 2010). Second, clinical services ought to recognize that engagement of at risk populations requires *good* caring relationships between the individual and mental health professionals. A growing body of qualitative research in mental healthcare strongly supports this claim (Brew et al., 2017; Lester et al., 2011; Stewart, 2013). Good caring relationships require that clinical services operate within a youth-friendly, low-stigma environment. Clinicians should try to establish relations based on *trust* with at-risk individuals (Corsico et al., 2018). Again, family involvement may be vital to this process, particularly in the case of young people. Ultimately, good caring relationships will be essential to improve clinical outcomes and to promote hope and social recovery.

Mental health legislation and the logic of risk

How does the logic of risk in the clinical management of psychotic illness interact with the logic of risk in mental health legislation? This is a further area of ethical concern. The history of hospital admission of people with schizophrenia warns us of the potentially dangerous consequences of an excessive medicalization of mental illness and of an uncritical reception of reductive biomedical conceptions of mental illness (Read and Dillon, 2013). Recently, scholars have highlighted a shift toward a "logic of risk" in mental health legislation, which may have dangerous repercussions on how this logic is implemented in clinical practice (Szmukler and Rose, 2013). Even though the clinical concept of "psychosis risk" and the legal concept of "risk" in mental health law are different in nature, the uncritical reception of potential overlaps between the two concepts may have harmful consequences for at-risk individuals (Rose, 1998). Here, we shall refer to England and Wales, though observations drawn with reference to this jurisdiction may well be relevant for other legal jurisdictions.

In England and Wales, mental health law is governed by the Mental Health Act (MHA) 1983 as amended by the MHA 2007 (Parliament of the United Kingdom, 2007) and by the Mental Capacity Act 2005 (Parliament of the United Kingdom, 2005). In particular, the MHA governs the process of compulsory admission to hospital, for assessment (Section 2) or treatment (Section 3), hence the term "being *sectioned*" to refer to compulsory admission to hospital of individuals with a mental disorder. Compulsory admission is often referred to as "civil detention under the MHA" (Bartlett and Sandland, 2014). The legislation is extremely complex, and projects of reform have been ongoing for decades. Thus, the UK Department of Health has issued a Code of Practice (COP) to the MHA to serve as a decision-making tool for professionals working in mental healthcare (Department of Health of the UK Government, 2015b). Criteria of admission for treatment (Section 3) under the MHA 2007 are defined as follows:

(a) [The patient] is suffering from mental disorder of a nature or degree which makes it appropriate to receive medical treatment in a hospital; and
(b) It is necessary for the health or safety of the patient or for the protection of other persons that he [sic] should receive such treatment and it cannot be provided unless he is detained under this section; and
(c) Appropriate medical treatment is available for him.

Criteria of admission for *assessment* (Section 2) are the same in nature as (a) and (b), but criterion (c) is omitted. None of these criteria make an explicit reference to the concept of *risk*. However, the logic of risk becomes apparent in the COP, which is the document that the majority of professionals refer to in everyday practice. The COP substantiates criterion (b) of the MHA, by listing the factors to consider with regard to the health and safety of the patient. According to the COP, when deciding whether a patient should be detained, professionals should consider if he is *at risk* of suicide, self-harm, self-neglect, jeopardizing his own health and safety accidently, or "otherwise putting their health or safety at risk" (Department of Health of the UK Government, 2015a, p. 114). The COP also states that "it is not always possible to differentiate risk of harm to the patient from the risk of harm to others" (Department of Health of the UK Government, 2015a, p. 115). In England and Wales, risk of harm—broadly defined—to self or others is thus identified as a condition for compulsory admission to a psychiatric hospital and, indeed, to preventive detention in a forensic context (Forrester, 2002).

Risk of harm and psychosis risk are different in nature. Yet, there is a possibility that the two definitions may become blurred in the practice of risk assessment in mental healthcare (Szmukler and Rose, 2013). Indeed, risk assessment is often carried out not by clinical researchers or mental health barristers, but by mental health professionals who operate within budget and time constraints. In addition, English Law identifies as criterion (a) for compulsory admission the presence of a "mental disorder of a nature *or* degree which makes it appropriate to receive medical treatment in a hospital." In this regard, APS or other risk syndromes might not meet the requirement of "degree," given their focus on *attenuated* psychotic symptoms. Yet, it is unclear whether they would meet the criterion of "nature of mental disorder." Here, it is the disjunctive "*or*" in criterion (a) that could create some problems. Lastly, criterion (c) could easily be met by individuals at risk of psychosis. Again, the reification of psychosis risk in a diagnosable syndrome could have dangerous consequences for patients' care.

At present, it seems very unlikely that individuals at risk of psychosis may be sectioned based on their psychosis risk. Yet, there were 37,065 compulsory admissions of adults under Section 2 and Section 3 of the MHA just in 2010–11, and just in England, with a rise of 44% over the previous 5 years (Bartlett and Sandland, 2014). Compulsory admission, even just for assessment, of individuals at risk of psychosis should not be ruled out as a mere theoretical possibility. Claims about psychosis risk, combined with powers of compulsory hospitalization, were used to successfully and coercively detain American black civil rights activists in the 1960s and 1970s (Metzl, 2010). Activists were then frequently labeled and treated against their will. These injustices took place in a different political and cultural space and time; nevertheless, the national investment of resources and placement in the diagnostic roster in DSM-5 represents the fact that psychosis risk is firmly a part of our contemporary prevention paradigm in mental health. Therefore, clinicians and researchers must learn from past injustices and also accept the active moral duty to promote ethical awareness while trying practically to avoid the pitfalls of an uncritical reception of the logic of risk.

Conclusion

The attempt to identify and treat psychosis risk ought not to be understood as a purely clinical endeavor. In this chapter, we have presented an overview of key ethical concerns that arise in the context of this endeavor. The clinical staging model is grounded in the ethical principle of beneficence. However, extending this model to individuals who have not (yet) suffered from a psychotic episode raises concerns in terms of nonmaleficence and respect of the autonomy of individuals involved in preclinical interventions. Some recommendations may be drawn from the overview presented in this chapter. First, disclosure of psychosis risk must be considered very carefully, in both research and clinical settings. Second, based on current evidence, prescription of antipsychotic medications to individuals based on their risk of developing psychosis should be avoided. Third, bad medicalization of mental distress and adverse life experiences is clinically inappropriate, societally burdensome, and personally harmful to individuals. Fourth, careful consideration will have to be given to how the logic of risk in psychosis prevention overlaps with the logic of risk in mental health legislation.

Managing risk is about managing *time* and *uncertainty*. We may be able to develop strategies to intervene as early as possible and better tools to establish psychosis risk in help-seeking or even in healthy individuals. Yet, we will not be able to defeat uncertainty or to avoid the influence of time on any clinical strategy; time and uncertainty are intrinsic components of any definition of risk. As good identification of psychosis risk improves, the ethical and social implications of identification and intervention are likely to become clearer and potentially to multiply. On the one hand, there are good reasons to strive to normalize mental illness: recognition that we are *all* at risk of suffering from mental distress at some point in our lives can lessen the burden of self and social stigma. Eleanor Longden has pointed to the importance of empathy and understanding in relation to dignity for persons experiencing mental health challenges:

> You cannot humiliate the person who feels pride. You cannot oppress the people who are not afraid anymore. For me, the achievements of the Hearing Voices Movement are a reminder that empathy, fellowship, justice and respect are more than words; they are convictions and beliefs, and that beliefs can change the world. ***Longden (2013).***

On the other hand, the drive to intervene ever earlier into "risk" may mean that research attention to, public understanding of, and resources for people who fully express cognitive, behavioral, and/or physical traits associated with psychosis and related mental illnesses progressively diminish. The Hearing Voices Movement has changed public perception and self-understanding of people who hear voices. This paradigm shift has been possible only because voice hearing has remained a visible dimension of psychosis experience. Moreover, it is important to note that some people who hear voices experience them as a positive part of "self" and would not want to "cure" them (Johns et al., 2014). Therefore, a crucial feature of an ethical analysis of intervention into psychosis risk must be awareness of the assumptions that guide the intervention paradigm, alongside an analysis of whether these assumptions can be said to achieve sufficient normative force that a national program of intervention can be justified in the longer term.

List of acronyms and abbreviations

APS Attenuated psychosis syndrome
ARMS At-risk mental state
CAARMS Comprehensive Assessment of At-Risk Mental State
CAMHS Child and Adolescent Mental Health Services
CBT Cognitive behavioral therapy
COP Mental Health Act 1983: Code of Practice
DSM-5 *Diagnostic and Statistical Manual of mental disorders,* 5th *Edition*
DUP Duration of untreated psychosis
EI Early intervention
FEP First episode of psychosis
HR High risk
IEPA International early psychosis association
IFs Incidental findings
MCA Mental Capacity Act
MHA Mental Health Act
MRI Magnetic resonance imaging
NICE National Institute of Health and Care Excellence
PGT Predictive genetic testing
SIPS Structured Interview for Psychosis-Risk Syndrome
UHR Ultra high risk

References

Addington, J., Stowkowy, J., Weiser, M., 2015. Screening tools for clinical high risk for psychosis. Early Interv. Psychiatry 9 (5), 345–356.

American Psychiatric Association, 2013a. Diagnostic and Statistical Manual of Mental Disorders: DSM-5. Washington DC.

American Psychiatric Association, 2013b. Attenuated psychosis syndrome. In: Diagnostic and Statistical Manual of Mental Disorders: DSM-5, pp. 783–786. Washington DC.

Appelbaum, P.S., 2015. Ethical challenges in the primary prevention of schizophrenia. Schizophr. Bull. 41 (4), 773–775.

Bartlett, P., Sandland, R., 2014. Civil Detention under the Mental Health Act 1983. In: Mental Health Law. Policy and Practice, fourth ed., ch. 6. Oxford University Press, Oxford, pp. 236–277.

Beauchamp, T.L., Childress, J.F., 2013. Principles of Biomedical Ethics, seventh ed. Oxford University Press, New York.

Bechdolf, A., Wagner, M., Ruhrmann, S., Harrigan, S., Putzfeld, V., Pukrop, R., Brockhaus-Dumke, A., Berning, J., Janssen, B., Decker, P., Bottlender, R., Maurer, K., Moller, H.J., Gaebel, W., Hafner, H., Maier, W., Klosterkotter, J., 2012. Preventing progression to first-episode psychosis in early initial prodromal states. Br. J. Psychiatry 200 (1), 22–29.

Bluhm, R., Raczek, G., Broome, M., Wall, M.B., 2015. Ethical issues in brain imaging in psychiatry. In: Sadler, J.Z., Fulford, K.W.M., Van Staden, C.W. (Eds.), Oxford Handbook of Psychiatric Ethics, ch. 75. Oxford University Press, Oxford, pp. 1109–1126.

Brew, B., Shannon, C., Storey, L., Boyd, A., Mulholland, C., 2017. A qualitative phenomenological analysis of the subjective experience and understanding of the at risk mental state. Int. J. Qual. Stud. Health Well-Being 12 (1), 1342504.

Broome, M., Fusar-Poli, P., 2012. Philosophical issues in the prodromal phase of psychosis. Curr. Pharmaceut. Des. 18 (4), 596–605.

Broome, M., Fusar-Poli, P., Wuyts, P., 2013. Conceptual and ethical issues in the prodromal phase of psychosis. In: Fulford, K.W.M., Davies, M., Gipps, R.G.T., Graham, G., Sadler, J.Z., Stanghellini, G., Thornton, T. (Eds.), The Oxford Handbook of Philosophy and Psychiatry, ch. 46. Oxford University Press, Oxford, pp. 779–802.

Cassetta, B.D., Goghari, V.M., 2015. Ethical considerations of screening and early intervention for clinical high-risk psychosis. Ethics Behav. 25 (1), 1–20.

Corsico, P., Griffin-Doyle, M., Singh, I., 2018. What constitutes 'good practice' in early intervention for psychosis? Analysis of clinical guidelines. Child Adolesc. Ment. Health 23 (3), 185–193.

Department of Health of the UK Government, 2015a. Applications for detention in hospital. In: Mental Health Act 1983: Code of Practice, ch. 14, pp. 113–135.

Department of Health of the UK Government, 2015b. Mental Health Act 1983: Code of Practice.

Forrester, A., 2002. Preventive detention, public protection and mental health. J. Forensic Psychiatry 13 (2), 329–344.

Francey, S.M., Nelson, B., Thompson, A., Parker, A.G., Kerr, M., Macneil, C., Fraser, R., Hughes, F., Crisp, K., Harrigan, S., Wood, S.J., Berk, M., McGorry, P.D., 2010. Who needs antipsychotic medication in the earliest stages of psychosis? A reconsideration of benefits, risks, neurobiology and ethics in the era of early intervention. Schizophr. Res. 119 (1–3), 1–10.

Fusar-Poli, P., Bonoldi, I., Yung, A.R., Borgwardt, S., Kempton, M.J., Valmaggia, L., Barale, F., Caverzasi, E., McGuire, P., 2012a. Predicting psychosis: meta-analysis of transition outcomes in individuals at high clinical risk. Arch. Gen. Psychiatr. 69 (3), 220–229.

Fusar-Poli, P., Borgwardt, S.J., McGuire, P., 2012b. Vulnerability to Psychosis. From Neuroscience to Psychopathology. Psychology Press, Hove and New York.

Fusar-Poli, P., Borgwardt, S., Bechdolf, A., Addington, J., Riecher-Rossler, A., Schultze-Lutter, F., Keshavan, M., Wood, S., Ruhrmann, S., Seidman, L.J., Valmaggia, L., Cannon, T., Velthorst, E., De Haan, L., Cornblatt, B., Bonoldi, I., Birchwood, M., McGlashan, T., Carpenter, W., McGorry, P., Klosterkotter, J., McGuire, P., Yung, A., 2013a. The psychosis high-risk state: a comprehensive state-of-the-art review. JAMA Psychiatry 70 (1), 107–120.

Fusar-Poli, P., Byrne, M., Badger, S., Valmaggia, L.R., McGuire, P.K., 2013b. Outreach and support in south London (OASIS), 2001–2011: ten years of early diagnosis and treatment for young individuals at high clinical risk for psychosis. Eur. Psychiatry 28 (5), 315–326.

Fusar-Poli, P., Nelson, B., Valmaggia, L., Yung, A.R., McGuire, P.K., 2014. Comorbid depressive and anxiety disorders in 509 individuals with an at-risk mental state: impact on psychopathology and transition to psychosis. Schizophr. Bull. 40 (1), 120–131.

Gifford, G., Crossley, N., Fusar-Poli, P., Schnack, H.G., Kahn, R.S., Koutsouleris, N., Cannon, T.D., McGuire, P., 2017. Using neuroimaging to help predict the onset of psychosis. Neuroimage 145 (Pt B), 209–217.

Glover-Thomas, N., 2011. The age of risk: risk perception and determination following the Mental Health Act 2007. Med. Law Rev. 19 (4), 581–605.

Goncalves, A.M., de Rosalmeida Dantas, C., Banzato, C.E.M., 2016. Values and DSM-5: looking at the debate on attenuated psychosis syndrome. BMC Med. Ethics 17 (1), 7.

Haslam, N., 2011. Genetic essentialism, neuroessentialism, and stigma: commentary on Dar-Nimrod and Heine (2011). Psychol. Bull. 137 (5), 819–824.

Hegelstad, W.T., Larsen, T.K., Auestad, B., Evensen, J., Haahr, U., Joa, I., Johannesen, J.O., Langeveld, J., Melle, I., Opjordsmoen, S., Rossberg, J.I., Rund, B.R., Simonsen, E., Sundet, K., Vaglum, P., Friis, S., McGlashan, T., 2012. Long-term follow-up of the TIPS early detection in psychosis study: effects on 10-year outcome. Am. J. Psychiatry 169 (4), 374–380.

Illes, J., Lombera, S., Rosenberg, J., Arnow, B., 2008. In the mind's eye: provider and patient attitudes on functional brain imaging. J. Psychiatr. Res. 43 (2), 107–114.

International Early Psychosis Association Writing Group, 2005. International clinical practice guidelines for early psychosis. Br. J. Psychiatry 187 (48), s120–s124.

Johannessen, J.O., McGorry, P., 2010. DSM-5 and the 'Psychosis Risk Syndrome': the need for a broader perspective. Psychosis 2 (2), 93–96.

Johns, L.C., Kompus, K., Connell, M., Humpston, C., Lincoln, T.M., Longden, E., Preti, A., Alderson-Day, B., Badcock, J.C., Cella, M., Fernyhough, C., McCarthy-Jones, S., Peters, E., Raballo, A., Scott, J., Siddi, S., Sommer, I.E., Laroi, F., 2014. Auditory verbal hallucinations in persons with and without a need for care. Schizophr. Bull. 40 (Suppl. 4), S255–S264.

Jorm, A.F., 2012. Ethics of giving antipsychotic medication to at-risk young people. Aust. N. Z. J. Psychiatr. 46 (9), 908–909.

Koutsouleris, N., Riecher-Rossler, A., Meisenzahl, E.M., Smieskova, R., Studerus, E., Kambeitz-Ilankovic, L., von Saldern, S., Cabral, C., Reiser, M., Falkai, P., Borgwardt, S., 2015. Detecting the psychosis prodrome across high-risk populations using neuroanatomical biomarkers. Schizophr. Bull. 41 (2), 471–482.

Lester, H., Marshall, M., Jones, P., Fowler, D., Amos, T., Khan, N., Birchwood, M., 2011. Views of young people in early intervention services for first-episode psychosis in England. Psychiatr. Serv. 62 (8), 882–887.

Lester, H., Khan, N., Jones, P., Marshall, M., Fowler, D., Amos, T., Birchwood, M., 2012. Service users' views of moving on from early intervention services for psychosis: a longitudinal qualitative study in primary care. Br. J. Gen. Pract. 62 (596), e183–e190.

Longden, E., TED 2013. The Voices in My Head (TED Talk) [Online]. Available at:. https://www.ted.com/talks/eleanor_longden_the_voices_in_my_head/transcript?language=en. Accessed 8 January 2020, to watch the full talk, visit TED.com.

Mand, C., Gillam, L., Delatycki, M.B., Duncan, R.E., 2012. Predictive genetic testing in minors for late-onset conditions: a chronological and analytical review of the ethical arguments. J. Med. Ethics 38 (9), 519–524.

Marshall, M., Rathbone, J., 2011. Early intervention for psychosis. Schizophr. Bull. 37 (6), 1111–1114.

Marwaha, S., Thompson, A., Upthegrove, R., Broome, M.R., 2016. Fifteen years on – early intervention for a new generation. Br. J. Psychiatry 209 (3), 186–188.

McCrone, P., Craig, T.K., Power, P., Garety, P.A., 2010. Cost-effectiveness of an early intervention service for people with psychosis. Br. J. Psychiatry 196 (5), 377–382.

McGorry, P., Keshavan, M., Goldstone, S., Amminger, P., Allott, K., Berk, M., Lavoie, S., Pantelis, C., Yung, A., Wood, S., Hickie, I., 2014. Biomarkers and clinical staging in psychiatry. World Psychiatry 13 (3), 211–223.

McGorry, P.D., 2010. Risk syndromes, clinical staging and DSM V: new diagnostic infrastructure for early intervention in psychiatry. Schizophr. Res. 120 (1–3), 49–53.

Mechelli, A., Riecher-Rossler, A., Meisenzahl, E.M., Tognin, S., Wood, S.J., Borgwardt, S.J., Koutsouleris, N., Yung, A.R., Stone, J.M., Phillips, L.J., McGorry, P.D., Valli, I., Velakoulis, D., Woolley, J., Pantelis, C., McGuire, P., 2011. Neuroanatomical abnormalities that predate the onset of psychosis: a multicenter study. Arch. Gen. Psychiatr. 68 (5), 489–495.

Metzl, J.M., 2010. The Protest Psychosis: How Schizophrenia Became a Black Disease. Beacon Press, Boston.

Mittal, V.A., Dean, D.J., Mittal, J., Saks, E.R., 2015. Ethical, legal, and clinical considerations when disclosing a high-risk syndrome for psychosis. Bioethics 29 (8), 543–556.

Mittelstadt, B.D., Allo, P., Taddeo, M., Wachter, S., Floridi, L., 2016. The ethics of algorithms: mapping the debate. Big Data Soc. 3 (2), 1–21.

Morris, S., Heinssen, R., 2014. Informed consent in the psychosis prodrome: ethical, procedural and cultural considerations. Philos. Ethics Humanit. Med. 9, 19.

NHS England, National Institute for Health and Care Excellence (NICE) & National Collaborating Centre for Mental Health, 2016. Implementing the Early Intervention in Psychosis Access and Waiting Time Standard: Guidance.

NICE, 2013. Psychosis and Schizophrenia in Children and Young People: Recognition and Management (CG155). United Kingdom.

NICE, 2014. Psychosis and Schizophrenia in Adults: Prevention and Management (CG178). United Kingdom.

Parliament of the United Kingdom, 2005. Mental Capacity Act 2005.

Parliament of the United Kingdom, 2007. Mental Health Act 2007.

Pettersson-Yeo, W., Benetti, S., Marquand, A.F., Dell'acqua, F., Williams, S.C., Allen, P., Prata, D., McGuire, P., Mechelli, A., 2013. Using genetic, cognitive and multi-modal neuroimaging data to identify ultra-high-risk and first-episode psychosis at the individual level. Psychol. Med. 43 (12), 2547–2562.

Racine, E., Illes, J., 2007. Emerging ethical challenges in advanced neuroimaging research: review, recommendations and research agenda. J. Empir. Res. Hum. Res. Ethics 2 (2), 1–10.

Raven, M., Stuart, G.W., Jureidini, J., 2012. 'Prodromal' diagnosis of psychosis: ethical problems in research and clinical practice. Aust. N. Z. J. Psychiatr. 46 (1), 64–65.

Read, J., Dillon, J., 2013. Models of Madness. Psychological, Social and Biological Approaches to Psychosis. Routledge, London & New York.

Rose, N., 1998. Governing risky individuals: the role of psychiatry in new regimes of control. Psychiatry Psychol. Law 5 (2), 177–195.

Sejnowski, T.J., Churchland, P.S., Movshon, J.A., 2014. Putting big data to good use in neuroscience. Nat. Neurosci. 17 (11), 1440–1441.

Shanmugaratnam, S.H., Edwards, A., 2016. Ethical aspects of risk communication. Medicine 44 (10), 607–610.

Stewart, K.D., 2013. Factors contributing to engagement during the initial stages of treatment for psychosis. Qual. Health Res. 23 (3), 336–347.

Szmukler, G., Rose, N., 2013. Risk assessment in mental health care: values and costs. Behav. Sci. Law 31 (1), 125–140.

Valli, I., Marquand, A.F., Mechelli, A., Raffin, M., Allen, P., Seal, M.L., McGuire, P., 2016. Identifying individuals at high risk of psychosis: predictive utility of support vector machine using structural and functional MRI data. Front. Psychiatry 7, 52.

van der Gaag, M., Smit, F., Bechdolf, A., French, P., Linszen, D.H., Yung, A.R., McGorry, P., Cuijpers, P., 2013. Preventing a first episode of psychosis: meta-analysis of randomized controlled prevention trials of 12 month and longer-term follow-ups. Schizophr. Res. 149 (1–3), 56–62.

Warner, R., 2005. Problems with early and very early intervention in psychosis. Br. J. Psychiatry 48, s104–s107.

Weiser, M., 2011. Early intervention for schizophrenia: the risk-benefit ratio of antipsychotic treatment in the prodromal phase. Am. J. Psychiatry 168 (8), 761–763.

Yang, L.H., Wonpat-Borja, A.J., Opler, M.G., Compton, M.T., Kelly, M., Purdie-Vaughns, V., Corcoran, C.M., 2012. Stigma in early stages of psychotic illness: connections with cognitive neuroscience. In: Fusar-Poli, P., Borgwardt, S., Mc Guire, P. (Eds.), Vulnerability to Psychosis. From Neurosciences to Psychopathology, ch. 12. Psychology Press, Hove and New York, pp. 159–176.

Yung, A.R., McGorry, P.D., McFarlane, C.A., Jackson, H.J., Patton, G.C., Rakkar, A., 1996. Monitoring and care of young people at incipient risk of psychosis. Schizophr. Bull. 22 (2), 283–303.

Yung, A.R., Yuen, H.P., McGorry, P.D., Phillips, L.J., Kelly, D., Dell'Olio, M., Francey, S.M., Cosgrave, E.M., Killackey, E., Stanford, C., Godfrey, K., Buckby, J., 2005. Mapping the onset of psychosis: the comprehensive assessment of at-risk mental states. Aust. N. Z. J. Psychiatr. 39 (11–12), 964–971.

Yung, A.R., McGorry, P.D., 2007. Prediction of psychosis: setting the stage. Br. J. Psychiatry 191 (Suppl. 51), s1–8.

Yung, A.R., Nelson, B., Thompson, A.D., Wood, S.J., 2010. Should a "risk syndrome for psychosis" be included in the DSMV? Schizophr. Res. 120 (1–3), 7–15.

Yung, A.R., 2012. Early intervention in psychosis: evidence, evidence gaps, criticism, and confusion. Aust. N. Z. J. Psychiatr. 46 (1), 7–9.

CHAPTER

18

Indicated prevention in psychosis risk—psychological approaches

Jean Addington, Danijela Piskulic, Daniel J. Devoe, Olga Santesteban-Echarri, Jacqueline Stowkowy

Hotchkiss Brain Institute, Department of Psychiatry, University of Calgary, Calgary, AB, Canada

Introduction

In 2004, Paul French and Tony Morrison from the University of Manchester published the first text related to psychosocial treatments for those at clinical high risk (CHR) of developing psychosis, "Early Detection and Cognitive Therapy for People at High Risk of Developing Psychosis." The book described in detail their cognitive therapy approach to the treatment of CHR youth (French and Morrison, 2004). This was followed in 2006 with a handbook that addressed a wide range of psychosocial treatments for CHR, "Working with People at High Risk of Developing Psychosis" (Addington et al., 2006). Much of the work in this handbook was based on clinical work with CHR youth that had been developed specifically for this at-risk group or had been modified from successful treatments with those in the early phases of schizophrenia and other psychotic disorders. Interestingly, at this time, only three trials, one of which was a medication-only trial, had been published: the Early Detection and Intervention Evaluation (EDIE) cognitive therapy trial in the United Kingdom that used the French and Morrison Manual (Morrison et al., 2004); the McGorry trial of risperidone and cognitive behavioral therapy (CBT) (McGorry et al., 2002); and the olanzapine trial by McGlashan et al. (2006). The main outcome for these trials was to prevent transition to a full-blown psychosis, defined as the development of either an affective or non affective psychotic disorder or even as attenuated psychotic symptoms (APS) reaching a psychotic level. However, it has been reported that for approximately half of those who do not go on to develop a psychotic illness, these young people continue to have other issues that may include continued APS, negative symptoms, and other diagnoses (Addington et al., 2011a). Furthermore, those at risk for psychosis often exhibit persistent difficulties in social functioning (Cornblatt et al., 2012), and poorer social functioning may be a predictor of transition to a psychotic disorder (Addington et al., 2017a). The aim of this chapter is first to review all the trials to date that examined psychosocial treatments for those at CHR and to present in more detail the treatments used. In our review of these

Risk Factors for Psychosis
https://doi.org/10.1016/B978-0-12-813201-2.00018-1

studies, we will consider a range of outcomes that include APS, negative symptoms, cognition, and social and role functioning.

Literature review

A literature search was conducted in several electronic databases (i.e., Medline, PsycINFO, Embase, and CINAHL). Two reviewers autonomously screened titles and abstracts, and the full text of any treatment study in high-risk samples reporting on a range of outcomes including transition, APS, negative symptoms, social functioning, and cognition was retrieved for a comprehensive review. The clinical trials registry (i.e., ClinicalTrials.gov) was searched using the keywords "psychosis" and "risk" and references of included articles were searched for suitable studies. Greater details about the searches are available in the respective papers (Devoe et al., 2018, 2019a, 2019b). A total of 57 treatment studies were found. These are presented in Table 18.1. Trials using a psychosocial treatment were the most common ($N = 25$) followed by trials using antipsychotics ($N = 16$). Several other types of interventions were identified including omega-3, glutamate modulators (i.e., glycine, D-serine, N-acetylcysteine), and antidepressants. However, this chapter will only focus on the 25 trials with a psychosocial treatment component. The previous chapter by Burkhardt et al. focuses on pharmacological interventions.

Characteristics of the 25 psychosocial treatment studies are outlined in Table 18.2. The number of high-risk participants ranged from 6 to 288, for a total of 2319 participants. The mean age was 20.3 years and 1310 (56.5%) were male. Participants in these trials usually met high-risk criteria based on the Structured Interview for Psychosis Risk (SIPS) (McGlashan et al., 2010) or an At-Risk Mental State (ARMS) based on the Comprehensive Assessment of At Risk Mental State (CAARMS) (Yung et al., 2006). Some participants may have met criteria based on the Schizophrenia Proneness Instrument, Adult version (SPI-A) (Schultze-Lutter et al., 2006). Some of the early trials used the Positive and Negative Syndrome Scale (PANSS) (Kay et al., 1987). The psychosocial treatments varied, CBT trials were the most common

TABLE 18.1 Treatment modalities.

Treatment modalities	Randomized controlled studies	Other study designs	Total
Cognitive behavioral therapy (CBT)	6	2	8
Family therapy	1	4	5
Cognitive remediation	4	3	7
Integrated psychological therapy	3	0	3
Antipsychotics	3	13	16
Omega-3	3	0	3
Glycine, D-serine, N-acetylcysteine	2	2	4
Antidepressants (AD)	0	2	2
Antidepressants and antipsychotics (AP)	0	4	4
Combinations (AP, AD, CBT, family)	2	3	5
Total	24	33	57

TABLE 18.2 Details of psychosocial treatment studies ($N = 25$).

Author, year	Country	Study design	Intervention	Control	Treatment duration (weeks)	CHR patients N	Age $M \pm SD$	Male N (%)
Cognitive behavioral therapy (n = 8)								
Addington et al. (2011b)	Canada	RCT	CBT	Supportive therapy	24	51	CBT: 20.8 ± 4.5 Supportive: 21.1 ± 3.7	36 (71)
Bechdolf et al. (2017)	Germany	RCT	CBT	Aripiprazole: 5−15 mg/day + CM/placebo + CM	52	280	24.4 ± 5.1	143 (66)
Fusar-Poli et al. (2015)	United Kingdom	Naturalistic	CBT	CBT + AP/CBT + AD/CBT + AP + AD	NR	258	22.9 ± 4.5	146 (57)
Kim et al. (2011)	Korea	Open label	CBT	None	10	22	19.4 ± 4.4	11 (50)
Morrison et al. (2004)	United Kingdom	RCT	CBT	TAU	24	58	22 ± 4.5	40 (69)
Morrison et al. (2012b)	United Kingdom	RCT	CBT + monitoring	Monitoring	24	288	20.7 ± 4.3	180 (63)
Stain et al. (2016)	Australia, New Zealand	RCT	CBT + TAU	NDRL + TAU	24	57	CBT: 16.2 ± 2.7 NDRL: 16.5 ± 3.2	23 (40)
Van Der Gaag et al. (2012a)	The Netherlands	RCT	CBT + TAU	TAU	24	201	CBT: 22.9 ± 5.6 TAU: 22.6 ± 5.5	99 (49)
Family therapy (n = 5)								
Grano et al. (2016)	Finland	Naturalistic	Family- and community-oriented integrative treatment model	TAU	52	56	FCTM: 15.5 ± 1.6 TAU: 16.3 ± 0.8	18 (32)
Landa et al. (2016)	USA	Open label	Group and family-based CBT	None	15	6	19.5 ± 1.5	2 (33)
McFarlane et al. (2015)	USA	Regression discontinuity design	Family-aided assertive community treatment	None	104	205	16.4 ± 3.3	116 (57)
Miklowitz et al. (2014)	USA + Canada	RCT	Family-focused therapy	Enhanced care	24	129	17.4 ± 4.1	74 (57)

(Continued)

TABLE 18.2 Details of psychosocial treatment studies ($N = 25$).—cont'd

Author, year	Country	Study design	Intervention	Control	Treatment duration (weeks)	N	CHR patients: Age M±SD	CHR patients: Male N (%)
O'Brien et al. (2007)	USA	Open label	Psychoeducational multifamily group	None	36	16	15.7 ± NR	8 (50)
Cognitive remediation (n = 7)								
Choi et al. (2017)	USA	RCT	CR	Active control	8	62	CR: 18.2 ± 3.8 Active control: 18.5 ± 3.7	30 (48)
Hooker et al. (2014)	USA	Open label	CR	Computer games	8	14	21.9 ± 4.2	7 (50)
Loewy et al. (2016)	USA	RCT	CR	Computer games	8	83	CR: 17.8 ± 3.1 Control: 18.7 ± 4.6	42 (51)
Nemoto et al. (2010)	Japan	Open label	CR	None	24	6	19.50 ± 2.5	4 (66)
Piskulic et al. (2015)	Canada	RCT	CR	Computer games	12	43	CR: 19.7 ± 5.7 Games: 17.5 ± 3.5	NR
Rauchensteiner et al. (2011)	Germany	Open label	CR	None	4	10	27.2 ± 5.3	7 (70)
Urben et al. (2012)	Switzerland	RCT	CR	Computer games	8	12	NR	NR
Integrated psychological therapy (n = 3)								
Albert et al. (2016)	Denmark	RCT	IPT	TAU	104	83	26.6 ± 4.4	38 (46)
Nordentoft et al. (2006)	Denmark	RCT	IPT	TAU	104	79	24.9 ± 4.9	53 (67)
Wessels et al. (2015)	Germany	RCT	IPT	Supportive therapy	52	128	IPT: 25.2 ± 5.4 Supportive: 26.8 ± 6.2	81 (63)
Risperidone plus CBT (n = 2)								
McGorry et al. (2002)	Australia	RCT	Risperidone: 1−2 mg/day + CBT	NBI	24	59	20 ± 4.0	34 (58)
McGorry et al. (2013)	Australia	RCT	Risperidone: 0.5−2 mg/day + CBT or CBT + placebo	Supportive therapy + placebo or monitoring	52	193	18.1 ± 3.0	76 (39)

AD, antidepressants; *AP*, antipsychotics; *CBT*, cognitive behavioral therapy; *CM*, case management; *CR*, cognitive remediation; *IPT*, integrated psychological therapy; *NBI*, needs-based intervention; *NDRL*, nondirective reflective listening; *NR*, not reported; *RCT*, randomized controlled trial; *TAU*, treatment as usual.

(N = 8), followed by cognitive remediation (CR) (N = 7), family-based treatments (N = 5), integrated psychological therapy (IPT) (N = 3), and risperidone plus CBT (N = 2). The mean treatment duration was 37.3 weeks. Study design varied as well, including 16 randomized controlled trials (RCTs), 6 open-label, 2 naturalistic, and 1 regression discontinuity design. Most studies were performed in North America (N = 9), followed by Europe (N = 11), Australia (N = 3), and Asia (N = 2).

The outcomes under study in the published meta-analyses were APS, negative symptoms, and social functioning. For each outcome, a random-effect pairwise meta-analysis was performed comparing active treatment to controls, followed by a network meta-analysis (NMA) for APS and negative symptoms. A NMA allows for comparisons between treatment arms that have not been compared before in existing RCTs (e.g., CBT to family therapy) that used a common comparator (e.g., supportive therapy). In this instance, several random-effect multivariate NMAs assuming consistency and a common heterogeneity across comparisons were utilized to analyze each outcome (Caldwell et al., 2005; Dias et al., 2013). In-depth details about the pairwise and NMA models are available in the individual papers (Devoe et al., 2018, 2019a, 2019b).

Transition to a psychotic disorder was the main outcome in most studies. Several studies suggest that CBT may have an impact on transition to psychosis. These studies are discussed in detail later in this chapter. Reducing APS was the most common secondary aim of the majority of psychosocial studies. In the pairwise analyses, CBT interventions were not significant at reducing APS compared to control treatments at 6-, 12-, 18-, or 24 months (Devoe et al., 2019b). Risperidone plus CBT interventions did not significantly reduce APS at 6- or 12-month follow-up nor did CR. In both the 6- and 12-month NMAs, no psychosocial intervention was significantly more effective at reducing APS compared to any other intervention.

Another common outcome reported in high-risk treatment studies was negative symptoms. In a pairwise analysis, CBT interventions were not associated with a significant reduction in negative symptoms (Devoe et al., 2018). Risperidone plus CBT interventions did not significantly reduce negative symptoms at 6- or 12-month follow-up. CR did not significantly reduce negative symptoms compared to computer games. At treatment end point, psychosocial interventions were not significantly more effective at reducing negative symptoms compared to any other intervention in the NMA.

One systematic review reported that none of the eight psychosocial treatment studies that they reviewed had any impact on functioning in CHR samples (Cotter et al., 2014). A more recent systematic review and meta-analysis of 19 studies supported this (Devoe et al., 2019a). This systematic review and meta-analysis demonstrated that no treatment significantly improved social functioning in youth at CHR. Interestingly, although several of the studies addressed social isolation, social skills training, and social relationships (Bechdolf et al., 2012; Stain et al., 2016), most had no impact on social functioning with the exception of one CR study which demonstrated improvements in social functioning (Choi et al., 2017).

In summary, there is some possibility that CBT has an impact on reducing transition to psychosis. However, in general, the impact of psychosocial interventions has had little to no impact on APS, negative symptoms, and social functioning.

Cognitive behavioral therapy

CBT has established its efficacy among different therapeutic approaches and is the treatment of choice for anxiety and depression disorders in youth (Weersing et al., 2017). Furthermore, CBT has shown effectiveness in reducing and coping with psychotic symptoms and the associated distress among patients with schizophrenia (Tarrier, 2010). Given the high prevalence of substance use, mood, anxiety, and APS among CHR youth (Addington et al., 2017b; Fusar-Poli et al., 2014), CBT could be especially appropriate to address CHR concerns and symptoms, teach stress management skills, and potentially delay or prevent transition to psychosis.

The first RCT for CHR population implemented an intervention that combined CBT and medication (McGorry et al., 2002), making it difficult to ascertain the contribution of the therapy in the change of outcome. However, later trials focused on assessing the impact of CBT versus monitoring or a supportive therapy.

Some CBT adaptations have been made for the CHR population. The majority of studies follow the manualized CBT treatment approach for people at CHR developed by French and Morrison (2004). This CBT model is based on Beck (1976) principles but was adapted to the target population. The authors conceptualize APS from an integrative model of delusions and hallucinations being on a continuum with normal beliefs (Morrison, 2001), where normalization of the experience is the primary goal. Behavioral experiments are developed to test, generate, and evaluate different explanations or interpretations of each individual's experiences. However, if other symptoms are more prominent (i.e., depression or anxiety symptoms), other models are followed (Clark and Wells, 1995; Salkovskis et al., 1998; Wells, 1995). Importantly, all these models share common processes, affective and behavioral components, which provide the foundation for a transdiagnostic generalization of results along different symptoms and problems. Some of the shared components that French and Morrison (2004) proposed for the formulation are (1) assessment, experimentation, and generation of alternatives to safety behaviors (i.e., strategies people use to avoid the feared symptoms or situations); (2) recognition and assessment of dysfunctional (both positive and negative) beliefs and metacognitions; (3) identification and modification of core beliefs; (4) improvement of social, community and vocational functioning, and identification of support; (5) stress management; (6) increase insight into the anomalous experiences; and (7) discussion of relapse prevention.

Morrison and colleagues tested their model through the Early Detection and Intervention Evaluation (EDIE) RCT, where participants were allocated to the CBT and monitoring group (CBT+M; up to 26 sessions; N = 35; mean age = 20.6) or the monitoring-only group (up to 13 sessions; N = 23; mean age = 21.5) over 6 months (Morrison et al., 2004). After adjustment for age, gender, family history, and baseline symptom ratings, results were that those in the CBT+M condition had 96% reduction in the odds of transitioning to psychosis compared to the monitoring group over 1 year. Moreover, the severity of APS was significantly reduced in the CBT+M group compared to the monitoring group, although there was no evidence of significant improvement of distress or functioning. In addition, CBT significantly reduced the likelihood of being prescribed antipsychotic medication and of meeting DSM-IV criteria for a psychotic disorder.

In a subsequent methodologically rigorous multisite RCT (EDIE-2) with a larger representative sample, CBT was delivered for people at CHR over a 6-month period (CBT+M: N = 144, mean age = 20.7; monitoring: N = 144, mean age = 20.8) (Morrison et al., 2011, 2012a). Results regarding the frequency and intensity of APS were similar to the original EDIE study, showing a significant reduction in the CBT+M condition compared to the monitoring group. Although distress, quality of life, depression, and anxiety symptoms improved, the improvement was not significantly different from the monitoring group. Finally, the authors could not replicate findings of their previous study regarding transition to psychosis. CBT did not impact the transition to psychosis over 12–24 months. However, the actual transition rate in the whole study was low, with only 17 out of 288 participants making the transition in the first 12 months and 6 in the next 12 months.

Interestingly, considering CBT components that may be associated with change in the EDIE-2, Flach et al. (2015) concluded that CBT had a larger effect on outcome if case formulation and homework were included in the intervention. Moreover, there was a trend toward significance suggesting that active change strategies improved CBT outcomes. Hence, fidelity in following the manual instructions and applying all therapeutic components can additionally reduce symptom severity.

Three other trials used the same model and in a recent meta-analysis (Stafford et al., 2013), when compiling data of two of these trials in addition to EDIE and EDIE-2 (Addington et al., 2011c; Morrison et al., 2004, 2011; van der Gaag et al., 2012b), CBT was associated with a greater decrease in APS over 6–12 months compared to the control treatments being offered. All these trials delivered the manualized, time-limited, and problem-focused CBT published by French and Morrison (2004). The Dutch EDIE trial (EDIE-NL) (Ising et al., 2016; van der Gaag et al., 2012b) demonstrated that the number of transitions to psychosis could be reduced by about 50% by using a CBT intervention, results which remained significant at 4-year follow-up. Both treatment groups received treatment as usual, and to enhance the CBT, this Dutch group enriched the CBT protocol with education on dopamine sensitivity, and how this affects thinking and perception and exercises on cognitive biases.

The two other RCTs employing the French and Morrison manual (Addington et al., 2011b; Stain et al., 2016) had different results. Addington and colleagues showed that there were no significant differences between supportive therapy (N = 24; mean age = 21.1) and CBT (N = 27; 20.8) (Addington et al., 2011b). However, transition rates were low with no transitions reported in the CBT group and three transitions in the supportive therapy group. Stain and colleagues showed that the nondirective reflective listening (NDRL) condition (N = 27; mean age 16.20) was superior to the CBT condition (N = 30; mean age = 16.2) in reducing distress associated with attenuated symptoms, with no differences between conditions for other outcomes (Stain et al., 2016). Interestingly, in this study, conversion was very low with only three transitions occurring, but this time in the CBT group. The only difference between the groups was that the NDRL group had significantly lower levels of distress about their APS. However, despite the good methodological designs, both studies were underpowered. Stain's sample was much younger and had better functioning and less APS than previously published trials.

Two other recent trials also reported positive effects of CBT. Bechdolf and colleagues in the PREVENT trial used a manualized version of CBT developed in Cologne that essentially contains typical CBT strategies, techniques, exercises, and homework tasks (Bechdolf et al., 2017). Up to 30 sessions of CBT were delivered over the course of 12 months. The CBT group had a trend toward fewer transitions compared to the aripiprazole plus case management or the placebo group (Bechdolf et al., 2011, 2017). The second trial was an open study in Korea (Kim et al., 2011), which used a 10-session modified version of French and Morrison's manual and that demonstrated significant improvement in APS, negative symptoms, and mood.

In summary, there is some evidence suggesting that CBT may have a positive effect on the reduction of APS and later transition to psychosis. However, other problems present in CHR youth such as negative symptoms (i.e., isolation and withdrawal), distress, and low social and role functioning do not seem to have benefited from CBT as much. Finally, although CBT is promising for this population, there is room for further work in larger RCTs. Questions that could still be answered include which components of CBT work better and for whom and what problems are being addressed. It may be that for some youth addressing anxiety or depression may be more important than APS (McAusland et al., 2015).

Family interventions

Literature with a focus on intervention with families of CHR youth is scarce, despite the well-known success of family intervention with individuals with an established schizophrenia illness as well as those experiencing a first-episode psychosis (Onwumere et al., 2011; Pharoah et al., 2010). However, as in psychosis (Butzlaff and Hooley, 1998; Hooley and Richters, 1995), the association of family environment with clinical and functional outcomes has been reported in CHR families. In one study, it was observed that a positive family environment predicted a decrease in symptoms and improved social functioning (O'Brien et al., 2006). In a second study, young CHR participants who were living in what was described as a critical family environment (i.e., high criticism, disapproval, and complains) presented with more severe APS at a 6-month follow-up relative to youth who were living in a much less critical family environment (Schlosser et al., 2010).

Up until 2006, the literature on the intervention with families of CHR youth was practically non existent. Collins and Addington (2006) proposed a theoretical framework to work with CHR youth families based on the Calgary and Toronto recovery stage model for first-episode psychosis (Addington et al., 2005). The authors suggested that work with families should cover (1) management of first signs (i.e., isolation, academic impairment, withdrawal, APS); (2) monitoring the youth to identify a possible transition to psychosis; and (3) self-care of family members (i.e., maintenance of functioning and well-being). Although not all the CHR youth and their families will progress in a linear trajectory, the authors proposed four stages to help frame the intervention with these families.

After the first contact with a mental health team about their young CHR family member, the main goals of Stage 1 are to (i) manage the concerns and any initial confusion of the family; (ii) establish a therapeutic alliance with the family; (iii) evaluate the family needs and family beliefs regarding causes, maintenance, and treatment of CHR state; and (iv) provide an initial explanation for the current problems through psychoeducation. After the decrease of the initial confusion expressed by families, the intervention should focus on consolidating Stage 1 goals, particularly because the development of an agreed explanatory model for what has occurred may help the family to move to a sense of recovery. The primary objectives of Stage 2 are to (i) evaluate family functioning and (ii) provide information of early warning signs of developing psychosis. Stage 3 assists with the consolidation of the learning from previous stages. Finally, and only if it appears to be necessary, Stage 4 covers the action plan in case the youth transitions to psychosis by (i) coping with changes of expectations; (ii) seamless transition to specialized first-episode services, and (iii) developing an agreed model of long-term prognosis. Translation of the intervention with families of CHR from the theory to research has only been developed in the last decade.

The first study involving an intervention with families of CHR youth (O'Brien et al., 2007) examined the feasibility and acceptance of a psychoeducational multifamily group (McFarlane, 2002) for CHR youth ($N = 16$, mean age $= 15.7$). After a one-session psychoeducational workshop, 90-minute group meetings were held biweekly over 9 months (18 sessions). The content of the sessions focused on symptom management, communication, social, and problem-solving skills. Youth improved in attenuated psychotic and general symptoms and social and role functioning. Moreover, family adaptability and cohesion and constructive coping improved after treatment. However, this was a single, non blinded, group study, and several interventions were delivered conjunctly (psychoeducational multifamily group, school intervention, case management, and psychiatric treatment) making it difficult to determine what mainly contributed to the changes.

The first family RCT was completed by Miklowitz et al. (2014), who adapted the psychoeducational family-focused therapy (FFT) (Miklowitz et al., 2013; Miklowitz and Scott, 2009), originally developed for bipolar youth, for CHR youth and their families. The FFT-CHR (Schlosser et al., 2012) is a manualized intervention that comprises 12 weekly and 6 biweekly 50-minute sessions over 6 months. Briefly, the FFT-CHR sessions are organized in three modules of six sessions each: (1) psychoeducation on stress management and the development of a prevention action plan; (2) interpersonal communication skills, and (3) problem-solving skills. Families are expected to do behavioral activation practice between modules. The control treatment was enhanced care (EC), which was three 1-hour sessions of brief psychoeducational treatment addressing a similar content of module 1 of FFT-CHR including the development of a prevention action plan. If needed, individual crisis management sessions were offered to all participants.

Findings from the FFT-CHR RCT (Miklowitz et al., 2014) (N = 129, mean age = 17.4) demonstrated that over 6 months, there was a decrease in APS in the FFT-CHR condition compared to EC. There was no difference in transition rates and only six participants (5.9%) made a transition to psychosis, one in the FFT-CHR group and five in the EC group. Interestingly, participants in both treatment conditions showed a reduction in negative symptoms. The behavioral activation component (i.e., participation in pleasant activities), present in both psychosocial treatments, could have played a role in the improvements in negative symptoms (Dimidjian et al., 2006). Both groups improved in social and role functioning over the 6 months of the trial. Furthermore, in terms of working with families, the age of the participant may have had a moderating effect on functional recovery and skill performance. Only participants 20 years and older in the FFT-CHR condition and those aged 16–19 years in the EC condition improved in social and role functioning (i.e., school or work performance). Young adults in the FFT-CHR group may have been more able than adolescents to generalize problem-solving and communication skills to other contexts such as school and work, whereas for adolescents, the brief coping strategies for stressful interpersonal situations provided in the EC group may be more useful to manage these circumstances. A subsequent report on a subsample of the Miklowitz et al. (2014) study ($N = 66$, mean age 16.9) reported that over 6 months, CHR youth and their families improved in constructive communication (i.e., calm communication and active listening) and reduced family conflict behaviors (i.e., anger, criticism, complaints, and irritability) compared to those in the EC group (O'Brien et al., 2014).

The Finish group of Grano et al. (2016) recently tested an adapted Family- and Community-Orientated Integrative Treatment Model (Grano et al., 2009) in CHR youth ($N = 28$, mean age = 15.5) compared with treatment as usual ($N = 28$, mean age = 16.3). This family treatment model included an adapted component of family therapy for psychosis (Alanen et al., 1991), open dialogue (Seikkula and Olson, 2003), and CBT. The therapy was delivered at the community level or at home. Main goals were to reduce stressors, involve and provide support for the family throughout the process, and reduce family conflict. Findings from the study showed that there was a significant reduction for self-reported depression and hopeless and an improvement in functioning in the experimental treatment group compared to treatment as usual. No significant differences were found between groups in APS or anxiety.

In summary, CHR youth may benefit from family treatments that focus on psychoeducation and skills training (Miklowitz et al., 2014). APS generally improved in several of the studies involving family members. The majority of family interventions focused on symptoms and psychoeducation, and some studies also targeted family communication and problem-solving skills. However, social and role functioning, negative symptoms, and other psychiatric symptoms were only improved in single-case studies with methodological limitations. Compared to working individually with youth, including family members in the intervention may provide a reinforcement of learned skills, which may be a unique resource to maintain gains, well-being, and prevent transition. Nevertheless, further research involving CHR family members is essential to understand family dynamics and to test hypotheses regarding the role of the family in the delay or prevention of transition to psychosis and improvement in functioning.

Cognitive remediation

Psychotic illness, such as schizophrenia, is marked by a wide range of cognitive impairments of varying magnitudes that are often present in the early phase of the illness (Fioravanti et al., 2012). Cognitive impairments tend to remain relatively stable over time and are said to precede the onset of the illness as is the case for CHR youth (Seidman et al., 2016). Results of prospective research studying youth at CHR of psychosis suggest that these young people exhibit cognitive functioning that falls between those of healthy controls and those with a first-episode psychosis (Hauser et al., 2017), suggesting a presence of some impairment. Cognitive impairments in CHR have been observed in global cognition and across different domains of cognitive function, most notably in verbal learning and memory, speed of processing, and attention (Addington and Barbato, 2012; Hauser et al., 2017).

Cognitive impairments are related to poor community functioning in schizophrenia (Fett et al., 2011) and to a lesser extent in CHR samples (Carrion et al., 2013). As the onset of psychosis typically occurs in late adolescence and early adulthood, cognitive deterioration at this time could be particularly disruptive. This phase of the lifespan is a critical period for learning and establishing new social relationships and is, as such, viewed as the fundamental period for early intervention. CR interventions employed to improve cognition in individuals with psychosis (Wykes et al., 2011) have been adapted for those at CHR, such as drill and practice-computerized approaches including Brain Fitness and CogPack. In the field of psychosis research, CR refers to a range of treatments aimed at improving cognitive abilities with the ultimate goal of improving community functioning. Given the notion that CR in CHR is implemented preemptively, often before discernible cognitive deficits, to preserve and perhaps recover cognitive skills that are essential for everyday functioning, all of the approaches used in CHR to date have been restorative (i.e., aimed at reinforcing and regenerating neuroanatomical connections that are purported to underpin cognitive functions).

Although CR approaches tend to differ in regard to structure, intensity, and duration of treatment, according to Best and Bowie (2017), they are typically based on combinations of three overarching principles: training, strategy monitoring, and generalization. Training typically involves repetitive practice, usually with computerized cognitive exercises. Strategy monitoring is a means to help the participants understand the training approaches being used. Lastly, generalization refers to the ability of the participant to be able to use the skills they have learned in the computerized training to their daily life. Another point of difference regarding CR approaches involves therapeutic targets and therapeutic modalities (Best and Bowie, 2017). Therapeutic targets could either be what is known as "bottom-up" processes such as basic perceptual skills or "top-down" processes such as executive functioning skills. CR approaches that target bottom-up processes aim to improve perceptual skills that have been identified as impaired to improve higher-order cognitive abilities. Alternatively, approaches that target top-down processes have a goal of training executive functioning directly and, hopefully, lead to a series of improvements in basic cognitive abilities and community functioning. It is recommended that the therapeutic modalities are chosen once a therapeutic target has been identified, and these will vary from stand alone or independent computerized training to intensive psychotherapy used in rehabilitation settings (Best and Bowie, 2017).

In the seven studies that assessed the efficacy of CR conducted in CHR populations to date (see Table 18.2), the training principle was examined in all of them while the generalization principle using pre- and postmeasures of community functioning was addressed in four out of seven studies. None of the studies used the strategy monitoring principle. Furthermore, the studies used a mixture of therapeutic targets and independent computerized training to do so. Five out of the seven studies used top-down CR approaches (Choi et al., 2017; Holzer et al., 2014; Hooker et al., 2014; Nemoto et al., 2010; Rauchensteiner et al., 2011) such as CogPack, Lumosity, Divergent Thinking, Captains Log, and Processing Speed Training, respectively, and the remaining two studies used a bottom-up CR strategy, Brain Fitness (Loewy et al., 2016; Piskulic et al., 2015). The majority of studies identified multiple cognitive domains as therapeutic targets of their CR training initiatives, with the exception of Choi et al. (2017) whose CR training was specifically designed and used to address processing speed abilities.

It is important to emphasize that all of the studies of CR in CHR reviewed to date have been significantly underpowered either at recruitment or because of a high attrition rate which, with one exception (Choi et al., 2017), ranged from 28% to 49%. A recently published CR study protocol had suggested a sample size of 63 CHR participants per group to detect a difference between CR and a control condition (Glenthøj et al., 2015). Low recruitment and high attrition rates may be reflective of a number of issues ranging from a choice of therapeutic targets and modalities to treatment relevance (participants' level of cognitive impairment, functional difficulties, and, therefore, intrinsic motivation). In fact, in the CR studies in CHR, the selection of participants was not based on any predetermined cognitive difficulties, which means that CHR samples were likely very heterogeneous with regard to their cognitive function. Efforts to remediate a cognitive impairment are inherently different from approaches aimed at improving intact cognitive function. Therefore, if the aim is to increase therapeutic generalizability to real-world outcomes, a CR that has demonstrated an efficacy to improve both proximal (cognitive function) and distal outcomes (community functioning) in CHR should be used in conjunction with other psychotherapeutic approaches (Best and Bowie, 2017). For this to be both effective and efficacious in CHR samples, however, there first has to be a clear evidence of a cognitive deficit (i.e., therapeutic target) and impairment in real-world functioning (e.g., reduced grades), which will then inform a choice of the most appropriate therapeutic modality (i.e., independent computerized training, intensive psychotherapeutic approaches, or a combination of the two). Approaching CR in this way would be to deliver it as an intensive psychotherapy, with a therapist at the forefront to aid with the development of new problem-solving strategies and the facilitation of real-world behavioral changes (Best and Bowie, 2017).

Social cognitive training

In psychosis research, the National Institute of Mental Health (NIMH) Workshop of Social Cognition in Schizophrenia has defined social cognition as a function that involves the perception, interpretation, and processing of information that underlies social interactions (Green et al., 2008). There has been a surge of studies examining social cognition in CHR populations compared to healthy controls and patients with psychosis, with the majority of studies reporting quantifiable deficits in social cognition in CHR populations relative to the comparison groups (Green et al., 2012; Thompson et al., 2011). Two recent meta-analyses of social cognition in CHR reported impairments in all domains of social cognition, with largest cumulative impairments in attributional bias and theory of mind (ToM) and somewhat smaller impairments in emotion perception and social perception (Lee et al., 2015; van Donkersgoed et al., 2015). Social cognition impacts community functioning in psychosis, both directly (Green et al., 2015) and through the mediation of cognitive functioning (Fett et al., 2011; Schmidt et al., 2011). In CHR, social cognition does not seem to mediate the relationship between cognition and functioning as is the case in the established illness (Barbato et al., 2013), but it is strongly associated with both cognition and community functioning (Glenthoj et al., 2016). It is, therefore, reasonable that social cognition could be a target of remediation efforts to improve real-world functioning. While there have been several social cognitive intervention studies to date, all of them were conducted in the established psychotic illness. To our knowledge, only one pilot study that included a social CR training in CHR participants in conjunction with CR has been published to date (Hooker et al., 2014). However, the focus of the study was on cognitive gains and their association with community functioning with no discussion as to the role of social cognitive training.

Unlike CR treatment, social cognitive intervention is a relatively new, although expanding, area of psychosocial rehabilitation research in psychosis. While viewed as a distinct type of therapy that emphasizes interpreting others' social cues, such as thoughts and emotions, social cognitive training shares features with the established interventions such as CR, skills training, and CBT (Fiszdon and Reddy, 2012). Social cognitive interventions tend to be classified into three overlapping categories: targeted, comprehensive, or broad-based. Targeted interventions focus on a single domain of social cognition (e.g., ToM or emotion perception); comprehensive interventions aim to address a broad range of social cognitive impairments using a single treatment modality; and broad-based interventions integrate training in multiple domains of social cognition within the context of other types of psychosocial treatment (Fiszdon and Reddy, 2012). However, one common feature of these interventions is that they reportedly affect basic or proximal social cognitive processes, such as emotion perception, while having a minimal effect on more distal, higher-order processes, such as attributional bias or social perception. The prevailing explanation for this outcome is that higher-order processes require that social cognitive skills are regularly practiced, both in therapy and in the real world, until they are completely integrated and practically reflexive. Another caveat is that cognitive impairments need to be addressed before social cognitive training. In other words, it is rare in psychosis that social cognition is intact in the presence of cognitive impairments, especially because social cognition mediates the effect of cognition on community functioning (Fett et al., 2011; Schmidt et al., 2011). Therefore, it appears that a combination of CR and social cognitive training could be used to ensure the optimal conditions for learning of new social cognitive skills and only in conjunction with repetitive practice, such as social skills training. Unlike CR, there are very few stand alone social cognition computerized interventions (e.g., SocialVille) presumably because social cognition develops through social interactions within social environments. This highlights the mammoth task of either adapting existing social cognitive interventions to the needs of CHR individuals or developing new, ecologically valid interventions. New research efforts are starting to explore the combination of existing social cognitive interventions and new technologies such as simulation techniques, or virtual reality, to mimic complex, dynamic, and interactive training environments (Peyroux and Franck, 2014).

Integrated psychological therapy

There have been a few studies that examined the effectiveness of an integrated model of psychological treatments. One of these trials offered individual case management, social skills training, and multiple family group psychoeducation; however, the participants were young adults with schizotypal personality disorder (Albert et al., 2016; Nordentoft et al., 2006). At the 24-month follow-up, the IPT postponed or reduced transition to psychosis compared to treatment as usual (Nordentoft et al., 2006); however, this difference was no longer apparent at the 3.5-year follow-up (Albert et al., 2016; Wessels et al., 2015). One of the more comprehensive IPTs was a German RCT (Bechdolf et al., 2012) to test the efficacy of the manualized Integrated Psychological Intervention (IPI) compared to 30 sessions of supportive counseling (SC) (IPI: N = 63; mean age 25.2; and SC: N = 65; 26.8), which consisted of (i) 25-weekly individual sessions of CBT successively complemented with the other modules; (ii) 12 sessions of CR; (iii) 15 group sessions of social skills training and problem-solving; and (iv) 3 sessions of multifamily education over 12 months. Participants met criteria for CHR based on the SIPS and some participants who were considered to be in an early prodromal phase met the criteria on the SPI-A. This study demonstrated significant differences in transition rates between groups at the 12-month follow-up (3.2% for IPI and 16.9% for SC) and these differences were maintained at the 24-month follow-up. However, as several modules were delivered conjunctly, it is difficult to ascertain the contribution of each therapy component to the changes in outcome.

Conclusion

This chapter has reviewed, to the best of our knowledge, all psychological treatment studies as of March 2018. Summarizing the literature to date, it seems as if CBT may have an impact on preventing transition and on improving APS. However, in a couple of studies, a clearly defined supportive therapy also appears to help these young people. In fact, CBT studies that offered a specific control treatment versus monitoring or treatment as usual reported that participants in those treatment conditions also demonstrated symptomatic improvement. Family interventions are promising, although there is only one RCT to date, which suggests that family intervention may improve APS. On one hand, there is little to suggest that negative symptoms and social and role functioning are helped by psychological treatments, but on the other hand, none of the treatments were specifically designed to address negative symptoms and functioning, and assessment of these was typically a secondary aim or required data to be sought from the investigators.

There are some important caveats. There is only a small number of studies representing each type of treatment, especially for RCTs. Studies have occurred over approximately a 15–20-year period, and in that time, rates of transition have been seen to decline. Although studies report on several outcome variables, the treatments were not necessarily chosen to target those outcomes. Furthermore, it has been reported in naturalistic studies that APS typically decline over time in longitudinal studies, and CHR samples include individuals who make a complete remission within the first few months (Addington et al., 2011a, 2015). Finally, CHR samples were not usually specifically selected as needing the specific treatment being tested as may be the case in schizophrenia studies. For example, participants in CR studies were not selected for having specific if any cognitive deficits.

Whether it is part of a clinical trial or a clinical service, implementation of treatment needs consideration. Because most of the reports are clinical trials, it is difficult to gauge how much psychological treatment is being delivered to CHR youth. However, a recent publication in the United Kingdom suggested that current provision for treatment for those at CHR did not match the National Institute for Health and Care Excellence (NICE) clinical guidelines with approximately 50% offering CBT and 42% offering family intervention (Stain et al., 2019). It is noticeable that out of the 82% of CHR help-seeking participants who received psychosocial treatment in the North American Prodrome Longitudinal Study (NAPLS-2; $N = 764$), only 11.4% had received CBT at any point in their life compared to supportive, school counseling, or case management (52.5%, 17.4%, and 13.0%, respectively) (Woodberry et al., 2016). Implementation needs to consider what treatment to use and who should be delivering that treatment. In our review it was evident that successful studies tended to use manualized treatment programs and include demonstration of fidelity to the model being used. For the majority of the CBT trials and the family intervention study, therapy is typically delivered by a psychologist or psychiatrist or at a minimum master level clinicians or therapists and nurses who have received specialized training in CBT (Addington et al., 2011b; Morrison et al., 2004, 2012c). The clinical training is comprehensive and typically requires weekly or biweekly close supervision and monitoring by experts, which may not be available in regular services.

Moreover, monitoring attrition may be important in the delivery of treatment. In the CBT studies where, typically, participants could have had 18–26 sessions over a period of 6 months, the mean number of sessions in many of the studies ranged from 9 to 12. It is not known whether this attrition from research treatment would be the same in a clinical setting. Thus, engagement may be an important issue to address in treatment implementation.

Finally, there is the question of which treatment may be the most beneficial for these young people. As we have seen with the studies already conducted, the usual outcome under study is prevention of transition to psychosis and improvement of APS. Yet clinically we know that these young people present with APS that have a wide range of severity, many have poor social and role functioning, and there is a high degree of comorbidity of other disorders, in particular depression and anxiety. Studies examining integrated psychological treatments were few but generally the outcome was promising. Offering a package of interventions such as family education, CBT, social skills training, and CRT may be an effective early intervention that would allow a wide range of symptoms such as attenuated psychotic and negative symptoms, anxiety, mood, and poor social and role functioning to be addressed. Alternatively, a modular-based treatment program similar to that used in the RAISE project (Kane et al., 2015, 2016) may be effective for those at CHR of psychosis. In RAISE, through shared decision-making, participants could select from a range of programs that included medication management, family education, and individual resiliency training that included CBT and supported education and employment.

Thus, good progress has already been made in testing relevant treatments for this population. The next steps in research may be to consider integrated treatments with clear identification of which problems are being addressed with which therapy.

References

Addington, J., Barbato, M., 2012. The role of cognitive functioning in the outcome of those at clinical high risk for developing psychosis. Epidemiol. Psychiatr. Sci. 21, 335–342.

Addington, J., Collins, A., Mccleery, A., Addington, D., 2005. The role of family work in early psychosis. Schizophr. Res. 79, 77–83.

Addington, J., Cornblatt, B.A., Cadenhead, K.S., Cannon, T.D., Mcglashan, T.H., Perkins, D.O., Seidman, L.J., Tsuang, M.T., Walker, E.F., Woods, S.W., Heinssen, R., 2011a. At clinical high risk for psychosis: outcome for nonconverters. Am. J. Psychiatry 168, 800–805.

Addington, J., Epstein, I., Liu, L., French, P., Boydell, K.M., Zipursky, R.B., 2011b. A randomized controlled trial of cognitive behavioral therapy for individuals at clinical high risk of psychosis. Schizophr. Res. 125, 54–61.

Addington, J., Francey, S.H., Morrison, A.P., 2006. Working with People at High Risk of Developing Psychosis. Wiley, Chichester, England.

Addington, J., Liu, L., Buchy, L., Cadenhead, K.S., Cannon, T.D., Cornblatt, B.A., Perkins, D.O., Seidman, L.J., Tsuang, M.T., Walker, E.F., Woods, S.W., Bearden, C.E., Mathalon, D.H., Mcglashan, T.H., 2015. North American Prodrome longitudinal study (NAPLS 2): the prodromal symptoms. J. Nerv. Ment. Dis. 203, 328–335.

Addington, J., Liu, L., Perkins, D.O., Carrion, R.E., Keefe, R.S.E., Woods, S.W., 2017a. The role of cognition and social functioning as predictors in the transition to psychosis for youth with attenuated psychotic symptoms. Schizophr. Bull. 43, 57–63.

Addington, J., Mancuso, E., Haarmans, M., 2011c. Cognitive behaviour therapy and early intervention. In: Turkington, D., Hagen, R., Berge, T., Gråwe, R.W. (Eds.), CBT for Psychosis: A Symptom-Based Approach Routledge.

Addington, J., Piskulic, D., Liu, L., Lockwood, J., Cadenhead, K.S., Cannon, T.D., Cornblatt, B.A., Mcglashan, T.H., Perkins, D.O., Seidman, L.J., Tsuang, M.T., Walker, E.F., Bearden, C.E., Mathalon, D.H., Woods, S.W., 2017b. Comorbid diagnoses for youth at clinical high risk of psychosis. Schizophr. Res.

Alanen, Y.O., Lehtinen, V., Räkköläinen, V., Aaltonen, J., 1991. Need-adapted treatment of new schizophrenic patients: experiences and results of the Turku project. Acta Psychiatr. Scand. 83, 363–372.

Albert, N., Glenthøj, L.B., Melau, M., Jensen, H., Hjorthøj, C., Nordentoft, M., 2016. Course of illness in a sample of patients diagnosed with a schizotypal disorder and treated in a specialized early intervention setting. Findings from the 3.5 year follow-up of the OPUS II study. Schizophr. Res. 182, 24–30.

Barbato, M., Liu, L., Penn, D.L., Keefe, R.S., Perkins, D.O., Woods, S.W., Addington, J., 2013. Social cognition as a mediator between neurocognition and functional outcome in individuals at clinical high risk for psychosis. Schizophr. Res. 150, 542–546.

Bechdolf, A., Müller, H., Stützer, H., Lambert, M., Karow, A., Zink, M., Lautenschlager, M., Heinz, A., De Millas, W., Janssen, B., Gaebel, W., Schneider, F., Juckel, G., Krüger-Özgürdal, S., Wobrock, T., Wagner, M., Maier, W., Klosterkötter, J., 2017. 108. PREVENT: a randomized controlled trial for the prevention of first-episode psychosis comparing cognitive–behavior therapy (CBT), clinical management, and aripiprazole combined and clinical management and placebo combined. Schizophr. Bull. 43, S56–S57.

Bechdolf, A., Muller, H., Stutzer, H., Wagner, M., Maier, W., Lautenschlager, M., Heinz, A., De Millas, W., Janssen, B., Gaebel, W., Michel, T.M., Schneider, F., Lambert, M., Naber, D., Brune, M., Kruger-Ozgurdal, S., Wobrock, T., Riedel, M., Klosterkotter, J., 2011. Rationale and baseline characteristics of PREVENT: a second-generation intervention trial in subjects at-risk (prodromal) of developing first-episode psychosis evaluating cognitive behavior therapy, aripiprazole, and placebo for the prevention of psychosis. Schizophr. Bull. 37, S111–S121.

Bechdolf, A., Wagner, M., Ruhrmann, S., Harrigan, S., Putzfeld, V., Pukrop, R., Brockhaus-Dumke, A., Berning, J., Janssen, B., Decker, P., Bottlender, R., Maurer, K., Moller, H.J., Gaebel, W., Hafner, H., Maier, W., Klosterkotter, J., 2012. Preventing progression to first-episode psychosis in early initial prodromal states. Br. J. Psychiatry 200, 22–29.

Beck, A.T., 1976. Cognitive Therapy and the Emotional Disorders. International Universities Press, New York.

Best, M.W., Bowie, C.R., 2017. A review of cognitive remediation approaches for schizophrenia: from top-down to bottom-up, brain training to psychotherapy. Expert Rev. Neurother. 17, 713–723.

Butzlaff, R.L., Hooley, J.M., 1998. Expressed emotion and psychiatric relapse. Arch. Gen. Psychiatr. 55, 547–552.

Caldwell, D.M., Ades, A.E., Higgins, J.P.T., 2005. Simultaneous comparison of multiple treatments: combining direct and indirect evidence. BMJ 331, 897–900.

Carrion, R.E., Mclaughlin, D., Goldberg, T.E., Auther, A.M., Olsen, R.H., Olvet, D.M., Correll, C.U., Cornblatt, B.A., 2013. Prediction of functional outcome in individuals at clinical high risk for psychosis. JAMA Psychiatry 70, 1133–1142.

Choi, J., Corcoran, C.M., Fiszdon, J.M., Stevens, M., Javitt, D.C., Deasy, M., Haber, L.C., Dewberry, M.J., Pearlson, G.D., 2017. Pupillometer-based neurofeedback cognitive training to improve processing speed and social functioning in individuals at clinical high risk for psychosis. Psychiatr. Rehabil. J. 40, 33–42.

Clark, D.M., Wells, A., 1995. A cognitive model of social phobia. In: Heimberg, R.G., Liebowitz, M.R. (Eds.), Social Phobia: Diagnosis, Assessment, and Treatment. Guilford Press, New York.

Collins, A.A., Addington, J., 2006. Working with families following the diagnosis of an at risk mental state. In: Addington, J.M., Francey, M.S., Morrison, A.P. (Eds.), Working with People at High Risk of Developing Psychosis. Wiley, Chichester, United Kingdom.

Cornblatt, B.A., Carrion, R.E., Addington, J., Seidman, L., Walker, E.F., Cannon, T.D., Cadenhead, K.S., Mcglashan, T.H., Perkins, D.O., Tsuang, M.T., Woods, S.W., Heinssen, R., Lencz, T., 2012. Risk factors for psychosis: impaired social and role functioning. Schizophr. Bull. 38, 1247–1257.

Cotter, J., Drake, R.J., Bucci, S., Firth, J., Edge, D., Yung, A.R., 2014. What drives poor functioning in the at-risk mental state? A systematic review. Schizophr. Res. 159, 267–277.

Devoe, D.J., Farris, M., Townes, P., Addington, J., 2019a. Interventions and social functioning in youth at risk of psychosis: a systematic review and meta-analysis. Early Interv. Psychiatry 13, 169–180.

Devoe, D.J., Farris, M.S., Townes, P., Addington, J., 2019b. Attenuated psychotic symptom interventions in youth at risk of psychosis: a systematic review and meta-analysis. Early Interv. Psychiatry 13, 3–17.

Devoe, D.J., Peterson, A., Addington, J., 2018. Negative symptom interventions in youth at risk of psychosis: a systematic review and network meta-analysis. Schizophr. Bull. 44, 807–823.

Dias, S., Sutton, A.J., Ades, A.E., Welton, N.J., 2013. Evidence synthesis for decision making 2: a generalized linear modeling framework for pairwise and network meta-analysis of randomized controlled trials. Med. Decis. Mak. 33, 607–617.

Dimidjian, S., Hollon, S.D., Dobson, K.S., Schmaling, K.B., Kohlenberg, R.J., Addis, M.E., Gallop, R., Mcglinchey, J.B., Markley, D.K., Gollan, J.K., Atkins, D.C., Dunner, D.L., Jacobson, N.S., 2006. Randomized trial of behavioral activation, cognitive therapy, and antidepressant medication in the acute treatment of adults with major depression. J. Consult. Clin. Psychol. 74, 658–670.

Fett, A.K., Viechtbauer, W., Dominguez, M.D., Penn, D.L., Van Os, J., Krabbendam, L., 2011. The relationship between neurocognition and social cognition with functional outcomes in schizophrenia: a meta-analysis. Neurosci. Biobehav. Rev. 35, 573–588.

Fioravanti, M., Bianchi, V., Cinti, M.E., 2012. Cognitive deficits in schizophrenia: an updated metanalysis of the scientific evidence. BMC Psychiatry 12, 64.

Fiszdon, J.M., Reddy, L.F., 2012. Review of social cognitive treatments for psychosis. Clin. Psychol. Rev. 32, 724–740.

Flach, C., French, P., Dunn, G., Fowler, D., Gumley, A.I., Birchwood, M., Stewart, S.L., Morrison, A.P., 2015. Components of therapy as mechanisms of change in cognitive therapy for people at risk of psychosis: analysis of the EDIE-2 trial. Br. J. Psychiatry 207, 123–129.

French, P., Morrison, A.P., 2004. Early Detection and Cognitive Therapy for People at High Risk of Developing Psychosis. Wiley, Chichester, United Kingdom.

Fusar-Poli, P., Frascarelli, M., Valmaggia, L., Byrne, M., Stahl, D., Rocchetti, M., Codjoe, L., Weinberg, L., Tognin, S., Xenaki, L., Mcguire, P., 2015. Antidepressant, antipsychotic and psychological interventions in subjects at high clinical risk for psychosis: OASIS 6-year naturalistic study. Psychol. Med. 45, 1327–1339.

Fusar-Poli, P., Nelson, B., Valmaggia, L., Yung, A.R., Mcguire, P.K., 2014. Comorbid depressive and anxiety disorders in 509 individuals with an at-risk mental state: impact on psychopathology and transition to psychosis. Schizophr. Bull. 40, 120–131.

Glenthoj, L.B., Fagerlund, B., Hjorthoj, C., Jepsen, J.R.M., Bak, N., Kristensen, T.D., Wenneberg, C., Krakauer, K., Roberts, D.L., Nordentoft, M., 2016. Social cognition in patients at ultra-high risk for psychosis: what is the relation to social skills and functioning? Schizophr. Res. Cogn. 5, 21–27.

Glenthøj, L.B., Fagerlund, B., Randers, L., Hjorthøj, C.R., Wenneberg, C., Krakauer, K., Vosgerau, A., Gluud, C., Medalia, A., Roberts, D.L., Nordentoft, M., 2015. The FOCUS trial: cognitive remediation plus standard treatment versus standard treatment for patients at ultra-high risk for psychosis: study protocol for a randomised controlled trial. Trials 16, 25-25.

Grano, N., Karjalainen, M., Anto, J., Itkonen, A., Edlund, V., Roine, M., 2009. Intervention to improve level of overall functioning and mental condition of adolescents at high risk of developing first-episode psychosis in Finland. Early Interv. Psychiatry 3, 94–98.

Grano, N., Karjalainen, M., Ranta, K., Lindgren, M., Roine, M., Therman, S., 2016. Community-oriented family-based intervention superior to standard treatment in improving depression, hopelessness and functioning among adolescents with any psychosis-risk symptoms. Psychiatry Res. 237, 9–16.

Green, M.F., Bearden, C.E., Cannon, T.D., Fiske, A.P., Hellemann, G.S., Horan, W.P., Kee, K., Kern, R.S., Lee, J., Sergi, M.J., Subotnik, K.L., Sugar, C.A., Ventura, J., Yee, C.M., Nuechterlein, K.H., 2012. Social cognition in schizophrenia, Part 1: performance across phase of illness. Schizophr. Bull. 38, 854–864.

Green, MF, Horan, WP, Lee, J, 2015. Social cognition in schizophrenia. Nat Rev Neurosci 16 (10), 620–631. Epub 2015 Sep 16. Review. PMID: 26373471.

Green, M.F., Penn, D.L., Bentall, R., Carpenter, W.T., Gaebel, W., Gur, R.C., Kring, A.M., Park, S., Silverstein, S.M., Heinssen, R., 2008. Social cognition in schizophrenia: an NIMH workshop on definitions, assessment, and research opportunities. Schizophr. Bull. 34, 1211–1220.

Hauser, M., Zhang, J.P., Sheridan, E.M., Burdick, K.E., Mogil, R., Kane, J.M., Auther, A., Carrion, R.E., Cornblatt, B.A., Correll, C.U., 2017. Neuropsychological test performance to enhance identification of subjects at clinical high risk for psychosis and to Be most promising for predictive algorithms for conversion to psychosis: a meta-analysis. J. Clin. Psychiatry 78, e28–e40.

Holzer, L., Urben, S., Passini, C.M., Jaugey, L., Herzog, M.H., Halfon, O., Pihet, S., 2014. A randomized controlled trial of the effectiveness of computer-assisted cognitive remediation (CACR) in adolescents with psychosis or at high risk of psychosis. Behav. Cognit. Psychother. 42, 421–434.

Hooker, C.I., Carol, E.E., Eisenstein, T.J., Yin, H., Lincoln, S.H., Tully, L.M., Dodell-Feder, D., Nahum, M., Keshavan, M.S., Seidman, L.J., 2014. A pilot study of cognitive training in clinical high risk for psychosis: initial evidence of cognitive benefit. Schizophr. Res. 157, 314–316.

Hooley, J.M., Richters, J.E., 1995. Expressed emotion: a developmental perspective. In: Cicchetti, D., Toth, S.L. (Eds.), Emotion, Cognition, and Representation. University of Rochester Press, Rochester, NY.

Ising, H.K., Kraan, T.C., Rietdijk, J., Dragt, S., Klaassen, R.M., Boonstra, N., Nieman, D.H., Willebrands-Mendrik, M., Van Den Berg, D.P., Linszen, D.H., Wunderink, L., Veling, W., Smit, F., Van Der Gaag, M., 2016. Four-year follow-up of cognitive behavioral therapy in persons at ultra-high risk for developing psychosis: the Dutch Early Detection Intervention Evaluation (EDIE-NL) trial. Schizophr. Bull. 42, 1243–1252.

Kane, J.M., Robinson, D.G., Schooler, N.R., Mueser, K.T., Penn, D.L., Rosenheck, R.A., Addington, J., Brunette, M.F., Correll, C.U., Estroff, S.E., Marcy, P., Robinson, J., Meyer-Kalos, P.S., Gottlieb, J.D., Glynn, S.M., Lynde, D.W., Pipes, R., Kurian, B.T., Miller, A.L., Azrin, S.T., Goldstein, A.B., Severe, J.B., Lin, H., Sint, K.J., John, M., Heinssen, R.K., 2016. Comprehensive versus usual community care for first-episode psychosis: 2-year outcomes from the NIMH RAISE early treatment program. Am. J. Psychiatry 173, 362–372.

Kane, J.M., Schooler, N.R., Marcy, P., Correll, C.U., Brunette, M.F., Mueser, K.T., Rosenheck, R.A., Addington, J., Estroff, S.E., Robinson, J., Penn, D.L., Robinson, D.G., 2015. The RAISE early treatment program for first-episode psychosis: background, rationale, and study design. J. Clin. Psychiatry 76, 240–246.

Kay, S.R., Fiszbein, A., Opler, L.A., 1987. The positive and negative syndrome scale (PANSS) for schizophrenia. Schizophr. Bull. 13, 261–276.

Kim, K.R., Lee, S.Y., Kang, J.I., Kim, B.R., Choi, S.-H., Park, J.Y., Lee, E., An, S.K., Kwon, J.S., 2011. Clinical efficacy of individual cognitive therapy in reducing psychiatric symptoms in people at ultra-high risk for psychosis. Early Interv. Psychiatry 5, 174–178.

Landa, Y., Mueser, K.T., Wyka, K.E., Shreck, E., Jespersen, R., Jacobs, M.A., Griffin, K.W., Van Der Gaag, M., Reyna, V.F., Beck, A.T., Silbersweig, D.A., Walkup, J.T., 2016. Development of a group and family-based cognitive behavioural therapy program for youth at risk for psychosis. Early Interv. Psychiatry 10, 511–521.

Lee, T.Y., Hong, S.B., Shin, N.Y., Kwon, J.S., 2015. Social cognitive functioning in prodromal psychosis: a meta-analysis. Schizophr. Res. 164, 28–34.

Loewy, R., Fisher, M., Schlosser, D.A., Biagianti, B., Stuart, B., Mathalon, D.H., Vinogradov, S., 2016. Intensive auditory cognitive training improves verbal memory in adolescents and young adults at clinical high risk for psychosis. Schizophr. Bull. 42, S118–S126.

McAusland, L., Buchy, L., Cadenhead, K.S., Cannon, T.D., Cornblatt, B.A., Heinssen, R., Mcglashan, T.H., Perkins, D.O., Seidman, L.J., Tsuang, M.T., Walker, E.F., Woods, S.W., Bearden, C.E., Mathalon, D.H., Addington, J., 2015. Anxiety in youth at clinical high risk for psychosis. Early Interv. Psychiatry 11, 480–487.

McFarlane, W.R., 2002. Multifamily Groups in the Treatment of Severe Psychiatric Disorders. Guilford Press, New York.

McFarlane, W.R., Levin, B., Travis, L., Lucas, F., Lynch, S., Verdi, M., Williams, D., Adelsheim, S., Calkins, R., Carter, C.S., Cornblatt, B., Taylor, S.F., Auther, A.M., Mcfarland, B., Melton, R., Migliorati, M., Niendam, T., Ragland, J., Sale, T., Salvador, M., Spring, E., 2015. Clinical and functional outcomes after 2 years in the early detection and intervention for the prevention of psychosis multisite effectiveness trial. Schizophr. Bull. 41, 30–43.

McGlashan, T., Walsh, B., Woods, S., 2010. The Psychosis-Risk Syndrome: Handbook for Diagnosis and Follow-Up. Oxford University Press, New York.

McGlashan, T.H., Zipursky, R.B., Perkins, D., Addington, J., Miller, T., Woods, S.W., Hawkins, K.A., Hoffman, R.E., Preda, A., Epstein, I., Addington, D., Lindborg, S., Trzaskoma, Q., Tohen, M., Breier, A., 2006. Randomized, double-blind trial of olanzapine versus placebo in patients prodromally symptomatic for psychosis. Am. J. Psychiatry 163, 790–799.

McGorry, P.D., Nelson, B., Phillips, L.J., Yuen, H.P., Francey, S.M., Thampi, A., Berger, G.E., Amminger, G., Simmons, M.B., Kelly, D., Thompson, A.D., Yung, A.R., 2013. Randomized controlled trial of interventions for young people at ultra-high risk of psychosis: twelve-month outcome. J. Clin. Psychiatry 74, 349–356.

McGorry, P.D., Yung, A.R., Phillips, L.J., Yuen, H.P., Francey, S., Cosgrave, E.M., Germano, D., Bravin, J., Mcdonald, T., Blair, A., Adlard, S., Jackson, H., 2002. Randomized controlled trial of interventions designed to reduce the risk of progression to first-episode psychosis in a clinical sample with subthreshold symptoms. Arch. Gen. Psychiatr. 59, 921–928.

Miklowitz, D.J., O'brien, M.P., Schlosser, D.A., Addington, J., Candan, K.A., Marshall, C., Domingues, I., Walsh, B.C., Zinberg, J.L., De Silva, S.D., Friedman-Yakoobian, M., Cannon, T.D., 2014. Family-focused treatment for adolescents and young adults at high risk for psychosis: results of a randomized trial. J. Am. Acad. Child Adolesc. Psychiatry 53, 848–858.

Miklowitz, D.J., Schneck, C.D., Singh, M.K., Taylor, D.O., George, E.L., Cosgrove, V.E., Howe, M.E., Dickinson, L.M., Garber, J., Chang, K.D., 2013. Early intervention for symptomatic youth at risk for bipolar disorder: a randomized trial of family-focused therapy. J. Am. Acad. Child Adolesc. Psychiatry 52, 121–131.

Miklowitz, D.J., Scott, J.P., 2009. Psychosocial treatments for bipolar disorder: cost-effectiveness, mediating mechanisms, and future directions. Bipolar Disord. 11, 110–122.

Morrison, A.P., 2001. The interprpretation of intrusions in psychosis: an integrative cognitive approach to hallucinations and delusions. Behav. Cognit. Psychother. 29, 257–276.

Morrison, A.P., French, P., Stewart, S.L., Birchwood, M., Fowler, D., Gumley, A.I., Jones, P.B., Bentall, R.P., Lewis, S.W., Murray, G.K., Patterson, P., Brunet, K., Conroy, J., Parker, S., Reilly, T., Byrne, R., Davies, L.M., Dunn, G., 2012a. Early detection and intervention evaluation for people at risk of psychosis: multisite randomised controlled trial. BMJ 344, e2233.

Morrison, A.P., French, P., Stewart, S.L.K., Birchwood, M., Fowler, D., Gumley, A.I., Jones, P.B., Bentall, R.P., Lewis, S.W., Murray, G.K., Patterson, P., Brunet, K., Conroy, J., Parker, S., Reilly, T., Byrne, R., Davies, L.M., Dunn, G., 2012b. Early detection and intervention evaluation for people at risk of psychosis: multisite randomised controlled trial. BMJ 344, e2233.

Morrison, A.P., French, P., Walford, L., Lewis, S.W., Kilcommons, A., Green, J., Parker, S., Bentall, R.P., 2004. Cognitive therapy for the prevention of psychosis in people at ultra-high risk: randomised controlled trial. Br. J. Psychiatry 185, 291–297.

Morrison, A.P., Hutton, P., Wardle, M., Spencer, H., Barratt, S., Brabban, A., Callcott, P., Christodoulides, T., Dudley, R., French, P., Lumley, V., Tai, S.J., Turkington, D., 2012c. Cognitive therapy for people with a schizophrenia spectrum diagnosis not taking antipsychotic medication: an exploratory trial. Psychol. Med. 42, 1049–1056.

Morrison, A.P., Stewart, S.L., French, P., Bentall, R.P., Birchwood, M., Byrne, R., Davies, L.M., Fowler, D., Gumley, A.I., Jones, P.B., Lewis, S.W., Murray, G.K., Patterson, P., Dunn, G., 2011. Early detection and intervention evaluation for people at high-risk of psychosis-2 (EDIE-2): trial rationale, design and baseline characteristics. Early Interv. Psychiatry 5, 24–32.

Nemoto, T., Takeshi, K., Togi, N., Haneda, M., Tsujino, N., Ito, S., Katagiri, N., Morita, K., Kobayashi, H., Mizuno, M., 2010. Divergent thinking as a potential target for cognitive intervention in early psychosis. Schizophr. Res. 117, 329.

Nordentoft, M., Thorup, A., Petersen, L., Ohlenschlaeger, J., Melau, M., Christensen, T.O., Krarup, G., Jorgensen, P., Jeppesen, P., 2006. Transition rates from schizotypal disorder to psychotic disorder for first-contact patients included in the OPUS trial. A randomized clinical trial of integrated treatment and standard treatment. Schizophr. Res. 83, 29–40.

O'Brien, M.P., Gordon, J.L., Bearden, C.E., Lopez, S.R., Kopelowicz, A., Cannon, T.D., 2006. Positive family environment predicts improvement in symptoms and social functioning among adolescents at imminent risk for onset of psychosis. Schizophr. Res. 81, 269–275.

O'brien, M.P., Miklowitz, D.J., Candan, K.A., Marshall, C., Domingues, I., Walsh, B.C., Zinberg, J.L., De Silva, S.D., Woodberry, K.A., Cannon, T.D., 2014. A randomized trial of family focused therapy with populations at clinical high risk for psychosis: effects on interactional behavior. J. Consultting Clin. Psychol. 82, 90–101.

O'Brien, M.P., Zinberg, J.L., Bearden, C.E., Daley, M., Niendam, T.A., Kopelowicz, A., Cannon, T.D., 2007. Psychoeducational multi-family group treatment with adolescents at high risk for developing psychosis. Early Interv. Psychiatry 1, 325–332.

Onwumere, J., Bebbington, P., Kuipers, E., 2011. Family interventions in early psychosis: specificity and effectiveness. Epidemiol. Psychiatr. Sci. 20, 113–119.

Peyroux, E., Franck, N., 2014. RC2S: a cognitive remediation program to improve social cognition in schizophrenia and related disorders. Front. Hum. Neurosci. 8, 400.

Pharoah, F., Mari, J., Rathbone, J., Wong, W., 2010. Family intervention for schizophrenia. Cochrane Database Syst. Rev. Cd000088.

Piskulic, D., Barbato, M., Liu, L., Addington, J., 2015. Pilot study of cognitive remediation therapy on cognition in young people at clinical high risk of psychosis. Psychiatry Res. 225, 93–98.

Rauchensteiner, S., Kawohl, W., Ozgurdal, S., Littmann, E., Gudlowski, Y., Witthaus, H., Heinz, A., Juckel, G., 2011. Test-performance after cognitive training in persons at risk mental state of schizophrenia and patients with schizophrenia. Psychiatry Res. 185, 334–339.

Salkovskis, P.M., Forresterter, E., Richards, H.C., Morrison, N., 1998. The devil is in the detail: conceptionalising and treating obsessional problems. In: Tarrier, N., Wells, A. (Eds.), Treating Complex Cases: The Cognitive Behavioural Therapy Approach. John WIley & Sons, New York, NY.

Schlosser, D.A., Miklowitz, D.J., O'brien, M.P., De Silva, S.D., Zinberg, J.L., Cannon, T.D., 2012. A randomized trial of family focused treatment for adolescents and young adults at risk for psychosis: study rationale, design and methods. Early Interv. Psychiatry 6, 283–291.

Schlosser, D.A., Zinberg, J.L., Loewy, R.L., Casey-Cannon, S., O'brien, M.P., Bearden, C.E., Vinogradov, S., Cannon, T.D., 2010. Predicting the longitudinal effects of the family environment on prodromal symptoms and functioning in patients at-risk for psychosis. Schizophr. Res. 118, 69–75.

Schmidt, S.J., Mueller, D.R., Roder, V., 2011. Social cognition as a mediator variable between neurocognition and functional outcome in schizophrenia: empirical review and new results by structural equation modeling. Schizophr. Bull. 37, S41–S54.

Schultze-Lutter, F., Ruhrmann, S., Picker, H., Klosterkötter, J., 2006. Development and Evaluation of the Schizophrenia Proneness Instrument, Adult Version (SPI-A).

Seidman, L.J., Shapiro, D.I., Stone, W.S., Woodberry, K.A., Ronzio, A., Cornblatt, B.A., Addington, J., Bearden, C.E., Cadenhead, K.S., Cannon, T.D., Mathalon, D.H., Mcglashan, T.H., Perkins, D.O., Tsuang, M.T., Walker, E.F., Woods, S.W., 2016. Association of neurocognition with transition to psychosis: baseline functioning in the second phase of the North American Prodrome Longitudinal Study. JAMA Psychiatry 73, 1239–1248.

Seikkula, J., Olson, M.E., 2003. The open dialogue approach to acute psychosis: its poetics and micropolitics. Fam. Process 42, 403–418.

Stafford, M.R., Jackson, H., Mayo-Wilson, E., Morrison, A.P., Kendall, T., 2013. Early interventions to prevent psychosis: systematic review and meta-analysis. BMJ 346, f185.

Stain, H.J., Bucci, S., Baker, A.L., Carr, V., Emsley, R., Halpin, S., Lewin, T., Schall, U., Clarke, V., Crittenden, K., Startup, M., 2016. A randomised controlled trial of cognitive behaviour therapy versus non-directive reflective listening for young people at ultra high risk of developing psychosis: the detection and evaluation of psychological therapy (DEPTh) trial. Schizophr. Res. 176, 212–219.

Stain, H.J., Mawn, L., Common, S., Pilton, M., Thompson, A., 2019. Research and practice for ultra-high risk for psychosis: a national survey of early intervention in psychosis services in England. Early Interv. Psychiatry 13 (1), 47–52.

Tarrier, N., 2010. Cognitive behavior therapy for schizophrenia and psychosis: current status and future directions. Clin. Schizophrenia Relat. Psychoses 4, 176–184.

Thompson, A.D., Bartholomeusz, C., Yung, A.R., 2011. Social cognition deficits and the 'ultra high risk' for psychosis population: a review of literature. Early Interv. Psychiatry 5, 192–202.

Urben, S., Pihet, S., Jaugey, L., Halfon, O., Holzer, L., 2012. Computer-assisted cognitive remediation in adolescents with psychosis or at risk for psychosis: a 6-month follow-up. Acta Neuropsychiatr. 24, 328–335.

Van Der Gaag, M., Nieman, D., Wunderink, L., Klaassen, R., Rietdijk, J., Dragt, S., Ising, H., Linszen, D., 2012a. The results of a specific CBT intervention in young help-seeking patients with social decline and an ultra-high risk for developing a first-episode of psychosis. Early Interv. Psychiatry 6, 12.

van der Gaag, M., Nieman, D.H., Rietdijk, J., Dragt, S., Ising, H.K., Klaassen, R.M., Koeter, M., Cuijpers, P., Wunderink, L., Linszen, D.H., 2012b. Cognitive behavioral therapy for subjects at ultrahigh risk for developing psychosis: a randomized controlled clinical trial. Schizophr. Bull. 38, 1180–1188.

van Donkersgoed, R.J., Wunderink, L., Nieboer, R., Aleman, A., Pijnenborg, G.H., 2015. Social cognition in individuals at ultra-high risk for psychosis: a meta-analysis. PLoS One 10, e0141075.

Weersing, V.R., Jeffreys, M., Do, M.T., Schwartz, K.T., Bolano, C., 2017. Evidence base update of psychosocial treatments for child and adolescent depression. J. Clin. Child Adolesc. Psychol. 53 46, 11–43.

Wells, A., 1995. Meta-cognition and worry: a cognitive model of generalised anxiety disorder. Behav. Cognit. Psychother. 23, 301–320.

Wessels, H., Wagner, M., Frommann, I., Berning, J., Putzfeld, V., Janssen, B., Bottlender, R., Maurer, K., Moller, H.-J., Gaebel, W., Hafner, H., Maier, W., Klosterkotter, J., Bechdolf, A., 2015. Neuropsychological functioning as a predictor of treatment response to psychoeducational, cognitive behavioral therapy in people at clinical high risk of first episode psychosis. Psychiatr. Prax. 42, 313–319.

Woodberry, K.A., Seidman, L.J., Bryant, C., Addington, J., Bearden, C.E., K, S.C., Cannon, T.D., Cornblatt, B.A., Mcglashan, T.H., Mathalon, D.H., Perkins, D.O., Tsuang, M.T., Walker, E.F., Woods, S.W., 2016. Treatment precedes positive symptoms in North American adolescent and young adult clinical high risk cohort. J. Clin. Child Adolesc. Psychol. 1–10.

Wykes, T., Huddy, V., Cellard, C., Mcgurk, S.R., Czobor, P., 2011. A meta-analysis of cognitive remediation for schizophrenia: methodology and effect sizes. Am. J. Psychiatry 168, 472–485.

Yung, A., Phillips, L., Simmons, M., Ward, J., Thompson, K., French, P., 2006. Comprehensive Assessment of At-Risk Mental States (CAARMS). University of Melbourne, Melbourne.

CHAPTER

19

Pharmacological intervention for people at risk of psychotic disorder

E. Burkhardt[1,2], *K. Leopold*[2,3], *A. Bechdolf*[1,2,3,4]

[1]Orygen, The National Centre of Excellence in Youth Mental Health, University of Melbourne, Melbourne, VIC, Australia; [2]Vivantes Klinikum Am Urban, Department of Psychiatry, Psychotherapy and Psychosomatic Medicine with Early intervention and recognition centre – FRITZ am Urban, Teaching hospital of Charité Medizin, Berlin, Germany; [3]Vivantes Klinikum im Friedrichshain, Department of Psychiatry, Psychotherapy and Psychosomatic Medicine, Teaching hospital of Charité Medizin, Berlin; [4]Universitätsklinikum Köln, Department of Psychiatry and Psychotherapy, Cologne, NRW, Germany

General considerations

According to international guidelines, treatment of psychotic disorders should include both pharmacological and psychotherapeutic interventions complemented by family support and vocational services (Galletly et al., 2016; NICE, 2014). For individuals with a high-risk status for psychosis, however, the situation is different. A stated ultra high risk (UHR) is not equivalent to a prodromal syndrome inevitably progressing to a full-blown psychotic episode. The likelihood of progression from UHR status to psychosis is as high as 22% within a year and up to 36% within 3 years (Fusar-Poli, 2012). However, individuals at UHR may also remain healthy or develop another psychiatric disorder (Bechdolf et al., 2015; Fusar-Poli et al., 2013). Against this background it is a difficult ethical question whether pharmacologic intervention with the aim of preventing the onset of psychosis is justifiable at all. As in other medical fields, a preventive treatment seems only conceivable if the intervention has proven good efficacy as well as low risk of side effects and alternative approaches with a better cost-effectiveness ratio are missing. On the other hand, the vast majority of individuals with UHR are distressed by their symptoms, already show a significant decline in psychosocial functioning, and experience poor quality of life (Fusar-Poli et al., 2015b). Therefore, symptom reduction and improvement of functioning and quality of life are possible targets of early intervention in a high-risk population too.

Studies have compared pharmacological intervention in UHR populations, especially the use of antipsychotic drugs, to psychotherapeutic treatment or the combination of both. A recent meta-analysis concluded that pharmacologic

Risk Factors for Psychosis
https://doi.org/10.1016/B978-0-12-813201-2.00019-3

intervention can significantly reduce the risk of developing psychosis at 6–48-month follow-up compared to control conditions (Schmidt et al., 2015). However, this effect was not superior to the results of psychotherapeutic approaches, especially cognitive behavioral therapy (CBT). Neither pharmacotherapy nor psychotherapy could show positive effects on the level of functioning when compared to control conditions (Schmidt et al., 2015).

Considering the lower risk of side effects and better acceptance among patients, international guidelines recommend favoring psychotherapeutic interventions over pharmacotherapy (Galletly et al., 2016; NICE, 2014; Schmidt et al., 2015). The recommendations for the use of antipsychotic medication in the treatment of people at risk differ considerably between the guidelines (see overview in Table 19.1). The British National Institute for Health and Care Excellence (NICE) guidelines for psychosis and schizophrenia in adults state explicitly that antipsychotic medication should not be offered to people at increased risk of developing psychosis or with the objective to prevent psychosis (NICE, 2014). The European Psychiatric Association (EPA) guidance on the early intervention in clinical high-risk states of psychoses suggests considering pharmacotherapy only if CBT is ineffective and severe symptoms or functional decline are present. It acknowledges that second-generation antipsychotics are able to improve the effectiveness of CBT and allows their use in a short-term treatment with the aim of symptom reduction in adult patients (Schmidt et al., 2015). The clinical practice guidelines for the management of schizophrenia and related disorders published by the Royal Australian and New Zealand College of Psychiatrists (Galletly et al., 2016) and the Australian Clinical Guidelines for Early Psychosis (Early Psychosis Guidelines Writing Group and EPPIC National Support Program, 2016) do not recommend the use of antipsychotics in young people at risk either. Exceptions are considered in case of self-harm or aggression associated with attenuated or brief positive symptoms or if CBT and other psychosocial treatments cannot reduce the distress and impairment caused by the symptoms sufficiently (Early Psychosis Guidelines Writing Group and EPPIC National Support Program, 2016; Galletly et al., 2016).

If individuals with a UHR status fulfill the criteria for other mental disorders such as substance use, anxiety, or affective disorders, these should be treated according to the respective guidelines. However, the majority of guidelines do not recommend the regular use of other

TABLE 19.1 International guidelines on treatment of individuals with ultra high risk status.

	Recommendation of antipsychotic medication for children and adolescents	Recommendation of antipsychotic medication for adults
NICE (NICE, 2014)	0[a]	–
EPA (Schmidt et al., 2015)	–	(+)[b]
RANZCP (Galletly et al., 2016)	(+)[c,d]	(+)[d]

– Not recommended.
+ Recommended.
[a] *No statement, guideline only targets adult patients.*
[b] *If CBT is ineffective and severe symptoms ± functional declines are present.*
[c] *Excluding childhood onset schizophrenia.*
[d] *If CBT and other psychosocial treatments are ineffective, ± acute risks of harm to self or others associated with positive symptoms are present.*

medication such as antidepressants or food supplements for the prevention of psychosis or treatment of subthreshold psychotic symptoms (Galletly et al., 2016; NICE, 2014; Schmidt et al., 2015). Only the Australian Clinical Guidelines for Early Psychosis mention that omega-3 fatty acids may delay or prevent the onset of psychosis (Early Psychosis Guidelines Writing Group and EPPIC National Support Program, 2016).

Use of specific antipsychotic agents in young people at risk

Several second-generation antipsychotic drugs have been tested in UHR populations (Table 19.2). However, the comparability of results is limited by substantial differences in study designs. Trials vary considerably in the number of participants, existence of control conditions, combination with other interventions (e.g., CBT), duration of treatment, follow-up period, outcome criteria, and assessment of side effects.

Amisulpride has been examined in a randomized controlled trial (RCT) targeting the treatment of people at risk with a relatively large sample size (n = 124) (Ruhrmann et al., 2007). Treatment and control groups both received a needs-focused intervention including psychoeducation, crisis intervention, family counseling, and support for difficulties with education or work. The control group did not receive placebo medication. Side effects were assessed in detail with questionnaires, e.g., the UKU (Udvalg for Kliniske Undersøgelser Side Effects Rating Scale for Clinicians) (Lingjærde et al., 1987), and clinical results (e.g., blood tests for prolactin levels). Despite the relatively low mean dosage of 170 mg per day, a number of side effects occurred (see Table 19.2). Prolactin-associated amenorrhea, galactorrhea, and sexual dysfunction led to drop out in three cases. On the other hand, treatment with amisulpride was associated with significant improvement of positive and negative symptoms and level of functioning. Transition rates to psychosis were not reported (Ruhrmann et al., 2007).

Aripiprazole has a relatively favorable side effect profile with a low risk of weight gain, prolactin increase, and sedation (Leucht et al., 2013) and is therefore frequently used in patients with first-episode psychosis (Morrison et al., 2018; Whale et al., 2016). An open-label trial of aripiprazole (mean dose 15 mg per day) in a small group of UHR individuals (n = 15) indicated a good efficacy in the reduction of positive and negative symptoms, an improvement of functioning, and a low level of side effects, except for a high rate (53%) of akathisia. Transition rates were not reported (Woods et al., 2007). The multicenter, randomized, and placebo-controlled trial PREVENT aimed to compare aripiprazole with placebo and with a CBT intervention in a larger sample (n = 156 at the time of publication of baseline characteristics) (Bechdolf et al., 2011), but results have not been published yet.

Olanzapine is an effective antipsychotic agent but is associated with a particular high risk of weight gain and metabolic syndrome that it is not even recommended as a first-line medication for patients with first-episode psychosis (Galletly et al., 2016). A randomized, placebo-controlled trial in a UHR population (n = 60) showed that olanzapine treatment (mean dose 9 mg per day) led to significant improvement of positive symptoms but was also associated with a substantial mean weight gain of 8.8 kg within 12 months. The study included 1 year of treatment with olanzapine or placebo and another year of observation without medication to examine whether the drug could prevent transition to psychosis or simply delay it. The transition rates during the year of intervention were 16% in the olanzapine group and 38% in the placebo group, but the difference ($p = 0.09$) did not meet the stated significance level ($p = 0.05$). During the observation period in the second year, 33% of the former olanzapine group developed

TABLE 19.2 Studies on pharmacological treatment of young people at risk of psychosis.

	Publication	Type/control conditions	Country	UHR definition	Participants (IG/CG)	Dosage range (mean at end point)	Duration of intervention	Benefit of medication compared to CG	Side effects (selected)
Amisulpride	Ruhrmann et al. (2007)	RCT NBI+AMI versus NBI	GER	APS/BLIPS (ERIraos)	(65/59)	50–800 mg (mean 170 mg)	12 weeks	↓Decreased positive symptoms ↓Decreased negative symptoms ↑Increased GAF	↑Increased prolactin level ↑Increased sexual dysfunction ↑Increased menstrual disorders (n.s.) ↑Increased galactorrhea (n.s.) ↑Increased weight (BMI +2.6%) ↑Increased sleep duration ↑Increased akathisia (7%)
Aripiprazole	Woods et al. (2007)	Open-label (no CG)	USA	APS/BLIPS/ GRD (SIPS/ SOPS)	(15/–)	5–30 mg (mean 15 mg)	8 weeks	↓Decreased positive symptoms ↓Decreased negative symptoms ↑Increased GAF	↑Increased akathisia (53%) ↑Increased irritability ↑Increased sedation ↑Increased weight (+1.2 kg)
Olanzapine	McGlashan et al. (2006)	RCT OLA versus PLA	USA	APS/BLIPS/ GRD (SIPS/ SOPS)	(31/29)	5–15 mg (mean 9 mg)	52 weeks	↓Decreased positive symptoms ↓Decreased transition (n.s.)	↑Increased fatigue ↑Increased weight (+8.8 kg)
Perospirone	Tsujino et al. (2013)	Open-label (no CG)	JAP	APS/BLIPS/ GRD (SIPS/ SOPS)	(11/–)	Flexible schedule (mean 10 mg)	26 weeks	↓Decreased positive symptoms	Not displayed

Risperidone	McGorry et al. (2002)	RCT NBI+CBT+RIS versus NBI	AUS	APS/BLIPS/ GRD (CAARMS)	(31/28)	1–2 mg (mean 1.3 mg)	26 weeks	↓Decreased transition	↑Increased minor sedation (↑)(increased) rigidity
	McGorry et al. (2013)	RCT CBT+RIS versus CBT+PLA versus ST+PLA	AUS	APS/BLIPS/ GRD (CAARMS)	(43/44/28)	0.5–2 mg (mean not reported)	52 weeks	↓Decreased transition (n.s.)	↑Increased fatigue (n.s.) ↑Increased depression (n.s.) ↑Increased concentration deficits (n.s.) ↑Increased orthostatic dizziness (n.s.)

AMI, amisulpride; *APS*, attenuated positive symptoms; *AUS*, Australia; *BLIPS*, brief limited intermittent symptoms; *BMI*, body mass index; *CAARMS*, Comprehensive Assessment of At-Risk Mental States; *CBT*, cognitive behavioral therapy; *CG*, control group; *ERIraos*, Early Recognition Inventory; *GAF*, Global Assessment of Functioning; *GER*, Germany; *GRD*, genetic risk and functional deterioration; *IG*, intervention group; *JAP*, Japan; *NBI*, needs-based intervention; *n.s.*, not significant; *OLA*, olanzapine; *PLA*, placebo; *RCT*, randomized controlled trial; *RIS*, risperidone; *SIPS*, Structured Interview for Prodromal Syndromes; *SOPS*, Scale Of Prodromal Symptoms; *ST*, supportive therapy; *UHR*, ultra high risk.

psychosis and 25% of the remaining patients in the control group converted. Perhaps due to the small number of participants remaining in the study for the second year (n = 17), differences did not prove significant (McGlashan et al., 2006; Woods et al., 2003).

Perospirone, a combined serotonin (5-HT) and dopamine antagonist and 5-HT1A receptor agonist, has been evaluated in a small group of high-risk patients (n = 11) in an open-label design without a control condition (Tsujino et al., 2013). Tolerability of the medication was examined using the scale of subjective well-being under neuroleptics, which addresses the overall physical and mental condition without distinguishing between effects of psychosis and medication (Naber et al., 2001). The only side effect assessed specifically was akathisia measured by the Barnes Akathisia Scale (Barnes, 1989). However, no statistical analysis is displayed for the changes of mean scores on the Barnes Akathisia Scale. The authors report a good tolerability and efficacy of perospirone (mean dose 10 mg per day) in the reduction of positive symptoms. Of the eight participants completing follow-up assessment after 26 weeks, none converted to psychosis (Tsujino et al., 2013).

Risperidone was tested in two randomized controlled Australian studies (McGorry et al., 2013, 2002). The first (n = 59) compared a combined intervention consisting of a needs-based therapy (counseling and support for social, vocational, and family issues, plus case management and family education and support), specific CBT, and low-dose risperidone (over 6 months) with needs-based therapy alone. The main outcome criterion was the transition rate to psychosis, which was significantly reduced in the intervention group during the 6 months of combined therapy (intervention group 10%, control group 36%). Toward the end of the observation period after 12 months, the transition rates of both groups converged (intervention group 19%, control group 36%). Symptomatic and functional measures showed improvement of both groups from baseline to follow-up but were not significantly different between intervention and control group. The authors concluded that low-dose risperidone (mean dose 1.3 mg per day) has a good tolerability (only four participants showed minor sedation or rigidity) and can, in combination with psychotherapy, delay the onset of psychosis (McGorry et al., 2002). A follow-up assessment 3–4 years after the 6-month intervention revealed no significant medium-term effect of the combined therapy on clinical symptoms or transition rates (Phillips et al., 2007).

The second study (n = 115) was designed with three arms, risperidone (doses between 0.5 and 2 mg per day, mean not reported) plus CBT, placebo plus CBT, and placebo plus supportive therapy. The intervention was delivered for 12 months followed by another 12 months of observation. Again, participants of all groups experienced comparable symptomatic and functional improvement as well as a similarly low level of side effects (fatigue, depression, concentration deficits, and orthostatic dizziness) during the 12 months of the intervention. Transition rates did not differ significantly (risperidone + CBT: 10.7%, placebo + CBT: 9.6%, placebo + supportive therapy: 21.8%). The authors conclude that treatment with risperidone does not substantially enhance the effect of psychotherapy (McGorry et al., 2013; Phillips et al., 2009; Yung et al., 2011).

In conclusion, RCTs showed evidence for a positive effect of amisulpride and olanzapine on clinical symptoms and psychosocial functioning in young people at risk for psychosis. However, both agents were associated with considerable side effects. Aripiprazole and perospirone showed good tolerability and promising effectiveness in small open-label studies but need to prove superiority over placebo or psychotherapeutic interventions. Risperidone appeared to reduce transition rates to psychosis in the short term in one study but the effect was not maintained in the medium to long term. When combined with CBT, risperidone could not add to the effect of CBT alone on symptoms, level of functioning, and transition to psychosis in RCTs. One study indicated that olanzapine treatment might delay the onset of psychosis. At present, no antipsychotic drug has been shown to prevent psychosis.

Other pharmacological interventions

From a clinician's perspective, the use of antipsychotics in the medical treatment of at-risk states is the most obvious choice, but a small number of studies also examined the effect of other psychotropic drugs, such as antidepressants and lithium, and of nutritional supplements.

Antidepressants have been found to be frequently prescribed to high-risk individuals in naturalistic studies. In the North American Prodromal Longitudinal Study, nearly half of the 191 participants were treated with selective serotonin reuptake inhibitors or other antidepressants (including bupropion, venlafaxine, trazodone, mirtazapine, and others), but no proof of a positive effect on symptom severity could be found (Walker et al., 2009). Another North American naturalistic study in adolescents (n = 48) observed that the adherence was significantly higher with antidepressant medication compared to antipsychotics and the transition rate in the antidepressant group was lower than in the antipsychotics group (in which 11 of 12 converters were non adherent to medication) (Cornblatt et al., 2007). A 6-year follow-up study (n = 258) in the early intervention setting of the Outreach and Support in South London (OASIS) revealed a lower transition rate in a group treated with antidepressants and CBT compared to the group receiving antipsychotic medication and CBT. The effect remained significant when corrections for baseline differences in symptom severity were made. However, transition rates of the antidepressant and CBT group were not significantly lower than those of participants receiving CBT alone (Fusar-Poli et al., 2015a). Unfortunately, the generalizability of all mentioned study results remains limited by the naturalistic design without randomization. An Australian RCT (STEP) currently investigates the effectiveness of fluoxetine as part of a staged intervention in people at UHR of psychosis, but results are not available yet (Nelson et al., 2018).

Lithium has been tested as a neuroprotective agent in UHR patients in an Australian MRI study. Eleven patients were treated with 450 mg per day of lithium carbonate while 10 individuals received treatment as usual. MRI scans showed a significant reduction of the hippocampal T2 relaxation time (a proxy for emerging brain pathology) in the lithium group. Clinical measures of symptoms did not improve significantly under lithium treatment. Transition rates were not reported (Berger et al., 2012).

Long-chain omega-3 polyunsaturated fatty acids (PUFAs) extracted from fish oil are a dietary supplementation that have been reported to have beneficial effects on people with schizophrenia due to possible neuroprotective properties (Amminger et al., 2010). An RCT in an Austrian UHR population (n = 81) revealed a significantly lower transition rate of the group receiving PUFAs (dose: 1.2g omega-3 PUFAs) for 12 weeks (5%) compared to the placebo group (28%) during 12 months of follow-up. PUFA treatment was also associated with a decline of positive, negative and general symptoms, and improvement of psychosocial functioning, without causing more side effects than placebo (Amminger et al., 2010). In a long-term follow-up after approximately 7 years, the group differences in transition rates (9.8% in the PUFA group vs. 40% in the placebo group), symptom severity, and psychosocial functioning remained significant (Amminger et al., 2015). Unfortunately, those promising results could not be replicated in a multicenter trial (n = 304) comparing one group undergoing PUFA treatment (dose: 1.4 g omega-3 PUFAs) for 26 weeks plus manualized cognitive behavioral case management (CBCM) to a group receiving placebo treatment plus CBCM. PUFA and placebo groups showed nearly identical transition rates (11.5% vs. 11.2% at 12 months) and similar improvement in symptoms and functioning, indicating that PUFA did not provide additional benefit to psychosocial interventions (McGorry et al., 2017). Several other international multicenter studies examining the use of PUFAs in UHR populations are currently being undertaken (Kahn, 2015; Qurashi et al., 2017).

NMDA receptor coagonists glycine and D-serine have been considered for the treatment of high-risk individuals based on the glutamate hypothesis of schizophrenia. If NMDA receptor hypoactivity was already present in prodromal stages, NMDAR coagonists could help to increase receptor activity and prevent the onset of psychosis (Kantrowitz et al., 2015; Woods et al., 2013). Based on this assumption, *glycine* has been tested in UHR individuals in two pilot studies. Descriptive analyses showed medium to large effect sizes of glycine (weight-adjusted dosage: 0.8 g/kg, up to 80 g per day, mean doses not reported) on positive and negative symptoms. Hypothesis testing was not performed due to the small sample sizes (10 participants in the first study, 8 in the second) (Woods et al., 2013). The use of *D-serine* in a UHR population was tested in a US-American multicenter RCT (n = 44) primarily examining the effect on negative symptoms. The group receiving D-serine (mean dose 4.2 g per day) for 16 weeks showed a significantly lower level of negative symptoms at follow-up than the placebo group. The small sample size did not allow comparisons of transition rates between the intervention groups (Kantrowitz et al., 2015).

In conclusion, of the other pharmacological interventions, only omega-3 PUFAs have been tested in RCTs with adequate sample sizes, but results of the two trials were contradictory. D-Serine showed a significant effect on negative symptoms in one RCT, but needs to be tested in a larger sample. Results of studies examining the effects of antidepressants, lithium, and other neuroprotective agents have been inconclusive because of study design or small sample sizes. Further research is necessary to identify alternative treatment options other than antipsychotics for young people at risk of developing psychosis.

Clinical recommendations

As outlined above, studies have failed to prove additional benefit of antipsychotic drugs and other pharmacological interventions on clinical symptoms and level of functioning when compared with or added to psychosocial interventions in people at risk for psychosis. In addition, there is no evidence for any drug in preventing the onset of psychosis. Therefore, the application of pharmacotherapy in individuals at risk, especially with the aim of preventing psychosis, remains questionable and is not recommended by clinical guidelines (Galletly et al., 2016; NICE, 2014; Schmidt et al., 2015).

Nevertheless, naturalistic data from the Netherlands show that even at the time of referral to a specialized center for early psychosis, 21% of the young people fulfilling UHR criteria have already been prescribed antipsychotic medication by previous services (Nieman et al., 2009). If the prescription of an antipsychotic drug is considered by the clinician, due to ineffectiveness of other therapeutic approaches, severe impairment caused by the symptoms, or if patients actively ask for it, advantages, and disadvantages need to be discussed openly according to the principles of shared decision-making. The patient needs to be broadly informed about the characteristics of the at-risk status, the possible course of the symptoms, and available treatment options. The prescription should be explicitly identified as an off-label use toward the patient. Based on the available study outcomes, no specific antipsychotic agent can be recommended over another, so choices should be made considering the side effect profile and patient preferences. The management of the pharmacological therapy, including physical assessments before the commencement of medication and during treatment, should be conducted in accordance with the recommendations for patients with first-episode psychosis. These recommendations include the preference for second-generation antipsychotics, usage of a low starting dose, and close monitoring of side effects (Galletly et al., 2016; NICE, 2014; Schmidt et al., 2015).

List of acronyms and abbreviations

CBT Cognitive behavioral
EPA European Psychiatric Association
NICE British National Institute for Health and Care Excellence
NMDAR N-methyl-D-aspartate-type glutamate receptor
PUFAs Polyunsaturated fatty acids
RCT Randomized controlled trial
UHR Ultra high risk

References

Amminger, G.P., Schäfer, M.R., Papageorgiou, K., Klier, C.M., Cotton, S.M., Harrigan, S.M., et al., 2010. Long-chain ω-3 fatty acids for indicated prevention of psychotic disorders: a randomized, placebo-controlled trial. Arch. Gen. Psychiatr. 67 (2), 146. https://doi.org/10.1001/archgenpsychiatry.2009.192.

Amminger, G.P., Schäfer, M.R., Schlögelhofer, M., Klier, C.M., McGorry, P.D., 2015. Longer-term outcome in the prevention of psychotic disorders by the Vienna omega-3 study. Nat. Commun. 6 https://doi.org/10.1038/ncomms8934.

Barnes, T.R., 1989. A rating scale for drug-induced akathisia. Br. J. Psychiatry J. Ment. Sci. 154, 672–676.

Bechdolf, A., Laier, S., Kallenbach, M., Biester, A., Leopold, K., 2015. Sollen bei Prodromalstadien der Schizophrenie Antipsychotika zur Psychoseprävention eingesetzt werden? Kontra. Der Nervenarzt 86 (11), 1424–1426. https://doi.org/10.1007/s00115-015-4428-0.

Bechdolf, A., Müller, H., Stützer, H., Wagner, M., Maier, W., Lautenschlager, M., et al., 2011. Rationale and baseline characteristics of PREVENT: a second-generation intervention trial in subjects at-risk (prodromal) of developing first-episode psychosis evaluating cognitive behavior therapy, aripiprazole, and placebo for the prevention of psychosis. Schizophr. Bull. 37 (Suppl. 2), S111–S121. https://doi.org/10.1093/schbul/sbr083.

Berger, G.E., Wood, S.J., Ross, M., Hamer, C.A., Wellard, R.M., Pell, G., et al., 2012. Neuroprotective effects of low-dose lithium in individuals at ultra-high risk for psychosis. A longitudinal MRI/MRS study. Curr. Pharmaceut. Des. 18 (4), 570–575.

Cornblatt, B.A., Lencz, T., Smith, C.W., Olsen, R., Auther, A.M., Nakayama, E., et al., 2007. Can antidepressants be used to treat the schizophrenia prodrome? Results of a prospective, naturalistic treatment study of adolescents. J. Clin. Psychiatry 68 (4), 546–557.

Early Psychosis Guidelines Writing Group and EPPIC National Support Program, 2016. Australian Clinical Guidelines for Early Psychosis, second ed. update. Orygen, The National Centre of Excellence in Youth Mental Health, Melbourne.

Fusar-Poli, P., Frascarelli, M., Valmaggia, L., Byrne, M., Stahl, D., Rocchetti, M., et al., 2015a. Antidepressant, antipsychotic and psychological interventions in subjects at high clinical risk for psychosis: OASIS 6-year naturalistic study. Psychol. Med. 45 (6), 1327–1339. https://doi.org/10.1017/S003329171400244X.

Fusar-Poli, P., 2012. Predicting psychosis: meta-analysis of transition outcomes in individuals at high clinical risk. Arch. Gen. Psychiatr. 69 (3), 220. https://doi.org/10.1001/archgenpsychiatry.2011.1472.

Fusar-Poli, P., Bechdolf, A., Taylor, M.J., Bonoldi, I., Carpenter, W.T., Yung, A.R., McGuire, P., 2013. At risk for schizophrenic or affective psychoses? A meta-analysis of DSM/ICD diagnostic outcomes in individuals at high clinical risk. Schizophr. Bull. 39 (4), 923–932.

Fusar-Poli, P., Rocchetti, M., Sardella, A., Avila, A., Brandizzi, M., Caverzasi, E., et al., 2015b. Disorder, not just state of risk: meta-analysis of functioning and quality of life in people at high risk of psychosis. Br. J. Psychiatry 207 (3), 198–206. https://doi.org/10.1192/bjp.bp.114.157115.

Galletly, C., Castle, D., Dark, F., Humberstone, V., Jablensky, A., Killackey, E., et al., 2016. Royal Australian and New Zealand College of Psychiatrists clinical practice guidelines for the management of schizophrenia and related disorders. Aust. N. Z. J. Psychiatr. 50 (5), 410–472. https://doi.org/10.1177/0004867416641195.

Kahn, R., 2015. Placebo-Controlled Trial in Subjects at Ultra-High Risk for Psychosis With Omega-3 Fatty Acids in Europe (PURPOSE). Retrieved May 12, 2018, from: https://clinicaltrials.gov/ct2/show/NCT02597439.

Kantrowitz, J.T., Woods, S.W., Petkova, E., Cornblatt, B., Corcoran, C.M., Chen, H., et al., 2015. D-serine for the treatment of negative symptoms in individuals at clinical high risk of schizophrenia: a pilot, double-blind, placebo-controlled, randomised parallel group mechanistic proof-of-concept trial. Lancet Psychiatry 2 (5), 403–412. https://doi.org/10.1016/S2215-0366(15)00098-X.

Leucht, S., Cipriani, A., Spineli, L., Mavridis, D., Orey, D., Richter, F., et al., 2013. Comparative efficacy and tolerability of 15 antipsychotic drugs in schizophrenia: a multiple-treatments meta-analysis. Lancet 382 (9896), 951–962. https://doi.org/10.1016/S0140-6736(13)60733-3.

Lingjærde, O., Ahlfors, U.G., Bech, P., Dencker, S.J., Elgen, K., 1987. The UKU side effect rating scale: a new comprehensive rating scale for psychotropic drugs and a cross-sectional study of side effects in neuroleptic-treated patients. Acta Psychiatr. Scand. 76 (s334), 1–100. https://doi.org/10.1111/j.1600-0447.1987.tb10566.x.

McGlashan, T.H., Zipursky, R.B., Perkins, D., Addington, J., Miller, T., Woods, S.W., et al., 2006. Randomized, double-blind trial of olanzapine versus placebo in patients prodromally symptomatic for psychosis. Am. J. Psychiatry 163 (5), 790–799. https://doi.org/10.1176/ajp.2006.163.5.790.

McGorry, P.D., Nelson, B., Markulev, C., Yuen, H.P., Schäfer, M.R., Mossaheb, N., et al., 2017. Effect of ω-3 polyunsaturated fatty acids in young people at ultrahigh risk for psychotic disorders: the NEURAPRO randomized clinical trial. JAMA Psychiatry 74 (1), 19–27. https://doi.org/10.1001/jamapsychiatry.2016.2902.

McGorry, P.D., Nelson, B., Phillips, L.J., Yuen, H.P., Francey, S.M., Thampi, A., et al., 2013. Randomized controlled trial of interventions for young people at ultra-high risk of psychosis: twelve-month outcome. J. Clin. Psychiatry 74 (4), 349–356. https://doi.org/10.4088/JCP.12m07785.

McGorry, P.D., Yung, A.R., Phillips, L.J., Yuen, H.P., Francey, S., Cosgrave, E.M., et al., 2002. Randomized controlled trial of interventions designed to reduce the risk of progression to first-episode psychosis in a clinical sample with subthreshold symptoms. Arch. Gen. Psychiatr. 59 (10), 921–928.

Morrison, A.P., Law, H., Carter, L., Sellers, R., Emsley, R., Pyle, M., et al., 2018. Antipsychotic drugs versus cognitive behavioural therapy versus a combination of both in people with psychosis: a randomised controlled pilot and feasibility study. Lancet Psychiatry 5 (5), 411–423. https://doi.org/10.1016/S2215-0366(18)30096-8.

Naber, D., Moritz, S., Lambert, M., Rajonk, F., Holzbach, R., Mass, R., et al., 2001. Improvement of schizophrenic patients' subjective well-being under atypical antipsychotic drugs. Schizophr. Res. 50 (1–2), 79–88. https://doi.org/10.1016/S0920-9964(00)00166-3.

Nelson, B., Amminger, G.P., Yuen, H.P., Wallis, N., Kerr, M.J., Dixon, L., et al., 2018. Staged treatment in early psychosis: a sequential multiple assignment randomised trial of interventions for ultra high risk of psychosis patients. Early Interv. Psychiatry 12 (3), 292–306. https://doi.org/10.1111/eip.12459.

NICE, 2014. Psychosis and schizophrenia in adults: prevention and management | 1-recommendations | Guidance and guidelines. NICE. Retrieved from: https://www.nice.org.uk/guidance/cg178/chapter/1-recommendations?unlid=214673181201631735658.

Nieman, D.H., Rike, W.H., Becker, H.E., Dingemans, P.M., van Amelsvoort, T.A., de Haan, L., et al., 2009. Prescription of antipsychotic medication to patients at ultra high risk of developing psychosis. Int. Clin. Psychopharmacol. 24 (4), 223–228. https://doi.org/10.1097/YIC.0b013e32832e0a76.

Phillips, L.J., McGorry, P.D., Yuen, H., Ward, J., Donovan, K., Kelly, D., et al., 2007. Medium term follow-up of a randomized controlled trial of interventions for young people at ultra high risk of psychosis. Schizophr. Res. 96 (1–3), 25–33. https://doi.org/10.1016/j.schres.2007.05.018.

Phillips, L.J., Nelson, B., Yuen, H.P., Francey, S.M., Simmons, M., Stanford, C., et al., 2009. Randomized controlled trial of interventions for young people at ultra-high risk of psychosis: study design and baseline characteristics. Aust. N. Z. J. Psychiatr. 43 (9), 818–829. https://doi.org/10.1080/00048670903107625.

Qurashi, I., Chaudhry, I.B., Khoso, A.B., Farooque, S., Lane, S., Husain, M.O., et al., 2017. A randomised, double-blind, placebo-controlled trial of minocycline and/or omega-3 fatty acids added to treatment as usual for at-risk mental states (NAYAB): study protocol. Trials 18 (1), 524. https://doi.org/10.1186/s13063-017-2275-y.

Ruhrmann, S., Bechdolf, A., Kühn, K.-U., Wagner, M., Schultze-Lutter, F., Janssen, B., et al., 2007. Acute effects of treatment for prodromal symptoms for people putatively in a late initial prodromal state of psychosis. Br. J. Psychiatry 191 (51), s88–s95. https://doi.org/10.1192/bjp.191.51.s88.

Schmidt, S.J., Schultze-Lutter, F., Schimmelmann, B.G., Maric, N.P., Salokangas, R.K.R., Riecher-Rössler, A., et al., 2015. EPA guidance on the early intervention in clinical high risk states of psychoses. Eur. Psychiatry 30 (3), 388–404.

Tsujino, N., Nemoto, T., Morita, K., Katagiri, N., Ito, S., Mizuno, M., 2013. Long-term efficacy and tolerability of perospirone for young help-seeking people at clinical high risk: a preliminary open trial. Clin. Psychopharmacol. Neurosci. 11 (3), 132–136. https://doi.org/10.9758/cpn.2013.11.3.132.

Walker, E.F., Cornblatt, B.A., Addington, J., Cadenhead, K.S., Cannon, T.D., McGlashan, T.H., et al., 2009. The relation of antipsychotic and antidepressant medication with baseline symptoms and symptom progression: a naturalistic study of the North American Prodrome Longitudinal Sample. Schizophr. Res. 115 (1), 50–57. https://doi.org/10.1016/j.schres.2009.07.023.

Whale, R., Harris, M., Kavanagh, G., Wickramasinghe, V., Jones, C.I., Marwaha, S., et al., 2016. Effectiveness of antipsychotics used in first-episode psychosis: a naturalistic cohort study. BJPsych Open 2 (5), 323–329. https://doi.org/10.1192/bjpo.bp.116.002766.

Woods, S.W., Breier, A., Zipursky, R.B., Perkins, D.O., Addington, J., Miller, T.J., et al., 2003. Randomized trial of olanzapine versus placebo in the symptomatic acute treatment of the schizophrenic prodrome. Biol. Psychiatry 54 (4), 453–464. https://doi.org/10.1016/S0006-3223(03)00321-4.

Woods, S.W., Tully, E.M., Walsh, B.C., Hawkins, K.A., Callahan, J.L., Cohen, S.J., et al., 2007. Aripiprazole in the treatment of the psychosis prodrome. Br. J. Psychiatry 191 (51), s96–s101. https://doi.org/10.1192/bjp.191.51.s96.

Woods, S.W., Walsh, B.C., Hawkins, K.A., Miller, T.J., Saksa, J.R., D'Souza, D.C., et al., 2013. Glycine treatment of the risk syndrome for psychosis: report of two pilot studies. Eur. Neuropsychopharmacol. 23 (8), 931–940. https://doi.org/10.1016/j.euroneuro.2012.09.008.

Yung, A.R., Phillips, L.J., Nelson, B., Francey, S.M., PanYuen, H., Simmons, M.B., et al., 2011. Randomized controlled trial of interventions for young people at ultra high risk for psychosis: 6-month analysis. J. Clin. Psychiatry 72 (4), 430–440. https://doi.org/10.4088/JCP.08m04979ora.

CHAPTER

20

International services for assessing and treating psychosis risk

Christy L.M. Hui, W.C. Chang, Sherry K.W. Chan, Edwin H.M. Lee, Y.N. Suen, Eric Y.H. Chen

Department of Psychiatry, University of Hong Kong, Hong Kong

Prevention and early intervention services for psychosis

Psychosis is best described as a collection of symptoms that emerge from an interaction between genetic vulnerability and an array of biological and psychosocial factors. As discussed in previous chapters, risk factors such as obstetric complications, substance use, early trauma, and social deficits commonly precede the onset of psychosis (Fusar-Poli et al., 2017). However, the presence of risk factors does not guarantee poor prognosis. International data indicate that much of these risk factors are changeable, as is the trajectory of the disorder (McGorry et al., 2008). Prompt attention to individuals at risk of developing psychosis could restrict or even reverse the damage.

The shift from a pessimistic outlook on the prognosis of psychosis to an optimistic one has inspired international efforts for the development of early intervention (EI) services (McGorry, 2007; Saraceno, 2007; Insel, 2007). Several EI programs established around the world have proved effective in ameliorating outcomes, namely in Australia (Carbone et al., 1999), the United Kingdom (Edwards and McGorry, 2002; Singh et al., 2003), Canada (Addington, 2003), Denmark (Nordentoft et al., 2002, 2003), Sweden (Cullberg et al., 2002), and Hong Kong (Chen et al., 2011; Chan et al., 2018b).

While EI services in different countries may operate according to distinct models, it is important to note that all EI programmes serve one or more of the following purposes: (1) to prevent the progression of prodromal cases into diagnosable psychosis, (2) to promote early identification of undetected cases, and (3) to improve outcomes for individuals with established first-episode psychosis (FEP; Murphy and Brewer, 2011). Of the three service aims, intervening with prodromal or at-risk cases is among the most controversial and challenging issues in EI. A key issue that underlies the challenge of prodromal intervention regards the difficulty of identifying true prodrome, due to the ambiguity of its definition (Singh and Fisher, 2004) and the nonspecificity of its symptoms (McGorry et al., 2010b). Even with true prodromal cases, up to

Risk Factors for Psychosis
https://doi.org/10.1016/B978-0-12-813201-2.00020-X

65%–80% do not develop symptoms for a full diagnosis of a psychotic disorder (Fusar-Poli et al., 2012), raising the question of whether the provision of treatment for all prodromal cases, including the false positives, can be justified.

Regardless of the challenges associated with identification and ethical implications of treatment (as discussed in Chapter 17), prodrome clinics do benefit its service users in multiple ways. Individuals with true prodrome detected by the services could receive intervention at an early stage, during which treatment produces an optimal response and prognosis (Yung, 2003), therefore reducing the need of and costs involved with later treatment for crystallized illness (McGorry et al., 2006, 2010a). In addition, prodrome clinics also detect patients at risk of developing psychiatric disorders other than psychosis, as the phenotypes of early psychotic and nonpsychotic disorders share overlapping similarities (McGorry et al., 2010b). Studies assessing the outcomes of prodromal interventions have also shown that individuals who attended clinical services before the onset of frank psychosis have a shorter duration of untreated psychosis (DUP) and a reduced need for hospitalization, compared to patients who only received treatment upon their first episode (Valmaggia et al., 2015). Another study gathered that compared to FEP patients presented at conventional services, those who attended services for individuals at high risk for psychosis had a shorter referral-to-diagnosis time, were less likely to receive compulsory hospital admission, spent fewer days in the hospital, and had a lower frequency of hospitalization (Fusar-Poli et al., 2016).

Despite the issues that continue to challenge the implementation of prevention services for psychosis, extant literature provides support for its benign effects on clinical outcomes for prodrome cases. In light of this evidence, recent years have witnessed an international expansion of prevention and EI services for psychosis. In the United Kingdom alone, up to 118 services were in operation in the mid-2000s (Tiffin and Glover, 2007) and there is now considerably more. This chapter presents an overview of the EI services worldwide, with a particular focus on recent programs established in Hong Kong. The service program rationales, designs, and outcomes of several major establishments around the world will be covered. The programs are the Personal Assessment and Crisis Evaluation (PACE) clinic in Melbourne, Outreach and Support in South London (OASIS), Support for Wellness Achievement Program (SWAP) in Singapore, Cambridgeshire and Peterborough Assessing, Managing and Enhancing Outcomes (CAMEO), Prevention through Risk Identification, Management and Education (PRIME) in Toronto, Prevention and Early Intervention Program for Psychoses (PEPP) in London, Ontario, Early Recognition and Intervention Center for mental crisis (FETZ) in Cologne, as well as the Early Assessment Service for Young People with Psychosis (EASY), and Jockey Club Mental Wellness Project for Women (JCWow) in Hong Kong.

A review of the prevention programs worldwide

Personal assessment and crisis evaluation

The very first established clinic specialized for individuals with incipient psychosis was PACE, which was founded in 1994 in Melbourne (Yung et al., 2004). The PACE clinic initially stated three main objectives: (i) to enhance understanding of the internal and external processes that take place during the prodromal phase and contribute to the onset of frank psychosis, (ii) to advance the existing psychological and pharmacological interventions by assessing their efficacy and safety for use at the prodromal phase, and (iii) to develop an evidence-based, accessible service designed for young individuals at risk of psychosis (Phillips et al., 2002).

The service identifies individuals who are highly likely to develop a psychotic disorder with the "ultra high risk" (UHR) criteria (Yung, 2003; Yung et al., 2004). According to the UHR criteria, appropriate intervention will be provided for young people between the ages of 14 and 25 years old, who also meets at least one of the following requirements: (i) patients have experienced subthreshold, attenuated positive psychotic symptoms in the past year, (ii) patients have experienced full psychotic symptoms which lasted no longer than a week and then have naturally subsided (i.e., brief limited intermittent psychotic symptoms [BLIPS]), or (iii) patients either have a schizotypal personality disorder or a first-degree relative with a psychotic disorder, *and* since 2003, it has been specified that this group should also have reported a marked, persistent decrease in functioning during the past year (Yung et al., 2003). On top of meeting at least one out of three of the aforementioned conditions, the UHR criteria also specify that service users must be between the ages of 14 and 30, reside in the Melbourne metropolitan area, and have no history of a previous psychotic episode (Phillips et al., 2002).

The stringent intake criteria set out by PACE add to the challenge of gathering suitable referrals. PACE has therefore implemented community education targeted at potential referrers such as school counselors, general practitioners, and local agencies that work with young people. Professional development meetings and case reviews were organized to educate potential referrers about PACE and its services.

Upon receiving referrals, PACE then provides case management for suitable service users, which aims to prevent the exacerbation of symptoms and their progression into a frank psychotic disorder. The clinic also provides both pharmacological and psychosocial treatment options. Extensive assessments of a client's pathology and physical health would be conducted before prescription of medication if necessary. The psychosocial interventions provided by the clinic are based upon the underlying notions of cognitive behavioral therapy (CBT), with adjustments made tailoring the treatment to manage difficulties that the young person faces (Phillips and Francey, 2004)—whether it be academic, employment, family, or housing related.

Outreach and support in South London

Following the success of the PACE services in Melbourne, limitations of the psychiatric services for young FEP patients in South London (Garety and Rigg, 2001) inspired the development of its own program specialized for people at UHR of psychosis—the OASIS. Similar to that of PACE, OASIS aims to reduce the risk of a deteriorating trajectory of illness and assist service users with getting their lives back on track.

The service was founded by Professor Philip McGuire in 2001, and it initially covers the single borough of Lambeth, working in close proximity with the Lambeth Early Onset (LEO) Community Team. With its high unemployment rates, high proportion of residing ethnic minorities, and proportion of homelessness, asylum seekers, and single-person households, Lambeth exhibits an incidence of psychosis that is almost four times the national average (Garety and Rigg, 2001). Taking into account the prevalence of psychosis in Lambeth and its surrounding areas (Power et al., 2007), the OASIS program has undergone service extension to cover the other boroughs under the South London and Maudsley NHS Trust.

This program in South London targets individuals with an "at-risk mental state" (ARMS). The ARMS criteria refer to the set of three clinical syndromes that define UHR (refer to the UHR criteria under the PACE section for details) and constitute risk for subsequent psychosis (Yung et al., 2003). To ascertain the presence of ARMS, the Comprehensive Assessment of the At-Risk Mental State (CAARMS) is used

(Phillips et al., 1999). Further details pertaining to the ARMS criteria are obtained using the family interview for genetic studies (FIGS), Structural Clinical Interview for the DSM-IV Axis I Disorders (SCID-1), DSM-IV Axis II Disorders (SCID-2), the Positive and Negative Syndrome Scale (PANSS), and the Hamilton depression and anxiety scale (First et al., 1997; Nurnberger et al., 1994).

Service users eligible for the service can self-refer via the phone, letter, e-mail, or fax. Referrals can also be initiated by the target individual's friends, family, and health professional(s) with whom they work (Broome et al., 2005). Referred clients then complete a preliminary assessment through the phone, which determines whether they fulfill the intake criteria, and are therefore suitable referrals. Upon acceptance into the OASIS program, an initial diagnostic assessment, constituted by two 1-h sessions, will then be performed by a psychiatrist along with a clinical psychologist. The assessment usually takes place either at the client's general practice or at their home. OASIS then provides case management as well as other treatment options for ARMS patients in the hope of preventing the exacerbation of symptoms and their progression into a frank psychotic disorder (McGlashan et al., 2007). The treatment options provided by OASIS are based upon the principles set out by the stress-vulnerability model of the development of psychosis—pharmacological and psychosocial interventions are incorporated according to the client's needs to optimize the program's ability in preventing a deteriorating prognosis (Yung et al., 2007).

Support for wellness achievement program

The Early Psychosis Intervention Program (EPIP) was established in 2001 in Singapore to promote early detection of and intervention with psychosis patients. Data gathered from the EPIP cohort indicated that the DUP of FEP patients was initially reduced from 12 to 4 months following its initiation (Chong et al., 2005), but examination of its more recent cohort revealed that the average DUP had now increased to 6 months (Verma et al., 2012). The lengthened DUP signaled for the need to enhance sensitivity and promptness of psychosis detection efforts. Noting the concerning figures, the SWAP was initiated with a primary focus to provide assessments and interventions for individuals with ARMS. The service also vows to distinguish itself from mainstream mental health clinics, so as to minimize the associated stigma (Rao et al., 2013). For instance, the name of the service made no reference to mental health services, and its clinics are located within a general polyclinic setting rather than a traditional mental health service setting.

The target service users of SWAP are young people between the ages of 16 and 30 years. Similar to OASIS, SWAP also offers its services for individuals with ARMS, who fulfills the criteria for at least one of the three previously mentioned groups, in addition to a 30% decline in functioning over the past year according to the Social and Occupational Functional Assessment Scale (SOFAS). However, individuals meeting one of the three conditions but not displaying significant deterioration in functioning are still considered eligible for SWAP services, as long as they express high levels of distress regarding the symptoms (Rao et al., 2013).

Referrals are accepted from school counselors, higher-level education institutions, general practitioners, family service centers, private and public clinics, and the Singapore Armed Forces. Following the diagnostic and eligibility assessments, individuals who fulfill the criteria are followed up by SWAP for at least 2 years. During this period, SWAP manages its service users with a multidisciplinary team approach—with guidance from a lead psychiatrist, psychotherapists, case managers, and occupational therapists collaborate to provide care regarding each client's biological, psychological, and social needs. However, antipsychotic medication is not usually prescribed unless the client expresses excessive concern over the intermittent or attenuated symptoms.

Case managers address their clients' occupational, social, and academic needs, while also working closely with their families, school counselors, and employers. The main focus of their supportive work is to equip patients with positive stress coping strategies, monitor symptomatic changes, and adjust their expectations. For patients with further comorbidities and psychological needs, case managers carry out solution-focused brief therapy in hopes of relieving them of their issues within a short amount of time. For additional psychological assistance, patients will be referred to psychotherapists; and for employment inquiries, patients will receive guidance and training from occupational therapists (Rao et al., 2013).

In addition to service provision, SWAP also incorporates public education and training to increase public awareness of their comprehensive services. Educational events are targeted at high-risk populations such as students at higher-level education institutions, as well as gatekeepers, namely school counselors, teachers, and general practitioners. Posters, leaflets, Website, videos, and talks have also been published to reach and network with relevant community organizations and potential referrers. Moreover, training specifically on the identification of psychosis and ARMS is also held by SWAP to enhance different agencies' sensitivity in detecting potential clients of their services.

Cambridgeshire and peterborough assessing, managing and enhancing outcomes

The CAMEO is one of the very first EI programs established in the United Kingdom (Cheng et al., 2011). It was originally designed for individuals between the age 14 and 35 years, experiencing symptoms of psychosis for the first time, and living in Cambridgeshire or Peterborough. Upon service extension, the CAMEO now accommodates for service users aged 14–65 years. They also cater for the needs of young persons aged 14–35 years with ARMS (Hui et al., 2013). CAMEO specifies that to be eligible for the ARMS services, one or more of the following criteria has to be met: (i) attenuated psychotic symptoms that interfere with daily life are experienced for more than once a week, (ii) a family history of psychosis or a biological vulnerability (e.g., head injury) is present, along with a notable recent decline in functioning, or (iii) transient psychotic symptoms. Referrals can be made by the person in need themselves, family or friends, general practitioners, or any health care professionals, via the phone, fax, e-mail, or letter (Cheng et al., 2011). Once a referral is received, a staff member from CAMEO will be in contact to conduct a comprehensive diagnostic assessment, which includes an hour-long neuropsychological battery (Barnett et al., 2005).

Following the initial clinical assessment and acceptance to the services, CAMEO then operates according to the Care Program Approach, which advocates for multidisciplinary caregiving, with particular emphasis on involvement of carers and service users themselves (Department of Health, 1995). CAMEO therefore ensures that any decision-making about subsequent intervention considers the service user's and the carer's opinions.

CAMEO offers a large variety of interventions to accommodate for the distinct needs that young persons with ARMS may have. These include case management, CBT, family therapy and focused groups, vocational services, social skills development, and medication if necessary.

The Toronto prevention through risk identification, management and education

The PRIME is a research-based clinic at the Center for Addiction and Mental Health (CAMH), an affiliated teaching hospital with the University of Toronto. It is a constituent of the First Episode Psychosis Program (FEPP), a service program designed for individuals with FEP (Addington et al., 2008). Through longitudinal research coupled with treatment provision for young people aged 14–30 years, PRIME aims to ultimately develop focused, all-inclusive, and effective treatments for prodromal clients, so as to lower the rate of conversion from prodrome to psychosis or to delay the onset of psychosis.

On top of its research-based nature, PRIME also distinguishes itself from other mainstream mental health services with the physical location and appearance of its clinic. Its clinic is situated on a street along with shops, restaurants, and university buildings, with no noticeable connection to CAMH, allowing for help-seeking in a nonstigmatizing environment.

As prodromal intervention has yet to be incorporated into Canada's health policy strategy, the PRIME program receives funding from research grants. Its operations are therefore supported by two ongoing longitudinal studies. One project aims to establish a model that will assist with the accurate identification of individuals with prodrome. Under this project, individuals who fulfill the prodromal criteria receive regular monitoring and assessments for 3–5 years. The other project is a randomized controlled trial (RCT) aimed at examining the effectiveness of CBT versus supportive therapy in the delay or prevention of psychosis onset. Participants under either study are assigned to one of PRIME's psychiatrists, as well as a case manager from FEPP if the participant expresses excessive distress or appears to be on the cusp of developing a full-blown psychotic disorder. Experienced therapists and well-trained clinical raters are also available at the service (Addington et al., 2008).

Prevention and early intervention program for psychoses

The PEPP is an EI program founded in 1997 in London, Ontario. The program incorporates enhancement of prodromal detection in the community and the provision of phase-specific interventions (Malla et al., 2003). In an attempt to enhance identification of prodrome cases, initially PEPP adopts an open referral policy, along with the promise of a prompt response (within 48 h) to referrals. They accomplished this by reaching out to all potential referral sources within their target area via the use of brochures explaining the nature and treatment of psychosis and services they provide. Upon appraisal of their progress, PEPP has adopted another set of tactics for reducing delays in treatment, as of 1999. The updated strategy aims to detect potential prodromal cases by reaching out to all members of the community. To maximize their influence, PEPP campaigns included the use of posters, pamphlets, bookmarks, calendars, radio, TV, and a variety of educational programs to disseminate about early signs of psychosis and benefits of EI and to introduce their own services.

Following initial contact with referrals, clinicians will then assess the potential service user to ensure that the help-seeking individual does present definite or attenuated psychotic symptoms. A full detailed clinical assessment will then be carried out by a psychiatrist to verify the presence of a prodrome diagnosis using the Structured Clinical Interview for DSM IV (SCID-IV; Malla et al., 2003).

Individuals with an FEP or those who present high risk for psychosis will then be accepted into the program. Much like other well-established EI services, PEPP also utilizes case management (Malla et al., 1999). The case manager develops a strong rapport with the patient and their families, while continuing to network with relevant educational institutions, community services, and potential employers. The follow-up period is generally restricted to 2 years, while the case manager provides psychosocial individual or group interventions according to the patient's needs, and regains normality in the service user's environment as much as possible. Upon the end of the 2 years, an additional 1–3 years of case management can be provided for patients who have yet to achieve a satisfactory level of independent functioning. In addition to case management and different therapies, medication treatment is also available at PEPP; however, they are generally only considered for patients with established psychosis.

The cologne early recognition and intervention center for mental crisis

One of the very first EI programs established in Europe is the FETZ. For many reasons, the service founded in 1997 is rather different to other institutes. For one, the clinic uses other criteria, on top of the UHR criteria, to detect individuals at an even earlier stage of psychosis (Klosterkötter et al., 2001; Schultze-Lutter and Ruhrmann, 2007). Furthermore, despite being part of the Department of Psychiatry and Psychotherapy of the University of Cologne, the FETZ clinic is built in isolation to the in- and outpatient parts of the department and functions in isolation from any FEP programs to offer more normalized access (Schultze-Lutter et al., 2009).

The FETZ employs a unique approach at identifying individuals at the prepsychotic phase. The condition is assessed based on two definitions of ARMS (Ruhrmann et al., 2003; Häfner et al., 2005; Bechdolf et al., 2005). Firstly, the early initial prodromal state is characterized by the occurrence of basic symptoms (i.e., self-reported subjective, minor disturbances that are self-perceived as a mental health issue; Klosterkötter et al., 2001) and/or by state trait factors such as a family history of psychosis or diagnosis of a schizotypal personality disorder (Yung et al., 1998). Secondly, the late initial prodromal state (LIPS) is also targeted by the FETZ services. The definition of LIPS is comparable to that of the attenuated psychosis syndrome (APS) and the BLIPS criteria as mentioned previously (Phillips et al., 2002; Yung et al., 2003; Morrison et al., 2004). Referrals are accepted from all sources and can be done through the phone, post, e-mail, or in person.

At initial contact with the service, at least two detailed diagnostic assessments which include clinical interviews are conducted, supplemented with medical examinations if necessary, to rule out physical factors being the cause of emotional distress. Self-report questionnaires are also included in the assessment process.

Interventions of FETZ are developed based on principles outlined by the vulnerability stress coping model of schizophrenia (Zubin and Spring 1977; Nuechterlein and Dawson, 1984) and that of the basic symptom concept (Huber and Gross, 1989). The model on which interventions are based assumes that biological, social, psychological, and vulnerability factors additively interplay to constitute a person's high-risk status for psychosis (Bechdolf et al., 2006). Acknowledging the interaction of said factors, FETZ then provides a variety of interventions for 12 months. This includes 30 sessions of one-to-one therapy, 15 sessions of group therapy, 12 sessions of cognitive training, and 3 sessions of psychoeducational multifamily therapy (Bechdolf et al., 2006). Each session is conducted following a standardized protocol to maintain quality of service delivery.

Prevention and intervention programs in Hong Kong

Psychiatric services in the territory

While the psychosis prevention and intervention services in Western countries have evidently been thriving, due to cultural beliefs held by Asian populations toward mental health problems, the acknowledgment and treatment of such issues are still highly stigmatized (Kung, 2004; Nguyen and Anderson, 2005; Lee et al., 2009; Masuda et al., 2009; Chan et al., 2016, 2017). However, the high prevalence of psychiatric issues in Asia (WHO, 2017) highlights the urgent need to address them. Out of the Asian regions, Hong Kong in particular has reported an escalated prevalence of mental health problems (Chang et al., 2017a).

Being one of the most densely populated cities in the world, Hong Kong has a population of over 7 million, with approximately 2.5% of Chinese adult population suffering from a lifetime psychotic disorder (Chang et al., 2017a). The population of Hong Kong consists of mostly Han Chinese (95%), and the primary spoken language is Cantonese. As of 2011, the gender ratio in the city is 876 males to 1000 females (Population Census Office, 2012), while that of the psychosis population has been reported to be more equal (Chen et al., 2005b, 2010; Chang et al., 2011; Hui et al., 2014). Another distinctive feature of the Hong Kong population is that majority of young people live with their family because of factors like its sky-rocketing property price and traditional Chinese culture. There is only 17% of the population living in a single-person household, a rate lower than in the United Kingdom (29%; Office for National Statistics, 2013), the United States (28%; United States Census Bureau, 2015), and Australia (24%; Australian Bureau of Statistics, 2011). Considering the risk that urbanicity, social exclusion, and expressed emotion pose on inducing psychosis (Heinz et al., 2013; Domínguez-Martínez et al., 2017), the high prevalence rate of psychosis in Hong Kong may be partially attributable to its population density and the common phenomenon of familial cohabitation.

A further notable feature of the city is its relatively lower rates of substance abuse compared to Western populations—a population-wide survey conducted in Hong Kong found that only 5% of males and 1.8% of females have used psychotropic substances in the past 12 months (Lau et al., 2005), whereas about 10% of the Canadian population revealed having long-term alcohol use issues and 3% with drug use problems (Rush et al., 2008). While substance use itself has been associated with the risk of developing psychotic symptoms (Arseneault et al., 2004), that in combination with urbanicity induces an additive risk for psychosis (Kuepper et al., 2011).

With the multitude of environmental risk factors interplaying with each other in Hong Kong, the question of whether individuals *at risk* of psychotic disorders are appropriately managed then becomes highly relevant. A government-subsidized public health care system is in place in Hong Kong to serve the various needs of the community. Local psychiatric health care depends heavily on this system, while being complemented by private services. In total, there are around 288 psychiatrists, 1880 psychiatric nurses (133 of whom are community nurses), and 197 medical social workers (Tang et al., 2010). They serve an approximate of 150,000 psychiatric patients. Needless to say, the service volume of the psychiatric health care system is extremely high, while government funding and available resources remain rather limited. On top of the insufficient supply of services to meet the demands, Hong Kong's psychiatric services are also largely burdened by social obstacles such as stigma, which is yet to be alleviated.

Owing to the various issues that strain the service, before the implementation of EI programs, patients with psychotic disorders in Hong Kong would have to wait for months, and often over a year, for a first consultation. The DUP was characteristically long, with an average of 513 days (Chen et al., 2005a). Even upon gaining contact with psychiatric services, patients are managed in busy outpatient clinics, where the consultation time is brief, averaging 5.6 min, and follow-ups are infrequent (Hui et al., 2008).

In addition to the issues of service load and access barriers, the social obstacle of stigma also contributes to the constraints of services. While the stigma for mental illnesses has long been identified as a significant barrier to the process of recovery (Sibitz et al., 2011), the issue is particularly concerning in Chinese populations compared to their Western counterparts (Knifton et al., 2010). Despite the enhanced understanding of psychotic disorders over the years, a phone interview study revealed that there was no significant change in the level of stigma toward psychosis among the general population in Hong Kong (Chan et al., 2017). The negative attitudes and stereotypes held against people with psychotic disorders impeded help-seeking behavior and treatment compliance (Chung and Wong, 2004). Acknowledging the multiple issues that prolong the DUP for patients, EI services for psychosis in Hong Kong were launched in 2001.

Early assessment service for young people with psychosis

The Hospital Authority founded the government-funded territory-wide program, EASY, in 2001. With evaluation of its services, the program has undergone extension since its establishment to accommodate for the needs of a wider range of people. EASY initially serves FEP patients between the ages of 15 and 25 years, in 4 out of 7 local hospital clusters. In 2011, upon receiving increased funding, it welcomes all adult FEP patients up to the age of 64 years, in all seven hospital clusters in the city. The service duration was also extended from 2 to 3 years. Patients are not eligible to receive EASY services if they are already receiving treatment; present moderate to severe intelligence deficiency; have significant neurological illnesses; or have substance-induced psychosis.

The EASY program has three primary goals: (1) to create an easily accessible channel for patients and reduce treatment delay; (2) to provide phase-specific care tailored to the needs of FEP patients during the critical period; and (3) to establish a quality, evidence-based, phase-specific intervention model by developing a team of professionals. A steering committee is in place to regulate the entire program, oversee quality control, and appraise service delivery.

Prodrome management and research

In light of evidence for the various risk factors for psychosis in Hong Kong, the EASY clinic also offers diagnostic and follow-up assessments for individuals with ARMS or who present with "prodromal" symptoms. Similar to OASIS, individuals' ARMS status would be verified using CAARMS (Yung et al., 2005). Individuals presenting with clinical high risk for psychosis, i.e., a putatively prodrome for psychotic disorder, including APS and BLIPS, are also enrolled and treated in the EASY program (Chen et al., 2013). Much like patients with an established psychotic disorder, prodromal cases would also receive case management and psychiatric follow-up with regular clinical monitoring, which has been introduced recently.

As EASY prodrome clinic continues to develop, research has been actively carried out in parallel, in hopes of optimizing subsequent service delivery. An ongoing prospective follow-up of Chinese ARMS subjects, with regular clinical monitoring, has been in place to better characterize the clinical course and cognitive profiles of subjects, as well as to examine psychosis transition and functional outcome (Chang et al., 2018a; Chu et al., 2018). For instance, our cross-sectional study comparing cognitive functioning across first-episode schizophrenia (FES) patients, their unaffected siblings (i.e., familial high risk or FHR), ARMS individuals, and healthy controls revealed that ARMS and FHR subjects displayed largely comparable cognitive impairment that was intermediate between FES and controls (Chu et al., 2018). Apart from the clinical, cognitive, and functional assessments, multimodal neuroimaging measurements including structural magnetic resonance imaging (MRI), diffusion tensor imaging, resting-state functional MRI (fMRI), and proton magnetic resonance spectroscopy have also been incorporated recently, hoping to build a more robust prediction model for psychosis transition and illness outcome.

Furthermore, in an attempt to relieve associated stigma, a recent study conducted by EASY research team explored preferences of health care professionals and laypersons for a Chinese term to describe people at risk for psychosis. The term "*yun-niang-qi*" (醞釀期; developing period) was preferred by majority of participants, regardless of their age, gender, level of education, occupation, and whether they have had contact with mental health services (Lee et al., 2017). As the preferred term implies uncertainty and possible reversibility, utilization of it instead of "*qian-qu-qi*" (前驅期; precursor period), "*zao-xian-qi*" (早顯期; early sign period), "*feng-xian-qi*" (風險期; risky period), or "*gao-wei-qi*" (高危期; high-risk period) may reduce stigmatization and bolster help-seeking behavior (Lee et al., 2017).

One key issue that impedes help-seeking behavior is the self-stigmatization associated with mental health disorders. Using vignettes, Lee et al. (2016) found that the general public held the most stigmatizing attitudes toward schizophrenia, which was followed by APS, depression, and psychosis-like experiences. While public stigma for APS is not as prevalent as that for schizophrenia, it still exerts negative influence on help-seeking and consequently patients' chances to recovery.

Case management

EASY uses comprehensive case management as a primary means of EI. Case managers, who are mostly nursing staff, deliver 3 years of clinical monitoring, psychosocial intervention, and psychoeducation. Along with intensive medical follow-up and other elaborate rehabilitation services, EASY's key component of case management is designed to provide phase-specific intervention for patients (Tang et al., 2010). However, the initial stage of management often coincides with the acute phase of psychosis, rather than the prodromal phase, when treatment brings about maximal response (Yung, 2003).

During the acute phase, patients usually receive second-generation antipsychotics as first-line treatment for their active psychotic symptoms. In addition, case managers begin to establish therapeutic alliances with patients as well as their families, to provide further support for more rapid and sustainable recovery. Following the acute phase of illness, intensive care and case management are continued during the maintenance phase. Beyond sustaining remission from positive symptoms, which is primarily handled by maintenance medication, case managers would then actively encourage and guide patients in evaluating their recovery progress. Particular focus is placed on independent living, symptom control, social network, and employment. To regulate the quality and consistency of intervention delivery, case managers oblige to a standardized module—the Psychological Intervention Programs in Early Psychosis I (PIPE-I; Chen et al., 2013), when delivering interventions to patients and their families. Patients who present more complex needs such as persisting residual symptoms and depressive symptoms would be referred to clinical psychologists, who would then intervene according to PIPE-II and PIPE-III.

Efficacy of EASY in comparison with standard care

Since EASY was established, it has proved efficient in managing patients in need. Each year on average, the program has responded to 3000 hotline inquiries, conducted 1000 diagnostic assessments, and accepted 600–700 new cases (Wong et al., 2012). Furthermore, the median waiting time between referral and diagnostic assessment averaged 5 days, which is notably shorter than the duration of waiting required in other public psychiatry systems in Hong Kong.

On top of its efficiency, EASY has also shown to effectively improve multifaceted outcomes for psychosis patients. A cohort study that compared the 3-year outcome of 700 EI patients against that of 700 patients who received standard care (SC) showed that the EI group had fewer positive and negative symptoms, better functional and occupational outcomes, reduced duration of hospitalization, and a lower rate of suicide (Chen et al., 2011). With the establishment of the public awareness campaign and the enhanced referral system, the DUP of the adult population has reduced by half (180 days to 93 days on average), although no significant change of DUP was seen among the younger population (age 15–25) (Chan et al., 2018a).

In addition to evaluating the short-term outcomes of EASY, long-term sustainability of the same cohort has also been explored. Chan et al. (2015) compared the 10-year outcomes of the aforementioned EASY cohort, against that of those who received SC treatment before EASY in Hong Kong. Medical records of all 148 matched pairs of patients were retrieved at 3-month intervals, and semistructured interviews were conducted at the 10-year time point. Findings revealed that users of EASY services had reduced rates of suicide and suicide attempts, fewer admissions to and shorter durations of hospitalization, and prolonged periods of employment compared to patients who received SC. However, between the two comparison groups, there were no differences in rates of psychotic symptoms, symptomatic remission, and functional recovery at 10 years. The overall long-term effects of EASY therefore support its continual execution, but the dilution of its initial short-term effects indicates the need to evaluate its optimal duration. More recently, it has been shown that patients who received EASY service have significantly lower suicide rate during the 12 years' follow-up than those who received SC (4.4% vs. 7.5%) (Chan et al., 2018b). Factors such as premorbid occupational impairment, number of relapses, and poor medication compliance in the initial 3 years of the disorders were important discerning indicators of later suicide between 4 and 12 years.

Furthermore, the cost-effectiveness of the EASY program has been examined as well. A comparison of the direct costs of this EI service and SC indicates that EI is associated with the reduction of outpatient contacts, inpatient hospitalization days, medication prescriptions, professionals' attendance, and accident and emergency department visits (Wong et al., 2011). Although EASY services involved additional costs regarding medication and professional contacts, these extra costs are counteracted by the reduced expenditures in other areas as mentioned.

In light of the updated evidence, the EASY program has been expanded (i.e., EASY+) in 2011 to cover patients from age range of 15–25 years to 15–64 years and from 2-year to 3-year phase-specific EI care, with lower caseload per case manager (1:50). Extension of its EI service duration has been empirically supported by our earlier randomized RCT conducted within the service context of EASY with an aim to compare 3-year versus 2-year EI care on Chinese FEP patients aged 15–25 years in Hong Kong (Chang et al., 2015). This is the first published RCT examining the effectiveness of a specialized EI program for FEP with its treatment period extended beyond 2 years. The study results demonstrated superiority of the 3-year EI over the standard 2-year care on improvement in negative symptoms, depressive symptoms and functional outcome, as well as on reduction in treatment default rate (Chang et al., 2015, 2016, 2017b). On the other hand, given that early and later-onset FEP samples likely have differential illness profiles and therapeutic needs, a territory-wide evaluation study comparing EASY+ with SC has recently been conducted in a large cohort of FEP patients (above 26 years) enrolled from all seven Hospital Authority clusters in Hong Kong. The study revealed the efficacy of EASY+ in treatment delay reduction, negative and depressive symptom alleviation, functional outcome and subjective quality of life improvement, and lowering of service default and disengagement rates in adult FEP patients (Chang et al., 2018b). Hence, the findings provide evidence supporting the recent expansion of EASY in Hong Kong.

Recent development of preventative intervention for women

While the EASY program is in place to address the needs of younger psychosis patients, there is also a need to tackle sex differences in psychiatric needs. The JCWow, led by the University of Hong Kong, is the first community project designed for women at risk of mental disorders, such as anxiety, depression, and psychosis. The project recognizes the effects of undertaking multiple roles, overworking, and the lack of personal space on the striking prevalence of common mental disorders among *women*, particularly those with socioeconomic deprivation in Hong Kong (Hui et al., 2018). It is funded by the Jockey Club Charities Fund and runs in collaboration with several local nongovernmental organizations (NGOs), namely the Caritas Hong Kong, Hong Kong Federation of Women's Centers (FWC), Early Psychosis Foundation (EPISO), and New Life Psychiatric Rehabilitation Association. Noting the need for a comprehensive, gender-specific program, the JCWow incorporates service provision, professional training, awareness-raising, and research. The project aims to enhance the detection of women at risk of mental disorders, to provide preventative interventions, and to improve outcomes for this particular group.

At present, the JCWow has three main projects in operation. One of these projects aims to deliver group coaching intervention, based on principles of positive psychology, for women at risk of common mental disorders. Another project provides aerobic exercise and yoga sessions for participants with established psychosis, in hopes of promoting improvement in their cognitive functions, such as working memory and attention. In addition, JCWow also runs a preventative program for women with incipient psychosis. This particular cohort is identified by the at-risk mental state criteria (CAARMS, Yung et al., 2005). Women who could potentially benefit from the program will then be offered CBT to enable healthy thinking habit and a strength-based program with tangible support to develop psychological resilience. The CBT will focus on modifying at-risk thinking habits and cognitive biases and flexibility and introduce alternative modes of thinking and behaving. The strength-based program is carried out mainly by NGOs to address issues of social relationship, self-compassion, and employment.

With their continued efforts, it is anticipated that the comprehensive programs of JCWow will contribute to the development of a preventative intervention model, improve mental illness outcomes, raise gender-specific awareness, and enhance NGOs' capability to identify and deliver low-intensity interventions for women in need in Hong Kong.

References

Addington, J., Epstein, I., Reynolds, A., Furimsky, I., Rudy, L., Mancini, B., et al., 2008. Early detection of psychosis: finding those at clinical high risk. Early Interv. Psychiatry 2 (3), 147–153.

Addington, J., 2003. The prodromal stage of psychotic illness: observation, detection or intervention? J. Psychiatry Neurosci. 28 (2), 93.

Arseneault, L., Cannon, M., Witton, J., Murray, R.M., 2004. Causal association between cannabis and psychosis: examination of the evidence. Br. J. Psychiatry 184 (2), 110–117.

Australian Bureau of Statistics, 2011. First Release Media Fact Sheets – Average Household Size.

Barnett, J.H., Sahakian, B.J., Werners, U., Hill, K.E., Brazil, R., Gallagher, O., et al., 2005. Visuospatial learning and executive function are independently impaired in first-episode psychosis. Psychol. Med. 35 (7), 1031–1041.

Bechdolf, A., Phillips, L.J., Francey, S.M., Leicester, S., Morrison, A.P., Veith, V., et al., 2006. Recent approaches to psychological interventions for people at risk of psychosis. Eur. Arch. Psychiatry Clin. Neurosci. 256 (3), 159–173.

Bechdolf, A., Pukrop, R., Köhn, D., Tschinkel, S., Veith, V., Schultze-Lutter, F., et al., 2005. Subjective quality of life in subjects at risk for a first episode of psychosis: a comparison with first episode schizophrenia patients and healthy controls. Schizophr. Res. 79 (1), 137–143.

Broome, M.R., Woolley, J.B., Johns, L.C., Valmaggia, L.R., Tabraham, P., Gafoor, R., et al., 2005. Outreach and support in south London (OASIS): implementation of a clinical service for prodromal psychosis and the at risk mental state. Eur. Psychiatry 20 (5–6), 372–378.

Carbone, S., Harrigan, S., McGorry, P.D., Curry, C., Elkins, K., 1999. Duration of untreated psychosis and 12-month outcome in first-episode psychosis: the impact of treatment approach. Acta Psychiatr. Scand. 100 (2), 96–104.

Chan, S.K., Chau, E.H., Hui, C.L., Chang, W.C., Lee, E.H., Chen, E.Y., 2018a. Long term effect of early intervention service on duration of untreated psychosis in youth and adult population in Hong Kong. Early Interv. Psychiatry 12 (3), 331–338.

Chan, S.K.W., Chan, S.W.Y., Pang, H.H., Yan, K.K., Hui, C.L.M., Chang, W.C., Lee, E.H.M., Chen, E.Y.H., 2018b. Association of an early intervention service for psychosis with suicide rate among patients with first-episode schizophrenia-spectrum disorders. JAMA Psychiatry 75 (5), 458–464.

Chan, S.K.W., Lee, K.W., Hui, C.L.M., Chang, W.C., Lee, E.H.M., Chen, E.Y.H., 2017. Gender effect on public stigma changes towards psychosis in the Hong Kong Chinese population: a comparison between population surveys of 2009 and 2014. Soc. Psychiatry Psychiatr. Epidemiol. 52 (3), 259–267.

Chan, S.K.W., Tam, W.W.Y., Lee, K.W., Hui, C.L.M., Chang, W.C., Lee, E.H.M., Chen, E.Y.H., 2016. A population study of public stigma about psychosis and its contributing factors among Chinese population in Hong Kong. Int. J. Soc. Psychiatry 62 (3), 205–213.

Chan, S.K.W., So, H.C., Hui, C.L.M., Chang, W.C., Lee, E.H.M., Chung, D.W.S., et al., 2015. 10-year outcome study of an early intervention program for psychosis compared with standard care service. Psychol. Med. 45 (6), 1181–1193.

Chang, W.C., Ho, C., Or, C.F., Liu, T.T., Lau, F.C., Chu, O.K., et al., 2018b. T252. Treatment delay and outcome comparison of extended early intervention service and standard psychiatric care for adults presenting with first-episode psychosis in Hong Kong. Schizophr. Bull. 44 (Suppl. 1), S215.

Chang, W.C., Lee, H.C., Chan, S.I., Chiu, S.Y., Lee, H.M., Chan, K.W., et al., 2018a. Negative symptom dimensions differentially impact on functioning in individuals at-risk for psychosis. Schizophr. Res. 202, 310.

Chang, W.C., Kwong, V.W.Y., Chan, G.H.K., Jim, O.T.T., Lau, E.S.K., Hui, C.L.M., et al., 2017b. Prediction of motivational impairment: 12-month follow-up of the randomized-controlled trial on extended early intervention for first-episode psychosis. Eur. Psychiatry 41, 37–41.

Chang, W.C., Wong, C.S.M., Chen, E.Y.H., Lam, L.C.W., Chan, W.C., Ng, R.M.K., et al., 2017a. Lifetime prevalence and correlates of schizophrenia-spectrum, affective, and other non-affective psychotic disorders in the Chinese adult population. Schizophr. Bull. 43 (6), 1280–1290.

Chang, W.C., Kwong, V.W.Y., Chan, G.H.K., Jim, O.T.T., Lau, E.S.K., Hui, C.L.M., et al., 2016. Prediction of functional remission in first-episode psychosis: 12-month follow-up of the randomized-controlled trial on extended early intervention in Hong Kong. Schizophr. Res. 173 (1–2), 79–83.

Chang, W.C., Chan, G.H.K., Jim, O.T.T., Lau, E.S.K., Hui, C.L.M., Chan, S.K.W., et al., 2015. Optimal duration of an early intervention programme for first-episode psychosis: randomised controlled trial. Br. J. Psychiatry 206 (6), 492–500.

Chang, W.C., Tang, J.Y., Hui, C.L., Chiu, C.P., Lam, M.M., Wong, G.H., et al., 2011. Gender differences in patients presenting with first-episode psychosis in Hong Kong: a three-year follow up study. Aust. N. Z. J. Psychiatr. 45 (3), 199–205.

Chen, E.Y.H., Chan, G.H.K., Wong, G.H.Y., Lee, H., 2013. Early Psychosis Intervention: A Culturally Adaptive Clinical Guide, vol. 1. Hong Kong University Press.

Chen, E.Y., Tang, J.Y., Hui, C.L., Chiu, C.P., Lam, M.M., Law, C.W., et al., 2011. Three-year outcome of phase-specific early intervention for first-episode psychosis: a cohort study in Hong Kong. Early Interv. Psychiatry 5 (4), 315–323.

Chen, E.Y., Hui, C.L., Lam, M.M., Chiu, C.P., Law, C.W., Chung, D.W., et al., 2010. Maintenance treatment with quetiapine versus discontinuation after one year of treatment in patients with remitted first episode psychosis: randomised controlled trial. Br. Med. J. 341, c4024.

Chen, E.Y.H., Dunn, E.L.W., Miao, M.Y.K., Yeung, W.S., Wong, C.K., Chan, W.F., et al., 2005a. The impact of family experience on the duration of untreated psychosis (DUP) in Hong Kong. Soc. Psychiatry Psychiatr. Epidemiol. 40 (5), 350–356.

Chen, E.Y.H., Hui, C.L.M., Dunn, E.L.W., Miao, M.Y.K., Yeung, W.S., Wong, C.K., et al., 2005b. A prospective 3-year longitudinal study of cognitive predictors of relapse in first-episode schizophrenic patients. Schizophr. Res. 77 (1), 99–104.

Cheng, F., Kirkbride, J.B., Lennox, B.R., Perez, J., Masson, K., Lawrence, K., et al., 2011. Administrative incidence of psychosis assessed in an early intervention service in England: first epidemiological evidence from a diverse, rural and urban setting. Psychol. Med. 41 (5), 949–958.

Chong, S.A., Mythily, S., Verma, S., 2005. Reducing the duration of untreated psychosis and changing help-seeking behaviour in Singapore. Soc. Psychiatry Psychiatr. Epidemiol. 40 (8), 619–621.

Chu, A.O.K., Chang, W.C., Chan, S.K.W., Lee, E.H.M., Hui, C.L.M., Chen, E.Y.H., 2018. Comparison of cognitive functions between first-episode schizophrenia patients, their unaffected siblings and individuals at clinical high-risk for psychosis. Psychol. Med. 1–8.

Chung, K.F., Wong, M.C., 2004. Experience of stigma among Chinese mental health patients in Hong Kong. Psychiatr. Bull. 28 (12), 451–454.

Cullberg, J., Levander, S., Holmqvist, R., Mattsson, M., Wieselgren, I.M., 2002. One-year outcome in first episode psychosis patients in the Swedish Parachute project. Acta Psychiatr. Scand. 106 (4), 276–285.

Department of Health, 1995. Building bridges. In: A Guide to Arrangements for Interagency Working for the Care and Protection of Severely Mentally Ill People. HMSO, London.

Domínguez-Martínez, T., Medina-Pradas, C., Kwapil, T.R., Barrantes-Vidal, N., 2017. Relatives' expressed emotion, distress and attributions in clinical high-risk and recent onset of psychosis. Psychiatry Res. 247, 323–329.

Edwards, J., McGorry, P.D., 2002. Implementing Early Intervention in Psychosis: A Guide to Establishing Psychosis Services. CRC Press.

First, M.B., Spitzer, R.L., Gibbon, M., Williams, J.B., 1997. User's Guide for the Structured Clinical Interview for DSM-IV axis I Disorders SCID-I: Clinician Version. American Psychiatric Pub.

Fusar-Poli, P., Bonoldi, I., Yung, A.R., Borgwardt, S., Kempton, M.J., Valmaggia, L., et al., 2012. Predicting psychosis: meta-analysis of transition outcomes in individuals at high clinical risk. Arch. Gen. Psychiatr. 69 (3), 220–229.

Fusar-Poli, P., Tantardini, M., De Simone, S., Ramella-Cravaro, V., Oliver, D., Kingdon, J., et al., 2017. Deconstructing vulnerability for psychosis: meta-analysis of environmental risk factors for psychosis in subjects at ultra high-risk. Eur. Psychiatry 40, 65–75.

Fusar-Poli, P., Diaz-Caneja, C.M., Patel, R., Valmaggia, L., Byrne, M., Garety, P., et al., 2016. Services for people at high risk improve outcomes in patients with first episode psychosis. Acta Psychiatr. Scand. 133 (1), 76–85.

Garety, P.A., Rigg, A., 2001. Early psychosis in the inner city: a survey to inform service planning. Soc. Psychiatry Psychiatr. Epidemiol. 36 (11), 537–544.

Häfner, H., Maurer, K., Trendler, G., an der Heiden, W., Schmidt, M., Könnecke, R., 2005. Schizophrenia and depression: challenging the paradigm of two separate diseases—a controlled study of schizophrenia, depression and healthy controls. Schizophr. Res. 77 (1), 11–24.

Heinz, A., Deserno, L., Reininghaus, U., 2013. Urbanicity, social adversity and psychosis. World Psychiatry 12 (3), 187–197.

Huber, G., Gross, G., 1989. The concept of basic symptoms in schizophrenic and schizoaffective psychoses. Recent. Prog. Med. 80 (12), 646–652.

Hui, C.L., Ko, W.T., Chang, W.C., Lee, E.H., Chan, S.K., Chen, E.Y., 2018. Clinical and functional correlates of financially deprived women with first-episode psychosis. Early Interv. Psychiatry 13 (3), 639–645.

Hui, C.L.M., Li, A.W.Y., Leung, C.M., Chang, W.C., Chan, S.K.W., Lee, E.H.M., Chen, E.Y.H., 2014. Comparing illness presentation, treatment and functioning between patients with adolescent-and adult-onset psychosis. Psychiatry Res. 220 (3), 797–802.

Hui, C., Morcillo, C., Russo, D.A., Stochl, J., Shelley, G.F., Painter, M., et al., 2013. Psychiatric morbidity, functioning and quality of life in young people at clinical high risk for psychosis. Schizophr. Res. 148 (1–3), 175–180.

Hui, C.L., Wong, G.H., Lam, C.Y., Chow, P.P., Chen, E.Y., 2008. Patient-clinician communication and needs identification for outpatients with schizophrenia in Hong Kong: role of the 2-COM instrument. Hong Kong J. Psychiatry 18 (2).

Insel, T.R., 2007. The arrival of preemptive psychiatry. Early Interv. Psychiatry 1 (1), 5–6.

Klosterkötter, J., Hellmich, M., Steinmeyer, E.M., Schultze-Lutter, F., 2001. Diagnosing schizophrenia in the initial prodromal phase. Arch. Gen. Psychiatr. 58 (2), 158–164.

Knifton, L., Gervais, M., Newbigging, K., Mirza, N., Quinn, N., Wilson, N., Hunkins-Hutchison, E., 2010. Community conversation: addressing mental health stigma with ethnic minority communities. Soc. Psychiatry Psychiatr. Epidemiol. 45 (4), 497–504.

Kuepper, R., Van Os, J., Lieb, R., Wittchen, H.U., Henquet, C., 2011. Do cannabis and urbanicity co-participate in causing psychosis? Evidence from a 10-year follow-up cohort study. Psychol. Med. 41 (10), 2121–2129.

Kung, W.W., 2004. Cultural and practical barriers to seeking mental health treatment for Chinese Americans. J. Community Psychol. 32 (1), 27–43.

Lau, J.T., Kim, J.H., Tsui, H.Y., 2005. Prevalence, health outcomes, and patterns of psychotropic substance use in a Chinese population in Hong Kong: a population-based study. Subst. Use Misuse 40 (2), 187–209.

Lee, E.H., Ching, E.Y., Hui, C.L., Lin, J.J., Chang, W.C., Chan, S.K., Chen, E.Y., 2017. Chinese label for people at risk for psychosis. Early Interv. Psychiatry 11 (3), 224–228.

Lee, E.H.M., Hui, C.L.M., Ching, E.Y.N., Lin, J., Chang, W.C., Chan, S.K.W., Chen, E.Y.H., 2016. Public stigma in China associated with schizophrenia, depression, attenuated psychosis syndrome, and psychosis-like experiences. Psychiatr. Serv. 67 (7), 766–770.

Lee, S., Juon, H.S., Martinez, G., Hsu, C.E., Robinson, E.S., Bawa, J., Ma, G.X., 2009. Model minority at risk: expressed needs of mental health by Asian American young adults. J. Community Health 34 (2), 144.

Malla, A., Malla, A., Norman, R., McLean, T., Scholten, D., Townsend, L., 2003. A Canadian programme for early intervention in non-affective psychotic disorders. Aust. N. Z. J. Psychiatr. 37 (4), 407–413.

Malla, A.K., Norman, R.M., Voruganti, L.P., 1999. Improving outcome in schizophrenia: the case for early intervention. CMAJ 160 (6), 843.

Masuda, A., Anderson, P.L., Twohig, M.P., Feinstein, A.B., Chou, Y.Y., Wendell, J.W., Stormo, A.R., 2009. Help-seeking experiences and attitudes among African American, Asian American, and European American college students. Int. J. Adv. Couns. 31 (3), 168–180.

McGlashan, T.H., Addington, J., Cannon, T., Heinimaa, M., McGorry, P., O'Brien, M., et al., 2007. Recruitment and treatment practices for help-seeking "prodromal" patients. Schizophr. Bull. 33 (3), 715–726.

McGorry, P., Johanessen, J.O., Lewis, S., Birchwood, M., Malla, A., Nordentoft, M., et al., 2010a. Early intervention in psychosis: keeping faith with evidence-based health care: a commentary on: 'Early intervention in psychotic disorders: faith before facts?' by Bosanac et al.(2009). Psychol. Med. 40 (3), 399–404.

McGorry, P.D., Nelson, B., Goldstone, S., Yung, A.R., 2010b. Clinical staging: a heuristic and practical strategy for new research and better health and social outcomes for psychotic and related mood disorders. Can. J. Psychiatr. 55 (8), 486–497.

McGorry, P.D., Killackey, E., Yung, A., 2008. Early intervention in psychosis: concepts, evidence and future directions. World Psychiatry 7 (3), 148–156.

McGorry, P., 2007. Welcome to early intervention in psychiatry. Early Interv. Psychiatry 1 (1), 1–2.

McGorry, P.D., Hickie, I.B., Yung, A.R., Pantelis, C., Jackson, H.J., 2006. Clinical staging of psychiatric disorders: a heuristic framework for choosing earlier, safer and more effective interventions. Aust. N. Z. J. Psychiatr. 40 (8), 616–622.

Morrison, A.P., French, P., Walford, L., Lewis, S.W., Kilcommons, A., Green, J., et al., 2004. Cognitive therapy for the prevention of psychosis in people at ultra-high risk: randomised controlled trial. Br. J. Psychiatry 185 (4), 291–297.

Murphy, B.P., Brewer, W.J., 2011. Early intervention in psychosis: strengths and limitations of services. Adv. Psychiatr. Treat. 17 (6), 401–407.

Nguyen, Q.C.X., Anderson, L.P., 2005. Vietnamese Americans' attitudes toward seeking mental health services: relation to cultural variables. J. Community Psychol. 33 (2), 213–231.

Nurnberger, J.I., Blehar, M.C., Kaufmann, C.A., York-Cooler, C., Simpson, S.G., Harkavy-Friedman, J., et al., 1994. Diagnostic interview for genetic studies: rationale, unique features, and training. Arch. Gen. Psychiatr. 51 (11), 849–859.

Nordentoft, M., Jeppesen, P., Petersen, L., Thorup, A., Abel, M., Øhlenschlæger, J.K., et al., 2003. OPUS project: a randomised controlled trial of integrated psychiatric treatment in first episode psychosis. Schizophr. Res. 1 (60), 297.

Nordentoft, M., Jeppesen, P., Abel, M., Kassow, P., Petersen, L., Thorup, A., et al., 2002. OPUS study: suicidal behaviour, suicidal ideation and hopelessness among patients with first-episode psychosis: one-year follow-up of a randomised controlled trial. Br. J. Psychiatry 181 (S43), s98–s106.

Nuechterlein, K.H., Dawson, M.E., 1984. A heuristic vulnerability/stress model of schizophrenic episodes. Schizophr. Bull. 10 (2), 300.

Office for National Statistics, 2013. Families and Households.

Phillips, L.J., Francey, S.M., 2004. Changing PACE: psychological interventions in the prepsychotic phase. In: Psychological Interventions in Early Psychosis: A Treatment Handbook, pp. 23–39.

Phillips, L.J., Leicester, S.B., O'DWYER, L.E., Francey, S.M., Koutsogiannis, J., Abdel-Baki, A., et al., 2002. The PACE clinic: identification and management of young people at "ultra" high risk of psychosis. J. Psychiatr. Pract. 8 (5), 255–269.

Phillips, L., Yung, A.R., Hearn, N., McFarlane, C., Hallgren, M., McGorry, P.D., 1999. Preventative mental health care: accessing the target population. Aust. N. Z. J. Psychiatr. 33 (6), 912–917.

Population Census Office, 2012. 2011 Population Census Summary Results. Retrieved from: http://www.censtatd.gov.hk/hkstat/sub/sp170.jsp?productCode=B1120055.

Power, P., McGuire, P., Iacoponi, E., Garety, P., Morris, E., Valmaggia, L., et al., 2007. Lambeth Early Onset (LEO) and outreach & support in south London (OASIS) service. Early Interv. Psychiatry 1 (1), 97–103.

Rao, S., Pariyasami, S., Tay, S.A., Lim, L.K., Yuen, S., Poon, L.Y., Lee, H., Verma, S., 2013. Support for Wellness Achievement Programme (SWAP): a service for individuals with at-risk mental state in Singapore. Ann. Acad. Med. Singapore 42 (10), 552–555.

Ruhrmann, S., Schultze-Lutter, F., Klosterkötter, J., 2003. Early detection and intervention in the initial prodromal phase of schizophrenia. Pharmacopsychiatry 36 (S 3), 162–167.

Rush, B., Urbanoski, K., Bassani, D., Castel, S., Wild, T.C., Strike, C., et al., 2008. Prevalence of co-occurring substance use and other mental disorders in the Canadian population. Can. J. Psychiatr. 53 (12), 800–809.

Saraceno, B., 2007. New knowledge and new hope to people with emerging mental disorders. Early Interv. Psychiatry 1 (1), 3–4.

Schultze-Lutter, F., Ruhrmann, S., 2007. Can schizophrenia be predicted phenomenologically? Evolving Psychosis 126–145.

Schultze-Lutter, F., Ruhrmann, S., Klosterkötter, J., 2009. Early detection of psychosis–establishing a service for persons at risk. Eur. Psychiatry 24 (1), 1–10.

Sibitz, I., Amering, M., Unger, A., Seyringer, M.E., Bachmann, A., Schrank, B., et al., 2011. The impact of the social network, stigma and empowerment on the quality of life in patients with schizophrenia. Eur. Psychiatry 26 (1), 28–33.

Singh, S.P., Fisher, H.L., 2004. Early intervention services. Psychiatry 3 (9), 17–21.

Singh, S., Wright, C., Joyce, E., Barnes, T., Burns, T., 2003. Developing early intervention services in the NHS: a survey to guide workforce and training needs. Psychiatr. Bull. 27 (7), 254–258.

Tang, J.Y., Wong, G.H., Hui, C.L., Lam, M.M., Chiu, C.P., Chan, S.K., et al., 2010. Early intervention for psychosis in Hong Kong–the EASY programme. Early Interv. Psychiatry 4 (3), 214–219.

Tiffin, P.A., Glover, G., 2007. From commitment to reality: early intervention in psychosis services in England. Early Interv. Psychiatry 1 (1), 104–107.

United States Census Bureau, 2015. Families and Living Arrangements. Table HH-4. Households by Size: 1960 to Present.

Valmaggia, L.R., Byrne, M., Day, F., Broome, M.R., Johns, L., Howes, O., et al., 2015. Duration of untreated psychosis and need for admission in patients who engage with mental health services in the prodromal phase. Br. J. Psychiatry 207 (2), 130–134.

Verma, S., Poon, L.Y., Subramaniam, M., Abdin, E., Chong, S.A., 2012. The Singapore early psychosis intervention programme (EPIP): a programme evaluation. Asian J. Psychiatry 5 (1), 63–67.

Wong, G.H., Hui, C.L., Tang, J.Y., Chang, W.C., Chan, S.K., Xu, J.Q., et al., 2012. Early intervention for psychotic disorders: real-life implementation in Hong Kong. Asian J. Psychiatry 5 (1), 68–72.

Wong, K.K., Chan, S.K., Lam, M.M., Hui, C.L., Hung, S.F., Tay, M., et al., 2011. Cost-effectiveness of an early assessment service for young people with early psychosis in Hong Kong. Aust. N. Z. J. Psychiatr. 45 (8), 673–680.

World Health Organization, 2017. Depression and Other Common Mental Disorders: Global Health Estimates (No. WHO/MSD/MER/2017.2). World Health Organization.

Yung, A.R., McGorry, P.D., McFarlane, C.A., Jackson, H.J., Patton, G.C., Rakkar, A., 2004. Monitoring and care of young people at incipient risk of psychosis. Focus 22 (1), 283-174.

Yung, A.R., Phillips, L.J., Yuen, H.P., Francey, S.M., McFarlane, C.A., Hallgren, M., McGorry, P.D., 2003. Psychosis prediction: 12-month follow up of a high-risk ("prodromal") group. Schizophr. Res. 60 (1), 21–32.

Yung, A.R., Yuen, H.P., Berger, G., Francey, S., Hung, T.C., Nelson, B., et al., 2007. Declining transition rate in ultra high risk (prodromal) services: dilution or reduction of risk? Schizophr. Bull. 33 (3), 673–681.

Yung, A.R., Yuen, H.P., McGorry, P.D., Phillips, L.J., Kelly, D., Dell'Olio, M., et al., 2005. Mapping the onset of psychosis: the comprehensive assessment of at-risk mental states. Aust. N. Z. J. Psychiatr. 39 (11–12), 964–971.

Yung, A.R., 2003. Commentary: the schizophrenia prodrome: a high-risk concept. Schizophr. Bull. 29 (4), 859–865.

Yung, A.R., Phillips, L.J., McGorry, P.D., McFarlane, C.A., Francey, S., Harrigan, S., et al., 1998. Prediction of psychosis: a step towards indicated prevention of schizophrenia. Br. J. Psychiatry 172 (S33), 14–20.

Zubin, J., Spring, B., 1977. Vulnerability: a new view of schizophrenia. J. Abnorm. Psychol. 86 (2), 103.

CHAPTER

21

New paradigms to study psychosis risk: clinical staging, pluripotency, and dynamic prediction

Rachael Spooner[1,2], Jessica A. Hartmann[1,2], Patrick D. McGorry[1,2], Barnaby Nelson[1,2]

[1]Orygen, Parkville, VIC, Australia; [2]Centre for Youth Mental Health, The University of Melbourne, Parkville, VIC, Australia

The ultra high risk for psychosis concept

Seminal work in the 1990s by McGorry and colleagues (Yung et al., 1996) introduced the concept of the "at-risk mental state" (ARMS) and operationalized ultra high risk (UHR) criteria (otherwise referred to as "clinical high risk" [CHR]) to identify help-seeking individuals at risk of progressing ("transitioning" or "converting") to a psychotic episode (see Chapter 3). This has been a powerful research paradigm for opening avenues for prospective etiological research and preventive intervention trials. Over the past 20 years, it is estimated that over 1500 studies have been published related to the concepts of UHR, CHR, and "transition" to psychotic disorder (van Os and Guloksuz, 2017). These well-validated criteria took a "close-in" approach to identify psychosis risk (see Chapter 3) by using state (attenuated or transient psychotic symptoms) and trait (family history or schizotypal personality disorder in combination with functional decline or chronic low functioning) characteristics of help-seeking people in the greatest age range for psychosis risk (15–25 years).

Increasingly, however, the call to revisit and possibly revise the ARMS concept has been growing for a number of reasons (Fusar-Poli et al., 2014b; Hartmann et al., 2017a; van Os and Guloksuz, 2017). Firstly, the reasonably low transition to psychosis rate in recent UHR cohorts has meant that intervention trials and prediction studies, particularly those with a short follow-up period, have suffered from low statistical power to investigate transition to psychosis as an outcome. For example, Morrison et al. (2012) reported an 8% transition rate in their CBT trial and McGorry et al. (2017) reported an 11% transition rate in their omega-3 fatty

Risk Factors for Psychosis
https://doi.org/10.1016/B978-0-12-813201-2.00021-1

acid trial. The issue of low statistical power is problematic in several ways. It reduces the chance of detecting a true effect, effect sizes are typically overestimated, and results are unlikely to be reproducible (Cuijpers, 2003). It also means that very large and most likely unachievable study samples are required to achieve adequate statistical power for a binary outcome (i.e., psychosis transition or not) (Cuijpers, 2003). Cuijpers (2003) outlines that one of the main ways to overcome the issues associated with statistical power in low prevalence disorders such as psychotic disorder or bipolar disorder is to target a *broader* range of disorders as outcomes of interest and not be limited to a single relatively low-incidence disorder.

It has also been argued that the dichotomous terms that typically define outcome in the "at risk" for psychosis literature (generally, "transition" to psychosis) are somewhat arbitrary—in other words, the line dividing the at-risk state from psychosis threshold (where, if crossed, an individual is deemed to have "transitioned" to an episode of psychosis) may not have robust validity markers (Yung et al., 2010) despite it having clinical implications for psychosocial intervention and antipsychotic treatment. Such categories (UHR/transition) can also obfuscate the diverse range of psychopathology that is often present in this patient population (Nelson and Yung, 2009; Nelson et al., 2013). Indeed, the use of these binary categories within statistical analyses can disguise the substantial diversity and often nonlinear progression of anomalous experiences along the continuum of psychotic symptomatology from sub- to full threshold distinctions (Fusar-Poli et al., 2014a) and also *within* both sub- and full threshold categories (Nelson et al., 2013; Nelson and Hartmann, 2018) and therefore blinker our exploration of the complexity of emerging psychosis (Nelson et al., 2018). New paradigms for studying psychosis risk and onset of mental disorder more broadly seek to address these complexities.

Recent developments

Comorbidity

There is some disagreement in the literature with regard to how useful the UHR criteria are for identifying risk for non-psychotic disorders. Recent literature has also investigated whether psychosis onset can be well-predicted *outside* the traditional UHR identification approach (Lee et al., 2018). A recent retrospective study by Shah et al. (2017) reported that 32% of their first-episode psychosis (FEP) sample did not undergo a period of subthreshold psychotic experiences before the onset of frank psychosis. This suggests that psychotic episodes may not always be preceded by a directly comparable subthreshold attenuated psychotic (*forme fruste*) state. Across the sample, the most prevalent early symptomatology reported was depression, anxiety, and low functioning (Shah et al., 2017). A recent systematic review by Lee et al. (2018) demonstrated that people at risk of non-psychotic disorders (identified through the presence of subthreshold nonpsychotic symptoms) were at elevated risk of *psychotic* disorder (3.87% 3-year incidence rate)—not as high as people meeting UHR criteria (24.63% 3-year incidence rate) but substantially higher (77.4-fold) than the general population. A good body of evidence also indicates a high rate (60%–80%) of non-psychotic disorders (primarily mood and anxiety disorders) and personality disorder (up to 75%) in the UHR population (Phillips et al., 2009; Wigman et al., 2012; Fusar-Poli et al., 2014a; Lim et al., 2015; Lin et al., 2015; Rutigliano et al., 2016; Sevilla-Llewellyn-Jones et al., 2018).

A long-term follow-up study by Lin et al. (2015) found high valence of the UHR criteria for *both* psychotic and non-psychotic disorders, a finding consistent with several other studies (Wigman et al., 2012; Lim et al., 2015). A substantial proportion of these samples had non-psychotic disorders at study entry (i.e., disorders were persistent rather than incident). A recent study by Sevilla-Llewellyn-Jones et al. (2018)

found that 75% of their UHR sample had personality disorders (most commonly depressive, borderline, or schizotypal) but concluded that personality disorders/traits may not signal an increase in "transition" risk but do, however, contribute to increased morbidity in UHR individuals. In contrast, several studies report that the UHR criteria are effective in identifying people at risk of psychotic disorders but not non-psychotic disorders (Webb et al., 2015; Fusar-Poli et al., 2017a; Woods et al., 2018). At the minimum, these studies collectively indicate that the UHR criteria certainly have utility for identifying those at risk of psychosis, albeit to a more modest degree than was suggested in the initial UHR studies, and *some* utility in identifying people with persistent or incident non-psychotic disorders. This suggests that the UHR criteria already identify help-seeking young people at risk of a *range* of disorders (incident psychosis, persistent, or incident nonpsychotic disorders) and could possibly be revised to increase their utility for pluripotent prediction.

One explanation for these findings is that attenuated psychotic symptoms may be a transdiagnostic phenomenon which can be predictive of, and co-occur with, a variety of non-psychotic disorders, lending support to the concept of phenomenological continuity (McGorry and Nelson, 2016; van Os and Reininghaus, 2016). van Os and Reininghaus (2016) propose that subthreshold psychotic phenomena can signify not only an explicit phenotypic expression of subclinical psychosis but also a set of core transphenotypic connections between various areas of psychopathology. As illustrated in Fig. 21.1, they suggest that a transdiagnostic psychosis phenotype could underlie both schizophrenia spectrum and bipolar disorder whereby positive, affective, negative, and disorganization symptoms interact in generating a clinical phenotype. The model is proposed to capture an extended phenotype in the general population as well. However, this model, while capturing an extended psychosis phenotype across (attenuated and full threshold) affective and non-affective psychoses, was not designed to account for nonpsychotic disorders. The growing body of evidence suggests that early clinical phenotypes are often highly undifferentiated and diffuse in symptomatology with a marked admixture of mood and anxiety symptoms, attenuated psychotic symptoms, personality disturbance, and functional impairment (Sawyer et al., 2000; Scott et al., 2012).

Pluripotency

Symptoms of mental disorder in young people (early adolescence to late 20s) generally defy clear diagnostic definitions and are a heterogeneous mix of clinical symptomatology (Sawyer et al., 2000; Scott et al., 2012), yet are still distressing and warranting clinical care (Rickwood et al., 2014, 2015). In some cases, these symptoms are self-limiting, while in others they represent emerging mental disorder that is more severe and impairing. Empirical studies suggest that diagnosable mental disorders share a range of overlapping symptoms during their prodromal periods. Therefore, an early clinical "at-risk" phenotype with pluripotentiality has been proposed (Johannessen and McGorry, 2010; McGorry 2013, 2014). An empirical study testing the theoretical robustness and clinical applicability of such an early pluripotent clinical phenotype is currently underway (Hartmann et al., 2019) (see below). This approach not only broadens the UHR criteria (the "input" or identification point) but also broadens the "output" point to include multiple targets or exit syndromes. Not only does this address the ongoing statistical concerns (Cuijpers, 2003; Cuijpers et al., 2005; Kane and Correll, 2017) regarding the use of a single relatively low-prevalence

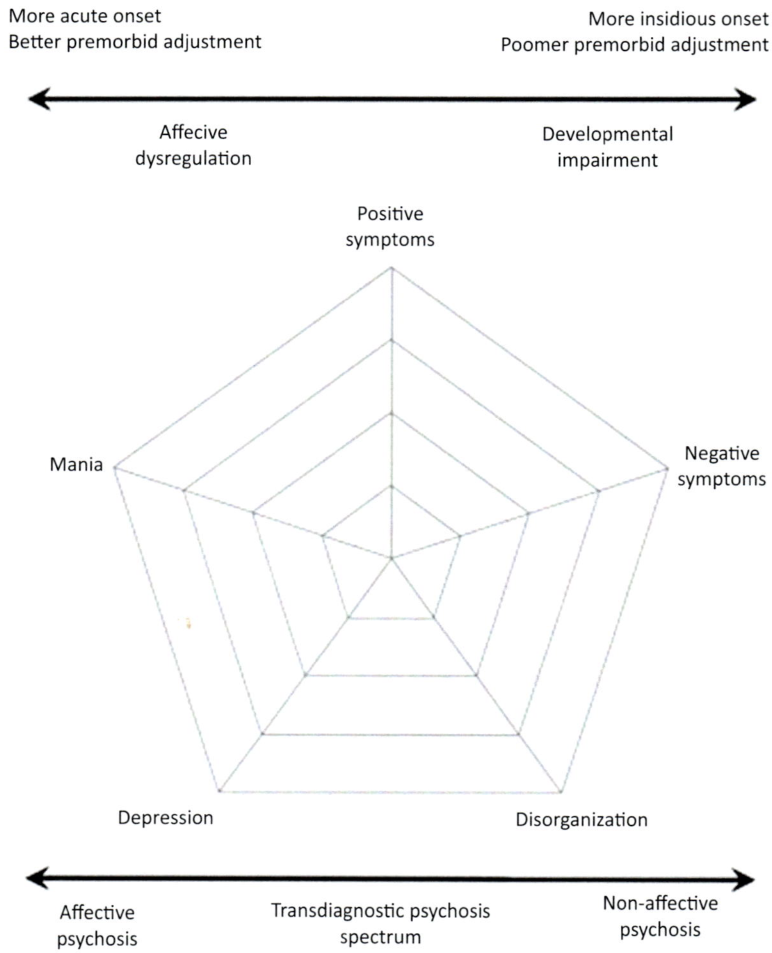

FIGURE 21.1 A transdiagnostic psychosis phenotype. A radar chart displaying the transdiagnostic psychosis continuum accounting for affective and non-affective psychotic symptomatology, which includes onset type and premorbid adjustment factors. *Reprinted from van Os, J., Reininghaus, U., 2016. Psychosis as a transdiagnostic and extended phenotype in the general population. World Psychiatry 15 (2), 118–124. Reprinted with permission.*

disorder as an outcome variable, but also the issue that early signs and symptoms of mental illness, including attenuated psychotic symptoms, may not necessarily signify the development for a *particular* type of disorder. Rather, it may signal risk for a broad range of psychiatric conditions (McGorry 2013, 2014), which is consistent with findings in the UHR population (Wigman et al., 2012; Fusar-Poli et al., 2014a; Lim et al., 2015; Lin et al., 2015; Rutigliano et al., 2016). A heuristic framework that supports this approach is clinical staging (McGorry et al., 2006; McGorry, 2007; Hickie et al., 2013).

Clinical staging

In recent years, owing to a growing disillusionment with the utility of traditional diagnostic approaches for early intervention and treatment and the limitations of existing diagnostic categories for researching etiology, a number of researchers have begun to investigate whether alternative models may be more fit for purpose (McGorry et al., 2006; Hyman, 2010; Hickie et al., 2013; Duffy, 2014; Verduijn et al., 2015). Adapted from general medicine, clinical staging in the field of psychiatry was first proposed by Fava and Kellner (1993). The clinical staging model attempts to identify the position of an individual along a continuum of illness defined according to stages (McGorry et al., 2006) and seeks to guide treatment interventions so that they are proportionate to clinical need, as well as organize research into psychosocial risk factors, neurocognitive variables, and biomarkers (both of current stage and risk for stage development) (McGorry et al., 2014). Emerging psychopathology can vary significantly over time and shift from one clinical picture to another, particularly in the early stages of disorder (McGorry and van Os, 2013). The clinical staging approach allows for an extension to the traditional clinical phenotype of mental illness progression and/or recovery (McGorry et al., 2006), broadly encompassing the following:

- Stage 0 = Asymptomatic. Nil mental ill-health symptoms.
- Stage 1a = Distressed and help-seeking. Does not meet diagnostic criteria for any mental health disorder.
- Stage 1b = Distressed and help-seeking. Displays attenuated (i.e., subthreshold) symptoms of a mental disorder.
- Stage 2: Clinically diagnosable mental illness.
- Stage 3: Persistent or recurrent mental illness.
- Stage 4: Chronic mental illness. Treatment-resistant.

The clinical staging model framework for psychotic and severe mood disorders proposed by McGorry et al. (2006) defines clinical stage (0–4) in terms of presenting symptomatology and functioning (increasing in severity as stage increases), target populations, appropriate psychosocial and pharmaceutical interventions, and stage-specific biological and endophenotypic markers. Given the widespread subdiagnostic and relatively nonspecific expression of early psychopathology in emerging psychosis (McGorry and Nelson, 2016; Hartmann et al., 2017a), proponents of this staging approach argue that the framework encompasses all dimensions of mental ill-health. A clinical staging approach to the study of psychosis allows us to broaden our understanding of emerging psychopathology and symptom trajectory outcomes outside the traditional conceptual structure provided by clinical diagnosis. The model includes those whose symptoms are transient or intermittent (attenuated psychotic experiences are transitory in about 80% of the general population (Linscott and Van Os, 2013)) as well as those who may be on a trajectory to more severe, enduring, and debilitating psychopathology. Preliminary neurocognitive and neuroimaging data also support the use of a staging model to investigate psychotic and affective disorders (Wood et al., 2011; Lin et al., 2013). Ellison-Wright et al. (2008), for example, found that decreases in gray matter were more profound and extensive in patients defined as meeting stage IV (schizophrenia diagnosis and low functioning) than in stage II (FEP). The lateral ventricles, an anatomical region commonly found to be abnormal in patients with schizophrenia, are less affected in stage II and commonly unaffected in stage 1b (UHR) (Wood et al., 2011). Overall, imaging data captured across the UHR (stage Ib) to FEP (stage II) transition consistently indicate gradual brain changes across the clinical staging threshold (Wood et al., 2011).

This approach represents a novel paradigm in which to search for stage-specific pathophysiological mechanisms as well as biological and cognitive markers. While highly consistent with theories of an extended phenotype for individual disorders (van Os and Linscott, 2012) and continuity with the general population (van Os et al., 2001), the staging model places strong diagnostic emphasis on where a person sits in the evolution of the clinical phenotype transdiagnostically. As such, one contentious issue is whether clinical staging paradigms should be disorder-specific or if a universal staging model for severe psychiatric disorders is possible. Proponents of a more universal approach base their case on the pluripotency and overlapping nature of early clinical phenotypes ("microphenotypes") (McGorry and van Os, 2013) and maintain that specific subtype trajectories can be studied *within* the broader, pragmatic pluripotent at-risk approach (Hartmann et al., 2019). In contrast, some argue that there is little evidence to support pluripotency of early clinical phenotypes and do not support a broad staging paradigm across psychiatric disorders because of the heterogeneous nature of such disorders (Duffy and Malhi, 2017; Duffy et al., 2017). Furthermore, it has been argued that such a universal framework is not well-suited to deal with mediating factors such as substance abuse (Scott and Henry, 2017). Although this debate is ongoing, considerations regarding transdiagnostic at-risk approaches and clinical staging have been hamstrung by a lack of cross-diagnostic discourse and empirical study design.

Our group recently commenced a study investigating a broad set of risk criteria with the aim of operationally defining the early clinical phenotypes (Stage 1b) for a range of exit syndromes (Stage 2) including psychosis, mania, severe depression, and borderline personality disorder (Hartmann et al., 2017a). Fig. 21.2 illustrates how this pluripotent approach is placed within the clinical staging paradigm. The term "clinical high at-risk mental state" (CHARMS,

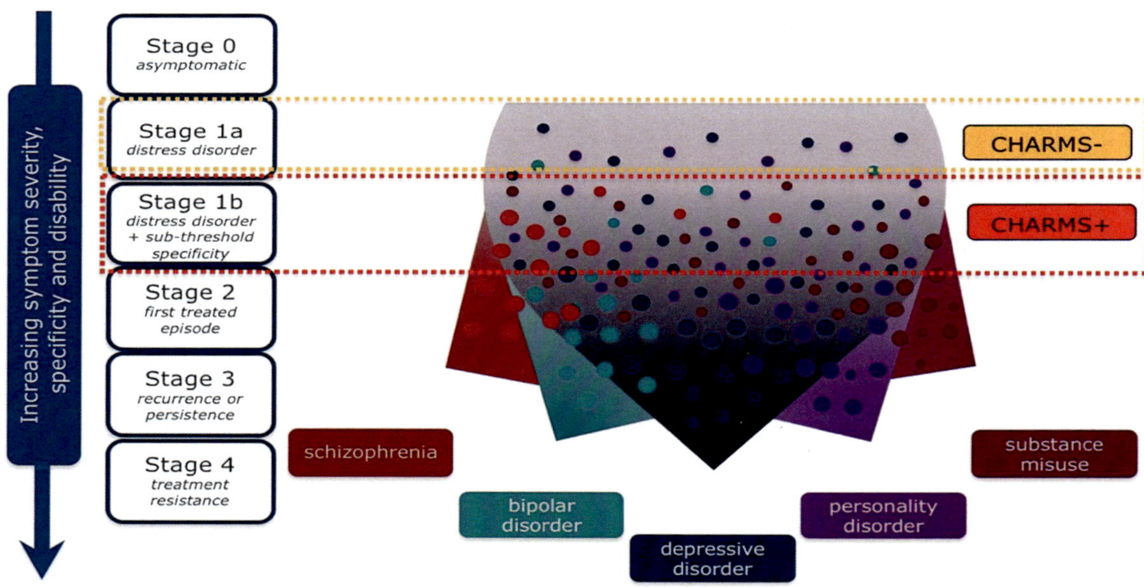

FIGURE 21.2 The clinical staging model, pluripotency, and the clinical high at-risk mental state (CHARMS) identification approach. A schematic representation of emerging psychopathology trajectories. Spots represent subthreshold symptoms (each color is referenced to a specific outcome disorder, i.e., red = schizophrenia) and blocked colors represent full threshold symptoms. Early psychopathology is shown to be broad and undifferentiated [i.e., CHARMS− (Stage 1a) and CHARMS+ (Stage 1b)]. The left panel represents clinical staging, and with increasing stage severity, the pluripotent risk for particular disorders is shown to crystallize over time as symptoms become more specific. This model illustrates that multiple (possibly overlapping) outcomes/mental disorders can be present in later disorder stages (i.e., Stage 3). *Adapted from McGorry, P., van Os, J., 2013. Redeeming diagnosis in psychiatry: timing versus specificity. Lancet 381 (9863), 343–345. Adapted with permission.*

an adaptation of the original "at-risk mental state" term (Yung et al., 1996)) is used to refer to this broad composite definition. The CHARMS approach aims to identify the subthreshold population at risk of developing severe mental ill health broadly defined, providing an operational definition of a pluripotent risk state, but who also warrant treatment in their own right on the basis of presenting symptoms. Our pilot CHARMS data indicate that evolution of symptoms may not necessarily follow a homotypic course (e.g., subthreshold psychosis evolving into threshold psychosis) but may be heterotypic (e.g., subthreshold psychosis evolving into severe depression), which is consistent with the findings of Lee et al. (2018). This does not present a problem for the CHARMS approach statistically or conceptually, as this broad at-risk approach is concerned with transition to and prediction of *any* Stage 2 "exit syndrome," rather than specific exit syndromes. Two clinical vignettes of participants in the CHARMS study are included below to illustrate these points.

CASE 1

Samantha is a 20-year-old help-seeking female attending a tertiary outpatient health care service (Orygen Youth Health Clinical Program). She presents with a first-degree relative with bipolar disorder with psychotic features, a recent drop in social and occupational functioning, attenuated psychotic symptoms (paranoid delusions, visual and auditory hallucinations), several borderline personality traits (affective instability, chronic emptiness, inappropriate anger), and comorbid moderate depression, anxiety (social and generalized), and alcohol abuse. The young person would be classified as Stage 1b in the clinical staging model and meets the CHARMS criteria for being at risk for progressing to a Stage 2 disorder. At the 6-month follow-up clinical interview, the young person has "transitioned" (meets criteria for Stage 2) to psychosis and severe depression, resulting in an inpatient admission. At the 12-month follow-up clinical interview, the young person has had a second psychotic episode and a second severe depressive episode, resulting in a second extended inpatient admission despite receiving ongoing pharmacological and psychological treatment throughout the follow-up period. Given that the psychopathology present as both persistent and recurrent, Samantha's symptoms may be classified as meeting Stage 3 according to the clinical staging model.

CASE 2

Tom is a 16-year-old help-seeking male attending a primary outpatient health care service (headspace youth mental health service). He presents with a first-degree relative with depression, sustained low social and occupational functioning, attenuated psychotic symptoms (visual and auditory hallucinations), anxiety (social and generalized), and cannabis abuse. Tom would be classified as 1b in the clinical staging model and meets the CHARMS criteria for

Continued

CASE 2 (*cont'd*)

being at risk for progressing to a Stage 2 disorder. At the 6-month follow-up clinical interview, the young person continues to be at risk for progressing to a Stage 2 disorder. At the 12-month follow-up clinical interview, the young person has "transitioned" (meets Stage 2 criteria) for severe depression. The young person was receiving ongoing pharmacological and psychological treatment throughout the study follow-up period.

These cases illustrate that a range of clinical microphenotypes can evolve into more specific macrophenotypic clinical pictures over time and that clinical symptomatology can be heterotypic and pluripotent in nature (e.g., Case 2's mix of attenuated psychotic symptoms, anxiety, substance abuse, functional decline evolving into Stage 2 depression). As such, nonspecific antecedents are consistent with a broad-based, transdiagnostic clinical staging model, at least in the initial stages of serious mental disorder. Using a pluripotent risk paradigm tackles, many of the shortcomings associated with the UHR for psychosis paradigm. Specifically, it addresses the low transition to psychosis rates, allowing for capturing a broad range of outcomes and therefore a higher "transition rate" to serious mental disorder generally, providing a clinical identification approach for broad target preventative intervention trials. Indeed, initial data from our CHARMS study indicate a 12-month transition rate to Stage 2 disorder of approximately 30% (Hartmann et al., 2019). It also places attenuated psychotic symptoms within the context of a range of multidimensional psychopathology, thereby addressing a major critique of the traditional UHR approach (van Os and Guloksuz, 2017).

Dynamic modeling approaches

In concert with the concepts of pluripotency and clinical staging, there has been increased recognition of the highly dynamic and changeable nature of mental disorder and risk, involving a complex combination of intraindividual and contextual factors interacting over time (Nelson et al., 2017b). Rather than being solely a search for a particular static factor that signals risk for future disorder (as in major *huntingtin* gene mutation in Huntington's disease), modeling risk for mental disorder may also require capturing factors (and their possible interaction) over time, i.e., must be dynamic in nature. As such, the development of predictive models with the capacity to incorporate such fluctuations in mental state (time-dependent variables) within the field of psychosis risk is increasing. The following sections will examine emerging psychiatric prediction models that can be applied to the UHR approach and integrated with the clinical staging model.

The traditional approach in psychosis prediction studies is to assess a range of variables (clinical, neurocognitive, neurobiological, genetic) upon entry to a mental health service and to determine whether these variables predict disorder onset (in the case of UHR research, FEP) or an increase/remission in symptom severity. This methodology adopts the notion that a single sampling of cross-sectional data can accurately predict the outcome of interest. However, a growing body of literature addresses the limited nature of such an approach, given the highly dynamic and changeable nature of psychopathology (van Os, 2013; McGorry and Nelson, 2016; Nelson et al., 2018). The heterogeneous nature of early psychotic experiences and symptom trajectories further supports the need to shift from static to more dynamic models of prediction (Nelson et al., 2017b; Polari et al., 2018). Models of dynamic change have predominantly emerged from disciplines outside of psychiatry. Several cross-disciplinary models of dynamic systems and how they may be conceptually and empirically applied to psychosis prediction research will be discussed.

With origins in mathematics and physics, dynamical systems theory (Scheffer, 2009) seeks to describe the behavior of complex dynamical systems such as the climate, ecosystems, and financial markets. Increasingly, mental health has been conceptualized as such a system: a system composed of many elements which interact with each other over time. The architecture of such a system reflects how it will change over time (Scheffer et al., 2012). For example, in a system with loosely connected elements, change occurs gradually over time in response to changing conditions (Fig. 21.3A). In a system with highly interconnected elements, it may initially resist change but then reach a critical threshold or "tipping point" that involves an abrupt shift to another state (Fig. 21.3B). The latter system is an example of a bistable state system with two "attractor" states. In the context of psychopathology, this has been hypothesized to correspond to a "healthy" and a "disordered state." Tipping points in bistable state systems are preceded by early warning signs or "resilience indicators," such as the phenomenon of "critical slowing down" (CSD; Fig. 21.3C). CSD indicates that the system takes increasingly longer to return to its previous state after a perturbation/stressor, the closer it is to the bifurcation (i.e., state transition) point. There is emerging evidence that transitions in mental health (at this stage, depression and bipolar disorder) are preceded by CSD (van de Leemput et al., 2014; Bayani et al., 2017).

The data required for this type of dynamic modeling require longitudinal within-person time series data. Using ecological momentary assessment, which involves the repeated sampling of variables such as behavior, contextual information, symptoms, and mood states in the course of daily life, several studies have applied dynamical system principles to the study of depression (van de Leemput et al., 2014; Wichers et al., 2016) and found evidence of CSD in the vulnerable person's response to stressors, serving as a predictor or early warning sign of a "tipping point" in mood state (from normal to depressed state or vice versa) (Wigman et al., 2013; van de Leemput et al., 2014; Wichers, 2014; Wichers et al., 2015; Wichers et al., 2016). Therefore, predictive modeling based on the principles of dynamical systems theory may be a useful tool to incorporate into psychosis risk paradigms and has already shown some utility in the area of depression (see below).

As stated above, complex dynamical systems are systems that consist of many interacting components, that is, a network of interacting elements. In the "network perspective" of psychopathology, which has gained traction in recent years, mental disorders are not conceptualized as the consequence of an underlying latent variable (a "common cause") but as a consequence of a dynamic interplay of symptoms (Fig. 21.4) (Borsboom and Cramer, 2013; van Borkulo et al., 2015; Isvoranu et al., 2016). That is, symptoms are actively influencing/causing each other, rather than being the passive expression of an underlying disease process.

Wigman et al. (2015) found that patients with a diagnosis of depression or psychotic disorder demonstrated stronger mental state cross-reactivity and long-term persistence of affective state than healthy controls. Isvoranu et al. (2017) recently demonstrated that the expression of certain psychopathology can link various types of trauma with psychotic symptoms. A network structure has also demonstrated that vulnerability factors such as substance use and urban environment are differentially

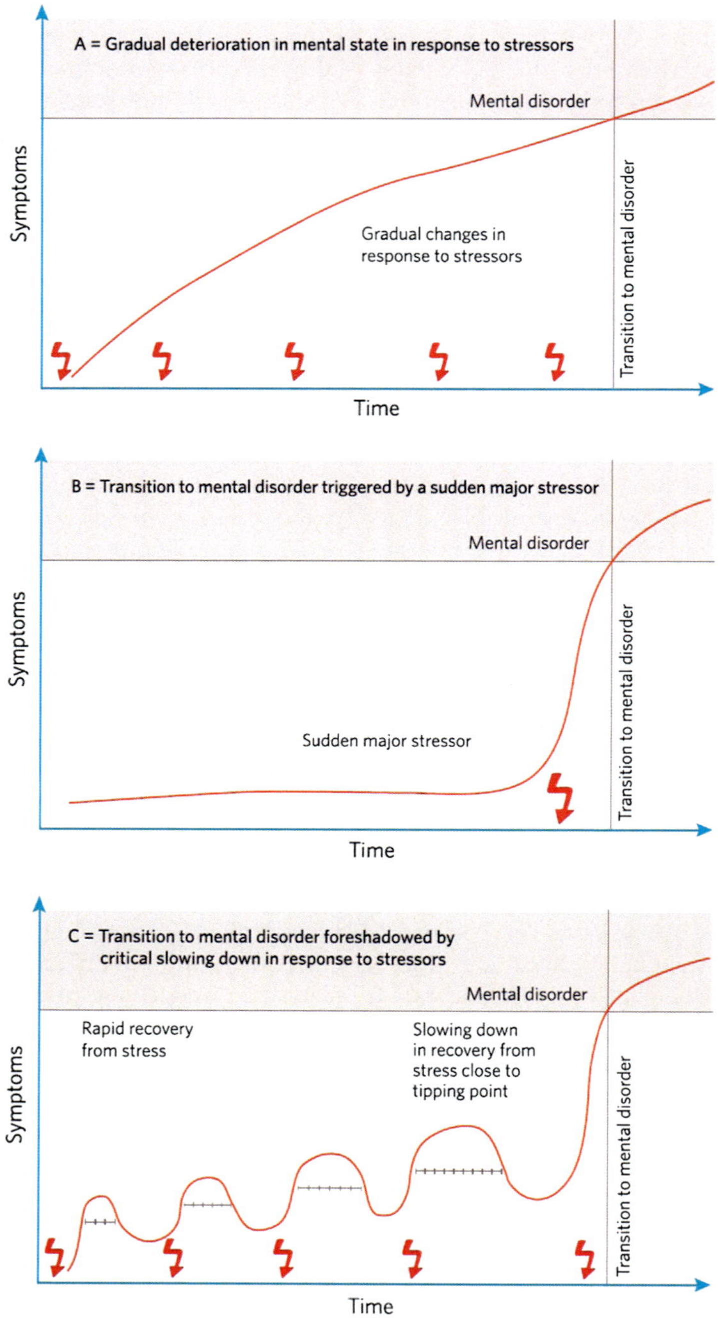

FIGURE 21.3 Modeling mental disorder onset using a dynamical systems approach. With stressors indicated by a *red arrow*, Fig. 21.3A–C collectively reflect how complex dynamical systems can allow us to differentially map changes in mental state across time. *Reprinted from Nelson, B., McGorry, P. D., Wichers, M., Wigman, J. T. W., Hartmann, J., 2017. Moving from static to dynamic models of the onset of mental disorder. JAMA Psychiatry 74 (5), 528–534. Reprinted with permission.*

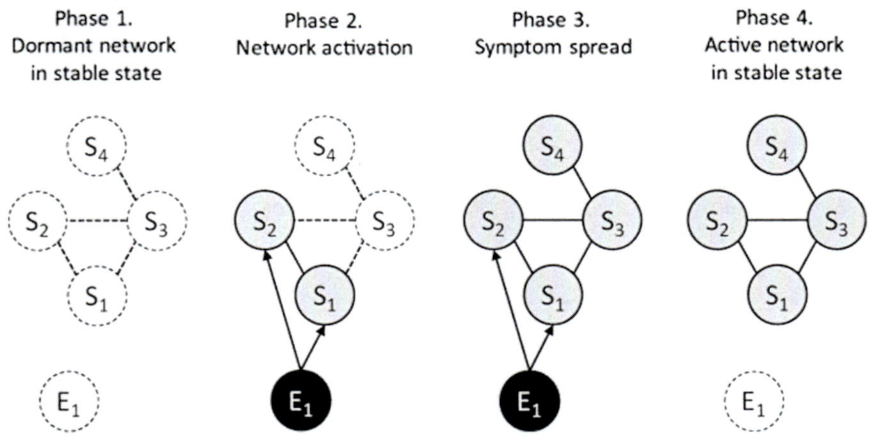

FIGURE 21.4 The onset of mental disorders within a network system model. Phase 1 represents an asymptomatic phase; Phase 2 represents the introduction of an external event (E1) to the network which activates some symptoms (S1, S2). In Phase 3, you see how this external event (E1) has then lead to the activation of several other interconnected symptoms (S3, S4). Phase 4 illustrates that if the network is strongly interconnected, the eradication of the previous external event (E1) does not lead to symptom remission as the network is self-sustaining and remains active. *Reprinted from Borsboom, D., 2017. A network theory of mental disorders. World Psychiatry 16 (1), 5–13. Reprinted with permission.*

connected to specific psychopathological dimensions (i.e., anxiety, depression) (Isvoranu et al., 2016). Such a model can demonstrate how various risk factors and symptoms of mental disorder are both differentiated and interrelated, possibly illuminating the mechanisms involved across the evolution of disorder. Recent work by Cramer et al. (2016) was able to demonstrate that system networks exhibit "early warning signals" (i.e., CSD; see above) before "tipping" to a depressed state. The authors conclude that such an approach could be applied to other mental disorders such as schizophrenia (Cramer et al., 2016). Within the context of pluripotency during early psychopathology, it has been proposed that the way in which networked symptoms influence each other during earlier stages of mental ill health may be less concentrated and stable than in later stages (Borsboom, 2017). Preliminary empirical work by Wigman et al. (2013) is consistent with this proposal, positioning network dynamics theory within the clinical staging framework and suggesting that with increased clinical stage severity, symptomatology becomes more specialized and differentiated, giving rise to diagnostic specificity associated with greater inter- and intramental state connection strength and greater inter- and intramental state connection variability. Empirical investigations into the predictive potential of dynamic symptom networks for the onset and progression of psychosis are currently underway (Booij et al., 2018).

Joint modeling

Another approach to modeling the predictive value of the dynamic nature of psychopathology is joint modeling. This approach is more agnostic with regard to theoretical principles. It is a statistical method that combines multilevel modeling (using repeated clinical assessments) with survival analysis (allowing for the time-to-event nature of determining outcome in psychopathology prediction studies) (Yuen and Mackinnon,

2016; Yuen et al. 2017, 2018). The approach allows for the identification of symptom trajectories (e.g., persistence of negative symptoms, intensification of general psychopathology) that predict outcome, taking into account censored data and time to follow-up (as in survival analysis). Initial work using this approach with data from a large UHR intervention trial (McGorry et al., 2017) shows that dynamic prediction using joint modeling produces much stronger predictive models, particularly positive predictive values, than using baseline data alone (Yuen et al., 2018).

Importantly, these dynamic prediction approaches link well with the process that occurs in real-world clinical decision-making (Chitty, 2005). Clinical decision-making regarding possible treatment changes and prognostic judgments is generally "adaptive" in nature, i.e., they react to and are updated based on the unfolding clinical symptomatology of the patient, rather than relying solely on the profile of the patient's first clinical presentation (Nelson et al., 2017a). Using the conceptual and analytic approaches outlined here may provide an empirically based and rigorous guide for clinical decision-making for modifying treatment in response to the evolution of a patient's clinical profile over time. In this way, they may be able to refine treatment decision-making and possibly be incorporated into adaptive clinical trial design, which are currently generally based purely on a category of response/nonresponse at the end of a prespecified time period (Bhatt and Mehta, 2016). For example, it is possible, using joint modeling and other strategies, to construct a risk calculator for transition and other outcomes that is *updated* at each assessment point, rather than being based solely on baseline variables (Cannon et al., 2016; Fusar-Poli et al., 2017b; Yuen et al., 2017; Yuen et al., 2018). While the latter provides a single risk estimate based on baseline measures, a dynamic calculator updated at each assessment point would mirror the unfolding clinical picture and thereby inform treatment decision-making (e.g., increase in risk of poor outcome might signal a need to intensify treatment, such as commencement or increase in medication).

A major consideration for dynamic approaches to prediction is the large volume of data required across multiple or extended time points. Study designs would be required to include repeat longitudinal assessment of relevant psychosocial or other variables (e.g., neurocognition, neuroimaging) achieved through either, or a combination of, micro- (momentary and day-to-day) or macro-level (month and year) assessments to move from a static to a more dynamic modeling approach (see Fig. 21.5). This may be a challenge in a population whose symptomatology (e.g., poor attention, low motivation) may interfere with such an intensive research study paradigm. A balance needs to be found between detail/richness of data collected and minimizing participant burden. For instance, while passive sensing, e.g., collecting data from mobile phone apps, allows ease of data collection and minimal participant burden, the data may be relatively superficial (Insel, 2017; Nelson et al., 2018). On the other hand, frequently repeated detailed in-person interviewing may not be feasible. The cost-effectiveness of dynamic approaches to data collection would also need to be considered. Simulation studies, as recently conducted by Cramer et al. (2016), may be a first logical step in exploring the utility of complex dynamic modeling in the field of psychosis research. However, the data available to date already indicate the superiority of these approaches over using baseline data alone, which would suggest that the challenges of collecting the data required are justified by the enhanced prediction models and resultant positive impact on clinical care.

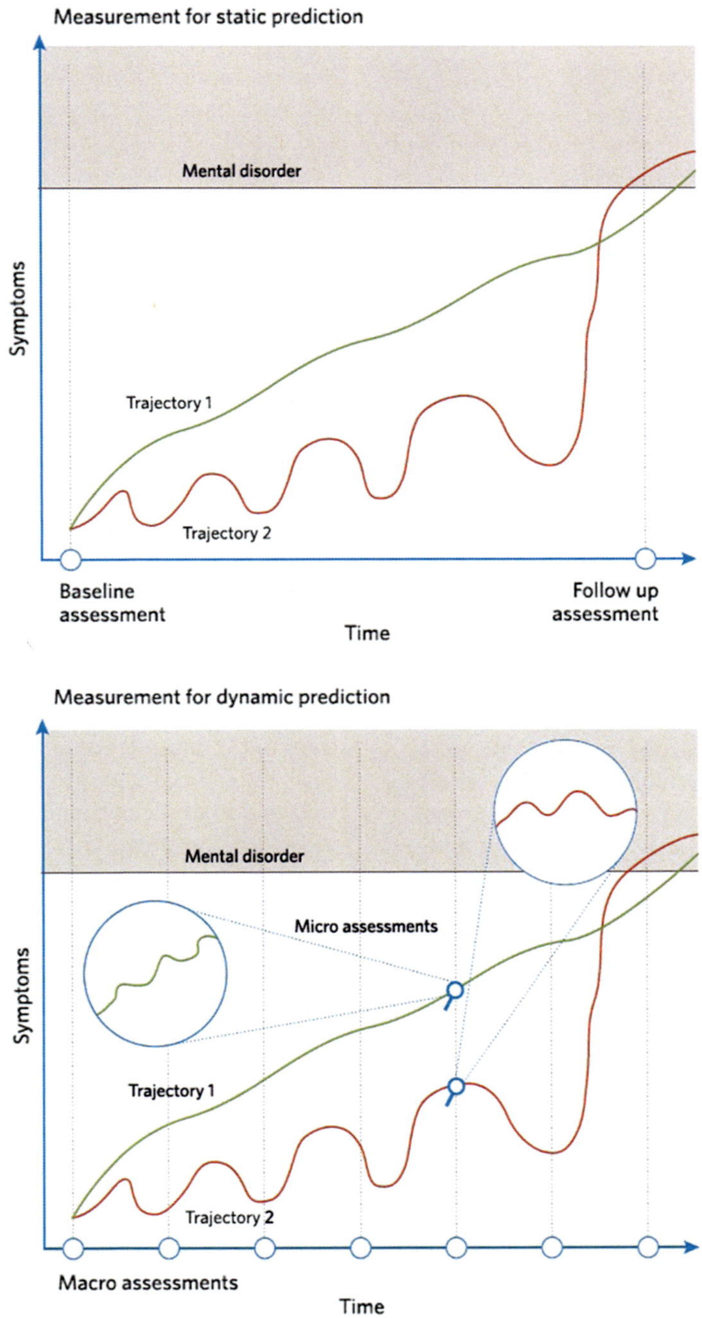

FIGURE 21.5 The macro- and micro-level assessments required for static versus dynamic predictive modeling. The *green and orange lines* represent two possible symptom trajectories leading to a full threshold mental disorder. The bottom panel represents how micro-level assessments (characterized by the magnifying glass symbol) may capture subtle fluctuations in mental state, which would not typically be captured by macro-level assessments (*blue circles* on the x-axis) before a system change (i.e., disorder onset). *Reprinted from Nelson, B., McGorry, P. D., Wichers, M., Wigman, J. T. W., Hartmann, J., 2017. Moving from static to dynamic models of the onset of mental disorder. JAMA Psychiatry 74 (5), 528–534. Reprinted with permission.*

Conclusion

The development of the UHR criteria for psychosis over 20 years ago created a new paradigm to investigate the prodrome of psychotic disorder (Yung et al., 1996). Although the UHR criteria capture elevated risk for subsequent psychotic disorders, attenuated psychotic symptoms are also cross-linked with other comorbid disorders and symptomatology (Wigman et al., 2012; Fusar-Poli et al., 2014a; Lim et al., 2015; Lin et al., 2015; Rutigliano et al., 2016). Diverse routes and a heterotypic course of emerging psychopathology are common (Sawyer et al., 2000; Scott et al., 2012). Sufficient study power to analyze predictors of psychosis onset and intervention strategies has also been an important consideration in the field of psychosis risk. Cuijpers (2003) introduced the value of including multiple outcomes or exit disorders when investigating a relatively low-incidence syndrome, such as schizophrenia spectrum or bipolar disorders.

To address these issues, we have proposed a new paradigm in which to study emerging mental disorder, including but not limited to onset of psychotic disorder, involving the identification of a pluripotent "at-risk" population with transdiagnostic targets (CHARMS (Hartmann et al., 2017a)). This shift in perspective creates a framework for new approaches to include broad risk factors and pathogenic mechanisms, which have traditionally been investigated in a disorder-specific fashion. It also lays the groundwork for the next generation of preventive intervention trials and overcomes the statistical issues of using a single low-incidence disorder as the single outcome target.

Clinical staging models provide the framework in which to move beyond the traditional dichotomous approach to psychiatric diagnosis by defining the extent of disease progression based on severity, duration, and course of symptoms (McGorry et al., 2006; McGorry, 2007; Hickie et al., 2013). This heuristic framework emphasizes that the developmental trajectory of psychosis is not linear or fixed. The clinical staging model also seeks to guide treatment interventions so that they are proportionate to clinical need, as well as organize research into psychosocial risk factors, neurocognitive variables, and biomarkers (both of current stage and risk for stage development). Although there is debate in the field about general versus disorder-specific staging models, our view, supported by our CHARMS study data, is that the literature increasingly supports a broad and universal staging model during the early stages of serious mental disorder, followed by more specific macrophenotypes as disorders progress and symptoms become more persistent and severe (McGorry et al., 2006; McGorry, 2007; Hickie et al., 2013; Hartmann et al., 2018).

The ability of predictive models to identify risk of evolving from "at-risk" state to full threshold disorder (whether for psychotic disorder or other disorders), both on the individual and group level, may benefit from incorporating dynamic, time-dependent variables (Nelson et al., 2017b). This requires different data collection and analytic strategies, as flagged above, to those that are most commonly used in the psychosis risk field. Specifically, dynamical systems theory and network modeling approaches may be of value (Scheffer, 2009; Borsboom, 2017), as well as joint modeling of symptom trajectories. These approaches may be able to capture fluctuations in mental state and contextual factors over time. Such paradigms may provide enriched knowledge of psychosis onset, points of vulnerability and resilience, and identify markers of impending clinical change, which has direct implications for adaptive preventive treatments.

Glossary

Clinical staging method used to stage illness severity and progression

Pluripotency can give rise to multiple different outcomes

Transdiagnostic across mental health diagnoses

List of acronyms and abbreviations

CHR Clinical high risk
FEP First-episode psychosis
UHR Ultra high risk

References

Bayani, A., Hadaeghi, F., Jafari, S., Murray, G., 2017. Critical slowing down as an early warning of transitions in episodes of bipolar disorder: a simulation study based on a computational model of circadian activity rhythms. Chronobiol. Int. 34 (2), 235–245.

Bhatt, D.L., Mehta, C., 2016. Adaptive designs for clinical trials. N. Engl. J. Med. 375 (1), 65–74.

Booij, S.H., Wichers, M., de Jonge, P., Sytema, S., van Os, J., Wunderink, L., Wigman, J.T., 2018. Study protocol for a prospective cohort study examining the predictive potential of dynamic symptom networks for the onset and progression of psychosis: the Mapping Individual Routes of Risk and Resilience (Mirorr) study. BMJ Open 8 (1), e019059.

Borsboom, D., 2017. A network theory of mental disorders. World Psychiatry 16 (1), 5–13.

Borsboom, D., Cramer, A.O., 2013. Network analysis: an integrative approach to the structure of psychopathology. Annu. Rev. Clin. Psychol. 9, 91–121.

Cannon, T.D., Yu, C., Addington, J., Bearden, C.E., Cadenhead, K.S., Cornblatt, B.A., Heinssen, R., Jeffries, C.D., Mathalon, D.H., McGlashan, T.H., Perkins, D.O., Seidman, L.J., Tsuang, M.T., Walker, E.F., Woods, S.W., Kattan, M.W., 2016. An individualized risk calculator for research in prodromal psychosis. Am. J. Psychiatry 173 (10), 980–988.

Chitty, R.N., 2005. Why clinicians are natural bayesians: is there a bayesian doctor in the house? BMJ 330 (7504), 1390.

Cramer, A.O., van Borkulo, C.D., Giltay, E.J., van der Maas, H.L., Kendler, K.S., Scheffer, M., Borsboom, D., 2016. Major depression as a complex dynamic system. PLoS One 11 (12), e0167490.

Cuijpers, P., 2003. Examining the effects of prevention programs on the incidence of new cases of mental disorders: the lack of statistical power. Am. J. Psychiatry 160 (8), 1385–1391.

Cuijpers, P., Van Straten, A., Smit, F., 2005. Preventing the incidence of new cases of mental disorders: a meta-analytic review. J. Nerv. Ment. Dis. 193 (2), 119–125.

Duffy, A., 2014. Toward a comprehensive clinical staging model for bipolar disorder: integrating the evidence. Can. J. Psychiatr. 59 (12), 659–666.

Duffy, A., Malhi, G.S., 2017. Mapping the Developmental Trajectory of Bipolar Disorder: Importance of Prerequisite Groundwork. SAGE Publications Sage UK, London, England.

Duffy, A., Malhi, G.S., Grof, P., 2017. Do the trajectories of bipolar disorder and schizophrenia follow a universal staging model? Can. J. Psychiatr. 62 (2), 115–122.

Ellison-Wright, I., Glahn, D.C., Laird, A.R., Thelen, S.M., Bullmore, E., 2008. The anatomy of first-episode and chronic schizophrenia: an anatomical likelihood estimation meta-analysis. Am. J. Psychiatry 165 (8), 1015–1023.

Fava, G.A., Kellner, R., 1993. Staging: a neglected dimension in psychiatric classification. Acta Psychiatr. Scand. 87 (4), 225–230.

Fusar-Poli, P., Nelson, B., Valmaggia, L., Yung, A.R., McGuire, P.K., 2014a. Comorbid depressive and anxiety disorders in 509 individuals with an at-risk mental state: impact on psychopathology and transition to psychosis. Schizophr. Bull. 40 (1), 120–131.

Fusar-Poli, P., Yung, A.R., McGorry, P., van Os, J., 2014b. Lessons learned from the psychosis high-risk state: towards a general staging model of prodromal intervention. Psychol. Med. 44 (1), 17–24.

Fusar-Poli, P., Rutigliano, G., Stahl, D., Davies, C., Bonoldi, I., Reilly, T., McGuire, P., 2017a. Development and validation of a clinically based risk calculator for the transdiagnostic prediction of psychosis. JAMA Psychiatry 74 (5), 493–500.

Fusar-Poli, P., Rutigliano, G., Stahl, D., Davies, C., De Micheli, A., Ramella-Cravaro, V., Bonoldi, I., McGuire, P., 2017b. Long-term validity of the at Risk Mental State (ARMS) for predicting psychotic and non-psychotic mental disorders. Eur. Psychiatry 42, 49–54.

Hartmann, J., Nelson, B., Ratheesh, A., Treen, D., McGorry, P.D., 2017a. An integrative review of clinical predictors of serious mental illness in at-risk young people. Psychol. Med. (in press).

Hartmann, J., Nelson, B., Spooner, R., Amminger, G.P., Chanen, A., Davey, C.G., McHugh, M., Ratheesh, A., Treen, D., Yuen, H.P., McGorry, P.D., 2019. A broad clinical high risk mental state (CHARMS): Methodology of a cohort study validating criteria for pluripotent risk. Early Intervention in Psychiatry 13 (3), 379–386.

Hartmann, J., Nelson, B., Ratheesh, A., Treen, D., McGorry, P., 2018. At-Risk Studies and Clinical Antecedents of Psychosis, Bipolar Disorder and Depression: A Scoping Review in the Context of Clinical Staging (Under review).

Hickie, I.B., Scott, E.M., Hermens, D.F., Naismith, S.L., Guastella, A.J., Kaur, M., Sidis, A., Whitwell, B., Glozier, N., Davenport, T., Pantelis, C., Wood, S.J., McGorry, P.D., 2013. Applying clinical staging to young people who present for mental health care. Early Interv. Psychiatry 7 (1), 31–43.

Hyman, S.E., 2010. The diagnosis of mental disorders: the problem of reification. Annu. Rev. Clin. Psychol. 6, 155–179.

Insel, T.R., 2017. Digital phenotyping: technology for a new science of behavior. JAMA 318 (13), 1215–1216.

Isvoranu, A.M., Borsboom, D., van Os, J., Guloksuz, S., 2016. A network approach to environmental impact in psychotic disorder: brief theoretical framework. Schizophr. Bull. 42 (4), 870–873.

Isvoranu, A.M., van Borkulo, C.D., Boyette, L.L., Wigman, J.T., Vinkers, C.H., Borsboom, D., I. Group, 2017. A network approach to psychosis: pathways between childhood trauma and psychotic symptoms. Schizophr. Bull. 43 (1), 187–196.

Johannessen, J.O., McGorry, P., 2010. DSM-5 and the 'Psychosis Risk Syndrome': the need for a broader perspective. Psychosis 2 (2), 93–96.

Kane, J.M., Correll, C.U., 2017. ω-3 polyunsaturated fatty acids to prevent psychosis: the importance of replication studies. JAMA Psychiatry 74 (1), 11–12.

Lee, T.Y., Lee, J., Kim, M., Choe, E., Kwon, J.S., 2018. Can we predict psychosis outside the clinical high-risk state? A systematic review of non-psychotic risk syndromes for mental disorders. Schizophr. Bull. 44 (2), 276–285.

Lim, J., Rekhi, G., Rapisarda, A., Lam, M., Kraus, M., Keefe, R.S., Lee, J., 2015. Impact of psychiatric comorbidity in individuals at ultra high risk of psychosis—findings from the longitudinal youth at risk study (LYRIKS). Schizophr. Res. 164 (1), 8–14.

Lin, A., Reniers, R.L., Wood, S.J., 2013. Clinical staging in severe mental disorder: evidence from neurocognition and neuroimaging. Br. J. Psychiatr. Suppl. 54, s11–17.

Lin, A., Wood, S.J., Nelson, B., Beavan, A., McGorry, P., Yung, A.R., 2015. Outcomes of nontransitioned cases in a sample at ultra-high risk for psychosis. Am. J. Psychiatry 172 (3), 249–258.

Linscott, R., Van Os, J., 2013. An updated and conservative systematic review and meta-analysis of epidemiological evidence on psychotic experiences in children and adults: on the pathway from proneness to persistence to dimensional expression across mental disorders. Psychol. Med. 43 (6), 1133–1149.

McGorry, P.D., 2007. Issues for DSM-V: clinical staging: a heuristic pathway to valid nosology and safer, more effective treatment in psychiatry. Am. J. Psychiatry 164 (6), 859–860.

McGorry, P., 2013. Early clinical phenotypes and risk for serious mental disorders in young people: need for care precedes traditional diagnoses in mood and psychotic disorders. Can. J. Psychiatr. 58 (1), 19–21.

McGorry, P.D., 2014. Beyond psychosis risk: early clinical phenotypes in mental disorder and the subthreshold pathway to safe, timely and effective care. Psychopathology 47 (5), 285–286.

McGorry, P., van Os, J., 2013. Redeeming diagnosis in psychiatry: timing versus specificity. Lancet 381 (9863), 343–345.

McGorry, P., Nelson, B., 2016. Why we need a transdiagnostic staging approach to emerging psychopathology, early diagnosis, and treatment. JAMA Psychiatry 1–2.

McGorry, P.D., Hickie, I.B., Yung, A.R., Pantelis, C., Jackson, H.J., 2006. Clinical staging of psychiatric disorders: a heuristic framework for choosing earlier, safer and more effective interventions. Aust. N. Z. J. Psychiatr. 40 (8), 616–622.

McGorry, P., Keshavan, M., Goldstone, S., Amminger, P., Allott, K., Berk, M., Lavoie, S., Pantelis, C., Yung, A., Wood, S., 2014. Biomarkers and clinical staging in psychiatry. World Psychiatry 13 (3), 211–223.

McGorry, P.D., Nelson, B., Markulev, C., Yuen, H.P., Schäfer, M.R., Mossaheb, N., Schlögelhofer, M., Smesny, S., Hickie, I.B., Berger, G.E., 2017. Effect of ω-3 polyunsaturated fatty acids in young people at ultrahigh risk for psychotic disorders: the NEURAPRO randomized clinical trial. JAMA Psychiatry 74 (1), 19–27.

Morrison, A.P., French, P., Stewart, S.L., Birchwood, M., Fowler, D., Gumley, A.I., Jones, P.B., Bentall, R.P., Lewis, S.W., Murray, G.K., 2012. Early detection and intervention evaluation for people at risk of psychosis: multisite randomised controlled trial. BMJ 344, e2233.

Nelson, B., Hartmann, J.A., 2018. Prediction and its limits (or why life can only be understood backwards but must be lived forwards). Early Intervention in Psychiatry 12 (5), 767–770.

Nelson, B., Hartmann, J., Parnas, J., 2018. Detail, dynamics, depth: useful correctives for some current research trends. Br J. Psychiatry 212, 262–264.

Nelson, B., Yung, A.R., 2009. Psychotic-like experiences as overdetermined phenomena: when do they increase risk for psychotic disorder? Schizophr. Res. 108 (1–3), 303–304.

Nelson, B., Thompson, A., Yung, A.R., 2013. Not all first-episode psychosis is the same: preliminary evidence of greater basic self-disturbance in schizophrenia spectrum cases. Early Interv. Psychiatry 7 (2), 200–204.

Nelson, B., Amminger, G.P., Yuen, H.P., Wallis, N., J Kerr, M., Dixon, L., Carter, C., Loewy, R., Niendam, T.A., Shumway, M., 2017a. Staged treatment in early psychosis: a sequential multiple assignment randomised trial of interventions for ultra high risk of psychosis patients. Early Interv. Psychiatry.

Nelson, B., McGorry, P.D., Wichers, M., Wigman, J.T.W., Hartmann, J., 2017b. Moving from static to dynamic models of the onset of mental disorder. JAMA Psychiatry 74 (5), 528–534.

Phillips, L.J., Nelson, B., Yuen, H.P., Francey, S.M., Simmons, M., Stanford, C., Ross, M., Kelly, D., Baker, K., Conus, P., 2009. Randomized controlled trial of interventions for young people at ultra-high risk of psychosis: study design and baseline characteristics. Aust. N. Z. J. Psychiatr. 43 (9), 818–829.

Polari, A., Lavoie, S., Yuen, H.P., Amminger, G.P., Berger, G., Chen, E., deHaan, L., Hartmann, J., Markulev, C., Melville, F., Nieman, D., Nordentoft, M., Riecher-Rössler, A., Smesny, S., Stratford, J., Verman, S., McGorry, P.D., Nelson, B., 2018. Clinical trajectories in the ultra-high risk for psychosis population. Schizophr. Res. (in press).

Rickwood, D.J., Telford, N.R., Parker, A.G., Tanti, C.J., McGorry, P.D., 2014. headspace—Australia's innovation in youth mental health: who are the clients and why are they presenting? Med. J. Aust. 200 (2), 108–111.

Rickwood, D.J., Mazzer, K.R., Telford, N.R., Parker, A.G., Tanti, C.J., McGorry, P.D., 2015. Changes in psychological distress and psychosocial functioning in young people visiting headspace centres for mental health problems. Med. J. Aust. 202 (10), 537–542.

Rutigliano, G., Valmaggia, L., Landi, P., Frascarelli, M., Cappucciati, M., Sear, V., Rocchetti, M., De Micheli, A., Jones, C., Palombini, E., McGuire, P., Fusar-Poli, P., 2016. Persistence or recurrence of non-psychotic comorbid mental disorders associated with 6-year poor functional outcomes in patients at ultra high risk for psychosis. J. Affect. Disord. 203, 101–110.

Sawyer, M., Arney, F., Baghurst, P., Clark, J., Graetz, B., Kosky, R., Nurcombe, B., Patton, G., Prior, M., Raphael, B., 2000. The mental health of young people in Australia. In: Canberra: Mental Health and Special Programs Branch. Commonwealth Department of Health and Aged Care.

Scheffer, M., 2009. Critical Transitions in Nature and Society. Princeton University Press.

Scheffer, M., Carpenter, S.R., Lenton, T.M., Bascompte, J., Brock, W., Dakos, V., van de Koppel, J., van de Leemput, I.A., Levin, S.A., van Nes, E.H., Pascual, M., Vandermeer, J., 2012. Anticipating critical transitions. Science 338 (6105), 344–348.

Scott, J., Henry, C., 2017. Clinical staging models: from general medicine to mental disorders. BJPsych Adv. 23 (5), 292–299.

Scott, E.M., Hermens, D.F., Glozier, N., Naismith, S.L., Guastella, A.J., Hickie, I.B., 2012. Targeted primary care-based mental health services for young Australians. Med. J. Aust. 196, 136–140.

Sevilla-Llewellyn-Jones, J., Camino, G., Russo, D.A., Painter, M., Montejo, A.L., Ochoa, S., Jones, P.B., Perez, J., 2018. Clinically significant personality traits in individuals at high risk of developing psychosis. Psychiatry Res. 261, 498–503.

Shah, J.L., Crawford, A., Mustafa, S.S., Iyer, S.N., Joober, R., Malla, A.K., 2017. Is the clinical high-risk state a valid concept? Retrospective examination in a first-episode psychosis sample. Psychiatr. Serv. 68 (10), 1046–1052.

van Borkulo, C., Boschloo, L., Borsboom, D., Penninx, B.W., Waldorp, L.J., Schoevers, R.A., 2015. Association of symptom network structure with the course of longitudinal depression. JAMA Psychiatry 72 (12), 1219–1226.

van de Leemput, I.A., Wichers, M., Cramer, A.O., Borsboom, D., Tuerlinckx, F., Kuppens, P., van Nes, E.H., Viechtbauer, W., Giltay, E.J., Aggen, S.H., Derom, C., Jacobs, N., Kendler, K.S., van der Maas, H.L., Neale, M.C., Peeters, F., Thiery, E., Zachar, P., Scheffer, M., 2014. Critical slowing down as early warning for the onset and termination of depression. Proc. Natl. Acad. Sci. USA 111 (1), 87–92.

van Os, J., 2013. The dynamics of subthreshold psychopathology: implications for diagnosis and treatment. Am. J. Psychiatry 170 (7), 695–698.

van Os, J., Guloksuz, S., 2017. A critique of the "ultra-high risk" and "transition" paradigm. World Psychiatry 16 (2), 200–206.

van Os, J., Linscott, R.J., 2012. Introduction: the extended psychosis phenotype—relationship with schizophrenia and with ultrahigh risk status for psychosis. Schizophr. Bull. 38 (2), 227–230.

van Os, J., Reininghaus, U., 2016. Psychosis as a transdiagnostic and extended phenotype in the general population. World Psychiatry 15 (2), 118–124.

van Os, J., Hanssen, M., Bijl, R.V., Vollebergh, W., 2001. Prevalence of psychotic disorder and community level of psychotic symptoms: an urban-rural comparison. Arch. Gen. Psychiatr. 58 (7), 663–668.

Verduijn, J., Milaneschi, Y., van Hemert, A.M., Schoevers, R.A., Hickie, I.B., Penninx, B.W., Beekman, A.T., 2015. Clinical staging of major depressive disorder: an empirical exploration. J. Clin. Psychiatry 76 (9), 1200–1208.

Webb, J.R., Addington, J., Perkins, D.O., Bearden, C.E., Cadenhead, K.S., Cannon, T.D., Cornblatt, B.A., Heinssen, R.K., Seidman, L.J., Tarbox, S.I., 2015. Specificity of incident diagnostic outcomes in patients at clinical high risk for psychosis. Schizophr. Bull. 41 (5), 1066–1075.

Wichers, M., 2014. The dynamic nature of depression: a new micro-level perspective of mental disorder that meets current challenges. Psychol. Med. 44 (7), 1349–1360.

Wichers, M., Wigman, J.T.W., Myin-Germeys, I., 2015. Micro-level affect dynamics in psychopathology viewed from complex dynamical system theory. Emotion Rev. 7 (4), 362–367.

Wichers, M., Groot, P.C., G, E.S.M., G, E.W.S., Psychosystems, 2016. Critical slowing down as a personalized early warning signal for depression. Psychother. Psychosom. 85 (2), 114–116.

Wigman, J.T., van Nierop, M., Vollebergh, W.A., Lieb, R., Beesdo-Baum, K., Wittchen, H.-U., van Os, J., 2012. Evidence that psychotic symptoms are prevalent in disorders of anxiety and depression, impacting on illness onset, risk, and severity—implications for diagnosis and ultra–high risk research. Schizophr. Bull. 38 (2), 247–257.

Wigman, J.T.W., van Os, J., Thiery, E., Derom, C., Collip, D., Jacobs, N., Wichers, M., 2013. Psychiatric diagnosis revisited: towards a system of staging and profiling combining nomothetic and idiographic parameters of momentary mental states. PLoS One 8 (3).

Wigman, J., Van Os, J., Borsboom, D., Wardenaar, K., Epskamp, S., Klippel, A., Viechtbauer, W., Myin-Germeys, I., Wichers, M., 2015. Exploring the underlying structure of mental disorders: cross-diagnostic differences and similarities from a network perspective using both a top-down and a bottom-up approach. Psychol. Med. 45 (11), 2375–2387.

Wood, S.J., Yung, A.R., McGorry, P.D., Pantelis, C., 2011. Neuroimaging and treatment evidence for clinical staging in psychotic disorders: from the at-risk mental state to chronic schizophrenia. Biol. Psychiatry 70 (7), 619–625.

Woods, S.W., Powers, A.R., Taylor, J.H., Davidson, C.A., Johannesen, J.K., Addington, J., Perkins, D.O., Bearden, C.E., Cadenhead, K.S., Cannon, T.D., 2018. Lack of diagnostic pluripotentiality in patients at clinical high risk for psychosis: specificity of comorbidity persistence and search for pluripotential subgroups. Schizophr. Bull. 44 (2), 254–263.

Yuen, H.P., Mackinnon, A., 2016. Performance of joint modelling of time-to-event data with time-dependent predictors: an assessment based on transition to psychosis data. PeerJ 4, e2582.

Yuen, H.P., Mackinnon, A., Nelson, B., 2017. A method for analysing transition to psychosis: joint modelling of time-to-event outcome with time-dependent predictors. Int. J. Methods Psychiatr. Res. (in press).

Yuen, H.P., Mackinnon, A., Hartmann, J., Amminger, G.P., Markulev, C., Lavoie, S., Schäfer, M.R., Mossaheb, N., Schlögelhofer, M., Smesny, S., Hickie, I.B., Berger, G.E., Chen, E.Y.H., de Haan, L., Nieman, D.H., Nordentoft, M., Riecher-Rössler, A., Verma, S., Thompson, A., Yung, A., McGorry, P., Nelson, B., 2018. A Novel Approach for Developing Prediction Models of Transition to Psychosis:Dynamic Prediction Using Joint Modelling (Under review).

Yung, A., Phillips, L., McGorry, P., Ward, J., Thompson, K., 1996. The Comprehensive Assessment of At-Risk Mental States (CAARMS) Manual. University of Melbourne, Melbourne, Victoria, Australia.

Yung, A.R., Nelson, B., Thompson, A., Wood, S.J., 2010. The psychosis threshold in Ultra High Risk (prodromal) research: is it valid? Schizophr. Res. 120 (1–3), 1–6.

CHAPTER

22

Future directions in risk research

Nikolai Albert[1], *Louise Birkedal Glenthøj*[1], *Merete Nordentoft*[1,2]

[1]**Mental Health Center Copenhagen, Copenhagen University Hospital, Copenhagen, Denmark;**
[2]**University of Copenhagen, Copenhagen, Denmark**

Introduction

This book has summarized several fields of risk research. In this chapter, we step back to review psychosis risk research, mostly within but also outside the ultra high risk (UHR) paradigm.

The concept of UHR for psychosis was conceived by Alison Yung and Patrick McGorry, who also used the term at-risk mental state (ARMS) to denote a clinical state that represented a putative psychosis prodrome. The UHR concept constitutes essential work in psychosis research, but it has become evident that, although help-seeking people fulfill criteria for UHR, only a minority develop psychotic disorders. In line with this, it could be argued that the UHR term could be replaced by a less stigmatizing and descriptive term, such as "Mild Disturbance of Thought and Perception" (MDTP), as this would more accurately reflect the situation of those who meet UHR criteria. The term can be seen in parallel to *mild cognitive impairment,* which is used in the field of dementia to characterize those who have cognitive disturbances but do not fulfill diagnostic criteria for dementia. It could be argued that a similar approach would be fruitful in the field of psychosis.

Genome-wide association studies (GWAS) have enabled assessment of the common genetic variation, which has been shown to play a major role in the etiology of human complex traits, including psychiatric disorders. Much of the heritability is captured by GWAS (single-nucleotide polymorphism heritability) and recent studies have identified several risk variants, most prominently in schizophrenia, and revealed substantial genetic overlap across disorders (mental traits as well as somatic traits) (Lee et al., 2013; Anttila et al., 2016). The finding of a substantial genetic overlap across disorders underlines the importance of keeping an open mind and not relying heavily on current diagnostic categories. Also, from a theoretical point of view, it has been argued that the psychiatric field has put too much emphasis on classification of disorders and that what matters in medicine is prediction, explanation, and identification of mechanisms and treatment (Bolton, 2012). Diagnostic categories are important for getting the science "off the ground," providing a common language of communication and relative confidence that we are studying the same phenomena at different places and times, so the importance of "getting a diagnosis right" (e.g., identifying a natural kind) might not be as essential. Given

Risk Factors for Psychosis
https://doi.org/10.1016/B978-0-12-813201-2.00022-3

the multifactorial etiology of psychiatric disorders, what matters is not so much the etiological validity of a diagnosis but its utility. Different framings of diagnosis serve different validities; selecting the diagnostic symptoms for a disorder in one way that might improve prediction but selecting them in a different way might improve treatment response. The point is a pragmatic one, suggesting that when the purpose of the diagnosis is determined, criteria can be selected to fit that purpose. This pragmatic approach to validity could be especially relevant for the UHR paradigm, where the etiology is thought to be even more heterogeneous than in more well-defined psychiatric conditions and where there is a large diversity in the symptomatology and illness trajectories of the patients.

In this chapter, we will first discuss the validity of the UHR, or the suggested MDTP, paradigm as a predictor of further illness course (especially in predicting psychosis). Second, we will discuss the finding that the majority of patients identified as UHR do not develop a psychotic disorder ("nontransitions") and question whether the UHR paradigm is meaningful for these patients. Third, we will review strategies for improving the validity of the UHR paradigm as a paradigm for the risk of psychosis. Fourth, we will examine treatment options, possible new opportunities for interventions, and some risk areas outside of the UHR field.

Prediction of psychosis

The UHR concept has been found to be effective as a tool for prediction of psychosis among help-seeking individuals. However, it has been argued that the concept is misguided and that it mostly catches young people with anxiety, depression, or substance use with a low risk of psychosis as an unspecific marker of poor prognosis (Os and Guloksuz, 2017). This argument seems erroneous, as the concept has been found to distinguish accurately between those who develop a psychosis (especially schizophrenia) and those who do not. Webb et al. (2015) compared UHR individuals to help-seeking individuals not fulfilling UHR criteria and found that 19.6% of those who fulfilled UHR criteria, compared to 1.8% of those who did not, developed schizophrenia. There were no significant differences between the rates of developing other mental disorders (Webb et al., 2015). This finding has been replicated in a larger sample (Fusar-Poli et al., 2017b). Danish register studies have shown that, among patients aged 14–17 years diagnosed with any child or adolescent psychiatric disorder, 8.74% develop schizophrenia within 8 years (Maibing et al., 2015). This elevated risk is low compared to UHR individuals who had a 22% –31.5% transition rate within 3 years (Fusar-Poli et al., 2012, 2016). Still, the critics raise an interesting point with regard to transition to psychosis. For instance, the Comprehensive Assessment of At-Risk Mental States (CAARMS) criteria define Brief Limited Intermittent Psychotic Symptoms (BLIPS) as a self-limiting psychotic symptom occurring for no more than 1 h per day for no more than a week. A higher frequency or duration score would be judged as a transition to psychosis. Intensity of symptoms is an important estimate, but as with any transformation of a continuous scale to a dichotomous outcome, there can be translational difficulties. Given the fluctuating nature of psychiatric symptoms, it is likely that there would be a proportion of false positive transitions in any UHR studies (e.g., patients who have not truly transitioned to a psychotic disorder, but where the psychotic-like symptoms have been somewhat more intense or of a longer duration, but with no other underlying change in symptomatology) (Yung et al., 2010; Hengartner et al., 2017).

As the original UHR concept was developed based on studies of the prodromal phase of schizophrenia and was established as a framework to prevent development of a psychotic disorder (e.g., schizophrenia), it is understandable that the outcome of most UHR studies has been development of psychotic disorders. However, there is an inherent paradox in the selection of this outcome. Patients with a schizotypal disorder are known to have a better prognosis than patients who are diagnosed with schizophrenia (often understood as two illnesses on a continuum, primarily separated by the presence or absence of psychotic symptoms) (Nelson et al., 2013b). What predicts the outcome of schizophrenia is most often not psychotic symptoms (e.g., hallucinations and delusions). Longitudinal studies of patients with schizophrenia have shown that the long-term outcome is primarily predicted by factors other than level of psychotic symptoms (e.g., age, negative symptoms, duration of untreated psychosis) (Marshall et al., 2005; Austin et al., 2013; Norman et al., 2017). Subsequently, it could be argued that psychotic symptoms are only important if they represent a meaningful distinction from the rest of the UHR group with regard to later outcome.

As most psychiatric diagnoses to some degree are based on arbitrary cutoffs, this is not a problem in itself, as long as the transitions assessed by the UHR instruments also predict development of a psychotic disorder with an outcome distinct from the UHR state. A metaanalysis ($n = 2182$) (Fusar-Poli et al., 2013) analyzed the *ICD/DSM* diagnostic outcome of patients who developed a psychotic disorder after having been identified as UHR individuals. They found that 560 (26%) developed a psychotic disorder and that 73% of those were within the schizophrenia spectrum. For years, there has been a debate regarding the utility of the UHR paradigm and transition rates (Carpenter, 2017). Transition rates vary widely between studies, ranging from 70% (in a study using the Schizophrenia Proneness instrument for adults (SPI-A) (Klosterkotter et al., 2001)) to as low as 9.11% (in studies using the CAARMS) (Nelson and Yung, 2010; McGorry et al., 2017). The reasons for this variety are manifold. First, the inclusion criteria in the studies were different; second, the duration of follow-up has a large impact. Furthermore, there have been reports of a long-term decrease in transition rates. This was most evident in the first UHR clinic established, the PACE clinic in Melbourne, Australia. They found that patients included from 2004 to 2006 had a lower risk for developing psychosis compared to those included from 1993 to 1997 (Nelson et al., 2013a). This tendency has been confirmed in a metaanalysis (Fusar-Poli et al., 2012).

Although the transition rates are lower now than in the 1990s, the same metaanalysis (Fusar-Poli et al., 2012) found the transition rate to be 31.5% within 3 years. This is a high proportion compared to the general population risk of 0.05% for developing schizophrenia within 3 years (McGrath et al., 2004; Lee et al., 2018). This comparison gives the impression that UHR clinics represent an effective early outreach program for patients on the path to psychosis. UHR clinics are not yet a part of standard mental health services in most countries; it has not been determined whether they could provide an early treatment option before development of psychosis. However, a study from London, where an established UHR clinic has been operating for almost 20 years, found that only 5.19% of patients who were diagnosed with a first-episode psychosis (FEP) came through the UHR clinics (Fusar-Poli et al., 2017a). This shows that the majority of patients who develop a psychotic disorder are not identified by the UHR clinic, either because the population or referrers are unaware of the service or because the UHR criteria apply to only a fraction of those patients who will develop a psychosis.

UHR clinics treat help-seeking individuals; there is therefore uncertainty whether the criteria could be used in the general population. An outreach to the general population could dilute the sample, resulting in lower transition rates. Among help-seeking individuals who are referred to UHR clinics, the UHR criteria can distinguish between people who are at high risk and those who are not. Outcome analyses of patients referred to clinics showed that patients who fulfilled UHR criteria had a 26% chance of developing a psychotic disorder, compared to 1.56% of those who were referred but did not fulfill the criteria (Fusar-Poli et al., 2015). The overall risk for transition within 38 months for all referred was 15%. This pretest risk is much higher than in the general population, but the pretest risk exerts a strong influence on the positive predictive value of the test. The predictive value of a test depends to a large degree on whether an illness is prevalent or nonprevalent in the tested population, regardless of high sensitivity and specificity. This is also true for the UHR population. One analysis of recruitment strategies for UHR studies found that clinics with a broad outreach and a high number of self-referrals had a lower rate of pretest transitions than clinics that were more restrictive in their referral practices (e.g., informed only health professionals) (Fusar-Poli et al., 2016). This effect has a strong influence on the usefulness of the UHR criteria for predicting psychosis and makes it unlikely that the instruments can be used for screening the general population. One study, screening all patients referred to secondary mental health services, found that the researchers could identify a large number of patients who fulfilled UHR criteria but were not referred to the UHR clinic and that those patients identified by the researchers had a higher transition rate over 12 months (21/78) compared to those referred to the UHR clinic (3/32) (Rietdijk et al., 2012). This study shows a possible way to expand recruitment efforts without diluting the usefulness of the test.

Nontransitioners

Rates of transition to a psychotic disorder vary depending on recruitment strategy, inclusion period, and instruments. Regardless of the approach used for inclusion of patients, the majority of UHR patients do not develop a psychotic disorder. As the studies included only help-seeking individuals with mental health problems, the outcome of the nonconverters is also of great importance. One recent study used data from two of the largest UHR cohorts and found that 62% of the patients developed a nonpsychotic axis-I disorder (McHugh et al., 2017). Follow-up studies have shown that, overall, UHR individuals improve in functional, cognitive, and symptom levels, but impairments over several domains remain for several years after they first met the UHR criteria (Lin et al., 2011; Velthorst et al., 2011; Hengartner et al., 2017).

There is a need to broaden the UHR concept from psychosis development to other equally relevant outcomes, as suggested in the CHARMS study (Hartmann et al., 2017) (see Chapter 21, new paradigms to study risk).

One such relevant outcome is functioning, which has received growing attention (Niendam et al., 2007; Yung et al., 2010; Lin et al., 2011; Lin et al., 2012; Brandizzi et al., 2015). Addington et al. (2011) investigated outcomes for UHR individuals who did not transition to psychosis within a 2.5-year period; they found that more than 50% of the nontransitioning individuals no longer presented any attenuated psychotic symptoms but remained at a significantly poor level of social and role functioning compared to healthy controls (Addington et al., 2011). The persistence of functional impairments is supported by evidence from a 6-year follow-up study that reported that nonconverting individuals displayed significant decrements in overall functioning compared to healthy controls (De Wit et al., 2014). In addition, metaanalytical evidence reveals UHR individuals to display large impairments in functioning compared to healthy controls but modest impairments compared to patients with established psychosis. Also, impairments in quality of life in UHR individuals equal those of patients with psychosis (Fusar-Poli et al., 2015). As the UHR paradigm captures a population that is at a maintained risk for mental disorders and significant and persistent functional decrements, there is a need for further research on predictors of the functional prognosis, as well as into interventions that target current mental disorders along with functional impairments, with the aim of improving long-term outcomes. Negative symptoms, which are currently not part of the UHR criteria, seem to be an area of special importance, as the level of negative symptoms has been found to be predictive of conversion to psychosis in UHR individuals and also predictive of their long-term functional impairment and impairments in quality of life (Bechdolf et al., 2005; Demjaha et al., 2012; Piskulic et al., 2012; Glenthøj et al., 2016). This underscores the importance of moving beyond the predominant focus on attenuated psychotic symptoms and merits further research into determinants of negative symptoms and their treatment.

In response to this criticism, there are endeavors to refocus the clinical and research field at a broader range of outcomes. One of these attempts is the establishment of the clinical high at-risk mental state (CHARMS) criteria. Including the CAARMS criteria for BLIPS and attenuated psychosis symptoms, the CHARMS expands inclusion criteria to include subthreshold symptoms for major depression, mania, and borderline personality disorder (Hartmann et al., 2017). This broadening of inclusion criteria will most likely lead to a higher number of patients being included. The outcome is not only psychosis but also major depression, bipolar disorder, and borderline personality disorder (all termed transitions). The study testing the CHARMS criteria is still in progress and no results are available. As a theoretical concept, the CHARMS criteria might be a valid category if the goal is to predict development of "any" psychiatric condition. It is, however, more questionable if treatment options that can be used uniformly for such a heterogeneous population can be developed, and there might be a need for subdivisions in valid treatment categories.

Increasing accuracy

While development of UHR criteria constitutes pivotal work in early psychosis research, there is a need to increase the specificity of the instruments used to identify UHR individuals. This merits further research into optimizing the UHR criteria to decrease the false positive rate. This book has in previous chapters discussed various biological and clinical areas of risk research (Chapters 9–16). We will briefly mention two further areas that might hold promise for further risk enrichment and look at the possibility for improving predictive accuracy by combining various biological and clinical measurements.

Initial evidence has shown that applying both the UHR criteria and the basic symptom criteria (cognitive disturbances) improves sensitivity and individual risk estimation. This may also translate into development of stage-targeted interventions (Schultze-Lutter et al., 2014). The Examination of Anomalies Self-Experiences (EASE) is an instrument developed for assessing self-disturbances in the "minimal self." The disturbances have been found to aggregate in schizophrenia spectrum disorders and to predict transition to psychosis in a UHR population (Nelson et al., 2012). Both of these studies show a possibility for increasing diagnostic prediction by drawing on the rich tradition of phenomenological research in psychiatry and by making more thorough clinical examination of UHR individuals.

Automated analysis of free speech, although using far newer technologies, has also shown promising results for psychosis prediction. A proof-of-principle study was published in 2015, using machine learning methods to analyze free speech in 35 high-risk individuals; the model predicted five transitions to psychosis with 100% accuracy (Bedi et al., 2015). A recent replication study also showed promising results (Corcoran et al., 2018). The studies indicated that baseline semantic coherence (the flow of meaning in speech) and syntactic complexity could predict later onset of psychosis. The later study discriminated with 83% accuracy between individuals who later transitioned to a psychotic disorder and those who did not. This study further tested the algorithm developed in the first group of participants in an independent UHR sample where they could predict transition with 79% accuracy. Although this finding should be replicated in a larger sample and in languages other than English, the method is promising and could be applied in other fields of psychiatry outside the UHR paradigm.

As mentioned, the UHR concept captures a population presenting with psychiatric symptoms and a need for intervention. Still, it is essential to decrease the false positive rate found in UHR populations to decrease the risk of stigma and potentially unnecessary medical treatment. As no one method at the moment holds the key to solving this problem, from a theoretical viewpoint, the combination of several tests should be more successful in predicting onset of psychosis. Using a three-stage model, Schmidt et al. explored this possibility (Schmidt et al., 2017). On evaluating clinical, biological, neurocognitive, combination, and environmental predictive models and testing 13 distinct three-stage approaches, they developed a theoretical model to predict psychosis onset with 98% accuracy in patients who tested positive on all three tests. The model is based on a UHR population and the highest positive predictive value was achieved by first using a combined test (clinical + EEG) (Nieman et al., 2014), followed by two biological markers (MRI) (Koutsouleris et al., 2015) and blood markers (Perkins et al., 2015). A positive result on all three tests would give a positive predicative value of 98%, compared to 71%–82% with two positive tests, 12%–21% with one positive complementary test, and 1% for no positive complementary tests. The model is based on data from several studies but is still theoretically viable and should be tested in a clinical sample. Also, health economic calculations are needed to assess the feasibility of conducting multiple tests on all UHR individuals. However, the basic principle of using multiple tests to construct a "risk ladder" in the UHR population is promising and should be developed further.

Treatments

Interventions in the UHR state have the aims of managing the current symptoms and causes of distress of UHR individuals and ideally prevent transition to psychosis. In addition, the beneficial effect of the intervention should outweigh the potential side effect. The psychotherapeutic treatments tested in the UHR population have predominantly been within a cognitive behavioral therapy (CBT) approach (Thompson et al., 2016).

CBT has been shown to be effective in patients with established psychosis in acute phases or in chronic and persistent psychotic symptoms (Tarrier et al., 1998; Sensky et al., 2000; Wykes et al., 2008) and with regard to relapse prevention (Birchwood, 1996). Metaanalytic findings, however, question a strong evidence for CBT for psychosis and point to more focused applications of CBT, rather than generic CBT, to achieve larger effect sizes (Turner et al., 2014). Given the selected findings of beneficial effects of CBT in the psychosis population, it seems to hold promise as an intervention for UHR individuals. In contrast to pharmacological treatments, it may be considered a less harmful intervention without potential detrimental side effects, although studies need to report on potential iatrogenic effects of CBT (e.g., self-stigma). The efficacy of CBT interventions, as a means to reduce transition to psychosis, has been tested in five randomized controlled trials within the UHR population, comparing CBT interventions with monitoring, supportive therapy, placebo, or treatment as usual. Metaanalytical evidence showed CBT to be associated with a pooled relative risk of transition to psychosis of 0.52 (95% CI = 0–79) (Van Der Gaag et al., 2013). A 4-year follow-up of one of the CBT trials included in the metaanalysis showed continued favorable effects of the CBT intervention in preventing transition to psychosis. Moreover, a significantly higher proportion of patients in the CBT group remitted from UHR status compared to the treatment as usual group (Ising et al., 2016). A more recent metaanalysis specifically addressing CBT interventions in the UHR population reviewed six trials and found that the relative risk for developing psychosis in those receiving CBT was reduced by at least 50% at 6-, 12-, and 18–24-month follow-up (Hutton and Taylor, 2014). The findings indicate robust effects of CBT interventions in preventing psychosis onset and in decreasing the level of attenuated psychotic symptoms. However, contrasting these positive findings for CBT interventions, a study by Stain et al. (2016) did not find any effectiveness of a CBT intervention, compared to nondirective reflective listening, in reducing transition rates, symptom severity, or improving functioning and quality of life in UHR individuals (Stain et al., 2016).

Early intervention is justified for preventing psychosis onset. The prominent and persistent mental and functional problems of UHR individuals justify them as a target for intervention to improve long-term prognosis. To date, the possibility of targeted interventions to reduce symptom severity and improve functioning and quality of life in the UHR population remains largely unexplored. It is paradoxical that, while it is well-known that negative symptoms impact not only psychosis development but also functional prognosis in UHR individuals (Piskulic et al., 2012; Glenthøj et al., 2016), the research field still lacks adequate treatment options for the negative symptom complex (Devoe et al., 2018). Also, cognitive deficits are known to have a strong impact on functional outcome in UHR individuals and on the long-term prognosis for patients with established psychosis (Niendam et al., 2006; Fusar-Poli et al., 2012; Meyer et al., 2014). Research within the psychosis population has shown cognitive remediation to have effect not only on cognitive performance but also on functional outcome (Wykes et al., 2011). Moreover, evidence indicates a positive effect of cognitive remediation on levels of negative symptoms in patients with schizophrenia (Cella et al., 2016). While the beneficial effect of cognitive remediation is examined quite extensively in the psychosis population, there is still a lack of knowledge regarding the effect of cognitive remediation in the UHR state of psychosis (Glenthøj et al., 2017). Cognitive remediation may be hypothesized to be more effective at this early stage of illness, with the potential of greater brain plasticity (Keshavan and Hogarty, 1999) than at more pronounced stages of illness. This calls for additional research into the effects of cognitive remediation in the UHR population aiming at improving functional outcome, quality of life, level of symptoms, and (potentially) the long-term prospect of preventing psychosis onset. Furthermore, social cognition is regarded as proximal to patients' real-life functioning and is critical for functional aspects such as community functioning, interpersonal relationships, and quality of life (Couture et al., 2006). Hence, the field needs vital knowledge on the effect of social cognitive remediation which can be hypothesized to significantly benefit the functional outcome of UHR individuals. Virtual reality exposure is another treatment approach that may hold promise in enhancing real-life social participation in patients with schizophrenia spectrum disorders (Fornells-Ambrojo et al., 2008; Veling et al., 2014). Virtual reality exposure has proven preliminary effectiveness in targeting persecutory beliefs and biases in thinking in UHR individuals (Valmaggia et al., 2007) and patients with psychosis (Fornells-Ambrojo et al., 2008; Pot-Kolder et al., 2018). Furthermore, virtual reality has the potential of using individually tailored exposure exercises to target the specific symptoms of each patient (e.g., anxiety or depressive symptoms), which may be a useful treatment approach in the UHR population, bearing in mind the heterogeneity of symptoms the UHR individuals experience. Hence, virtual reality exposure could potentially be a beneficial additive treatment component to CBT or cognitive remediation in UHR state.

Development of schizophrenia can be the start of a life-long illness with persisting psychotic symptoms and functional impairments; however, for most patients, the outcome is much more heterogeneous (Hegarty et al., 1994). A minority will experience recovery, understood as both functional and symptomatic remission (Jääskeläinen et al., 2013). Most patients who are diagnosed with schizophrenia experience negative symptoms and cognitive and functional problems, but the psychotic symptoms that distinguish between UHR and FEP are the ones for which there is an effective medical treatment. Only a minority of patients are treatment-resistant to antipsychotic medications. There are still no effective treatment options for cognitive and functional impairments and negative symptoms. These symptoms/deficits are shared by both UHR and FEP patients. Recognition of this pairing could lead to more approaches to treatment and research beyond the separation of psychosis. There might be patients among those diagnosed with schizophrenia who are too debilitated to be a part of psychosocial interventions with UHR individuals; however, most adequately medicated FEP patients present problems similar to those of UHR individuals. For example, persecutory or self-referring ideas are prevalent in both UHR individuals and post-psychotic FEP patients. These symptoms should be open to CBT-based interventions for both groups. The integration of treatment facilities could be a further development. Our clinical experience in the OPUS clinics in Denmark, where both patients with schizotypal disorder and patients with schizophrenia are treated, has been shown to be a viable approach for both patients and clinicians with no reported cases of increased stigma in the schizotypal group. A reservation about integrating the services has been the fear that the integration might lead to more liberal use of antipsychotic medications in the UHR population and increase stigma. There is no strong evidence that the use of antipsychotics for UHR individuals reduces transitions rates (McGlashan et al., 2006; McGorry et al., 2013). This reservation is well founded, as we have found a high rate of patients with schizotypal disorder who are treated with antipsychotic medications in the OPUS clinics (Albert et al., 2016). More research is needed into the clinical considerations for these prescriptions and if there are outcomes other than psychosis that are relevant to explore.

Other risk groups

Most people who currently develop psychosis have not been identified as UHR cases before the psychosis onset. This can be explained partly by the fact that UHR clinics are few in number and not sufficiently well-known to the public. However, this finding indicates that other initiatives to prevent psychosis can be relevant. Such initiatives could include intervening in other high-risk groups, such as children with parents with severe mental disorders (Rasic et al., 2014; Thorup et al., 2015). Metaanalytic evidence in studies following the offspring of parents with a severe mental illness (SMI) until adulthood has shown that the children have a 32% chance of developing an SMI. SMI includes schizophrenia, bipolar disorder, and major depression. However, in all mental disorders for the offspring of parents with a SMI, the rate for developing a disorder is 55%, and this is more likely to be an underestimation than an overestimation (Rasic et al., 2014). The relative risk of children of parents with schizophrenia of developing schizophrenia compared to healthy controls is 7.54. This elevated risk highlights the importance of developing monitoring or intervention programs. There is no need for healthy children to be enrolled in treatment programs, but development of family treatment and swift referral procedures when problems occur should be a minimum requirement. Trauma such as bullying is another factor that has been found to associate with the risk for psychosis development. Using data from a large nation-based survey, Catone and colleagues found that a lifetime history of bullying was associated with the presence of psychotic symptoms (persecutory ideations and hallucinations). Controlling for other traumas and sexual abuse did not affect this association (Catone et al., 2015). This finding underlines the importance of decreasing bullying, for example, by instigating national antibullying campaigns. It also calls for early detection of bullying and initiation of relevant interventions to ameliorate the psychological consequences of bullying, as this might aid in preventing psychosis development.

Another group at elevated risk for developing a psychotic disorder is people who have experienced psychosis in relation to use of drugs, especially use of cannabis (Hjorthøj et al., 2018). A drug-induced psychosis is clinically often considered a transitory state, with no need for further intervention. Using the Danish national registers, it has been shown in a large sample ($N = 6788$) that 32.2% of those who were treated for a drug-induced psychosis were later diagnosed with schizophrenia or bipolar disorder. The highest risk was observed for cannabis-induced psychosis, where 47.4% were diagnosed with a psychotic disorder within 20 years, with 50% occurring within the first 3.1 years (Starzer et al., 2017). This elevated risk calls for "watchful waiting" and not discharging patients without proper follow-up after the psychotic episode has remitted. More appropriate would be development of targeted interventions for patients who have experienced a drug-induced psychosis, potentially in collaboration with UHR clinics but with a special focus on drug and alcohol abuse. In addition, in a public health perspective, it points to an increased awareness on reducing cannabis abuse in the youth population.

Conclusion

The book has discussed a number of risk paradigms encompassing genetic high risk, including 22q11.2 deletion syndrome, schizotypal disorder, UHR, basic symptoms, and psychotic-like experiences. The development of the UHR concept was part of a wider movement in psychiatry, starting in the 1990s, which refocused the clinical and research field on the early course of psychotic illness. The development of specialized early services for FEP and the establishment of UHR clinics have transformed the treatment provided for young people who are suffering from mental illnesses. Treatment facilities for both UHR and FEP patients have been shown to be beneficial (Van Der Gaag et al., 2013; Correll et al., 2018). The research field has also yielded large long-term follow-up studies of both UHR and FEP patients (Nelson et al., 2013a; Secher et al., 2015). These studies have shown a wide variety in the outcomes of early mental disorders. Recognizing that the UHR concept has a low false negative rate and actually identifies individuals at heightened risk of psychosis, the outcome for most patients is not psychosis. In line with this, we propose renaming the UHR concept to "Mild Disturbances of Thought and Perception (MDTP)," as this term is more descriptive of the symptoms presented by patients and might help to reduce stigma.

Finally, and in line with the notion of a pragmatic validity, we believe that the UHR, or MDTP, paradigm should continue to balance a plurality of validities: *First*, psychosis prediction, the original *raison d'etre* for the UHR field, is still important and efforts should continue to increase its accuracy by improving the psychopathological evaluations, finding new biomarkers, or using machine learning. All of these approaches are promising, and combining them might produce the best results. *Second*, raising the predictive power of UHR paradigm must not lead to a neglect of the at-need population who do not develop a psychotic disorder. The identification of a large nondiagnosed impaired population is one major achievement of the UHR paradigm and the field should evolve the criteria and outcomes to reflect the pluripotent outcomes of early mental illnesses. *Third*, the most robust evidence for preventing transition to psychosis comes from CBT. The search for potential medical preventions should go on; however, as of now, no medical treatments have been robustly shown to be effective in preventing psychosis. Even in the absence of such treatment options, identification of an UHR individual and watchful waiting are important for minimizing the duration of untreated psychosis. Also, as the UHR field should refocus on outcomes other than psychosis, treatments should be developed to reflect this shift. Low doses of antipsychotic medications might not reduce transition rates for UHR patients, but they might alleviate other symptoms that are distressing to the patient. CBT might reduce transition rates; it might also be used transdiagnostically as a treatment for both UHR and FEP patients who are suffering from ideas of reference or social anxiety. *Fourth*, outside the UHR field, the clear majority of patients who develop a psychotic disorder are never identified as UHR individuals before the psychosis. How to identify more patients at risk without diluting the ability to predict outcome is a question of great importance and complexity which should be a major focus for the research field in the future.

List of acronyms and abbreviations

APS Attenuated psychosis symptoms
ARMS At-risk mental state
BLIPS Brief Limited Intermittent Psychotic Symptoms
CAARMS Comprehensive Assessment of At-Risk Mental State
CBT Cognitive behavioral therapy
CHARMS Clinical high at-risk mental state
CHR Clinical high risk
FEP First-episode psychosis
GWAS Genome-wide association studies
MDTP Mild Disturbance of Thought and Perception
SNP Single-nucleotide polymorphism
SMI Severe mental disorder
UHR Ultra high risk

References

Addington, J., et al., 2011. At clinical high risk for psychosis: outcome for nonconverters. Am. J. Psychiatry 168 (8), 800–805. https://doi.org/10.1176/appi.ajp.2011.10081191.

Albert, N., et al., 2016. Course of illness in a sample of patients diagnosed with a schizotypal disorder and treated in a specialized early intervention setting. Findings from the 3.5 year follow-up of the OPUS II study. Schizophr. Res. https://doi.org/10.1016/j.schres.2016.10.013.

Anttila, V., et al., 2016. Analysis of Shared Heritability in Common Disorders of the Brain, p. 048991. https://doi.org/10.1101/048991 bioRxiv.

Austin, S.F., et al., 2013. Predictors of recovery in first episode psychosis: the OPUS cohort at 10year follow-up. Schizophr. Res. 150 (1), 163–168. https://doi.org/10.1016/j.schres.2013.07.031.

Bechdolf, A., et al., 2005. Subjective Quality of Life in Subjects at Risk for a First Episode of Psychosis : A Comparison with First Episode Schizophrenia Patients and Healthy Controls, vol. 79, pp. 137–143. https://doi.org/10.1016/j.schres.2005.06.008.

Bedi, G., et al., 2015. 'Automated analysis of free speech predicts psychosis onset in high-risk youths'. NPJ Schizophr. 1 (1), 15030. https://doi.org/10.1038/npjschz.2015.30. Nature Publishing Group.

Birchwood, M., 1996. Early intervention in psychotic relapse: cognitive approaches to detection and management. In: Haddock, G., Slade, P.D. (Eds.), Cognitive-Behavioural Interventions with Psychotic Disorders. Routledge, pp. 171–211.

Bolton, D., 2012. Chapter 1 Classification and causal mechanisms: a deflationary approach to the classification problem. In: Kendler, K.S., Parnas, J. (Eds.), Philosophical Issues in Psychiatry II: Nosology, first ed. Oxford University Press, Oxford, pp. 6–11.

Brandizzi, M., et al., 2015. Predictors of functional outcome in individuals at high clinical risk for psychosis at six years follow-up. J. Psychiatr. Res. 65, 115–123. https://doi.org/10.1016/j.jpsychires.2015.03.005. Elsevier Ltd.

Carpenter, W.T., February 2017. Clinical high risk controversies and challenge for the experts. Schizophr. Bull. https://doi.org/10.1093/schbul/sbx182, pp. 2017–2019.

Catone, G., et al., 2015. Bullying victimisation and risk of psychotic phenomena: analyses of British national survey data. Lancet Psychiatry 2 (7), 618–624. https://doi.org/10.1016/S2215-0366(15)00055-3. Elsevier Ltd.

Cella, M., et al., 2016. Cognitive remediation for negative symptoms of schizophrenia: a network meta-analysis. Clin. Psychol. Rev. 52, 43–51. https://doi.org/10.1016/j.cpr.2016.11.009. Elsevier B.V.

Corcoran, C.M., et al., 2018. Prediction of psychosis across protocols and risk cohorts using automated language analysis. World Psychiatry 17 (1), 67–75. https://doi.org/10.1002/wps.20491.

Correll, C., et al., 2018. Effectiveness of early intervention services for early psychosis: systematic review, meta-analysis, and metaregression-analysis. JAMA Psychiatry 75 (6), 555–565. https://doi.org/10.1001/jamapsychiatry.2018.0623.

Couture, S.M., Penn, D.L., Roberts, D.L., 2006. The functional significance of social cognition in schizophrenia: a review. Schizophr. Bull. 32 (Suppl. 1), S44–S63. https://doi.org/10.1093/schbul/sbl029.

De Wit, S., et al., 2014. Adolescents at ultra-high risk for psychosis: long-term outcome of individuals who recover from their at-risk state. Eur. Neuropsychopharmacol. 24 (6), 865–873. https://doi.org/10.1016/j.euroneuro.2014.02.008. Elsevier.

Demjaha, A., et al., 2012. Disorganization/cognitive and negative symptom dimensions in the at-risk mental state predict subsequent transition to psychosis. Schizophr. Bull. 38 (2), 351–359. https://doi.org/10.1093/schbul/sbq088.

Devoe, D.J., Peterson, A., Addington, J., February 2018. Negative Symptom Interventions in Youth at Risk of Psychosis : A Systematic Review and Network Meta-Analysis, pp. 1–17. https://doi.org/10.1093/schbul/sbx139.

Fornells-Ambrojo, M., et al., 2008. Virtual reality and persecutory delusions: safety and feasibility. Schizophr. Res. 104 (1–3), 228–236. https://doi.org/10.1016/j.schres.2008.05.013.

Fusar-Poli, P., Bonoldi, I., et al., 2012. Predicting psychosis meta-analysis of transistion outcomes in individuals at high clinical risk. Arch. Gen. Psychiatr. 69 (3), 220–229. https://doi.org/10.1001/archgenpsychiatry.2011.1472.

Fusar-Poli, P., Cappucciati, M., et al., 2015. At risk or not at risk? A meta-analysis of the prognostic accuracy of psychometric interviews for psychosis prediction. World Psychiatry 14 (3), 322–332. https://doi.org/10.1002/wps.20250.

Fusar-Poli, P., Cappucciati, M., et al., 2016. Heterogeneity of psychosis risk within individuals at clinical high risk: a meta-analytical stratification. JAMA Psychiatry 73 (2), 113–120. https://doi.org/10.1001/jamapsychiatry.2015.2324.

Fusar-Poli, P., Deste, G., et al., 2012. Cognitive functioning in prodromal psychosis. Arch. Gen. Psychiatr. 69 (6), 562–571. https://doi.org/10.1001/archgenpsychiatry.2011.1592.

Fusar-Poli, P., Rocchetti, M., et al., 2015. Disorder, not just state of risk: meta-analysis of functioning and quality of life in people at high risk of psychosis. Br. J. Psychiatry 207 (3), 198–206. https://doi.org/10.1192/bjp.bp.114.157115.

Fusar-Poli, P., Schultze-Lutter, F., et al., 2016. The dark side of the moon: meta-analytical impact of recruitment strategies on risk enrichment in the clinical high risk state for psychosis. Schizophr. Bull. 42 (3), 732–743. https://doi.org/10.1093/schbul/sbv162.

Fusar-Poli, P., et al., 2013. At risk for schizophrenic or affective psychoses? A meta-analysis of DSM/ICD diagnostic outcomes in individuals at high clinical risk. Schizophr. Bull. 39 (4), 923–932. https://doi.org/10.1093/schbul/sbs060.

Fusar-Poli, P., Rutigliano, G., Stahl, D., Davies, C., Bonoldi, I., et al., 2017a. Development and validation of a clinically based risk calculator for the transdiagnostic prediction of psychosis. JAMA Psychiatry 74 (5), 493–500. https://doi.org/10.1001/jamapsychiatry.2017.0284.

Fusar-Poli, P., Rutigliano, G., Stahl, D., Davies, C., De Micheli, A., et al., 2017b. Long-term validity of the at Risk Mental State (ARMS) for predicting psychotic and non-psychotic mental disorders. Eur. Psychiatry 49–54. https://doi.org/10.1016/j.eurpsy.2016.11.010. The Author(s), 42(2017).

Glenthøj, L.B., et al., 2016. Social cognition in patients at ultra-high risk for psychosis: what is the relation to social skills and functioning? Schizophr. Res. Cogn. 5 https://doi.org/10.1016/j.scog.2016.06.004.

Glenthøj, L.B., et al., 2017. The effect of cognitive remediation in individuals at ultra-high risk for psychosis: a systematic review. NPJ Schizophr. 3 (1), 20. https://doi.org/10.1038/s41537-017-0021-9.

Hartmann, J.A., et al., June 2017. Broad clinical high-risk mental state (CHARMS): methodology of a cohort study validating criteria for pluripotent risk. Early Interv. Psychiatry 2, 1–8. https://doi.org/10.1111/eip.12483.

Hegarty, J., et al., 1994. 'One-hundred years of schizophrenia : a meta-analysis of the outcome literature'. Am. J. Psychiatry 151, 1409.

Hengartner, M.P., et al., 2017. Course of psychotic symptoms, depression and global functioning in persons at clinical high risk of psychosis: results of a longitudinal observation study over three years focusing on both converters and non-converters. Schizophr. Res. 189, 19–26. https://doi.org/10.1016/j.schres.2017.01.040. Elsevier B.V.

Hjorthøj, C., Albert, N., Nordentoft, M., 2018. Association of substance use disorders with conversion from schizotypal disorder to schizophrenia. JAMA Psychiatry 75 (7). https://doi.org/10.1001/jamapsychiatry.2018.0568.

Hutton, P., Taylor, P.J., 2014. Cognitive behavioural therapy for psychosis prevention : a systematic review and meta-analysis. Psychol. Med. 449–468. https://doi.org/10.1017/S0033291713000354.

Ising, H.K., et al., 2016. Four-Year follow-up of cognitive behavioral therapy in persons at ultra-high risk for developing psychosis: the Dutch early detection intervention evaluation (EDIE-NL) trial. Schizophr. Bull. 42 (5), 1243–1252. https://doi.org/10.1093/schbul/sbw018. United States.

Jääskeläinen, E., et al., 2013. A systematic review and meta-analysis of recovery in schizophrenia. Schizophr. Bull. 39 (6), 1296–1306. https://doi.org/10.1093/schbul/sbs130.

Keshavan, M.S., Hogarty, G.E., 1999. Brain maturational processes and delayed onset in schizophrenia. Dev. Psychopathol. 11 (3), 525–543.

Klosterkotter, J., et al., 2001. Diagnosing schizophrenia in the prodromal phase. Arch. Gen. Psychiatr. 58, 158–164. https://doi.org/10.1016/S0924-9338(99)80046-7.

Koutsouleris, N., et al., 2015. Detecting the psychosis prodrome across high-risk populations using neuroanatomical biomarkers. Schizophr. Bull. 41 (2), 471–482. https://doi.org/10.1093/schbul/sbu078.

Lee, S., et al., 2013. Genetic relationship between five psychiatric disorders estimated from genome-wide SNPs. Nat. Genet. 45 (9), 984–994.

Lee, T.Y., et al., February 2018. Can we predict psychosis outside the clinical high-risk state? A systematic review of non-psychotic risk syndromes for mental disorders. Schizophr. Bull. 1–10. https://doi.org/10.1093/schbul/sbx173.

Lin, A., et al., 2011. Neurocognitive predictors of functional outcome two to 13years after identification as ultra-high risk for psychosis. Schizophr. Res. 132 (1), 1–7. https://doi.org/10.1016/j.schres.2011.06.014. Elsevier B.V.

Lin, A., Nelson, B., Yung, A.R., 2012. "At-risk" for psychosis research: where are we heading? Epidemiol. Psychiatr. Sci. 21 (04), 329–334. https://doi.org/10.1017/S2045796012000388.

Maibing, C.F., et al., 2015. Risk of schizophrenia increases after all child and adolescent psychiatric disorders: a nationwide study. Schizophr. Bull. 41 (4), 963–970. https://doi.org/10.1093/schbul/sbu119.

Marshall, M., et al., 2005. Association between duration of untreated psychosis and outcome in cohorts of first-episode patients. Arch. Gen. Psychiatr. 62 (9), 975–983. https://doi.org/10.1001/archpsyc.62.9.975.

McGlashan, T.H., et al., 2006. Randomized, double-blind trial of olanzapine versus placebo in patients prodromally symptomatic for psychosis. Am. J. Psychiatry 163 (5), 790. https://doi.org/10.1176/appi.ajp.163.5.790.

McGorry, P.D., et al., 2013. Randomized controlled trial of interventions for young people at ultra-high risk of psychosis: twelve-month outcome. J. Clin. Psychiatry 74 (4), 349–356. https://doi.org/10.4088/JCP.12m07785.

McGorry, P.D., et al., 2017. Effect of ω-3 polyunsaturated fatty acids in young people at ultrahigh risk for psychotic disorders: the NEURAPRO randomized clinical trial. JAMA Psychiatry 74 (1), 19–27. https://doi.org/10.1001/jamapsychiatry.2016.2902.

McGrath, J., et al., 2004. A systematic review of the incidence of schizophrenia: the distribution of rates and the influence of sex, urbanicity, migrant status and methodology. BMC Med. 2 (1), 13. https://doi.org/10.1186/1741-7015-2-13.

McHugh, M.J., et al., 2017. The Ultra-High-Risk for psychosis groups: evidence to maintain the status quo. Schizophr. Res. https://doi.org/10.1016/j.schres.2017.09.003. Elsevier B.V.

Meyer, E.C., et al., 2014. The relationship of neurocognition and negative symptoms to social and role functioning over time in individuals at clinical high risk in the first phase of the North American Prodrome Longitudinal Study. Schizophr. Bull. 40 (6), 1452–1461. https://doi.org/10.1093/schbul/sbt235.

Nelson, B., Yung, A.R., 2010. Can clinicians predict psychosis in an ultra high risk group? Aust. N. Z. J. Psychiatr. 44 (7), 625–630. https://doi.org/10.3109/00048671003620210. SAGE Publications Ltd.

Nelson, B., Thompson, A., Yung, A.R., 2012. Basic self-disturbance predicts psychosis onset in the ultra high risk for psychosis "prodromal" population. Schizophr. Bull. 38 (6), 1277–1287. https://doi.org/10.1093/schbul/sbs007.

Nelson, B., et al., 2013a. Long-term follow-up of a group at ultra high risk ("Prodromal") for psychosis the PACE 400 study. JAMA Psychiatry 70 (8), 793–802. https://doi.org/10.1001/jamapsychiatry.2013.1270.

Nelson, M.T., et al., 2013b. Evidence of a dimensional relationship between schizotypy and schizophrenia: a systematic review. Neurosci. Biobehav. Rev. 37 (3), 317–327. https://doi.org/10.1016/j.neubiorev.2013.01.004. Elsevier Ltd.

Nieman, D.H., et al., 2014. Psychosis prediction: stratification of risk estimation with information-processing and premorbid functioning variables. Schizophr. Bull. 40 (6), 1482–1490. https://doi.org/10.1093/schbul/sbt145.

Niendam, T.A., et al., 2006. Neurocognitive performance and functional disability in the psychosis prodrome. Schizophr. Res. 84 (1), 100–111. https://doi.org/10.1016/j.schres.2006.02.005.

Niendam, T.A., et al., 2007. The course of neurocognition and social functioning in individuals at ultra high risk for psychosis. Schizophr. Bull. 33 (3), 772–781. https://doi.org/10.1093/schbul/sbm020.

Norman, R.M.G., et al., 2017. An examination of components of recovery after five years of treatment in an early intervention program for psychosis. Schizophr. Res. 6–11. https://doi.org/10.1016/j.schres.2017.08.054. Elsevier B.V.

Os, J. Van, Guloksuz, S., 2017. 'A critique of the "ultra-high risk" and "transition" paradigm'. World Psychiatry 16 (2), 200–206.

Perkins, D.O., et al., 2015. Towards a psychosis risk blood diagnostic for persons experiencing high-risk symptoms: preliminary results from the NAPLS project. Schizophr. Bull. 41 (2), 419–428. https://doi.org/10.1093/schbul/sbu099.

Piskulic, D., et al., 2012. Negative symptoms in individuals at clinical high risk of psychosis. Psychiatry Res. 196 (2–3), 220–224. https://doi.org/10.1016/j.psychres.2012.02.018. Elsevier Ireland Ltd.

Pot-Kolder, R.M.C.A., et al., 2018. Virtual-reality-based cognitive behavioural therapy versus waiting list control for paranoid ideation and social avoidance in patients with psychotic disorders: a single-blind randomised controlled trial. Lancet Psychiatry 5 (3), 217–226. https://doi.org/10.1016/S2215-0366(18)30053-1. Elsevier Ltd.

Rasic, D., et al., 2014. Risk of mental illness in offspring of parents with schizophrenia, bipolar disorder, and major depressive disorder: a meta-analysis of family high-risk studies. Schizophr. Bull. 40 (1), 28–38. https://doi.org/10.1093/schbul/sbt114. United States.

Rietdijk, J., et al., 2012. Detection of people at risk of developing a first psychosis: comparison of two recruitment strategies. Acta Psychiatr. Scand. 126 (1), 21–30. https://doi.org/10.1111/j.1600-0447.2012.01839.x.

Schmidt, A., et al., 2017. Improving prognostic accuracy in subjects at clinical high risk for psychosis: systematic review of predictive models and meta-analytical sequential testing simulation. Schizophr. Bull. 43 (2), 375–388. https://doi.org/10.1093/schbul/sbw098.

Schultze-Lutter, F., Klosterkötter, J., Ruhrmann, S., 2014. Improving the clinical prediction of psychosis by combining ultra-high risk criteria and cognitive basic symptoms. Schizophr. Res. 154 (1–3), 100–106. https://doi.org/10.1016/j.schres.2014.02.010.

Secher, R.G., et al., 2015. Ten-year follow-up of the OPUS specialized early intervention trial for patients with a first episode of psychosis. Schizophr. Bull. 41 (3), 617–626. https://doi.org/10.1093/schbul/sbu155. United States.

Sensky, T., et al., 2000. A randomized controlled trial of cognitive-behavioral therapy for persistent symptoms in schizophrenia resistant to medication. Arch. Gen. Psychiatr. 57 (165–72).

Stain, H.J., et al., 2016. 'A randomised controlled trial of cognitive behaviour therapy versus non-directive re fl ective listening for young people at ultra high risk of developing psychosis : the detection and evaluation of psychological therapy (DEPTh) trial'. Schizophr. Res. 176 (2–3), 212–219. https://doi.org/10.1016/j.schres.2016.08.008. Elsevier B.V.

Starzer, M.S.K., Nordentoft, M., Hjorthøj, C., 2017. Rates and predictors of conversion to schizophrenia or bipolar disorder following substance-induced psychosis. Am. J. Psychiatry (1). https://doi.org/10.1176/appi.ajp.2017.17020223 appi.ajp.2017.1.

Tarrier, N., et al., 1998. Randomised controlled trial of intensive cognitive behaviour therapy for patients with chronic schizophrenia. BMJ 317 (7154), 303–307. https://doi.org/10.1136/bmj.317.7154.303.

Thompson, A., Marwaha, S., Broome, M.R., 2016. At-risk mental state for psychosis: identification and current treatment approaches. BJPscyh Adv. 22 (3), 186–193. https://doi.org/10.1192/apt.bp.115.015487.

Thorup, A.A.E., et al., 2015. The Danish High Risk and Resilience Study – VIA 7 – a cohort study of 520 7-year-old children born of parents diagnosed with either schizophrenia, bipolar disorder or neither of these two mental disorders. BMC Psychiatry 15 (1), 233. https://doi.org/10.1186/s12888-015-0616-5.

Turner, D., et al., May 2014. 'Psychological Interventions for Psychosis : a meta-analysis of comparative outcome studies'. Am. J. Psychiatry 171, 523–538. https://doi.org/10.1176/appi.ajp.2013.13081159.

Valmaggia, L.R., et al., 2007. Virtual reality and paranoid ideations in people with an "at-risk mental state" for psychosis. Br. J. Psychiatry (Suppl. 51), s63–s68. https://doi.org/10.1192/bjp.191.51.s63.

Van Der Gaag, M., et al., 2013. Preventing a first episode of psychosis: meta-analysis of randomized controlled prevention trials of 12month and longer-term follow-ups. Schizophr. Res. 149 (1–3), 56–62. https://doi.org/10.1016/j.schres.2013.07.004. Elsevier B.V.

Veling, W., et al., 2014. Virtual reality experiments linking social environment and psychosis: a pilot study. Cyberpsychol. Behav. Soc. Netw. 17 (3), 191–195.

Velthorst, E., et al., 2011. Three-year course of clinical symptomatology in young people at ultra high risk for transition to psychosis. Acta Psychiatr. Scand. 123 (1), 36–42. https://doi.org/10.1111/j.1600-0447.2010.01593.x.

Webb, J.R., et al., 2015. Specificity of incident diagnostic outcomes in patients at clinical high risk for psychosis. Schizophr. Bull. 41 (5), 1066–1075. https://doi.org/10.1093/schbul/sbv091.

Wykes, T., et al., 2008. Cognitive behavior therapy for schizophrenia: effect sizes, clinical models, and methodological rigor. Schizophr. Bull. https://doi.org/10.1093/schbul/sbm114.

Wykes, T., et al., 2011. A meta-analysis of cognitive remediation for schizophrenia: methodology and effect sizes. Am. J. Psychiatry 168 (5), 472–485. https://doi.org/10.1176/appi.ajp.2010.10060855.

Yung, A.R., et al., 2010. The psychosis threshold in Ultra High Risk (prodromal) research: is it valid? Schizophr. Res. 120 (1–3), 1–6. https://doi.org/10.1016/j.schres.2010.03.014. Elsevier B.V.

Index

Note: 'Page numbers followed by "t" indicate tables, "f" indicate figures and "b" indicate boxes'.

E

F

R

T

FSC
www.fsc.org
MIX
Papier | Fördert gute Waldnutzung
FSC® C083411

Kolibri 360 GmbH
Ferdinand-Jühlke-Straße 7
99095 Erfurt, Deutschland
produktsicherheit@kolibri360.de